MATHEMATICAL PAPERS.

London: C. J. CLAY AND SONS,
CAMBRIDGE UNIVERSITY PRESS WAREHOUSE,
AVE MARIA LANE.
Glasgow: 263, ARGYLE STREET.

Leipzig: F. A. BROCKHAUS.
New York: THE MACMILLAN COMPANY.
Bombay: E. SEYMOUR HALE.

THE COLLECTED

MATHEMATICAL PAPERS

OF

ARTHUR CAYLEY, Sc.D., F.R.S.,

LATE SADLERIAN PROFESSOR OF PURE MATHEMATICS IN THE UNIVERSITY OF CAMBRIDGE.

VOL. XIII.

CAMBRIDGE:
AT THE UNIVERSITY PRESS.
1897.

CAMBRIDGE:

PRINTED BY J. AND C. F. CLAY,

AT THE UNIVERSITY PRESS.

ADVERTISEMENT.

THE present volume contains 80 papers, numbered 888 to 967, published for the most part in the years 1889 to 1895, the year in which Professor Cayley died. The volume completes the collection of his mathematical papers.

The table for the thirteen volumes is

Vol.	I.	Numbers	1	to	100,
,,	II.	,,	101	,,	158,
,,	III.	,,	159	,,	222,
,,	IV.	,,	223	,,	299,
,,	V.	,,	300	,,	383,
,,	VI.	,,	384	,,	416,
,,	VII.	,,	417	,,	485,
,,	VIII.	,,	486	,,	555,
,,	IX.	,,	556	,,	629,
,,	X.	,,	630	,,	705,
,,	XI.	,,	706	,,	798,
,,	XII.	,,	799	,,	887,
,,	XIII.	,,	888	,,	967.

A. R. FORSYTH.

30 *November*, 1897.

CONTENTS.

[An Asterisk means that the paper is not printed in full.]

CLASSIFICATION.

888.

ON A FORM OF QUARTIC SURFACE WITH TWELVE NODES.

[From the *British Association Report*, (1886), pp. 540, 541.]

USING throughout capital letters to denote homogeneous quadric functions of the coordinates (x, y, z, w), we have as a form of quartic surface with eight nodes $\Omega = (*\between U, V, W)^2 = 0$; viz. the nodes are here the octad of points, or eight points of intersection of the quadric surfaces $U=0$, $V=0$, $W=0$; the equation can, by a linear transformation on the functions U, V, W (that is, by substituting for the original functions U, V, W linear functions of these variables), be reduced to the form $\Omega = U^2 + V^2 + W^2 = 0$.

Suppose now that the function Ω can in a second manner be expressed in the like form $\Omega = P^2 + Q^2 + R^2$ (where P, Q, R are not linear functions of U, V, W); that is, suppose that we have identically $U^2 + V^2 + W^2 = P^2 + Q^2 + R^2$, this gives $U^2 - P^2 + V^2 - Q^2 + W^2 - R^2 = 0$; or, writing $U+P$, $V+Q$, $W+R = A$, B, C, and $U-P$, $V-Q$, $W-R = F$, G, H, the identity becomes $AF + BG + CH = 0$; and this identity being satisfied, the equation $\Omega = 0$ of the quartic surface may be written in the two forms

$$\Omega = (A+F)^2 + (B+G)^2 + (C+H)^2 = 0,$$

and

$$\Omega = (A-F)^2 + (B-G)^2 + (C-H)^2 = 0;$$

viz. the quartic surface has the nodes which are the intersections of the three quadric surfaces $A+F=0$, $B+G=0$, $C+H=0$, and also the nodes which are the intersections of the three quadric surfaces $A-F=0$, $B-G=0$, $C-H=0$. We may of course also write the equation of the surface in the form

$$\Omega = A^2 + B^2 + C^2 + F^2 + G^2 + H^2 = 0.$$

An easy way of satisfying the identity $AF + BG + CH = 0$ is to assume

$$A,\ B,\ C,\ F,\ G,\ H = ayz,\ bzx,\ cxy,\ fxw,\ gyw,\ hzw,$$

where the constants a, b, c, f, g, h satisfy the condition $af + bg + ch = 0$; this being so, the functions A, B, C, F, G, H, and consequently the functions $A+F$, $B+G$, $C+H$ and $A-F$, $B-G$, $C-H$ each of them vanish for the four points ($y=0$, $z=0$, $w=0$), ($z=0$, $x=0$, $w=0$), ($x=0$, $y=0$, $w=0$), ($x=0$, $y=0$, $z=0$), or say the points (1, 0, 0, 0), (0, 1, 0, 0), (0, 0, 1, 0), (0, 0, 0, 1). It hence appears that the quartic surface

$$\Omega = a^2y^2z^2 + b^2z^2x^2 + c^2x^2y^2 + f^2x^2w^2 + g^2y^2w^2 + h^2z^2w^2 = 0$$

is a quartic surface with twelve nodes; viz. it has as nodes the last-mentioned four points, the remaining four points of intersection of the surfaces

$$ayz + fxw = 0, \quad bzx + gyw = 0, \quad cxy + hzw = 0,$$

and the remaining four points of intersection of the surfaces

$$ayz - fxw = 0, \quad bzx - gyw = 0, \quad cxy - hzw = 0.$$

The above is the analytical theory of one of the two forms of quartic surface with twelve nodes recently established by Dr K. Rohn in a paper in the *Berichte ü. d. Verhandlungen der K. Sächsische Gesellschaft zu Leipzig*, (1884), pp. 52—60.

889.

ON A DIFFERENTIAL EQUATION AND THE CONSTRUCTION OF MILNER'S LAMP.

[From the *Proceedings of the Edinburgh Mathematical Society*, vol. v. (1887), pp. 99—101.]

WHAT sort of an equation is

$$b^3 \cos(\alpha+\theta) = a\cos\theta \int_\theta^\beta r^2 d\theta - \tfrac{2}{3}\left\{\cos\theta \int_\theta^\beta r^3 \cos\theta\, d\theta + \sin\theta \int_\theta^\beta r^3 \sin\theta\, d\theta\right\}\ ? \qquad\text{.......(1).}$$

Write

$$X = \int_\theta^\beta r^2 d\theta,\quad Y = \int_\theta^\beta r^3\cos\theta\, d\theta,\quad Z = \int_\theta^\beta r^3 \sin\theta\, d\theta \qquad\text{.......(2),}$$

and start with the equations

$$d\theta = \frac{dX}{-r^2} = \frac{dY}{-r^3\cos\theta} = \frac{dZ}{-r^3\sin\theta} \qquad\text{.......(3),}$$

$$\left(\frac{d^2}{d\theta^2}+1\right)\{a\cos\theta\,.\,X - \tfrac{2}{3}(Y\cos\theta + Z\sin\theta)\} = 0 \qquad\text{.......(4).}$$

This last gives

$$(r - a\cos\theta)\,dr + ar\sin\theta\,.\,d\theta = 0 \qquad\text{.......(5),}$$

and the system thus is

$$d\theta = \frac{dX}{-r^2} = \frac{dY}{-r^3\cos\theta} = \frac{dZ}{-r^3\sin\theta} = \frac{(r-a\cos\theta)\,dr}{-ar\sin\theta} \qquad\text{.......(6),}$$

viz. this is a system of ordinary differential equations between the five variables θ, r, X, Y, Z: the system can therefore be integrated with four arbitrary constants, and these may be so determined that for the value β of θ, X, Y, Z shall be each $=0$; and r shall have the value r_0.

But this being so, from the assumed equations (3) and (4) we have

$$X=\int_\theta^\beta r^2 d\theta,\quad Y=\int_\theta^\beta r^3\cos\theta d\theta,\quad Z=\int_\theta^\beta r^3\sin\theta d\theta,$$

and further, by integration of (4),

$$L\cos\theta+M\sin\theta=a\cos\theta\,.\,X-\tfrac{2}{3}(Y\cos\theta+Z\sin\theta).$$

Here L and M denote properly determined constants: viz. the conclusion is that r, X, Y, Z admit of being determined as functions of θ and of an arbitrary constant r_0, in such wise that

$$a\cos\theta\,.\,X-\tfrac{2}{3}(Y\cos\theta+Z\sin\theta)$$

shall be a function of θ, of the proper form $L\cos\theta+M\sin\theta$, but not so that it shall be the precise function $b^3\cos(\alpha+\theta)$. To make it have this value, we must have $L=b^3\cos\alpha$, $M=-b^3\sin\alpha$ (where L, M are given functions of a, β, r_0), i.e. we must have *two* given relations between a, b, α, β, r_0: or treating r_0 as a disposable constant, we must have *one* given relation between a, b, α, β.

The equation $d\theta=\dfrac{r-a\cos\theta}{-ar\sin\theta}dr$ gives $r^2-2ar\cos\theta=C$, where $C=r_0^2-2ar_0\cos\beta$. There would be considerable difficulty in working the question out with r_0 arbitrary, but we may do it easily enough for the particular value $r_0=0$ or $r_0=2a\cos\beta$, giving $C=0$ and therefore $r=2a\cos\theta$: and we ought in this case to be able to satisfy the given equation not in general but with *two* determinate relations between the constants a, b, α, β.

We have

$$\int\cos^2\theta d\theta=\tfrac{1}{2}\theta+\tfrac{1}{4}\sin 2\theta,$$

$$\int\cos^4\theta d\theta=\tfrac{3}{8}\theta+\tfrac{1}{4}\sin 2\theta+\tfrac{1}{32}\sin 4\theta,$$

$$\int\cos^3\theta\sin\theta d\theta=-\tfrac{1}{4}\cos^4\theta.$$

And thence

$$\begin{aligned}
&a\cos\theta\,.\,X-\tfrac{2}{3}(Y\cos\theta+Z\sin\theta)\\
&=\quad 4a^3\cos\theta\,\{\tfrac{1}{2}(\beta-\theta)+\tfrac{1}{4}(\sin 2\beta-\sin 2\theta)\}\\
&\quad-\tfrac{16}{3}a^3\cos\theta\,\{\tfrac{3}{8}(\beta-\theta)+\tfrac{1}{4}(\sin 2\beta-\sin 2\theta)+\tfrac{1}{32}(\sin 4\beta-\sin 4\theta)\}\\
&\quad-\tfrac{16}{3}a^3\sin\theta\,\{\qquad\qquad\qquad\qquad\qquad\qquad -\tfrac{1}{4}(\cos^4\beta-\cos^4\theta)\}\\
&=-\tfrac{1}{3}a^3\cos\theta(\sin 2\beta-\sin 2\theta)\\
&\quad-\tfrac{1}{6}a^3\cos\theta(\sin 4\beta-\sin 4\theta)\\
&\quad+\tfrac{4}{3}a^3\sin\theta(\cos^4\beta-\cos^4\theta),
\end{aligned}$$

where the terms containing β are readily reduced to $\frac{4}{3}a^3\cos^3\beta\sin(\theta-\beta)$; hence also the terms without β disappear of themselves: and we have

$$a\cos\theta\,.\,X-\tfrac{2}{3}(Y\cos\theta+Z\sin\theta)=\tfrac{4}{3}a^3\cos^3\beta\,.\,\sin(\theta-\beta),$$

which may be put

$$=b^3\cos(\theta+\alpha):$$

viz. this will be so if we have the *two* relations

$$\alpha=\tfrac{1}{2}\pi-\beta;\quad\text{and}\quad b^3=-\tfrac{4}{3}a^3\cos^3\beta.$$

I make (see figure) Milner's lamp, with a circular section, β arbitrary, but a

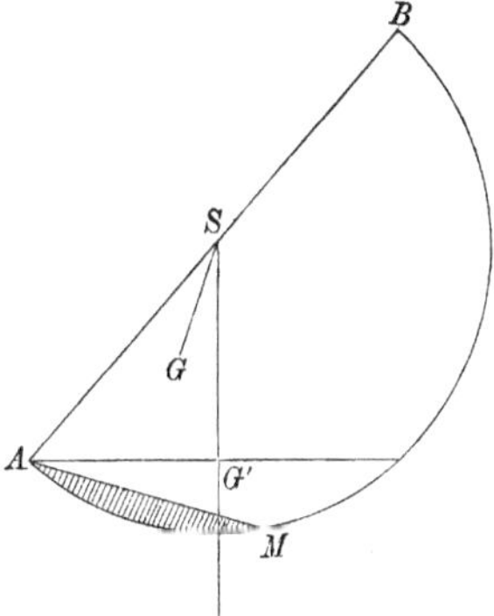

segment AM ($\angle SAM=\beta$) made solid. G in the line SG at right angles to AM is the C. G. of the lamp, and G' the C. G. of the oil.

And this seems to be the *only* form—for the pole of r must, it seems to me, be *on* the bounding circle—viz. in the equation $r^2-2ar\cos\theta=C$, we must have $C=0$.

890.

NOTE ON THE HYDRODYNAMICAL EQUATIONS.

[From the *Proceedings of the Royal Society of Edinburgh*, vol. XV. (1889), pp. 342—344.]

WRITING for shortness $D = \frac{d}{dt} + u\frac{d}{dx} + v\frac{d}{dy} + w\frac{d}{dz}$, then if from the hydrodynamical equations

$$Du = \frac{d}{dx}\left(V - \frac{p}{\rho}\right),\quad Dv = \frac{d}{dy}\left(V - \frac{p}{\rho}\right),\quad Dw = \frac{d}{dz}\left(V - \frac{p}{\rho}\right),$$

without the aid of the equation

$$\frac{du}{dx} + \frac{dv}{dy} + \frac{dw}{dz} = 0,$$

we eliminate $V - \frac{p}{\rho}$, we obtain equations not equivalent to those of Helmholtz,

$$D\xi = \left(\xi\frac{d}{dx} + \eta\frac{d}{dy} + \zeta\frac{d}{dz}\right)u, \quad = \xi\frac{du}{dx} + \eta\frac{dv}{dx} + \zeta\frac{dw}{dx}, \text{ \&c.,}$$

($2\xi,\ 2\eta,\ 2\zeta = \frac{dv}{dz} - \frac{dw}{dy},\ \frac{dw}{dx} - \frac{du}{dz},\ \frac{du}{dy} - \frac{dv}{dx}$, as usual), but which, transforming them by means of the omitted equation, agree as they should do with his equations. But the form of the equations obtained directly by elimination as above, is an interesting one, which it is worth while to give.

We have

$$\begin{aligned} D\left(\frac{dv}{dz} - \frac{dw}{dy}\right) &= D\left(\frac{dv}{dz} - \frac{dw}{dy}\right) - \frac{d}{dz}Dv + \frac{d}{dy}Dw, \\ &= \left(\frac{d}{dt} + u\frac{d}{dx} + v\frac{d}{dy} + w\frac{d}{dz}\right)\left(\frac{dv}{dz} - \frac{dw}{dy}\right) \\ &\quad - \frac{d}{dz}\left(\frac{dv}{dt} + u\frac{dv}{dx} + v\frac{dv}{dy} + w\frac{dv}{dz}\right) \\ &\quad + \frac{d}{dy}\left(\frac{dw}{dt} + u\frac{dw}{dx} + v\frac{dw}{dy} + w\frac{dw}{dz}\right), \end{aligned}$$

where the terms containing second derived functions disappear of themselves, and the expression on the right-hand is thus

$$\begin{aligned}&= -\frac{du}{dz}\frac{dv}{dx} - \frac{dv}{dz}\frac{dv}{dy} - \frac{dw}{dz}\frac{dv}{dz}\\ &\quad + \frac{du}{dy}\frac{dw}{dx} + \frac{dv}{dy}\frac{dw}{dy} + \frac{dw}{dy}\frac{dw}{dz}.\end{aligned}$$

Representing for shortness the Matrix

$$\left|\begin{matrix} \frac{du}{dx}, & \frac{du}{dy}, & \frac{du}{dz} \\ \frac{dv}{dx}, & \frac{dv}{dy}, & \frac{dv}{dz} \\ \frac{dw}{dx}, & \frac{dw}{dy}, & \frac{dw}{dz} \end{matrix}\right| \text{ by } \left|\begin{matrix} a\ , & b\ , & c \\ a', & b', & c' \\ a'', & b'', & c'' \end{matrix}\right|, \text{ and its square by } \left|\begin{matrix} A\ , & B\ , & C \\ A', & B', & C' \\ A'', & B'', & C'' \end{matrix}\right|,$$

we have

$$\left|\begin{matrix} A\ , & B\ , & C \\ A', & B', & C' \\ A'', & B'', & C'' \end{matrix}\right| = \begin{array}{c|ccc} & \left(\frac{du}{dx}, \frac{dv}{dx}, \frac{dw}{dx}\right), & \left(\frac{du}{dy}, \frac{dv}{dy}, \frac{dw}{dy}\right), & \left(\frac{du}{dz}, \frac{dv}{dz}, \frac{dw}{dz}\right) \\ \hline \frac{du}{dx}, \frac{du}{dy}, \frac{du}{dz} & ,, & ,, & ,, \\ \frac{dv}{dx}, \frac{dv}{dy}, \frac{dv}{dz} & ,, & ,, & ,, \\ \frac{dw}{dx}, \frac{dw}{dy}, \frac{dw}{dz} & ,, & ,, & ,, \end{array}$$

viz. the combinations which enter into the foregoing formula are

$$C' = \frac{dv}{dx}\frac{du}{dz} + \frac{dv}{dy}\frac{dv}{dz} + \frac{dv}{dz}\frac{dw}{dz},$$

and

$$B'' = \frac{dw}{dx}\frac{du}{dy} + \frac{dw}{dy}\frac{dv}{dy} + \frac{dw}{dz}\frac{dw}{dy},$$

and the equation thus is $D(c' - b'') + C' - B'' = 0$; viz. the three equations are

$$\begin{aligned} D(c' - b'') + C' - B'' &= 0,\\ D(a'' - c\ ) + A'' - C\ &= 0,\\ D(b\ - a') + B\ - A' &= 0,\end{aligned}$$

which are the equations in question.

Observe that we have

$$\begin{aligned} C' - B'' &= (a', b', c'')(c, c', c'') - (a'', b'', c'')(b, b', b'')\\ &= a'c' + b'c' + c'c'' \qquad - a''b - b'b'' - b''c'',\end{aligned}$$

and thence, writing

$$\begin{aligned}\rho &= a(c' - b'') + b(a'' - c) + c(b - a'),\\ &= ac' - ab'' + a''b - a'c,\end{aligned}$$

we have

$$C' - B'' + \rho = (a + b' + c'')(c' - b'') = 0,$$

if $a+b'+c''=0$; viz. this being so, $C'-B''=-\rho$, or the first equation is

$$D(c'-b'')=\rho, \;= a(c'-b'')+b(a''-c)+c(b-a'),$$

that is, $D\xi=\xi\frac{du}{dx}+\eta\frac{du}{dy}+\zeta\frac{du}{dz}$, the first equation of Helmholtz, and we thus have the equations of Helmholtz, if $a-b'+c''=0$, that is, if $\frac{du}{dx}+\frac{dv}{dy}+\frac{dw}{dz}=0$.

The foregoing three equations $D(c'-b'')+C'-B''=0$, &c., are the quaternion equation ($\sigma=iu+jv+kw$, $\nabla=i\frac{d}{dx}+j\frac{d}{dy}+k\frac{d}{dz}$, $\frac{d}{dt}=D$, denotes a complete differentiation),

$$\frac{d}{dt}V\nabla\sigma=V\nabla_1\sigma_2 S\sigma_1\nabla_2$$

of Mr M[c]Aulay's paper "Some General Theorems in Quaternion Integration," *Messenger of Mathematics*, vol. XIV. (1884), pp. 26—37; see p. 34.

891.

ON THE BINODAL QUARTIC AND THE GRAPHICAL REPRESENTATION OF THE ELLIPTIC FUNCTIONS.

[From the *Transactions of the Cambridge Philosophical Society*, vol. XIV. (1889), pp. 484—494. Read May 6, 1889.]

I APPROACH the subject from the question of the graphical representation of the elliptic functions: assuming as usual that the modulus is real, positive, and less than unity, and to fix the ideas considering the function sn (but the like considerations are applicable to the functions cn and dn), then the equation $x+iy = \text{sn}(x'+iy')$ establishes a (1, 1) correspondence between the xy infinite quarter plane, and the $x'y'$ rectangle (sides K and K'): viz. to any given point $x+iy$, x and y each positive, there corresponds a single point $x'+iy'$, x', y' each positive and less than K, K' respectively: and conversely to any such point $x'+iy'$, there corresponds a single point $x+iy$, x and y each positive.

I draw in the $x'y'$-figure the rectangle $A'B'C'D'$ (sides K and K'), and in the xy-figure, I take on the axis of x, the points B, C where $AB=1$, $AC=\frac{1}{k}$: and the

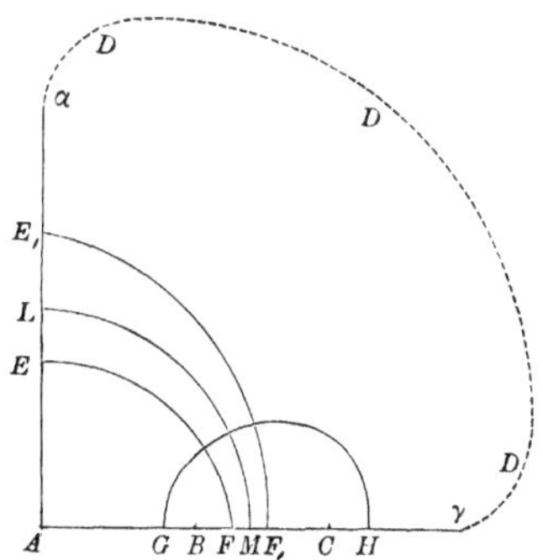

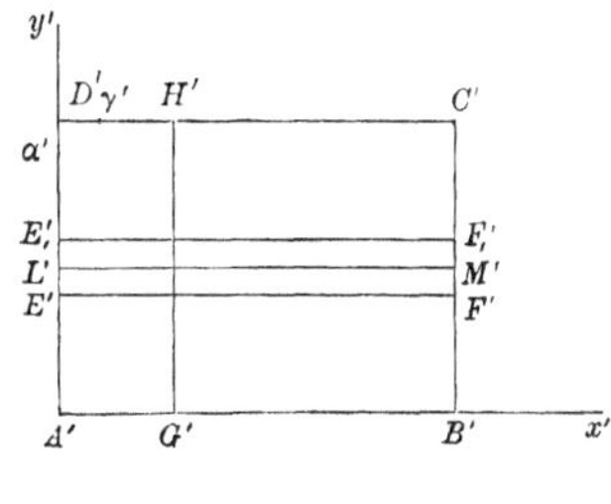

point D at infinity. We have thus in the $x'y'$-figure the closed curve or contour

$A'B'C'D'A'$: and corresponding hereto we have in the xy-figure the closed curve or contour $ABCDA$, viz. here D is the point at infinity considered as a line always at infinity, extending from the point at infinity on the positive part of the axis of x, to the point at infinity on the positive part of the axis of y, the contour being thus AB, BC, CD (D at infinity on the axis of x); and then D (at infinity on the axis of y) A. And thus to a point P' describing successively the lines $A'B'$, $B'C'$, $C'D'$, $D'A'$ there corresponds a point P describing successively the lines AB, BC, CD, DA: to P' at D' there corresponds P at D, viz. this is any point at infinity from D on the axis of x to D on the axis of y. There is no real breach of continuity: in further illustration, suppose that P', instead of actually coming to D', just cuts off the corner, viz. that it passes from a point γ' on $C'D'$ to a point α' on $D'A'$ (γ', α' each of them very near to D'): then P passes from a point γ very near D on the axis of x (that is, at a great distance from A) to a point α very near D on the axis of y (that is, at a great distance from A): and to the indefinitely small arc $\gamma'\alpha'$ described by P' there corresponds the indefinitely large arc $\gamma\alpha$ described by P.

We thus see that, if P' describe any arc $E'F'$ passing from a point E' of $A'D'$ to a point F' of $B'C'$, then P will describe an arc EF passing from a point E of AD to a point F of BC: and similarly, if P' describe any arc $G'H'$ passing from a point G' of $A'B'$ to a point H' of $C'D'$, then P will describe an arc GH passing from a point G of AB to a point H of CD.

Supposing $E'F'$ is a straight line parallel to $A'x'$, that is, cutting $A'D'$ and $B'C'$ each at right angles, then EF will be an arc cutting AD and BC each at right angles: and so if $G'H'$ is a straight line parallel to $A'y'$, that is, cutting $A'B'$ and $C'D'$ each at right angles, then GH will be an arc cutting AB and CD each at right angles: and moreover, since $E'F'$ and $G'H'$ cut each other at right angles, then also EF and GH cut each other at right angles.

Supposing, as above, that $E'F'$ and $G'H'$ are straight lines, we have EF and GH each of them the arc of a special bicircular quartic: the theory was in fact established in a very elegant manner in a memoir by Siebeck, "Ueber eine Gattung von Curven vierten Grades, welche mit den elliptischen Functionen zusammenhängen," *Crelle*, t. LVII. (1860), pp. 359—370, and t. LIX. (1861), pp. 173—184.

In particular, if P' describe the line $L'M'$ lying halfway between $A'B'$ and $D'C'$ (that is, if $A'L' = \frac{1}{2}K'$), then P will describe the circular quadrant LM, radius $\frac{1}{\sqrt{k}}$, viz. in this case the bicircular quartic degenerates into a circle twice repeated: and so if P' describe successively the lines $E'F'$ and $E_1'F_1'$ equidistant from $L'M'$ ($AE' = \frac{1}{2}K' - \eta$, $AE_1' = \frac{1}{2}K' + \eta$), then P will describe the arcs EF and E_1F_1 which are the images of each other in regard to the centre A and circular quadrant LM, and which together constitute the quadrant of one and the same bicircular quartic.

A bicircular quartic is of course a binodal quartic with the circular points at infinity for the two nodes: there is no real gain of generality in considering the binodal quartic rather than the bicircular quartic, but I have preferred to do so, and I have accordingly introduced the term Binodal Quartic into the title of the present Memoir. I present in a compendious form the properties of the general curve, and I show how the curve is to be particularised so as to obtain from it the special bicircular quartics which present themselves as above in the graphical representation of the elliptic functions.

A binodal quartic has the Plückerian numbers

$$\begin{array}{cccccc} m & n & \delta & \kappa & \tau & \iota \\ =4 & 8 & 2 & 0 & 8 & 12. \end{array}$$

The number of tangents to the curve which can be drawn from either of the nodes is $n-4$, $=4$; and the pencil of tangents from the one node is homographic with the pencil of tangents from the other node. Call the nodes I and J: and let the tangents from I be called (a, b, c, d) and those from J be called (a', b', c', d'), then if the tangents which correspond to (a', b', c', d') respectively are (a, b, c, d), they may also be taken to be (b, a, d, c), (c, d, a, b) or (d, c, b, a): and considering the intersections of corresponding tangents, we have thus four tetrads of points, say the f-points, such that the points of each tetrad lie in a conic through the two nodes: and we have consequently four conics each passing through the two nodes, say these are the f-conics.

Starting as above with the correspondence (a, b, c, d), (a', b', c', d'), if the intersections of a and a', b and b', c and c', d and d' are called A, B, C, D respectively, then we have A, B, C, D for a tetrad of f-points, lying on the f-conic $(ABCD)$.

Writing AB for the two points, the intersections of IA, JB and of IB, JA respectively, and so in other cases, then the remaining three tetrads of f-points are

AB, CD	lying on the f-conic	(AB, CD),
AC, BD	„ „	(AC, BD),
AD, BC	„ „	(AD, BC).

The two points AB may be spoken of as the antipoints of A, B: and so in other cases.

Any two of the f-conics have in common the nodes I, J, and they therefore intersect in two points besides: at each of these the tangents to the two conics, and the lines to I, J respectively, form a harmonic pencil.

Consider the two tangents at I and the two tangents at J: we have a conic touching these four lines and passing through the tetrad of f-points, or what is the same thing, intersecting the f-conic in the four f-points: say this is a c-conic. There are thus four c-conics corresponding to the four f-conics respectively.

We may consider a variable conic, passing through the points I and J; and such that the tangents thereto at these points respectively meet on a point of a

c-conic: the variable conic and the corresponding f-conic each pass through the points I, J and they meet in two points besides: and the variable conic may be such that at each of these the tangents to the variable conic and the f-conic form with the lines drawn to the points I and J a harmonic pencil. The variable conic, as thus defined, contains a single variable parameter; and it has for its envelope the binodal quartic: the binodal quartic is thus in four different ways the envelope of a variable conic. This is of course Casey's Theorem for the fourfold generation of the bicircular quartic as the envelope of a variable circle.

One of the f-conics, say $(ABCD)$, may break up into the line IJ and a line, say the axis $(ABCD)$: we have then the four f-points A, B, C, D on this line. The other f-conics say (AB, CD), (AC, BD), (AD, BC) are proper conics as before: any one of these meets the line $(ABCD)$ in two points: and at each of these the line and the tangent to the conic form with the lines drawn to the points I and J a harmonic pencil. The binodal quartic is in this case said to have an axis.

But a second f-conic, say (AD, BC), may break up into the line IJ and a line, say the axis (AD, BC): we have then the four f-points AD, BC on this line. Writing moreover A', D' for the two points AD, or say $(AI, DJ) = A'$, $(AJ, DI) = D'$; and similarly B', C' for the two points BC, or say $(BI, CJ) = B'$, $(BJ, CI) = C'$; then the four f-points are A', B', C', D', and the axis (AD, BC) may be called $(A'B'C'D')$.

The relation of the f-points A, B, C, D and A', B', C', D' and of the two axes is as shown in the figure: taking X for the intersection of IJ with $(ABCD)$ and Y for

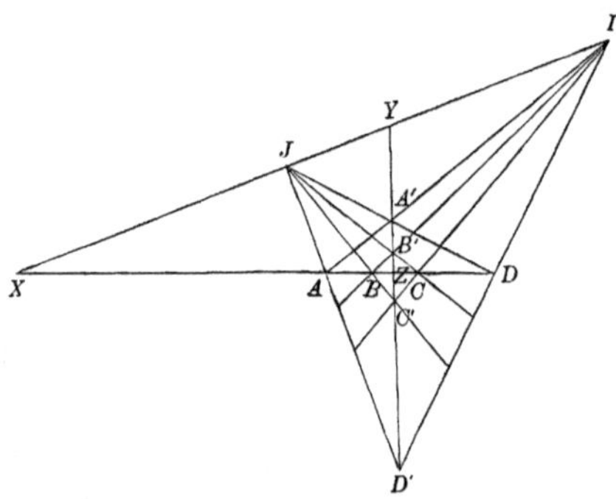

that of IJ with $(A'B'C'D')$, also Z for the intersection of the two axes, then X, Z are the sibiconjugate points of the involution AD, BC and Y, Z the sibiconjugate points of the involution $A'D'$, $B'C'$. The two axes intersect in Z, and form with the lines ZI, ZJ a pencil in involution.

The two remaining f-conics (AB, CD) and (AC, BD), or as they might also be called $(A'B', C'D')$ and $(A'C', B'D')$, are proper conics as before: they touch each other at the points I, J, and have for their common tangents at these points the lines ZI,

ZJ respectively. Each conic meets each axis in two points; and at each of these points the axis and the tangent to the conic form with the lines to I, J a harmonic pencil. The binodal quartic is in this case said to be biaxial. The point Z, which is the intersection of the two axes, may be called the centre.

In the case where an f-conic breaks up into the line IJ, and a line containing four f-points, say an f-line, the corresponding c-conic coincides with the f-conic, viz. it also breaks up into the line IJ and the f-line: the variable conic is a conic through the points I, J such that the tangents thereto at these points respectively meet on the f-line. Moreover, the variable conic must be such that at each of its intersections with the f-conic, that is, the f-line, the tangent to the variable conic and the f-line must be harmonics in regard to the lines drawn from the point to the points I, J respectively: but this condition is satisfied *ipso facto* for each of the intersections of the variable conic and the f-line. This depends on the theorem that, taking on a conic any three points P, I, J, then the tangent at P and the line drawn from P to the pole of IJ are harmonics in regard to the lines PI, PJ. Thus we have only three conditions for the variable conic, or, as above defined, it would in the case in question (of four f-points in a line) depend upon two variable parameters. There is really another condition—but what this in general is I have not ascertained: and this being so the variable conic in the case in question (of the four f-points in a line) depends upon a single variable parameter, and we have as before the bicircular quartic as the envelope. The foregoing is the axial case; in the biaxial case, the same thing happens in regard to the variable conics belonging to the two axes respectively. Thus in every case we have the fourfold generation of the curve as the envelope of a variable conic: only in the axial case, the variable conics belonging to the axis, and in the biaxial case the variable conics belonging to the two axes respectively, are not by the foregoing definitions completely defined. It will be seen further on how, in the case of the biaxial bicircular quartic, we complete the definition of the variable circles belonging to the two axes respectively.

Taking the points I, J to be the circular points at infinity, we have a bicircular quartic. The f-points are the foci, and the f-conics are circles, viz. we have 16 foci situate in fours upon four focal circles. The harmonic relation of two lines to the lines through I, J means of course that the lines cut at right angles; hence the focal circles cut each other at right angles: this must certainly be a known property, but it is not mentioned in Salmon's *Higher Plane Curves*, Ed. 3, Dublin, 1879, and I cannot find it in Darboux or Casey: it is given No. 81 in Lachlan's Memoir "On Systems of Circles and Spheres," *Phil. Trans.*, vol. CLXXVII. (1886), and I find it as a question in the *Educational Times*, March 1, 1889, 10034 (Prof. Morley). "Prove that, of the four focal circles of a circular cubic or bicircular quartic, any two are orthogonal, and the radii are connected by the relation $\Sigma(\mu^{-2}) = 0$." The theorem is not as well known as it should be.

The c-conics are confocal conics having for their real foci the so-called double-foci of the quartic (more accurately, the common foci are the four quadruple foci of the quartic); we have thus four conics corresponding to the four focal circles respectively, each conic intersecting the corresponding circle in the four foci upon this circle. And we have then the quartic as the envelope of a variable circle having its centre upon one of these conics and cutting at right angles the corresponding focal circle: the bicircular quartic is thus generated in four different ways.

Instead of one of the focal circles, we may have a line or axis, and the quartic is then said to be axial: the foci on the axis may be any four points; and for a real curve they may be all real, or two real and two imaginary, or all four imaginary. The remaining focal circles are real or imaginary circles, cut by the axis at right angles, that is, having their centres on the axis, and cutting each other at right angles.

But instead of another of the focal circles, we may have a line or axis, and the quartic is then said to be biaxial: the two axes cut at right angles at a point which may be called the centre of the curve. The foci on each axis form pairs of points situate symmetrically in regard to the centre. If on one of the axes the foci are real, then on the other axis they form two imaginary conjugate pairs; and conversely: but if on one of the axes the foci are two of them real and the other two conjugate imaginaries, then this is so for the other axis also. There are thus only the two cases: 1°, foci on the one axis real, and on the other conjugate imaginaries; 2°, foci on each axis two of them real and the other two conjugate imaginaries: there is however a limiting case where on each axis two foci are united at the centre, the other two foci being real on the one axis and conjugate imaginaries on the other. The remaining two focal circles are real or imaginary circles, cutting each axis at right angles, that is, having their centres at the centre; and cutting each other at right angles, that is, having the sum of the squares of their radii $=0$.

The biaxial form of bicircular quartic is, in fact, that which presents itself in the theory of the representation of the elliptic functions.

I consider for a moment the case of a variable circle having its centre upon a given line, and cutting at right angles a given circle. The variable circles pass all of them through two fixed points, the antipoints of the intersections of the given line and circle, and which are thus real or imaginary according as the intersections of the given line and circle are imaginary or real. Hence, considering any one variable circle and the consecutive variable circle, these intersect in two real points, when the given line does not meet the given circle (meets it in two imaginary points); but when the given line meets the given circle in two real points, then the two variable circles intersect in two imaginary points: if however the given line touches the given circle, then the two variable circles touch each other. Taking now the curve of centres to be any given curve whatever, and considering one of the variable circles, and the consecutive variable circle, it at once appears that, if the tangent to the curve of centres at the centre of the variable circle does not meet the given circle, then the two variable circles intersect in two real points (which, if the tangent touch the given circle, unite in a single real point): but if the tangent to the curve of centres meets the given circle, then the two variable circles do not intersect. It hence appears that the real portions of the envelope arise exclusively from those portions of the curve of centres which are such that at any point thereof the tangent to the curve of centres does not meet the given circle. In particular, if the given circle be real, and the curve of centres is a real ellipse enclosing the given circle, then the real portion of the envelope arises from the whole ellipse: but if the curve of centres be a real ellipse cutting the given circle in four real points, then drawing the four common tangents of the ellipse and circle, it is at once seen that there are on the ellipse two detached portions such that, at any point of either portion, the tangent to the ellipse does not meet the circle: and the real portions of the envelope arise exclusively from these portions of the ellipse.

In the case just referred to, there are on the ellipse four portions each lying outside the circle and terminating in the four intersections respectively of the ellipse and circle, such that at a point of any one of these portions the tangent to the ellipse meets the circle in two real points. Starting from the extremity of one of these portions of the ellipse and proceeding to the other extremity on the circle, the corresponding variable circles do not intersect each other, but each of them is a circle lying wholly inside that which immediately precedes it; and the variable circle becomes ultimately a point, viz. this point is a focus of the curve: this agrees with the foregoing statement that the f-conic intersects the circle in the four foci upon this circle. For the two portions of the ellipse which lie inside the circle, the variable circle is of course always imaginary. The like considerations apply to the case where the locus of the centre of the variable circle is a hyperbola or parabola. The foregoing remarks illustrate the actual generation of a bicircular quartic as the envelope of the variable circle.

Starting now from the equation

$$x + iy = \text{sn}\,(x' + iy'),$$

we have

$$x = \frac{\text{sn}\,x'\,\text{cn}\,iy'\,\text{dn}\,iy'}{1 - k^2\,\text{sn}^2 x'\,\text{sn}^2 iy'}, \quad iy = \frac{\text{sn}\,iy'\,\text{cn}\,x'\,\text{dn}\,x'}{1 - k^2\,\text{sn}^2 x'\,\text{sn}^2 iy'},$$

or putting

$$\text{sn}\,x' = p, \quad \text{sn}\,iy' = iq\,;$$

these equations are

$$x = \frac{p\sqrt{1+q^2\,.\,1+k^2q^2}}{1+k^2p^2q^2}, \quad y = \frac{q\sqrt{1-p^2\,.\,1-k^2p^2}}{1+k^2p^2q^2}\,;$$

whence also

$$x^2 + y^2 = \frac{p^2+q^2}{1+k^2p^2q^2} = r^2, \quad (\text{if } x^2+y^2 \text{ be put } = r^2).$$

These equations, considering therein q as a given constant, and p as a variable parameter, determine the curve EF: and considering p as a given constant, and q as a variable parameter, they determine the curve GH. But the eliminations are easily effected; we have

$$p^2(1 - k^2q^2r^2) = r^2 - q^2, \quad q^2(1 - k^2p^2r^2) = r^2 - p^2.$$

Hence, for EF,

$$p^2 = \frac{r^2 - q^2}{1 - k^2r^2q^2}, \qquad 1 - p^2 = \frac{1 + q^2 - (1 + k^2q^2)\, r^2}{1 - k^2r^2q^2},$$

$$1 - k^2p^2 = \frac{1 + k^2q^2 - k^2(1 + q^2)\, r^2}{1 - k^2r^2q^2}, \quad 1 + k^2p^2q^2 = \frac{1 - k^2q^4}{1 - k^2r^2q^2},$$

and consequently

$$x = \frac{\sqrt{1 + q^2 \,.\, 1 + k^2q^2}}{1 - k^2q^4} \sqrt{r^2 - q^2 \,.\, 1 - k^2r^2q^2},$$

$$y = \frac{q\sqrt{1 + q^2 - (1 + k^2q^2)\, r^2}\, \sqrt{1 + k^2q^2 - k^2(1 + q^2)\, r^2}}{1 - k^2q^4},$$

giving x and y each of them in terms of r. And from the first of these we at once derive

$$(x^2 + y^2)^2 - 2Ax^2 - 2By^2 + \frac{1}{k^2} = 0,$$

where

$$2A = \frac{1 + q^2}{1 + k^2q^2} + \frac{1}{k^2}\, \frac{1 + k^2q^2}{1 + q^2}, \qquad 2B = q^2 + \frac{1}{k^2q^2}.$$

Similarly, for GH,

$$q^2 = \frac{r^2 - p^2}{1 - k^2r^2p^2}, \qquad 1 + q^2 = \frac{(1 - p^2) + r^2(1 - k^2p^2)}{1 - k^2r^2p^2},$$

$$1 + k^2q^2 = \frac{1 - k^2p^2 + k^2(1 - p^2)\, r^2}{1 - k^2r^2p^2}, \quad 1 + k^2p^2q^2 = \frac{1 - k^2p^4}{1 - k^2r^2p^2};$$

and consequently

$$x = \frac{p\sqrt{(1 - p^2) + (1 - k^2p^2)\, r^2}\, \sqrt{1 - k^2p^2 + k^2(1 - p^2)\, r^2}}{1 - k^2p^4},$$

$$y = \frac{\sqrt{1 - p^2 \,.\, 1 - k^2p^2}}{1 - k^2p^4} \sqrt{r^2 - p^2 \,.\, 1 - k^2r^2p^2},$$

giving x, y each of them in terms of r. And from the second of them we at once derive

$$(x^2 + y^2)^2 - 2Ax^2 - 2By^2 + \frac{1}{k^2} = 0,$$

where

$$2A = p^2 + \frac{1}{k^2p^2}, \qquad 2B = -\frac{1 - p^2}{1 - k^2p^2} - \frac{1}{k^2}\, \frac{1 - k^2p^2}{1 - p^2}.$$

Consider, in particular, the case where the line EF is the midway line LM: here $y' = \tfrac{1}{2}K'$, and thence $iq = \operatorname{sn} iy' = \operatorname{sn} \tfrac{1}{2}iK' = \dfrac{i}{\sqrt{k}}$, that is, $q = \dfrac{1}{\sqrt{k}}$: and we thence obtain $A = B = \dfrac{1}{k}$; the equation of the bicircular quartic is

$$(x^2 + y^2)^2 - \frac{2}{k}(x^2 + y^2) + \frac{1}{k^2} = 0,$$

viz. this is the circle $x^2+y^2-\frac{1}{k}=0$ twice repeated. As a direct verification, observe that we have here

$$x+iy=\operatorname{sn}(x'+\tfrac{1}{2}iK')=\frac{\frac{1+k}{\sqrt{k}}\operatorname{sn}x'+\frac{i}{\sqrt{k}}\operatorname{cn}x'\operatorname{dn}x'}{1+k^2.\frac{1}{k}\operatorname{sn}^2x'}=\frac{(1+k)\,p+i\sqrt{1-p^2\,.\,1-k^2p^2}}{\sqrt{k}\,(1+kp^2)},$$

and hence

$$x^2+y^2=\frac{(1+k)^2p^2+1-(1+k^2)\,p^2+k^2p^4}{k\,(1+2kp^2+k^2p^4)}=\frac{1}{k},$$

as it should be.

Reverting to the equation

$$x+iy=\operatorname{sn}(x'+iy'),$$

I write successively

$$y'=\tfrac{1}{2}K'-z',\quad \operatorname{sn}iy'=iq_1=\operatorname{sn}i\,(\tfrac{1}{2}K'-z'),$$

and

$$y'=\tfrac{1}{2}K'+z',\quad \operatorname{sn}iy'=iq_2=\operatorname{sn}i\,(\tfrac{1}{2}K'+z');$$

we then have

$$iq_1\,.\,iq_2=\operatorname{sn}i\,(\tfrac{1}{2}K'-z')\operatorname{sn}i\,(\tfrac{1}{2}K'+z'),\quad =-\frac{1}{k},$$

that is, $q_1q_2=\frac{1}{k}$; hence for q writing q_1 or q_2, we have in each case the same values of A and B; that is, we have the same bicircular quartic corresponding to the lines $E'F'$ and $E_1'F_1'$, equidistant from the line $L'M'$: but to one of these lines there corresponds the quadrant lying inside, to the other that lying outside, the circular quadrant

$$x^2+y^2-\frac{1}{k}=0.$$

The curve

$$(x^2+y^2)^2-2Ax^2-2By^2+\frac{1}{k^2}=0$$

is in four different ways the envelope of a variable circle: viz. the circle may have its centre on a conic $\alpha x^2+\beta y^2-1=0$, and cut at right angles one of the circles

$$x^2+y^2\pm\frac{1}{k}=0;$$

or it may have its centre on the axis of x, or on the axis of y. The circle, having its centre on either axis, cuts this axis at right angles; but this condition being *ipso facto* satisfied, we do not thereby determine the radius of the circle having for its centre a given point on the axis: the expression for the radius must be sought for independently.

Write for shortness

$$\lambda^2=\frac{B-\frac{1}{k^2}}{A-B},$$

and consider the circle

$$x^2 + y^2 - 2\alpha x = \lambda \sqrt{A - B - 2\alpha^2} + B,$$

where α is a variable parameter. Differentiating, we have

$$x = \frac{\lambda\alpha}{\sqrt{A - B - 2\alpha^2}}, \text{ giving } \alpha = \frac{x\sqrt{A-B}}{\sqrt{2x^2+\lambda^2}}:$$

the equation of the circle then gives

$$x^2 + y^2 - B = \frac{\alpha}{x}(2x^2 + \lambda^2), \ = \sqrt{A-B}\sqrt{2x^2+\lambda^2};$$

that is,

$$(x^2+y^2)^2 - 2Ax^2 - 2By^2 = B^2 + \lambda^2(A - B), \ = -\frac{1}{k^2};$$

or we have

$$(x^2+y^2)^2 - 2Ax^2 - 2By^2 + \frac{1}{k^2} = 0$$

as the envelope of the variable circle

$$x^2 + y^2 - 2\alpha x = \frac{\sqrt{B - \frac{1}{k^2}}}{\sqrt{A-B}}\sqrt{A - B - 2\alpha^2} + B,$$

having its centre on the axis of x.

And similarly, the same quartic curve is the envelope of the variable circle

$$x^2 + y^2 - 2\beta y = \frac{\sqrt{A - \frac{1}{k^2}}}{\sqrt{B-A}}\sqrt{B - A - 2\beta^2} + A,$$

having its centre on the axis of y.

If we have in like manner the equation $x + iy = \text{cn}(x' + iy')$, then writing, as before,

$$\text{sn}\, x' = p, \quad \text{sn}\, iy' = iq,$$

we find

$$x^2 = \frac{(1-p^2)(1+q^2)}{(1+k^2p^2q^2)^2}, \quad y^2 = \frac{p^2q^2(1-k^2p^2)(1+k^2q^2)}{(1+k^2p^2q^2)^2},$$

and hence

$$r^2 = \frac{1 - p^2 + q^2 - k^2p^2q^2}{1 + k^2p^2q^2}, \text{ where } r^2 = x^2 + y^2.$$

For the curve EF corresponding to a line $E'F'$ parallel to the axis of x', we have to eliminate p from these equations; the expression for r^2 gives

$$p^2 = \frac{1 + q^2 - r^2}{1 + k^2q^2 + k^2q^2r^2}, \qquad 1 - p^2 = \frac{-(1-k^2)q^2 + r^2(1+k^2q^2)}{1 + k^2q^2 + k^2q^2r^2},$$

$$1 - k^2p^2 = \frac{(1-k^2) + k^2r^2(1+q^2)}{1 + k^2q^2 + k^2q^2r^2}, \qquad 1 + k^2p^2q^2 = \frac{1 + 2k^2q^2 + k^2q^4}{1 + k^2q^2 + k^2q^2r^2},$$

and thence

$$x = \frac{\sqrt{1+q^2}\sqrt{-(1-k^2)q^2 + r^2(1+k^2q^2)}\sqrt{1+k^2q^2+k^2q^2r^2}}{1+2k^2q^2+k^2q^4},$$

$$y = \frac{q\sqrt{1+q^2-r^2}\sqrt{1+k^2q^2}\sqrt{1-k^2+k^2r^2(1+q^2)}}{1+2k^2q^2+k^2q^4},$$

from which we deduce

$$(x^2+y^2)^2 - 2Ax^2 - 2By^2 - \frac{k'^2}{k^2} = 0,$$

where

$$2A = \frac{-1+2k^2+2k^2q^2+k^2q^4}{k^2(1+q^2)}, \quad 2B = \frac{-1-2k^2q^2+k^2(1-2k^2)q^4}{k^2q^2(1+k^2q^2)},$$

which is a bicircular quartic of the foregoing form.

Similarly, for the curve GH corresponding to a line $G'H'$ parallel to the axis of y', we have to eliminate q: the expression for r^2 gives

$$q^2 = \frac{-1+p^2+r^2}{1-k^2p^2-k^2p^2r^2}, \qquad 1+q^2 = \frac{(1-k^2)p^2+r^2(1-k^2p^2)}{1-k^2p^2-k^2p^2r^2},$$

$$1+k^2q^2 = \frac{(1-k^2)+k^2r^2(1-p^2)}{1-k^2p^2-k^2p^2r^2}, \quad 1+k^2p^2q^2 = \frac{1-2k^2p^2+k^2p^4}{1-k^2p^2-k^2p^2r^2}.$$

Hence

$$x = \frac{\sqrt{1-p^2}\sqrt{(1-k^2)p^2+r^2(1-k^2p^2)}\sqrt{1-k^2p^2-k^2p^2r^2}}{1-2k^2p^2+k^2p^4},$$

$$y = \frac{p\sqrt{-1+p^2+r^2}\sqrt{1-k^2p^2}\sqrt{1-k^2+k^2r^2(1-p^2)}}{1-2k^2p^2+k^2p^4},$$

from which we deduce

$$(x^2+y^2)^2 - 2Ax^2 - 2By^2 - \frac{k'^2}{k^2} = 0,$$

where

$$2A = \frac{-1+2k^2-2k^2p^2+k^2p^4}{k^2(1-p^2)}, \quad 2B = \frac{1-2k^2p^2-k^2(1-2k^2)p^4}{k^2p^2(1-k^2p^2)},$$

which is again a bicircular quartic of the foregoing form. And we have the like results for the equation

$$x+iy = \text{dn}(x'+iy');$$

so that, for the sn, cn, and dn, the curve EF or GH is in each case a biaxial bicircular quartic of the form

$$(x^2+y^2)^2 - 2Ax^2 - 2By^2 + C = 0.$$

892.

NOTE ON THE ORTHOMORPHIC TRANSFORMATION OF A CIRCLE INTO ITSELF.

[From the *Proceedings of the Edinburgh Mathematical Society*, vol. VIII. (1890), pp. 91, 92.]

THE following is, of course, substantially well known, but it strikes me as rather pretty:—to find the orthomorphic transformation of the circle

$$x^2+y^2-1=0$$

into itself. Assume this to be

$$x_1+iy_1=\frac{A(x+iy)+B}{1+C(x+iy)}.$$

Then, writing A', B', C' for the conjugates of A, B, C, we have

$$x_1-iy_1=\frac{A'(x-iy)+B'}{1+C'(x-iy)};$$

and then

$$x_1^2+y_1^2=\frac{AA'(x^2+y^2)+AB'(x+iy)+A'B(x-iy)+BB'}{1+C(x+iy)+C'(x-iy)+CC'(x^2+y^2)},$$

which should be an identity for $x^2+y^2=1$, $x_1^2+y_1^2=1$.

Evidently $C=AB'$, whence $C'=A'B$; and the equation then is

$$1+AA'BB'=AA'+BB',$$

that is,

$$(1-AA')(1-BB')=0.$$

But $BB'=1$ gives the illusory result

$$x_1+iy_1=B,$$

therefore

$$1-AA'=0;$$

and the required solution thus is

$$x_1+iy_1=\frac{A(x+iy)+B}{1+AB'(x+iy)};$$

where A is a unit-vector (say $A=\cos\lambda+i\sin\lambda$) and B, B' are conjugate vectors. Or, writing $B=b+i\beta$, $B'=b-i\beta$, the constants are λ, b, β; three constants as it should be.

893.

THE BITANGENTS OF THE QUINTIC.

(LETTER FROM PROFESSOR CAYLEY TO MR HEAL.)

[From the *Annals of Mathematics*, vol. v. (1890), pp. 109, 110.]

DEAR SIR,

I have to thank you very much for your letter concerning the bitangents of the quintic, and am glad you have obtained more simple numerical coefficients than you had at first. I do not see my way to verifying your result, but assuming it to be correct, it is a very interesting and valuable one. The relation to Salmon's *Higher Plane Curves* (p. 351) is rather curious, viz. the result there given is of the deg-order 18-54, with an extraneous factor $(\alpha x+\beta y+\gamma z)^6$, the rejection of which would reduce it to the proper form, deg-order 18-48; whereas yours is of deg-order 24-66, with an extraneous factor H^2, the rejection of which would reduce it to the same proper form, deg-order 18-48.

But there is another known solution, not by any means a handy one, but which has no extraneous factor*, viz. if

$$\Omega = D^3H - 4D^3H_1 + 6D^3H_2,$$

then the required equation is

$$F\Omega = 0,$$

where the facients of the reciprocant are the first derived functions,

$$\partial_x U,\quad \partial_y U,\quad \partial_z U.$$

Say this is

$$(B^2C^2 - \ldots)(\partial_x U)^6 = 0\,;$$

the coefficients A, B, C, &c., are those of Ω, viz. these are of deg-order 3-6, and $\partial_x U$, &c., are of deg-order 1-4; so that the deg-order is

$$4\,(3\text{-}6) + 6\,(1\text{-}4) = (12\text{-}24) + (6\text{-}24) = (18\text{-}48),$$

as it should be. It might be worth while to further consider this form, but I doubt whether, practically, the whole question is not too difficult to be worth working at.

I remain, dear sir, yours very sincerely,

A. CAYLEY.

Cambridge, January 17, 1890.

* See *Phil. Trans.*, vol. CXLIX. pp. 193—212, [260].

894.

THE INVESTIGATION BY WALLIS OF HIS EXPRESSION FOR π.

[From the *Quarterly Journal of Pure and Applied Mathematics,* vol. XXIII. (1889), pp. 165—169.]

THE following is in effect the investigation by Wallis in the *Arithmetica Infinitorum* (Oxford, 1656) of his well-known expression for π. He obtains the series of equations which in modern notation are

$$\int_0^1 (x-x^2)^0\,dx = 1,$$

$$\int_0^1 (x-x^2)^1\,dx = \tfrac{1}{2}\cdot\tfrac{1}{3},$$

$$\int_0^1 (x-x^2)^2\,dx = \tfrac{1}{6}\cdot\tfrac{1}{5},$$

$$\int_0^1 (x-x^2)^3\,dx = \tfrac{1}{20}\cdot\tfrac{1}{7},$$

or say, in general,

$$\int_0^1 (x-x^2)^n\,dx = \frac{1}{\phi(n)}\,\frac{1}{2n+1}.$$

In the case $n=\frac{1}{2}$, the integral gives the area of a semicircle, diameter $=1$, viz. this is $=\frac{1}{8}\pi$, or we have $\frac{1}{8}\pi=\dfrac{1}{2\phi(\frac{1}{2})}$; or writing with him $\square=\dfrac{4}{\pi}$, this is $\square=\phi(\frac{1}{2})$, where $\phi(0)$, $\phi(1)$, $\phi(2)$, ... are the series of numbers 1, 2, 6, 20, ..., which are in fact the middle binomial coefficients of the several even powers, and are the diagonal numbers in the table which he subsequently considers. As regards the form in which Wallis exhibits these results, his theorem CXXXIII. is:

Si exponatur series Primanorum mulctata serie secundanorum, Residuorum Quadrata, Cubi, Biquadrata, &c. ad seriem Æqualium rationem habebunt cognitam.

And then, after an algebraical calculation,

$$\frac{1}{2}-\frac{1}{3}=\frac{1}{6} \quad \Bigg| \quad \frac{1}{3}-\frac{2}{4}+\frac{1}{5}=\frac{1}{30} \quad \Bigg| \quad \frac{1}{4}-\frac{3}{5}+\frac{3}{6}-\frac{1}{7}=\frac{1}{140}$$

$$\frac{1}{2\,.\,3}=\frac{1}{6} \quad \Bigg| \quad \frac{1\,.\,2}{3\,.\,4\,.\,5}=\frac{1}{30} \quad \Bigg| \quad \frac{1\,.\,2\,.\,3}{4\,.\,5\,.\,6\,.\,7}=\frac{1}{140}$$

$$\frac{1}{2\,.\,3}=\frac{1}{6} \quad \Bigg| \quad \frac{1}{2\,.\,3}\cdot\frac{4}{4\,.\,5}=\frac{1}{30} \quad \Bigg| \quad \frac{1}{2\,.\,3}\cdot\frac{4}{4\,.\,5}\cdot\frac{9}{6\,.\,7}=\frac{1}{140}$$

...et sic deinceps; continue multiplicando Numeratores per numeros quadratos et denominatores per binos continue sequentes arithmetice proportionales—viz. he thus explains the law of his calculated numbers.

He then forms the table

	1		$\frac{1}{2}$		$\frac{3}{8}$		$\frac{15}{48}$		$\frac{105}{384}$	$-\frac{1}{2}$
1	1	1	1	1	1	1	1	1	1	0
	1		$\frac{3}{2}$		$\frac{15}{8}$		$\frac{105}{48}$		$\frac{945}{384}$	$\frac{1}{2}$
$\frac{1}{2}$	1	$\frac{3}{2}$	2	$\frac{5}{2}$	3	$\frac{7}{2}$	4	$\frac{9}{2}$	5	1
	1		$\frac{5}{2}$		$\frac{35}{8}$		$\frac{315}{48}$		$\frac{3465}{384}$	$\frac{3}{2}$
$\frac{3}{8}$	1	$\frac{15}{8}$	3	$\frac{35}{8}$	6	$\frac{63}{8}$	10	$\frac{99}{8}$	15	2
	1		$\frac{7}{2}$		$\frac{63}{8}$		$\frac{693}{48}$		$\frac{9009}{384}$	$\frac{5}{2}$
$\frac{15}{48}$	1	$\frac{105}{48}$	4	$\frac{315}{48}$	10	$\frac{693}{48}$	20	$\frac{1287}{48}$	35	3
	1		$\frac{9}{2}$		$\frac{99}{8}$		$\frac{1287}{48}$		$\frac{19305}{384}$	$\frac{7}{2}$
$\frac{105}{384}$	1	$\frac{945}{384}$	5	$\frac{3465}{384}$	15	$\frac{9009}{384}$	35	$\frac{19305}{384}$	70	4
$-\frac{1}{2}$	0	$\frac{1}{2}$	1	$\frac{3}{2}$	2	$\frac{5}{2}$	3	$\frac{7}{2}$	4	

where the figures in the second column are $=1$, those in the fourth column are $=\frac{n+1}{1}$, those of the sixth column are $=\frac{(n+1)(n+2)}{1\,.\,2}$, and so on: n being in each case the rank in the column as shown by the right-hand outside column, viz. these ranks are $-\frac{1}{2}$, 0, $\frac{1}{2}$, 1, ... at intervals of $\frac{1}{2}$. And the several even lines contain the same figures as the several even columns respectively.

Wallis then remarks that, if in any line the first and second terms respectively are A and 1, then the even lines are as follows, viz. the second line is

$$A,\ 1,\ A\frac{1}{1},\ 1.\frac{2}{2},\ A\frac{1\,.\,3}{1\,.\,3},\ 1.\frac{2\,.\,4}{2\,.\,4},\ A\frac{1\,.\,3\,.\,5}{1\,.\,3\,.\,5},\ 1.\frac{2\,.\,4\,.\,6}{2\,.\,4\,.\,6},\ \ldots,$$

the fourth line is

$$A,\ 1,\ A\frac{3}{1},\ 1.\frac{4}{2},\ A\frac{3\,.\,5}{1\,.\,3},\ 1.\frac{4\,.\,6}{2\,.\,4},\ A\frac{3\,.\,5\,.\,7}{1\,.\,3\,.\,5},\ 1.\frac{4\,.\,6\,.\,8}{2\,.\,4\,.\,6},\ \ldots,$$

the sixth line is

$$A,\ 1,\ A\frac{5}{1},\ 1.\frac{6}{2},\ A\frac{5.7}{1.3},\ 1.\frac{6.8}{2.4},\ A\frac{5.7.9}{1.3.5},\ 1.\frac{6.8.10}{2.4.6},\ \ldots;$$

and this being so, he completes the table by inserting a term $\square$ in the diagonal line between the 1 and 2, and calculating the odd lines as follows: viz. if in each case A is the first term and B the second term of the line, then the complete rule is:

the first line is

$$A,\ 1,\ A\frac{0}{1},\ 1.\frac{1}{2},\ A\frac{0.2}{1.3},\ 1.\frac{1.3}{2.4},\ A\frac{0.2.4}{1.3.5},\ 1.\frac{1.3.5}{2.4.6},$$

the second line is

$$A,\ 1,\ A\frac{1}{1},\ 1.\frac{2}{2},\ A\frac{1.3}{1.3},\ 1.\frac{2.4}{2.4},\ A\frac{1.3.5}{1.3.5},\ 1.\frac{2.4.6}{2.4.6},$$

the third line is

$$A,\ 1,\ A\frac{2}{1},\ 1.\frac{3}{2},\ A\frac{2.4}{1.3},\ 1.\frac{3.5}{2.4},\ A\frac{2.4.6}{1.3.5},\ 1.\frac{3.5.7}{2.4.6},$$

the fourth line is

$$A,\ 1,\ A\frac{3}{1},\ 1.\frac{4}{2},\ A\frac{3.5}{1.3},\ 1.\frac{4.6}{2.4},\ A\frac{3.5.7}{1.3.5},\ 1.\frac{4.6.8}{2.4.6},$$

the fifth line is

$$A,\ 1,\ A\frac{4}{1},\ 1.\frac{5}{2},\ A\frac{4.6}{1.3},\ 1.\frac{5.7}{2.4},\ A\frac{4.6.8}{1.3.5},\ 1.\frac{5.7.9}{2.4.6};$$

and so on, the rule for the odd lines being thus an interpolation from that for the even lines. Observe that, in the first line A is $=\infty$, and that, in the several odd terms after the first, $A\frac{0}{1}$ is $=1$.

The table thus calculated (I have given here the fractions in their least terms) is

∞	1	$\frac{1}{2}\square$	$\frac{1}{2}$	$\frac{1}{3}\square$	$\frac{3}{8}$	$\frac{4}{15}\square$	$\frac{5}{16}$	$\frac{8}{35}\square$	$\frac{35}{128}$	$-\frac{1}{2}$
1	1	1	1	1	1	1	1	1	1	0
$\frac{1}{2}\square$	1	$\square$	$\frac{3}{2}$	$\frac{4}{3}\square$	$\frac{15}{8}$	$\frac{8}{5}\square$	$\frac{35}{16}$	$\frac{64}{35}\square$	$\frac{315}{128}$	$\frac{1}{2}$
$\frac{1}{2}$	1	$\frac{3}{2}$	2	$\frac{5}{2}$	3	$\frac{7}{2}$	4	$\frac{9}{2}$	5	1
$\frac{1}{3}\square$	1	$\frac{4}{3}\square$	$\frac{5}{2}$	$\frac{8}{3}\square$	$\frac{35}{8}$	$\frac{64}{15}\square$	$\frac{105}{16}$	$\frac{128}{21}\square$	$\frac{1155}{128}$	$\frac{3}{2}$
$\frac{3}{8}$	1	$\frac{15}{8}$	3	$\frac{35}{8}$	6	$\frac{63}{8}$	10	$\frac{99}{8}$	15	2
$\frac{4}{15}\square$	1	$\frac{8}{5}\square$	$\frac{7}{2}$	$\frac{64}{15}\square$	$\frac{63}{8}$	$\frac{128}{15}\square$	$\frac{231}{16}$	$\frac{512}{35}\square$	$\frac{3003}{128}$	$\frac{5}{2}$
$\frac{5}{16}$	1	$\frac{35}{16}$	4	$\frac{105}{16}$	10	$\frac{231}{16}$	20	$\frac{429}{16}$	35	3
$\frac{8}{35}\square$	1	$\frac{64}{35}\square$	$\frac{9}{2}$	$\frac{128}{21}\square$	$\frac{99}{8}$	$\frac{512}{35}\square$	$\frac{429}{16}$	$\frac{1024}{35}\square$	$\frac{6435}{128}$	$\frac{7}{2}$
$\frac{35}{128}$	1	$\frac{315}{128}$	5	$\frac{1155}{128}$	15	$\frac{3003}{128}$	35	$\frac{6435}{128}$	70	4
$-\frac{1}{2}$	0	$\frac{1}{2}$	1	$\frac{3}{2}$	2	$\frac{5}{2}$	3	$\frac{7}{2}$	4	

in which table □, as the term interpolated between the diagonal terms 1, 2, denotes the value $\frac{4}{\pi}$ as before.

The third line of the table is

$$\frac{1}{2}\square,\quad 1,\quad \square,\quad \frac{3}{2},\quad \frac{4}{3}\square,\quad \frac{3.5}{2.4},\quad \frac{4.6}{3.5}\square,\quad \frac{3.5.7}{2.4.6},\quad \frac{4.6.8}{3.5.7}\square,\quad \frac{3.5.7.9}{2.4.6.8},\ \ldots.$$

The successive even terms continually increase but tend to equality, and in like manner the successive odd terms continually increase but tend to equality; it seems to have been assumed that, for any three consecutive terms x, y, z, we have $\frac{y}{x} > \frac{z}{y}$, that is, $y^2 > xz$. Taking this to be so, we have

$$1 > \frac{1}{2}\square^2,\quad \square^2 > \frac{3}{2},\quad \frac{3^2}{2^2} > \frac{4}{3}\square^2,\quad \frac{4^2}{3^2}\square^2 > \frac{3^2.5}{2^2.4},\quad \frac{3^2.5^2}{2^2.4^2} > \frac{4^2.6}{3^2.5}\square^2,\ \ldots,$$

and these equations give □

less than	greater than
$\sqrt{\frac{2}{1}}$,	$\sqrt{\frac{3}{2}}$,
$\frac{3^2}{2.4}\sqrt{\frac{4}{3}}$,	$\frac{3^2}{2.4}\sqrt{\frac{5}{4}}$,
$\frac{3^2.5^2}{2.4^2.6}\sqrt{\frac{6}{5}}$,	$\frac{3^2.5^2}{2.4^2.6}\sqrt{\frac{7}{6}}$,
$\frac{3^2.5^2.7^2}{2.4^2.6^2.8}\sqrt{\frac{8}{7}}$,	$\frac{3^2.5^2.7^2}{2.4^2.6^2.8}\sqrt{\frac{9}{8}}$,

limits which tend continually to equality. We thus have

$$\square = \frac{4}{\pi},\quad = \frac{3.3.5.5.7.7\ldots}{2.4.4.6.6.8\ldots},$$

the number of factors in the numerator being always equal to the number in the denominator, and the accuracy of the approximation increasing with the number of factors.

It is to be remarked that for a square; row m and column n; m or $n = -\frac{1}{2}$, 0, $\frac{1}{2}$, 1, $\frac{3}{2}$, ... as before; the term of the square is in general $\Pi(m+n) \div \Pi(m)\Pi(n)$; thus $m = n = \frac{1}{2}$, the term is $\Pi(1) \div \{\Pi(\frac{1}{2})\}^2$, $= 1 \div (\frac{1}{2}\pi)^2 = \frac{4}{\pi}$, $= \square$; $m = 3$, $n = \frac{1}{2}$ it is $\Pi(\frac{7}{2}) \div \Pi(3)\Pi(\frac{1}{2})$, $= \frac{7}{2}.\frac{5}{2}.\frac{3}{2} \div 6$, $= \frac{105}{48}$; and so in any other case.

895.

A THEOREM ON TREES.

[From the *Quarterly Journal of Pure and Applied Mathematics*, vol. XXIII. (1889), pp. 376—378.]

THE number of trees which can be formed with $n+1$ given knots $\alpha, \beta, \gamma, \ldots$ is $=(n+1)^{n-1}$; for instance $n=3$, the number of trees with the 4 given knots $\alpha, \beta, \gamma, \delta$ is $4^2=16$, for in the first form shown in the figure the $\alpha, \beta, \gamma, \delta$ may be arranged

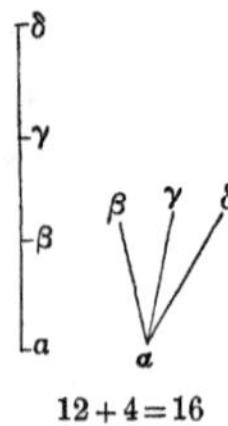

12+4=16

in 12 different orders ($\alpha\beta\gamma\delta$ being regarded as equivalent to $\delta\gamma\beta\alpha$), and in the second form any one of the 4 knots $\alpha, \beta, \gamma, \delta$ may be in the place occupied by the α: the whole number is thus $12+4, =16$.

Considering for greater clearness a larger value of n, say $n=5$, I state the particular case of the theorem as follows:

No. of trees $(\alpha, \beta, \gamma, \delta, \epsilon, \zeta)=$ No. of terms of $(\alpha+\beta+\gamma+\delta+\epsilon+\zeta)^4\,\alpha\beta\gamma\delta\epsilon\zeta, =6^4, =1296$, and it will be at once seen that the proof given for this particular case is applicable for any value whatever of n.

I use for any tree whatever the following notation: for instance, in the first of the forms shown in the figure, the branches are $\alpha\beta$, $\beta\gamma$, $\gamma\delta$; and the tree is said to be $\alpha\beta^2\gamma^2\delta$ (viz. the knots α, δ occur each once, but β, γ each twice); similarly in the second of the same forms, the branches are $\alpha\beta$, $\alpha\gamma$, $\alpha\delta$, and the tree is said to be $\alpha^3\beta\gamma\delta$ (viz. the knot α occurs three times, and the knots β, γ, δ each once). And so in other cases.

This being so, I write

$$(\alpha+\beta+\gamma+\delta+\epsilon+\zeta)^4\,\alpha\beta\gamma\delta\epsilon\zeta = \left.\begin{array}{rll} 1 & \alpha^4 & 6 \\ +\ 4 & \alpha^3\beta & 30 \\ +\ 6 & \alpha^2\beta^2 & 15 \\ +12 & \alpha^2\beta\gamma & 60 \\ +24 & \alpha\beta\gamma\delta & 15 \end{array}\right\}\alpha\beta\gamma\delta\epsilon\zeta;\quad \begin{array}{r} 6 \\ 120 \\ 90 \\ 720 \\ \underline{360} \\ 1296, \end{array}$$

where the numbers of the left-hand column are the polynomial coefficients for the index 4; and the numbers of the right-hand column are the numbers of terms of the several types, 6 terms α^4, 30 terms $\alpha^3\beta$, 15 terms $\alpha^2\beta^2$, &c.: the products of the corresponding terms of the two columns give the outside column 6, 120, 90, &c.; and the sum of these numbers is of course 6^4, $=1296$.

It is to be shown that we have

1 tree $\alpha^4 . \alpha\beta\gamma\delta\epsilon\zeta$ $(=\alpha^5\beta\gamma\delta\epsilon\zeta)$; 4 trees $\alpha^3\beta . \alpha\beta\gamma\delta\epsilon\zeta$ $(=\alpha^4\beta^2\gamma\delta\epsilon\zeta)$, ...,

24 trees $\alpha\beta\gamma\delta . \alpha\beta\gamma\delta\epsilon\zeta$ $(=\alpha^2\beta^2\gamma^2\delta^2\epsilon\zeta)$:

for this being so, then by the mere interchange of letters, the numbers 1, 4, 6, ... of the left-hand column have to be multiplied by the numbers 6, 30, 15, ... of the right-hand column, and we have the numbers in the outside column, the sum of which is $=1296$ as above.

Start with the last term $\alpha\beta\gamma\delta . \alpha\beta\gamma\delta\epsilon\zeta$, $=\alpha^2\beta^2\gamma^2\delta^2\epsilon\zeta$. We have the trees

$$\epsilon\alpha\beta\gamma\delta\zeta\ (=\epsilon\alpha . \alpha\beta . \beta\gamma . \gamma\delta . \delta\zeta),$$

where the α, β, γ, δ may be written in any one of the 24 orders, and the number of such trees is thus $=24$. If we consider only the 12 orders ($\alpha\beta\gamma\delta$ being regarded as equivalent to $\delta\gamma\beta\alpha$), then the ϵ, ζ may be interchanged; and the number is thus 2×12, $=24$ as before.

Now for the δ of $\alpha\beta\gamma\delta$ substitute α, or consider the form $\alpha\beta\gamma\alpha . \alpha\beta\gamma\delta\epsilon\zeta$, $=\alpha^3\beta^2\gamma^2\delta\epsilon\zeta$. We see at once in the form $\epsilon\alpha . \alpha\beta . \beta\gamma . \gamma\delta . \delta\zeta$, which one it is of the two δ's that must be changed into α: in fact, changing the first δ, we should have $\epsilon\alpha . \alpha\beta . \beta\gamma . \gamma\alpha . \delta\zeta$ which contains a circuit $\alpha\beta\gamma$, and a detached branch $\delta\zeta$, and is thus *not* a tree: changing the second δ, we have $\epsilon\alpha . \alpha\beta . \beta\gamma . \gamma\delta . \alpha\zeta$ which is a tree $\alpha^3\beta^2\gamma^2\delta\epsilon\zeta$, $=\alpha\zeta . \alpha\epsilon . \alpha\beta . \beta\gamma . \gamma\delta$. And similarly for any other order of the $\alpha\beta\gamma\delta$, there is in each case only one of the δ's which can be changed into α; and thus from each of the 24 forms we obtain a tree $\alpha^3\beta^2\gamma^2\delta\epsilon\zeta$. But dividing the 24 forms into the $12+12$ forms corresponding to the interchange of the letters ϵ, ζ, then the first 12 forms, and the second 12 forms, give each of them the same trees $\alpha^3\beta^2\gamma^2\delta\epsilon\zeta$; and the number of these trees is thus $\frac{1}{2} . 24$, $=12$.

And in like manner reducing the $\alpha\beta\gamma\delta$ to $\alpha^2\beta^2$, $\alpha^3\beta$ or α^4, we obtain in each case the number of trees equal to the proper sub-multiple of 24, viz. 6, 4, 1 in the three cases respectively (for the last case this is obvious, viz. there is 1 tree $\alpha^5\beta\gamma\delta\epsilon\zeta$, $=\alpha\beta\,.\,\alpha\gamma\,.\,\alpha\delta\,.\,\alpha\epsilon\,.\,\alpha\zeta$); and the subsidiary theorem is thus proved. Hence the original theorem is true: as already remarked, it is easy to see that the proof is perfectly general.

The theorem is one of a set as follows:

Let $(\lambda, \alpha, \beta, \gamma, \ldots)$ denote as above the trees with the given knots $\lambda, \alpha, \beta, \gamma, \ldots$; $(\lambda+\mu, \alpha, \beta, \gamma, \ldots)$ the pairs of trees with the given knots $\lambda, \mu, \alpha, \beta, \gamma, \ldots$, the knots λ, μ belonging always to different trees; $(\lambda+\mu+\nu, \alpha, \beta, \gamma, \ldots)$ the triads of trees with the given knots $\lambda, \mu, \nu, \alpha, \beta, \gamma, \ldots$, the knots λ, μ, ν always belonging to different trees; and so on: then if $i+1$ be the number of the knots $\lambda, \mu, \nu, \ldots$, and n the number of the knots $\alpha, \beta, \gamma, \ldots$, the number of trees or pairs, or triads, &c., of trees is $=(i+1)(i+n+1)^{n-1}$. In particular, if $i=0$, then n being the number of knots $\alpha, \beta, \gamma, \ldots$, and therefore $n+1$ the whole number of knots $\lambda, \alpha, \beta, \gamma, \ldots$, the number of trees is $=(n+1)^{n-1}$ as before.

As a simple example, consider the pairs $(\lambda+\mu, \alpha, \beta)$: here $i=1$, $n=2$, and we have $(i+1)(i+n+1)^{n-1}=2\,.\,4, =8$: in fact, the pairs of trees are

$$(\lambda\alpha,\ \alpha\beta,\ \mu),\ (\lambda\beta,\ \beta\alpha,\ \mu),\ (\lambda\alpha,\ \lambda\beta,\ \mu),$$
$$(\mu\alpha,\ \alpha\beta,\ \lambda),\ (\mu\beta,\ \beta\alpha,\ \lambda),\ (\mu\alpha,\ \mu\beta,\ \lambda);\ (\lambda\alpha,\ \mu\beta),\ (\lambda\beta,\ \mu\alpha).$$

We may arrange the trees $(\alpha, \beta, \gamma, \delta, \epsilon)$ as follows:

$$\begin{array}{llll}
(\alpha, \beta, \gamma, \delta, \epsilon) = \alpha\beta & (\beta,\ \gamma,\ \delta,\ \epsilon); & 125 = \ \ 4\times 1\,.\,4^2 = & 64 \\
\quad + \alpha\beta\,.\,\alpha\gamma & (\beta+\gamma,\ \delta,\ \epsilon) & \quad + 6\times 2\,.\,4^1 & 48 \\
\quad + \alpha\beta\,.\,\alpha\gamma\,.\,\alpha\delta & (\beta+\gamma+\delta,\ \epsilon) & \quad + 4\times 3\,.\,4^0 & 12 \\
\quad + \alpha\beta\,.\,\alpha\gamma\,.\,\alpha\delta\,.\,\alpha\epsilon & & \quad + 1 & 1 \\
\hline
 & & & 125,
\end{array}$$

viz. to obtain the trees $(\alpha, \beta, \gamma, \delta, \epsilon)$, we join on the branch $\alpha\beta$ to any tree $(\beta, \gamma, \delta, \epsilon)$: the branches $\alpha\beta$, $\alpha\gamma$ to any pair of trees $(\beta+\gamma, \delta, \epsilon)$; the branches $\alpha\beta$, $\alpha\gamma$, $\alpha\delta$ to any triad of trees $(\beta+\gamma+\delta, \epsilon)$; and take lastly the tree $\alpha\beta\,.\,\alpha\gamma\,.\,\alpha\delta\,.\,\alpha\epsilon$: the knots $\beta, \gamma, \delta, \epsilon$ being then interchanged in every possible manner. The whole number of trees 125 is thus obtained as $=64+48+12+1$; the theorem is of course perfectly general.

The foregoing theory in effect presents itself in a paper by Borchardt, "Ueber eine der Interpolation entsprechende Darstellung der Eliminations-Resultante," *Crelle*, t. LVII. (1860), pp. 111—121, viz. Borchardt there considers a certain determinant, composed of the elements 10, 12, ..., $1n$, 20, 21, 23, ..., $2n$, ..., $n0$, $n1$, ..., $nn-1$, and represented by means of the trees $(0, 1, 2, \ldots, n)$; the branches of the tree being the aforesaid elements, and the tree being regarded as equal to the product of the several branches: the number of terms of the determinant is thus $=(n+1)^{n-1}$ as above.

896.

A TRANSFORMATION IN ELLIPTIC FUNCTIONS.

[From the *Quarterly Journal of Pure and Applied Mathematics*, vol. XXIV. (1890), pp. 259—262.]

THE formula in question is given in Klein's Memoir "Ueber hyperelliptische Sigmafunctionen," *Math. Ann.* t. XXVII. (1886), pp. 431—464, see p. 454, in the form

$$u=\int_y^x \frac{(\varpi d\varpi)}{\sqrt{\{f(z)\}}},\quad \wp(u)=\frac{\sqrt{\{f(x)\}}\sqrt{\{f(y)\}}+F(x,\ y)}{2(xy)^2},$$

(the discovery of it being ascribed to Weierstrass); and it is also given in Halphen's *Traité des Fonctions Elliptiques*, t. II. (1888), p. 357.

The algebraic foundation of the theorem is as follows: writing

$$z=\tfrac{1}{2}\{\sqrt{(ae)}+c\},$$

and as usual I, J for the two invariants of the quartic function $(a,\ b,\ c,\ d,\ e\between x,\ y)^4$, we have identically, as is easily verified,

$$4z^3-Iz-J=\{d\sqrt{(a)}+b\sqrt{(e)}\}^2,$$

or say

$$\sqrt{(4z^3-Iz-J)}=d\sqrt{(a)}+b\sqrt{(e)}.$$

Hence putting

$$A=(a,\ b,\ c,\ d,\ e\between x_1,\ x_2)^4,$$

$$B=(a,\ \ldots\between x_1,\ x_2)^3(y_1,\ y_2),$$

$$\vdots$$

$$E=(a,\ \ldots\between y_1,\ y_2)^4;$$

and

$$\lambda=x_1y_2-x_2y_1,$$

so that

$$\lambda^4 I = AE - 4BD + 3C^2,$$

$$\lambda^6 J = ACE - AD^2 - B^2E + 2BCD - C^3,$$

then writing

$$z = \frac{1}{2\lambda^2}\{\sqrt{(AE)} + C\},$$

so that $\lambda^2 z$ is a bipartite irrational quadric function of the variables (x_1, x_2) and (y_1, y_2), we have

$$\sqrt{(4z^3 - Iz - J)} = \frac{1}{\lambda^3}\{D\sqrt{(A)} + B\sqrt{(E)}\},$$

and we thence infer the existence of a transformation

$$\frac{dz}{\sqrt{(4z^3 - Iz - J)}} = M\left\{\frac{x_1dx_2 - x_2dx_1}{\sqrt{(A)}} + \frac{y_1dy_2 - y_2dy_1}{\sqrt{(E)}}\right\}.$$

But for the verification hereof and a direct determination of the value of M, observe that we have

$$dz = \frac{1}{2\lambda^2}\left\{\sqrt{(E)}\frac{dA}{2\sqrt{(A)}} + dC - \{\sqrt{(AE)} + C\}\frac{2}{\lambda}d\lambda\right\},$$

where, attending only to the terms in dx_1, dx_2,

$$\begin{aligned} dA &= 4\{(a, b, c, d \between x_1, x_2)^3 dx_1 + (b, c, d, e \between x_1, x_2)^3 dx_2\}, \\ &= 4(A_1dx_1 + A_2dx_2), \text{ suppose;} \\ C &= (a, b, c \between y_1, y_2)^2 x_1^2 + 2(b, c, d \between y_1, y_2)^2 x_1x_2 + (c, d, e \between y_1, y_2)^2 x_2^2, \\ &= E_1x_1^2 + 2E_2x_1x_2 + E_3x_2^2, \text{ suppose;} \end{aligned}$$

and therefore

$$dC = 2\{(E_1x_1 + E_2x_2)dx_1 + (E_2x_1 + E_3x_2)dx_2\}.$$

Hence

$$\begin{aligned} dz = \frac{1}{2\lambda^2}\bigg\{&\frac{1}{2}\frac{\sqrt{(E)}}{\sqrt{(A)}}4(A_1dx_1 + A_2dx_2) + 2(E_1x_1 + E_2x_2)dx_1 \\ &+ 2(E_2x_1 + E_3x_2)dx_2 - \{\sqrt{(AE)} + C\}\frac{2}{\lambda}(y_2dx_1 - y_1dx_2)\bigg\}. \end{aligned}$$

The whole coefficient herein of dx_1 is

$$= \frac{1}{\lambda^2}\left\{\frac{\sqrt{(E)}}{\sqrt{(A)}}A_1 + (E_1x_1 + E_2x_2) - \frac{\sqrt{(AE)} + C}{\lambda}y_2\right\},$$

which is

$$= \frac{1}{\lambda^2\sqrt{(A)}}\left\{A_1\sqrt{(E)} + (E_1x_1 + E_2x_2)\sqrt{(A)} - \frac{1}{\lambda}\{A\sqrt{(E)} + C\sqrt{(A)}\}y_2\right\}$$

$$= \frac{1}{\lambda^3\sqrt{(A)}}\{[-Cy_2 + \lambda(E_1x_1 + E_2x_2)]\sqrt{(A)} + [A_1\lambda - Ay_2]\sqrt{(E)}\},$$

which is

$$=\frac{1}{\lambda^3\sqrt{(A)}}(-x_2)\{D\sqrt{(A)}+B\sqrt{(E)}\};$$

and similarly the whole coefficient of dx_2 is

$$=\frac{1}{\lambda^3\sqrt{(A)}}\quad(x_1)\{D\sqrt{(A)}+B\sqrt{(E)}\}.$$

Hence the terms in dx_1, dx_2 are together

$$=\frac{1}{\sqrt{(A)}}(x_1dx_2-x_2dx_1)\frac{D\sqrt{(A)}+B\sqrt{(E)}}{\lambda^3};$$

and since the terms in dy_1, dy_2 are of the like form, we have

$$dz=\left\{\frac{1}{\sqrt{(A)}}(x_1dx_2-x_2dx_1)+\frac{1}{\sqrt{(E)}}(y_1dy_2-y_2dy_1)\right\}\times\frac{D\sqrt{(A)}+E\sqrt{(B)}}{\lambda^3},$$

and combining herewith the foregoing value

$$\sqrt{(4z^3-Iz-J)}=\frac{D\sqrt{(A)}+E\sqrt{(B)}}{\lambda^3},$$

we have the required formula

$$\frac{dz}{\sqrt{(4z^3-Iz-J)}}=\left\{\frac{x_1dx_2-x_2dx_1}{\sqrt{(A)}}+\frac{y_1dy_2-y_2dy_1}{\sqrt{(B)}}\right\}.$$

As a very simple verification, suppose $a=1$, $b=c=d=e=0$; then

$$(A,\ B,\ C,\ D,\ E)=(x_1^4,\ x_1^3y_1,\ x_1^2y_1^2,\ x_1y_1^3,\ y_1^4),$$

and if $\lambda=x_1y_2-x_2y_1$ as before, then

$$z=\frac{x_1^2y_1^2}{(x_1y_2-x_2y_1)^2}=\theta^2,$$

where

$$\theta=\frac{-x_1y_1}{x_1y_2-x_2y_1},\quad\text{or}\quad\frac{1}{\theta}=\frac{x_2}{x_1}-\frac{y_2}{y_1}.$$

Also $I=0$, $J=0$, and consequently

$$\frac{dz}{\sqrt{(4z^3-Iz-J)}}=\frac{dz}{2\sqrt{(z^3)}}=\frac{d\theta}{\theta^2}=\frac{x_1dx_2-x_2dx_1}{x_1^2}-\frac{y_1dy_2-y_2dy_1}{y_1^2},$$

which, in virtue of the foregoing values of A, E, is

$$=\frac{x_1dx_2-x_2dx_1}{\sqrt{(A)}}-\frac{y_1dy_2-y_2dy_1}{\sqrt{(E)}}.$$

I remark that an even more simple transformation from the general quartic radical to the Weierstrassian cubic radical was obtained by Hermite; this is alluded to in my "Note sur les Covariants, &c.," *Crelle*, t. L. (1855), pp. 285—287, [135], and is given with a demonstration in my paper "Sur quelques formules pour la transformation des intégrales elliptiques," *Crelle*, t. LV. (1858), pp. 15—24, [235], see No. IV.; viz. from the identical relation $JU^3 - IU^2H + 4H^3 = -\Phi^2$, which connects the covariants of a quartic function, it at once follows that if

$$U = (a,\ b,\ c,\ d,\ e \between x,\ 1)^4,$$

and H is the Hessian hereof,

$$H = (ac - b^2,\quad (ad - bc),\quad ae + 2bd - 3c^2,\quad 2(be - cd),\quad ce - d^2 \between x,\ 1)^4;$$

then writing

$$z = \frac{H}{U},$$

we have

$$\frac{dz}{\sqrt{(-4z^3 + zI - J)}} = \frac{2dx}{\sqrt{\{(a,\ b,\ c,\ d,\ e \between x,\ 1)^4\}}},$$

which is the transformation in question.

897.

SUR LES RACINES D'UNE ÉQUATION ALGÉBRIQUE.

[From the *Comptes Rendus de l'Académie des Sciences de Paris*, t. CX. (*Janvier—Juin*, 1890), pp. 174—176, 215—218.]

SOIT $f(u)$ une fonction rationnelle et entière avec des coefficients réels ou imaginaires, de l'ordre n; en supposant que l'équation $f'(u) = 0$, de l'ordre $n-1$, ait $n-1$ racines, je démontre que l'équation $f(u) = 0$ aura n racines. Pour cela, soit $f(u) = f(x+iy) = P + iQ$: je suppose que c dénote une quantité positive donnée, et je considère la surface $c - z = P^2 + Q^2$, en attribuant à la coordonnée z des valeurs positives; c'est seulement pour avoir des maxima au lieu de minima, et pour faciliter ainsi l'exposition, que je prends cette surface au lieu de $z = P^2 + Q^2$. On peut donner à c une valeur si grande que la courbe $c = P^2 + Q^2$ soit une courbe fermée qui ne se coupe pas, c'est-à-dire un contour simple: cela étant, on peut se figurer ce contour comme la ligne de rivage d'une île montagneuse; la valeur de z est au plus $= c$, et, en donnant à z une valeur plus petite, $= b$, on a le contour qui correspond à l'altitude b: évidemment, les contours qui correspondent à des altitudes différentes ne se coupent pas. Il s'agit de prouver que l'île a précisément n sommets, chacun de l'altitude c.

J'écris

$$\frac{dP}{dx} = X, \quad \frac{dP}{dy} = Y;$$

donc

$$\frac{dQ}{dx} = -Y, \quad \frac{dQ}{dy} = X;$$

et, de plus,

$$\frac{dX}{dx} = a, \quad \frac{dX}{dy} = \frac{dY}{dx} = h, \quad \frac{dY}{dy} = -a,$$

et de là

$$\frac{d^2P}{dx^2} = \quad a, \qquad \frac{d^2Q}{dx^2} = -h,$$

$$\frac{d^2P}{dx\,dy} = \quad h, \qquad \frac{d^2Q}{dx\,dy} = \quad a,$$

$$\frac{d^2P}{dy^2} = -a, \qquad \frac{d^2Q}{dy^2} = \quad h.$$

Cela étant, nous avons

$$-\tfrac{1}{2}\frac{dz}{dx} = PX - QY, \qquad -\tfrac{1}{2}\frac{dz}{dy} = PY + QX;$$

on aura un plan tangent horizontal pour $P=0$, $Q=0$, ou pour $X=0$, $Y=0$; les valeurs $P=0$, $Q=0$, appartiennent à la valeur c de z et correspondent à des sommets de montagne de cette altitude c; les valeurs $X=0$, $Y=0$ correspondent à des sommets de col. En effet, nous avons

$$-\tfrac{1}{2}\,\frac{d^2z}{dx^2} = \quad Pa - Qh + X^2 + Y^2,$$

$$-\tfrac{1}{2}\,\frac{d^2z}{dx\,dy} = \quad Ph + Qa,$$

$$-\tfrac{1}{2}\,\frac{d^2z}{dy^2} = -Pa + Qh + X^2 + Y^2;$$

donc

$$\tfrac{1}{4}\left[\frac{d^2z}{dx^2}\frac{d^2z}{dy^2} - \left(\frac{d^2z}{dx\,dy}\right)^2\right] = (X^2+Y^2)^2 - (Pa-Qh)^2 - (Ph+Qa)^2$$
$$= (X^2+Y^2)^2 - (P^2+Q^2)(a^2+h^2),$$

valeur positive pour $P=0$, $Q=0$; négative pour $X=0$, $Y=0$.

A présent, nous avons $f'(x+iy) = X - iY$; donc, en supposant que l'équation $f'(x+iy)=0$ ait $n-1$ racines, il y aura $n-1$ systèmes de valeurs réelles de x, y qui satisfont aux équations $X=0$, $Y=0$; pour chaque système, il y aura des valeurs déterminées de P, Q et de là aussi de z; ces valeurs de z seront en général différentes. Ainsi il y aura dans l'île $n-1$ cols, dont les altitudes seront en général différentes: soient $c_1, c_2, \ldots, c_{n-1}$ ces altitudes, commençant avec la plus petite.

Pour $z=0$, nous avons un contour simple, et de même pour une valeur quelconque plus petite que c_1; mais, pour $z=c_1$, nous avons un col; le contour est une courbe, figure de 8; en donnant à z une valeur un peu plus grande, le contour se divise en deux courbes fermées, ou bien contours simples, extérieurs l'un à l'autre. Pour $z=c_2$, nous avons encore un col; l'un des contours simples s'est changé en figure de 8, et, pour une valeur un peu plus grande, le contour se divise en trois contours simples, chacun extérieur aux autres. En continuant de cette manière, on a, pour c_{n-1}, le col le plus haut; le contour est composé de $n-2$ contours simples et d'une figure de 8; et, en donnant à z une valeur un peu plus grande, on obtient un contour composé de n contours simples, chacun extérieur aux autres. Enfin, en faisant croître z, chacun des contours simples doit se réduire à un point, c'est-à-dire qu'il doit y avoir précisément n sommets de montagne; mais il n'y a pas de sommet de montagne, sinon pour la valeur $z=c$; donc il y a précisément n sommets de montagne, chacun de l'altitude c. On suppose toujours que l'équation $f'(x+iy)=0$ n'ait pas de racines égales, mais il peut bien arriver que deux ou plusieurs des valeurs $c_1, c_2, \ldots, c_n$ deviennent égales; la démonstration est très peu changée, en donnant à z une valeur un peu plus grande que celle qui correspond à l'altitude des cols d'altitude égale; le contour se divise toujours en contours simples, extérieurs chacun aux autres.

Il va sans dire que cette démonstration repose sur les mêmes principes que celles de Gauss et de Cauchy.

Je reprends la théorie des racines de l'équation $f(u)=0$; au lieu de la surface $c-z=P^2+Q^2$, il convient de considérer la surface $(c-z)^2=P^2+Q^2$, en faisant attention seulement aux valeurs de z positives et pas plus grandes que c. La théorie est très peu changée; les contours sont les mêmes qu'auparavant, mais ils appartiennent à des altitudes différentes; et, au lieu de maxima $z=c$ pour $P=0$, $Q=0$, on a des points coniques, c'est-à-dire, dans l'île montagneuse, au lieu d'un sommet arrondi de montagne, on a un cône ou un pic.

Mais avec la nouvelle surface, on construit graphiquement l'approximation de Newton: partant d'une valeur réelle ou imaginaire approximative u, on obtient la nouvelle valeur

$$u_1 = u + h = u - \frac{f(u)}{f'(u)}.$$

Je représente u par le point (x, y, z) de la surface $(c-z)^2=P^2+Q^2$, ou le point (x, y) du plan des sommets $z-c$; et, de même, u_1 par le point (x_1, y_1, z_1) de la surface, ou (x_1, y_1) du plan des sommets: cela étant, si, par le point (x, y, z) de la surface, on mène la droite de plus grande pente (droite tangente à la surface et perpendiculaire au contour), cette droite rencontrera le plan des sommets en un point (x_1, y_1), et l'on obtient ainsi le point (x_1, y_1, z_1) de la surface, qui représente la valeur cherchée u_1. En particulier, si les coefficients de $f(u)$ sont réels, on a

$$Q=0;$$

l'équation $(c-z)^2=P^2+Q^2$ devient

$$(c-z)^2=P^2,$$

c'est-à-dire

$$c-z=\pm P,$$

ou enfin

$$c-z=\pm f(x),$$

et la section verticale de l'île est formée par des parties de ces deux courbes symétriques: pour la théorie géométrique, on peut évidemment y substituer la seule courbe $c-z=f(x)$.

J'ai proposé, il y a plus de dix ans (*Amer. Math. Journ.*, t. II., 1879, [694]), le problème que je nomme "The Newton-Fourier imaginary Problem," et dans une Note (*Quart. Math. Journ.*, t. XVI., 1879, [736]), "Application of the Newton-Fourier method to an imaginary root of an equation," j'ai considéré le cas d'une équation quadratique. Pour l'équation $u^2 - 1 = 0$, on a

$$u_1 = u - \frac{u^2 - 1}{2u} = \tfrac{1}{2}\left(u + \frac{1}{u}\right);$$

cela donne

$$u_1 - 1 = \frac{1}{2u}(u-1)^2, \qquad u_1 + 1 = \frac{1}{2u}(u+1)^2,$$

et de là

$$\frac{u_1 - 1}{u_1 + 1} = \left(\frac{u-1}{u+1}\right)^2.$$

Cette dernière équation a rapport aux deux racines $+1$ et -1, et, quoiqu'elle donne les résultats les plus élégants, cependant, en vue de la théorie générale, il vaut mieux considérer l'équation $u_1 - 1 = \dfrac{1}{2u}(u-1)^2$ qui se rapporte à la seule racine $+1$.

Je remarque d'abord que la formule originale $u_1 = \tfrac{1}{2}\left(u + \dfrac{1}{u}\right)$ donne

$$x_1 = \tfrac{1}{2}x\left(1 + \frac{1}{x^2 + y^2}\right), \qquad y_1 = \tfrac{1}{2}y\left(1 - \frac{1}{x^2 + y^2}\right);$$

donc les valeurs de x et x_1 seront à la fois positives ou négatives, et ainsi l'on peut ne faire attention qu'aux valeurs positives. Cela étant, nous avons

$$x_1 + iy_1 - 1 = \frac{(x + iy - 1)^2}{2(x + iy)}.$$

Désignons par A le point (0, 1), par B le point (0, -1), par 0 le point (0, 0); et aussi par P le point (x, y), et de même par P_1 le point (x_1, y_1); écrivons aussi $x + iy = se^{i\phi}$, $x + iy - 1 = re^{i\theta}$, $x_1 + iy_1 = r_1 e^{i\theta}$; l'équation est

$$r_1 e^{i\theta_1} = \tfrac{1}{2}\frac{r^2 e^{2i\theta}}{se^{i\phi}};$$

donc

$$r_1 = \tfrac{1}{2}\frac{r^2}{s}, \quad \left(\text{c'est-à-dire } AP_1 = \frac{AP^2}{2OP}\right),$$

et

$$\theta_1 = 2\theta - \phi, \quad (\text{c'est-à-dire } \widehat{AP_1x} = \widehat{2APx} - \widehat{OPx}).$$

Je remarque que, dans la géométrie des vecteurs, la seule équation $AP_1 = \dfrac{AP^2}{2OP}$ dénote l'équation en $x_1 + iy_1$, $x + iy$, c'est-à-dire les deux équations que je viens de trouver.

Partant d'un point quelconque P, on obtient une suite de points $P_1, P_2, P_3, \ldots$; et, si le point P est sur l'axe des $y\,(x = 0)$, tous les autres points seront aussi sur l'axe de y, et l'on n'approche ni du point A ni du point B. Mais, si la coordonnée x a une valeur positive si petite que l'on veut, on arrive enfin infiniment près du point A, et l'on peut même (dans un sens qui sera expliqué plus bas, mais qui n'est pas le sens le plus naturel) dire que l'approximation est régulière. En effet, on n'a pas toujours $AP_1 < AP$, et ainsi, dans le sens le plus naturel, l'approximation n'est pas toujours régulière. Pour étudier cela, j'écris $AP_1 = AP$; cela donne $AP = 2OP$, ou, ce qui est la même chose, $x^2 + y^2 + \frac{2}{3}x = \frac{1}{3}$, c'est-à-dire que le point P sera situé sur le cercle, centre $x = -\frac{1}{3}$ et rayon $= \frac{2}{3}$; en ne faisant attention qu'aux valeurs positives de x, on a un segment compris entre l'axe des y et un arc par les points $\left(x = 0,\ y = \pm\frac{1}{\sqrt{3}}\right)$, $(x = \frac{1}{3},\ y = 0)$. Si le point P est sur l'arc, on aura $AP_1 = AP$; si P est en dedans du segment, alors $AP_1 > AP$; si P est en dehors du segment, $AP_1 < AP$.

Mais, en supposant P en dehors du segment, et ainsi $AP_1 < AP$, il peut bien arriver que P_1 soit en dedans du segment, et, cela étant, on aura $AP_2 > AP_1$, et l'approximation ne sera pas régulière. Mais, en considérant le cercle $x^2 + y^2 - \frac{2}{3}x = \frac{1}{3}$, lequel est le cercle, centre A et rayon $\frac{2}{3}$, qui touche le segment au point $(x = \frac{1}{3},\ y = 0)$, alors, en supposant que le point P soit en dedans de ce cercle, on aura $AP_1 < AP$, le point P_1 sera aussi en dedans du cercle, et ainsi en dehors du segment; et les points successifs P, P_1, P_2, ... approcheront continuellement du point A; l'approximation sera dans ce cas régulière.

Il y a ainsi trois régions: le segment, le cercle $x^2 + y^2 - \frac{2}{3}x = \frac{1}{3}$ et le résidu du demi-plan; on pourrait les nommer régions *noire*, *blanche* et *grise* respectivement. C'est seulement pour un point P à l'intérieur de la région blanche que l'approximation est certainement régulière.

Nous venons de considérer en effet les cercles qui ont pour centre le point A; $AP_1 < AP$ veut dire que le point P est situé sur un cercle plus grand, et P_1 sur un cercle plus petit; mais, au lieu de ces cercles concentriques, considérons des cercles quelconques qui entourent le point A, sans se couper les uns les autres; et convenons de dire que c'est un bon pas quand on passe du point P sur un cercle plus grand à un point P_1 sur un cercle plus petit: avec cette convention on aura, en général, trois régions, lesquelles cependant ne seront pas les mêmes comme auparavant. En particulier, si nous considérons les cercles $AP = kBP$ (k une constante quelconque plus petite que l'unité), alors il n'y a pas de région noire, ou, si l'on veut, la région noire se réduit à la seule droite $y = 0$; donc il n'y a pas non plus de région grise, et le demi-plan entier est région blanche, c'est-à-dire, dans le sens que je viens d'expliquer, l'approximation est toujours régulière. En effet, c'est là la théorie à laquelle on est conduit au moyen de l'équation $\dfrac{u_1 - 1}{u_1 + 1} = \left(\dfrac{u - 1}{u + 1}\right)^2$ ci-dessus mentionnée.

En parlant de cercles, j'ai fait une restriction qui n'est nullement nécessaire; j'aurais pu parler d'ovales, de forme quelconque, qui entourent le point A sans se couper les uns les autres.

J'espère appliquer cette théorie au cas d'une équation cubique, mais les calculs sont beaucoup plus difficiles.

898.

SUR L'ÉQUATION MODULAIRE POUR LA TRANSFORMATION DE L'ORDRE 11.

[From the *Comptes Rendus de l'Académie des Sciences de Paris*, t. CXI. (*Juillet—Décembre*, 1890), pp. 447—449.]

L'ÉQUATION en u, v, en y écrivant $u=x$, $v=y$, est

$$\left.\begin{array}{ll} y^{12} & \\ +y^{11} & (32x^{11}-22x^{3}) \\ +y^{10}\,. & 44x^{6} \\ +y^{9} & (88x^{9}+22x\,) \\ +y^{8}\,. & 165x^{4} \\ +y^{7}\,. & 132x^{7} \\ +y^{6} & (-\;44x^{10}+44x^{2}) \\ +y^{5} & -132x^{5} \\ +y^{4} & -165x^{8} \\ +y^{3} & (-\;22x^{11}-88x^{3}) \\ +y^{2} & -\;\;44x^{6} \\ +y & (22x^{9}-22x\,) \\ +1\,. & -\;\;x^{12} \end{array}\right\} = 0.$$

Selon un résultat trouvé par H. J. S. Smith, pour la transformation de l'ordre p, la courbe est de l'ordre $2p$, et il y a à l'origine un point double, à l'infini deux points singuliers équivalents chacun à $\frac{1}{2}(p-1)(p-2)$ points doubles, et de plus $(p-1)(p-3)$ points doubles. Au cas $p=11$, le nombre de ces derniers points doubles est donc $=80$. Cela s'accorde avec l'expression

$$D=x^{12}(1-x^{8})^{10}(16x^{16}-31x^{8}+16)^{2}(x^{64}-301960x^{56}+\ldots+1)^{2},$$

trouvée par M. Hermite pour le discriminant de la fonction; et l'on voit ainsi que les valeurs de x, qui correspondent aux quatre-vingts points doubles, sont données par les équations

$$16x^{16} - 31x^8 + 16 = 0,$$

$$x^{64} - 301960x^{56} + \ldots + 1 = 0.$$

Je ne considère que les seize points doubles donnés par la première équation. Cette équation donne

$$x^8 = \tfrac{1}{32}(31 + 3i\sqrt{7}), \quad x^4 = \tfrac{1}{8}(i + 3\sqrt{7}), \quad x^2 = \frac{1+i}{4\sqrt{2}}(-3 + i\sqrt{7});$$

il y a ainsi quatre points doubles, pour lesquels les valeurs de x sont

$$x = \pm\sqrt{\frac{1+i}{4\sqrt{2}}(-3 + i\sqrt{7})}, \quad x = \pm\sqrt{\frac{1+i}{4\sqrt{2}}(-3 - i\sqrt{7})};$$

je trouve que les valeurs correspondantes de y sont $y = \dfrac{1+i}{\sqrt{2}}x$, savoir que les quatre points sont situés sur la droite $y = \dfrac{1+i}{\sqrt{2}}x$, et, en changeant successivement les signes de i et $\sqrt{2}$, on voit ainsi que les seize points sont situés, quatre à quatre, sur les droites

$$y = \frac{1+i}{\sqrt{2}}x, \quad y = \frac{1-i}{\sqrt{2}}x, \quad y = -\frac{1+i}{\sqrt{2}}x, \quad y = -\frac{1-i}{\sqrt{2}}x.$$

J'écris, pour abréger,

$$m = \frac{1+i}{\sqrt{2}}, \text{ (donc } m^4 = -1),$$

$$p = \tfrac{1}{4}(-3 + i\sqrt{7}),$$

donc

$$2p^2 + 3p + 2 = 0.$$

En écrivant $y = mx$ dans l'équation et en rejetant le facteur x^2, puis en écrivant $x^2 = mp$, l'équation se présente sous la forme

$$\left.\begin{array}{ll} m^{10}p^{10}\,. & 32m^{11} \\ +\, m^8p^8\,. & 88m^9 \\ +\, m^7p^7\,. & 44m^{10} - 44m^6 \\ +\, m^6p^6\,. & -22m^{11} + 132m^7 - 22m^3 \\ +\, m^5p^5\,. & m^{12} + 165m^8 - 165m^4 - 1 \\ +\, m^4p^4\,. & 22m^9 - 132m^5 + 22m \\ +\, m^3p^3\,. & 44m^6 - 44m^2 \\ +\, m^2p^2\,. & -88m^3 \\ +\, 1\,. & -32m \end{array}\right\} = 0,$$

où les coefficients ne contiennent que les puissances m^{21}, m^{17}, m^{13}, m^9, m^5, m^1 de m et se réduisent ainsi à des multiples de m; il y a aussi un facteur numérique 8, et, en divisant par $-8m$, l'équation devient

$$4p^{10}-11p^8-11p^7+22p^6+41p^5+22p^4-11p^3-11p^2+4=0;$$

cette équation est de la forme

$$(2p^2+3p+2)^2(p^6-3p^5+2p^4+p^3+2p^2-3p+1)=0.$$

La droite $y=mx$ a donc, avec la courbe, quatre intersections doubles

$$p=\tfrac{1}{4}(-3\pm i\sqrt{7}),$$

c'est-à-dire

$$x^2=\frac{1+i}{4\sqrt{2}}(-3\pm i\sqrt{7}):$$

on démontre sans peine que la droite n'est pas une tangente, et ces valeurs correspondent ainsi à des points doubles de la courbe, c'est-à-dire qu'il y a sur la droite $y=\dfrac{1+i}{\sqrt{2}}x$ quatre points doubles. Réciproquement, cette valeur de x^2 conduit au facteur $(16x^{16}-31x^8+16)^2$ du déterminant de l'équation modulaire.

899.

SUR LES SURFACES MINIMA.

[From the *Comptes Rendus de l'Académie des Sciences de Paris*, t. CXI. (*Juillet—Décembre*, 1890), pp. 953, 954.]

On peut généraliser tant la définition que la construction de ces surfaces, en substituant pour le cercle imaginaire à l'infini une conique ou même une surface quadrique quelconque.

Je rappelle que, dans la théorie ordinaire, une surface minima est une surface telle qu'un point quelconque de la surface est situé à mi-chemin entre les deux centres de courbure, et qu'une telle surface est le lieu des points à mi-chemin entre les deux points situés respectivement sur deux courbes de longueur nulle quelconques. Or on peut rattacher la notion d'une courbe de longueur nulle à celle d'une courbe de poursuite. Pour expliquer cela, j'observe que, dans le plan, en supposant comme à l'ordinaire que le lièvre coure selon une droite et que le chien et le lièvre courent avec des vitesses uniformes, la courbe de poursuite est une courbe déterminée; mais si les vitesses varient arbitrairement, alors la définition exprime seulement que la courbe est une courbe plane. Mais, dans l'espace, si au lieu d'une droite on considère une courbe plane ou à double courbure (disons une directrice) quelconque, alors, quelles que soient les vitesses, la définition précise toujours la courbe, savoir: on a toujours pour courbe de poursuite une courbe dont chaque tangente rencontre la courbe directrice. Au lieu d'une courbe, on peut avoir une surface directrice; dans ce cas, le nom est moins applicable, néanmoins je le retiens, et je dis que la courbe de poursuite est une courbe dont chaque tangente touche la surface directrice. Nous avons, de cette manière, la définition d'une courbe de poursuite par rapport à une courbe ou surface directrice quelconque.

A présent, au lieu du cercle imaginaire à l'infini, considérons une conique quelconque, l'absolue: on établit, comme on sait, par rapport à cette conique, la notion de la perpendicularité, et ainsi les notions d'une normale et des centres de courbure ne cessent pas de subsister. On peut donc considérer une surface telle que chaque point de la surface soit l'harmonicale par rapport aux deux centres de courbure du point de rencontre de la normale avec le plan de l'absolue: on a ainsi ce que je nomme une surface *quasi-minima*. Il va sans dire qu'il faut modifier convenablement la notion métrique d'une aire minima pour qu'elle soit applicable à cette nouvelle surface.

Pour construire la surface, on prend par rapport à l'absolue deux courbes de poursuite quelconques, et puis sur la droite, menée par deux points quelconques de ces courbes respectivement, l'harmonicale par rapport à ces deux points du point de rencontre de la droite avec le plan de l'absolue: le lieu de ce point harmonical sera la surface quasi-minima.

Il paraît permis de substituer pour la conique absolue une surface quadrique quelconque, que je nomme aussi l'*absolue*: on a, comme on sait, les notions de la normale et des centres de courbure. Pour la surface quasi-minima, le point sur la surface sera l'un des points doubles (foyers) de l'involution formée par les deux centres de courbure et les deux points de rencontre de la normale avec l'absolue; et de même pour la construction de la surface, il faut prendre sur la droite menée par deux points quelconques des deux courbes de poursuite respectivement les points doubles (foyers) de l'involution formée par ces deux points et les deux points de rencontre de la droite avec l'absolue.

900.

JAMES JOSEPH SYLVESTER.

[(*Scientific Worthies in Nature*, xxv.); *Nature*, vol. xxxix. (1889), pp. 217—219.]

James Joseph Sylvester, born in London on September 3, 1814, is the sixth and youngest son of the late Abraham Joseph Sylvester, formerly of Liverpool*. He was educated at two private schools in London, and at the Royal Institution, Liverpool, whence he proceeded in due course of time to St John's College, Cambridge. In these early days he manifested considerable aptitude for mathematics, and so it was not matter for surprise that he came out in the Tripos Examination of 1837 as Second Wrangler; being incapacitated, by the fact of his Jewish origin, from *taking* his degree, he was not able to compete for either of the Smith's Prizes. In more enlightened times (1872), he had the degrees of B.A. and M.A., by accumulation, conferred upon him, and received therewith the honour of a Latin speech from the Public Orator. He himself says: "I am perhaps the only man in England who am a full (voting) Master of Arts for the three Universities of Dublin, Cambridge, and Oxford, having received that degree from these Universities in the order above given: from Dublin, by *ad eundem;* from Cambridge, *ob merita;* from Oxford, by decree." He is now D.C.L. of Oxford, LL.D. of Dublin and Edinburgh, and Hon. Fellow of St John's College, Cambridge. It is still open for him to receive yet higher recognition from his own *alma mater***.

Prof. Sylvester became a student of the Inner Temple, July 29, 1846, and was called to the Bar on November 22, 1850†. He has been Professor of Natural Philosophy at University College, London; of Mathematics at the University of Virginia, U.S.A.‡; then ten years later Professor at the Royal Military Academy,

* Foster's *Hand-book of Men at the Bar.*

[** The honorary degree Sc.D. was conferred upon him by Cambridge in 1890.]

† Foster, *l.c.*

‡ The late Prof. Key, of University College and School, was the first occupant of the Chair, founded by Mr Jefferson, once President of the United States, in 1824.

Woolwich; and again, after a five years' interval, Professor of Mathematics at the Johns Hopkins University, Baltimore, U.S.A., from its foundation in 1877. Finally, in December 1883, he was elected Savilian Professor of Geometry at Oxford, in succession to Prof. Henry Smith*. His first printed paper was on Fresnel's optical theory (in the *Phil. Mag.*, 1837).

We can here only briefly allude to a communication which was accompanied by many important results: we refer to the Friday evening address (January 23, 1874) to the Royal Institution, "On Recent Discoveries in Mechanical Conversion of Motion." He says:—"It would be difficult to quote any other discovery which opens out such vast and varied horizons as this of Peaucellier's,—in one direction, descending to the wants of the workshop, the simplification of the steam-engine, the revolutionising of the mill-wright's trade, the amelioration of garden-pumps, and other domestic conveniences (the sun of science glorifies all it shines upon); and in the other, soaring to the sublimest heights of the most advanced doctrines of modern analysis, lending aid to, and throwing light from a totally unexpected quarter on the researches of such men as Abel, Riemann, Clebsch, Grassmann, and Cayley. Its head towers above the clouds, while its feet plunge into the bowels of the earth."

The only works that Prof. Sylvester has published, we believe, are: (1) "A Probationary Lecture on Geometry, delivered before the Gresham Committee and the Members of the Common Council of the City of London, December 4, 1854," a slight thing which had to be written and delivered at a few hours' notice; (2) "Laws of Verse," 1870; (3) several short poems, sonnets, and translations, which have appeared in our columns and elsewhere.

Our notice would be incomplete without some record of the honours that have been conferred upon Dr Sylvester. He was elected a Fellow of the Royal Society on April 25, 1839; has received a Royal Medal (1860) and the Copley Medal (1880), this latter rarely awarded, we believe, to a pure mathematician. On this last occasion, Mr Spottiswoode accompanied the presentation with the words, "His extensive and profound researches in pure mathematics, especially his contributions to the theory of invariants and covariants, to the theory of numbers and to modern geometry, may be regarded as fully establishing Mr Sylvester's claim to the award." He is a Fellow of New College, Oxford; Foreign Associate of the United States National Academy of Sciences; Foreign Member of the Royal Academy of Sciences, Göttingen, of the Royal Academy of Sciences of Naples, and of the Academy of Sciences of Boston; Corresponding Member of the Institute of France, of the Imperial Academy of Science of St Petersburg, of the Royal Academy of Science of Berlin, of the Lyncei of Rome, of the Istituto Lombardo, and of the Société Philomathique. He has been long connected with the editorial staff of the *Quarterly Journal of Mathematics* (under

* He commences his Oxford lecture (*Nature*, vol. XXXIII. p. 222), of date December 12, 1885, with the words: "It is now two years and seven days since a message by the Atlantic cable containing the single word 'elected' reached me in Baltimore informing me that I had been appointed Savilian Professor of Geometry in Oxford, so that for three weeks I was in the unique position of filling the post and drawing the pay of Professor of Mathematics in each of two Universities."

one or another of its titles), and was the first editor of, and is a considerable contributor to, the *American Journal of Mathematics*; and he was at one time Examiner in Mathematics and Natural Philosophy in the University of London. He was not an original member of the London Mathematical Society (founded January 16, 1865), but was elected a member on June 19, 1865, Vice-President on January 15, 1866, and succeeded Prof. De Morgan as the second President on November 8, 1866. The Society showed its recognition of his great services to them and to mathematical science generally by awarding him its De Morgan Gold Medal in November 1887. Wherever Dr Sylvester goes, there is sure to be mathematical activity; and the latest proof of this is the formation, during the last term at Oxford, of a Mathematical Society, which promises, we hear without surprise, to do much for the advancement of mathematical science there.

The writings of Sylvester date from the year 1837; the number of them in the Royal Society Index up to the year 1863 is 112, in the next ten years 38, and in the volume for the next ten years 81, making 231 for the years 1837 to 1883: the number of more recent papers is also considerable. They relate chiefly to finite analysis, and cover by their subjects a large part of it: algebra, determinants, elimination, the theory of equations, partitions, tactic, the theory of forms, matrices, reciprocants, the Hamiltonian numbers, &c.; analytical and pure geometry occupy a less prominent position; and mechanics, optics, and astronomy are not absent. A leading feature is the power which is shown of originating a theory or of developing it from a small beginning; there is a breadth of treatment and determination to make the most of a subject, an appreciation of its capabilities, and real enjoyment of it. There is not unfrequently an adornment or enthusiasm of language which one admires, or is amused with: we have a motto from Milton, or Shakespeare; a memoir is a trilogy divided into three parts, each of which has its action complete within itself, but the same general cycle of ideas pervades all three, and weaves them into a sort of complex unity; the apology for an unsymmetrical solution is—symmetry, like the grace of an eastern robe, has not unfrequently to be purchased at the expense of some sacrifice of freedom and rapidity of action; and, he remarks, may not music be described as the mathematic of sense, mathematic as the music of the reason? the soul of each the same! &c. It is to be mentioned that there is always a generous and cordial recognition of the merit of others, his fellow-workers in the science.

It would be in the case of any first-rate mathematician—and certainly as much so in this as in any other case—extremely interesting to go carefully through the whole of a long list of memoirs, tracing out as well their connexion with each other, and the several leading ideas on which they depend, as also their influence on the development of the theories to which they relate; but for doing this properly, or at all, space and time, and a great amount of labour, are required. Short of doing so, one can only notice particular theorems—and there are, in the case of Sylvester, many of these, "beautiful exceedingly," which, for their own sakes, one is tempted to refer to—or one can give titles, which, to those familiar with the memoirs themselves, will recall the rich stores of investigation and theory contained therein.

A considerable number of papers, including some of the earliest ones, relate to the question of the reality of the roots of a numerical equation: in the several connexions thereof with Sturm's theorem, Newton's rule for the number of imaginary roots, and the theory of invariants. Sylvester obtained for the Sturmian functions, divested of square factors, or say for the reduced Sturmian functions, singularly elegant expressions in terms of the roots, viz. these were

$$f_2(x) = \Sigma (a-b)^2 (x-c)(x-d) \ldots, \quad f_3(x) = \Sigma (a-b)^2 (a-c)^2 (b-c)^2 (x-d) \ldots, \text{ \&c.};$$

but not only this: applying the Sturmian process of the greatest common measure (not to $f(x)$, $f'(x)$, but instead) to two independent functions $f(x)$, $\phi(x)$, he obtained for the several resulting functions expressions involving products of differences between the roots of the one and the other equation, $f(x) = 0$, $\phi(x) = 0$; the question then arose, what is the meaning of these functions? The answer is given by his theory of *intercalations:* they are signaletic functions, indicating in what manner (when the real roots of the two equations are arranged in order of magnitude) the roots of the one equation are intercalated among those of the other. The investigations in regard to Newton's rule (not previously demonstrated) are very important and valuable: the principle of Sturm's demonstration is applied to this wholly different question: viz. x is made to vary continuously, and the consequent gain or loss of changes of sign is inquired into. The third question is that of the determination of the character of the roots of a quintic equation by means of invariants. In connexion with it, we have the noteworthy idea of *facultative* points; viz. treating as the coordinates of a point in n-dimensional space those functions of the coefficients which serve as criteria for the reality of the roots, a point is facultative or non-facultative according as there is, or is not, corresponding thereto any equation with real coefficients: the determination of the characters of the roots depends (and, it would seem, depends only) on the bounding surface or surfaces of the facultative regions, and on a surface depending on the discriminant. Relating to these theories there are two elaborate memoirs, "On the Syzygetic Relations &c." and "Algebraical Researches &c." in the *Philosophical Transactions* for the years 1853 and 1864 respectively; but as regards Newton's rule later papers must also be consulted.

In the years 1851—54, we have various papers on homogeneous functions, the calculus of forms, &c. (*Camb. and Dub. Math. Journal*, vols. VI. to IX.), and the separate work "On Canonical Forms" (London, 1851). These contain crowds of ideas, embodied in the new words, *cogredient, contragredient, concomitant, covariant, contravariant, invariant, emanant, combinant, commutant, canonical form, plexus,* &c., ranging over and vastly extending the then so-called theories of linear transformations and hyperdeterminants. In particular, we have the introduction into the theory of the very important idea of *continuous* or *infinitesimal* variation: say that a function, which (whatever are the values of the parameters on which it depends) is invariant for an infinitesimal change of the parameters, is absolutely invariant.

There is, in 1844, in the *Philosophical Magazine*, a valuable paper, "Elementary Researches in the Analysis of Combinatorial Aggregation," and the titles of two other papers, 1865 and 1866, may be mentioned: "Astronomical Prolusions; commencing with the instantaneous proof of Lambert's and Euler's theorems, and modulating through the construction of the orbit of a heavenly body from two heliocentric distances, the subtended chord, and the periodic time, and the focal theory of Cartesian ovals, into a discussion of motion in a circle and its relation to planetary motion"; and the sequel thereto, "Note on the periodic changes of orbit under certain circumstances of a particle acted upon by a central force, and on vectorial coordinates, &c., together with a new theory of the analogues of the Cartesian ovals in space."

Many of the later papers are published in the *American Mathematical Journal*, founded, in 1878, under the auspices of the Johns Hopkins University, and for the first six volumes of which Sylvester was editor-in-chief. We have, in vol. I., a somewhat speculative paper entitled "An application of the new atomic theory to the graphical representation of the invariants and covariants of binary quantics," followed by appendices and notes relating to various special points of the theory; and in the same and subsequent volumes various memoirs on binary and ternary quantics, including papers (by himself, with the aid of Franklin) containing tables of the numerical generating functions for binary quantics of the first ten orders, and for simultaneous binary quantics of the first four orders, &c. The memoir (vols. II. and III.) on "Ternary cubic-form equations" is connected with some early papers relating to the theory of numbers. We have in it the theory of residuation on a cubic curve, and the beautiful chain-rule of rational derivation; viz. from an arbitrary point 1 on the curve it is possible to derive the singly infinite series of points $(1, 2, 4, 5, \ldots, 3p \pm 1)$ such that the chord through any two points, m, n, again meets the curve in a point $m+n$, $m \sim n$ (whichever number is not divisible by 3) of the series; moreover, the coordinates of any point m are rational and integral functions of the degree m^2 of those of the point 1.

There is in vol. V. the memoir, "A Constructive Theory of Partitions arranged in three acts, an Interact in two parts, and an Exodion," and in vol. VI. we have "Lectures on the Principles of Universal Algebra," (referring to a course of lectures on multinomial quantity, in the year 1881). The memoir is incomplete, but the general theories of nullity and vacuity, and of the corpus formed by two independent matrices of the same order, are sketched out; and there are, in the *Comptes rendus* of the French Academy, later papers containing developments of various points of the theory,—the conception of "nivellators" may be referred to.

The last-mentioned paper in the *American Mathematical Journal* was published subsequently to Sylvester's return to England on his appointment as Savilian Professor of Mathematics at Oxford. In December 1886, he gave there a public lecture containing an outline of his new theory of reciprocants (reported in *Nature*, January 7, 1887), and the lectures since delivered are published under the title, "Lectures on the Theory of Reciprocants" (reported by J. Hammond), same *Journal*, vols. VIII. to X.; thirty-three lectures actually delivered, entire or in abstract, in the course of three terms, to a class in the University, with a concluding so-called lecture 34, which is due to Hammond. The subject, as is well known, is that of the functions of a dependent variable, y, and its differential coefficients, y', y'', ..., in regard to x (or, rather, the functions of y', y'', ...), which remain unaltered by the interchange of the variables x and y: this is a less stringent condition than that imposed by Halphen ("Thèse," 1878) on his differential invariants, and the theory is accordingly a more extensive one. A passage may be quoted:—"One is surprised to reflect on the change which is come over Algebra in the last quarter of a century. It is now possible to enlarge to an almost unlimited extent on any branch of it. These thirty lectures, embracing only a fragment of the theory of reciprocants, might be compared to an unfinished epic in thirty cantos. Does it not seem as if Algebra had attained to the dignity of a fine art, in which the workman has a free hand to develop his conceptions, as in a musical theme or a subject for painting? Formerly, it consisted in detached theorems, but nowadays it has reached a point in which every properly-developed algebraical composition, like a skilful landscape, is expected to suggest the notion of an infinite distance lying beyond the limits of the canvas." And, indeed, the theory has already spread itself out far and wide, not only in these lectures by its founder, but in various papers by auditors of them, and others,—Elliott, Hammond, Leudesdorf, Rogers, Macmahon, Berry, Forsyth.

Sylvester's latest important investigations relate to the Hamiltonian numbers: there is a memoir, *Crelle*, t. C. (1887), and, by Sylvester and Hammond jointly, two memoirs in the *Philosophical Transactions*. The subject is that of the series of numbers 2, 3, 5, 11, 47, 923, calculated thus far by Sir W. R. Hamilton in his well-known Report to the British Association, on Jerrard's method. A formula for the independent calculation of any term of the series was obtained by Sylvester, but the remarkable law by means of a generating function was discovered by Hammond, viz. E_0, E_1, E_2, ..., being the series 3, 4, 6, ... of the foregoing numbers, each increased by unity; then these are calculated by the formula

$$(1-t)^{E_0} + t(1-t)^{E_1} + t^2(1-t)^{E_2} + \ldots = 1 - 2t,$$

equating the powers of t on the two sides respectively: observe the paradox, $t = \frac{1}{2}$, then the formula gives $0 =$ sum of a series of positive powers of $\frac{1}{2}$.

Enough has been said to call to mind some of Sylvester's achievements in mathematical science. Nothing further has been attempted in the foregoing very imperfect sketch.

901.

NOTE ON THE SUMS OF TWO SERIES.

[From the *Messenger of Mathematics*, vol. XIX. (1890), pp. 29—31.]

I CONSIDER the two series

$$S = \frac{1}{1+e^{\pi\alpha}} + \frac{1}{3(1+e^{3\pi\alpha})} + \frac{1}{5(1+e^{5\pi\alpha})} + \dots,$$

and

$$S_1 = \frac{1}{2+\pi\alpha} + \frac{1}{3(2+3\pi\alpha)} + \frac{1}{5(2+5\pi\alpha)} + \dots,$$

where α is real, positive, and indefinitely small; these would at first sight appear to be equal to each other, but this is not in fact the case.

Taking first the series S_1, putting therein $\pi\alpha = 2x$, this is

$$2S_1 = \frac{1}{1+x} + \frac{1}{3(1+3x)} + \frac{1}{5(1+5x)} + \dots.$$

Now we have, (Legendre, *Théorie des Fonctions Elliptiques*, t. II. p. 438),

$$\frac{y}{1+y} + \frac{y}{2(2+y)} + \frac{y}{3(3+y)} + \dots = C + \frac{d}{dy}\log\Gamma(y+1),$$

where C is Euler's constant, $= {\cdot}577\dots$; and if y be real, positive, and very large, then

$$\Gamma(y+1) = \sqrt{(2\pi)}\, y^{y+\frac{1}{2}} e^{-y+\frac{1}{12y}+\dots};$$

whence, differentiating the logarithm and neglecting the terms which contain negative powers of y, then the value is $= C + \log y$; hence, writing $y = \frac{1}{x}$, we obtain

$$\frac{1}{1+x} + \frac{1}{2(1+2x)} + \frac{1}{3(1+3x)} + \dots = C - \log x.$$

Writing herein $2x$ for x, and dividing by 2, we have

$$\frac{1}{2(1+2x)}+\frac{1}{4(1+4x)}+\ldots=\tfrac{1}{2}C-\tfrac{1}{2}\log 2x,\ =\tfrac{1}{2}(C-\log 2)-\tfrac{1}{2}\log x;$$

or, subtracting,

$$\frac{1}{1+x}+\frac{1}{3(1+3x)}+\frac{1}{5(1+5x)}+\ldots=\tfrac{1}{2}(C+\log 2)-\tfrac{1}{2}\log x.$$

Hence, writing for x its value, $=\tfrac{1}{2}\pi\alpha$, we have

$$S_1=\tfrac{1}{4}(C+\log 2)-\tfrac{1}{4}\log\tfrac{1}{2}\pi\alpha,\ =\tfrac{1}{4}(C+2\log 2-\log\pi)-\tfrac{1}{4}\log\alpha.$$

For the series S, we have (*Fundamenta Nova*, p. 103*) the formula

$$\tfrac{1}{4}\log\frac{2K}{\pi}=\frac{1}{1+q^{-1}}+\frac{1}{3(1+q^{-3})}+\frac{1}{5(1+q^{-5})}+\ldots;$$

or, putting herein $\alpha=\dfrac{K'}{K}$, then $q=e^{-\frac{\pi K'}{K}}=e^{-\pi\alpha}$, and thence

$$S=\frac{1}{1+e^{\pi\alpha}}+\frac{1}{2(1+e^{2\pi\alpha})}+\frac{1}{3(1+e^{3\pi\alpha})}+\ldots,\ =\tfrac{1}{4}\log\frac{2K}{\pi}.$$

We have $q=e^{-\pi\alpha}$, which is real, positive, and less than but indefinitely near to 1; hence also k is real, positive, and less than but indefinitely near to 1, say the value is $=1-\beta$; thence $k'=\sqrt{(2\beta)}$, and $K=\log\dfrac{4}{k'}$, $=\log\dfrac{2\sqrt{2}}{\sqrt{\beta}}$; also $K'=\tfrac{1}{2}\pi$, and therefore $\alpha=\dfrac{K'}{K}=\tfrac{1}{2}\pi\div\log\dfrac{2\sqrt{2}}{\sqrt{\beta}}$, whence $\log\dfrac{2\sqrt{2}}{\sqrt{\beta}}=\dfrac{\pi}{2\alpha}$, which is the relation between α and β; and we thus have $\dfrac{2K}{\pi}=\dfrac{2}{\pi}\log\dfrac{2\sqrt{2}}{\sqrt{\beta}}$, $=\dfrac{1}{\alpha}$; and consequently $S=\tfrac{1}{4}\log\dfrac{2K}{\pi}$, $=-\tfrac{1}{4}\log\alpha$. The two values thus are

$$S=-\tfrac{1}{4}\log\alpha,\quad S_1=\tfrac{1}{4}(C+2\log 2-\log\pi)-\tfrac{1}{4}\log\alpha,$$

each depending on $\log\alpha$, and having for this term the same coefficient $-\tfrac{1}{4}$; but there is in S_1 a constant term $\tfrac{1}{4}(C+2\log 2-\log\pi)$, where C is the constant ·577....

It is easy to see why the series S is not reducible to S_1; however small α may be in the general term $\dfrac{1}{n(1+e^{n\pi\alpha})}$, then taking n sufficiently large, not only $n\pi\alpha$ is not indefinitely small, but it in fact becomes indefinitely large; the general term of the first series thus approximates to $\dfrac{1}{ne^{n\pi\alpha}}$, or the terms diminish somewhat more rapidly than in a geometric series with the ratio $e^{-\pi\alpha}$ (a small positive value less than but very near to 1), whereas in the second series the general term approximates to $\dfrac{1}{n^2\pi\alpha}$, or the convergence is ultimately that of the series $\dfrac{1}{n^2}+\dfrac{1}{(n+1)^2}+\ldots$.

[* Jacobi's *Gesammelte Werke*, t. I, p. 159.]

902.

ON THE FOCALS OF A QUADRIC SURFACE.

[From the *Messenger of Mathematics*, vol. XIX. (1890), pp. 113—117.]

IN plane geometry, the focus of a curve is the node of the circumscribed line-system of the curve and the circular points at infinity; and so, in solid geometry, the focal of a surface or curve is the nodal line of the circumscribed developable of the surface or curve and the circle at infinity. And as in plane geometry the circular points at infinity may be regarded as an indefinitely thin conic, so in solid geometry the circle at infinity may be regarded as an indefinitely thin quadric surface.

In plane geometry, let it be proposed to find the circumscribed line-system (common tangents) of the two quadrics

$$\frac{x^2}{a}+\frac{y^2}{b}+\frac{z^2}{c}=0,\quad \frac{x^2}{a'}+\frac{y^2}{b'}+\frac{z^2}{c'}=0;$$

if a common tangent be

$$\xi x+\eta y+\zeta z=0,$$

then we have

$$a\xi^2+b\eta^2+c\zeta^2=0,$$

$$a'\xi^2+b'\eta^2+c'\zeta^2=0;$$

here writing for shortness

$$f,\ g,\ h=bc'-b'c,\ ca'-c'a,\ ab'-a'b,$$

we have

$$\xi^2:\eta^2:\zeta^2=f:g:h;$$

and thence the tangent is

$$x\sqrt{(f)}\pm y\sqrt{(g)}\pm z\sqrt{(h)}=0,$$

viz. we have thus four tangents, and the rationalised form is of course

$$f^2x^4+g^2y^4+h^2z^4-2ghy^2z^2-2hfz^2x^2-2fgx^2y^2=0.$$

In connexion with the corresponding question in solid geometry, I obtain this equation in a different manner. We investigate the envelope of the line $\xi x + \eta y + \zeta z = 0$, considering ξ, η, ζ as parameters connected by the foregoing two equations; by the ordinary process of indeterminate multipliers, we have

$$x + (\lambda a + \mu a')\,\xi = 0, \quad y + (\lambda b + \mu b')\,\eta = 0, \quad z + (\lambda c + \mu c')\,\zeta = 0,$$

and thence, eliminating ξ, η, ζ, we obtain

$$\frac{x^2}{\lambda a + \mu a'} + \frac{y^2}{\lambda b + \mu b'} + \frac{z^2}{\lambda c + \mu c'} = 0,$$

$$\frac{ax^2}{(\lambda a + \mu a')^2} + \frac{by^2}{(\lambda b + \mu b')^2} + \frac{cz^2}{(\lambda c + \mu c')^2} = 0,$$

$$\frac{a'x^2}{(\lambda a + \mu a')^2} + \frac{b'y^2}{(\lambda b + \mu b')^2} + \frac{c'z^2}{(\lambda c + \mu c')^2} = 0,$$

equations equivalent to two equations, from which λ and μ are to be eliminated. The second and third equations are the derived functions of the first equation in regard to λ, μ respectively; and hence, expressing the first equation in an integral form, the result is

$$\text{Disct. } \{x^2 (\lambda b + \mu b')(\lambda c + \mu c') + \&\text{c.}\} = 0;$$

viz. this is

$$\{(bc' + b'c)\,x^2 + (ca' + c'a)\,y^2 + (ab' + a'b)\,z^2\}^2 - 4\,(bcx^2 + cay^2 + abz^2)\,(b'c'x^2 + c'a'y^2 + a'b'z^2) = 0,$$

or developing and reducing, we have the foregoing result, the coefficients entering through the combinations

$$f,\ g,\ h = bc' - b'c,\ ca' - c'a,\ ab' - a'b.$$

Writing $c = -1$, also $a' = b' = 1$, $c' = 0$: the equation of the first conic is

$$\frac{x^2}{a} + \frac{y^2}{b} - z^2 = 0,$$

and that of the second may be replaced by the two equations $x^2 + y^2 = 0$, $z = 0$, viz. these give the circular points at infinity: we have $f, g, h = 1, -1, a - b$, and the equation of the line-system is

$$x^4 + 2x^2y^2 + y^4 - 2h\,(x^2 - y^2)\,z^2 + h^2z^4 = 0.$$

If finally, $z = 1$, then for the tangents from the circular points at infinity to the quadric

$$\frac{x^2}{a} + \frac{y^2}{b} - 1 = 0,$$

the equation is

$$x^4 + 2x^2y^2 + y^4 - 2h\,(x^2 - y^2) + h^2 = 0,$$

where, as before, $h = a - b$; the four tangents intersect in pairs in the two circular points at infinity, and in four other points which are the foci of the quadric.

Passing now to the problem in solid geometry, we consider the two quadric surfaces

$$\frac{x^2}{a}+\frac{y^2}{b}+\frac{z^2}{c}+\frac{w^2}{d}=0,$$

$$\frac{x^2}{a'}+\frac{y^2}{b'}+\frac{z^2}{c'}+\frac{w^2}{d'}=0;$$

if a common tangent plane hereof be

$$\xi x + \eta y + \zeta z + \omega w = 0,$$

then we have

$$a\xi^2 + b\eta^2 + c\zeta^2 + d\omega^2 = 0,$$

$$a'\xi^2 + b'\eta^2 + c'\zeta^2 + d'\omega^2 = 0;$$

and the circumscribed developable is the envelope of the plane, considering in the equation thereof ξ, η, ζ, ω as variable parameters connected by the last two equations. By what precedes, it is at once seen that the resulting equation is

$$\text{Disct.}\ \{x^2(b\lambda+b'\mu)(c\lambda+c'\mu)(d\lambda+d'\mu)+\&\text{c.}\}=0,$$

viz. this is a quadric equation in (x^2, y^2, z^2, w^2), the coefficients a, b, c, d, a', b', c', d', entering therein through the combinations

$$\mathrm{a},\ \mathrm{b},\ \mathrm{c} = bc' - b'c,\ \ ca' - c'a,\ \ ab' - a'b;$$

$$\mathrm{f},\ \mathrm{g},\ \mathrm{h} = ad' - a'd,\ \ bd' - b'd,\ \ cd' - c'd.$$

The developed result is given in my paper "On the developable surfaces which arise from two surfaces of the second order," *Camb. and Dub. Math. Jour.*, t. v. (1850), pp. 45—53, [84]. We require here, for the particular case, $d=-1$, $a'=b'=c'=1$, $d'=0$, viz. one of the surfaces is taken to be

$$\frac{x^2}{a}+\frac{y^2}{b}+\frac{z^2}{c}-w^2=0,$$

and the other is the circle at infinity $x^2+y^2+z^2=0$, $w=0$; we thus have a, b, c = 1, 1, 1; f, g, h = $b-c$, $c-a$, $a-b$, or now using the italic letters f, g, h to signify these values, we have $f+g+h=0$. The equation is

$$\begin{aligned}
& f^2x^8+g^2y^8+h^2z^8+f^2g^2h^2w^8 \\
& + 2g(g-h)y^6z^2 + 2h(h-f)z^6x^2 + 2f(f-g)x^6y^2 \\
& - 2h(g-h)y^2z^6 - 2f(h-f)z^2x^6 - 2g(f-g)x^2y^6 \\
& + 2f^2(g-h)x^6w^2 + 2g^2(h-f)y^6w^2 + 2h^2(f-g)z^6w^2 \\
& - 2f^2gh(g-h)x^2w^6 - 2fg^2h(h-f)y^2w^6 - 2fgh^2(f-g)z^2w^6 \\
& + f^2(f^2-6gh)x^4w^4 + g^2(g^2-6hf)y^4w^4 + h^2(h^2-6fg)z^4w^4 \\
& + (f^2-6gh)y^4z^4 + (g^2-6hf)z^4x^4 + (h^2-6fg)x^4y^4 \\
& + 2gh(gh-3f^2)w^4y^2z^2 + 2hf(hf-3g^2)w^4z^2x^2 + 2fg(fg-3h^2)w^4x^2y^2 \\
& + 2h(gh-3f^2)z^4x^2w^2 + 2f(hf-3g^2)x^4y^2w^2 + 2g(fg-3h^2)y^4z^2w^2 \\
& - 2g(gh-3f^2)y^4x^2w^2 - 2h(hf-3g^2)z^4y^2w^2 - 2f(fg-3h^2)x^4z^2w^2 \\
& + 2(gh-3f^2)x^4y^2z^2 - 2(hf-3g^2)x^2y^4z^2 - 2(fg-3h^2)x^2y^2z^4 \\
& + 2(g-h)(h-f)(f-g)x^2y^2z^2w^2 = 0.
\end{aligned}$$

The equation may be written in the following four forms:

$$x^2\Theta_1 + (gy^2 - hz^2 + ghw^2)^2 \{y^4 + 2y^2z^2 + z^4 - 2f(y^2 - z^2)w^2 + f^2w^4\} = 0,$$

$$y^2\Theta_1 + (hz^2 - fx^2 + hfw^2)^2 \{z^4 + 2z^2x^2 + x^4 - 2g(z^2 - x^2)w^2 + g^2w^4\} = 0,$$

$$z^2\Theta_1 + (fx^2 - gy^2 + fgw^2)^2 \{x^4 + 2x^2y^2 + y^4 - 2h(x^2 - y^2)w^2 + h^2w^4\} = 0,$$

$$w^2\Theta_1 + (x^2 + y^2 + z^2)^2 \times \{f^2x^4 + g^2y^4 + h^2z^4 - 2ghy^2z^2 - 2hfz^2x^2 - 2fgx^2y^2\} = 0;$$

the last of these shows that the circle at infinity $x^2 + y^2 + z^2 = 0$, $w = 0$, is a nodal line on the surface; this, however, is not regarded as a focal. The other three show that we have also, as nodal lines, three conics, which are the focals of the given surface; viz. now writing $w = 1$, we have, for the quadric surface

$$\frac{x^2}{a} + \frac{y^2}{b} + \frac{z^2}{c} - 1 = 0,$$

the three focal conics

$$x = 0, \quad -\frac{y^2}{h} + \frac{z^2}{g} - 1 = 0,$$

$$y = 0, \quad \frac{x^2}{h} \quad - \frac{z^2}{f} - 1 = 0,$$

$$z = 0, \quad -\frac{x^2}{g} + \frac{y^2}{f} \quad - 1 = 0,$$

where, as before, f, g, $h = b - c$, $c - a$, $a - b$ respectively. If, as usual, a, b, c are positive and in order of decreasing magnitude, then for the ellipsoid

$$\frac{x^2}{a} + \frac{y^2}{b} + \frac{z^2}{c} - 1 = 0,$$

we have the focal conics

$$x = 0, \quad -\frac{y^2}{a - b} + \frac{z^2}{a - c} - 1 = 0,$$

$$y = 0, \quad \frac{x^2}{a - b} \quad - \frac{z^2}{b - c} - 1 = 0,$$

$$z = 0, \quad \frac{x^2}{a - c} + \frac{y^2}{b - c} \quad - 1 = 0;$$

viz. these are an imaginary conic, the focal hyperbola, and the focal ellipse, in the coordinate planes $x = 0$, $y = 0$, $z = 0$ respectively.

903.

ON LATIN SQUARES.

[From the *Messenger of Mathematics*, vol. XIX. (1890), pp. 135—137.]

IF in each line of a square of n^2 compartments the same n letters $a, b, c, \ldots$ are arranged so that no letter occurs twice in the same column, we have what was termed by Euler "a Latin square." Supposing that in one of the lines the letters are arranged in the natural order $abcde\ldots$, then in the remaining lines there must be arrangements beginning with b, c, d, e, &c., respectively, and we may consider the case in which the bottom line has the arrangement $abcde\ldots$, and in the other lines, reckoning from the bottom one in order, the arrangements begin with b, c, d, e, &c., respectively: if the number of such squares be $=N$, then, obviously, the whole number of squares which can be formed with the same n arrangements is $=N\,[n]^n$.

Starting with the bottom line as above, then it is a well-known problem to determine the number of arrangements for the second line; viz. this number is

$$=1.2.3\ldots n\left\{1-\frac{1}{1}+\frac{1}{1.2}-\ldots\pm\frac{1}{1.2\ldots n}\right\};$$

and if we assume, as above, that the second line begins with b, then the whole number of arrangements is this number divided by $(n-1)$, the quotient being of course integral. For instance, when $n=5$, the number is $=120-120+60-20+5-1$, $=44$, which is divisible by 4, and the number of arrangements for the second line is thus $=11$.

But the number of arrangements for the third line will be different according to the arrangement selected for the second line, and it is not easy to see how in general the whole number of arrangements for the third line is to be calculated, and the difficulty of course increases for the next following lines; it may be remarked that, when all the lines are filled up except the top-line, the top-line is completely determined.

Imagine the square completed: we may write down the substitutions by which we pass from the bottom line to itself (this is of course the substitution 1) and to each of the other lines respectively; we have thus a set of n substitutions, which may form *a group*; and when this is so, we may conversely from the group construct the Latin square. But it is not every Latin square which is thus connected with a group of n substitutions.

In the cases $n = 2, 3, 4$ there is no difficulty; the squares are

$$\begin{array}{cccccc} b\,a, & c\,a\,b, & d\,c\,a\,b, & d\,c\,b\,a, & d\,c\,b\,a, & d\,a\,b\,c, \\ a\,b, & b\,c\,a, & c\,d\,b\,a, & c\,d\,a\,b, & c\,a\,d\,b, & c\,d\,a\,b, \\ & a\,b\,c, & b\,a\,d\,c, & b\,a\,d\,c, & b\,d\,a\,c, & b\,c\,d\,a, \\ & & a\,b\,c\,d, & a\,b\,c\,d, & a\,b\,c\,d, & a\,b\,c\,d; \end{array}$$

viz. when $n = 2$ the number is 1, when $n = 3$ it is 1, when $n = 4$ it is 4; in this last case, the arrangement *badc* for the second line gives two squares, but each of the other arrangements only one square.

In each of the squares of 4, we have a group, viz. for the four squares respectively, these are

$$\begin{array}{lllll} 1^\circ, & 1, & (ab)\,(cd), & (acbd)\;, & (adbc)\;, \\ 2^\circ & 1, & (ab)\,(cd), & (ac)\,(bd), & (ad)\,(bc), \\ 3^\circ & 1, & (abdc)\;, & (acdb)\;, & (ad)\,(bc), \\ 4^\circ & 1, & (abcd)\;, & (ac)\,(bd), & (adcb)\;. \end{array}$$

1°, 3°, 4° are the cyclical groups of (*acbd*), (*abdc*), and (*abcd*), respectively; 2° is a different kind of group.

In the case when $n = 5$, the whole number of squares is 56; viz. there are five arrangements of the second line each giving four squares, and six arrangements each giving six squares, $5 \,.\, 4 + 6 \,.\, 6 = 56$. The five arrangements are

$$\begin{array}{ccccc} b\,a\,e\,c\,d, & b\,a\,d\,e\,c, & b\,c\,a\,e\,d, & b\,d\,e\,a\,c, & b\,e\,d\,c\,a, \\ a\,b\,c\,d\,e, & a\,b\,c\,d\,e, & a\,b\,c\,d\,e, & a\,b\,c\,d\,e, & a\,b\,c\,d\,e, \end{array}$$

viz. in these cases, the substitutions for passing to the second line are (*ab*) (*ced*) (*ab*) (*cde*), (*abc*) (*de*), (*abd*) (*ce*), (*abe*) (*cd*), respectively.

The six arrangements are

$$\begin{array}{cccccc} b\,d\,a\,e\,c, & b\,e\,a\,c\,d, & b\,c\,e\,a\,d, & b\,e\,d\,a\,c, & b\,c\,d\,e\,a, & b\,d\,e\,c\,a, \\ a\,b\,c\,d\,e, & a\,b\,c\,d\,e, & a\,b\,c\,d\,e, & a\,b\,c\,d\,e, & a\,b\,c\,d\,e, & a\,b\,c\,d\,e; \end{array}$$

viz. in these cases, the substitutions for passing to the second line are (*abdec*), (*abedc*), (*abced*), (*abecd*), (*abcde*), (*abdce*), respectively.

A set of four squares is

e c d b a,	*e d a b c*,	*e c d a b*,	*e d b a c*,
d e b a c,	*d c b e a*,	*d e a b c*,	*d c a e b*,
c d a e b,	*c e d a b*,	*c d b e a*,	*c e d b a*,
b a e c d,	*b a e c d*,	*b a e c d*,	*b a e c d*,
a b c d e,	*a b c d e*,	*a b c d e*,	*a b c d e*;

and a set of six squares is

e c d a b,	*e a d c b*,	*e c d b a*,	*e a b c d*,	*e a b c d*,	*e c b a d*,
d e b c a,	*d c e b a*,	*d a e c b*,	*d c e b a*,	*d c e a b*,	*d a e c b*,
c a e b d,	*c e b a d*,	*c e b a d*,	*c e d a b*,	*c e d b a*,	*c e d b a*,
b d a e c,	*b d a e c*,	*b d a e c*,	*b d a e c*,	*b d e a c*,	*b d a e c*,
a b c d e,	*a b c d e*,	*a b c d e*,	*a b c d e*,	*a b c d e*,	*a b c d e*.

In a square belonging to a set of four, the substitutions for obtaining from any one line all the other lines are of a form such as 1, $(ab)(ced)$, $(ac)(bde)$, $(ad)(bec)$, $(ae)(bcd)$, which are not a group. In the case of a set of six squares, there is one square of the set (in the foregoing instance the first square) where the substitutions are of a form such as 1, $(abdec)$, $(acedb)$, $(adceb)$, $(aebcd)$, and which thus form a cyclical group of five substitutions.

904.

NOTE ON RECIPROCAL LINES.

[From the *Messenger of Mathematics*, vol. XIX. (1890), pp. 174, 175.]

IF two lines are reciprocal in regard to a quadric surface, then any point on the one line and any point on the other line are harmonics in regard to the surface, viz. the two points and the intersections of their line of junction with the surface form a harmonic range. This is obvious: the polar plane of the first point passes through the second line, and thus the second point is a point in the polar plane of the first point, that is, the two points are harmonics.

But it is worth while to look at the theorem in a different point of view. If the first line meets the surface in the points A, C and the second line meets the surface in the points B, D (in order to the four points being real, the surface must of course be a skew hyperboloid), then AB, BC, CD, DA are lines on the surface,

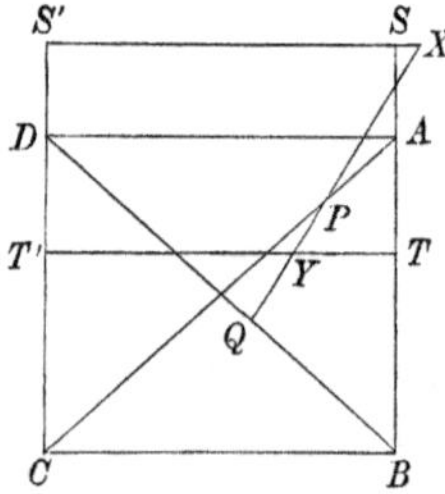

say AB, CD are directrices, and BC, AD are generatrices, the two reciprocal lines being the diagonals AC and BD of the skew quadrilateral. Taking on AC a point P and on BD a point Q, the theorem is that, if PQ meets the surface in the points

X, Y, then the four points P, Q, X, Y are harmonics. Let the generatrix through X meet AB, CD in the points S, S' respectively, and the generatrix through Y meet the same lines in the points T, T' respectively. Then the four lines CB, DA, $T'T$, $S'S$, are all met by an infinite series of lines, and in particular they are met by the lines CD, BA in the points C, D, T', S' and B, A, T, S respectively. Consequently, writing AH for anharmonic ratio, we have

$$AH\,(C,\ D,\ T',\ S') = AH\,(B,\ A,\ T,\ S).$$

Consider now the four lines CA, DB, $T'T$, $S'S$: these are each of them met by the lines CD and BA, and also by the line PQ; therefore, by an infinity of lines (that is, they are directrices of a hyperboloid). They are met by the last-mentioned three lines in the points C, D, T', S'; A, B, T, S; and P, Q, Y, X respectively; hence

$$AH\,(C,\ D,\ T',\ S') = AH\,(A,\ B,\ T,\ S) = AH\,(P,\ Q,\ Y,\ X).$$

Comparing with the foregoing equation, it appears that

$$AH\,(B,\ A,\ T,\ S) = AH\,(A,\ B,\ T,\ S),$$

viz. the anharmonic ratio is not altered by the interchange of the two points A, B: this implies that the points A, B, T, S are harmonics, viz. the pairs A, B and S, T are harmonics; and, this being so, the equation

$$AH(A,\ B,\ T,\ S) = AH\,(P,\ Q,\ Y,\ X)$$

shows that P, Q, Y, X are harmonics, viz. that the pairs P, Q and X, Y are harmonics; or say P, Q are harmonics in regard to the quadric surface, which is the theorem in question.

905.

ON THE EQUATION $x^{17}-1=0$.

[From the *Messenger of Mathematics*, vol. XIX. (1890), pp. 184—188.]

WRITING $\rho=\cos\frac{2\pi}{17}+i\sin\frac{2\pi}{17}$, I carry the solution up to the determination of the periods each of two roots, $\rho+\rho^{16}$, $=2\cos\frac{2\pi}{17}$, &c. The expressions contain the radicals

$$a=\surd(17),\quad b=\surd\{2(17-a)\},\quad c=\surd\{4(17+3a)-2(3+a)b\},$$

where a, b, c are taken to be positive ($a=4{\cdot}12$, $b=5{\cdot}07$, $c=6{\cdot}72$). Taking for a moment r to be any imaginary seventeenth root, $r=\rho^\theta$, then the algebraical expression for the period P_1 of eight roots is $P_1=\frac{1}{2}(-1\pm a)$, but I assume the value to be $P_1=\frac{1}{2}(-1+a)$, and thus determine θ to denote some one of the values 1, 2, 4, 8, 9, 13, 15, 16; similarly, I assume the value of Q_1 to be $=\frac{1}{4}(-1+a+b)$; and thus further determine θ to denote some one of the values 1, 4, 13, 16: and, again, I assume the value of R_1 to be $=\frac{1}{8}(-1+a+b+c)$, and thus further determine θ to denote one of the values 1 and 16. As regards the values of the periods R, it is obviously indifferent which value is taken, and I assume therefore $\theta=1$. This comes to saying that the signs of the radicals are determined in suchwise that r shall denote the root $\cos\frac{2\pi}{17}+i\sin\frac{2\pi}{17}$; and it is to be understood that r has this value.

I write now

$$\begin{aligned}
P_1&=r+r^9+r^{13}+r^{15}+r^{16}+r^8+r^4+r^2,\\
P_2&=r^3+r^{10}+r^5+r^{11}+r^{14}+r^7+r^{12}+r^6,\\
Q_1&=r+r^{13}+r^{16}+r^4,\\
Q_2&=r^9+r^{15}+r^8+r^2,\\
Q_3&=r^3+r^5+r^{14}+r^{12},\\
Q_4&=r^{10}+r^{11}+r^7+r^6,\\
R_1&=r+r^{16},\\
R_2&=r^{13}+r^4,\\
R_3&=r^9+r^8,
\end{aligned}$$

$$R_4 = r^{15} + r^2,$$
$$R_5 = r^3 + r^{14},$$
$$R_6 = r^5 + r^{12},$$
$$R_7 = r^{10} + r^7,$$
$$R_8 = r^{11} + r^6,$$

and moreover

$$\begin{aligned} a &= P_1 - P_2, \\ b &= 2\,(Q_1 - Q_2), \\ +b_1 &= 2\,(Q_3 - Q_4), \\ c &= 4\,(R_1 - R_2), \\ -c_1 &= 4\,(R_3 - R_4), \\ +c_2 &= 4\,(R_5 - R_6), \\ -c_3 &= 4\,(R_7 - R_8). \end{aligned}$$

It will appear by what follows that a is determined by the quadric equation $a^2 = 17$, but I have assumed that a denotes the positive root $a = \sqrt{(17)}$; similarly, b is determined by the quadric equation $b^2 = 2\,(17 - a)$, but it is assumed that b denotes the positive root, $b = \sqrt{\{2\,(17-a)\}}$; and c is determined by the quadric equation $c^2 = 4\,(17+3a) - 2\,(3+a)\,b$, but it is assumed that c denotes the positive root,

$$c = \sqrt{\{4\,(17+3a) - 2\,(3+a)\,b\}}.$$

If in the equations I had written b_1, c_1, c_2, c_3, instead of $+b_1$, $-c_1$, $+c_2$, $-c_3$, then b_1 comes out rationally in terms of a, b; and c_1, c_2, c_3 come out rationally in terms of a, b, c; the signs were attached to them *à posteriori*, in suchwise that the values of b_1, c_1, c_2, c_3 might be each of them positive; for their independent determination, we have, in fact, for b_1^2 an expression such as that for b^2; and for c_1^2, c_2^2, c_3^2 expressions such as that for c^2; and taking as above for each of them the positive value of the square root, we have

$$\begin{aligned} a &= \sqrt{(17)}, && (= 4{\cdot}12), \\ b &= \sqrt{\{2\,(17 - a)\}}, && (= 5{\cdot}07), \\ b_1 &= \sqrt{\{2\,(17 + a)\}}, && (= 6{\cdot}49), \\ c &= \sqrt{\{4\,(17+3a) - 2\,(3+a)\,b\}}, && (= 6{\cdot}72), \\ c_1 &= \sqrt{\{4\,(17+3a) + 2\,(3+a)\,b\}}, && (= 13{\cdot}77), \\ c_2 &= \sqrt{\{4\,(17-3a) + 2\,(-3+a)\,b_1\}}, && (= 5{\cdot}75), \\ c_3 &= \sqrt{\{4\,(17-3a) - 2\,(-3+a)\,b_1\}}, && (= 2{\cdot}02). \end{aligned}$$

The relations between the periods P are

$$P_1 + P_2 = -1,$$

	P_1	P_2
P_1	$-P_1+4$	-4
P_2		$-P_2+4$

	a
a	17

that is,

$$P_1^2 = -P_1 + 4, \text{ \&c.}; \; a^2 = 17.$$

Here for the second square, we have

$$a^2 = (P_1 - P_2)^2 = P_1^2 + P_2^2 - 2P_1P_2 = (-P_1+4) + (-P_2+4) + 8, \; = -P_1 - P_2 + 16, \; = 17;$$

and similarly in the other cases which follow.

For the periods Q, we have

$$Q_1 + Q_2 = P_1,$$
$$Q_3 + Q_4 = P_2,$$

	Q_1	Q_2	Q_3	Q_4
Q_1	Q_2+2Q_3+4	-1	$2Q_1+Q_2+Q_4$	$Q_2+Q_3+2Q_4$
Q_2		Q_1+2Q_4+4	Q_1+2Q_3+4	$Q_1+2Q_2+Q_3$
Q_3			$2Q_2+Q_4+4$	-1
Q_4				$2Q_1+Q_3+4$

	b	b_1
b	$34-2a$	$8a$
b_1		$34+2a$

so that $bb_1 = 8a$, or b_1 is given rationally in terms of a, b.

And for the periods R, we have

$$R_1 + R_2 = Q_1,$$
$$R_3 + R_4 = Q_2,$$
$$R_5 + R_6 = Q_3,$$
$$R_7 + R_8 = Q_4,$$

	R_1	R_2	R_3	R_4	R_5	R_6	R_7	R_8
R_1	R_4+2	$\overline{R_5+R_6}$	R_3+R_7	R_1+R_5	R_2+R_4	R_2+R_8	R_3+R_8	R_6+R_7
R_2		R_3+2	R_2+R_6	R_4+R_8	R_1+R_7	R_1+R_3	R_5+R_8	R_4+R_7
R_3			R_1+2	$\overline{R_7+R_8}$	R_6+R_8	R_2+R_5	R_1+R_4	R_4+R_5
R_4				R_2+2	R_1+R_6	R_5+R_7	R_3+R_6	R_2+R_3
R_5					R_8+2	$\overline{R_3+R_4}$	R_2+R_7	R_3+R_5
R_6						R_7+2	R_4+R_6	R_1+R_8
R_7							R_5+2	$\overline{R_1+R_2}$
R_8								R_3+2

	c	c_1	c_2	c_3
c	$4(17+3a)-2(3+a)b$	$8(b+b_1)$	$4(2a-b+b_1)$	$4(-2a+b+b_1)$
c_1		$4(17+3a)+2(3+a)b$	$4(2a+b+b_1)$	$4(2a+b-b_1)$
c_2			$4(17-3a)+2(-3+a)b_1$	$8(-b+b_1)$
c_3				$4(17-3a)-2(-3+a)b_1$

where observe that the overlined terms R_5+R_6, R_7+R_8, R_3+R_4, and R_1+R_2, have the values Q_4, Q_3, Q_2, Q_1, respectively, and in the last table b_1 may be considered as denoting its value $=\frac{8a}{b}$, so that c_1, c_2, c_3 are each given rationally in terms of a, b, c.

And from the foregoing results, we have

$$
\begin{aligned}
P_1 &= \tfrac{1}{2}(-1+a), &&= 1{\cdot}56,\\
P_2 &= \tfrac{1}{2}(-1-a), &&= -2{\cdot}56,\\[4pt]
Q_1 &= \tfrac{1}{4}(-1+a+b), &&= 2{\cdot}05,\\
Q_2 &= \tfrac{1}{4}(-1+a-b), &&= -0{\cdot}49,\\
Q_3 &= \tfrac{1}{4}(-1-a+b_1), &&= 0{\cdot}34,\\
Q_4 &= \tfrac{1}{4}(-1-a-b_1), &&= -2{\cdot}90,\\[4pt]
R_1 &= \tfrac{1}{8}(-1+a+b\ +c), &&= 1{\cdot}87,\\
R_2 &= \tfrac{1}{8}(-1+a+b\ -c), &&= 0{\cdot}18,\\
R_3 &= \tfrac{1}{8}(-1+a-b\ -c_1), &&= -1{\cdot}96,\\
R_4 &= \tfrac{1}{8}(-1+a-b\ +c_1), &&= 1{\cdot}47,\\
R_5 &= \tfrac{1}{8}(-1-a+b_1+c_2), &&= 0{\cdot}89,\\
R_6 &= \tfrac{1}{8}(-1-a+b_1-c_2), &&= -0{\cdot}55,\\
R_7 &= \tfrac{1}{8}(-1-a-b_1-c_3), &&= -1{\cdot}70,\\
R_8 &= \tfrac{1}{8}(-1-a-b_1+c_3), &&= -1{\cdot}20.
\end{aligned}
$$

The approximate numerical values have been given throughout only for the purpose of showing that the signs of the square roots have been rightly determined.

906.

NOTE ON SCHLAEFLI'S MODULAR EQUATION FOR THE CUBIC TRANSFORMATION; WITH A CORRECTION.

[From the *Messenger of Mathematics*, vol. XX. (1891), pp. 59, 60; 120.]

THE equation in question, *Crelle*, t. LXXII. (1870), p. 369, is

$$S^4 + T^4 + 8ST - S^3T^3 = 0,$$

where

$$S = \frac{\sqrt{2}}{\sqrt[4]{(kk')}}, \quad T = \frac{\sqrt{2}}{\sqrt[4]{(\lambda\lambda')}};$$

which must be of course equivalent to the ordinary modular equation

$$u^4 - v^4 + 2uv - 2u^3v^3 = 0,$$

where

$$u = \sqrt[4]{k}, \quad v = \sqrt[4]{\lambda};$$

the resemblance in form between the two equations, with such different meanings of the S and T in the one case, and the u and v in the other, is very noticeable.

Schlaefli's equation is

$$\frac{4}{kk'} + \frac{4}{\lambda\lambda'} + \frac{16}{(kk'\lambda\lambda')^{\frac{1}{4}}} - \frac{8}{(kk'\lambda\lambda')^{\frac{3}{4}}} = 0,$$

that is,

$$kk' + \lambda\lambda' = 2\,(kk'\lambda\lambda')^{\frac{1}{4}} - 4\,(kk'\lambda\lambda')^{\frac{3}{4}},$$

or say

$$k^2k'^2 + \lambda^2\lambda'^2 = 4\,(kk'\lambda\lambda')^{\frac{1}{2}} - 18\,(kk'\lambda\lambda') + 16\,(kk'\lambda\lambda')^{\frac{3}{2}}.$$

To deduce this from the uv-modular equation, we have (Jacobi's *Fund. Nova*, p. 68, *Ges. Werke*, t. I., p. 124),

$$(1 - u^8)(1 - v^8) = (1 - u^2v^2)^4,$$

or, multiplying each side by u^8v^8, and extracting the fourth root, we have

$$\sqrt{(kk'\lambda\lambda')} = u^2v^2(1-u^2v^2) = x - x^2,$$

if for shortness we write $x = u^2v^2$.

The equation to be proved thus is

$$u^8(1-u^8) + v^8(1-v^8) = 4(x-x^2) - 18(x-x^2)^2 + 16(x-x^2)^3.$$

But from the foregoing equation

$$(1-u^8)(1-v^8) = (1-u^2v^2)^4,$$

we have

$$u^8 + v^8 = 4u^2v^2 - 6u^4v^4 + 4u^6v^6, \quad = 4x - 6x^2 + 4x^3,$$

and thence

$$u^{16} + v^{16} = (4x - 6x^2 + 4x^3)^2 - 2x^4;$$

and the equation to be proved thus becomes

$$(4x - 6x^2 + 4x^3) - (4x - 6x^2 + 4x^3)^2 + 2x^4 = 4(x-x^2) - 18(x-x^2)^2 + 16(x-x^2)^3,$$

which is in fact an identity, each side being

$$= 4x - 22x^2 + 52x^3 - 66x^4 + 48x^5 - 16x^6.$$

CORRECTION, p. 120.

I find that I misquoted Schlaefli's equation, viz. in effect, I took it to be

$$S_1^4 + T_1^4 + 8S_1T_1 - S_1^3T_1^3 = 0, \text{ where } S_1 = \frac{\sqrt{2}}{\sqrt[4]{(kk')}}, \quad T_1 = \frac{\sqrt{2}}{\sqrt[4]{(\lambda\lambda')}}:$$

whereas his equation really is

$$S^4 + T^4 - 8ST + S^3T^3 = 0, \text{ where } S = 2\sqrt[4]{(kk')}, \quad T = 2\sqrt[4]{(\lambda\lambda')}.$$

The change is only a change of form, for writing $S_1 = \frac{2\sqrt{2}}{S}$ and $T_1 = \frac{2\sqrt{2}}{T}$, the equation in (S_1, T_1) is converted into that in (S, T); but it was quite an unnecessary one, and I cannot account for having made it, as the paper in *Crelle* must have been before me.

907.

NOTE ON THE NINTH ROOTS OF UNITY.

[From the *Messenger of Mathematics*, vol. XX. (1891), p. 63.]

LET θ be a prime ninth root of unity, so that $\theta^6 + \theta^3 + 1 = 0$; and write

$$a = \theta + \theta^8,$$
$$b = \theta^2 + \theta^7,$$
$$c = \theta^4 + \theta^5;$$

then

$$a + b + c + \theta^3 + \theta^6 = \frac{1 - \theta^9}{1 - \theta} - 1, = -1,$$

that is,

$$a + b + c = 0.$$

Also

$$a^2 = b + 2, \quad bc = b - 1,$$
$$b^2 = c + 2, \quad ca = c - 1,$$
$$c^2 = a + 2, \quad ab = a - 1,$$

whence

$$ab + ac + bc = -3,$$
$$abc = -1;$$

and a, b, c are thus the roots of the equation $x^3 - 3x + 1 = 0$. We have

$$a^2b + b^2c + c^2a = a^2 + b^2 + c^2 = 6,$$
$$ab^2 + bc^2 + ca^2 = bc + ca + ab = -3,$$

and thence

$$-(a^2b + b^2c + c^2a) + (ab^2 + bc^2 + ca^2) = (b - c)(c - a)(a - b) = -6 - 3, = -9.$$

The equation $x^3 - 3x + 1 = 0$ is thus such that $a^2b + b^2c + c^2a$, and consequently any rational function whatever of a, b, c, invariable by the cyclical interchange (abc) of the roots, has a rational value.

908.

ON TWO INVARIANTS OF A QUADRIQUADRIC FUNCTION.

[From the *Messenger of Mathematics*, vol. XX. (1891), pp. 68, 69.]

THE quadriquadric function

$$\begin{aligned} &z^2\ (ax^2 + 2hxy + g'y^2) \\ +\ &2zw\ (h'x^2 + 2bxy + fy^2) \\ +\ &w^2\,(gx^2 + 2f'xy + cy^2), \end{aligned}$$

considered successively as a function of (z, w) and of (x, y), has the discriminants U, V, equal to

$$(ax^2 + 2hxy + g'y^2)(gx^2 + 2f'xy + cy^2) - (h'x^2 + 2bxy + fy^2)^2,$$

$$(az^2 + 2h'zw + gw^2)(g'z^2 + 2fzw + cw^2) - (hz^2 + 2bzw + f'w^2)^2,$$

respectively. As is well known, these quartic functions have each of them the same quadrinvariant and the same cubinvariant; these are the invariants in question of the quadriquadric function.

The quadrinvariant has been calculated in a different notation, but I am not aware that the cubinvariant has been before calculated; the two values are as follows:

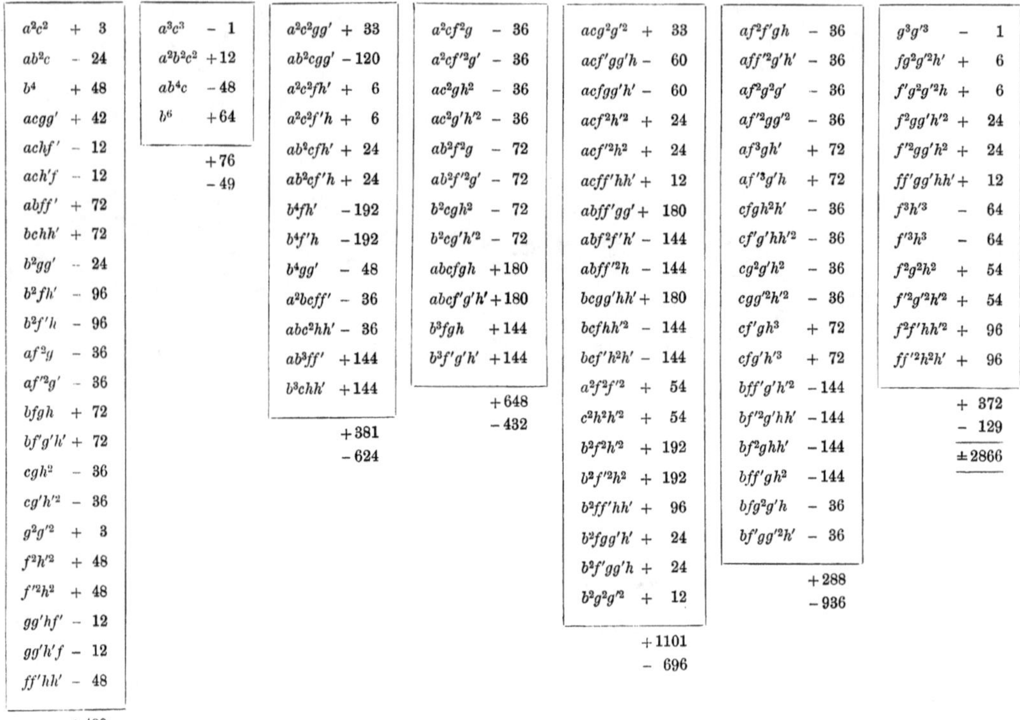

Quadrinvariant is

a^2c^2	$+\ 3$
ab^2c	$-\ 24$
b^4	$+\ 48$
$acgg'$	$+\ 42$
$achf'$	$-\ 12$
$ach'f$	$-\ 12$
$abff'$	$+\ 72$
$bchh'$	$+\ 72$
b^2gg'	$-\ 24$
b^2fh'	$-\ 96$
$b^2f'h$	$-\ 96$
af^2g	$-\ 36$
af'^2g'	$-\ 36$
$bfgh$	$+\ 72$
$bf'g'h'$	$+\ 72$
cgh^2	$-\ 36$
$cg'h'^2$	$-\ 36$
$g^2g'^2$	$+\ 3$
$f^2h'^2$	$+\ 48$
f'^2h^2	$+\ 48$
$gg'hf'$	$-\ 12$
$gg'h'f$	$-\ 12$
$ff'hh'$	$-\ 48$
	± 480

Cubinvariant is

a^3c^3	$-\ 1$
$a^2b^2c^2$	$+12$
ab^4c	-48
b^6	$+64$
	$+76$
	-49

a^2c^2gg'	$+\ 33$
ab^2cgg'	-120
a^2c^2fh'	$+\ 6$
$a^2c^2f'h$	$+\ 6$
ab^2cfh'	$+\ 24$
$ab^2cf'h$	$+\ 24$
b^4fh'	-192
$b^4f'h$	-192
b^4gg'	$-\ 48$
a^2bcff'	$-\ 36$
abc^2hh'	$-\ 36$
ab^3ff'	$+144$
b^3chh'	$+144$
	$+381$
	-624

a^2cf^2g	$-\ 36$
$a^2cf'^2g'$	$-\ 36$
ac^2gh^2	$-\ 36$
$ac^2g'h'^2$	$-\ 36$
ab^2f^2g	$-\ 72$
$ab^2f'^2g'$	$-\ 72$
b^2cgh^2	$-\ 72$
$b^2cg'h'^2$	$-\ 72$
$abcfgh$	$+180$
$abcf'g'h'$	$+180$
b^3fgh	$+144$
$b^3f'g'h'$	$+144$
	$+648$
	-432

$acg^2g'^2$	$+\ 33$
$acf'gg'h$	$-\ 60$
$acfgg'h'$	$-\ 60$
$acf^2h'^2$	$+\ 24$
acf'^2h^2	$+\ 24$
$acff'hh'$	$+\ 12$
$abff'gg'$	$+\ 180$
$abf^2f'h'$	$-\ 144$
$abff'^2h$	$-\ 144$
$bcgg'hh'$	$+\ 180$
$bcfhh'^2$	$-\ 144$
$bcf'h^2h'$	$-\ 144$
$a^2f^2f'^2$	$+\ 54$
$c^2h^2h'^2$	$+\ 54$
$b^2f^2h'^2$	$+\ 192$
$b^2f'^2h^2$	$+\ 192$
$b^2ff'hh'$	$+\ 96$
$b^2fgg'h'$	$+\ 24$
$b^2f'gg'h$	$+\ 24$
$b^2g^2g'^2$	$+\ 12$
	$+1101$
	$-\ 696$

$af^2f'gh$	$-\ 36$
$aff'^2g'h'$	$-\ 36$
af^2g^2g'	$-\ 36$
$af'^2gg'^2$	$-\ 36$
af^3gh'	$+\ 72$
$af'^3g'h$	$+\ 72$
$cfgh^2h'$	$-\ 36$
$cf'g'hh'^2$	$-\ 36$
$cg^2g'h^2$	$-\ 36$
$cgg'^2h'^2$	$-\ 36$
$cf'gh^3$	$+\ 72$
$cfg'h'^3$	$+\ 72$
$bff'g'h'^2$	-144
$bf'^2g'hh'$	-144
bf^2ghh'	-144
$bff'gh^2$	-144
$bfg^2g'h$	$-\ 36$
$bf'gg'^2h'$	$-\ 36$
	$+288$
	-936

$g^3g'^3$	$-\ 1$
$fg^2g'^2h'$	$+\ 6$
$f'g^2g'^2h$	$+\ 6$
$f^2gg'h'^2$	$+\ 24$
$f'^2gg'h^2$	$+\ 24$
$ff'gg'hh'$	$+\ 12$
$f^3h'^3$	$-\ 64$
f'^3h^3	$-\ 64$
$f^2g^2h^2$	$+\ 54$
$f'^2g'^2h'^2$	$+\ 54$
$f^2f'hh'^2$	$+\ 96$
ff'^2h^2h'	$+\ 96$
	$+\ 372$
	$-\ 129$
	± 2866

By writing herein $f', g', h' = f, g, h$, we obtain of course the two invariants of the symmetrical quadriquadric function.

909.

ON A PARTICULAR CASE OF KUMMER'S DIFFERENTIAL EQUATION OF THE THIRD ORDER.

[From the *Messenger of Mathematics*, vol. XX. (1891), pp. 75—79.]

THE general form of equation in question is

$$\frac{x'''}{x'} - \frac{3}{2}\left(\frac{x''}{x'}\right)^2 + x'^2\left\{\frac{A}{(x-1)^2} + \frac{B}{x(x-1)} + \frac{C}{x^2}\right\} - \left\{\frac{A'}{(t-1)^2} + \frac{B'}{t(t-1)} + \frac{C'}{t^2}\right\} = 0,$$

here x is a function of t; and A, B, C, A', B', C' are numerical constants. For various given values of A, B, C, and values determined thereby of A', B', C', the equation admits of a solution in the form $x =$ rational function of t; the theory in reference to the cases considered by Schwarz is considered in my paper "On the Schwarzian Derivative and the Polyhedral Functions," *Camb. Phil. Trans.*, t. XIII. (1883), pp. 5—68, [744]. But the theory is considered in a more general and exhaustive manner in Goursat's memoir, "Recherches sur l'équation de Kummer," *Acta Soc. Sci. Fennicæ*, t. XV. (1888), pp. 47—127. I consider here one of the solutions given by him, viz. writing

$$\begin{aligned} P &= 4t - 5 , & X &= t^2P^3, \\ Q &= 5t - 4 , & Y &= Q^3, \\ R &= 8t^2 - 11t + 8, & Z &= -(t-1)^2 R^2, \end{aligned}$$

so that, identically, $X + Y + Z = 0$; then the solution is expressed by either of the equivalent equations

$$x = -\frac{X}{Z} = \frac{t^2P^3}{(t-1)^2 R^2},$$

$$x - 1 = \frac{Y}{Z} = -\frac{Q^3}{(t-1)^2 R^2}.$$

The values of the constants to which this solution belongs are

$$A=\tfrac{4}{9},\ B=-\tfrac{37}{32},\ C=\tfrac{4}{9};\ A'=\tfrac{3}{8},\ B'=\tfrac{131}{144},\ C'=\tfrac{5}{18}.$$

But instead of assuming these values in the first instance, I leave the values indeterminate; and starting from the foregoing expression for x, I substitute this in

$$\Omega=\frac{x'''}{x'}-\frac{3}{2}\left(\frac{x''}{x'}\right)^2+x'^2\left\{\frac{A}{(x-1)^2}+\frac{B}{x(x-1)}+\frac{C}{x^2}\right\}-\left\{\frac{A'}{(t-1)^2}+\frac{B'}{t(t-1)}+\frac{C'}{t^2}\right\},$$

thus obtaining Ω as a function of t which, as will appear, vanishes identically when A, B, C, A', B', C' have the foregoing values.

I remark that this is, in effect, doing in a somewhat different form for the particular case what Goursat does for the general case, viz. starting from

$$\Omega_1=\frac{x'''}{x'}-\frac{3}{2}\left(\frac{x''}{x'}\right)^2+x'^2\left\{\frac{A}{(x-1)^2}+\frac{B}{x(x-1)}+\frac{C}{x^2}\right\},$$

with values of A, B, C which belong to the solution considered, he shows that this is a function of t having no infinities other than $(0, 1, \infty)$; that ∞ is not an infinity of the function or of the function multiplied into t, and that 0 and 1 are each of them a twofold infinity; that is, that the function is of the form

$$\frac{Lt^2+Mt+N}{t^2(t-1)^2}\quad\text{or}\quad\frac{A'}{(t-1)^2}+\frac{B'}{t(t-1)}+\frac{C'}{t^2}.$$

Proceeding to carry out the process, we have

$$\frac{x'}{x-1}=\quad-\frac{1}{t-1}+\frac{3Q'}{Q}-\frac{2R'}{R},$$

$$\frac{x'}{x}\quad=\frac{2}{t}-\frac{1}{t-1}+\frac{3P'}{P}-\frac{2R'}{R},$$

and from either of these equations, collecting and reducing,

$$x'\quad=\frac{5tP^2Q^2}{(t-1)^2R^3},$$

where observe that, from the values of x and $x-1$ respectively, it appears *à priori* that tP^2 and Q^2 must be factors in the numerator of x'. From this value of x', we have

$$\frac{x''}{x'}\quad=\frac{1}{t}-\frac{2}{t-1}+\frac{2P'}{P}+\frac{2Q'}{Q}-\frac{3R'}{R};$$

and hence, P' and Q' being mere constants,

$$\frac{x'''}{x'}-\left(\frac{x''}{x'}\right)^2=-\frac{1}{t^2}+\frac{2}{(t-1)^2}-\frac{3R''}{R}-2\left(\frac{P'}{P}\right)^2-2\left(\frac{Q'}{Q}\right)^2+3\left(\frac{R'}{R}\right)^2,$$

and consequently

$$\begin{aligned}\Omega = & -\frac{1}{t^2}+\frac{2}{(t-1)^2}-\frac{3R''}{R}+2\left(\frac{P'}{P}\right)^2-2\left(\frac{Q'}{Q}\right)^2+3\left(\frac{R'}{R}\right)^2\\ & -\tfrac{1}{2}\left(\frac{1}{t}-\frac{2}{t-1}+\frac{2P'}{P}+\frac{2Q'}{Q}-\frac{3R'}{R}\right)^2\\ & +A\left(-\frac{1}{t-1}+\frac{3Q'}{Q}-2\frac{R'}{R}\right)^2\\ & +B\left(-\frac{1}{t-1}+\frac{3Q'}{Q}-2\frac{R'}{R}\right)\left(\frac{2}{t}-\frac{1}{t-1}+\frac{3P'}{P}-\frac{2R'}{R}\right)\\ & +C\left(\frac{2}{t}-\frac{1}{t-1}+\frac{3P'}{P}-\frac{2R'}{R}\right)^2\\ & -\frac{C'}{(t-1)^2}-\frac{B'}{t(t-1)}-\frac{A'}{t^2}.\end{aligned}$$

Putting for shortness

$$\frac{1}{t}=\alpha,\ \frac{1}{t-1}=\beta,\ \frac{P'}{P}=p,\ \frac{Q'}{Q}=q,\ \frac{R'}{R}=r,$$

this equation gives

$$\begin{aligned}\Omega = & -\alpha^2+2\beta^2-\frac{3R''}{R}-2p^2-2q^2+3r^2\\ & -\tfrac{1}{2}(\alpha-2\beta+2p+2q-3r)^2\\ & +A(-\beta+3q-2r)^2\\ & +B(-\beta+3q-2r)(2\alpha-\beta+3p-2r)\\ & +C(2\alpha-\beta+3p-2r)^2\\ & -C'\alpha^2-B'\alpha\beta-A'\beta^2,\end{aligned}$$

which is

$$\begin{aligned} = \ & \alpha^2(-\tfrac{3}{2}+4C-C')-\frac{3R''}{R}: \text{ say it is } = && L\alpha^2-\frac{48}{R}\\ & +\alpha\beta(2-2B-4C-B') && +M\alpha\beta\\ & +\beta^2\ (A+B+C-A') && +N\beta^2\\ & +\alpha p(-2+12C) && +F\alpha p\\ & +\alpha q(-2+\ 6B) && +G\alpha q\\ & +\alpha r\ (3-4B-8C) && +H\alpha r\\ & +\beta p(4-3B-6C) && +F'\beta p\\ & +\beta q(4-6A-3B) && +G'\beta q\\ & +\beta r(-6+4A+4B+4C) && +H'\beta r\\ & +p^2\ (-4+9C) && +A''p^2\\ & +q^2\ (-4+9A) && +B''q^2\\ & +r^2\ (-\tfrac{3}{2}+4A+4B+4C) && +C''r^2\\ & +qr\ (6-12A-\ 6B) && +F''qr\\ & +rp\,(6-\ 6B-12C) && +G''rp\\ & +pq(-4+9B) && +H''pq.\end{aligned}$$

By decomposing $\alpha\beta$, αp, &c., into simple fractions, this becomes

$$\begin{aligned}\Omega = \quad & L\alpha^2 - \frac{48}{R} \\ & + M(-\alpha+\beta) \\ & + N\beta^2 \\ & + F(-\tfrac{4}{5}\alpha \ + \tfrac{4}{5}p) \\ & + G(-\tfrac{5}{4}\alpha \ + \tfrac{5}{4}q) \\ & + H\left(-\tfrac{11}{8}\alpha + \frac{11t+\frac{7}{8}}{R}\right) \\ & + F'\ (-4\beta+4p) \\ & + G'\ (\ \ 5\beta-5q) \\ & + H'\left(\ \ \beta + \frac{-8t+19}{R}\right) \\ & + A''p^2 \\ & + B''q^2 \\ & + C''r^2 \\ & + F''\tfrac{5}{12}\left(q - \frac{8t-43}{R}\right) \\ & + G''\tfrac{4}{3}\left(p - \frac{8t-13}{R}\right) \\ & + H''\tfrac{20}{9}(p-q).\end{aligned}$$

This is

$$\begin{aligned} = \quad & \alpha^2 L \\ & + \alpha(-M - \tfrac{4}{5}F - \tfrac{5}{4}G - \tfrac{11}{8}H) \\ & + \beta^2 N \\ & + \beta(M - 4F' + 5G' + H') \\ & + p^2 A'' \\ & + p(4F' + \tfrac{4}{3}G'' + \tfrac{20}{9}H'' + \tfrac{4}{5}F) \\ & + q^2 B'' \\ & + q(-5G' - \tfrac{5}{12}F'' - \tfrac{20}{9}H'' + \tfrac{5}{4}G) \\ & + r^2 C'' \\ & + \frac{1}{R}\{-48 + H(11t+\tfrac{7}{8}) + H'(-8t+19) - \tfrac{5}{12}F''(8t-43) - \tfrac{4}{3}G''(8t-13)\}.\end{aligned}$$

This should be identically $=0$; making $A''=0$, $B''=0$, $C''=0$, we find

$$A = \tfrac{4}{9},\ B = -\tfrac{37}{32},\ C = \tfrac{4}{9};\ (A+B+C=\tfrac{3}{8}).$$

and thence

$$F = \tfrac{10}{3},\ G = -\tfrac{61}{12},\ H = \tfrac{3}{2};\ F' = \tfrac{23}{8},\ G' = \tfrac{23}{8},\ H' = -\tfrac{9}{2};$$

$$F'' = \tfrac{15}{4},\ G'' = \tfrac{15}{4},\ H'' = -\tfrac{68}{9}.$$

These values make the coefficients of p and q to be each $=0$; and they make the coefficient of R to be identically $=0$, viz. we have

$$0 = -48 + \tfrac{7}{8}H + 19H' + \tfrac{215}{12}F'' + \tfrac{52}{3}G'',$$

and

$$0 = 11H - 8H' - \tfrac{10}{3}F'' - \tfrac{32}{3}G''.$$

We have, moreover,

$$L = \tfrac{5}{18} - C',\ M = \tfrac{365}{144} - B',\ N = \tfrac{3}{8} - A';$$

and the coefficients of α and β are $= -M + \frac{13}{8}$ and $M - \frac{13}{8}$ respectively; hence the coefficients of α^2, β^2, α and β will all vanish if only $L=0$, $M=\frac{13}{8}$, $N=0$, that is,

$$A' = \tfrac{3}{8},\ B' = \tfrac{131}{144},\ C' = \tfrac{5}{8};$$

and we have thus identically $\Omega = 0$, if only A, B, C, A', B', C' have the above-mentioned values.

910.

NOTE ON THE INVOLUTANT OF TWO BINARY MATRICES.

[From the *Messenger of Mathematics*, vol. XX. (1891), pp. 136, 137.]

CONSIDER the two matrices

$$M = \begin{pmatrix} a, & b \\ c, & d \end{pmatrix}, \quad M' = \begin{pmatrix} a', & b' \\ c', & d' \end{pmatrix},$$

and their product in one or the other order

$$MM' = \begin{pmatrix} A, & B \\ C, & D \end{pmatrix}, \quad M'M = \begin{pmatrix} A_1, & B_1 \\ C_1, & D_1 \end{pmatrix}.$$

Then the Involutant is by definition = either of the determinants

$$I = \begin{vmatrix} 1, & a, & a', & A \\ 0, & b, & b', & B \\ 0, & c, & c', & C \\ 1, & d, & d', & D \end{vmatrix}, \qquad I_1 = \begin{vmatrix} 1, & a', & a, & A_1 \\ 0, & b', & b, & B_1 \\ 0, & c', & c, & C_1 \\ 1, & d', & d, & D_1 \end{vmatrix};$$

viz. it is to be shown that these two values are in fact equal.

We have

$$MM' = \begin{array}{c} (a,\ b) \\ (c,\ d) \end{array} \left| \begin{array}{cc} (a',\ c') & (b',\ d') \\ \hline ,, & ,, \\ ,, & ,, \end{array} \right. = \begin{pmatrix} A, & B \\ C, & D \end{pmatrix},$$

that is,

$$A = aa' + bc', \quad B = ab' + bd',$$

$$C = ca' + dc', \quad D = cb' + dd',$$

and similarly

$$M'M = \begin{matrix} \\ (a',\ b') \\ (c',\ d') \end{matrix} \begin{vmatrix} (a,\ c) & (b,\ d) \\ ,, & ,, \\ ,, & ,, \end{vmatrix} = \begin{pmatrix} A_1, & B_1 \\ C_1, & D_1 \end{pmatrix},$$

that is,

$$A_1 = aa' + cb', \quad B_1 = ba' + db',$$

$$C_1 = ac' + cd', \quad D_1 = bc' + dd',$$

viz. A_1, B_1, C_1, D_1 are obtained from A, B, C, D by the interchange of the accented and unaccented letters.

We have then, from the first expression for the Involutant,

$$I = A \begin{vmatrix} 0, & b, & b' \\ 0, & c, & c' \\ 1, & d, & d' \end{vmatrix} - B \begin{vmatrix} 0, & c, & c' \\ 1, & d, & d' \\ 0, & b, & b' \end{vmatrix} + C \begin{vmatrix} 1, & d, & d' \\ 1, & a, & a' \\ 0, & b, & b' \end{vmatrix} - D \begin{vmatrix} 1, & a, & a' \\ 0, & b, & b' \\ 0, & c, & c' \end{vmatrix},$$

$$= A(bc' - b'c) - B(ac' - a'c + cd' - c'd) + C(ab' - a'b + bd' - b'd) - D(bc' - b'c),$$

or substituting for $A - D$, B and C their values, this is

$$(aa' - dd' + bc' - b'c)(bc' - b'c) - (ab' + bd')(ac' - a'c + cd' - c'd)$$
$$+ (ca' + dc')(ab' - a'b + bd' - b'd);$$

and multiplying out and grouping together the terms in bc, $b'c'$, bc' and $b'c$, this is found to be

$$= -(a' - d')^2 bc + (a - d)(a' - d')(bc' + b'c) - (a - d)^2 b'c' + (bc' - b'c)^2,$$

which is

$$= -\{(a - d)b' - (a' - d')b\}\{(a - d)c' - (a' - d')c\} + (bc' - b'c)^2.$$

Hence, writing

$$\mathrm{a} = bc' - b'c, \quad \mathrm{f} = ad' - a'd,$$

$$\mathrm{b} = ca' - c'a, \quad \mathrm{g} = bd' - b'd,$$

$$\mathrm{c} = ab' - a'b, \quad \mathrm{h} = cd' - c'd,$$

we have

$$I = -(\mathrm{c} + \mathrm{g})(-\mathrm{b} + \mathrm{h}) + \mathrm{a}^2,$$

that is,

$$I = \mathrm{bc} - \mathrm{ch} + \mathrm{bg} - \mathrm{gh} + \mathrm{a}^2.$$

To obtain the value of I_1, we must interchange the accented and unaccented letters, that is, change the signs of the several quantities a, b, c, f, g, h; but I, being a quadric function of the six quantities, is not altered by the change; that is, we have $I = I_1$.

911.

ON AN ALGEBRAICAL IDENTITY RELATING TO THE SIX COORDINATES OF A LINE.

[From the *Messenger of Mathematics*, vol. XX. (1891), pp. 138—140.]

THE following identity may be verified without difficulty; but it is interesting in regard to the analytical theory of the line.

The identity is

$$\begin{aligned}0 = & \begin{vmatrix} B, & C, & F \\ b\,, & c\,, & f \\ b', & c', & f' \end{vmatrix} \begin{vmatrix} A, & B, & C \\ a\,, & b\,, & c \\ a', & b', & c' \end{vmatrix} \\ & - (bc' - b'c)^2 (AF + BG + CH) \\ & + (bc' - b'c)\,(bC - cB)\,(Af' + Bg' + Ch' + Fa' + Gb' + Hc') \\ & - (bc' - b'c)\,(b'C - c'B)\,(Af + Bg + Ch + Fa + Gb + Hc) \\ & - (bC - cB)^2 \qquad (a'f' + b'g' + c'h') \\ & + (bC - cB)\,(b'C - c'B)\,(af' + a'f + bg' + b'g + ch' + c'h) \\ & - (b'C - c'B)^2 \qquad (af + bg + ch),\end{aligned}$$

where all the letters have arbitrary values.

It follows that, if

$$\begin{aligned}& af + bg + ch = 0, \\ & a'f' + b'g' + c'h' = 0, \\ & af' + a'f + bg' + b'g + ch' + c'h = 0, \\ & AF + BG + CH = 0, \\ & Af + Bg + Ch + Fa + Gb + Hc = 0, \\ & Af' + Bg' + Ch' + Fa' + Gb' + Hc' = 0,\end{aligned}$$

then either

$$\begin{vmatrix} B, & C, & F \\ b\,, & c\,, & f \\ b', & c', & f' \end{vmatrix} = 0,$$

or else

$$\begin{vmatrix} A, & B, & C \\ a\,, & b\,, & c \\ a', & b', & c' \end{vmatrix} = 0.$$

Supposing the first three of the six equations are satisfied, then (a, b, c, f, g, h) and (a', b', c', f', g', h') are the coordinates of two intersecting lines; and supposing that the last three equations are also satisfied, then (A, B, C, F, G, H) will be the coordinates of a line meeting each of the intersecting lines. The third line is thus either in the plane of the intersecting lines, or else passes through their point of intersection; and in fact, the first of the two determinant equations is the condition in order that the line may be in the plane of the two intersecting lines, and the second determinant equation is the condition in order that it may pass through their point of intersection. Each equation is in fact one out of four equivalent forms, viz. we may have in the first equation (B, C, F), (C, A, G), (A, B, H), or (F, G, H); and in the second equation (A, B, C), (A, H, G), (H, B, F), or (G, F, C).

The analytical theory may be presented in a complete form by means of the formulæ of my memoir "On the six coordinates of a line," (1867), *Camb. Phil. Trans.*, t. XI. pp. 290—323, [435]. Considering the two intersecting lines (a, b, c, f, g, h) and (a', b', c', f', g', h'), the coordinates of the plane through these two lines (that is, the coefficients ξ, η, ζ, ω of the equation $\xi x + \eta y + \zeta z + \omega w = 0$ of the plane) are there given (see p. 295*) in a fourfold form; and if we thence form the condition in order that the line (A, B, C, F, G, H) may lie in this plane, we have

$$\left(\begin{array}{cccc} 0, & C, & -B, & F \\ -C, & 0, & A, & G \\ B, & -A, & 0\,, & H \\ -F, & -G, & -H, & 0 \end{array}\right)$$

$$\left(\begin{array}{cccc} af'+b'g+c'h, & bf'-b'f\;, & cf'-c'f\;, & -(bc'-b'c) \\ ag'-a'g\;, & a'f+bg'+c'h, & cg'-c'g\;, & -(ca'-c'a) \\ ah'-a'h\;, & bh'-b'h\;, & a'f+b'g+ch', & -(ab'-a'b) \\ gh'-g'h\;, & hf'-h'f\;, & fg'-f'g\;, & af'+bg'+ch' \end{array}\right) = 0;$$

viz. the condition is expressible in any one of the 16 forms obtained by combining a line of the first matrix with a line of the second matrix; thus one form is

$$0\,(af'+b'g+c'h) + C\,(bf'-b'f) - B\,(cf'-c'f) - F\,(bc'-b'c) = 0,$$

[* This Collection, vol. VII., p. 71.]

that is,

$$\begin{vmatrix} B, & C, & F \\ b, & c, & f \\ b', & c', & f' \end{vmatrix} = 0,$$

the foregoing first determinant equation, which thus belongs to the case where the line lies in the plane of the two intersecting lines.

Again we have (see p. 296*) an expression, in a fourfold form, for the coordinates of the point of intersection of the two intersecting lines; and thence for the condition, in order that the line (A, B, C, F, G, H) may pass through this point, we have

$$\begin{pmatrix} 0, & H, & -G, & A \\ -F, & 0, & F, & B \\ G, & -H, & 0, & C \\ -A, & -B, & -C, & 0 \end{pmatrix}$$

$$\begin{pmatrix} af'+b'g+c'h, & ag'-a'g, & ah'-a'h, & gh'-g'h \\ bf'-b'f, & a'f+bg'+c'h, & bh'-b'h, & hf'-h'f \\ cf'-c'f, & cg'-c'g, & a'f+b'g+ch', & fg'-f'g \\ -(bc'-b'c), & -(ca'-c'a), & -(ab'-a'b), & af'+bg'+ch' \end{pmatrix} = 0,$$

viz. the condition is expressible in any one of the 16 forms obtained by combining a line of the first matrix with a line of the second matrix; thus one form is

$$A\,(bc'-b'c)+B\,(ca'-c'a)+C\,(ab'-a'b)+0\,(af'+bg'+ch')=0,$$

that is,

$$\begin{vmatrix} A, & B, & C \\ a, & b, & c \\ a', & b', & c' \end{vmatrix} = 0,$$

the foregoing second determinant equation, which thus belongs to the case where the line passes through the point of intersection of the two intersecting lines.

I remark that the original identity may be written in the very compendious symbolical form

$$\begin{pmatrix} A, & a, & a' \\ F, & f, & f' \end{pmatrix} \begin{pmatrix} B, & b, & b' \\ C, & c, & c' \end{pmatrix}^2 = \begin{vmatrix} A, & a, & a' \\ B, & b, & b' \\ C, & c, & c' \end{vmatrix} \begin{vmatrix} F, & f, & f' \\ B, & b, & b' \\ C, & c, & c' \end{vmatrix},$$

viz. here on the left-hand side the second factor denotes the three determinants $bc'-b'c$, $b'C-c'B$, $Bc-Cb$: the whole is a quadric function of these, the coefficients being

$$AF+BG+CH, \quad af+bg+ch,$$
$$a'f'+b'g'+c'h', \quad af'+bg'+ch'+fa'+gb'+hc',$$
$$a'F+b'G+c'H+f'A+g'B+h'C,$$
$$Af+Bg+Ch+Fa+Gb+Hc,$$

respectively; and the right-hand side is simply the product of two determinants.

[* *Loc. cit.*, p. 72.]

912.

ON THE NOTION OF A PLANE CURVE OF A GIVEN ORDER.

[From the *Messenger of Mathematics*, vol. XX. (1891), pp. 148—150.]

WE have a complete geometrical notion of a curve of a given order, viz. a curve of the order n is a curve which is met by any line whatever in n points and no more; but starting with this definition, how do we know that there exists a curve of the order n? and, further, how do we know that it depends linearly on $\frac{1}{2}n(n+3)$ parameters, or, what is the same thing, that there is one and only one curve which can be drawn through $\frac{1}{2}n(n+3)$ given points?

The last-mentioned property does not by itself constitute a definition of a curve of the order n; thus $n=2$, we cannot define a curve of the second order as a curve which is uniquely determined by the condition of passing through 5 given points; for a cubic passing through 4 given points is a curve uniquely determined by the condition in question; but we may differentiate between these two solutions by adding the further condition that, when 3 of the 5 points are in a line, the curve of the second order shall include as part of itself this line. And we are thus led to the definition: A curve of the order n is a curve which is uniquely determined by the condition of passing through $\frac{1}{2}n(n+3)$ given points; and of being moreover such that, when $n+1$ of these points lie on a line, it includes as part of itself this line.

Starting from the foregoing definition, the first property is, I think, demonstrable, viz. the property that a curve of the order n is met by any line whatever in n points and no more. Thus $n=2$: start with a line chosen at pleasure, and on it take 2 points which are regarded as indeterminate points: to fix the ideas, let one of these be regarded as depending on a parameter λ and the other on a parameter μ, (so that when λ has a determinate value assigned to it, the first point becomes a determinate point, and similarly when μ has a determinate value assigned to it, the second point becomes a determinate point; and consequently, when λ and μ have determinate values, the 2 points are determinate points on the line). Take now any other 3 points not on the line; then, for the moment regarding the 2 points on the line as determinate points, we can through the 5 points draw a curve of the second order; this is the general curve of the second order through the 3 points; for it is a curve of the second order through the 3 points, and which, when the parameters λ and μ are regarded as undetermined and arbitrary, or choosable at pleasure, might be made to pass through any other two points whatever. But by hypothesis, the curve meets the line in question, that is, *any line*, in two points: and the conic cannot meet the line in more than two points; for in like manner, starting with the given line, and upon it 3 points (which may be considered as depending on the parameters λ, μ and ν respectively), and taking any two points not on the line, we have through the 5 points a conic; and this conic, regarding the parameters λ, μ, ν as undetermined and arbitrary, or choosable at pleasure, will be the general conic through the two points; for by a proper determination of the parameters, it might be made to pass through any other 3 points whatever. But, by hypothesis, the conic contains as part of itself the line; that is, the general conic through the 2 points contains as part of itself any line whatever, which is absurd.

So again for the cubic, $n=3$: here starting with a line taken at pleasure, we take on it 3 points, which may be regarded as depending on the parameters λ, μ, ν respectively, and we take any other 6 points not on the line. We have through the 9 points a cubic; and this is the general cubic through the 6 points, for it depends on the parameters λ, μ, ν, which might be determined so as to make the curve pass through any other 3 points whatever; and by hypothesis the cubic meets the line, that is, any line whatever, in 3 points. And it cannot meet it in more than 3 points: for starting with the same line and upon it 4 points depending on the parameters λ, μ, ν, ρ respectively, and taking any other 5 points not upon the line, we then have through the 9 points a cubic which will be the general cubic through the 5 points (for it depends on the 4 parameters λ, μ, ν, ρ). But by hypothesis, the cubic contains as part of itself the line, viz. the general cubic through the 5 points contains as part of itself any line whatever, which is absurd. The reasoning is quite general; and applying to a curve of the order n, the conclusion is that such a curve meets any line in n points and no more.

913.

ON THE EPITROCHOID.

[From the *Messenger of Mathematics*, vol. XX. (1891), pp. 150—158.]

IF we have a curve C_1 rolling on a fixed curve C, and consider the epitrochoid described by a point P, attached to and carried along with the curve C_1, then there is a known construction for the radius of curvature at any point of the epitrochoid; the construction is probably much older, but I refer to Mannheim, *Géométrie Descriptive*, Paris, 1886, pp. 177 and 194. In fact, if Q be a position of the point of contact, P_1 the corresponding position of the describing point, R_1 and R the radii of curvature at Q of the two curves respectively, ρ the distance QP_1, and ϕ the inclination of this distance to the common normal at Q, then the radius of curvature x at the point P_1 of the epitrochoid is given by the formula

$$\frac{x}{x-\rho} = \frac{\rho}{\cos\phi}\left(\frac{1}{R}+\frac{1}{R_1}\right).$$

I prove this as follows: take Q, Q' consecutive positions of the point of contact, P_1, P_1' consecutive positions of P_1; the centre of curvature is the intersection of the lines P_1Q, $P_1'Q'$; hence, if for a moment M and N are the perpendicular distances between these two lines at the points P_1 and Q respectively, we have $\dfrac{x}{x-\rho}=\dfrac{M}{N}$. Let ds be the element QQ' considered as belonging to each of the two curves respectively, θ and θ_1 the angles which this element subtends at the two centres of curvature respectively; we have $ds = R\theta = R_1\theta_1$, whence

$$\theta+\theta_1 = ds\left(\frac{1}{R}+\frac{1}{R_1}\right).$$

The instantaneous centre is P_1, and hence

$$M = P_1P_1' = \rho(\theta+\theta_1); \text{ also } N = ds\cos\phi;$$

we thus have

$$\frac{x}{x-\rho}=\frac{\rho(\theta+\theta_1)}{ds\cos\phi};$$

or, substituting for $\theta+\theta_1$ the foregoing value, we have the required expression

$$\frac{x}{x-\rho}=\frac{\rho}{\cos\phi}\left(\frac{1}{R}+\frac{1}{R_1}\right).$$

If the curve C_1 roll on a straight line, $R=\infty$ and therefore

$$\frac{x}{x-\rho}=\frac{\rho}{R_1\cos\phi}.$$

Suppose that at any instant we have on the epitrochoid an inflexion, then $x=\infty$, and we have $\rho=R_1\cos\phi$; this denotes that P_1 is a point of the circle described on the radius of curvature at Q as its diameter, and we have thus the theorem:

If at any instant we consider the circle, which has for its diameter the radius of curvature of the rolling curve at its point of contact with the right line, then each point on this circle is an inflexion on the epitrochoid described by the same point as the curve rolls on the straight line.

The circle in question for any point Q of the rolling curve C_1 is of course a circle, having its centre on the normal at Q and its radius equal to one-half of the radius of curvature; we may call it the circle of double curvature. Regarding Q as a given point on the curve C_1, and supposing that the circle of double curvature intercepts the curve C_1 in a point X, then taking P_1 at X, we have the theorem:

If the curve C_1 roll on a straight line, then the epitrochoid described by the point X has an inflexion corresponding to the position of the rolling curve for which the point Q comes into contact with the straight line.

Consider the curve C_1 referred to axes fixed in the plane of the curve and moveable with it; take x, y as the coordinates of the point Q; α, β as the coordinates of the corresponding centre of curvature, and X, Y as current coordinates; the equation of the circle of double curvature is

$$(X-x)(X-\alpha)+(Y-y)(Y-\beta)=0,$$

viz. this is the equation of a circle having for a diameter the two points (x, y) and (α, β).

Suppose that C_1 is a conic; the circle of double curvature at any point Q meets it in the point Q counting twice, and in two other points, say P_1, P_2, which are determined as the intersections of the conic by the line P_1P_2 which joins them. I propose to determine the equation of this line. The investigation would be more symmetric for the ellipse, but I nevertheless prefer to consider for the curve C_1 the hyperbola $b^2X^2-a^2Y^2-a^2b^2=0$.

Representing for a moment the equation of the circle by $U=0$, and that of the hyperbola by $V=0$, it must be possible to determine the ratio of the coefficients p, q, in such wise that $pU+qV=0$ shall break up into a pair of lines, one of which is the tangent at Q of the hyperbola; and then the other of them will be the required line P_1P_2.

The equation of the tangent is $b^2xX-a^2yY-a^2b^2=0$; we then see that the equation of the line P_1P_2 must be of the form $b^2xX+a^2yY+\Omega=0$; and we have to find p, q, Ω so as to verify the identity

$$(p+qb^2)X^2+(p-qa^2)Y^2-p(\alpha+x)X-p(\beta+y)Y+p(\alpha x+\beta y)-q\alpha^2\beta^2$$
$$=(b^2xX-a^2yY-a^2b^2)(b^2xX+a^2yY+\Omega);$$

that is, we ought to have

$$p+qb^2=b^4x^2,$$
$$p-qa^2=a^4y^2,$$
$$p(x+\alpha)=(\Omega-a^2b^2)\,b^2x,$$
$$p(y+\beta)=(\Omega+a^2b^2)\,a^2y,$$
$$qa^2b^2-p(\alpha x+\beta y)=\Omega a^2b^2.$$

It will be recollected that α, β denote the coordinates of the centre of curvature corresponding to the point (x, y) of the hyperbola; their values thus are

$$\alpha=\ \ (a^2+b^2)\frac{x^3}{a^4},$$
$$\beta=-(a^2+b^2)\frac{y^3}{b^4}.$$

There is no difficulty in finding the values

$$p=\frac{a^4b^4}{a^2+b^2},\quad q=\frac{b^4x^2+a^4y^2}{a^2+b^2},$$

whence also

$$q-a^2b^2=\frac{a^4b^2}{a^2+b^2}\left\{\left(\frac{1}{a^2}+\frac{1}{b^2}\right)y^2-1\right\},$$
$$q+a^2b^2=\frac{a^2b^4}{a^2+b^2}\left\{\left(\frac{1}{a^2}+\frac{1}{b^2}\right)x^2+1\right\},$$
$$\Omega=-a^2b^2\frac{x^2+y^2}{a^2+b^2};$$

these satisfy the identities in question.

The equation of the line P_1P_2 is thus

$$b^2xX+a^2yY-\frac{a^2b^2}{a^2+b^2}(x^2+y^2)=0,$$

and combining this with the equation

$$b^2X^2-a^2Y^2-a^2b^2=0$$

of the hyperbola, we find

$$\left\{xX - \frac{a^2}{a^2+b^2}(x^2+y^2)\right\}^2 = \frac{a^4y^2Y^2}{b^4}, \; = \frac{a^2y^2}{b^2}(X^2-a^2),$$

that is,

$$X^2\left(x^2 - \frac{a^2y^2}{b^2}\right) - \frac{2a^2x}{a^2+b^2}(x^2+y^2)X + \frac{a^4}{(a^2+b^2)^2}(x^2+y^2)^2 + \frac{a^4y^2}{b^2} = 0,$$

viz.

$$X^2 - 2\frac{x^2+y^2}{a^2+b^2}xX + a^2\left\{\left(\frac{x^2+y^2}{a^2+b^2}\right)^2 + \frac{y^2}{b^2}\right\} = 0,$$

and we thence obtain

$$X = \frac{x^2+y^2}{a^2+b^2}x \pm \frac{ay}{b}\sqrt{\left\{\left(\frac{x^2+y^2}{a^2+b^2}\right)^2 - 1\right\}},$$

$$Y = -\frac{x^2+y^2}{a^2+b^2}y \pm \frac{bx}{a}\sqrt{\left\{\left(\frac{x^2+y^2}{a^2+b^2}\right)^2 - 1\right\}},$$

values belonging to the points P_1, P_2 respectively: these points are thus real or imaginary, according as x^2+y^2 is greater or less than a^2+b^2. In the limiting case $x^2+y^2=a^2+b^2$, we have $X=x$, $Y=-y$, viz. here the circle of double curvature has its centre on the axis of x, and has double contact with the hyperbola at the points (x, y) and $(x, -y)$ respectively.

Suppose, to fix the ideas, x and y are both positive, i.e. that Q is on the right-hand upper half branch of the hyperbola; and that $x^2+y^2=$ or $>a^2+b^2$, so that P_1, P_2 are both real. The two values of X are both positive, those of Y are originally both negative, and as Q moves along the half branch of the hyperbola one of these values increases negatively, the other increases positively until it becomes $=0$, after which it becomes positive and continually increases: say the points P_1, P_2 are originally on the right-hand lower half branch, but one of them, say P_1, moves up to the vertex, and we afterwards have P_1 on the upper half branch and P_2 on the lower half branch. The limiting case is when P_1 is at the vertex; here $\frac{x^2+y^2}{a^2+b^2}=\frac{x}{a}$, giving for (x, y) a real point, which is easily constructed on the upper half branch. There is thus on the half branch a real position of Q, such that the corresponding circle of double curvature passes through the vertex of the same half branch.

Hence in the epitrochoid described by the vertex of a hyperbola rolling on a right line there is always an inflexion.

Imagine the branch $B'AB$ of a hyperbola rolling on the line $y=0$, where A is the vertex, B', B the points at infinity, and so that as x increases positively (that is, towards the right) the point of contact Q passes from A towards B; and consider the epitrochoid described by the vertex A. Supposing that the point of contact is at first at A; the point A is a cusp on the epitrochoid, the motion of A is at first vertically upwards, and towards the right, the curve being concave to the axis; we then come to a position of the point of contact for which there is an inflexion; the motion of A is still upwards and towards the right, but the curve has now become convex to the axis, and the motion continues in this manner as the point of contact moves along the hyperbola towards B; when the point of contact is at B, i.e. when the asymptote has come to coincide with the axis, then the motion of A is again vertically upwards; the other half branch of the hyperbola now comes into contact with the axis; the point A still moves upwards but towards the left, the curve being concave to the axis, and this continues until the point of contact arrives at the other vertex (i.e. the vertex on the other branch), when the motion of A is horizontal and towards the left, viz. the coordinate y has here a maximum value, equal transverse axis of the hyperbola; the motion of A is thenceforth towards the left and downwards, until the other asymptote comes to coincide with the curve, after which it is downwards and towards the right (the point of contact being now on the half $B'A$ of the original branch $B'AB$), and ultimately the point of contact is again at A, and we have $y=0$, a cusp; the curve is symmetrical in regard to the before-mentioned maximum ordinate: it is to be noticed that the distance between the two cusps is equal to four times the difference between the length of the half asymptote and the half branch of the hyperbola (this difference being a finite value expressible by elliptic functions), and that, as already mentioned, the maximum ordinate is equal to the transverse axis. The above described portion of the curve is of course repeated continually right and left *ad infinitum*; two consecutive portions intersect each other, and there is thus a real node vertically above each cusp of the infinite curve. We have, in what precedes, a complete explanation of the motion of a hyperbola rolling on a straight line, and in particular our explanation of the form of the epitrochoid described by a vertex; the epitrochoid described by any other point of the curve is of a similar character but must be non-symmetrical in regard to the maximum ordinate; and it would not be difficult to explain the form of the curve for other positions of the describing point; but as regards the analytical theory, I prefer to consider the case of the ellipse.

In the case of the ellipse $b^2X^2+a^2Y^2-a^2b^2=0$ rolling on the line $y=0$ (see figure), take as origin the point O which comes in contact with the extremity B of

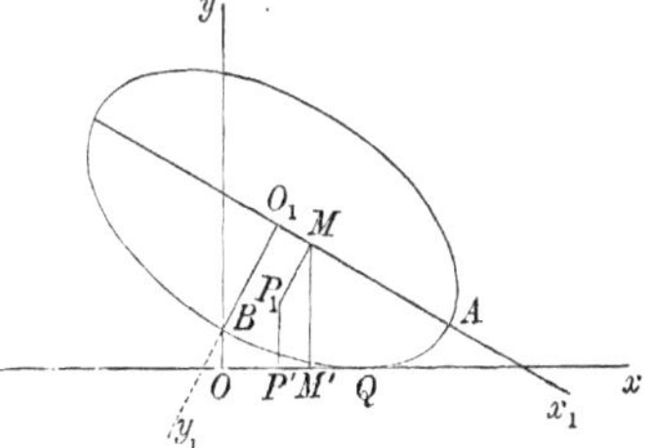

the minor axis; let x_1, y_1 be the coordinates of the describing point P_1, and for the point Q, taking ϕ as the complement of the eccentric angle, let the coordinates referred to the axes of the ellipse be

$$x_1 = a\sin\phi,\quad y_1 = b\cos\phi;$$

the arc BQ is then

$$= \int_0^\phi \sqrt{(a^2 \cos^2 \phi + b^2 \sin^2 \phi)}\, d\phi,$$

which is $= aE\phi$, the modulus being the eccentricity e of the ellipse; we then have for the coordinates of P_1

$$x = OP' = OQ - QP', \quad = BQ - QP',$$
$$y \qquad\qquad = \qquad P_1P',$$

where QP' and P_1P' are the projections of P_1Q in the direction of, and at right angles to, the tangent QO at Q; we find the values without difficulty, and we thus have

$$x = \int_0^\phi \sqrt{(a^2 \cos^2 \phi + b^2 \sin^2 \phi)}\, d\phi - \frac{(a^2 - b^2) \sin\phi \cos\phi - ax_1 \cos\phi + by_1 \sin\phi}{\sqrt{(a^2 \cos^2 \phi + b^2 \sin^2 \phi)}},$$
$$y = \frac{ab - bx_1 \sin\phi - ay_1 \cos\phi}{\sqrt{(a^2 \cos^2 \phi + b^2 \sin^2 \phi)}};$$

which are the expressions for x, y in terms of a variable parameter ϕ.

In particular, for the point B at the extremity of the minor axis, $x_1 = 0$, $y_1 = b$, and the equations become

$$x = \int_0^\phi \sqrt{(a^2 \cos^2 \phi + b^2 \sin^2 \phi)}\, d\phi - \frac{(a^2 - b^2) \sin\phi \cos\phi + b^2 \sin\phi}{\sqrt{(a^2 \cos^2 \phi + b^2 \sin^2 \phi)}},$$
$$y = \frac{ab\,(1 - \cos\phi)}{\sqrt{(a^2 \cos^2 \phi + b^2 \sin^2 \phi)}},$$

which, for $a = b$, become

$$x = a\,(\phi - \sin\phi), \quad y = a\,(1 - \cos\phi),$$

the ordinary formulæ for the cycloid.

In the case of the parabola $Y^2 = 4aX$ (see figure), taking x_1, y_1 for the coordinates

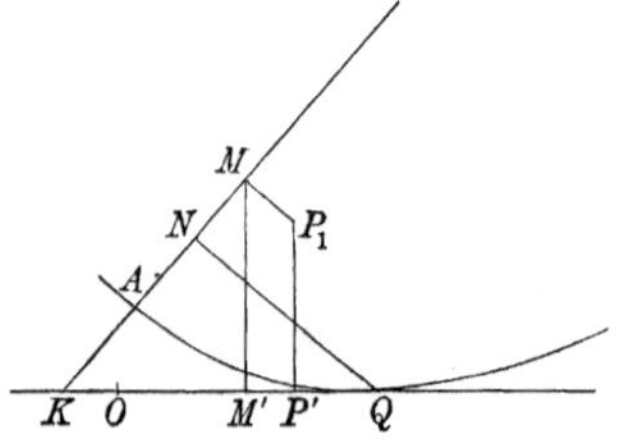

of the describing point P_1, and representing those of the point of contact Q by means of the parameter ϕ, $X = a\phi^2$, $Y = 2a\phi$, then the arc AQ is

$$= a\,[\phi \sqrt{(\phi^2 + 1)} + \log \{\phi + \sqrt{(\phi^2 + 1)}\}],$$

so that, taking the origin at the point O, which comes into contact with the vertex A, this is also the value of OQ, and we then have

$$x = OQ - QP',$$
$$y = P_1P'.$$

Producing the axis of the parabola to meet the line $y = 0$ in k, we have

$$AK = AN = a\phi^2,$$

and thence $KM = a\phi^2 + x_1$; and then, if θ be the inclination AKQ of the axis of the parabola,

$$KM' = (a\phi^2 + x_1)\cos\theta, \quad MM' = (a\phi^2 + x_1)\sin\theta;$$

and thence also

$$KP' = (a\phi^2 + x_1)\cos\theta + y_1\sin\theta, \quad P_1P' = (a\phi^2 + x_1)\sin\theta - y_1\cos\theta;$$

also

$$KQ = 2a\phi^2\sec\theta,$$

and thence

$$QP' = 2a\phi^2\sec\theta - (a\phi^2 + x_1)\cos\theta - y_1\sin\theta.$$

We have $\tan\theta = \dfrac{1}{\phi}$, and thence

$$\sin\theta = \frac{1}{\sqrt{(\phi^2+1)}}, \quad \cos\theta = \frac{\phi}{\sqrt{(\phi^2+1)}}, \quad \sec\theta = \frac{\sqrt{(\phi^2+1)}}{\phi};$$

hence

$$QP' = 2a\phi\sqrt{(\phi^2+1)} - \frac{(a\phi^2 + x_1)\phi + y_1}{\sqrt{(\phi^2+1)}} = \frac{a(\phi^3 + 2\phi) - x_1\phi - y_1}{\sqrt{(\phi^2+1)}},$$

$$P_1P' = \frac{a\phi^2 + x_1 - y_1\phi}{\sqrt{(\phi^2+1)}},$$

and the values of x, y thus become

$$x = a[\phi\sqrt{(\phi^2+1)} + \log\{\phi + \sqrt{(\phi^2+1)}\}] - \frac{a(\phi^3 + 2\phi) - x_1\phi - y_1}{\sqrt{(\phi^2+1)}},$$

that is,

$$x = a\log\{\phi + \sqrt{(\phi^2+1)}\} + \frac{(x_1 - a)\phi + y_1}{\sqrt{(\phi^2+1)}},$$

$$y = \frac{a\phi^2 + x_1 - \phi y_1}{\sqrt{(\phi^2+1)}}.$$

In particular, for the focus of the parabola, $x_1 = a$, $y_1 = 0$, and the equations become

$$x = a\log\{\phi + \sqrt{(\phi^2+1)}\},$$

$$y = a\sqrt{(\phi^2+1)},$$

that is,

$$\frac{x}{a} = \log\left\{\frac{y}{a} + \sqrt{\left(\frac{y^2}{a^2} - 1\right)}\right\},$$

or say

$$\exp.\left(\frac{x}{a}\right) = \frac{y}{a} + \sqrt{\left(\frac{y^2}{a^2} - 1\right)};$$

this gives

$$\exp.\left(-\frac{x}{a}\right) = \frac{y}{a} - \sqrt{\left(\frac{y^2}{a^2} - 1\right)};$$

and we have therefore

$$y = \tfrac{1}{2}a\left\{\exp.\left(\frac{x}{a}\right) + \exp.\left(-\frac{x}{a}\right)\right\},$$

viz. as is well known, the epitrochoid described by the focus of the parabola is the catenary.

914.

ON A SOLUBLE QUINTIC EQUATION.

[From the *American Journal of Mathematics*, vol. XIII. (1891), pp. 53—58.]

MR YOUNG, in his paper, "Solvable Quintic Equations with Commensurable Coefficients," *American Journal of Mathematics*, X. (1888), pp. 99—130, has given, in illustration of his general theory of the solution of soluble quintic equations (founded upon a short note by Abel), no less than twenty instances of the solution of a quintic equation with purely numerical coefficients, having a solution of the form $\sqrt[5]{A}+\sqrt[5]{B}+\sqrt[5]{C}+\sqrt[5]{D}$, where A, B, C, D are numerical expressions involving only square roots. But the solutions are not presented in their most simple form: thus in example 1, $x^5+3x^2+2x-1=0$, the expression involves a radical

$$\sqrt{\tfrac{47}{8}(21125+9439\sqrt{5})}:$$

here

$$(21125+9439)\sqrt{5},\ =\sqrt{5}\,(9439+4225\sqrt{5}),\ =\sqrt{5}\,.\tfrac{1}{2}(18+5\sqrt{5})^2(1+\sqrt{5})^2(2+\sqrt{5}),$$

so that, taking out the roots of the squared factors, we have as the proper form of the radical the very much more simple form $\sqrt{47(2+\sqrt{5})\sqrt{5}}$; where observe that $(2+\sqrt{5})(2-\sqrt{5})=-1$, and thence $(2+\sqrt{5})\sqrt{-47(2-\sqrt{5})\sqrt{5}}=\sqrt{47(2+\sqrt{5})\sqrt{5}}$, viz. the conjugate radicals $\sqrt{-47(2-\sqrt{5})\sqrt{5}}$ and $\sqrt{47(2+\sqrt{5})\sqrt{5}}$ differ only by a factor $2+\sqrt{5}$ which is rational in 1 and $\sqrt{5}$. To avoid fractions I consider the foregoing equation under the form

$$x^5+3000x^2+20000x-100000=0,$$

and I will presently give the solution; but first I consider the general theory. Writing

$$\begin{aligned} A&=\alpha^5, & A'&=\alpha^2\gamma, & A''&=\alpha^3\beta,\\ B&=\beta^5, & B'&=\alpha\beta^2, & B''&=\beta^3\delta,\\ C&=\gamma^5, & C'&=\gamma^2\delta, & C''&=\alpha\gamma^3,\\ D&=\delta^5, & D'&=\beta\delta^2, & D''&=\gamma\delta^3,\end{aligned}$$

we have $A'D'=\alpha^2\delta^2\beta\gamma$, $B'C'=\alpha\delta\beta^2\gamma^2$. Also

$$A''=\frac{A'B'}{\beta\gamma},\quad B''=\frac{B'D'}{\alpha\delta},\quad C''=\frac{A'C'}{\alpha\delta},\quad D''=\frac{C'D'}{\beta\gamma},$$

which determine A'', B'', C'', D'' in terms of A', B', C', D', $\alpha\delta$, $\beta\gamma$; and then

$$A=\frac{A'A''}{\beta\gamma},\quad B=\frac{B'B''}{\alpha\delta},\quad C=\frac{C'C''}{\alpha\delta},\quad D=\frac{D'D''}{\beta\gamma},$$

which give A, B, C, D.

If now we assume $x=\alpha+\beta+\gamma+\delta$, and regard A, B, C, D, A', B', C', D', A'', B'', C'', D'', $\alpha\delta$, $\beta\gamma$ each as a rational function, we may express x, x^2, x^3, x^5 each of them by means of rational functions or of rational functions multiplied into α, β, γ, δ respectively: thus,

$$x=\alpha+\beta+\gamma+\delta \qquad =\alpha+\beta+\gamma+\delta,$$

$$x^2=\alpha^2+\beta^2+\gamma^2+\delta^2 \qquad =\frac{A'\beta}{\beta\gamma}+\frac{B'\delta}{\alpha\delta}+\frac{C'\alpha}{\alpha\delta}+\frac{D'\gamma}{\beta\gamma},$$

$$+2\alpha\beta+2\alpha\gamma+2\alpha\delta+2\beta\gamma+2\beta\delta+2\gamma\delta \qquad +\frac{2B'\gamma}{\beta\gamma}+\frac{2A'\delta}{\alpha\delta}+2\alpha\delta+2\beta\gamma+\frac{2D'\alpha}{\alpha\delta}+\frac{2C'\beta}{\beta\gamma},$$

&c.; and we thus obtain

$$\begin{aligned}
&x^5+qx^3+rx^2+sx+t\\
&=A+B+C+D\\
&\quad+(20\alpha\delta+30\beta\gamma)(A'+D')+30(\alpha\delta+20\beta\gamma)(B'+C')\\
&\qquad\qquad+3q(A'+B'+C'+D')+2r(\alpha\delta+\beta\gamma)+t\\
&\quad+\alpha\left\{5A''+5\frac{C'^2}{\alpha\delta}+\frac{5B''\beta\gamma}{\alpha\delta}+10C''+\frac{10D'^2}{\alpha\delta}\right.\\
&\qquad\qquad+10\alpha^2\delta^2+20B''+20D''+30\beta^2\gamma^2+30D''\frac{\beta\gamma}{\alpha\delta}+60\alpha\beta\gamma\delta\\
&\qquad\qquad\left.+q\left(\frac{B''}{\alpha\delta}+\frac{3D''}{\alpha\delta}+3\alpha\delta+6\beta\gamma\right)+r\left(\frac{C'}{\alpha\delta}+\frac{2D'}{\alpha\delta}\right)+s\right\}\\
&\quad+\beta\left\{5B''+\frac{5A'^2}{\beta\gamma}+5\frac{D''\alpha\delta}{\beta\gamma}+10A''+\frac{10C'^2}{\beta\gamma}\right.\\
&\qquad\qquad+10\beta^2\gamma^2+20D''+20C''+30\alpha^2\delta^2+30C''\frac{\alpha\delta}{\beta\gamma}+60\alpha\beta\gamma\delta\\
&\qquad\qquad\left.+q\left(\frac{D''}{\beta\gamma}+\frac{3C''}{\beta\gamma}+3\beta\gamma+6\alpha\delta\right)+r\left(\frac{A'}{\beta\gamma}+\frac{2C'}{\beta\gamma}\right)+s\right\}\\
&\quad+\gamma\left\{5C''+\frac{5D'^2}{\beta\gamma}+5\frac{A''\alpha\delta}{\beta\gamma}+10D''+\frac{10B'^2}{\beta\gamma}\right.\\
&\qquad\qquad+10\beta^2\gamma^2+20A''+20B''+30\alpha^2\delta^2+30B''\frac{\alpha\delta}{\beta\gamma}+60\alpha\beta\gamma\delta\\
&\qquad\qquad\left.+q\left(\frac{A''}{\beta\gamma}+\frac{3B''}{\beta\gamma}+3\beta\gamma+6\alpha\delta\right)+r\left(\frac{D'}{\beta\gamma}+\frac{2B'}{\beta\gamma}\right)+s\right\}\\
&\quad+\delta\left\{5D''+\frac{5B'^2}{\alpha\delta}+5\frac{C''\beta\gamma}{\alpha\delta}+10B''+\frac{10A'^2}{\alpha\delta}\right.\\
&\qquad\qquad+10\alpha^2\delta^2+20C''+20A''+30\beta^2\gamma^2+30A''\frac{\beta\gamma}{\alpha\delta}+60\alpha\beta\gamma\delta\\
&\qquad\qquad\left.+q\left(\frac{C''}{\alpha\delta}+\frac{3A''}{\alpha\delta}+3\alpha\delta+6\beta\gamma\right)+r\left(\frac{B'}{\alpha\delta}+\frac{2A'}{\alpha\delta}\right)+s\right\}.
\end{aligned}$$

If, then, $x^5 + qx^3 + rx^2 + sx + t = 0$, we have the rational term $=0$, and the coefficients of α, β, γ, δ each $=0$; in the class of equations under consideration, these last equations differ only in the signs of the radicals contained therein, so that one of them being satisfied identically, the others will be also satisfied. In particular, if $q=0$, then $\alpha\delta+\beta\gamma=0$: the rational term gives

$$A+B+C+D-10\alpha\delta(A'+D'-B'-C')+t=0,$$

and the term in α gives

$$5A''+15B''+10C''-10D''+\frac{5}{\alpha\delta}(C'^2+2D'^2)+\frac{r}{\alpha\delta}(C'+2D')-20\alpha^2\delta^2+s=0.$$

For the equation $x^5+3000x^2+20000x-100000=0$, the expression for the root is $x=\sqrt[5]{A}+\sqrt[5]{B}+\sqrt[5]{C}+\sqrt[5]{D}$, where

$$\begin{aligned}
A &= 39000+18200\sqrt{5}+(1720+920\sqrt{5})\sqrt{235+94\sqrt{5}},\\
D &= 39000+18200\sqrt{5}+(-1720-920\sqrt{5})\sqrt{235+94\sqrt{5}},\\
B &= 39000-18200\sqrt{5}+(-1720+920\sqrt{5})\sqrt{235-94\sqrt{5}},\\
C &= 39000-18200\sqrt{5}+(1720-920\sqrt{5})\sqrt{235-94\sqrt{5}},
\end{aligned}$$

and where also

$$\begin{aligned}
A' &= -150-70\sqrt{5}+(-10-2\sqrt{5})\sqrt{235+94\sqrt{5}},\\
D' &= -150-70\sqrt{5}+(10+2\sqrt{5})\sqrt{235+94\sqrt{5}},\\
B' &= -150+70\sqrt{5}+(10-2\sqrt{5})\sqrt{235-94\sqrt{5}},\\
C' &= -150+70\sqrt{5}+(-10+2\sqrt{5})\sqrt{235-94\sqrt{5}},
\end{aligned}$$

and

$$\begin{aligned}
A'' &= -940-100\sqrt{5}+(-100+20\sqrt{5})\sqrt{235+94\sqrt{5}},\\
D'' &= -940-100\sqrt{5}+(100-20\sqrt{5})\sqrt{235+94\sqrt{5}},\\
B'' &= -940+100\sqrt{5}+(100+20\sqrt{5})\sqrt{235-94\sqrt{5}},\\
C'' &= -940+100\sqrt{5}+(-100-20\sqrt{5})\sqrt{235-94\sqrt{5}}.
\end{aligned}$$

The foregoing forms are in some respects the most convenient; but it is to be observed that we have

$$\begin{aligned}
A &= 2600\sqrt{5}(1+\sqrt{5})(2+\sqrt{5})+40(1+\sqrt{5})(18+5\sqrt{5})\sqrt{47\sqrt{5}(2+\sqrt{5})}, \text{ \&c.,}\\
A' &= -10\sqrt{5}(1+\sqrt{5})(2+\sqrt{5})-2\sqrt{5}(1+\sqrt{5})\sqrt{47\sqrt{5}(2+\sqrt{5})}, \text{ \&c.,}\\
A'' &= 20(1-\sqrt{5})(18+13\sqrt{5})+20\sqrt{5}(1-\sqrt{5})\sqrt{47\sqrt{5}(2+\sqrt{5})}, \text{ \&c.,}
\end{aligned}$$

or putting for shortness

$$\sqrt{Q}=\sqrt{47\sqrt{5}(2+\sqrt{5})},\quad \sqrt{Q_1}=\sqrt{-47\sqrt{5}(2-\sqrt{5})},$$

(so that, according to a foregoing remark, we have $(2+\sqrt{5})\sqrt{Q}=\sqrt{Q_1}$), then we have

$$\begin{aligned} A\ &= 40(1+\sqrt{5})\{65\sqrt{5}(2+\sqrt{5})+(18+5\sqrt{5})\sqrt{Q}\}, \text{ \&c.},\\ A' &= -2\sqrt{5}(1+\sqrt{5})\{5(2+\sqrt{5})+\sqrt{Q}\}, \text{ \&c.},\\ A'' &= 20(1-\sqrt{5})\{18+13\sqrt{5}+\sqrt{5}\sqrt{Q}\}, \text{ \&c.},\end{aligned}$$

where observe that the term $2+\sqrt{5}$ is a factor of Q.

Starting from the values of A', B', C', D', we have

$$\begin{aligned} A' &= -2\sqrt{5}(1+\sqrt{5})\{5(2+\sqrt{5})+\sqrt{Q}\},\\ D' &= -2\sqrt{5}(1+\sqrt{5})\{5(2+\sqrt{5})-\sqrt{Q}\},\end{aligned}$$

and therefore

$$A'D'=20(1+\sqrt{5})^2(2+\sqrt{5})\{25(2+\sqrt{5})-47\sqrt{5}\},$$

where the last factor is

$$=50-22\sqrt{5},\ =-2\sqrt{5}(11-5\sqrt{5}),\ =-\sqrt{5}(1-\sqrt{5})^2(2-\sqrt{5}).$$

Hence

$$A'D'=-\sqrt{5}\,.\,20(-4)^2(-1)=320\sqrt{5},$$

that is,

$$A'D'=(\alpha\delta)^2\beta\gamma=320\sqrt{5}, \text{ and similarly } B'C'=\alpha\delta(\beta\gamma)^2=-320\sqrt{5},$$

whence

$$\alpha\delta=-4\sqrt{5},\quad \beta\gamma=4\sqrt{5}, \text{ and } \alpha\delta+\beta\gamma=0, \text{ as above.}$$

We have, moreover,

$$\begin{aligned} A' &= -2\sqrt{5}(1+\sqrt{5})\{5(2+\sqrt{5})+\sqrt{Q}\},\\ B' &= 2\sqrt{5}(1-\sqrt{5})\{5(2-\sqrt{5})-\sqrt{Q_1}\},\end{aligned}$$

and thence

$$\begin{aligned} A'B' &= 80\{-25-\sqrt{QQ_1}+5(2-\sqrt{5})\sqrt{Q}-5(2+\sqrt{5})\sqrt{Q_1}\},\\ &= 80\{-25-47\sqrt{5}+(5(2-\sqrt{5})-5)\sqrt{Q}\},\end{aligned}$$

that is,

$$\begin{aligned} A'B'\div\beta\gamma &= 4\sqrt{5}\{-25-47\sqrt{5}+5(1-\sqrt{5})\sqrt{Q}\}\\ &= -20(47+5\sqrt{5})+20\sqrt{5}(1-\sqrt{5})\sqrt{Q},\ =A'';\end{aligned}$$

and similarly we verify the values of B'', C'' and D''.

We have next

$$A'A''=160\sqrt{5}\{(10+5\sqrt{5}+\sqrt{Q})(18+13\sqrt{5}+\sqrt{5}\sqrt{Q})\},$$

or observing that $Q\sqrt{5}$ is $=235(2+\sqrt{5})$, the whole term in $\{\ \}$ is

$$\begin{aligned} &=(505+220\sqrt{5})+(470+235\sqrt{5})+(18+13\sqrt{5}+25+10\sqrt{5})\sqrt{Q},\\ &=975+455\sqrt{5}+(43+23\sqrt{5})\sqrt{Q}=65\sqrt{5}(7+3\sqrt{5})+(43+23\sqrt{5})\sqrt{Q};\end{aligned}$$

or we have

$$\begin{aligned} A'A'' &= 160\sqrt{5}\{65\sqrt{5}(7+3\sqrt{5})+(43+23\sqrt{5})\sqrt{Q}\},\\ &= 160\sqrt{5}(1+\sqrt{5})\{65\sqrt{5}(2+\sqrt{5})+(18+5\sqrt{5})\sqrt{Q}\},\end{aligned}$$

and consequently

$$A'A''\div\beta\gamma=40(1+\sqrt{5})\{65\sqrt{5}(2+\sqrt{5})+(18+5\sqrt{5})\sqrt{Q}\},\ =A;$$

and similarly we verify the values of B, C, D.

In the proposed equation $x^5+3000x^2+20000x-100000=0$, we have $r=3000$, $s=20000$, $t=-100000$, $\alpha\delta=-4\sqrt{5}$; the two equations to be verified thus are

$$A+B+C+D+40\sqrt{5}(A'+D'-B'-C')-100000=0,$$

and

$$5A''+15B''+10C''-10D''-\frac{\sqrt{5}}{4}(C'^2+2D'^2)-150\sqrt{5}(C'+2D')-1600+20000=0.$$

As to the first of these, we have $A+B+C+D=156000$, $A'+D'-B'-C'=-280\sqrt{5}$, and the equation thus is

$$156000+40\sqrt{5}(-280\sqrt{5})-100000=0,$$

which is right.

For the second equation, if in the calculation we keep the radicals in the first instance distinct, we have

$$5A''+15B''+10C''-10D''=-18800+3000\sqrt{5}+(-1500+300\sqrt{5})\sqrt{Q}+(500+100\sqrt{5})\sqrt{Q_1},$$

$$-150\sqrt{5}(C'+2D')=\{-450-70\sqrt{5}$$
$$+(20+4+\sqrt{5})\sqrt{Q}+(-10+2\sqrt{5})\sqrt{Q_1}\}(-150\sqrt{5})$$

$$-1600+20000 \quad =18400$$

$$-\frac{\sqrt{5}}{4}(C'^2+2D'^2) \quad =-\frac{\sqrt{5}}{4}\{282000+416800\sqrt{5}$$
$$+(-8800-4000\sqrt{5})\sqrt{Q}+(4400-2000\sqrt{5})\sqrt{Q_1}\}.$$

Substituting in the equation, we ought to have

$$\begin{aligned}0=&-18800+3000\sqrt{5}+(-1500+300\sqrt{5})\sqrt{Q}+(500+100\sqrt{5})\sqrt{Q_1}\\&+52500+67500\sqrt{5}+(-3000-3000\sqrt{5})\sqrt{Q}+(-1500+1500\sqrt{5})\sqrt{Q_1}\\&+18400\\&-52100-70500\sqrt{5}+(5000+2200\sqrt{5})\sqrt{Q}+(2500-1100\sqrt{5})\sqrt{Q_1},\end{aligned}$$

that is,

$$0=(500-500\sqrt{5})\sqrt{Q}+(1500+500\sqrt{5})\sqrt{Q_1},$$

viz.

$$0=500\{(1-\sqrt{5})\sqrt{Q}+(3+\sqrt{5})\sqrt{Q_1}\},$$

which is satisfied in virtue of $\sqrt{Q}=(2+\sqrt{5})\sqrt{Q_1}$: this completes the verification.

915.

ON THE PARTITIONS OF A POLYGON.

[From the *Proceedings of the London Mathematical Society*, vol. XXII. (1891), pp. 237—262. Read March 12, 1891.]

1. THE partitions are made by non-intersecting diagonals; the problems which have been successively considered are (1) to find the number of partitions of an r-gon into triangles, (2) to find the number of partitions of an r-gon into k parts, and (3) to find the number of partitions of an r-gon into p-gons, r of the form

$$n(p-2)+2.$$

Problem (1) is a particular case of (2); and it is also a particular case of (3); but the problems (2) and (3) are outside each other; for problem (3) a very elegant solution, which will be here reproduced, is given in the paper, H. M. Taylor and R. C. Rowe, "Note on a Geometrical Theorem," *Proc. Lond. Math. Soc.*, t. XIII. (1882), pp. 102—106, and this same paper gives the history of the solution of (1).

The solution of (2) is given in the memoir, Kirkman "On the k-partitions of the r-gon and r-ace," *Phil. Trans.*, t. CXLVII. (for 1857), p. 225; viz. he there gives for the number of partitions of the r-gon into k parts (or, what is the same thing, by means of $k-1$ non-intersecting diagonals) the expression

$$\frac{[r+k-2]^{k-1}\,[r-3]^{k-1}}{[k]^{k-1}\,[k-1]^{k-1}};$$

but there is no complete demonstration of this result.

If $k=r-2$, we have the solution of the problem (1); viz. the number of partitions of the r-gon into triangles is

$$=[2r-4]^{r-3}\div[r-2]^{r-3}.$$

The present paper relates chiefly to the foregoing problem (2), the determination of the number of partitions of the r-gon into k parts, or, what is the same thing, by means of $k-1$ non-intersecting diagonals.

2. Assuming for the moment the foregoing result, then for $k=1$, the number of partitions is

$$=1,$$

for $k=2$ it is

$$=\frac{r\,.\,r-3}{2},$$

for $k=3$ it is

$$=\frac{r+1\,.\,r\,.\,r-3\,.\,r-4}{12},$$

and so on. As a simple verification, $k=2$, the number of partitions is equal to the number of diagonals, viz. this is number of pairs of summits *less* number of sides, that is,

$$\tfrac{1}{2}r(r-1)-r, \quad =\tfrac{1}{2}r(r-3).$$

For convenience, I give the first Table on the next page, which is a tabulation of the functions

$$U_1 = x^3 + x^4 + x^5 + x^6 + \ldots + x^r,$$

$$U_2 = 2x^4 + 5x^5 + 9x^6 + \ldots + \frac{r\,.\,r-3}{2\,.\,1}\,x^r,$$

$$U_3 = 5x^5 + 21x^6 + \ldots + \frac{r+1\,.\,r\,.\,r-3\,.\,r-4}{3\,.\,2\,.\,2\,.\,1}\,x^r,$$

$$U_4 = 14x^6 + \ldots + \frac{r+2\,.\,r+1\,.\,r\,.\,r-3\,.\,r-4\,.\,r-5}{4\,.\,3\,.\,2\,.\,3\,.\,2\,.\,1}\,x^r,$$

$$U_5 = \ldots + \frac{r+3\,.\,r+2\,.\,r+1\,.\,r\,.\,r-3\,.\,r-4\,.\,r-5\,.\,r-6}{5\,.\,4\,.\,3\,.\,2\,.\,4\,.\,3\,.\,2\,.\,1}\,x^r,$$

&c.

3. And in connexion herewith, I give the second Table on the next page, which is a tabulation of the functions

$$V_2 = x^6 + 2x^7 + 3x^8 + 4x^9 + \ldots + 1\,\frac{r-3}{1}\,x^{r+2},$$

$$V_3 = 4x^7 + 14x^8 + 32x^9 + \ldots + 2\,\frac{r+1\,.\,r-3\,.\,r-4}{3\,.\,2\,.\,1}\,x^{r+2},$$

$$V_4 = 14x^8 + 72x^9 + \ldots + 3\,\frac{r+2\,.\,r+1\,.\,r-3\,.\,r-4\,.\,r-5}{4\,.\,3\,.\,3\,.\,2\,.\,1}\,x^{r+2},$$

$$V_5 = 48x^9 + \ldots + 4\,\frac{r+3\,.\,r+2\,.\,r+1\,.\,r-3\,.\,r-4\,.\,r-5\,.\,r-6}{5\,.\,4\,.\,3\,.\,4\,.\,3\,.\,2\,.\,1}\,x^{r+2},$$

&c.

$r=$	$k=1$	2	3	4	5	6	7	8	9	10	11	12	13
3	1												
4	1	2											
5	1	5	5										
6	1	9	21	14									
7	1	14	56	84	42								
8	1	20	120	300	330	132							
9	1	27	225	825	1485	1287	429						
10	1	35	385	1925	5005	7007	5005	1430					
11	1	44	616	4004	14014	28028	32032	19448	4862				
12	1	54	936	7644	34398	91728	148512	143208	75582	16796			
13	1	65	1365	13650	76440	259896	556920	755820	629850	293930	58786		
14	1	77	1925	23100	157080	659736	1790712	3197700	3730650	2735810	1144066	208012	
15	1	90	2640	37400	302940	1534896	5116320	11511720	17587350	17978180	11767536	4457400	742900

$r=$	$k=2$	3	4	5	6	7	8	9	10	11	12	13
4	1											
5	2	4										
6	3	14	14									
7	4	32	72	48								
8	5	60	225	330	165							
9	6	100	550	1320	1430	572						
10	7	154	1155	4004	7007	6006	2002					
11	8	224	2184	10192	25480	34944	24052	7072				
12	9	312	3822	22932	76440	148512	167076	100776	25194			
13	10	420	6300	47040	199920	514080	813960	775200	396800	90440		
14	11	550	9900	89760	476240	1534996	3197700	4263600	3517470	1634380	326876	
15	12	704	14960	161568	1023264	4093056	10744272	18759840	18573816	15690048	6547520	1188640

4. These functions, U and V, are particular values satisfying the equation

$$(V_2 + V_3 y + V_4 y^2 + \ldots) = (U_1 + U_2 y + U_3 y^2 + U_4 y^3 + \ldots)^2;$$

that this is so will appear from the following general investigation.

5. Taking x, y as independent variables, denoting by X an arbitrary function of x, and using accents to denote differentiations in regard to x, we require the following identity:

$$\frac{2}{1.2} X^2 + \frac{4y}{1.2.3}(X^3)' + \frac{6y^2}{1.2.3.4}(X^4)'' + \ldots = \left\{X + \frac{y}{1.2}(X^2)' + \frac{y^2}{1.2.3}(X^3)'' + \ldots\right\}^2,$$

which I prove as follows. Writing U to denote the same function of u which X is of x, I start from the equation

$$u = x + yU,$$

which determines u as a function of the independent variables x, y. We have

$$\frac{du}{dy}(1 - yU') = U, \quad \frac{du}{dx}(1 - yU') = 1,$$

where the accent denotes differentiation in regard to u; hence

$$\frac{du}{dy} = U\frac{du}{dx} = \frac{u-x}{y}\frac{du}{dx},$$

or say

$$y\frac{du}{dy} = (u-x)\frac{du}{dx}.$$

Writing

$$u_1 = \int u\,dx,$$

and therefore

$$\frac{du_1}{dx} = u,$$

this equation may be written

$$y\frac{d^2u_1}{dxdy} - \frac{du_1}{dx} = u\frac{du}{dx} - u - x\frac{du}{dx};$$

or, integrating with respect to x, we have

$$y\frac{du_1}{dy} - u_1 = \tfrac{1}{2}u^2 - ux,$$

or say

$$\frac{2}{y}\frac{du_1}{dy} - \frac{2(u_1 - \frac{1}{2}x^2)}{y^2} = \frac{(u-x)^2}{y^2},$$

that is,

$$2\frac{d}{dy}\left(\frac{u_1 - \frac{1}{2}x^2}{y}\right) = \frac{(u-x)^2}{y^2}.$$

6. But, from the equation

$$u = x + yU,$$

we have

$$u = x + yX + \frac{y^2}{1.2}(X^2)' + \frac{y^3}{1.2.3}(X^3)'' + \ldots,$$

and thence

$$u_1 = \tfrac{1}{2}x^2 + yX_1 + \frac{y^2}{1.2}X^2 + \frac{y^3}{1.2.3}(X^3)' + \ldots,$$

if for a moment X_1 is written for $\int X\,dx$. And hence, from the relation obtained above, we have the required identity

$$\frac{2}{1.2}X^2 + \frac{4y}{1.2.3}(X^3)' + \frac{6y^2}{1.2.3.4}(X^4)'' + \ldots = \left\{X + \frac{y}{1.2}(X^2)' + \frac{y^2}{1.2.3}(X^3)'' + \ldots\right\}^2.$$

This of course gives the series of identities

$$\frac{2}{1.2}X^2 = X^2,$$

$$\frac{4}{1.2.3}(X^3)' = \frac{2}{1.2}X(X^2),$$

$$\frac{6}{1.2.3.4}(X^4)'' = \frac{2}{1.2.3}X(X^3)'' + \left\{\frac{1}{1.2}(X^2)'\right\}^2;$$

$$\vdots$$

or say

$$X^2 = X^2,$$

$$(X^3)' = \tfrac{3}{2}X(X^2)',$$

$$(X^4)'' = \tfrac{4}{3}X(X^3)'' + \{(X^2)'\}^2,$$

$$\vdots$$

all of which may be easily verified.

7. I multiply each side of the identity by x^2, and write

$$U_1 = x.X, \qquad V_2 = x^2\frac{2}{1.2}X^2,$$

$$U_2 = x\frac{1}{1.2}(X^2)', \qquad V_3 = x^2\frac{4}{1.2.3}(X^3)',$$

$$U_3 = x\frac{1}{1.2.3}(X^3)'', \qquad V_4 = x^2\frac{6}{1.2.3.4}(X^4)'',$$

$$U_4 = x\frac{1}{1.2.3.4}(X^4)''', \qquad V_5 = x^2\frac{8}{1.2.3.4.5}(X^5)''',$$

$$\vdots \qquad\qquad \vdots$$

We thus obtain two sets of functions U and V, satisfying the before-mentioned equation. We have

$$(V_2 + yV_3 + y^2V_4 + \ldots) = (U_1 + yU_2 + y^2U_3 + \ldots)^2;$$

and it will be observed that we have, moreover, the relations

$$U_2 = \tfrac{1}{2}x(x^{-2}V_2)', \quad U_3 = \tfrac{1}{4}x(x^{-2}V_3)', \quad U_4 = \tfrac{1}{6}x(x^{-2}V_4)', \ \ldots.$$

8. In particular, if

$$X = \frac{x^2}{1-x},$$

then the general term

in X is x^{r-1}, the first term occurring when $r = 3$,

in X^2 is $(r-3)x^r$, ,, ,, $r = 4$,

in X^3 is $\dfrac{r-3.r-4}{1.2}x^{r+1}$, ,, ,, $r = 5$,

in X^4 is $\dfrac{r-3.r-4.r-5}{1.2.3}x^{r+2}$, ,, ,, $r = 6$,

$\vdots$

from which it appears that, for this value of X, U_1, U_2, U_3, U_4, &c., have the before-mentioned values (No. 2), and further that V_2, V_3, V_4, V_5, &c., have also the before-mentioned values (No. 3).

9. We do not absolutely require, but it is interesting to obtain, the finite expressions of these functions. We have

$$(1-x)\ U_1 = x^3(1),$$
$$(1-x)^3\ U_2 = x^4(2-x),$$
$$(1-x)^5\ U_3 = x^5(5-4x+x^2),$$
$$(1-x)^7\ U_4 = x^6(14-14x+6x^2-x^3),$$
$$(1-x)^9\ U_5 = x^7(42-48x+27x^2-8x^3+x^4),$$
$$(1-x)^{11}\ U_6 = x^8(132-165x+110x^2-44x^3+10x^4-x^5);$$
$$\vdots$$
$$(1-x)^2\ V_2 = x^6\ (1),$$
$$(1-x)^4\ V_3 = x^7\ (4-2x),$$
$$(1-x)^6\ V_4 = x^8\ (14-12x+3x^2),$$
$$(1-x)^8\ V_5 = x^9\ (48-54x+24x^2-4x^3),$$
$$(1-x)^{10}\ V_6 = x^{10}(165-220x+132x^2-40x^3+5x^4),$$
$$\vdots$$

and here the factors in ( ) satisfy the series of relations

$$1 = 1^2,$$
$$4-2x = 2(2-x),$$
$$14-12x+3x^2 = 2.1(5-4x+x^2)+(2-x)^2,$$
$$48-54x+24x^2-4x^3 = 2.1(14-14x+6x^2-x^3)+2(2-x)(5-4x+x^2),$$
$$\vdots$$

corresponding to

$$V_2 = U_1^2,\quad V_3 = 2U_1U_2,\ \&\text{c.},$$

given by the before-mentioned equation (No. 7), between the functions V and U.

10. It is to be shown that, taking U_1, U_2, U_3, ... for the functions which belong to the partitions of the r-gon (assumed to be unknown functions of r and the suffixes), and connecting them with a set of functions V_2, V_3, V_4, ... by the relations

$$U_2 = \tfrac{1}{2}x(x^{-2}V_2)',\quad U_3 = \tfrac{1}{4}x(x^{-2}V_3)',\quad U_4 = \tfrac{1}{6}x(x^{-2}V_4)',\ \&\text{c.},$$

then we have the foregoing identical equation

$$(V_2+V_3y+V_4y^2+\ldots) = (U_1+U_2y+U_3y^2+U_4y^3+\ldots)^2.$$

This implies the relations

$$V_2 = U_1^2,$$
$$V_3 = 2U_1U_2,$$
$$V_4 = 2U_1U_3 + U_2^2,$$
$$V_5 = 2U_1U_4 + 2U_2U_3,$$
&c.

Thus, if U_1 is known, the equation

$$V_2 = U_1^2$$

determines V_2, and then the equation

$$U_2 = \tfrac{1}{2}x(x^{-2}V_2)'$$

determines U_2, so that U_1, U_2 are known; and we thence in the same way find successively U_3 and V_3, U_4 and V_4, and so on; that is, assuming only that U_1 has the before-mentioned value,

$$U_1 = x^3 + x^4 + x^5 + \ldots + x^r + \ldots,$$

it follows that all the remaining functions U and V must have their before-mentioned values. But the function

$$U_1 = x^3 + x^4 + x^5 + \ldots,$$

where each coefficient is $=1$, is evidently the proper expression for the generating function of the number of partitions of the r-gon into a single part; and we thus arrive at the proof that the remaining functions U, which are the generating functions for the number of partitions of the r-gon into 2, 3, 4, ..., k, parts, have their before-mentioned values.

11. Considering, then, the partition problem from the point of view just referred to, I write A_r, B_r, C_r, ... for the number of partitions of an r-gon into 1 part, 2 parts, 3 parts, &c., and form therewith the generating functions

$$U_1 = A_3x^3 + A_4x^4 + \quad \ldots + \quad A_rx^r \quad + \ldots,$$
$$U_2 = \qquad B_4x^4 + \quad \ldots + \quad B_rx^r \quad + \ldots,$$
$$U_3 = \qquad\qquad C_5x^5 + \ldots + \quad C_rx^r \quad + \ldots,$$
$$\vdots$$

and also the functions

$$V_2 = \qquad \frac{2}{4}B_4x^6 + \quad \ldots + \frac{2}{r}\,B_rx^{r+2} + \ldots,$$
$$V_3 = \qquad\qquad \frac{4}{5}C_5x^7 + \ldots + \frac{4}{r}\,C_rx^{r+2} + \ldots,$$
$$\vdots$$

where observe that the functions U, V are such that

$$U_2 = \tfrac{1}{2}x(x^{-2}V_2)', \quad U_3 = \tfrac{1}{4}x(x^{-2}V_3)', \quad U_4 = \tfrac{1}{6}x(x^{-2}V_4)', \text{ \&c.}$$

To fix the ideas, consider an r-gon which is to be divided into six parts. Choosing any particular summit, and from this summit drawing a diagonal successively to each of the non-adjacent $r-3$ summits, we divide the r-gon into two parts in $r-3$ different ways; viz. the two parts are

$$\begin{array}{lll} \text{a } 3\text{-gon} & \text{and} & (r-1)\text{-gon}, \\ 4\text{-gon} & \text{and} & (r-2)\text{-gon}, \\ \vdots & & \\ (r-1)\text{-gon} & \text{and} & 3\text{-gon}; \end{array}$$

say any one of these ways is

$$\text{an } \alpha\text{-gon and } \beta\text{-gon}, \quad \alpha+\beta=r+2.$$

Next, writing

$$a+b=6,$$

that is,

$$\begin{aligned} a, b = 1&, 5, \\ 2&, 4, \\ 3&, 3, \\ 4&, 2, \\ 5&, 1, \end{aligned}$$

we divide in every possible way the α-gon into a parts, and the β-gon into b parts (so dividing the r-gon into six parts). Observing that A, B, C, D, E, F are the letters belonging to the numbers 1, 2, 3, 4, 5, 6, respectively, the number of parts which we thus obtain (corresponding to the different values of a, b) are

$$A_\alpha E_\beta + B_\alpha D_\beta + C_\alpha C_\beta + D_\alpha B_\beta + E_\alpha A_\beta,$$

and summing for the different values of α, β $(\alpha+\beta=r+2)$, the whole number of parts is

$$= \text{coeff. } x^{r+2} \text{ in } (U_1U_5 + U_2U_4 + U_3U_3 + U_4U_2 + U_5U_1),$$

that is,

$$= \text{coeff. } x^{r+2} \text{ in } (2U_1U_5 + 2U_2U_4 + U_3^2).$$

12. To obtain the whole number of the partitions of the r-gon into six parts, we must perform the foregoing process successively with each summit of the r-gon as the summit from which is drawn the diagonal which divides the r-gon into two parts; that is, the number found as above is to be multiplied by r. We thus obtain all the partitions repeated a certain number of times, viz. each partition into six parts is a partition by means of five diagonals, and is thus obtainable by the foregoing process, taking any one of the ten extremities of these diagonals as the point from which is drawn the diagonal which divides the r-gon into two parts; that is, we have to divide the foregoing product by 10. The final result thus is

$$\frac{10}{r} F_r = \text{coeff. } x^{r+2} \text{ in } (2U_1U_5 + 2U_2U_4 + U_3^2),$$

where

$$\frac{10}{r} F_r \text{ is } = \text{coeff. } x^{r+2} \text{ in } V_6;$$

we thus have

$$V_6 = 2U_1U_5 + 2U_2U_4 + U_3^2.$$

13. The reasoning is perfectly general; and applying it successively to the partitions into two parts, three parts, &c., we have

$$\begin{aligned} V_2 &= U_1^2, \\ V_3 &= 2U_1U_2, \\ V_4 &= 2U_1U_3 + U_2^2, \\ V_5 &= 2U_1U_4 + 2U_2U_3, \\ &\vdots \end{aligned}$$

where any function V is related to the corresponding function U as above. The value of U_1 is obviously

$$U_1 = x^3 + x^4 + x^5 + \ldots = \frac{x^3}{1-x};$$

and hence the several functions U and V have the values above written down; the general term of U_k is

$$\frac{[r+k-2]^{k-1}\,[r-3]^{k-1}}{[k]^{k-1}\,[k-1]^{k-1}}\,x^r;$$

and the number of partitions of the r-gon into k parts is equal to the coefficient of x^r in this general term.

14. In the investigations which next follow, I consider, without using the method of generating functions, the problem of the partition of the r-gon into 2, 3, 4 or 5 parts; it will be convenient to state the results as follows:

Number of Partitions.

2 parts, $\frac{r}{2}\ A$,

3 parts, $\frac{r}{4}\ 2A$,

4 parts, $\frac{r}{6}\,(3A+2B)$,

5 parts, $\frac{r}{8}\,(4A+8B+2C)$;

where the capital letters refer to different "diagonal-types," thus:

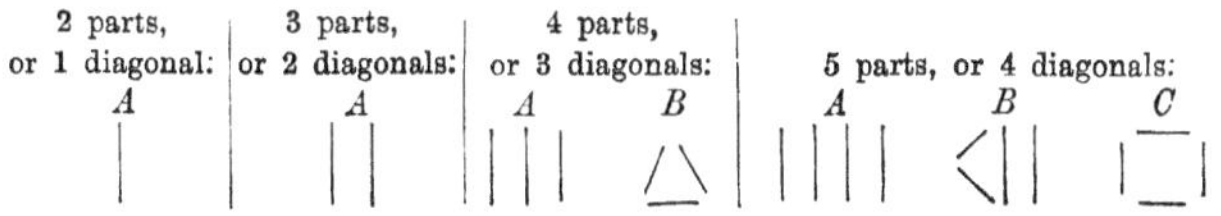

2 parts, or 1 diagonal:	3 parts, or 2 diagonals:	4 parts, or 3 diagonals:		5 parts, or 4 diagonals:		
A	A	A	B	A	B	C

viz. if, in a polygon divided into k parts by means of $k-1$ diagonals, we delete all the sides of the polygon, leaving only the diagonals, then these will present themselves under distinct forms, which are what I call "diagonal-types"; for instance, when

$$k = 4,$$

there are the two types A and B shown in the above diagram for four parts.

15. It is to be observed that we have sub-types corresponding to the coalescence of the terminal points of different diagonals; thus, suppose

$$k = 4.$$

Writing now A° and B° to denote the forms without coalescences, we have the sub-types A°, A', A'' and B°, B', B'', B''', as follows:

4 parts, or 3 diagonals:

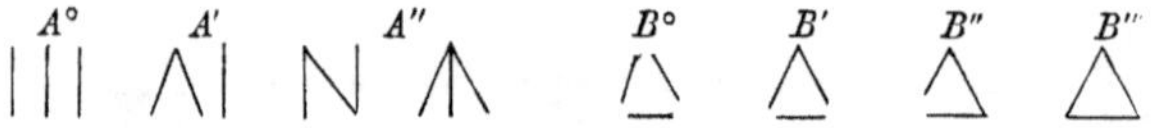

A° A' A'' B° B' B'' B'''

where observe that under A'' are included two distinct forms, which, nevertheless, by reason that there is in each of them the same number $(= 2)$ of coalescences, are reckoned as belonging to the same sub-type.

16. The numbers called A, B, C, &c., have values which may be directly determined. I write down as follows:

1 diagonal, $A = \dfrac{r-3}{1}$.

2 diagonals, $A = \dfrac{r-3\,.\,r-2\,.\,r-1}{6} - \dfrac{r-3}{1} = \dfrac{r-3\,.\,r-4}{6}(r+1)$;

where the calculation is

$$\begin{array}{rr} r-2\,.\,r-1 = & r^2-3r+2 \\ -6 & -6 \\ \hline & r^2-3r-4 \\ & = r-4\,.\,r+1. \end{array}$$

$$\begin{aligned} \text{3 diagonals, } A &= \frac{r-3\,.\,r-2\,.\,r-1\,.\,r\,.\,r+1}{120} - 2\left(\frac{r-3\,.\,r-2\,.\,r-1}{6} - \frac{r-3}{1}\right) - \frac{r-3}{1} \\ &= \frac{r-3\,.\,r-2\,.\,r-1\,.\,r\,.\,r+1}{120} - 2\,\frac{r-3\,.\,r-2\,.\,r-1}{6} + 1\,\frac{r-3}{1} \\ &= \frac{r-3\,.\,r-4\,.\,r-5}{120}(r^2+7r+2); \end{aligned}$$

where the calculation is

$$\begin{array}{rr} r-2\,.\,r-1\,.\,r\,.\,r+1 = & r^4-2r^3-r^2+2r \\ -40\,.\,r-2\,.\,r-1 & -40r^2+120r-80 \\ +120 & +120 \\ \hline & r^4-2r^3-41r^2+122r+40 \\ & = r-4\,.\,r-5\,.\,r^2+7r+2. \end{array}$$

$$\begin{aligned} B &= \frac{r-5\,.\,r-4\,.\,r-3\,.\,r-2\,.\,r-1}{120} \\ &= \frac{r-3\,.\,r-4\,.\,r-5}{120}(r-1\,.\,r-2). \end{aligned}$$

4 diagonals, $A = \frac{r-3 \,.\, r-2 \,.\, r-1 \,.\, r \,.\, r+1 \,.\, r+2 \,.\, r+3}{5040}$

$$- 3\,\frac{r-3 \,.\, r-2 \,.\, r-1 \,.\, r \,.\, r+1}{120} + 3\,\frac{r-3 \,.\, r-2 \,.\, r-1}{6} - 1\,\frac{r-3}{1}$$

$$= \frac{r-3 \,.\, r-4 \,.\, r-5 \,.\, r-6}{5040}\,(r^3 + 18r^2 + 65r);$$

where the calculation is

$$\begin{array}{rr} r-2 \,.\, r-1 \,.\, r \,.\, r+1 \,.\, r+2 \,.\, r+3 = & r^6 + 3r^5 - 5r^4 - 15r^3 + 4r^2 + 12r \\ -126 \,.\, r-2 \,.\, r-1 \,.\, r \,.\, r+1 & - 126r^4 + 252r^3 + 126r^2 - 252r \\ +2520 \,.\, r-2 \,.\, r-1 & + 2520r^2 - 7560r + 5040 \\ -5040 & - 5040 \\ \hline & r^6 + 3r^5 - 131r^4 + 237r^3 + 2650r^2 - 7800r \\ & = r-4 \,.\, r-5 \,.\, r-6 \,.\, r^3 + 18r^2 + 65r. \end{array}$$

Also

$$B = \frac{r-5 \,.\, r-4 \,.\, r-3 \,.\, r-2 \,.\, r-1 \,.\, r \,.\, r+1}{5040} - \frac{r-5 \,.\, r-4 \,.\, r-3 \,.\, r-2 \,.\, r-1}{120}$$

$$= \frac{r-3 \,.\, r-4 \,.\, r-5 \,.\, r-6}{5040}\,(r-1 \,.\, r-2 \,.\, r+7),$$

$$C = \frac{r-7 \,.\, r-6 \,.\, r-5 \,.\, r-4 \,.\, r-3 \,.\, r-2 \,.\, r-1}{5040}$$

$$= \frac{r-3 \,.\, r-4 \,.\, r-5 \,.\, r-6}{5040}\,(r-1 \,.\, r-2 \,.\, r-7);$$

where the calculation is

$$\begin{array}{rr} r \,.\, r+1 & = r^2 + r \\ -42 & -42 \\ \hline & r^2 + r - 42 \\ & = r-6 \,.\, r+7. \end{array}$$

17. To explain the formation of these expressions, observe that:

One diagonal.—There must be on each side of the diagonal, or say in each of the two "intervals" formed by the diagonal, two sides; there remain $r-4$ sides which may be distributed at pleasure between the two intervals, and the number of ways in which this can be done is

$$= \frac{r-3}{1}.$$

Two diagonals.—There must be on each side of the two diagonals, or say in two of the four intervals formed by the diagonals, two sides; there remain $r-4$ sides to be distributed between the same four intervals, and the number of ways in which this can be done is

$$= \frac{r-3 \,.\, r-2 \,.\, r-1}{6}.$$

But we must exclude the distributions where there is no side in the one interval and no side in the other interval between the two diagonals; the number of these is that for the case of the coalescence of the two diagonals into a single diagonal, viz. it is

$$=\frac{r-3}{1};$$

and thus the number required is

$$\frac{r-3\,.\,r-2\,.\,r-1}{6}-\frac{r-3}{1}.$$

18. Three diagonals, *A*.—There must be on each side of the three diagonals, that is, in two of the six intervals formed by the diagonals, two sides; there remain $r-4$ sides to be distributed between the same six intervals, and the number of ways in which this can be done is

$$=\frac{r-3\,.\,r-2\,.\,r-1\,.\,r\,.\,r+1}{120}.$$

But we must exclude distributions which would permit the coalescence of the first and second, or of the second and third, or of all three of the diagonals. For the coalescence of the first and second diagonals (the third diagonal not coalescing), the term to be subtracted is

$$\frac{r-3\,.\,r-2\,.\,r-1}{6}-\frac{r-3}{1};$$

and the same number for the coalescence of the second and third diagonals (the first diagonal not coalescing); that is, the last-mentioned number is to be multiplied by 2; and for the coalescence of all three diagonals the number to be subtracted is

$$=\frac{r-3}{1};$$

we have thus the foregoing value

$$\frac{r-3\,.\,r-2\,.\,r-1\,.\,r\,.\,r+1}{120}-2\,.\,\frac{r-3\,.\,r-2\,.\,r-1}{6}+1\,\frac{r-3}{1},$$

where it will be observed that we have the binomial coefficients 1, 2, 1 with the signs +, −, +.

Three diagonals, *B*.—There must be outside each of the three diagonals, that is, in three of the six intervals formed by the diagonals, two sides; and there remain $r-6$ sides to be distributed between the six intervals; the number of ways in which this can be done is

$$=\frac{r-5\,.\,r-4\,.\,r-3\,.\,r-2\,.\,r-1}{120};$$

and there is here no coalescence of diagonals, so that this is the number required.

19. Four diagonals, *A*.—There must be on each side of the four diagonals, that is, in two of the eight intervals formed by the diagonals, two sides; there remain

$r-4$ sides to be distributed between the eight intervals, and the number of ways in which this can be done is

$$\frac{r-3\,.\,r-2\,.\,r-1\,.\,r\,.\,r+1\,.\,r+2\,.\,r+3}{5040}.$$

But this number requires to be corrected for coalescences, as in the case Three diagonals, A; and the required number is thus found to be

$$\frac{r-3\,.\,r-2\,.\,r-1\,.\,r\,.\,r+1\,.\,r+2\,.\,r+3}{5040}-3\,\frac{r-3\,.\,r-2\,.\,r-1\,.\,r\,.\,r+1}{120}$$
$$+3\,\frac{r-3\,.\,r-2\,.\,r-1}{6}-1\,\frac{r-3}{1}.$$

Four diagonals, B.—There must be outside of three of the diagonals, that is, in each of three of the eight intervals formed by the diagonals, two sides; there remain $r-6$ sides to be distributed between the eight intervals, and the number of ways in which this can be done is

$$\frac{r-5\,.\,r-4\,.\,r-3\,.\,r-2\,.\,r-1\,.\,r\,.\,r+1}{5040}.$$

There is a correction for the coalescence of two of the diagonals, giving rise to a form such as Three diagonals, B; and consequently there is a term

$$-\frac{r-5\,.\,r-4\,.\,r-3\,.\,r-2\,.\,r-1}{120},$$

which, with the first-mentioned term, gives the required number.

Four diagonals, C.—There must be outside of each of the diagonals, that is, in each of four of the eight intervals formed by the diagonals, two sides; there remain $r-8$ sides to be distributed between the eight intervals, and the number of ways in which this can be done is

$$\frac{r-7\,.\,r-6\,.\,r-5\,.\,r-4\,.\,r-3\,.\,r-2\,.\,r-1}{5040},$$

which is the required number.

20. In the expressions of No. 14, A, $2A$, $3A+2B$, $4A+8B+2C$, if we regard the terminals of the diagonals as given points, then (1) we have two summits, which can be joined in one way only, giving rise to the diagonal-type A; (2) we have four summits, which can be joined in two ways only, so as to give rise to the diagonal-type A; (3) we have six summits, which can be joined in three ways so as to give rise to a diagonal-type A, and in two ways so as to give rise to a diagonal-type B; and (4) we have eight summits, which can be joined in four ways so as to give rise to a diagonal-type A, in eight ways so as to give rise to a diagonal-type B, and in two ways so as to give rise to a diagonal-type C; we have thus the linear forms in question. To obtain the number of partitions, we have in each case to multiply by r. To explain this, after the polygon is drawn, imagine

the summits to be numbered 1, 2, 3, ..., r in succession (the numbering may begin at any one of the r summits); regarding each of these numberings as giving a different partition, we should have the factor r. But, in fact, the partitions so obtained are not all of them distinct, but we have in each case a system of partitions repeated as many times as there are summits of the diagonals, that is, a number of times equal to twice the number of the diagonals; and we have thus, after the multiplication by r, to divide by the numbers 2, 4, 6, 8, in the four cases respectively.

21. We hence have immediately:—

Two parts, the number of partitions

$$= \frac{r}{2} A = \frac{r \,.\, r-3}{2 \,.\, 1};$$

Three parts, the number of partitions

$$= \frac{r}{2} A = \frac{r+1 \,.\, r \,.\, r-3 \,.\, r-4}{3 \,.\, 2 \,.\, 2 \,.\, 1};$$

Four parts, the number of partitions

$$= \frac{r}{6}(3A + 2B) = \frac{r+2 \,.\, r+1 \,.\, r \,.\, r-3 \,.\, r-4 \,.\, r-5}{4 \,.\, 3 \,.\, 2 \,.\, 3 \,.\, 2 \,.\, 1},$$

the calculation being

$$\begin{array}{rr} 3(r^2 + 7r + 2) = & 3r^2 + 21r + 6 \\ + 2 \,.\, r-1 \,.\, r-2 & + 2r^2 - 6r + 4 \\ \hline & 5r^2 + 15r + 10 \\ = & 5 \,.\, r+1 \,.\, r+2; \end{array}$$

Five parts, the number of partitions

$$= \frac{r}{8}(4A + 8B + 2C) = \frac{r+3 \,.\, r+2 \,.\, r+1 \,.\, r \,.\, r-3 \,.\, r-4 \,.\, r-5 \,.\, r-6}{5 \,.\, 4 \,.\, 3 \,.\, 2 \,.\, 4 \,.\, 3 \,.\, 2 \,.\, 1},$$

the calculation being

$$\begin{array}{rr} 4(r^3 + 18r^2 + 65r) = & 4r^3 + 72r^2 + 260r \\ + 8 \,.\, r-1 \,.\, r-2 \,.\, r+7 & + 8r^3 + 32r^2 - 152r + 112 \\ + 2 \,.\, r-1 \,.\, r-2 \,.\, r-7 & + 2r^3 - 20r^2 + 46r - 28 \\ \hline = & 14r^3 + 84r^2 + 154r + 84 \\ = & 14(r^3 + 6r^2 + 11r + 6) \\ = & 14 \,.\, r+1 \,.\, r+2 \,.\, r+3. \end{array}$$

To complete the theory, it would be in the first instance necessary to find for any given number of diagonals, $k-1$, whatever, the number and form of the diagonal-types, A, B, C, &c.; this is itself an interesting question in the Theory of Partitions, but I have not considered it.

22. Although the foregoing process (which, it will be observed, deals with the diagonal-types, without any consideration of the sub-types) is the most simple for the determination of the numbers A, B, C, &c., yet it is interesting to give a second process. Considering the several cases in order:

One diagonal, A.—The diagonal has two summits; we must have on each side of it one summit, and there remain $r-4$ summits which may be distributed between the two intervals formed by the diagonals. This can be done in $\frac{r-3}{1}$ ways, or we have, as before,

$$A=\frac{r-3}{1}.$$

Two diagonals, A.—The diagonals have four summits; we must have outside each diagonal one summit, and there remain $r-6$ summits to be distributed between the four intervals formed by the diagonals; this can be done in $\frac{r-5 \,.\, r-4 \,.\, r-3}{6}$ ways, or we have this value for A°. But the two top summits of the diagonals, or the two bottom summits, may coalesce; in either case, the diagonals have three summits. We must have outside each diagonal one summit, and there remain $r-5$ summits to be distributed between the three intervals formed by the diagonals; the number of ways in which this can be done is

$$=\frac{r-4 \,.\, r-3}{2},$$

say this is the value of A'. And we then have $A=A^\circ+2A'$,

$$=\frac{r+1 \,.\, r-3 \,.\, r-4}{6},$$

as before. The calculation is

$$r-5+6=r+1.$$

23. Three diagonals, A.—See No. 15 for the figures of the sub-types. We have

$$A=A^\circ+4A'+4A'',$$

where the coefficients, 4 and 4, are the number of ways in which A' and A'' respectively can be derived from A° by coalescences of summits. For A°, the diagonals have six summits, and there must be outside of two diagonals one summit; there remain $r-8$ summits to be distributed between the six intervals formed by the diagonals, and we have

$$A^\circ=\frac{r-7 \,.\, r-6 \,.\, r-5 \,.\, r-4 \,.\, r-3}{120}.$$

For A', the diagonals have five summits, and we must have outside of each of two diagonals, one summit; there remain $r-7$ summits to be distributed between the five intervals formed by the diagonals; we thus have

$$A'=\frac{r-6 \,.\, r-5 \,.\, r-4 \,.\, r-3}{24}.$$

For A'', the diagonals have four summits; there must be outside of each of two diagonals one summit, and there remain $r-6$ summits to be distributed between the four intervals formed by the diagonals; we thus have

$$A'' = \frac{r-5 \, . \, r-4 \, . \, r-3}{6}.$$

The foregoing values give

$$A = \frac{r-3 \, . \, r-4 \, . \, r-5}{120} (r^2 + 7r + 2),$$

as before. The calculation is

$$\begin{array}{rr} r-6 \, . \, r-7 = & r^2 - 13r + \ \ 42 \\ +\, 20 \, . \, r-6 \ \ & +\, 20r - 120 \\ +\, 80 \ \ & +\ \ 80 \\ \hline & r^2 + \ \ 7r + \ \ \ 2. \end{array}$$

Three diagonals, B.—See No. 15 for the figures of the sub-types. We have

$$B = B^\circ + 3B' + 3B'' + B'''.$$

For B°, the diagonals have six summits, and there must be outside each of the three diagonals one summit; there remain $r-9$ summits to be distributed between the six intervals formed by the diagonals. We thus have

$$B^\circ = \frac{r-8 \, . \, r-7 \, . \, r-6 \, . \, r-5 \, . \, r-4}{120}.$$

Similarly,

$$B' = \frac{r-7 \, . \, r-6 \, . \, r-5 \, . \, r-4}{24}, \quad B'' = \frac{r-6 \, . \, r-5 \, . \, r-4}{6}, \quad B''' = \frac{r-5 \, . \, r-4}{2}.$$

Hence

$$B = \frac{r-5 \, . \, r-4 \, . \, r-3 \, . \, r-2 \, . \, r-1}{120},$$

as before. The calculation is

$$\begin{array}{rr} r-6 \, . \, r-7 \, . \, r-8 = & r^3 - 21r^2 + 146r - 336 \\ +\, 15 \, . \, r-6 \, . \, r-7 \ \ & +\, 15r^2 - 195r + 630 \\ +\, 60 \, . \, r-6 \ \ & +\ \ 60r - 360 \\ +\, 60 \ \ & +\ \ 60 \\ \hline & r^3 - \ \ 6r^2 + \ \ 11r - \ \ \ 6 \\ & = r-1 \, . \, r-2 \, . \, r-3. \end{array}$$

24. Four diagonals, A.—The figures of the sub-types of A, B, C can be supplied without difficulty. We have

$$A = A^\circ + 6A' + 12A'' + 8A''',$$

where I remark that the numerical coefficients 1, 6, 12, 8 are the terms of $(1, 2)^3$. We have

$$A^\circ = \frac{r-9\,.\,r-8\,.\,r-7\,.\,r-6\,.\,r-5\,.\,r-4\,.\,r-3}{5040},$$

$$A' = \frac{r-8\,.\,r-7\,.\,r-6\,.\,r-5\,.\,r-4\,.\,r-3}{720},$$

$$A'' = \frac{r-7\,.\,r-6\,.\,r-5\,.\,r-4\,.\,r-3}{120},$$

$$A''' = \frac{r-6\,.\,r-5\,.\,r-4\,.\,r-3}{24},$$

and thence

$$A = \frac{r-3\,.\,r-4\,.\,r-5\,.\,r-6}{5040}(r^3+18r^2+65r),$$

as before. The calculation is

$$\begin{array}{rr} r-9\,.\,r-8\,.\,r-7 = & r^3 - 24r^2 + 191r - 504 \\ +\,42\,.\,r-8\,.\,r-7 & +\,42r^2 - 630r + 2352 \\ +\,504\,.\,r-7 & +\,504r - 3528 \\ +\,1680 & +\,1680 \\ \hline & r^3 + 18r^2 + 65r . \end{array}$$

Four diagonals, *B*.—We have

$$B = B^\circ + 5B' + 9B'' + 7B''' + 2B^{\text{iv}},$$

where the coefficients, 1, 5, 9, 7, 2, are the terms of $(1, 1)^3(1, 2)$. We have

$$B^\circ = \frac{r-10\,.\,r-9\,.\,r-8\,.\,r-7\,.\,r-6\,.\,r-5\,.\,r-4}{5040},$$

$$B' = \frac{r-9\,.\,r-8\,.\,r-7\,.\,r-6\,.\,r-5\,.\,r-4}{720},$$

$$B'' = \frac{r-8\,.\,r-7\,.\,r-6\,.\,r-5\,.\,r-4}{120},$$

$$B''' = \frac{r-7\,.\,r-6\,.\,r-5\,.\,r-4}{24},$$

$$B^{\text{iv}} = \frac{r-6\,.\,r-5\,.\,r-4}{6},$$

and thence

$$B = \frac{r-3\,.\,r-4\,.\,r-5\,.\,r-6}{5040}\,r-1\,.\,r-2\,.\,r+7,$$

as before. The calculation is

$$\begin{aligned}
r-10\,.\,r-9\,.\,r-8\,.\,r-7 &= r^4-34r^3+431r^2-2414r+5040\\
+35\,.\,r-9\,.\,r-8\,.\,r-7 &\quad +35r^3-840r^2+6685r-17640\\
+378\,.\,r-8\,.\,r-7 &\quad +378r^2-5670r+21168\\
+1470\,.\,r-7 &\quad +1470r-10290\\
+1680 &\quad +1680\\
\hline
& r^4+r^3-31r^2+71r-42\\
& =r-3\,.\,r-2\,.\,r-1\,.\,r+7.
\end{aligned}$$

Four diagonals, C.—We have

$$C=C^{\circ}+4C'+6C''+4C'''+1C^{\text{iv}},$$

where the coefficients are the terms of $(1,\ 1)^4$. We have

$$C^{\circ}=\frac{r-11\,.\,r-10\,.\,r-9\,.\,r-8\,.\,r-7\,.\,r-6\,.\,r-5}{5040},$$

$$C'=\frac{r-10\,.\,r-9\,.\,r-8\,.\,r-7\,.\,r-6\,.\,r-5}{720},$$

$$C''=\frac{r-9\,.\,r-8\,.\,r-7\,.\,r-6\,.\,r-5}{120},$$

$$C'''=\frac{r-8\,.\,r-7\,.\,r-6\,.\,r-5}{24},$$

$$C^{\text{iv}}=\frac{r-7\,.\,r-6\,.\,r-5}{6},$$

and thence

$$C=\frac{r-7\,.\,r-6\,.\,r-5\,.\,r-4\,.\,r-3\,.\,r-2\,.\,r-1}{5040}.$$

I omit the calculation, as the equation is at once seen to be a particular case of a known factorial formula.

25. We may analyse the partitions of an r-gon into a given number of parts, according to the nature of the parts, that is, the numbers of the sides of the several component polygons. It is for this purpose convenient to introduce the notion of "weight"; say a triangle has the weight 1, then a quadrangle, as divisible into two triangles, has the weight 2, a pentagon, as divisible into three triangles, has the weight 3, ..., and generally an r-gon, as divisible into $r-2$ triangles, has the weight $r-2$. It at once follows that, if

$$W=w+w',\ \text{or}\ =w+w'+w'',\ \&\text{c.},$$

then a polygon of weight W is divisible into two polygons of the weights w, w', or into three polygons of the weights w, w', w'' respectively; and so on. Thus the 2-partitions of an 8-gon (weight = 6) are 15, 24, and 33; the 3-partitions are 114,

123, 222, and so on. Of course the number of the partitions 15, 24, 33, is equal to the whole number of the 2-partitions of the 8-gon, that is, $=20$; the number of the partitions 114, 123, 222, is equal to the whole number of the 3-partitions of the 8-gon, that is, it is $=120$; and so in other cases. It is easy to derive in order one from the other the numbers of the partitions of each several kind of the polygons of the several weights 2, 3, 4, 5, 6, &c.; and I write down the accompanying Table (No. 26), facing page 112, the process for the construction being as follows:

27. The first column (2 parts) is at once obtained. For a polygon of an odd number of sides, for instance the 9-gon (weight $=7$), imagining the summits numbered in order 1, 2, ..., 9, we divide this into a triangle and octagon, or obtain the partitions 16, by drawing the diagonals 13, 24, ..., 81, 92: viz. the number is $=9$. In the Table this is written, $16=9$; and so in other cases. Similarly we divide it into a quadrangle and heptagon, or obtain the partitions 25, by drawing the diagonals 14, 25, ..., 82, 93: viz. the number is again $=9$; and we divide it into a pentagon and a hexagon, or obtain the partitions 34, by drawing the diagonals 15, 26, ..., 83, 94: viz. the number is $=9$, and here

$$9+9+9=27,$$

the whole number of 2-partitions of the 9-gon. For a polygon of an even number of sides, for instance the 10-gon (weight $=8$), the process is a similar one, the only difference being that for the division into two hexagons, (that is, for the partitions 44), each partition is thus obtained twice, or the number of such partitions is $\frac{1}{2}10$, $=5$; the numbers for the partitions 17, 26, 35, 44, thus are 10, 10, 10, 5; and we have

$$10+10+10+5=35,$$

the whole number of the 2-partitions of the 10-gon.

28. To obtain the second column (3 parts)—suppose, for instance, the 3-partitions of the 9-gon; these are 115, 124, 133, 223. We obtain the number of the partitions 115 from the terms

$$16=9 \text{ and } 25=9$$

of the first column: viz. in 16, changing the 6 into 15, that is, dividing the polygon of weight 6 into two parts of weights 1 and 5 respectively: this can be done in eight ways (see, higher up, $15=8$ in the first column); and we thus obtain the number of partitions

$$9\times 8=72;$$

and again, in 25, changing the 2 into 11, that is, dividing the polygon of weight 2 into two parts each of weight 1: this can be done in two ways (see, higher up, $11=2$ in the first column); and we thus obtain the number of partitions

$$9\times 2=18;$$

we should thus have, for the number of partitions 115, the sum

$$72+18=90,$$

only, as it is easy to see, each partition is obtained twice, and the number of the partitions 115 is the half of this, $=45$. And by the like process it is found that the numbers of the partitions 124, 133, 223 are equal to 90, 45, 45 respectively; and then, as a verification, we have

$$45+90+45+45=225,$$

the whole number of the 3-partitions of the 9-gon.

29. The third column (4 parts) is derived in like manner from the second column by aid of the first column; and so in general, each column is derived in like manner from the column which immediately precedes it, by aid of the first column. And we have for the numbers in each compartment of any column the verification that the sum of these numbers is equal to the whole number (for the proper values of k and r) of the k-partitions of the r-gon.

It might be possible, by an application of the method of generating functions, to find a law for the numbers in any compartment of a column of the table; but I have not attempted to make this investigation.

30. In the table in No. 2, the numbers 1, 2, 5, 14, 42, &c., of the diagonal line show the number of partitions of the triangle, the quadrangle, the 5-gon, ..., r-gon into triangles: viz. these numbers show the number of partitions of the r-gon into $r-2$ parts, that is, into triangles; and, for the r-gon, writing

$$k=r-2,$$

the number is

$$=\frac{[2r-4]^{r-3}}{[r-2]^{r-3}}.$$

If, as above, taking the weight of the triangle to be 1, we write

$$r-2=w,$$

then the number is

$$=\frac{[2w]^{w-1}}{[w]^{w-1}},$$

viz. this is the expression for the number of partitions of the polygon of weight w, or $(w+2)$-gon, into triangles.

31. The question considered by Taylor and Rowe, in the paper referred to in No. 1, is that of the partition of the r-gon into p-gons, for p, a given number >3; this implies a restriction on the form of r, viz. we must have $r-2$ divisible by $p-2$. In fact, generalizing the definition of w, if we attribute to a p-gon the weight 1, and accordingly to a polygon divisible into w p-gons the weight w, then, r being the number of summits, we must have

$$r=(p-2)\,w+2.$$

In particular, if $p=4$, so that the r-gon is to be divided into quadrangles, then r is necessarily even, and for the values

$$w=1,\ 2,\ 3,\ \ldots,$$

we have

$$r=4,\ 6,\ 8,\ \ldots.$$

26. THE TABLE REFERRED TO (p. 111).

[*To face page* 112.

r	W	2 PARTS	3 PARTS	4 PARTS	5 PARTS	6 PARTS	7 PARTS	8 PARTS
3	1							
4	2	11 = 2 2						
5	3	12 = 5 5	111 5 × 2 = 10 2) 10 = 5					
6	4	13 = 6 22 = 3 9	112 6 × 5 = 30 3 × 4 = 12 2) 42 = 21	1111 21 × 2 = 42 3) 42 = 14				
7	5	14 = 7 23 = 7 14	113 7 × 6 = 42 7 × 2 = 14 2) 56 = 28 122 7 × 3 = 21 7 × 5 = 35 2) 56 = 28 56	1112 28 × 5 = 140 28 × 4 = 112 3) 252 = 84	11111 84 × 2 = 168 4) 168 = 42			
8	6	15 = 8 24 = 8 33 = 4 20	114 8 × 7 = 56 8 × 2 = 16 2) 72 = 36 123 8 × 7 = 56 8 × 6 = 48 4 × 10 = 40 2) 144 = 72 222 8 × 3 = 24 2) 24 = 12 120	1113 36 × 6 = 216 72 × 2 = 144 3) 360 = 120 1122 36 × 3 = 108 72 × 5 = 360 12 × 6 = 72 3) 540 = 180 300	11112 120 × 5 = 600 180 × 4 = 720 4) 1320 = 330	111111 330 × 2 = 660 5) 660 = 132		
9	7	16 = 9 25 = 9 34 = 9 27	115 9 × 8 = 72 9 × 2 = 18 2) 90 = 45 124 9 × 8 = 72 9 × 7 = 63 9 × 5 = 45 2) 180 = 90 133 9 × 4 = 36 9 × 6 = 54 2) 90 = 45 223 9 × 7 = 63 9 × 3 = 27 2) 90 = 45 225	1114 45 × 7 = 315 90 × 2 = 180 3) 495 = 165 1123 45 × 7 = 315 90 × 6 = 540 45 × 10 = 450 45 × 4 = 180 3) 1485 = 495 1222 90 × 3 = 270 45 × 5 = 225 3) 495 = 165 825	11113 165 × 6 = 990 495 × 2 = 990 4) 1980 = 495 11122 165 × 3 = 495 495 × 5 = 2475 165 × 6 = 990 4) 3960 = 990 1485	111112 495 × 5 = 2475 990 × 4 = 3960 5) 6435 = 1287	1111111 1287 × 2 = 2574 6) 2574 = 429	
10	8	17 = 10 26 = 10 35 = 10 44 = 5 35	116 10 × 9 = 90 10 × 2 = 20 2) 110 = 55 125 10 × 9 = 90 10 × 8 = 80 10 × 5 = 50 2) 220 = 110 134 10 × 9 = 90 10 × 7 = 70 5 × 12 = 60 2) 220 = 110 224 10 × 8 = 80 5 × 6 = 30 2) 110 = 55 233 10 × 4 = 40 10 × 7 = 70 2) 110 = 55 385	1115 55 × 8 = 440 110 × 2 = 220 3) 660 = 220 1124 55 × 8 = 440 110 × 7 = 770 110 × 5 = 550 55 × 4 = 220 3) 1980 = 660 1133 55 × 4 = 220 110 × 6 = 660 55 × 2 = 110 3) 990 = 330 1223 110 × 7 = 770 110 × 3 = 330 55 × 6 = 330 55 × 10 = 550 3) 1980 = 660 2222 55 × 3 = 165 3) 165 = 55 1925	11114 220 × 7 = 1540 660 × 2 = 1320 4) 2860 = 715 11123 220 × 7 = 1540 660 × 6 = 3960 330 × 10 = 3300 660 × 4 = 2640 4) 11440 = 2860 11222 660 × 3 = 1980 660 × 5 = 3300 55 × 8 = 440 4) 5720 = 1430 5005	111113 715 × 6 = 4290 2860 × 2 = 5720 5) 10010 = 2002 111122 715 × 3 = 2145 2860 × 5 = 14300 1430 × 6 = 8580 5) 25025 = 5005 7007	1111112 2002 × 5 = 10010 5005 × 4 = 20020 6) 30030 = 5005	11111111 5005 × 2 = 10010 7) 10010 = 1430

32. To fix the ideas, I assume $p=4$, and thus consider the problem of the division of the $(2w+2)$-gon into quadrangles. Writing

$$w-1=a+b+c,$$

we take at pleasure any one side of the $(2w+2)$-gon, making this the first side of a quadrangle, the second, third, and fourth sides being diagonals of the polygon such that outside the second side we have a polygon of weight a, outside the third side a polygon of weight b, and outside the fourth side a polygon of weight c. Any one or more of the numbers a, b, c may be $=0$; they cannot be each of them $=0$ except in the case $w=1$. The meaning is that the corresponding side of the quadrangle, instead of being a diagonal, is a side of the $(2w+2)$-gon, viz. there is no polygon outside such side. Suppose, in general, that P_w is the number of ways in which a polygon of weight w can be divided in quadrangles, and let each of the polygons of weights a, b, c respectively, be divided into quadrangles: the number of ways in which this can be done is $P_aP_bP_c$; and it is to be noticed that, if for instance $a=0$, then the number is $=P_bP_c$, viz. the formula remains true if only we assume $P_0=1$. The number of partitions thus obtained is $\Sigma P_aP_bP_c$, where the summation extends to all the partitions of $w-1$ into the parts a, b, c (zeros admissible and the order of the parts being attended to). And we thus obtain *all* the partitions of the $(2w+2)$-gon into w parts; for first, the partitions so obtained are all distinct from each other, and next every partition of the $(2w+2)$-gon into w parts is a partition in which the selected side of the $(2w+2)$-gon is a side of some one of the quadrangles. That is, we have

$$P_w=\Sigma P_aP_bP_c \text{ } (a,\ b,\ c \text{ as above});$$

and it hence appears that, considering the generating function

$$f=1+P_1x+P_2x^2+P_3x^3+\dots,$$

we have

$$f=1+xf^3.$$

The reasoning is precisely the same if, instead of a division into quadrangles, we have a division into p-gons; the only difference is that instead of the three parts a, b, c, we have the $p-1$ parts a, b, c, ..., and the equation for f thus is

$$f=1+xf^{p-1}.$$

33. Writing for a moment

$$f=u+xf^{p-1},$$

and expanding by Lagrange's theorem, we have

$$f=u+\frac{x}{1}(u^{p-1})+\frac{x^2}{1\,.\,2}(u^{2(p-1)})'+\dots+\frac{x^w}{1\,.\,2\dots w}(u^{w(p-1)})'^{\dots(w-1)}+\dots,$$

viz. after the differentiation, writing $u=1$, we have

$$P_w=\frac{[(p-1)w]^{w-1}}{[w]^{w-1}},$$

where it will be recollected that, for the number of sides of the polygon, we have

$$r=(p-2)w+2.$$

In the case of the partition into triangles, $p=3$, and we have the before-mentioned value

$$[2w]^{w-1}\div[w]^{w-1},\quad w=r-2.$$

916.

[NOTE ON A THEOREM IN MATRICES.]

[From the *Proceedings of the London Mathematical Society*, vol. XXII. (1891), p. 458.]

PROF. CAYLEY remarks that a "simple instance [of the theorem] is that, if the real symmetric matrix

$$\begin{pmatrix} a, & h, & g \\ h, & b, & f \\ g, & f, & c \end{pmatrix}$$

has two latent roots each $=0$, and therefore a vacuity $=2$, then it has also a nullity $=2$ [which may be shown as follows], viz. the conditions for a vacuity $=2$ are

$$\begin{vmatrix} a, & h, & g \\ h, & b, & f \\ g, & f, & c \end{vmatrix} = 0, \quad bc + ca + ab - f^2 - g^2 - h^2 = 0,$$

or, if as usual the determinant is called K, and if

$$(A,\ B,\ C,\ F,\ G,\ H) = (bc - f^2,\ ac - g^2,\ \ldots),$$

then, if

$$K = 0, \quad A + B + C = 0,$$

these equations give

$$BC - F^2 = Ka = 0, \quad AC - G^2 = Kb = 0, \quad AB - H^2 = Kc = 0,$$

i.e.

$$BC = F^2, \quad AC = G^2, \quad AB = H^2;$$

and therefore

$$A\,(A + B + C) = A^2 + H^2 + G^2,$$
$$B\,(A + B + C) = H^2 + B^2 + F^2,$$
$$C\,(A + B + C) = G^2 + F^2 + C^2;$$

or, if $A + B + C = 0$, then for real values

$$A = B = C = F = G = H = 0,$$

i.e. nullity $= 2$."

917.

[NOTE ON THE THEORY OF RATIONAL TRANSFORMATION.]

[From the *Proceedings of the London Mathematical Society*, vol. XXII. (1891), pp. 475, 476.]

IN my paper, "Note on the Theory of the Rational Transformation between Two Planes, and on Special Systems of Points," *Proc. Lond. Math. Soc.* t. III. (1870), pp. 106—108, [450], I notice a difficulty which presents itself in the theory. The transformation is given by the equations

$$x' : y' : z' = X : Y : Z,$$

where X, Y, Z are functions $(*\between x, y, z)^n$, such that $X=0$, $Y=0$, $Z=0$ are curves in the first plane passing through α_1 points each once, α_2 points each twice (that is, having each of the α_2 points for a double point), α_3 points each 3 times, and so on. We have as the condition of a single variable point of intersection,

$$\alpha_1 + 4\alpha_2 + 9\alpha_3 + \ldots = n^2 - 1,$$

and as the condition in order that each of the curves $X=0$, $Y=0$, $Z=0$, or say the curve $aX+bY+cZ=0$, may be unicursal,

$$\alpha_2 + 3\alpha_3 + \ldots = \tfrac{1}{2}(n-1)(n-2);$$

and we thence deduce

$$\alpha_1 + 3\alpha_2 + 6\alpha_3 + \ldots = \tfrac{1}{2}n(n+3) - 2;$$

viz. the postulation of the fixed points *quoad* a curve of the order n is less by 2 than the postulandum (or, as I prefer to call it, the capacity) $\tfrac{1}{2}n(n+3)$ of the curve of the order n; that is, there are precisely the three asyzygetic curves $X=0$, $Y=0$, $Z=0$. This is as it should be, assuming that the $(\alpha_1, \alpha_2, \alpha_3, \ldots)$ points are an ordinary system of points: but what if they form a special system having a postulation less than $\alpha_1+3\alpha_2+6\alpha_3+\ldots$? If, for instance, the postulation is $=\alpha_1+3\alpha_2+6\alpha_3+\ldots-1$, then this would be $=\frac{1}{2}n(n+3)-3$, and there would be four asyzygetic curves $X=0$, $Y=0$, $Z=0$, $W=0$. I believe this to be impossible; but the only proof which I can offer rests upon a remark in regard to the form of the tables*, pp. 148, 149, of my paper "On the Rational Transformation between Two Spaces," *Proc. Lond. Math. Soc.*, t. III. (1870), pp. 127—180, [447]. I recall that the Jacobian curve $J(X, Y, Z)=0$ consists of α_1' lines, α_2' conics, α_3' cubics, ..., &c., each passing a certain number of times through the $(\alpha_1, \alpha_2, \alpha_3, \ldots)$ points, and that the number of times of passage is shown by these tables; thus, *loc. cit.*, $n=5$, $\alpha_1=8$, $\alpha_4=1$: the Jacobian consists of eight lines and a quartic, and we have the table $(\alpha_1'=8,\ \alpha_4'=1)$,

	$a_1 = 8$	$a_2 = 0$	$a_3 = 0$	$a_4 = 1$
$a_1'=8$	1			1
$a_2'=0$				
$a_3'=0$				
$a_4'=1$	8			1^3

showing that the quartic passes through the eight points α_1, and through the point α_4 three times (has α_4 for a triple point). Imagine a new function W. Then in like manner $J(X, Y, W)=0$ consists of eight lines and a quartic, and this quartic passes through the eight points α_1 and the point α_4 three times; that is, the two quartics intersect in $8+3.3$, $=17$ points; and thus the two quartics must be one and the same curve; this implies a syzygy between X, Y, Z, W, viz. W is a mere linear function of X, Y, Z. The general remark is that, if in the tables m^p is reckoned as mp^2, then in the table for the several lines (exclusive of those for which the outside accented letter is $=0$, and therefore the tabular numbers of the line are each $=0$), i.e. for the lines which correspond to a line, a conic, a cubic, a quartic, &c., respectively, the sums of the tabular numbers are 1^2+1, 2^2+1, 3^2+1, 4^2+1, &c., respectively. This is, in fact, the case for each of the eleven tables (*loc. cit.*).

[* This Collection, vol. VII., pp. 208, 209.]

918.

ON THE SUBSTITUTION GROUPS FOR TWO, THREE, FOUR, FIVE, SIX, SEVEN, AND EIGHT LETTERS.

[From the *Quarterly Journal of Pure and Applied Mathematics*, vol. XXV. (1891), pp. 71—88, 137—155.]

THE substitution groups for two, three, four, and five letters were obtained by Serret: those for six, seven and eight letters have recently been obtained by Mr Askwith. I wish to reproduce these results in a condensed form.

The following table shows for the several cases respectively, the orders of the several groups, and for any order the number of distinct groups. As regards the case of eight letters, the numbers mentioned do not exactly agree with Mr Askwith: he gives a few non-existent groups, and omits some which I have supplied (see *post*, the list of the groups for eight letters); and it is possible that there are other omissions: the several numbers in the column and the sum total of 155 are given subject to correction.

2		3		4		5		6		7		8	
1	2	1	3	1	2	1	5	1	2	2	6	1	2
		1	6	3	4	2	6	1	3	1	7	8	4
				1	8	1	10	4	4	2	10	4	6
				1	12	1	12	3	6	9	12	25	8
				1	24	1	20	6	8	1	14	6	12
						1	60	1	9	2	20	1	15
						1	120	1	12	1	21	22	16
								1	16	7	24	3	18
								3	18	1	36	4	24
								4	24	1	40	3	30
								2	36	1	42	12	32
								2	48	1	48	5	36
								1	60	3	72	14	48
								1	72	2	120	3	60
								1	120	1	144	3	64
								1	360	1	240	3	72
								1	720	1	2520	10	96
										1	5040	3	120
												2	144
												1	168
												1	180
												3	192
												1	240
												3	288
												1	336
												3	360
												1	384
												2	576
												3	720
												1	1152
												1	1440
												1	20160
												1	40320
1		2		7		8		34		38		155	

Here the top line shows the numbers of letters; each second column shows the order of the groups, and each first column the number of groups of the several orders: the sums at the foot of the first columns show therefore the whole number of groups, viz.

$$\frac{\text{No. of letters} = 2,\ 3,\ 4,\ 5,\ 6,\ 7,\ 8}{\text{No. of groups} = 1,\ 2,\ 7,\ 8,\ 34,\ 38,\ 155}.$$

In the enumeration of the groups, I use some notations which must be explained. For greater simplicity I omit parentheses, and write *ab*, *abc*, *ab . cd*, *abc . def*, &c., to denote substitutions, viz. *ab* is the interchange of *a* and *b*; *abc* the cyclical change *a* into *b*, *b* into *c*, *c* into *a*; *ab . cd* the combined interchange of *a* and *b* and of *c* and *d*; *abc . def* the combined cyclical changes *a* into *b*, *b* into *c*, *c* into *a*, and *d* into *e*, *e* into *f*, *f* into *d*; and so in other cases.

Again, (*abc*) all, means the complete group of all the substitutions

(1, *abc*, *acb*, *bc*, *ca*, *ab*)

upon the three letters; and so in other cases. In the case, however, of two letters, I write simply (ab) to denote the complete group $(1, ab)$ of substitutions, and so also for any substitution such as $ab.cd$, where the complete group is $(1, ab.cd)$, I denote the group by $(ab.cd)$. Moreover, (abc) cyc., denotes the group of cyclical substitutions $(1, abc, acb)$ upon the three letters; and so in other cases.

A group will in general contain positive and negative substitutions, and, when this is so, the positive substitutions will form a group which is denoted by the symbol, pos.; the negative substitutions (which of course do not form a group) are denoted in like manner by the symbol, neg. Thus, (abc) all, pos., or for shortness, (abc) pos., will denote the group formed by the positive substitutions of (abc) all; (abc) pos. is thus the same thing as (abc) cyc., but obviously $(abcd)$ pos. and $(abcd)$ cyc. have quite different meanings. It is to be noticed that, for any odd number of letters, the substitutions of a group (abc) cyc. are all positive, and thus (abc) cyc. pos. is the original group: for any even number of letters, the group $(abcd)$ cyc. or $(abcdef)$ cyc. contains positive and negative substitutions, but the positive substitutions thereof form a cyclical group, thus

$$(abcd) \text{ cyc. pos.} = (ab.cd) \text{ cyc.}, \quad (abcdef) \text{ cyc. pos.} = (ace.bdf) \text{ cyc.},$$

and the notation, ( ) cyc. pos., is thus unnecessary, and it will not be used.

Substitutions or groups which have no letter in common are said to be independent. The product of two independent groups is of course a group, and the components may be called independent factors of the resultant group: we use for such a product the notation $A.B$, and call the group a composite group. If each of the groups A and B contain positive and negative substitutions, we thence derive a new group, $(A.B)$ pos., viz. the substitutions hereof are the products of a positive substitution of A and a positive substitution of B, and the products of a negative substitution of A and a negative substitution of B, say

$$(A.B) \text{ pos.} = (A \text{ pos. } B \text{ pos.}) + (A \text{ neg.})(B \text{ neg.}):$$

obviously the number of substitutions or order of the group $(A.B)$ pos. is one half of that of the group $(A.B)$.

A more general, but not perfectly definite, notation is that of $(A.B)$ dim., the dimidiate of the group $A.B$. Suppose, for instance, that A is a group of substitutions of the letters (a, b, c, d); and that B is the group (ef). Here if A is composed of two equal sets $1, P, Q, \ldots; R, S, T, \ldots$, where the first set $1, P, Q, \ldots$, is a group, then we have a group $1, P, Q, \ldots, ef.R, ef.S, ef.T, \ldots$, which is a group of the form in question, $(A.B)$ dim. But in some cases the group A can be in more than one way divided into two equal sets the first of which forms a group, and a further explanation of the notation is required. I do not give at present a more complete explanation, nor explain the analogous notions trisection (tris.), &c.

Substitutions or groups having a letter or letters in common are non-independent. If A, B are such groups, then the substitutions of A are not commutable with those of B, and we have not in general such a group as $B.A$ or $A.B$. It may happen that, although the individual substitutions of A are not commutable with those of B, yet that the groups are commutable, say we have $A.B = B.A$, viz. here the substitutions of $A.B$ are in a different order identical with those of $B.A$. We have in this case a group $A.B$; this is not a composite group; and the notation will never be employed without explanation.

I consider, in particular, the two groups $(abc.def)$ cyc. and (abc) cyc. (def) cyc.; in each of them, the six letters (a, b, c, d, e, f) are divided into two sets (a, b, c), (d, e, f), and the substitutions are the product of an abc-substitution into a def-substitution: viz. in the first group the substitutions are

$$1,\quad abc.def,\quad acb.dfe,$$

and in the second group they are

$$\begin{array}{ll} 1, & abc, \quad abc.def, \\ & acb, \quad acb.def, \\ & def, \quad abc.dfe, \\ & dfe, \quad acb.dfe. \end{array}$$

We can from each of the groups, introducing substitutions which interchange (a, b, c) with (d, e, f), or say by combination with the group $(ad.be.cf)$, derive a group of double the order; and it is worth while to consider the two cases in detail. First, for the group

$$(ad.be.cf)(abc.def) \text{ cyc.}$$

We have

Arrangements.	Substitutions.
abcdef,	1,
bcaefd,	*abc.def,*
cabfde,	*acb.dfe,*
defabc,	*ad.be.cf,*
efdbca,	*aecdbf,*
fdecab,	*afbdce,*

say

$$(ad.be.cf)(abc.def) \text{ cyc.} = (abcdef)_6.$$

This group of the order 6 is, as it happens, the group $(aecdbf)$ cyc.

It may be remarked that if, instead of $(abc.def)$ cyc., we consider the precisely similar group $(abc.dfe)$ cyc., then operating upon it with the same group $(ad.be.cf)$, we obtain quite a different form of group $(abcdef)_6$. In fact, for $(ad.be.cf)(abc.dfe)$ cyc., we have

Arrangements.	Substitutions.
abcdef,	1,
bcafde,	*abc . dfe*,
cabefd,	*acb . def*,
defabc,	*ad . be . cf*,
efdcab,	*ae . bf . cd*,
fdebca;	*af . bd . ce*;

which is not the cyclical group. See *post*, six letters, ord. 6_3.

Next, for the group $(ad\,.\,be\,.\,cf)\,(abc)$ cyc. (def) cyc.; we have here

Arrangements.	Substitutions.
abcdef,	1,
bcadef,	*abc*,
cabdef,	*acb*,
abcefd,	*def*,
bcaefd,	*abc . def*,
cabefd,	*acb . def*,
abcfde,	*dfe*,
bcafde,	*abc . dfe*,
cabfde,	*acb . dfe*,
defabc,	*ad . be . cf*,
efdabc,	*aebfcd*,
fdeabc,	*afcebd*,
defbca,	*adbecf*,
efdbca,	*aecdbf*,
fdebca,	*af . bd . ce*,
defcab,	*adcfbe*,
efdcab,	*ae . bf . cd*,
fdecab;	*afbdce*;

say

$$(ad\,.\,be\,.\,cf)\,\{(abc)\text{ cyc. }(def)\text{ cyc.}\} = (abcdef)_{18}.$$

See *post*, six letters, ord. 18.

The groups thus obtained, with substitutions which interchange the two sets of letters, are said to be "woven" groups.

By means of the foregoing notations a large number, but by no means all, of the groups belonging to the several cases of 2, 3, 4, 5, 6, 7, and 8 letters can be represented in a very compendious form. It is, even in the case of 4 letters, proper to introduce special notations: thus for the order 4, we have $(ab)(cd)$, (or as I generally write it $(ac)(bd)$), and $(abcd)$ cyc. (which it is convenient thus to represent), and another group $(1,\ ab.cd,\ ac.bd,\ ad.bc)$: this is a woven group $(ac.bd)(ab)(cd)$, but in thus representing it, we fail to exhibit the symmetrical character of the group, and I prefer to represent it as $(abcd)_4$. Again for the order 8, there is a single group, which I write $(abcd)_8$: this can be, in regard to dimidiation, divided in three different ways into two sets of four, and to distinguish them I write

$$(abcd)_8 \text{ com.} = (1,\ ac,\ bd,\ ac.bd;\ abcd,\ adbc,\ ab.cd,\ ad.bc),$$

$$(abcd)_8 \text{ cyc.} = (1,\ abcd,\ ac.bd,\ adbc;\ ac,\ bd,\ ab.cd,\ ad.bc),$$

$$(abcd)_8 \text{ pos.} = (1,\ ab.cd,\ ac.bd,\ ad.bc;\ ac,\ bd,\ abcd,\ adcb);$$

viz. com. denotes that the first set of four is the composite group $(ac)(bd)$; cyc. that it is the cyclical group $(abcd)$ cyc.; and pos. that it is the positive group $(abcd)_4$. We have thus, in the case of 6 letters, the three dimidiation forms,

$$\{(abcd)_8 \text{ com.} (ef)\} \text{ dim.} = (ac)(bd) + ef(abcd,\ adbc,\ ab.cd,\ ad.bc),$$

$$\{(abcd)_8 \text{ cyc.} (ef)\} \text{ dim.} = (abcd) \text{ cyc.} + ef(ac,\ bd,\ ab.cd;\ ad.bc),$$

$$\{(abcd)_8 \text{ pos.} (ef)\} \text{ dim.} = (abcd)_4 \quad + ef(ac,\ bd,\quad abcd,\ adcb),$$

the last of which may be more simply written as $\{(abcd)_8(ef)\}$ pos.

In the cases of 2, 3, 4, and 5 letters, the groups are as follows:

2 letters ord.		3 letters ord.		4 letters ord.		5 letters ord.	
2	(ab)	3	(abc) cyc.	2	$(ab.cd)$	5	$(abcde)$ cyc.
		6	(abc) all	4	$(ab)(cd)$	6	(abc) cyc. (de)
				,,	$(abcd)$ cyc.	,,	$\{(abc)$ all $(de)\}$ pos.
				,,	$(abcd)_4$	10	$(abcde)_{10}$
				8	$(abcd)_8$	12	(abc) all (de)
				12	$(abcd)$ pos.	20	$(abcde)_{20}$
				24	$(abcd)$ all	60	$(abcde)$ pos.
						120	$(abcde)$ all

The nature of the group is in most cases at once intelligible from the foregoing explanations: thus (ab) denotes the group $(1,\ ab)$; (abc) cyc., the cyclical group $(1,\ abc,\ acb)$: (abc) all, the group of the six substitutions $(1,\ abc,\ acb,\ ab,\ ac,\ bc)$, and so in other cases: but, when a further explanation is required, the nature of the group is merely indicated by writing down the order as a suffix: thus $(abcd)_4$ means a group of four substitutions, $(abcd)_8$ a group of 8 substitutions, and so in other cases. In the case of four letters, the necessary explanations have already been given: it may be remarked that the group $(abcd)_8$ is that of the substitutions which leave unaltered the three-valued function $ac + bd$.

Five letters. Explanations. Five letters.

ord. 10. $(abcde)_{10}$. This may be written $\{(abcde)_{20}\}$ pos., viz. it consists of the 10 positive substitutions out of the next-mentioned group of 20 substitutions. Referring to that group, and writing $\sigma = S^2 = ac \,.\, bd$, and $T = aecdb$, the present group is $(1, \sigma)(1, T, T^2, T^3, T^4)$, where $\sigma^2 = 1$, $T^5 = 1$, $\sigma T = T^4\sigma$, $\sigma T^4 = T\sigma$, $\sigma T^3 = T^2\sigma$, $\sigma T^2 = T^3\sigma$: or, retaining S^2 instead of σ, say the group is $(1, S^2)(1, T, T^2, T^3, T^4)$.

ord. 20. $(abcde)_{20}$. The substitutions, distinguishing the positive and the negative ones, are

+	+	+	−
1,	$ab \,.\, de$,	$abdce$,	$abcd$,
	$ac \,.\, bd$,	$acbed$,	$abec$,
	$ad \,.\, ce$,	$adebc$,	$acde$,
	$ae \,.\, bc$,	$aecdb$,	$aceb$,
	$be \,.\, cd$,		$adbe$,
			$adcb$,
			$aebd$,
			$aedc$,
			$bced$,
			$bdec$;

and here, if $S = abcd$, $T = aecdb$, then the group is $(1, S, S^2, S^3)(1, T, T^2, T^3, T^4)$, where $S^4 = 1$, $T^5 = 1$, and generally $S^\alpha T^\beta = T^{\beta \,.\, 2^\alpha} S^\alpha$; that is,

	1	T	T^2	T^3	T^4
1	1	T	T^2	T^3	T^4
S	S	T^2S	T^4S	TS	T^3S
S^2	S^2	T^4S^2	T^3S^2	T^2S^2	TS^2
S^3	S^3	T^3S^3	TS^3	T^4S^3	T^2S^3

read

$$ST = T^2S,$$
$$ST^2 = T^4S, \text{ \&c.}$$

In the cases of 6, 7, and 8 letters, where the number of groups is larger, I introduce current numbers for the groups of the same order: thus, *infra*, 6 letters order 4, we have the groups $4._1$, $4._2$, $4._3$, $4._4$; and so in other cases, where there is more than one group for the same order.

Six letters.

Six letters: the groups are

ord.		
2.		$(ab\,.\,cd\,.\,ef)$,
3.		$(abc\,.\,def)$ cyc.
4.	1	$(ab\,.\,cd)\,(ef)$,
	2	$\{(ab)\,(cd)\,(ef)\}$ pos.
	3	$\{(abcd)$ cyc. $(ef)\}$ pos.
	4	$\{(abcd)_4\,(ef)\}$ dim.
6.	1	$(abcdef)$ cyc.
	2	$(abc\,.\,def)$ all,
	3	$(ad\,.\,bf\,.\,ce)\,(abc\,.\,def)$ cyc.
8.	1	$(ab)\,(cd)\,(ef)$,
	2	$(abcd)$ cyc. (ef),
	3	$(abcd)_4\,(ef)$,
	4	$\{(abcd)_8$ com. $(ef)\}$ dim.
	5	$\{(abcd)_8$ cyc. $(ef)\}$ dim.
	6	$\{(abcd)_8$ pos. $(ef)\}$ dim. $=\{(abcd)_8\,(ef)\}$ pos.
9.		(abc) cyc. (def) cyc.
12.		$(abcdef)_{12}$,
16.		$(abcd)_8\,(ef)$,
18.	1	(abc) all (def) cyc.,
	2	$\{(abc)$ all (def) all$\}$ pos.,
	3	$(ad\,.\,be\,.\,cf)\,\{(abc)$ cyc. (def) cyc.$\}$,
24.	1	$(abcd)$ pos. (ef),
	2	$\{(abcd)$ all $(ef)\}$ pos.,
	4	$(+\,abcdef)_{24}$,
	3	$(\pm\,abcdef)_{24}$,
36.	1	(abc) all (def) all,
	2	$(abcdef)_{36}$,
48.	1	$(abcd)$ all (ef),
	2	$(abcdef)_{48}$,
60.		$(abcdef)_{60}$,
72.		$(abcdef)_{72}$,
120.		$(abcdef)_{120}$,
360.		$(abcdef)$ pos.,
720.		$(abcdef)$ all.

The $\pm$ and $+$ are used for distinction, the $\pm$ showing that the group is one with positive and negative substitutions, the $+$ that it is a group with positive substitutions only.

Six letters. Explanations. Six letters.

ord. 6. 2. $(abc.def)$ all; the substitutions are those of (abc) all each, compounded with the corresponding substitution of (def) all; viz. they are

$$\begin{array}{lll} 1, & ab.de, & abc.def, \\ & ac.df, & acb.dfe, \\ & bc.ef. & \end{array}$$

ord. 6. 3. $(ad.bf.ce)(abc.def)$ cyc. This is, in fact, the group $(ad.be.cf)(abc.dfe)$ cyc. explained in the introductory paragraphs; only for convenience the letters e, f have been interchanged. Making this interchange, the substitutions of the group $(ad.bf.ce)(abc.def)$ cyc. are

$$\begin{array}{lll} 1, & abc.def, & ad.bf.ce, \\ & acb.dfe; & ae.bd.cf, \\ & & af.be.cd. \end{array}$$

It may be added that, writing $\beta = ad.bf.ce$ and $\theta = abc.def$, the group is $(1, \beta)(1, \theta, \theta^2)$, where $\beta^2 = 1$, $\theta^3 = 1$, and $\beta\theta = \theta^2\beta$, $\beta\theta^2 = \theta\beta$. If $U = adbecf$, then $U^2 = abc.def = \theta$, and the form is $(1, \beta)(1, U^2, U^4)$: see $(abcdef)_{12}$, *infra*.

ord. 8. 4, 5, 6; the nature of these groups is explained in the introductory paragraphs.

ord. 12. $(abcdef)_{12}$. The substitutions are

$$\begin{array}{lllll} 1, & ab.ef, & ad.bf.ce, & abc.def, & adbecf, \\ & ac.de, & af.be.cd, & acb.dfe, & afcebd; \\ & bc.df, & ae.bd.cf, & & \\ & & ae.bf.cd, & & \end{array}$$

writing $U = adbecf$ and $\beta = ad.bf.ce$, this is

$$(1, \beta)(1, U, U^2, U^3, U^4, U^5),$$

where

$$U^6 = 1, \quad \beta^2 = 1, \quad \beta U = U^5\beta,$$

and thence

$$\beta(1, U, U^2, U^3, U^4, U^5) = (1, U^5, U^4, U^3, U^2, U)\beta.$$

ord. 18. $(ad.be.cf)\{(abc)$ cyc. (def) cyc.$\}$; this has been explained in the introductory paragraphs. Arranging them in a more convenient order, the substitutions are

$$\begin{array}{lllll} 1, & abc, & abc.def, & ad.be.cf, & adcfbe, \\ & acb, & abc.dfe, & ae.bf.cd, & adbecf, \\ & def, & acb.def, & af.bd.ce; & aebfcd, \\ & dfe; & acb.dfe; & & aecdbf, \\ & & & & afbdce, \\ & & & & afcebd. \end{array}$$

Six letters.

ord. 24. 3. $(\pm\, abcdef)_{24}$: the substitutions are

+	+	+	−	−
1,	*ac . bd*,	*abe . cdf*,	*abcd*,	*ab . cd . ef*,
	ac . ef,	*abf . cde*,	*adcb*,	*ac . be . df*,
	bd . ef,	*ade . bfc*,	*aecf*,	*ac . bf . de*,
		adf . bce,	*afce*,	*ad . bc . ef*,
		aeb . cfd,	*bedf*,	*ae . bd . fc*,
		aed . bcf,	*bfde*,	*af . bd . ce*,
		afb . ced,		
		afd . bce.		

This includes, as part of itself, the group

$$\{(abcd)_8 \text{ cyc. } (ef)\} \text{ dim.};$$

and it may be written

$$(ade\,.\,bcf) \text{ cyc. } \{(abcd)_8 \text{ cyc. } (ef)\} \text{ dim.}$$

24. 4. $(+\,abcdef)_{24}$; the substitutions are

1,	*ab . cd*,	*abe . cdf*,	*abcd . ef*,
	ac . bd,	*abf . cde*,	*adcb . ef*,
	ac . ef,	*ade . bfc*,	*aecf . bd*,
	ad . bc,	*adf . bec*,	*afce . bd*,
	ae . cf,	*aeb . cfd*,	*bedf . ac*,
	af . ce,	*aed . bcf*,	*bfde . ac*,
	bd . ef,	*afb . ced*,	
	be . df,	*afd . bce*,	
	bf . de.		

This includes, as part of itself, the group $\{(abcd)_8\,(ef)\}$ pos.; and it may be written $(ade\,.\,bfc)$ cyc. $\{(abcd)_8\,(ef)\}$ pos. It consists of the positive terms of $(abcdef)_{48}$, and might thus be written $(abcdef)_{48}$ pos.

ord. 36. $(abcdef)_{36}$; this consists of the positive terms of $(abcdef)_{72}$. The group consists of the substitutions which leave unaltered $b-c\,.\,c-a\,.\,a-b\,.\,e-f\,.\,f-d\,.\,d-e$.

ord. 48. $(abcdef)_{48}$. The substitutions are Six letters.

+	+	+	+	−	−	−	−
1, *ab . cd*,	*abe . cdf*,	*abcd . ef*,	*ac*,	*abcd*,	*ab . cd . ef*,	*abecfd*,	
ac . bd,	*abf . cde*,	*adcb . ef*,	*bd*,	*adcb*,	*ac . bd . ef*,	*abecdf*,	
ac . ef,	*ade . cbf*,	*aecf . bd*,	*ef*;	*aecf*,	*ac . be . df*,	*abfcde*,	
ad . bc,	*adf . cbe*,	*afce . bd*,		*afce*,	*ac . bf . de*,	*adecbf*,	
ae . cf,	*aeb . cfd*,	*bedf . ac*,		*bedf*,	*ad . bc . ef*,	*adfcbe*,	
af . ce,	*aed . cfb*,	*bfde . ac*;		*bfde*;	*ae . cf . bd*,	*aedcfb*,	
bd . ef,	*afb . ced*,				*af . ce . bd*;	*afbced*,	
be . df,	*afd . ceb*;					*afdceb*;	
bf . de.							

ord. 60. $(abcdef)_{60}$; this consists of the positive terms of $(abcdef)_{120}$.

ord. 72. $(abcdef)_{72}$. The substitutions are

+	+	+	+	+	−	−	−	−
1,	*abc*,	*abc . def*,	*ab . de*,	*ad . becf*,	*ab*,	*ab . def*,	*ad . be . cf*,	*adbecf*,
	acb,	*abc . dfe*,	*ab . df*,	*ad . bfce*,	*ac*,	*ab . dfe*,	*ad . bf . ce*,	*adbfce*,
	def,	*acb . def*,	*ab . ef*,	*ae . bdcf*,	*bc*,	*ac . def*,	*ae . bd . cf*,	*adcebf*,
	dfe;	*acb . dfe*;	*ac . de*,	*ae . bfcd*,	*de*,	*ac . dfe*,	*ae . bf . cd*,	*adcfbe*,
			ac . df,	*af . bdce*,	*df*,	*bc . def*,	*af . bd . ce*,	*aebdcf*,
			ae . cf,	*af . becd*,	*ef*;	*bc . dfe*,	*af . be . cd*;	*aebfcd*,
			bc . de,	*bd . aecf*,		*de . abc*,		*aecdbf*,
			bc . df,	*bd . afce*,		*de . acb*,		*aecfbd*,
			bc . ef;	*be . adcf*,		*df . abc*,		*afbdce*,
				be . afcd,		*df . acb*,		*afbecd*,
				bf . adce,		*ef . abc*,		*afcdbe*,
				bf . aecd,		*ef . acb*;		*afcebd*;
				cd . aebf,				
				cd . afbe,				
				ce . adbf,				
				ce . afbd,				
				cf . adbe,				
				cf . aebd.				

This contains, as part of itself, the group (abc) all (def) all; and it may be written $(ad . be . cf)$ $\{(abc)$ all (def) all$\}$. It leaves unaltered the function $abc + def$.

Six letters.

ord. 120. $(abcdef)_{120}$. The substitutions are

+	+	+	+	−	−	−
1,	*ab . cf*,	*abcef*,	*abc . dfe*,	*abcd*,	*abcfde*,	*ab . cd . ef*,
	ab . de,	*abdce*,	*abd . cfe*,	*abdf*,	*abdefc*,	*ab . ce . df*,
	ac . bd,	*abefd*,	*abe . cdf*,	*abec*,	*abedcf*,	*ac . be . df*,
	ac . ef,	*abfdc*,	*abf . cde*,	*abfe*,	*abfced*,	*ac . bf . de*,
	ad . bf,	*acbed*,	*acb . def*,	*acbf*,	*acbdfe*,	*ad . bc . ef*,
	ad . ce,	*acdfb*,	*acd . bfe*,	*acde*,	*acdbef*,	*ad . be . cf*,
	ae . bc,	*acedf*,	*ace . bfd*,	*aceb*,	*acefbd*,	*ae . bd . cf*,
	ae . df,	*acfbe*,	*acf . bde*,	*acfd*,	*acfedb*,	*ae . bf . cd*,
	af . be,	*adbcf*,	*adb . cef*,	*adbe*,	*adbfec*,	*af . bc . de*,
	af . cd,	*adbec*,	*adc . bef*,	*adcb*,	*adcebf*,	*af . bd . ce*;
	bc . df,	*adcfe*,	*ade . bfc*,	*adef*,	*adecfb*,	
	bd . ef,	*adfeb*,	*adf . bec*,	*adfc*,	*adfbce*,	
	be . cd,	*aebdf*,	*aeb . cfd*,	*aebd*,	*aebcdf*,	
	bf . ce,	*aebfc*,	*aec . bdf*,	*aecd*,	*aecbfd*,	
	cf . de;	*aefcd*,	*aed . bcf*,	*aecf*,	*aedfcb*,	
		aecdb,	*aef . bdc*,	*aefb*,	*aefdbc*,	
		afbde,	*afb . ced*,	*afbc*,	*afbecd*,	
		afcdb,	*afc . bed*,	*afce*,	*afcdeb*,	
		afdec,	*afd . bce*,	*afdb*,	*afdcbe*,	
		afecb,	*afe . bcd*;	*afed*,	*afebdc*;	
		bcdef,		*bced*,		
		bdfce,		*bcfe*,		
		becfd,		*bdcf*,		
		bfedc;		*bdec*,		
				bedf,		
				befc,		
				bfcd,		
				bfde,		
				cdfe,		
				cefd.		

This is the remarkable group giving rise to a six-valued function of six letters, not symmetrical in regard to five of the letters. Such a function is Six letters.

$$\begin{array}{l} ab \,.\, cf \,.\, de, \\ ac \,.\, db \,.\, ef, \\ ad \,.\, ec \,.\, fb, \\ ae \,.\, fd \,.\, bc, \\ af \,.\, be \,.\, cd; \end{array}$$

viz. this denotes any symmetrical function of the five functions $ab \,.\, cf \,.\, de$, &c., where $ab \,.\, cf \,.\, de$ is a symmetrical function of ab, cf, de, these denoting the same symmetrical functions of a and b, of c and f, and of d and e, respectively.

Writing $S = abcd$, $T = aecdb$, $U = abfced$, the group $(1, S, S^2, S^3)(1, T, T^2, T^3, T^4)$ is convertible with the group $(1, U, U^2, U^3, U^4, U^5)$, as may be verified by means of the diagram:

	1	U	U^2	U^3	U^4	U^5
1	1	$U \,.\, 1$	$U^2 \,.\, 1$	$U^3 \,.\, 1$	$U^4 \,.\, 1$	U^5
T	T	$U^3 \,.\, S^2$	$U \,.\, S^3T^2$	$U^5 \,.\, S^2T^2$	$U^2 \,.\, T^4$	$U^4 \,.\, ST^4$
T^2	T^2	$U^5 \,.\, T^3$	$U^3 \,.\, ST^2$	$U^4 \,.\, S^3T^3$	$U \,.\, S^3T$	$U^2 \,.\, ST$
T^3	T^3	$U^4 \,.\, ST^2$	$U^5 \,.\, S^3T^3$	$U^2 \,.\, S^3T$	$U^3 \,.\, ST$	$U \,.\, T^2$
T^4	T^4	$U^2 \,.\, ST^4$	$U^4 \,.\, T$	$U \,.\, S^2$	$U^5 \,.\, S^3T^2$	$U^3S^2 \,.\, T^2$
S	S	$U^3 \,.\, ST^3$	$U^2 \,.\, S^3T^4$	$U^4 \,.\, S^2T^3$	$U^5 \,.\, S^2T^4$	$U \,.\, S^3$
ST	ST	$U^4 \,.\, T^2$	$U^3 \,.\, T^3$	$U \,.\, ST^2$	$U^2 \,.\, S^3T^3$	$U^5 \,.\, S^3T$
ST^2	ST^2	$U \,.\, S^3T^3$	$U^4 \,.\, S^3T$	$U^5 \,.\, ST$	$U^3 \,.\, T^2$	$U^2 \,.\, T^3$
ST^3	ST^3	$U^5 \,.\, S^3T^4$	$U \,.\, S^2T^3$	$U^2 \,.\, S^2T^4$	$U^4 \,.\, S^3$	$U^3 \,.\, S$
ST^4	ST^4	$U^2 \,.\, T$	$U^5 \,.\, S^2$	$U^3 \,.\, S^3T^2$	$U \,.\, S^2T^2$	$U^4 \,.\, T^4$
S^2	S^2	$U^4 \,.\, S^3T^2$	$U^2 \,.\, S^2T^2$	$U^5 \,.\, T^4$	$U \,.\, ST^4$	$U^3 \,.\, T$
S^2T	S^2T	$U^5 \,.\, S^3T$	$U^4 \,.\, S^2T$	$U^3 \,.\, S^2T$	$U^2 \,.\, S^2T$	$U \,.\, S^2T$
S^2T^2	S^2T^2	$U^3 \,.\, T^4$	$U^5 \,.\, ST^4$	$U \,.\, T$	$U^4 \,.\, S^2$	$U^2 \,.\, S^3T^2$
S^2T^3	S^2T^3	$U \,.\, S^2T^4$	$U^3 \,.\, S^3$	$U^2 \,.\, S$	$U^5 \,.\, ST^3$	$U^4 \,.\, S^3T^4$
S^2T^4	S^2T^4	$U^2 \,.\, S^3$	$U \,.\, S$	$U^4 \,.\, ST^3$	$U^3 \,.\, S^3T^4$	$U^5 \,.\, S^2T^3$
S^3	S^3	$U^5 \,.\, S$	$U^2 \,.\, ST^3$	$U \,.\, S^3T^4$	$U^3 \,.\, S^2T^3$	$U^4 \,.\, S^2T^4$
S^3T	S^3T	$U \,.\, ST$	$U^5 \,.\, T^2$	$U^4 \,.\, T^3$	$U^2 \,.\, ST^2$	$U^3 \,.\, S^3T^3$
S^3T^2	S^3T^2	$U^4 \,.\, S^2T^2$	$U \,.\, T^4$	$U^3 \,.\, ST^4$	$U^5 \,.\, T$	$U^2 \,.\, S^2$
S^3T^3	S^3T^3	$U^3 \,.\, S^3T$	$U^4 \,.\, ST$	$U^2 \,.\, T^2$	$U \,.\, T^3$	$U^5 \,.\, ST^2$
S^3T^4	S^3T^4	$U^2 \,.\, S^2T^3$	$U^3 \,.\, S^2T^4$	$U^5 \,.\, S^3$	$U^4 \,.\, S$	$U \,.\, ST^3$

Six letters. and the group is thus

$$(1, \ldots, S^3)(1, \ldots, T^4)(1, \ldots, U^5) \text{ or } (1, \ldots, U^5)(1, \ldots, S^3)(1, \ldots, T^4),$$

or, since $(1, \ldots, S^3)(1, \ldots, T^4)$ is a group, we may invert the order of the factors $(1, \ldots, S^3)$ and $(1, \ldots, T^4)$.

But it is noticeable that we cannot write the group as

$$(1, \ldots, S^3)(1, \ldots, U^5)(1, \ldots, T^4) \text{ or } (1, \ldots, T^4)(1, \ldots, U^5)(1, \ldots, S^3).$$

The 120 substitutions of either of these products are not the 120 substitutions of the group in question; but some of these are missing altogether, and others of them occur twice repeated. And it is to be remarked also that

$$(1, \ldots, S^3)(1, \ldots, U^5) \text{ and } (1, \ldots, T^4)(1, \ldots, U^5)$$

are neither of them a group. To illustrate this, I give the following fragment of a table

	STU	*SUT*	do.	*TUS*	do.
abcdef	000	000		000	
abdfce	033	122		330,	412
abecfd	124	032,	251	343	
abfedc	105	150,	331	431,	013
acbfed	323	111		453	
acdefb	044	—		440	
adefbc	113	243		—	

viz. here the second line denotes that the substitution *abdfce*, which is $S^0T^3U^3$, is in one way *SUT*, viz. it is $S^1U^2T^2$: and in two ways *TUS*, viz. it is $T^3U^3S^0$ and also $T^4U^1S^2$. But *acdefb*, which is $S^0T^4U^4$, is not in any way *SUT*; and so *adefbc*, which is $S^1T^1U^3$, is not in any way *TUS*.

Seven letters. Seven letters.

ord.

6. 1 (*ac* . *bd*) (*efg*) cyc.
   2 {(*ac* . *bd*) (*efg*) all} dim.

7. (*abcdefg*) cyc.

10. 1 (*abcde*) cyc. (*fg*),
    2 {$(abcde)_{10}$ (*fg*)} dim.

12. 1 (*ac* . *bd*) (*efg*) all,
    2 (*ab*) (*cd*) (*efg*) cyc.
    3 (*abcd*) cyc. (*efg*) cyc.
    4 $(abcd)_4$ (*efg*) cyc.
    5 {(*ab*) (*cd*) (*efg*) all} pos.
    6 {(*abcd*) cyc. (*efg*) all} pos.
    7 {$(abcd)_4$ (*efg*) all} dim.
    8 {(*abc*) all (*de*)} pos. (*fg*),
    9 {(*abcd*) pos. (*efg*) cyc.} tris.

Seven letters.

14. $(abcdefg)_{14}$,

20. 1 $(abcde)_{10}\,(fg)$,

2 $\{(abcde)_{20}\,(fg)\}$ pos.

21. $(abcdefg)_{21}$,

24. 1 $(ab)\,(cd)\,(efg)$ all,

2 $(abcd)$ cyc. (efg) all,

3 $(abcd)_4\,(efg)$ all,

4 $(abcd)_8\,(efg)$ cyc.

5 $\{(abcd)_8$ com. (efg) all$\}$ dim.

6 $\{(abcd)_8$ cyc. (efg) all$\}$ dim.

7 $\{(abcd)_8$ pos. (efg) all$\}$ dim.

36. $(abcd)$ pos. (efg) cyc.

40. $(abcde)_{20}\,(fg)$,

42. $(abcdefg)_{42}$,

48. $(abcd)_8\,(efg)$ all,

72. 1 $(abcd)$ all (efg) cyc.

2 $(abcd)$ pos. (efg) all,

3 $\{(abcd)$ all (efg) all$\}$ pos.

120. 1 $(abcde)$ pos. (fg),

2 $\{(abcde)$ all $(fg)\}$ pos.

144. $(abcd)$ all (efg) all,

240. $(abcde)$ all (fg),

2520. $(abcdefg)$ pos.

5040. $(abcdefg)$ all.

Seven letters. Explanations.

ord. 12. 9. $\{(abcd)$ pos. (efg) cyc.$\}$ tris. The substitutions are

$$
\begin{array}{lll}
1, & ab\,.\,cd, & abc\,.\,efg, \\
 & ac\,.\,bd, & acd\,.\,efg, \\
 & ad\,.\,bc, & adb\,.\,efg, \\
 & & bdc\,.\,efg, \\
 & & acb\,.\,egf, \\
 & & adc\,.\,egf, \\
 & & abd\,.\,egf, \\
 & & bcd\,.\,egf;
\end{array}
$$

Seven letters. viz. the 12 substitutions of $(abcd)$ pos. are divided into fours, which are connected with the substitutions 1, efg, egf respectively.

ord. 14. $(abcdefg)_{14}$: this is $(1, P^3)(1, Q, Q^2, Q^3, Q^4, Q^5, Q^6)$, where $P^3 = ad\,.\,be\,.\,cf$, $Q = agdecfb$, $(P^3)^2 = 1$, $Q^7 = 1$, $P^3Q = Q^6P^3$, see *infra*, ord. 42. The substitutions are

1,	$abfcedg$,	$ab\,.\,cd\,.\,fg$,
	$acgfdbe$,	$ac\,.\,bf\,.\,eg$,
	$adcbgef$,	$ad\,.\,be\,.\,cf$,
	$aebdfgc$,	$ae\,.\,bc\,.\,dg$,
	$afegbcd$,	$af\,.\,cg\,.\,de$,
	$agdecfb$,	$ag\,.\,bd\,.\,ef$,
		$bg\,.\,ce\,.\,df$.

ord. 21. $(abcdefg)_{21}$; this is $(1, P^2, P^4)(1, Q, Q^2, Q^3, Q^4, Q^5, Q^6)$, where

$$P^2 = acb\,.\,dfe,\ Q = agdecfb,\ (P^2)^3 = 1,\ Q^7 = 1,\ P^2Q = Q^2P^2;$$

see *infra* order 42. The substitutions are the positive substitutions of $(abcdefg)_{42}$.

ord. 24. The last three groups correspond to the before-mentioned three modes of division of $(abcd)_8$ into two sets of four, viz. we have,

24. 5; $\{(abcd)_8$ com. (efg) all$\}$ dim. $= \{(ac)(bd)\}(efg)$ pos.
$+ (abcd,\ adbc,\ ab\,.\,cd,\ ad\,.\,bc)(efg)$ neg.

24. 6; $\{(abcd)_8$ cyc. (efg) all$\}$ dim. $= (abcd)$ cyc. (efg) pos.
$+ (\ ac\ ,\ bd\ ,\ ab\,.\,cd,\ ad\,.\,bc)(efg)$ neg.

24. 7; $\{(abcd)_8$ pos. (efg) all$\}$ dim. $= (abcd)_4\ (efg)$ pos.
$+ (\ ac\ ,\ bd\ ,\ abcd\ ,\ adbc\ )(efg)$ neg.

ord. 42. $(abcdefg)_{42}$: this is

$$(1, P, P^2, P^3, P^4, P^5)(1, Q, Q^2, Q^3, Q^4, Q^5, Q^6),$$

where $P = aecdbf$, $Q = agdecfb$, $P^6 = 1$, $Q^7 = 1$, and $PQ = Q^5P$. The substitutions are

+	+	+	−	−
1,	$abc\,.\,def$,	$abfcedg$,	$ab\,.\,cd\,.\,fg$,	$abegdf$,
	$abd\,.\,cge$,	$acgfdbe$,	$ac\,.\,bf\,.\,eg$,	$abgcfe$,
	$acb\,.\,dfe$,	$adcbgef$,	$ad\,.\,be\,.\,cf$,	$acdebg$,
	$acf\,.\,bdg$,	$aebdfgc$,	$ae\,.\,bc\,.\,dg$,	$acefgd$,
	$adb\,.\,ceg$,	$afegbcd$,	$af\,.\,cg\,.\,de$,	$adfbcg$,
	$ade\,.\,bfg$,	$agdecfb$,	$ag\,.\,bd\,.\,ef$,	$adgfec$,
	$aed\,.\,bgf$,		$bg\,.\,ce\,.\,df$,	$aecdbf$,
	$aeg\,.\,cfd$,			$aefcgb$,

Seven letters.

afc . bgd,	*afbdce*,
afg . bec,	*afdgeb*,
age . cdf,	*agbedc*,
agf . bce,	*agcbfd*,
bef . cdg,	*bcfged*,
bfe . cgd,	*bdegfc*.

The table for the combination of the powers of P and Q is

	1	Q	Q^2	Q^3	Q^4	Q^5	Q^6	
1	$(1,$	$Q,$	$Q^2,$	$Q^3,$	$Q^4,$	$Q^5,$	$Q^6)$	$P,$
P	$(1,$	$Q^5,$	$Q^3,$	$Q,$	$Q^6,$	$Q^4,$	$Q^2)$	$P,$
P^2	$(1,$	$Q^4,$	$Q,$	$Q^5,$	$Q^2,$	$Q^6,$	$Q^3)$	$P^2,$
P^3	$(1,$	$Q^6,$	$Q^5,$	$Q^4,$	$Q^3,$	$Q^2,$	$Q)$	$P^3,$
P^4	$(1,$	$Q^2,$	$Q^4,$	$Q^6,$	$Q,$	$Q^3,$	$Q^5)$	$P^4,$
P^5	$(1,$	$Q^3,$	$Q^6,$	$Q^2,$	$Q^5,$	$Q,$	$Q^4)$	$P^5;$

read

$$PQ = Q^5P, \quad PQ^2 = Q^3P, \text{ \&c.}$$

This completes the explanations as to the groups of seven letters.

I proceed to consider the substitution groups of eight letters. The abbreviation A. refers to Mr Askwith's paper (*Quart. Journal of Math.*, vol. XXIV. (1890), pp. 263—331). The list is

Eight letters.

ord.

2. 1 $(ab\,.\,cd\,.\,ef\,.\,gh)$.

4. 1 $(ab\,.\,cd\,.\,ef)\,(gh)$,
   2 $(ab\,.\,cd)\,(ef\,.\,gh)$,
   3 $\{(ab)\,(cd)\,(ef\,.\,gh)\}$ dim.,
   4 $\{(abcd)$ cyc. $(ef\,.\,gh)\}$ dim.,
   5 $\{(abcd)_4\,(ef\,.\,gh)\}$ dim.,
   6 $(abcd\,.\,efgh)$ cyc.,
   7 $\{(abcd)_4\,(ef\,.\,gh)\}$ dim.,
   8 $(abcdefgh)_4$. Substitutions are

   1, *ae . bf . cg . dh*,
   ac . eg . bd . fh,
   ag . ce . bh . df.

6. 1 $(abc\,.\,def)$ cyc. (gh),
   2 $\{(abcdef)$ cyc. $(gh)\}$ pos.
   3 $\{(abc\,.\,def)$ all $(gh)\}$ dim. [not in A.],
   4 $\{(abcdef)_6\,(gh)\}$ dim.

Eight letters.

8\. 1 $(ab \,.\, cd)\,(ef)\,(gh)$,

2 $\{(ab)\,(cd)\,(ef)\}$ pos. (gh),

3 $\{(abcd)$ cyc. $(ef)\}$ pos. (gh),

4 $\{(abcd)_4\,(ef)\}$ dim. (gh),

5 $(abcd)$ cyc. $(ef \,.\, gh)$,

6 $(abcd)_4\,(ef \,.\, gh)$,

7 $(1,\ abcd,\ ac \,.\, bd,\ adcb) + ef \,.\, gh\,(ac,\ bd,\ ab \,.\, cd,\ ad \,.\, bc)$,

8 $(1,\ ab \,.\, cd,\ ac \,.\, bd,\ ad \,.\, bc) + ef \,.\, gh\,(ac,\ bd,\ abcd,\ adcb)$,

9 $(1,\ ac,\ bd,\ ac \,.\, bd) + ef \,.\, gh\,(abcd,\ adbc,\ ab \,.\, cd,\ ad \,.\, bc)$ [Not in A.],

10 $(1,\ ac \,.\, bd)\,(1,\ eg \,.\, fh) + (ac,\ bd)\,(eg,\ fh) = A'\,(abcd \,.\, efgh)_8$,

11 $(1,\ ac \,.\, bd)\,(1,\ eg \,.\, fh) + (abcd,\ adcb)\,(efgh,\ ehfg) = B'\,(abcd \,.\, efgh)_8$,

12 $(1,\ ac \,.\, bd)\,(1,\ eg \,.\, fh) + (ab \,.\, cd,\ ad \,.\, bc)\,(ef \,.\, gh,\ eh \,.\, fg) = C'\,(abcd \,.\, efgh)_8$,

13 $(1,\ ac \,.\, bd)\,(1,\ eg \,.\, fh) + (ac,\ bd)\,(efgh,\ ehgf) = D'\,(abcd \,.\, efgh)_8$,

14 $(1,\ ac \,.\, bd)\,(1,\ eg \,.\, fh) + (ac,\ bd)\,(ef \,.\, gh,\ eh \,.\, fg) = E'\,(abcd \,.\, efgh)_8$,

15 $(1,\ ac \,.\, bd)\,(1,\ eg \,.\, fh) + (ab \,.\, cd,\ ad \,.\, bc)\,(efgh,\ ehgf) = F'\,(abcd \,.\, efgh)_8$,

16 $(ae \,.\, bf \,.\, cg \,.\, dh)\,(ac \,.\, bd)\,(eg \,.\, fh)$. Substitutions are

1, $ac \,.\, bd$, $ac \,.\, bd \,.\, eg \,.\, fh$, $agce \,.\, bhdf$,
$eg \,.\, fh$, $ae \,.\, bf \,.\, cg \,.\, dh$, $aecg \,.\, bfdh$.
$ag \,.\, bh \,.\, ce \,.\, df$,

17 $(1,\ ac \,.\, bd,\ ac \,.\, ef,\ bd \,.\, ef) + gh\,(abcd,\ adcb,\ ab \,.\, cd \,.\, ef,\ ad \,.\, bc \,.\, ef)$,

18 $(1,\ ac \,.\, bd,\ ac \,.\, ef,\ bd \,.\, ef) + gh\,(ab \,.\, cd,\ ad \,.\, bc,\ abcd \,.\, ef,\ adcb \,.\, ef)$,

19 $(1,\ ac \,.\, bd,\ ab \,.\, cd \,.\, ef,\ ad \,.\, bc \,.\, ef) + gh\,(abcd,\ adcb,\ ac \,.\, ef,\ bd \,.\, ef)$,

20 $(abcdefgh)_8$. Substitutions are

1, $ac \,.\, be \,.\, dg$, $ac \,.\, bd \,.\, eg \,.\, fh$, $afch \,.\, bgde$,
$bd \,.\, af \,.\, ch$, $ahcf \,.\, bedg$.
$eg \,.\, ah \,.\, cf$,
$fh \,.\, bg \,.\, de$,

21 $(abcdefgh)$ cyc.,

22 $A\,(abcdefgh)_8$,

23 $B\,(abcdefgh)_8$,

24 $C\,(abcdefgh)_8$,

25 $D\,(abcdefgh)_8$.

Eight letters.

12. 1 $(abcdef)$ cyc. (gh),

2 $(abc\,.\,def)$ all (gh),

3 $(abcdef)_6\,(gh)$,

4 $(1,\ abcdef,\ ace\,.\,bdf,\ ad\,.\,be\,.\,cf,\ aec\,.\,bfd,\ afedcb)$
$+\,gh\,(af\,.\,be\,.\,cd,\ ab\,.\,cf\,.\,de,\ ad\,.\,bc\,.\,ef,\ bf\,.\,ce,\ ac\,.\,df,\ ae\,.\,bd)$,

5 $(1,\ bf\,.\,ce,\ ac\,.\,df,\ ae\,.\,bd,\ ace\,.\,bdf,\ aec\,.\,bfd)$
$+\,gh\,(ab\,.\,cf\,.\,de,\ ad\,.\,bc\,.\,ef,\ ad\,.\,bf\,.\,ce,\ af\,.\,be\,.\,cd,\ abcdef,\ afedcb)$,

6 $(1,\ ab\,.\,cf\,.\,de,\ ad\,.\,bc\,.\,ef,\ af\,.\,be\,.\,cd,\ ace\,.\,bdf,\ aec\,.\,bfd)$
$+\,gh\,(bf\,.\,ce,\ ac\,.\,df,\ ae\,.\,bd,\ ad\,.\,be\,.\,cf,\ abcdef,\ afedcb)$.

15. 1 $(abcde)$ cyc. (fgh) cyc.

16. 1 $(ab)\,(cd)\,(ef)\,(gh)$,

2 $(abcd)$ cyc. $(ef)\,(gh)$,

3 $(abcd)_4\,(ef)\,(gh)$,

4 $\{(abcd)_8$ com. $(ef)\}$ dim. (gh),

5 $\{(abcd)_8$ cyc. $(ef)\}$ dim. (gh),

6 $\{(abcd)_8$ pos. $(ef)\}$ dim. (gh),

7 $(abcd)$ cyc. $(efgh)$ cyc.,

8 $(abcd)_4\,(efgh)$ cyc.,

9 $(abcd)_4\,(efgh)_4$,

10 $(acbd)_8\,(ef\,.\,gh)$,

11 $(1,\ ac\,.\,bd)\,(1,\ eg,\ fh,\ eg\,.\,fh) + (ac,\ bd)\,(ef\,.\,gh,\ eh\,.\,fg,\ efgh,\ ehgf)$,

12 $(1,\ ac\,.\,bd)\,(1,\ eg,\ fh,\ eg\,.\,fh) + (ab\,.\,cd,\ ad\,.\,bc)\,(ef\,.\,gh,\ eh\,.\,fg,\ efgh,\ ehgf)$,

13 $(1,\ ac\,.\,bd)\,(1,\ eg,\ fh,\ eg\,.\,fh) + (abcd,\ adcb)\,(ef\,.\,gh,\ eh\,.\,fg,\ efgh,\ ehfg)$,

14 $(1,\ ac\,.\,bd)\,(1,\ eg\,.\,fh,\ efgh,\ ehgf) + (abcd,\ adcb)\,(eg,\ fh,\ ef\,.\,gh,\ eh\,.\,fg)$,

15 $(1,\ ac\,.\,bd)\,(1,\ eg\,.\,fh,\ efgh,\ ehgf) + (ab\,.\,cd,\ ad\,.\,bc)\,(eg\,.\,fh,\ ef\,.\,gh,\ eh\,.\,fg)$,

16 $(1,\ ac\,.\,bd)\,(1,\ ef\,.\,gh,\ eg\,.\,fh,\ eh\,.\,fg) + (abcd,\ adcb)\,(eg,\ fh,\ efgh,\ ehfg)$,

17 $(1,\ ac\,.\,bd)\,(1,\ ef\,.\,gh,\ eg\,.\,fh,\ eh\,.\,fg) + (ab\,.\,cd,\ ad\,.\,bc)\,(eg,\ fh,\ efgh,\ ehgf)$,

18 $(1,\ ac\,.\,bd)\,(1,\ efgh,\ eg\,.\,fh,\ ehgf) + (ac,\ bd)\,(eg,\ fh,\ ef\,.\,gh,\ eh\,.\,fg)$,

19 $(1,\ ac\,.\,bd)\,(1,\ ef\,.\,gh,\ eg\,.\,fh,\ eh\,.\,fg) + (ac,\ bd)\,(eg,\ fh,\ efgh,\ ehgf)$,

20 $(ae\,.\,bf\,.\,cg\,.\,dh)\,B'\,(abcd\,.\,efgh)_8$,

21 $(ae\,.\,bf\,.\,cg\,.\,dh)\,C'\,(abcd\,.\,efgh)_8$,

22 $(aebf\,.\,cgdh)$ cyc. $(ac\,.\,bd)\,(eg\,.\,fh)$,

23 $(aebfcgdh)$ cyc. $(1,\ ac\,.\,bd)$.

Eight letters.

18\. 1 (*abc*) cyc. (*def*) cyc. (*gh*),

2 {(*abc*) all (*de*)} pos. (*fgh*),

3 {(*abc*) pos. (*def*) pos.} + *gh* {(*abc*) neg. (*def*) neg.} = $A\,(abcdefgh)_{18}$,

4 $B\,(abcdefgh)_{18}$.

[A. gives, *loc. cit.*, p. 310, a form which is not a group.]

24\. 1 $(abcdef)_{12}\,(gh)$,

2 (*abcd*) pos. (*ef* . *gh*),

3 {(*abcd*) all (*ef* . *gh*)} dim.

4 $(abcdefgh)_{24}$.

30\. 1 (*abcde*) cyc. (*fgh*) all,

2 $(abcde)_{10}\,(fgh)$ cyc.

3 $(abcdefgh)_{30}$.

32\. 1 $(abcd)_8\,(ef)\,(gh)$,

2 $(abcd)_8\,(efgh)$ cyc.

3 $(abcd)_8\,(efgh)_4$,

4 $L\,\{(abcd)_8\,(efgh)_8\}$ dim.

5 $M\,\{(abcd)_8\,(efgh)_8\}$ dim.

6 $N\,\{(abcd)_8\,(efgh)_8\}$ dim.

7 $P\,\{(abcd)_8\,(efgh)_8\}$ dim.

8 $Q\,\{(abcd)_8\,(efgh)_8\}$ dim.

9 $R\,\{(abcd)_8\,(efgh)_8\}$ dim.

10 (*ae* . *bf* . *cg* . *dh*) (*abcd*) cyc. (*efgh*) cyc.

11 $(ae\,.\,bf\,.\,cg\,.\,dh)\,(abcd)_4\,(efgh)_4$.

[A. has, *loc. cit.*, p. 291 and p. 295, two forms which are identical, and, p. 275, there is a single form which is, by mistake, counted as two.]

36\. 1 {(*abc*) all (*de*)} pos. (*fgh*),

2 {(*abc*) all (*def*) all} pos. (*gh*),

3 {(*abc*) all (*def*) (*gh*)} pos.

4 $(abcdef)_{18}\,(gh)$,

5 $(ad\,.\,be\,.\,cf)\,A\,(abcdefgh)_{18}$ is $= \{(abcdef)_{36}\,(gh)\}$ dim.

[The group, A., *loc. cit.*, p. 312, is derived from the non-existent group of 18, p. 310, and is thus non-existent.]

Eight letters.

48. 1 $(abcd)$ all $(ef\,.\,gh)$,

2 $(abcd)$ pos. $(ef)\,(gh)$,

3 $(abcd)$ pos. $(efgh)$ cyc.

4 $(abcd)$ pos. $(efgh)_4$,

5 $\{(abcd)$ all $(ef)\}$ pos. (gh),

6 $(\pm\ abcdef)_{24}\,(gh)$,

7 $(+\ abcdef)_{24}\,(gh)$,

8 $\{(abcd)$ all $(efgh)$ cyc.$\}$ dim.

9 $\{(abcd)$ all $(ef)\,(gh)\}$ dim.

10 $\{(abcd)$ all $(efgh)_4\}$ dim.

11 $\{(abcdef)_{48}\,(gh)\}$ dim.

12 $\{(abcdef)_{48}\,(gh)\}$ pos.

[A. *loc. cit.*, p. 272 is not a group.]

13 $(abcdefgh)_{48}$,

14 $(abcd)_{12}\,(efgh)_{12}$ tris.

60. 1 $(abcde)_{10}\,(fgh)$ all,

2 $(abcde)_{20}\,(fgh)$ cyc.

3 $\{(abcde)_{20}\,(fgh)$ all$\}$ dim.

64. 1 $(abcd)_8\,(efgh)_8$,

2 $(ae\,.\,bf\,.\,cg\,.\,dh)\,M\,\{(abcd)_8\,(efgh)_8\}$ dim.

3 $(ae\,.\,bf\,.\,cg\,.\,dh)\,N\,\{(abcd)_8\,(efgh)_8\}$ dim.

72. 1 (abc) all (def) all (gh),

2 $(abcdef)_{36}\,(gh)$,

3 $\{(abcdef)_{72}\,(gh)\}$ dim.

4 $(abcdefgh)_{72}$.

96. 1 $(abcd)$ all $(ef)\,(gh)$,

2 $(abcd)$ all $(efgh)$ cyc.

3 $(abcd)$ all $(efgh)_4$,

4 $(abcdef)_{48}\,(gh)$,

5 $(abcd)$ pos. $(efgh)_8$,

6 $\{(abcd)$ all $(efgh)_8$ com.$\}$ dim.

7 $\{(abcd)$ all $(efgh)_8$ cyc.$\}$ dim.

Eight letters.

96. 8 $\{(abcd)$ all $(efgh)_8$ pos.$\}$ dim.

9 $(ae\,.\,bf\,.\,cg\,.\,dh)\,\{(abcd)_{12}\,(efgh)_{12}\}$ tris.

10 $(ae\,.\,bg\,.\,cf\,.\,dh)\,\{(abcd)_{12}\,(efgh)_{12}\}$ tris.

120. 1 $(abcde)_{20}\,(fgh)$ all,

2 $(abcdef)_{60}\,(gh)$,

3 $\{(abcdef)_{120}\,(gh)\}$ pos.

144. 1 $(abcd)$ pos. $(efgh)$ pos.

2 $(ad\,.\,be\,.\,cf)\,\{(abc)$ all (def) all $(gh)\}$.

168. $(abcdefgh)_{168}$.

180. $(abcde)$ pos. (fgh) cyc.

192. 1 $(abcd)$ all $(efgh)_8$,

2 $(1,\ agch,\ bgdh,\ egfh)\,\{(abcdef)_{48}\,(gh)\}$ dim.

3 $(1,\ bd\,.\,agch,\ efbgdh,\ ac\,.\,egfh)\,\{(abcdef)_{48}\,(gh)\}$ pos.

240. $(abcdef)_{120}\,(gh)$.

288. 1 $(abcd)$ all $(efgh)$ pos.

2 $\{(abcd)$ all $(efgh)$ all$\}$ pos.

3 $(ae\,.\,bf\,.\,cg\,.\,dh)\,(abcd)$ pos. $(efgh)$ pos.

336. $(abcdefgh)_{336}$.

360. 1 $(abcde)$ all (fgh) cyc.

2 $(abcde)$ pos. (fgh) all,

3 $\{(abcde)$ all (fgh) all$\}$ pos.

384. $(1,\ ag\,.\,bh,\ cf\,.\,dh,\ eg\,.\,fh)\,\{(abcdef)_{48}\,(gh)\}$.

576. 1 $(abcd)$ all $(efgh)$ all,

2 $(ae\,.\,bf\,.\,cg\,.\,dh)\,\{(abcd)$ all $(efgh)$ all$\}$ pos.

720. 1 $(abcde)$ all (fgh) all,

2 $(abcdef)$ pos. (gh),

3 $\{(abcdef)$ all $(gh)\}$ pos.

1152. $(ae\,.\,bf\,.\,cg\,.\,dh)\,\{(abcd)$ all $(efgh)$ all$\}$.

1440. $(abcdef)$ all (gh).

20160. $(abcdefgh)$ pos.

40320. $(abcdefgh)$ all.

Eight letters, explanations. Eight letters. Explanations.

ord. 8. Ord. 8.

$A\,(abcdefgh)_8$. The substitutions are

$$\begin{array}{l} 1,\ ab\,.\,cd\,.\,ef\,.\,gh, \\ ac\,.\,bd\,.\,eg\,.\,fh, \\ ad\,.\,bc\,.\,eh\,.\,fg, \\ ae\,.\,bf\,.\,cg\,.\,dh, \\ af\,.\,be\,.\,ch\,.\,dg, \\ ag\,.\,bh\,.\,ce\,.\,df, \\ ah\,.\,bg\,.\,cf\,.\,de. \end{array}$$

$B\,(abcdefgh)_8$. The substitutions are

$$\begin{array}{ll} 1,\ ab\,.\,ch\,.\,dg\,.\,ef, & aceg\,.\,bdfh, \\ ad\,.\,bc\,.\,eh\,.\,fg, & agec\,.\,bhdf. \\ ae\,.\,bf\,.\,cg\,.\,dh, & \\ af\,.\,be\,.\,cd\,.\,gh, & \\ ah\,.\,bg\,.\,cf\,.\,de, & \end{array}$$

$C\,(abcdefgh)_8$. The substitutions are

$$\begin{array}{ll} 1,\ ad\,.\,bg\,.\,cf\,.\,eh, & aceg\,.\,bdfh, \\ ae\,.\,bf\,.\,cg\,.\,dh, & agec\,.\,bhfd, \\ ah\,.\,bc\,.\,de\,.\,fg, & abef\,.\,cdgh, \\ & afeb\,.\,chgd. \end{array}$$

$D\,(abcdefgh)_8$. The substitutions are

$$\begin{array}{ll} 1,\ ae\,.\,bf\,.\,cg\,.\,dh, & aceg\,.\,bdfh, \\ & agec\,.\,bhfd, \\ & abef\,.\,chgd, \\ & afeb\,.\,cdgh, \\ & adeh\,.\,bgfc\,, \\ & ahed\,.\,bcfg\,. \end{array}$$

ord. 12. The last three groups, the substitutions of which are given in full, are each of them of the form $\{(abcdef)_{12}\,(gh)\}$ dim. They may be written: Ord. 12.

$\{(abcdef)_{12}$ cyc. $(gh)\}$ dim.; viz. here the six substitutions of $(abcdef)_{12}$ combined with 1 are those of $(abcdef)$ cyc.

$\{(abcdef)_{12}$ pos. $(gh)\}$ dim.; viz. here the six substitutions of $(abcdef)_{12}$ combined with 1 are the positive substitutions of $(abcdef)_{12}$.

$\{(abcdef)_{12} \pm (gh)\}$ dim.; viz. here the six substitutions of $(abcdef)_{12}$ combined with 1 are three positive and three negative substitutions.

Eight letters.
Ord. 16.

ord. 16. I have not thought it necessary to devise any notation for the set of groups 11 to 19, the substitutions of which are given in full.

For the remaining groups 20 to 23:

(*ae . bf . cg . dh*) *B′* (*abcd . efgh*)$_8$. The substitutions are

1,	*ac . bd*,	*ac . bd . ef . gh*,	*abcd . efgh*,
	eg . fh,	*ae . bf . cg . dh*,	*abcd . ehgf*,
		af . bg . ch . de,	*adcb . efgh*,
		ag . bh . ce . df,	*adcb . ehgf*,
		ah . be . cf . dg,	*agce . bhdf*,
			aecg . bfdh,
			afch . bgde,
			ahcf . bedg.

(*ae . bf . cg . dh*) *C′* (*abcd . efgh*)$_8$. The substitutions are

1,	*ac . bd*,	*ac . bd . eg . fh*,	*aceg . bfdh*,
	eg . fh,	*ab . cd . ef . gh*,	*agce . bhdf*,
		ab . cd . eh . fg,	*afch . bedg*,
		ad . bc . ef . gh,	*ahcf . bgde*.
		ad . bc . eh . fg,	
		ae . bf . cg . dh,	
		af . bg . ch . de,	
		ag . bh . ce . df,	
		ah . bg . cf . de,	

(*aebf . cdgh*) cyc. (*ac . bd*)(*ef . gh*). The substitutions are

1,	*ac . bd*,	*ac . bd . eg . fh*,	*aebf . cgdh*,
	eg . fh,	*ab . cd . ef . gh*,	*aedh . bfcg*,
		ab . cd . eh . fg,	*afdg . bech*,
		ad . bc . ef . gh,	*agbh . cedf*,
		ad . bc . eh . fg,	*afbe . chdg*,
			ahde . bgcf,
			agdf . bhce,
			ahbg . cfde.

$(aebfcgdh)$ cyc. $(1,\ ac . bd)$. The substitutions are Eight letters. Ord. 16.

1,	$ac . bd$,	$ac . bd . eg . fh$,	$abcd . efgh$,	$aebfcgdh$,
	$eg . fh$,		$abcd . ehgf$,	$afdechbg$,
			$adcb . efgh$,	$agbhcedf$,
			$adcb . ehgf$,	$ahdgcfbe$,
				$aedhcgbf$,
				$afbgchde$,
				$agdfcebh$,
				$ahbecfdg$.

ord. 18. $B\,(abcdefgh)_{18}$. This group, communicated to me by Mr Askwith, might be written Ord. 18.

$$[((ad . be . cf)\,\{(abc)\text{ cyc. }(def)\text{ cyc.}\})\,gh]\text{ pos.},$$

or for shortness

$$[(abcdef)_{18}\,gh]\text{ pos.}$$

ord. 24. $(abcdefgh)_{24}$. The substitutions are Ord. 24.

1,	$ac . dg$,	$ace . bdf$,	$ac . bh . df . eg$,	$aceg . bhdf$,
	$ae . bd$,	$aec . bfd$,	$ae . bd . cg . fh$,	$agec . bfdh$,
	$ag . bh$,	$acg . dfh$,	$ag . bf . ce . dh$,	$acge . bdfh$,
	$bf . ce$,	$agc . dhf$,		$aegc . bhfd$,
	$cg . fh$,	$aeg . bhd$,		$aecg . bfhd$,
	$eg . bh$,	$age . bdh$,		$agce . bdhf$.
		$bfh . cge$,		
		$bhf . ceg$,		

ord. 30. $(abcdefgh)_{30}$. The substitutions are Ord. 30.

1,	$abcde$,	fgh,	$abcde . fgh$,	$ab . ce . fg$,
	$acebd$,	fhg,	$abcde . fhg$,	$ac . de . fg$,
	$adbec$,		$acebd . fgh$,	$ad . bc . fg$,
	$aedcb$,		$acebd . fhg$,	$ae . bd . fg$,
			$adbec . fgh$,	$be . cd . fg$,
			$adbec . fhg$,	$ab . ce . fh$,
			$aedcb . fgh$,	$ac . de . fh$,
			$aedcb . fhg$,	$ad . bc . fh$,
				$ae . bd . fh$,
				$be . cd . fh$,
				$ab . ce . gh$,
				$ac . de . gh$,
				$ad . bc . gh$,
				$ae . bd . gh$,
				$be . cd . gh$.

Eight letters. Ord. 32.

ord. 32. Six groups $L, M, N, P, Q, R\,\{(abcd)_8\,(efgh)_8\}$ dim., the two groups $(\quad)_8$ may be dimidiated

$(abcd)_8$ into	$(efgh)_8$ into
$A,\ A'$,	$E,\ E'$,
$B,\ B'$,	$F,\ F'$,
$C,\ C'$,	$G,\ G'$;

where

$$A = (1,\ ac\ ,\ bd\ ,\ ac\,.\,bd),\quad E = (1,\ eg\ ,\ fh\ ,\ eg\,.\,fh),$$

$$B = (1,\ abcd\ ,\ ac\,.\,bd,\ adcb\ ),\quad F = (1,\ efgh\ ,\ eg\,.\,fh,\ ehgf\ ),$$

$$C = (1,\ ab\,.\,cd,\ ac\,.\,bd,\ ad\,.\,bc),\quad G = (1,\ ef\,.\,gh,\ eg\,.\,fh,\ eh\,.\,fg\ ),$$

and A', B', C', E', F', G' are the tails of the groups $(abcd)_8$ and $(efgh)_8$ respectively.

We have then

$$L\,\{(abcd)_8\,(efgh)_8\}\ \text{dim.} = AE + A'E',$$

viz. the substitutions are

$$(1,\ ac,\ bd,\ ac\,.\,bd)\,(1,\ eg,\ fh,\ eg\,.\,fh)$$

$$+\,(ab\,.\,cd,\ ad\,.\,bc,\ abcd,\ adbc)\,(ef\,.\,gh,\ eh\,.\,fg,\ efgh,\ ehfg),$$

$$M \quad " \quad = BF + B'F',$$

$$N \quad " \quad = CG + C'G',$$

$$P \quad " \quad = BG + B'G'\ (\text{or } CF + C'F'),$$

$$Q \quad " \quad = CE + C'E'\ (\text{or } AG + A'G'),$$

$$R \quad " \quad = AF + A'F'\ (\text{or } BE + B'E').$$

For the remaining groups, we have

$(ae\,.\,bf\,.\,cg\,.\,dh)\,(abcd)$ cyc. $(efgh)$ cyc. The substitutions are

+	+	−	−	+	+	−
1,	*ac . bd*,	*abcd*,	*ac . bd . efgh*,	*ac . bd . ef . gh*,	*abcd . efgh*,	*aebfcgdh*,
	ef . gh;	*adcb*,	*ac . bd . ehgf*,	*ae . bf . cg . dh*,	*abcd . ehgf*,	*aedhcgbf*,
		efgh,	*eg . fh . abcd*,	*af . bg . ch . de*,	*adcb . efgh*,	*afbgchde*,
		ehfg;	*eg . fh . adcb*;	*ag . bh . ce . df*,	*adcb . ehgf*,	*afdechbg*,
				ah . be . cf . dg;	*aecg . bfdh*,	*agbhcedf*,
					agce . bhdf,	*agdfcebh*,
					afch . bgde,	*ahbecfdg*,
					ahcf . bedg;	*ahdgcfbe*.

$(ae\,.\,bf\,.\,cg\,.\,dh)(abcd)_4(efgh)_4$. The substitutions are

Eight letters. Ord. 32.

+	+	+	+
1,	*ab . cd,*	*ab . cd . ef . gh,*	*aebf . cdgh,*
	ac . bd,	*ab . cd . eg . fh,*	*aecg . bfdh,*
	ad . bc,	*ab . cd . eh . fg,*	*aedh . bfcg,*
	ef . gh,	*ac . bd . ef . gh,*	*afbe . chdg,*
	eg . fh,	*ac . bd . eg . fh,*	*afch . bedg,*
	eh . fg;	*ac . bd . eh . fg,*	*afdg . bech,*
		ad . bc . ef . gh,	*agbh . cedf,*
		ad . bc . eg . fh,	*agce . bhdf,*
		ad . bc . eh . fg,	*agdf . bhce,*
		ae . bf . cg . dh,	*ahbg . cfde,*
		af . be . ch . dg,	*ahcf . bgde,*
		ag . bh . ce . df,	*ahde . bgcf.*
		ah . bg . cf . de;	

ord. 48. $\{(abcdef)_{48}(gh)\}$ dim. The substitutions are

Ord. 48.

					|				
1, *abcd,*	*ac . bd,*	*abe . cdf,*	*ab . cd . ef,*	$+gh$	*ac,*	*ab . cd,*	*ac . bedf,*	*ac . bd . ef;*	*abfcde,*
adcb,	*ac . ef,*	*aeb . cfd,*	*ac . be . df,*		*bd,*	*ad . bc,*	*ac . bfde,*		*abecdf,*
aecf,	*bd . ef;*	*abf . cde,*	*ac . bf . de,*		*ef;*	*ae . cf,*	*bd . aecf,*		*adecbf,*
afce,		*afb . ced,*	*ad . bc . ef,*			*af . ce,*	*bd . afce,*		*adfcbe,*
bedf,		*ade . bfc,*	*ae . bd . cf,*			*be . df,*	*ef . abcd,*		*aebcfd,*
bfde;		*aed . bcf,*	*af . bd . ce;*			*bf . de;*	*ef . adcb;*		*aedcfb,*
		adf . bec,							*afbced,*
		afd . bce;							*afdceb;*

$\{(abcdef)_{48}(gh)\}$ pos. The substitutions are

					|			
1,	*ab . cd,*	*ac . bedf,*	*abe . cdf,*	$+gh$	*ac,*	*abcd,*	*ab . cd . ef,*	*abecdf,*
	ac . bd,	*ac . bfde,*	*abf . cde,*		*bd,*	*adcb,*	*ac . bd . ef,*	*abfcde,*
	ac . ef,	*bd . aecf,*	*ade . bfc,*		*ef;*	*aecf,*	*ac . be . df,*	*adecbf,*
	ad . bc,	*bd . afce,*	*adf . cbe,*			*afce,*	*ac . bf . de,*	*adfcbe,*
	ae . cf,	*ef . abcd,*	*aeb . cfd,*			*bedf,*	*ad . bc . ef,*	*aebcfd,*
	af . ce,	*ef . adcb;*	*aed . bcf,*			*bfde;*	*ae . bd . cf,*	*aedcfb,*
	bd . ef,		*afb . ced,*				*af . bd . ce,*	*afbced,*
	be . df,		*afd . ceb;*					*afdceb;*
	bf . de;							

Eight letters. Ord. 48.

$(abcdefgh)_{48}$. The substitutions are

1,	ac . df,	ace . bdf,	ab . cf . de . gh,	ad . bchefg,	abch . defg,
	ae . bd,	acg . bdh,	ab . ch . de . fg,	ad . bgfehc,	abgf . cdeh,
	ag . bh,	aec . bfd,	ac . df . eg . bh,	be . afgdch,	aceg . bhdf,
	bf . ce,	aeg . bhd,	ad . bc . ef . gh,	be . ahcdgf,	acge . bdfh,
	bh . eg,	agc . bhd,	ad . be . cf . gh,	cf . abgdeh,	aecg . bfhd,
	cg . fh,	age . bdh,	ad . be . ch . fg,	cf . ahedgb,	aegc . bhfd,
		bhf . ceg,	ad . bg . ch . ef,	gh . abcdef,	afeh . bgdc,
		bfh . cge,	ae . bd . cg . fh,	gh . afedcb,	afgb . ched,
			af . be . cd . gh,		agce . bdhf,
			af . bg . cd . eh,		agec . bfdh,
			ag . bf . ce . dh,		ahcb . dgfe,
			ah . bc . dg . ef,		ahef . bcdg.
			ah . be . cf . dg,		

$(abcd)_{12}(efgh)_{12}$ tris. The substitutions are

1	1,	+ abc	efg,	+ acb	egf,
ab . cd	ef . gh,	adb	ehf,	abd	efh,
ac . bd	eg . fh,	bdc	fhg,	bcd	fgh,
ad . bc	eh . fg,	acd	egh,	adc	ehg.

Ord. 64.

ord. 64. $(ae . bf . cg . dh) M \{(abcd)_8 (efgh)_8\}$ dim. The substitutions are

+	−	+	+	−	−	+	−
1,	abcd,	ac . bd,	abcd . efgh,	ac . ef . gh,	ac . bd . efgh,	ac . bd . ef . gh,	aebfcgdh,
	adbc,	eg . fh,	adbc . efgh,	ac . eh . fg,	ac . bd . ehfg,	ab . cd . ef . gh,	aebhcgdf,
	efgh,	ac . eg,	abcd . ehgf,	bd . ef . gh,	eg . fh . abcd,	ab . cd . eh . fg,	aedfcgbh,
	ehgf;	ac . fh,	adbc . ehgf,	bd . eh . fg,	eg . fh . adbc;	ad . bc . ef . gh,	aedhcgbf,
		bd . eg,	aecg . bhdf,	ab . cd . eg,		ad . bc . eh . fg,	afbechdg,
		bd . fh;	agce . bhdf,	ab . cd . fh,		ae . bf . cg . dh,	afbgchde,
			aecg . bfdh,	ad . bc . eg,		ae . bh . cg . df,	afdechbg,
			agce . bfdh,	ad . bc . fh;		af . be . ch . dg,	afdgchbe,
			afch . bgde,			af . bg . ch . de,	agbfcedh,
			ahcf . bgde,			ag . bf . ce . dh,	agbhcedf,
			afch . bedg,			ag . bh . ce . df,	agdfcebh,
			ahcf . bedg;			ah . be . cf . dg,	agdhcebf,
						ah . bg . cf . de;	ahbecfdg,
							ahbgcfde,
							ahdecfbg,
							ahdgcfbe;

$(ae\,.\,bf\,.\,cg\,.\,dh)\ N\{(abcd)_8\,(efgh)_8\}$ dim. Eight letters. Ord. 64.

The substitutions (all positive) are

1,	$ab\,.\,cd\,.\,ef\,.\,gh$,	$ac\,.\,efgh$,	$abcd\,.\,efgh$,
$ab\,.\,cd$,	$ab\,.\,cd\,.\,eg\,.\,fh$,	$ac\,.\,ehgf$,	$abcd\,.\,ehgf$,
$ac\,.\,bd$,	$ab\,.\,cd\,.\,eh\,.\,fg$,	$bd\,.\,efgh$,	$adcb\,.\,efgh$,
$ad\,.\,bc$,	$ac\,.\,bd\,.\,ef\,.\,gh$,	$bd\,.\,ehgf$,	$adcb\,.\,ehgf$,
$ef\,.\,gh$,	$ac\,.\,bd\,.\,eg\,.\,fh$,	$eg\,.\,abcd$,	$aebf\,.\,cgdh$,
$eg\,.\,fh$,	$ac\,.\,bd\,.\,eh\,.\,fg$,	$eg\,.\,adbc$,	$aebh\,.\,cgdf$,
$eh\,.\,fg$,	$ad\,.\,bc\,.\,ef\,.\,gh$,	$fh\,.\,abcd$,	$aecg\,.\,bfdh$,
$ac\,.\,eg$,	$ad\,.\,bc\,.\,eg\,.\,fh$,	$fh\,.\,adcb$,	$aecg\,.\,bhdf$,
$ac\,.\,fh$,	$ad\,.\,bc\,.\,eh\,.\,fg$,		$aedf\,.\,bhcg$,
$bd\,.\,eg$,	$ae\,.\,bf\,.\,cg\,.\,dh$,		$aedh\,.\,bfcg$,
$bd\,.\,fh$,	$ae\,.\,bh\,.\,cg\,.\,df$,		$afbe\,.\,chdg$,
	$af\,.\,be\,.\,ch\,.\,dg$,		$afbg\,.\,chde$,
	$af\,.\,bg\,.\,ch\,.\,de$,		$afch\,.\,bedg$,
	$ag\,.\,bf\,.\,ce\,.\,dh$,		$afch\,.\,bgde$,
	$ag\,.\,bh\,.\,ce\,.\,df$,		$afde\,.\,bgch$,
	$ah\,.\,be\,.\,cf\,.\,dg$,		$afdg\,.\,bech$,
	$ah\,.\,bg\,.\,cf\,.\,de$,		$agbf\,.\,cedh$,
			$agbh\,.\,cedf$,
			$agce\,.\,bfdh$,
			$agce\,.\,bhdf$,
			$agdf\,.\,bhce$,
			$agdh\,.\,bfce$,
			$ahbe\,.\,cfdg$,
			$ahbg\,.\,cfde$,
			$ahcf\,.\,begd$,
			$ahcf\,.\,bgde$,
			$ahde\,.\,bgcf$,
			$ahdg\,.\,becf$.

ord. 72. $(abcdefgh)_{72}$. This group, communicated to me by Mr Askwith, may be written Ord. 72.

$$\{(abcdef)_{72}\,gh\}\ \text{pos.}$$

ord. 96. $\{(abcd)$ all $(efgh)_8$ com.$\}$ dim. This means Ord. 96.

$$(abcd)\ \text{pos.}\ (1,\ eg,\ fh,\ eg\,.\,fh) + (abcd)\ \text{neg. residue.}$$

Eight letters. Ord. 96.

$\{(abcd)$ all $(efgh)_8$ cyc.$\}$ dim. This means

$$(abcd) \text{ pos. } (1,\ efgh,\ eg\,.\,fh,\ ehgf) + (abcd) \text{ neg. residue.}$$

$\{(abcd)$ all $(efgh)_8$ pos.$\}$ dim. This means

$$(abcd) \text{ pos. } (1,\ ef\,.\,gh,\ eg\,.\,fh,\ eh\,.\,fg) + (abcd) \text{ neg. residue;}$$

viz. each form $(efgh)_8$ is divided into two sets, which are combined with $(abcd)$ pos. and $(abcd)$ neg. respectively.

ord. 96. $(ae\,.\,bf\,.\,cg\,.\,dh)\{(abcd)_{12}(efgh)_{12}\}$ tris.

The substitutions (all positive) are those of

$\{(abcd)_{12}(efgh)_{12}\}$ tris., viz.

1	1 ,	+ *abc*	*efg* ,	+ *acb*	*egf* ,
ab . cd	*ef . gh*,	*adb*	*ehf* ,	*abd*	*efh* ,
ac . bd	*eg . fh*,	*bdc*	*fhg*,	*bcd*	*fgh*,
ad . bc	*eh . fg*,	*acd*	*egh* ,	*adc*	*ehg* ,

together with the following 48 substitutions

ae . bf . cg . dh,	*aebf . cgdh*,	*ae . bgdfch*,
af . be . ch . dg,	*aecg . bfdh*,	*ae . bhcfdg*,
ag . bh . ce . df,	*aedh . bfcg*,	*af . bgcedh*,
ah . bg . cf . de;	*afbe . chdg*,	*af . bhdecg*,
	afch . bedg,	*ag . bedhcf*,
	afdg . bech,	*ag . bfchde*,
	agbh . cedf,	*ah . becgdf*,
	agce . bhdf,	*ah . bfdgce*,
	agdf . bhce,	*be . agcfdh*,
	ahbg . cfde,	*be . ahdfcg*,
	ahcf . bgde,	*bf . agdech*,
	ahde . bgcf;	*bf . ahcedg*,
		bg . aechdf,
		bg . afdhce,
		bh . aedgcf,
		bh . afcgde,
		ce . afbgdh,
		ce . ahdgbf,
		cf . aebhdg,

Eight letters. Ord. 96.

cf . agdhbe,
cg . afdebh,
cg . ahbedf,
ch . aedfbg,
ch . agbfde,
de . afbhcg,
de . agchbf,
df . aebgch,
df . ahcgeb,
dg . aecfbh,
dg . ahbfce,
dh . afcebg,
dh . agbecf.

$(ae \, . \, bg \, . \, cf \, . \, dh)$ $\{(abcd)_{12}\,(efgh)_{12}\}$ tris. The substitutions (all positive) are those of $\{(abcd)_{12}\,(efgh)_{12}\}$ tris., viz.

1	1 ,	+ *abc*	*efg* ,	+ *acb*	*egf* ,
ab . cd	*ef . gh,*	*adb*	*ehf* ,	*abd*	*efh* ,
ac . bd	*eg . fh,*	*bdc*	*fhg,*	*bcd*	*fgh* ,
ad . bc	*eh . fg,*	*acd*	*egh* ,	*adc*	*ehg* ,

together with the following 48 substitutions

ae . bf . ch . dg,	*aebf . chdg ,*
ae . bg . cf . dh,	*aebg . cfdh ,*
ae . bh . cg . df,	*aebh . cgdf ,*
af . be . cg . dh,	*aecf . bgdh,*
af . bg . ch . de,	*aecg . bhdf,*
af . bh . ce . dg,	*aech . bfdg,*
ag . be . ch . df,	*aedf . bhcg ,*
ag . bf . ce . dh,	*aedg . bfch ,*
ag . bh . cf . de,	*aedh . bgcf ,*
ah . be . cf . dg,	*afbe . cgdh ,*
ah . bf . cg . de,	*afbg . chde ,*
ah . bg . ce . df ;	*afbh . cedg ,*
	afce . bhdg ,
	afcg . bedh ,
	afch . bgde ,

Eight letters. Ord. 96.

afde . bgch ,
afdg . bhce ,
afdh . becg ,
agbe . chdf ,
agbf . cedh ,
agbh . cfde ,
agce . bfdh ,
agcf . bhde ,
agch . bedf .
agde . bhcf ,
agdf . bech ,
agdh . bfce ,
ahbe . cfdg ,
ahbf . cgde ,
ahbg . cedf ,
ahce . bgdf ,
ahcf . bedg ,
ahcg . bfde ,
ahde . bfcg ,
ahdf . bgce ,
ahdg . becf .

Ord. 168. ord. 168. $(abcdefgh)_{168}$.

This is $(1, S^2, S^4)(1, T, \ldots, T^6)(1, U, \ldots, U^7)$, where $S = bdcgef$, $T = abcdefg$, $U = ahbfgecd$; it is a derivative of $(abcdefg)_{21} = (1, S^2, S^4)(1, T, \ldots, T^6)$, see post $(abcdefgh)_{336}$.

Ord. 192. ord. 192. [Two forms not examined.]

Ord. 288. ord. 288. $(ae \,.\, bf \,.\, cg \,.\, dh)(abcd)$ pos. $(efgh)$ pos.

The substitutions are those of $(abcd)$ pos. $(efgh)$ pos., viz. the 144 substitutions

1	1,
ab . cd	*ef . gh,*
ac . bd	*eg . fh,*
ad . bc	*eh . fg,*
abc	*efg,*
acb	*egf,*
abd	*efh,*
adb	*ehf,*
acd	*egh,*
adc	*ehg,*
bcd	*fgh,*
bdc	*fhg,*

together with the following 144 substitutions, viz. **Eight letters. Ord. 288.**

ae . bf . cg . dh,	*aebf . cgdh,*	*afbe . chdg,*	*agbe . cfdh,*	*ahbe . cgdf,*
ae . bg . ch . df,	*aebg . chdf,*	*afbg . cedh,*	*agbf . chde,*	*ahbf . cedg,*
ae . bh . cf . dg,	*aebh . cfdg,*	*afbh . cgde,*	*agbh . cedf,*	*ahbg . cfde,*
af . be . ch . dg,	*aecf . bhdg,*	*afce . bgdh,*	*agce . bhdf,*	*ahce . bfdg,*
af . bg . ce . dh,	*aecg . bfdh,*	*afcg . bhde,*	*agcf . bedh,*	*ahcf . bgde,*
af . bh . cg . de,	*aech . bgdf,*	*afch . bedg,*	*agch . bfde,*	*ahcg . bedf,*
ag . bh . ce . df,	*aedf . bgch,*	*afde . bhcg,*	*agde . bfch,*	*ahde . bgcf,*
ag . be . cf . dh,	*aedg . bhcf,*	*afdg . bech,*	*agdf . bhce,*	*ahdf . becg,*
ag . bf . ch . de,	*aedh . bfcg,*	*afdh . bgce,*	*agdh . becf,*	*ahdg . bfce;*
ah . bg . cf . de,				
ah . be . cg . df,				
ah . bf . ce . dg;				

ae . bfcgdh,	*af . bechdg,*	*ag . becfdh,*	*ah . becgdf*	*ce . afbgdh,*	*cf . aebhdg,*	*cg . aebfdh,*	*ch . aebgdf,*
„ *bfdhcg,*	„ *bedgch,*	„ *bedhcf,*	„ *bedfcg*	„ *afdhbg,*	„ *aedgbh,*	„ *aedhbf,*	„ *aedfbg,*
„ *bgchdf,*	„ *bgcedh,*	„ *bfchde,*	„ *bfcedg*	„ *agbhdf,*	„ *agbedh,*	„ *afbhde,*	„ *afbedg,*
„ *bgdfch,*	„ *bgdhce,*	„ *bfdech,*	„ *bfdgce*	„ *agdfbh,*	„ *agdhbe,*	„ *afdebh,*	„ *afdgbe,*
„ *bhcfdg,*	„ *bhcdge,*	„ *bhcedf,*	„ *bgcfde*	„ *ahbfdg,*	„ *ahbgde,*	„ *ahbedf,*	„ *agbfde,*
„ *bhdgcf;*	„ *bhdecg;*	„ *bhdfce;*	„ *bgdecf;*	„ *ahdgbf;*	„ *ahdebg;*	„ *ahdfbe;*	„ *agdebf;*

be . afchdg,	*bf . aecgdh,*	*bg . aechdf,*	*bh . aecfdg*	*de . afbhcg,*	*df . aebgch,*	*dg . aebhcf,*	*dh . aebfcg,*
„ *afdgch,*	„ *aedhcg,*	„ *aedfch,*	„ *aedgcf*	„ *afcgbh,*	„ *aechbg,*	„ *aecfbh,*	„ *aecgbf,*
„ *agcfdh,*	„ *agchde,*	„ *afcedh,*	„ *afcgde*	„ *agbfch,*	„ *agbhce,*	„ *afbech,*	„ *afbgce,*
„ *agdhcf,*	„ *agdech,*	„ *afdhce,*	„ *afdecg*	„ *agchbf,*	„ *agcebh,*	„ *afchbe,*	„ *afcebg,*
„ *ahcgdf,*	„ *ahcedg,*	„ *ahcfde,*	„ *agcedf*	„ *ahbgcf,*	„ *ahbecg,*	„ *ahbfce,*	„ *agbecf,*
„ *ahdfcg;*	„ *ahdgce;*	„ *ahdecf;*	„ *agdfce;*	„ *ahcfbg;*	„ *ahcgbe;*	„ *ahcebf;*	„ *agcfbe.*

ord. 336. $(abcdefgh)_{336}$. This is **Ord. 336.**

$$(1,\ S,\ S^2,\ S^3,\ S^4,\ S^5)(1,\ T,\ \ldots,\ T^6)(1,\ U,\ \ldots,\ U^7),$$

where $S = bdcgef$, $T = abcdefg$, $U = ahbfgecd$; it is a derivative of

$$(abcdefg)_{42} = (1,\ S,\ \ldots,\ S^5)(1,\ T,\ \ldots,\ T^6).$$

It will be noticed that there are some alterations in the numbers of the groups of the several orders as stated in the Table, p. 118, and that the number here obtained for the total number of the groups of eight letters (instead of 155 as in the table) is 157. Some of the groups, in particular those of the order 192, require further explanation.

919.

ON THE PROBLEM OF TACTIONS.

[From the *Quarterly Journal of Pure and Applied Mathematics*, vol. XXV. (1891), pp. 104—127.]

1. I REMARK that the problem "to draw a circle touching each of three given circles" is not properly a problem with eight solutions, but it is a set of four problems each with two solutions: viz. if a, b, c are the radii of the given circles, ϑ the radius of the tangent circle, and r, s, t the distances of its centre from the centres of the given circles respectively, then in the four problems respectively we have

$$\begin{array}{llll} r=a+\vartheta, & r=\ \ a+\vartheta, & r=\ \ a+\vartheta, & r=\ \ a+\vartheta, \\ s=b+\vartheta, & s=-b+\vartheta, & s=\ \ b+\vartheta, & s=-b+\vartheta, \\ t=c+\vartheta, & t=\ \ c+\vartheta, & t=-c+\vartheta, & t=-c+\vartheta; \end{array}$$

and thence also

$$\begin{array}{llll} s-t=b-c, & s-t=-b-c, & s-t=\ \ b+c, & s-t=-b+c, \\ t-r=c-a, & t-r=\ \ c-a, & t-r=-c-a, & t-r=-c-a, \\ r-s=a-b, & r-s=\ \ a+b, & r-s=\ \ a-b, & r-s=\ \ a+b, \end{array}$$

where a, b, c may be regarded each of them as positive; but the sign of each distance r, s, t, and of the radius ϑ is not assumable at pleasure, but analytically it comes out as a result in the solution, or it may be found by geometrical considerations. Thus, if the given circles are external to each other, then in the first problem we have two solutions, a first tangent circle touched externally by each of the given circles, and a second tangent circle touched internally by each of the given circles; and taking r, s, t, ϑ_1, ϑ_2, each of them as positive, the signs in the two solutions respectively are

$$\begin{array}{lll} r=a+\vartheta_1, & r=-a+\vartheta_2, \text{ or say } & -r=a-\vartheta_2, \\ s=b+\vartheta_1, & s=-b+\vartheta_2, & -s=b-\vartheta_2, \\ t=c+\vartheta_1, & t=-c+\vartheta_2, & -t=c-\vartheta_2, \end{array}$$

and so in other cases. The second, third, and fourth problems are, it is clear, derived from the first problem by the change of (b, c) into $(-b, c)$, $(b, -c)$, $(-b, -c)$ respectively: so that only the first problem need be considered, viz. this is as above

$$r = a + \vartheta, \quad s = b + \vartheta, \quad t = c + \vartheta,$$

whence also

$$s - t = b - c, \quad t - r = c - a, \quad r - s = a - b,$$

where $b - c$, $c - a$, $a - b$ are given magnitudes the algebraical sum of which is $= 0$, viz. they are one of them positive, and the other two each negative, or else two of them each positive, and the remaining one negative.

2. The most simple and straightforward geometrical solution is that given in the *Principia*, Book I. Lemma XVI.; I reproduce this as given in Motte's translation (*The Mathematical Principles of Natural Philosophy*, by Sir Isaac Newton, translated into English by Andrew Motte, 8°, 2 vols., London, 1729).

"Lemma XVI. *From three given points to draw to a fourth point which is not given three right lines whose differences shall be either given or none at all.*

Case 1. Let the given points be A, B, C (see figure) and Z the fourth point which we are to find: because of the given difference of the line AZ, BZ, the locus

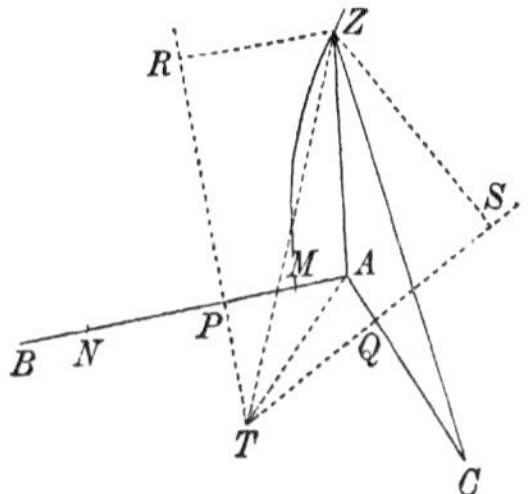

of the point Z will be a hyperbola whose foci are A and B and whose principal axe is the given difference. Let that axe be MN. Taking PM to MA as MN is to AB, erect PR perpendicular to AB, and let fall ZR perpendicular to PR; then from the nature of the hyperbola ZR will be to AZ as MN is to AB. And by the like argument the locus of the point Z will be another hyperbola whose foci are A, C, and whose principal axe is the difference between AZ and CZ; and QS a perpendicular on AC may be drawn to which (QS), if from any point Z of this hyperbola a perpendicular ZS is let fall, this (ZS) shall be to AZ as the difference between AZ and CZ is to AC. Wherefore the ratios of ZR and ZS to AZ are given and consequently the ratio of ZR to ZS one to the other: and therefore if the right lines RP, QS meet in T, and TZ and TA are drawn, the figure $TRZS$ will be given in specie, and the right line TZ, in which the point Z is somewhere placed, will be given in position. There will be given also the right line TA and the angle ATZ; and because the ratios of AZ and TZ to ZS are given, their ratio to each other is given also; and thence will be given also the triangle ATZ whose vertex is the point Z. Q. E. I.

Case 2. If two of the three lines, for instance AZ and BZ, are equal, &c.

Case 3. If all the three are equal, &c.

This problematic lemma is likewise resolved in Apollonius's Book of Tactions restored by Vieta."

3. Newton, in fact, considers the hyperbolas AB and AC, each of given axis, having the foci (A, B) and (A, C) respectively, and having PR, QS for the directrices which in the two hyperbolas respectively belong to the common focus A. The required point Z thus lies on a given line through the intersection T of these two directrices; and its position on this line is determined by the condition that the distances AZ, TZ shall be in a given ratio: the locus of the points which satisfy this last condition is of course a circle; and the position of Z is thus determined as the intersection of the given line by a given circle (which I will call a Newton-circle); there are two intersections giving points Z_1, Z_2, which are the centres of the two tangent circles respectively: and the line as a *locus in quo* of these two points is of course a determinate line, but Newton's circle is only one of a singly infinite series of circles through the two points: any other solution of the problem gives therefore the same line, but not in general the same circle.

4. In what immediately follows, I use for convenience the letter F in place of the foregoing letter T.

Effecting Newton's construction, first as above, with the points $A(B, C)$; and then in like manner, secondly with the points $B(C, A)$, and thirdly with the points $C(A, B)$; and in regard to a hyperbola AB or BA, writing AB for the directrix which belongs to the focus A, and BA for the directrix which belongs to the focus B; then

For hyperbolas AB, AC, we have intersection of directrices AB, AC is a point F; for hyperbolas BC, BA, we have intersection of directrices BC, BA is a point G; for hyperbolas CA, CB, we have intersection of directrices CA, CB is a point H.

Hence these three points F, G, H lie in a line, which is the line containing the required points Z_1, Z_2; or say it is the line Z_1Z_2.

The points Z_1, Z_2 are determined as the intersections of this line by a circle which is the locus of the points whose distances A, F are in a given ratio; similarly they are determined as the intersections by a circle which is the locus of the points whose distances from B, G are in a given ratio; and they are determined as the intersections by a circle which is the locus of the points whose distances from C, H are in a given ratio. We have thus three Newton-circles: if the centres of these are F', G', H' respectively, then clearly these points lie on a line $F'G'H'$, which bisects at right angles the line (or chord) Z_1Z_2; the points A, F, F' are obviously in a line, as are also the points B, G, G', and the points C, H, H'; or (what is the same thing) considering for a moment the line $F'G'H'$ as a given line bisecting Z_1Z_2 at right angles, then the centres F', G', H' would be found as the intersections of this line $F'G'H'$ with the lines AF, BG, CH respectively.

The points Z_1, Z_2 being determined as above, then the points of contact α_1, β_1, γ_1 of the circle Z_1 with the circles A, B, C respectively are points of intersection of the lines Z_1A, Z_1B, Z_1C with these circles respectively; and similarly the points of contact α_2, β_2, γ_2 of the circle Z_2 with the same circles respectively are points of intersection of the lines Z_2A, Z_2B, Z_2C with these circles respectively.

5. I compare with Newton's the construction in which the centres Z_1, Z_2 are determined by means of the points of contact with the given circles: I may refer to Prop. 10, pp. 118—120 of Casey's *Sequel to Euclid* (12°, Dublin, 1881). We have here the line Z_1Z_2 determined as the line through the radical centre Ω of the three circles A, B, C, perpendicular to an axis of symmetry (say the axis containing the three centres of direct symmetry) of the same circles: this point Ω is the centre of the orthotomic circle. And if the common chords of the orthotomic circle and the circles A, B, C respectively meet the axis of symmetry in the points a, b, c; then we have α_1, α_2 as the points of contact of the tangents from a to the circle A; β_1, β_2 as the points of contact of the tangents from b to the circle B; and γ_1, γ_2 as the points of contact of the tangents from c to the circle C: the suffixes 1 and 2 can and must be so applied that the three lines $A\alpha_1$, $B\beta_1$, $C\gamma_1$ meet in a point Z_1 of the line Z_1Z_2, and the three lines $A\alpha_2$, $B\beta_2$, $C\gamma_2$ in a point Z_2 of the same line. We thus obtain the required points Z_1 and Z_2.

6. Taking the equations of the three circles to be

$$(x-\alpha)^2+(y-\alpha_1)^2=a^2,$$
$$(x-\beta)^2+(y-\beta_1)^2=b^2,$$
$$(x-\gamma)^2+(y-\gamma_1)^2=c^2,$$

I wish to obtain the equations of the line Z_1Z_2 and of the three Newton-circles; but I will first find, by a separate analytical investigation, an expression for the length of the chord Z_1Z_2.

Writing f, g, h for the distances BC, CA, AB of the points A, B, C from each other; r_1, s_1, t_1 for the distances of Z_1 from these points respectively, and r_2, s_2, t_2 for the distances of Z_2 from these points respectively; we have a triangle whose sides are f, g, h, and two points Z_1, Z_2 whose distances from the vertices are r_1, s_1, t_1 and r_2, s_2, t_2 respectively; and we can in terms of these data find an expression for the distance x of the points Z_1, Z_2 from each other. In fact, considering any four points 1, 2, 3, 4 and any other four points 1′, 2′, 3′, 4′, then if 11′, 12′, &c., denote the squared distances of the points 1 and 1′ from each other, of the points 1 and 2′ from each other, &c., we have between the several distances the relation

$$\begin{vmatrix} 0, & 1\ , & 1\ , & 1\ , & 1 \\ 1, & 11', & 12', & 13', & 14' \\ 1, & 21', & 22', & 23', & 24' \\ 1, & 31', & 32', & 33', & 34' \\ 1, & 41', & 42', & 43', & 44' \end{vmatrix} = 0,$$

and thence, taking the points 1, 2, 3 and also the points 1′, 2′, 3′ to be the points A, B, C respectively, and the points 4 and 4′ to be the points Z_1 and Z_2 respectively, we have the required relation

$$\begin{vmatrix} 0, & 1, & 1, & 1, & 1 \\ 1, & 0, & h^2, & g^2, & r_1^2 \\ 1, & h^2, & 0, & f^2, & s_1^2 \\ 1, & g^2, & f^2, & 0, & t_1^2 \\ 1, & r^2, & s^2, & t^2, & x^2 \end{vmatrix} = 0,$$

viz. putting for shortness

$$\Delta = f^4 + g^4 + h^4 - 2g^2h^2 - 2h^2f^2 - 2f^2g^2,$$

this equation is

$$\Delta x^2 + \begin{vmatrix} 0, & 1, & 1, & 1, & 1 \\ 1, & 0, & h^2, & g^2, & r_1^2 \\ 1, & h^2, & 0, & f^2, & s_1^2 \\ 1, & g^2, & f^2, & 0, & t_1^2 \\ 1, & r^2, & s^2, & t^2, & 0 \end{vmatrix} = 0,$$

where the determinant has the value

$$\begin{aligned} &(r^2 + r_1^2) f^2 (-f^2 + g^2 + h^2) + f^2 \{(t_1^2 - r_1^2)(r_2^2 - s_2^2) + (t_2^2 - r_2^2)(r_1^2 - s_1^2)\} \\ &+ (s^2 + s_1^2)\, g^2 (\ \ f^2 - g^2 + h^2) + g^2 \{(r_1^2 - s_1^2)(s_2^2 - t_2^2) + (r_2^2 - s_2^2)(s_1^2 - t_1^2)\} \\ &+ (t_1^2 + t_2^2)\, h^2 (\ \ f^2 + g^2 - h^2) + h^2 \{(s_1^2 - t_1^2)(t_2^2 - r_2^2) + (s_2^2 - t_2^2)(t_1^2 - r_1^2)\} \\ &- 2f^2g^2h^2. \end{aligned}$$

7. For the points Z_1, Z_2, the distances r_1, s_1, t_1 and r_2, s_2, t_2 have the values $a + \vartheta_1$, $b + \vartheta_1$, $c + \vartheta_1$ and $a + \vartheta_2$, $b + \vartheta_2$, $c + \vartheta_2$ respectively, where ϑ_1, ϑ_2 are the radii of the tangent circles; substituting these values, we find

$$\Delta x^2 + 2\mathfrak{A} + 2\mathfrak{B}(\vartheta_1 + \vartheta_2) + \mathfrak{D}(\vartheta_1^2 + \vartheta_2^2) + 2\mathfrak{E}\vartheta_1\vartheta_2 = 0,$$

where

$$\begin{aligned} \mathfrak{A} = \ & a^2f^2(-f^2 + g^2 + h^2) + f^2(c^2 - a^2)(a^2 - b^2) - f^2g^2h^2 \\ & + b^2g^2(\ \ f^2 - g^2 + h^2) + g^2(a^2 - b^2)(b^2 - c^2) \\ & + c^2h^2(\ \ f^2 + g^2 - h^2) + h^2(b^2 - c^2)(c^2 - a^2), \\ \mathfrak{B} = \ & af^2(-f^2 + g^2 + h^2) + f^2(c - a)(a - b)(2a + b + c) \\ & + bg^2(\ \ f^2 - g^2 + h^2) + g^2(a - b)(b - c)(a + 2b + c) \\ & + ch^2(\ \ f^2 + g^2 - h^2) + h^2(b - c)(c - a)(a + b + 2c), \\ \mathfrak{D} = \ & f^2(-f^2 + g^2 + h^2) \\ & + g^2(\ \ f^2 - g^2 + h^2) \\ & + h^2(\ \ f^2 + g^2 - h^2), = -\Delta, \\ \mathfrak{E} = \ & 4f^2(c - a)(a - b) \\ & + 4g^2(a - b)(b - c) \\ & + 4h^2(b - c)(c - a). \end{aligned}$$

But ϑ_1, ϑ_2 are the roots of an equation in ϑ, which is at once obtained from the foregoing equation by putting therein $x=0$, $\vartheta_1=\vartheta_2$, $=\vartheta$: viz. if, for shortness,

$$\mathfrak{C}=\mathfrak{D}+\mathfrak{E},\ =-\Delta+4\{f^2(c-a)(a-b)+g^2(a-b)(b-c)+h^2(b-c)(c-a)\},$$

then the equation in ϑ is

$$\mathfrak{A}+2\mathfrak{B}\vartheta+\mathfrak{C}\vartheta^2=0\,;$$

we have therefore

$$\vartheta_1+\vartheta_2=-\frac{2\mathfrak{B}}{\mathfrak{C}},\quad \vartheta_1\vartheta_2=\frac{\mathfrak{A}}{\mathfrak{C}}\,;$$

and consequently

$$\Delta x^2+2\mathfrak{A}-\frac{4\mathfrak{B}^2}{\mathfrak{C}}+\mathfrak{D}\,\frac{4\mathfrak{B}^2-2\mathfrak{A}\mathfrak{C}}{\mathfrak{C}^2}+\frac{2\mathfrak{A}\mathfrak{E}}{\mathfrak{C}}=0,$$

that is,

$$\Delta\mathfrak{C}^2x^2+2\mathfrak{A}\mathfrak{C}^2-4\mathfrak{B}^2\mathfrak{C}+\mathfrak{D}(4\mathfrak{B}^2-2\mathfrak{A}\mathfrak{C})+2\mathfrak{A}\mathfrak{C}(\mathfrak{C}-\mathfrak{D})=0,$$

or, reducing,

$$\Delta\mathfrak{C}^2x^2+4(\mathfrak{B}^2-\mathfrak{A}\mathfrak{C})(\mathfrak{D}-\mathfrak{C})=0,$$

that is,

$$\Delta\mathfrak{C}^2x^2=4(\mathfrak{B}^2-\mathfrak{A}\mathfrak{C})\,\mathfrak{E},$$

where Δ, $\mathfrak{B}$, $\mathfrak{C}$, $\mathfrak{E}$ have the values given above; we hence find

$$\begin{aligned}\mathfrak{B}^2-\mathfrak{A}\mathfrak{C}&=-\{f^2-(b-c)^2\}\{g^2-(c-a)^2\}\{h^2-(a-b)^2\}\times(f^4+g^4+h^4-2g^2h^2-2h^2f^2-2f^2g^2)\\&=-\{f^2-(b-c)^2\}\{g^2-(c-a)^2\}\{h^2-(a-b)^2\}\,\Delta\,;\end{aligned}$$

and, putting for x its value ZZ_1, the equation thus reduces itself to

$$\mathfrak{C}^2(ZZ_1)^2=-4\{f^2-(b-c)^2\}\{g^2-(c-a)^2\}\{h^2-(a-b)^2\}\,\mathfrak{E}.$$

8. The denominator factor $\mathfrak{C}^2$ and the several numerator factors of $(ZZ_1)^2$ may be accounted for. It is to be observed that $(ZZ_1)^2$ does not contain the factor Δ of $\mathfrak{B}^2-\mathfrak{A}\mathfrak{C}$. If $\Delta=0$, the centres of the circles A, B, C are in a line; the two tangent circles are circles situate symmetrically in regard to the line ABC, that is, their radii are equal, and the line through their centres is bisected at right angles by the line ABC; the radii are equal, and thus $\mathfrak{B}^2-\mathfrak{A}\mathfrak{C}=0$, but the centres are not coincident, and thus ZZ_1 is not $=0$. The expression for $(ZZ_1)^2$ assumes a more simple form, for we have $\mathfrak{C}=-\Delta+\mathfrak{E}=\mathfrak{E}$, and the formula thus becomes

$$\mathfrak{E}(ZZ_1)^2=-4\{f^2-(b-c)^2\}\{g^2-(c-a)^2\}\{h^2-(a-b)^2\}.$$

If in this formula $\mathfrak{E}=0$, then we have $ZZ_1=\infty$; in fact, the circles A, B, C have here a pair of common tangents, and the tangent circles are these common tangents: each of the centres is thus a point at infinity, and the distance of the two centres is to be regarded as $=$ infinity. In verification observe that, if the circles have a pair of common tangents, then, taking the intersection of these for the origin, if P, Q, R be the distances of the centres from this origin, we have

$$P:Q:R=a:b:c\,;$$

and therefore

$$f:g:h=b-c:c-a:a-b\,;$$

whence

$$\mathfrak{E},\ =(c-a)(a-b)f^2+(a-b)(b-c)g^2+(b-c)(c-a)h^2,$$

contains a factor $(b-c)+(c-a)+(a-b)$, and is thus $=0$.

Supposing Δ not $=0$, then $\mathfrak{B}^2-\mathfrak{A}\mathfrak{C}$ is $=0$, that is, the two radii are equal only if one of the factors $f^2-(b-c)^2$, $g^2-(c-a)^2$, $h^2-(a-b)^2$ is $=0$, and in this case we have also $ZZ_1=0$; the two tangent circles are here coincident. The equation $f^2-(b-c)^2=0$ signifies that the circles B, C touch each other internally, or, what is the same thing (if, as was initially assumed, the radii b, c be taken to be each of them positive), the point of contact is a direct centre of symmetry; hence, when $f^2-(b-c)^2=0$, the two tangent circles are coincident. And similarly if $g^2-(c-a)^2=0$, or if $h^2-(a-b)^2=0$.

9. If in the general formula we have $\mathfrak{C}=0$, then $ZZ_1=\infty$; the circles A, B, C have here a common tangent and one of the tangent circles becomes this common tangent; we have thus a centre at infinity, and the distance of the two centres is thus also infinite.

To show that $\mathfrak{C}=0$ is the condition of a common tangent, suppose that the three circles have a common tangent, and let P, Q, R denote the distances of the points of contact from any fixed point on this tangent: we have

$$f^2=(Q-R)^2+(b-c)^2,$$
$$g^2=(R-P)^2+(c-a)^2,$$
$$h^2=(P-Q)^2+(a-b)^2,$$

equations which I represent by f^2, g^2, $h^2=\alpha^2+l^2$, β^2+m^2, γ^2+n^2, where $\alpha+\beta+\gamma=0$, $l+m+n=0$. We have

$$-f^2+g^2+h^2=-\alpha^2+\beta^2+\gamma^2-l^2+m^2+n^2, =-2\beta\gamma-2mn,$$

and forming the corresponding values of $f^2-g^2+h^2$, $f^2+g^2-h^2$, and multiplying by f^2, g^2, $h^2=\alpha^2+l^2$, β^2+m^2, γ^2+n^2 respectively, we find

$$-\Delta=-2\alpha^2mn-2\beta^2nl-2\gamma^2lm-2l^2\beta\gamma-2m^2\gamma\alpha-2n^2\alpha\beta.$$

But the last three terms hereof are

$$=l^2(-\alpha^2+\beta^2+\gamma^2)+m^2(-\beta^2+\gamma^2+\alpha^2)+n^2(-\gamma^2+\alpha^2+\beta^2),$$

which is

$$=\alpha^2(m^2+n^2-l^2)+\beta^2(n^2+l^2-m^2)+\gamma^2(l^2+m^2-n^2),$$

and this is

$$=-2\alpha^2mn-2\beta^2nl-2\gamma^2lm.$$

Hence we have

$$-\Delta=-4\alpha^2mn-4\beta^2nl-4\gamma^2lm,$$

that is,

$$=-4f^2mn-4g^2nl-4h^2lm,$$

or replacing l, m, n by their values, we find

$$-\Delta=-4\{f^2(c-a)(a-b)+g^2(a-b)(b-c)+h^2(b-c)(c-a)\},$$

which is the required condition $\mathfrak{C}=0$.

10. It would at first sight appear that the distance ZZ_1 of the two centres

would always vanish if $\mathfrak{C}=0$. But if A, B, C are real circles, this condition $\mathfrak{C}=0$, implies

$$\frac{f^2}{(b-c)^2}=\frac{g^2}{(c-a)^2}=\frac{h^2}{(a-b)^2},$$

whence $\Delta=0$, and this being so we have $\mathfrak{C}=-\Delta+\mathfrak{C}$, $=0$, and the value of ZZ_1 instead of being $=0$, is or appears to be infinite. In proof, take for a moment the origin at A and the line AB for the axis of x; we have thus $(0, h)$ for the coordinates of B, and taking (x, y) for the coordinates of $\mathfrak{C}$, we have $g^2=x^2+y^2$; $f^2=(h-x)^2+y^2$. Writing as before l, m, n, to denote $b-c$, $c-a$, $a-b$ respectively, we have

$$\begin{aligned}\tfrac{1}{4}\mathfrak{C} &= mn\{(h-x)^2+y^2\}+nl(x^2+y^2)+lmh^2,\\ &= m(n+l)h^2+n(l+m)(x^2+y^2)-2mnhx,\\ &=-m^2h^2-n^2(x^2+y^2)-2mnhx,\\ &=-(mh+nx)^2-n^2y^2,\end{aligned}$$

and thus, for real values, $\mathfrak{C}$ can only vanish for $y=0$, $x=-\frac{mh}{n}$; these values of x, y give $f^2=\frac{l^2h^2}{n^2}$, $g^2=\frac{m^2h^2}{n^2}$, that is, $\frac{f^2}{l^2}=\frac{g^2}{m^2}=\frac{h^2}{n^2}$, or writing for l, m, n their values, they give

$$\frac{f^2}{(b-c)^2}=\frac{g^2}{(c-a)^2}=\frac{h^2}{(a-b)^2}.$$

11. But for imaginary circles the condition $\mathfrak{C}=0$ does not imply $\Delta=0$, and supposing $\mathfrak{C}=0$, the distance Z_1Z_2 is $=0$; the equation $\mathfrak{B}^2-\mathfrak{A}\mathfrak{C}=0$, is not satisfied, and thus the two radii are unequal; it would seem that we have *concentric* circles Z_1, Z_2 each touching the three given circles A, B, C, and this would imply that the radii a, b, c were equal to each other: this cannot be the case, for the only relation is that given by the foregoing condition $\mathfrak{C}=0$. The explanation of this paradox is that the two circles Z_1, Z_2 are not really concentric, but it is only the distance Z_1Z_2 of the centres which is $=0$, viz. the centres are points on an imaginary line $x-\alpha\pm i(y-\beta)=0$.

In verification hereof, I start from two circles Z_1, Z_2,

$$(x+1)^2+(y+i)^2=m^2,$$
$$(x-1)^2+(y-i)^2=n^2,$$

having for centres the two points $(-1, -i)$, $(1, i)$ the distance of which two points from each other is $=0$. Consider for a moment a conic having these two imaginary points for its foci; viz. writing ξ, η for the coordinates of a point of the conic, the equation is

$$\surd\{(\xi+1)^2+(\eta+i)^2\}-\surd\{(\xi-1)^2+(\eta-i)^2\}=m-n;$$

we thence obtain

$$(\xi+1)^2+(\eta+i)^2=(m-n)^2+2(m-n)\surd\{(\xi-1)^2+(\eta-i)^2\}+(\xi-1)^2+(\eta-i)^2,$$

that is,

$$4(\xi+i\eta)-(m-n)^2=2(m-n)\surd\{(\xi-1)^2+(\eta-i)^2\},$$

or putting $m - n = 2k$, we have

$$\xi + i\eta - k^2 = k\sqrt{\{(\xi - 1)^2 + (\eta - i)^2\}},$$

and thence

$$k^2(\xi^2 + \eta^2) - (\xi + i\eta)^2 = k^4,$$

for the equation of the conic. The last preceding equation gives

$$\sqrt{\{(\xi - 1)^2 + (\eta - i)^2\}} = -k + \frac{\xi + i\eta}{k}, \; = -\tfrac{1}{2}(m - n) + \frac{\xi + i\eta}{k},$$

or say

$$\sqrt{\{(\xi - 1)^2 + (\eta - i)^2\}} - n = -\tfrac{1}{2}(m + n) + \frac{\xi + i\eta}{k};$$

and we have similarly

$$\sqrt{\{(\xi + 1)^2 + (\eta + i)^2\}} - m = -\tfrac{1}{2}(m + n) + \frac{\xi + i\eta}{k}.$$

This being so, it at once appears that, if (ξ, η) are coordinates of a point on the conic, then the circle

$$(x - \xi)^2 + (y - \eta)^2 = a^2,$$

where

$$a = -\tfrac{1}{2}(m + n) + \frac{\xi + i\eta}{k},$$

is a circle touching each of the given circles Z_1, Z_2. In fact, the distance of the centre from the point Z_1 is $\sqrt{\{(\xi + 1)^2 + (\eta + i)^2\}}$, which is $= a + m$, the sum of the two radii; and similarly the distance of the centre from the point Z_2 is $\sqrt{\{(\xi - 1)^2 + (\eta - i)^2\}}$, which is $= a + n$, the sum of the two radii.

Hence if (ξ', η'), (ξ'', η'') belong to any other two points on the conic, and we write

$$a = -\tfrac{1}{2}(m + n) + \frac{\xi + i\eta}{k},$$

$$b = -\tfrac{1}{2}(m + n) + \frac{\xi' + i\eta'}{k},$$

$$c = -\tfrac{1}{2}(m + n) + \frac{\xi'' + i\eta''}{k},$$

we have

$$(x - \xi)^2 + (y - \eta)^2 = a^2,$$

$$(x - \xi')^2 + (y - \eta')^2 = b^2,$$

$$(x - \xi'')^2 + (y - \eta'')^2 = c^2,$$

for the equations of three circles A, B, C each touching the two circles Z_1, Z_2. Writing as before f, g, h for the mutual distances BC, CA, AB of the centres of these circles, then

$$f^2 = (\xi' - \xi'')^2 + (\eta' - \eta'')^2,$$

and similarly for g^2 and h^2. But we have

$$b - c = \frac{1}{k}\{(\xi' - \xi'') + i(\eta' - \eta'')\},$$

and therefore

$$\frac{f^2}{b-c} = k\{\xi' - \xi'' - i(\eta' - \eta'')\};$$

and similarly

$$\frac{g^2}{c-a} = k\{\xi'' - \xi - i(\eta'' - \eta)\},$$

$$\frac{h^2}{a-b} = k\{\xi - \xi' - i(\eta - \eta')\},$$

and hence

$$\frac{f^2}{b-c} + \frac{g^2}{c-a} + \frac{h^2}{a-b} = 0, \text{ that is, } \mathfrak{G} = 0,$$

viz. it thus appears that the condition $\mathfrak{G} = 0$ applies to a pair of circles Z_1, Z_2 which are not concentric, but which have for their centres two imaginary points the distance of which from each other is $= 0$.

This completes the explanation of the denominator and numerator factors in the expression for the distance Z_1Z_2 between the centres of the two tangent circles.

12. I consider now the analytical solution: the equations of the given circles A, B, C are

$$(X-\alpha)^2 + (Y-\alpha_1)^2 - a^2 = 0,$$

$$(X-\beta)^2 + (Y-\beta_1)^2 - b^2 = 0,$$

$$(X-\gamma)^2 + (Y-\gamma_1)^2 - c^2 = 0,$$

viz. (α, α_1), (β, β_1), and (γ, γ_1) are the coordinates of the centres and (a, b, c) are the radii. Taking (x, y) for the coordinates of the centre of the tangent circle and ϑ for its radius, the equation of the tangent circle is

$$(X-x)^2 + (Y-y)^2 - \vartheta^2 = 0;$$

and if we write r, s, t for the distances of this centre from the points A, B, C respectively, that is,

$$r = \sqrt{\{(x-\alpha)^2 + (y-\alpha_1)^2\}},$$

$$s = \sqrt{\{(x-\beta)^2 + (y-\beta_1)^2\}},$$

$$t = \sqrt{\{(x-\gamma)^2 + (y-\gamma_1)^2\}};$$

then for the determination of the unknown quantities x, y, ϑ we have the three equations

$$r = a + \vartheta, \quad s = b + \vartheta, \quad t = c + \vartheta,$$

or eliminating ϑ, the centre is determined by means of the hyperbolas

$$s - t = b - c, \quad t - r = c - a, \quad r - s = a - b;$$

these three hyperbolas have, in fact, two common intersections which are the two centres Z_1, Z_2.

In all that follows, I write, as before, $b-c$, $c-a$, $a-b=l$, m, n; the last-mentioned equations are therefore

$$s-t=l,\quad t-r=m,\quad r-s=n,$$

and we deduce

$$r=\tfrac{1}{2}n+\frac{r^2-s^2}{2n}=-\tfrac{1}{2}m+\frac{t^2-r^2}{2m},$$

$$s=\tfrac{1}{2}l+\frac{s^2-t^2}{2l}=-\tfrac{1}{2}n+\frac{r^2-s^2}{2n},$$

$$t=\tfrac{1}{2}m+\frac{t^2-r^2}{2m}=-\tfrac{1}{2}l+\frac{s^2-t^2}{2l},$$

viz. writing

$$R=\frac{s^2-t^2}{2l},\quad S=\frac{t^2-r^2}{2m},\quad T=\frac{r^2-s^2}{2n};$$

and therefore

$$lR+mS+nT=0,$$

these equations are

$$r=S-\tfrac{1}{2}m,\quad s=T-\tfrac{1}{2}n,\quad t=R-\tfrac{1}{2}l,$$

$$=T+\tfrac{1}{2}n,\quad =R+\tfrac{1}{2}l,\quad =S+\tfrac{1}{2}m;$$

R, S, T are each of them a linear function of the coordinates (x, y), say we have

$$R=\lambda x+\lambda_1 y+\lambda_2,$$

$$S=\mu x+\mu_1 y+\mu_2,$$

$$T=\nu x+\nu_1 y+\nu_2,$$

where

$$\lambda,\ \lambda_1,\ \lambda_2=-\frac{\beta-\gamma}{l},\quad -\frac{\beta_1-\gamma_1}{l},\quad \frac{\beta^2+\beta_1^2-\gamma^2-\gamma_1^2}{2l},$$

$$\mu,\ \mu_1,\ \mu_2=-\frac{\gamma-\alpha}{m},\quad -\frac{\gamma_1-\alpha_1}{m},\quad \frac{\gamma^2+\gamma_1^2-\alpha^2-\alpha_1^2}{2m},$$

$$\nu,\ \nu_1,\ \nu_2=-\frac{\alpha-\beta}{n},\quad -\frac{\alpha_1-\beta_1}{n},\quad \frac{\alpha^2+\alpha_1^2-\beta^2-\beta_1^2}{2n}.$$

13. From the two equations $r=S-\frac{1}{2}m=T+\frac{1}{2}n$, we deduce the equations of a line and a circle.

The line is $S-T-\frac{1}{2}(m+n)=0$, viz. substituting for S and T their values, this is

$$\frac{t^2-r^2}{m}-\frac{r^2-s^2}{n}-(m+n)=0,$$

that is,

$$n(t^2-r^2)-m(r^2-s^2)-mn(m+n)=0,$$

or, since $l+m+n=0$, the equation is

$$lr^2+ms^2+nt^2+lmn=0,$$

which is symmetrical in regard to the three circles. The equation may be written

$$l(r^2 - a^2) + m(s^2 - b^2) + n(t^2 - c^2) = 0;$$

and it thus appears that the line passes through the radical centre of the three circles.

We have

$$(\nu - \mu) r = \nu (S - \tfrac{1}{2}m) - \mu (T + \tfrac{1}{2}n) = -(\mu\nu_1 - \mu_1\nu) y + \nu\mu_2 - \mu\nu_2 - \tfrac{1}{2}(m\nu + n\mu),$$

$$(\nu_1 - \mu_1) r = \nu_1 (S - \tfrac{1}{2}m) - \mu_1 (T + \tfrac{1}{2}n) = (\mu\nu_1 - \mu_1\nu) x + \nu\mu_2 - \mu\nu_2 - \tfrac{1}{2}(m\nu_1 + n\mu_1),$$

and thence

$$\{(\nu - \mu)^2 + (\nu_1 - \mu_1)^2\} r^2 = \{\nu (S - \tfrac{1}{2}m) - \mu (T + \tfrac{1}{2}n)\}^2 + \{\nu_1 (S - \tfrac{1}{2}m) - \mu_1 (T + \tfrac{1}{2}n)\}^2,$$

which is the equation of a circle; in fact, on the left-hand side and right-hand side the only terms of the second order in (x, y) are $\{(\nu - \mu)^2 + (\nu_1 - \mu_1)^2\}(x^2 + y^2)$ and $(\mu\nu_1 - \mu_1\nu)^2(x^2 + y^2)$ respectively. We have thus the equation of the Newton-circle F; but I reduce the form by substituting for $\mu, \mu_1, \mu_2, \nu, \nu_1, \nu_2$ their values. Writing for shortness

$$l\alpha + m\beta + n\gamma = K,$$

$$l\alpha_1 + m\beta_1 + n\gamma_1 = K_1,$$

$$\beta\gamma_1 - \beta_1\gamma + \gamma\alpha_1 - \gamma_1\alpha + \alpha\beta_1 - \alpha_1\beta = \Omega,$$

$$(\beta - \gamma)(\alpha^2 + \alpha_1^2) + (\gamma - \alpha)(\beta^2 + \beta_1^2) + (\alpha - \beta)(\gamma^2 + \gamma_1^2) = \Pi,$$

$$(\beta_1 - \gamma_1)(\alpha^2 + \alpha_1^2) + (\gamma_1 - \alpha_1)(\beta^2 + \beta_1^2) + (\alpha_1 - \beta_1)(\gamma^2 + \gamma_1^2) = \Pi_1,$$

after some easy reductions the equation is found to be

$$\begin{aligned} 4(K^2 + K_1^2)(x^2 + y^2 - 2\alpha x - 2\alpha_1 y + \alpha^2 + \alpha_1^2) \\ = \{2\Omega y + \Pi + (\beta - \gamma) mn + (m - n) K\}^2 \\ + \{-2\Omega x + \Pi_1 + (\beta_1 - \gamma_1) mn + (m - n) K_1\}^2. \end{aligned}$$

14. To further abbreviate, I write

$$(\beta - \gamma) mn + (m - n) K = F, \quad (\beta_1 - \gamma_1) mn + (m - n) K_1 = F_1,$$

$$(\gamma - \alpha) nl + (n - l) K = G, \quad (\gamma_1 - \alpha_1) nl + (n - l) K_1 = G_1,$$

$$(\alpha - \beta) lm + (l - m) K = H, \quad (\alpha_1 - \beta_1) lm + (l - m) K_1 = H_1;$$

also

$$l(\alpha^2 + \alpha_1^2) + m(\beta^2 + \beta_1^2) + n(\gamma^2 + \gamma_1^2) = \Theta;$$

and then writing down the three equations, we have

$$4(K^2 + K_1^2) r^2 = (-2\Omega x + \Pi_1 + F_1)^2 + (2\Omega y + \Pi + F)^2,$$

$$4(K^2 + K_1^2) s^2 = (-2\Omega x + \Pi_1 + G_1)^2 + (2\Omega y + \Pi + G)^2,$$

$$4(K^2 + K_1^2) t^2 = (-2\Omega x + \Pi_1 + H_1)^2 + (2\Omega y + \Pi + H)^2,$$

which are the equations of the three Newton-circles, each meeting the chord

$$lr^2 + ms^2 + nt^2 + lmn = 0,$$

or, say

$$-2Kx - 2K_1 y + \Theta + lmn = 0,$$

in the points Z_1, Z_2.

15. The first of these equations is

$$\begin{aligned}&4(K^2+K_1^2-\Omega^2)(x^2+y^2)\\ &\qquad -2\{4(K^2+K_1^2)\alpha-2\Omega(\Pi_1+F_1)\}\,x\\ &\qquad -2\{4(K^2+K_1^2)\alpha_1+2\Omega(\Pi+F)\}\,y\\ &\qquad +4(K^2+K_1^2)(\alpha^2+\alpha_1^2)-(\Pi_1+F_1)^2-(\Pi+F)^2=0,\end{aligned}$$

that is,

$$\begin{aligned}&\{4(K^2+K_1^2-\Omega^2)x-4(K^2+K_1^2)\alpha+2\Omega(\Pi_1+F_1)\}^2\\ &+\{4(K^2+K_1^2-\Omega^2)y-4(K^2+K_1^2)\alpha_1-2\Omega(\Pi+F)\}^2\\ &+4(K^2+K_1^2-\Omega^2)\{4(K^2+K_1^2)(\alpha^2+\alpha_1^2)-(\Pi_1+F_1)^2-(\Pi+F)^2\}\\ &-\{4(K^2+K_1^2)\alpha-2\Omega(\Pi_1+F_1)\}^2\\ &-\{4(K^2+K_1^2)\alpha_1+2\Omega(\Pi+F)\}^2=0,\end{aligned}$$

where the last term is

$$\begin{aligned}=\;&16(K^2+K_1^2)^2(\alpha^2+\alpha_1^2)\\ &-4(K^2+K_1^2)(\Pi_1+F_1)^2\\ &-4(K^2+K_1^2)(\Pi+F)^2\\ &-16(K^2+K_1^2)(\alpha^2+\alpha_1^2)\Omega^2\\ &-16(K^2+K_1^2)^2(\alpha^2+\alpha_1^2)\\ &+16(K^2+K_1^2)\alpha\Omega(\Pi_1+F_1)\\ &-16(K^2+K_1^2)\alpha_1\Omega(\Pi+F)\\ =\;&(K^2+K_1^2)\{-4(\Pi_1+F_1)^2-4(\Pi+F)^2\\ &\qquad -16(\alpha^2+\alpha_1^2)\Omega^2+16\alpha\Omega(\Pi_1+F_1)-16\alpha_1\Omega(\Pi+F)\}.\end{aligned}$$

It thus appears that the equation of the Newton-circle F is

$$\begin{aligned}&4(K^2+K_1^2-\Omega^2)^2\{(x-\mathrm{f})^2+(y-\mathrm{f}_1)^2\}\\ &\quad=(K^2+K_1^2)\{(\Pi_1+F_1)^2+(\Pi+F)^2\\ &\qquad\quad -4\alpha\Omega(\Pi_1+F_1)+4\alpha_1\Omega(\Pi+F)+4(\alpha^2+\alpha_1^2)\Omega^2\}\\ &\quad=(K^2+K_1^2)\{(\Pi_1+F_1-2\alpha\Omega)^2+(\Pi+F+2\alpha_1\Omega)^2\},\end{aligned}$$

where the coordinates of the centre are

$$\mathrm{f}=\frac{2(K^2+K_1^2)\alpha-\Omega(\Pi_1+F_1)}{2(K^2+K_1^2-\Omega^2)},$$

$$\mathrm{f}_1=\frac{2(K^2+K_1^2)\alpha_1+\Omega(\Pi+F)}{2(K^2+K_1^2-\Omega^2)},$$

and

$$\mathrm{rad}^2=\frac{K^2+K_1^2}{4(K^2+K_1^2-\Omega^2)^2}\{(\Pi_1+F_1-2\alpha\Omega)^2+(\Pi+F+2\alpha_1\Omega)^2\};$$

and similarly for the Newton-circles G and H.

16. The centres are in a line at right angles to

$$-2Kx-2K_1y+\Theta+lmn=0,$$

say the equation of this line is

$$K_1x-\ Ky\ +\Psi=0;$$

then we ought to have

$$\frac{K_1\{2(K^2+K_1^2)\alpha-\Omega(\Pi_1+F_1)\}-K\{2(K^2+K_1^2)\alpha_1+\Omega(\Pi+F)\}}{2(K^2+K_1^2-\Omega^2)}+\Psi=0,$$

that is,

$$2(K^2+K_1^2)(K_1\alpha\ -K\alpha_1)-(K_1\Pi_1+K\Pi)\Omega-(K_1F_1+KF)\Omega+2(K^2+K_1^2-\Omega^2)\Psi-0.$$

This should agree with

$$2(K^2+K_1^2)(K_1\beta-K\beta_1)-(K_1\Pi_1+K\Pi)\Omega-(K_1G_1+KG)\Omega+2(K^2+K_1^2-\Omega^2)\Psi=0,$$

viz. we ought to have

$$2(K^2+K_1^2)\{K_1(\alpha-\beta)-K(\alpha_1-\beta_1)\}-\{K_1(F_1-G_1)+K(F-G)\}\Omega=0.$$

This can be true only if $K_1(\alpha-\beta)-K(\alpha_1-\beta_1)$ is a multiple of Ω; and, in fact,

$$\begin{aligned}&K_1(\alpha-\beta)-K(\alpha_1-\beta_1)\\&\quad=(\alpha-\beta)(l\alpha_1+m\beta_1+n\gamma_1)-(\alpha_1-\beta_1)(l\alpha+m\beta+n\gamma)\\&\quad=l(\alpha\beta_1-\alpha_1\beta)+m(\alpha\beta_1-\alpha_1\beta)+n\{-(\beta\gamma_1-\beta_1\gamma)-(\gamma\alpha_1-\gamma_1\alpha)\}\\&\quad=-n(\beta\gamma_1-\beta_1\gamma+\gamma\alpha_1-\gamma_1\alpha+\alpha\beta_1-\alpha_1\beta),\ =-n\Omega.\end{aligned}$$

The equation to be verified thus is

$$2n(K^2+K_1^2)-K_1(F_1-G_1)-K(F-G)=0,$$

and here

$$\begin{aligned}F-G&=(\beta-\gamma)mn-(\gamma-\alpha)ln+(l+m-2n)K\\&=\alpha ln+\beta mn-\gamma(l+m)n-3nK=(n-3n)K=-2nK.\end{aligned}$$

Hence

$$-K\ (F\ -G\ )=2nK^2,$$

and similarly

$$-K_1(F_1-G_1)=2nK_1^2;$$

and thus the equation is verified.

17. Writing for shortness

$$\beta\gamma_1-\beta_1\gamma,\quad\gamma\alpha_1-\gamma_1\alpha,\quad\alpha\beta_1-\alpha_1\beta=X,\ Y,\ Z,$$

and therefore $\Omega=X+Y+Z$; we have

$$\begin{aligned}&2(K^2+K_1^2)(K_1\alpha-K\alpha_1)-(KF+K_1F_1)\Omega\\&\quad=\ K\{mn(\beta-\gamma)(\ X-Y-Z)\\&\quad+\quad nl\ (\gamma-\alpha)(-X+Y-Z)\\&\quad+\quad lm\ (\alpha-\beta)(-X-Y+Z)\}\\&\quad+K_1\{mn(\beta_1-\gamma_1)(\ X-Y-Z)\\&\quad+\quad nl\ (\gamma_1-\alpha_1)(-X+Y-Z)\\&\quad+\quad lm\ (\alpha_1-\beta_1)(-X-Y+Z)\}.\end{aligned}$$

To verify this, observe that the left-hand side is

$$2(K^2+K_1^2)(mZ-nY)-\{(\beta-\gamma)mnK+(\beta_1-\gamma_1)mnK_1+(m-n)(K^2+K_1^2)\Omega\},$$

or putting herein $X+Y+Z$ for Ω, this is

$$(K^2+K_1^2)\{(n-m)X+lY-lZ\}-mn\{(\beta-\gamma)K+(\beta_1-\gamma_1)K_1\}(X+Y+Z),$$

which is thus

$$\begin{aligned}= \;& mn\{(\beta-\gamma)K+(\beta_1-\gamma_1)K_1\}(\;X-Y-Z)\\ &+nl\;\{(\gamma-\alpha)K+(\gamma_1-\alpha_1)K_1\}(-X+Y-Z)\\ &+lm\;\{(\alpha-\beta)K+(\alpha_1-\beta_1)K_1\}(-X-Y+Z).\end{aligned}$$

The equation to be verified thus becomes

$$\begin{aligned}&(K^2+K_1^2)\{(n-m)X+lY-lZ\}\\ &\quad= mn\{(\beta-\gamma)K+(\beta_1-\gamma_1)K_1\}\,2X\\ &\quad\quad+nl\;\{(\gamma-\alpha)K+(\gamma_1-\alpha_1)K_1\}(-X+Y-Z)\\ &\quad\quad+lm\;\{(\alpha-\beta)K+(\alpha_1-\beta_1)K_1\}(-X-Y+Z).\end{aligned}$$

This breaks up into two equations,

$$\begin{aligned}K\{(n-m)X+lY-lZ\}= \;& mn(\beta-\gamma)\,2X\\ &+nl\;(\gamma-\alpha)(-X+Y-Z)\\ &+lm\,(\alpha-\beta)(-X-Y+Z);\end{aligned}$$

and a like equation with the suffixed letters. And the equation just written down, observing that each side is a linear function of X and $Y-Z$, again breaks up into the two equations

$$\begin{aligned}(n-m)K&=2mn(\beta-\gamma)-nl(\gamma-\alpha)-lm(\alpha-\beta),\\ lK&=\qquad\qquad\quad\;\; -nl(\gamma-\alpha)-lm(\alpha-\beta),\end{aligned}$$

which are at once verified: in fact, for the first equation the right-hand side is

$$\begin{aligned}&=(nl-lm)\;\alpha+(2mn+lm)\beta+(-2mn-nl)\gamma,\\ &=(n-m)\;l\alpha+\;(2n+l)m\beta+(-2m\;-l)n\gamma,\\ &=(n-m)(l\alpha+\qquad m\beta+\qquad n\gamma),\;=(n-m)K;\end{aligned}$$

and similarly in the second equation the right-hand side is

$$(-nl-lm)\alpha+lm\beta+ln\gamma,\quad=l(l\alpha+m\beta+n\gamma),\quad=lK.$$

Writing then

$$\begin{aligned}\Phi=\;& K\;\{mn(\beta\;-\;\gamma)(X-Y-Z)\\ &+nl(\gamma\;-\alpha\;)(-X+Y-Z)+lm(\alpha\;-\beta\;)(-X-Y+Z)\}\\ &+K_1\{mn(\beta_1-\gamma_1)(X-Y-Z)\\ &+nl(\gamma_1-\alpha_1)(-X+Y-Z)+lm(\alpha_1-\beta_1)(-X-Y+Z)\},\end{aligned}$$

we have

$$\Phi=(K_1\Pi_1+K\Pi)\Omega-2(K^2+K_1^2-\Omega^2)\Psi.$$

This equation determines Φ, and thus the equation of the line of centres is

$$2(K^2+K_1^2-\Omega^2)(K_1x-Ky)+(K\Pi+K_1\Pi_1)\Omega-\Phi=0.$$

18. This line meets

$$-2Kx-2K_1y+\Theta+lmn=0,$$

in the mid-point of the chord Z_1Z_2. We thus have for the coordinates x, y of this mid-point

$$2(K^2+K_1^2-\Omega^2)(K^2+K_1^2)x+K_1\{(K\Pi+K_1\Pi_1)\Omega-\Phi\}-K\,(K^2+K_1^2-\Omega^2)(\Theta+lmn)=0,$$

$$2(K^2+K_1^2-\Omega^2)(K^2+K_1^2)y-K\,\{(K\Pi+K_1\Pi_1)\Omega-\Phi\}-K_1(K^2+K_1^2-\Omega^2)(\Theta+lmn)=0.$$

The perpendicular distance of the centre of the circle F from the chord is

$$=\frac{-2Kf-2K_1f_1+\Theta+lmn}{2\sqrt{(K^2+K_1^2)}};$$

here

$$2(Kf+K_1f_1)=\frac{1}{K^2+K_1^2-\Omega^2}[2(K^2+K_1^2)(K\alpha+K_1\alpha_1)-\Omega\{K(\Pi_1+F_1)-K_1(\Pi+F)\}],$$

$$\begin{aligned}K\Pi_1-K_1\Pi=&\ (\alpha^2+\alpha_1^2)\{K(\beta_1-\gamma_1)-K_1(\beta-\gamma)\}\\&+(\beta^2+\beta_1^2)\{K(\gamma_1-\alpha_1)-K_1(\gamma-\alpha)\}\\&+(\gamma^2+\gamma_1^2)\{K(\alpha_1-\beta_1)-K_1(\alpha-\beta)\}\\=&\ \Omega\{l(\alpha^2+\alpha_1^2)+m(\beta^2+\beta_1^2)+n(\gamma^2+\gamma_1^2)\}=\Omega\Theta,\\KF_1-K_1F=&\ mn\{K(\beta_1-\gamma_1)-K_1(\beta-\gamma)\}=lmn\Omega.\end{aligned}$$

Thus

$$2(Kf+K_1f_1)=\frac{1}{K^2+K_1^2-\Omega^2}\{2(K^2+K_1^2)(K\alpha+K_1\alpha_1)-\Omega^2(\Theta+lmn)\},$$

and hence the numerator of the fraction is

$$\frac{1}{K^2+K_1^2-\Omega^2}\{-2(K^2+K_1^2)(K\alpha+K_1\alpha_1)+\Omega^2(\Theta+lmn)+(K^2+K_1^2-\Omega^2)(\Theta+lmn)\}$$

$$=\frac{1}{K^2+K_1^2-\Omega^2}(K^2+K_1^2)\{-2(K\alpha+K_1\alpha_1)+\Theta+lmn\}.$$

Thus the perpendicular distance of the centre of the circle F from the chord Z_1Z_2 is

$$=\frac{\sqrt{(K^2+K_1^2)}}{2(K^2+K_1^2-\Omega^2)}\{-2(K\alpha+K_1\alpha_1)+\Theta+lmn\};$$

moreover, by what precedes, we have

$$\text{Radius}=\frac{\sqrt{(K^2+K_1^2)}}{2(K^2+K_1^2-\Omega^2)}\sqrt{\{(\Pi+F+2\alpha_1\Omega)^2+(\Pi_1+F_1-2\alpha\Omega)^2\}}.$$

19. Hence also

$$(Z_1Z_2)^2=\frac{K^2+K_1^2}{(K^2+K_1^2-\Omega^2)^2}[(\Pi+F+2\alpha_1\Omega)^2+(\Pi_1+F_1-2\alpha\Omega)^2-\{\Theta+lmn-2(K\alpha+K_1\alpha_1)\}^2].$$

We have

$$K^2+K_1^2-\Omega^2=(l\alpha+m\beta+n\gamma)^2+(l\alpha_1+m\beta_1+n\gamma_1)^2-(\beta\gamma_1-\beta_1\gamma+\gamma\alpha_1-\gamma_1\alpha+\alpha\beta_1-\alpha_1\beta)^2;$$

but

$$f^2=(\beta-\gamma)^2+(\beta_1-\gamma_1)^2,$$
$$g^2=(\gamma-\alpha)^2+(\gamma_1-\alpha_1)^2,$$
$$h^2=(\alpha-\beta)^2+(\alpha_1-\beta_1)^2,$$

and hence

$$-f^2+g^2+h^2=-2(\gamma-\alpha)(\alpha-\beta)-2(\gamma_1-\alpha_1)(\alpha_1-\beta_1),$$

whence

$$-f^2(-f^2+g^2+h^2)=\{2(\gamma-\alpha)(\alpha-\beta)+2(\gamma_1-\alpha_1)(\alpha_1-\beta_1)\}\{(\beta-\gamma)^2+(\beta_1-\gamma_1)^2\};$$

or forming the like values of $-g^2(f^2-g^2+h^2)$ and $-h^2(f^2+g^2-h^2)$ and adding, we have

$$\begin{aligned}\Delta&=f^4+g^4+h^4-2g^2h^2-2h^2f^2-2f^2g^2\\&=\ \ 2(\gamma-\alpha)(\alpha-\beta)(\beta_1-\gamma_1)^2+2(\gamma_1-\alpha_1)(\alpha_1-\beta_1)(\beta-\gamma)^2\\&\quad+2(\alpha-\beta)(\beta-\gamma)(\gamma_1-\alpha_1)^2+2(\alpha_1-\beta_1)(\beta_1-\gamma_1)(\gamma-\alpha)^2\\&\quad+2(\beta-\gamma)(\gamma-\alpha)(\alpha_1-\beta_1)^2+2(\beta_1-\gamma_1)(\gamma_1-\alpha_1)(\alpha-\beta)^2\\&=-4\{\alpha_1(\beta-\gamma)+\beta_1(\gamma-\alpha)+\gamma_1(\alpha-\beta)\}^2=-4\Omega^2.\end{aligned}$$

But

$$\begin{aligned}mnf^2+nlg^2+lmh^2=&\ \ mn\{\beta^2+\beta_1^2+\gamma^2+\gamma_1^2-2(\beta\gamma+\beta_1\gamma_1)\}\\&+nl\ \{\gamma^2+\gamma_1^2+\alpha^2+\alpha_1^2-2(\gamma\alpha+\gamma_1\alpha_1)\}\\&+lm\ \{\alpha^2+\alpha_1^2+\beta^2+\beta_1^2-2(\alpha\beta+\alpha_1\beta_1)\},\\=&-l^2(\alpha^2+\alpha_1^2)-m^2(\beta^2+\beta_1^2)-n^2(\gamma^2+\gamma_1^2)\\&-2mn(\beta\gamma_1+\beta_1\gamma)-2nl(\gamma\alpha_1+\gamma_1\alpha)-2lm(\alpha\beta_1+\alpha_1\beta),\\=&-(l\alpha+m\beta+n\gamma)^2-(l\alpha_1+m\beta_1+n\gamma_1)^2,\\=&-K^2-K_1^2.\end{aligned}$$

Thus

$$4(K^2+K_1^2)=-4(mnf^2+nlg^2+lmh^2),\quad -4\Omega^2=\Delta,$$

and therefore

$$4(K^2+K_1^2-\Omega^2)=\Delta-4(mnf^2+nlg^2+lmh^2).$$

But

$$\mathfrak{S}=\qquad 4(mnf^2+nlg^2+lmh^2),$$
$$\mathfrak{C}=-\Delta+4(mnf^2+nlg^2+lmh^2),$$

and thus

$$4(K^2+K_1^2)\qquad=-\mathfrak{S},$$
$$4(K^2+K_1^2-\Omega^2)=-\mathfrak{C},$$

hence

$$\frac{K^2+K_1^2}{(K^2+K_1^2-\Omega^2)^2}=\frac{-4\mathfrak{S}}{\mathfrak{C}^2},$$

and we have

$$(Z_1Z_2)^2 = \frac{-4\mathfrak{C}}{\mathfrak{G}^2}[(\Pi + F + 2\alpha_1\Omega)^2 + (\Pi_1 + F_1 - 2\alpha\Omega)^2 - \{\Theta + lmn - 2(K\alpha + K_1\alpha_1)\}^2].$$

20. This should agree with the expression in No. 7, that is, we ought to have

$$(\Pi + F + 2\alpha_1\Omega)^2 + (\Pi_1 + F_1 - 2\alpha\Omega)^2 - \{\Theta + lmn - 2(K\alpha + K_1\alpha_1)\}^2 = (f^2 - l^2)(g^2 - m^2)(h^2 - n^2),$$

and this breaks up into the equations

$$\begin{aligned} (\Pi + 2\alpha_1\Omega) + (\Pi_1 - 2\alpha\Omega)^2 &= f^2g^2h^2 \\ 2F(\Pi + 2\alpha_1\Omega) + 2F_1(\Pi_1 - 2\alpha\Omega) - \{\Theta - 2(K\alpha + K_1\alpha_1)\}^2 &= -g^2h^2l^2 - h^2f^2m^2 - f^2g^2n^2 \\ F^2 + F_1^2 - 2lmn\{\Theta - 2(K\alpha + K_1\alpha_1)\} &= f^2m^2n^2 + g^2n^2l^2 + h^2l^2m^2 \\ -l^2m^2n^2 &= -l^2m^2n^2, \end{aligned}$$

which may be separately verified.

21. In fact, we have

$$\begin{aligned} \Pi + 2\alpha_1\Omega = {} & (\beta - \gamma)(\alpha^2 + \alpha_1^2) + (\gamma - \alpha)(\beta^2 + \beta_1^2) + (\alpha - \beta)(\gamma^2 + \gamma_1^2) \\ & + 2\alpha_1\{-(\beta - \gamma)\alpha_1 - (\gamma - \alpha)\beta_1 - (\alpha - \beta)\gamma_1\}, \\ = {} & (\beta - \gamma)(\alpha^2 - \alpha_1^2) + (\gamma - \alpha)\{\beta^2 - \alpha_1^2 + (\alpha_1 - \beta_1)^2\} + (\alpha - \beta)\{\gamma^2 - \alpha_1^2 + (\gamma_1 - \alpha_1)^2\}, \\ = {} & \alpha^2(\beta - \gamma) + \beta^2(\gamma - \alpha) + \gamma^2(\alpha - \beta) + (\gamma - \alpha)(\alpha_1 - \beta_1)^2 + (\alpha - \beta)(\gamma_1 - \alpha_1)^2, \\ = {} & -(\beta - \gamma)(\gamma - \alpha)(\alpha - \beta) + (\gamma - \alpha)(\alpha_1 - \beta_1)^2 + (\alpha - \beta)(\gamma_1 - \alpha_1)^2, \end{aligned}$$

or, putting for shortness

$$\beta - \gamma,\ \gamma - \alpha,\ \alpha - \beta = \lambda,\ \mu,\ \nu;\quad \beta_1 - \gamma_1,\ \gamma_1 - \alpha_1,\ \alpha_1 - \beta_1 = \lambda_1,\ \mu_1,\ \nu_1,$$

(where the letters λ, μ, ν, λ_1, μ_1, ν_1 have a meaning different from that assigned to them in No. 7), this is

$$\Pi + 2\alpha_1\Omega = -\lambda\mu\nu + \mu\nu_1^2 + \nu\mu_1^2,$$

and similarly

$$\Pi_1 - 2\alpha\Omega = -\lambda_1\mu_1\nu_1 + \mu_1\nu^2 + \nu_1\mu^2.$$

Also

$$f^2,\ g^2,\ h^2 = \lambda^2 + \lambda_1^2,\ \mu^2 + \mu_1^2,\ \nu^2 + \nu_1^2,$$

and the first equation thus becomes

$$(-\lambda\mu\nu + \mu\nu_1^2 + \nu\mu_1^2)^2 + (-\lambda_1\mu_1\nu_1 + \mu_1\nu^2 + \nu_1\mu^2)^2 = (\lambda^2 + \lambda_1^2)(\mu^2 + \mu_1^2)(\nu^2 + \nu_1^2),$$

or if on the left-hand we write $\lambda = -\mu - \nu$, $\lambda_1 = -\mu_1 - \nu_1$, this is

$$\{\nu(\mu^2 + \mu_1^2) + \mu(\nu^2 + \nu_1^2)\}^2 + \{\nu_1(\mu^2 + \mu_1^2) + \mu_1(\nu^2 + \nu_1^2)\}^2,$$

which is

$$\begin{aligned} &= (\mu^2 + \mu_1^2)(\nu^2 + \nu_1^2)(\mu^2 + 2\mu\nu + \nu^2 + \mu_1^2 + 2\mu_1\nu_1 + \nu_1^2) \\ &= (\mu^2 + \mu_1^2)(\nu^2 + \nu_1^2)(\lambda^2 + \lambda_1^2), \end{aligned}$$

which is right.

22. For the second equation, we use the values

$$\Pi\ + 2\alpha_1\Omega = \nu\ (\mu^2 + \mu_1^2) + \mu\ (\nu^2 + \nu_1^2),$$

$$\Pi_1 - 2\alpha\ \Omega = \nu_1 (\mu^2 + \mu_1^2) + \mu_1 (\nu^2 + \nu_1^2)\ ;$$

and the equation to be verified thus is

$$\begin{aligned} 2F\ \{\nu\ (\mu^2 + \mu_1^2) + \mu\ (\nu^2 + \nu_1^2)\} &= - l^2\ (\mu^2 + \mu_1^2)(\nu^2 + \nu_1^2) \\ + 2F_1 \{\nu_1 (\mu^2 + \mu_1^2) + \mu_1 (\nu^2 + \nu_1^2)\} &\quad - m^2 (\nu^2 + \nu_1^2)(\lambda^2 + \lambda_1^2) \\ - \{\Theta - 2(K\alpha + K_1\alpha_1)\}^2 &\quad - n^2\ (\lambda^2 + \lambda_1^2)(\mu^2 + \mu_1^2)\ ; \end{aligned}$$

we have

$$F = mn\lambda + (m - n)(l\alpha + m\beta + n\gamma) = mn\lambda + (m - n)(- m\nu + n\mu),$$

that is,

$$F = - m^2\nu\ - n^2\mu\ ;$$

and similarly

$$F_1 = - m^2\nu_1 - n^2\mu_1.$$

Also

$$\begin{aligned} &\Theta - 2(K\alpha + K_1\alpha_1) \\ &= \ \ l(\alpha^2 + \alpha_1^2) + m(\beta^2 + \beta_1^2) + n(\gamma^2 + \gamma_1^2) - 2\{l(\alpha^2 + \alpha_1^2) + m(\alpha\beta + \alpha_1\beta_1) + n(\alpha\gamma + \alpha_1\gamma_1)\}, \\ &= - l(\alpha^2 + \alpha_1^2) + m(\beta^2 + \beta_1^2 - 2\alpha\beta - 2\alpha_1\beta_1) + n(\gamma^2 + \gamma_1^2 - 2\alpha\gamma - 2\alpha_1\gamma_1), \\ &= \ \ m\{(\alpha - \beta)^2 + (\alpha_1 - \beta_1)^2\} + n\{(\gamma - \alpha)^2 + (\gamma_1 - \alpha_1)^2\}, \\ &= \ \ m(\nu^2 + \nu_1^2) + n(\mu^2 + \mu_1^2). \end{aligned}$$

The left-hand side is

$$\begin{aligned} &- 2(m^2\nu\ + n^2\mu\ )\{\nu\ (\mu^2 + \mu_1^2) + \mu\ (\nu^2 + \nu_1^2)\} \\ &- 2(m^2\nu_1 + n^2\mu_1)\{\nu_1 (\mu^2 + \mu_1^2) + \mu_1 (\nu^2 + \nu_1^2)\} \\ &- \{m(\nu^2 + \nu_1^2) + n(\mu^2 + \mu_1^2)\}^2, \end{aligned}$$

which is

$$\begin{aligned} &= \ \ m^2\{- 2(\nu^2 + \nu_1^2)(\mu^2 + \mu_1^2) - 2(\mu\nu + \mu_1\nu_1)(\nu^2 + \nu_1^2) - (\nu^2 + \nu_1^2)^2\} \\ &\quad + n^2\{- 2(\mu\nu + \mu_1\nu_1)(\mu^2 + \mu_1^2) - 2(\mu^2 + \mu_1^2)(\nu^2 + \nu_1^2) - (\mu^2 + \mu_1^2)^2\} \\ &\quad + \ \ (- l^2 + m^2 + n^2)(\mu^2 + \mu_1^2)(\nu^2 + \nu_1^2), \\ &= - l^2(\mu^2 + \mu_1^2)(\nu^2 + \nu_1^2) \\ &\quad - m^2(\nu^2 + \nu_1^2)(\mu^2 + 2\mu\nu + \nu^2 + \mu_1^2 + 2\mu_1\nu_1 + \nu_1^2) \\ &\quad - n^2(\mu^2 + \mu_1^2)(\mu^2 + 2\mu\nu + \nu^2 + \mu_1^2 + 2\mu_1\nu_1 + \nu_1^2), \end{aligned}$$

which is equal to the right-hand side.

23. For the third equation we have as above

$$F = - m^2\nu - n^2\mu, \quad F_1 = - m^2\nu_1 - n^2\mu_1,$$

$$\Theta - 2(K\alpha + K_1\alpha_1) = m(\nu^2 + \nu_1^2) + n(\mu^2 + \mu_1^2),$$

and the equation thus is

$$(m^2\nu + n^2\mu)^2 + (m^2\nu_1 + n^2\mu_1)^2 - 2lmn\{m(\nu^2+\nu_1^2) + n(\mu^2+\mu_1^2)\}$$
$$= (\lambda^2+\lambda_1^2)\,m^2n^2 + (\mu^2+\mu_1^2)\,n^2l^2 + (\nu^2+\nu_1^2)\,l^2m^2;$$

here the left-hand side is

$$\begin{aligned} = \;& (m^4 - 2lm^2n)(\nu^2+\nu_1^2) \\ & + (n^4 - 2lmn^2)(\mu^2+\mu_1^2) \\ & + m^2n^2(\lambda^2+\lambda_1^2-\mu^2-\mu_1^2-\nu^2-\nu_1^2), \end{aligned}$$

which is

$$\begin{aligned} = \;& (\lambda^2+\lambda_1^2)\,m^2n^2 \\ & + (\mu^2+\mu_1^2)\,n^2(n^2 - 2lm - m^2) \\ & + (\nu^2+\nu_1^2)\,m^2(m^2 - 2ln - n^2), \end{aligned}$$

which is equal to the right-hand side.

24. The fourth equation is the identity $-l^2m^2n^2 = -l^2m^2n^2$, and the whole equation is thus verified: viz. the analytical solution leads to the expression

$$\mathfrak{G}^2(Z_1Z_2)^2 = -4\{f^2-(b-c)^2\}\{g^2-(c-a)^2\}\{h^2-(a-b)^2\}\,\mathfrak{C},$$

obtained by an independent process in No. 7 for the squared distance of the two centres.

920.

ON ORTHOMORPHOSIS.

[From the *Quarterly Journal of Pure and Applied Mathematics*, vol. XXV. (1891), pp. 203—226.]

1. THE equation $x_1 + iy_1 = \phi(x + iy)$, where ϕ is in general an imaginary function (that is, a function with imaginary coefficients), and where (x, y), (x_1, y_1) are rectangular coordinates, determines x_1, y_1, each of them as a function of (x, y), and hence, first eliminating y, we have a set of curves depending on the parameter x, and secondly eliminating x, we have a set of curves depending on the parameter y; we have thus two sets of curves, or say trajectories, which are orthomorphoses of the two sets of lines $x = \text{const.}$ and $y = \text{const.}$ respectively. They thus cut everywhere at right angles, and are moreover such that, giving to the parameters values at equal infinitesimal intervals (the same for x and y respectively), they form a double system of infinitesimal squares; say that we have two systems of square trajectories S and T. We may assume at pleasure a trajectory S, and also the consecutive trajectory S'; but it is at once seen geometrically that the entire system of trajectories S and T is then completely determined. For at any point t of S drawing a normal to meet S', and on S the element tt' equal to the normal distance at t, then at t' drawing a normal to meet S', and on S the element $t't''$ equal to the normal distance at t', and so on, we divide the strip between S and S' into infinitesimal squares: at the successive points of S' drawing normals, and on these measuring off distances equal to the successive elements of S', we construct a new curve S'', at the same time dividing the strip between S' and S'' into infinitesimal squares; and proceeding in this manner, we obtain the series of curves S, S', S'', S''', &c., and at the same time the series of curves T, T', T'', T''', &c., proceeding from the points t, t', t'', t''', &c., respectively, and forming with the first set of curves the double system of infinitesimal squares.

2. But to translate this into Analysis, and to obtain the equations of the two sets of curves, we proceed as follows*.

Suppose that the curve S is given in the form

$$x_1 = p, \quad y_1 = q,$$

where p, q are given real functions of a variable parameter θ: we obtain a series of square trajectories S and T, one of the first set being the given curve S, by forming from these equations the equation

$$x_1 + iy_1 = p + iq,$$

and writing therein $\theta = x + iy$. In fact, for the value $y = 0$, this equation becomes $x_1 = p$, $y_1 = q$, where p, q are now the same functions of x which they were originally of θ, and the elimination of x leads therefore to the equation of the given curve S. More generally, we may take $\theta = f(w)$, an arbitrary real function of w, and then assume $w = x + iy$: in this case, for $y = 0$, we have as before $x_1 = p$, $y_1 = q$, where p and q are now the same functions of $f(x)$ which they originally were of θ, and the elimination of x from these equations thus leads as before to the given curve S.

We have next to determine $f(w)$ in suchwise that the curve corresponding to an infinitesimal real value ϵ of y shall be the given consecutive curve S'. We assume that this curve S' is given by the equations

$$x_1 = p + \gamma P, \quad y_1 = q + \gamma Q,$$

where p, q are as before and P, Q are given real functions of θ; γ is a real infinitesimal. The expression for w as a function of θ is then determined by the condition

$$\gamma = C dw = \frac{(p'^2 + q'^2)\, d\theta}{p'Q - q'P},$$

where p', q' are the derived functions of p, q in regard to θ: C is a real constant, the value of which may be assumed at pleasure. Say we have

$$Cw = \int \frac{(p'^2 + q'^2)\, d\theta}{p'Q - q'P},$$

where the integral may be regarded as containing a real or imaginary constant of integration $x_0 + iy_0$; but this being so, we ultimately have

$$w + x_0 + iy_0 = x + iy, \quad \text{or} \quad w = (x - x_0) + i(y - y_0),$$

which is the same thing as $w = x + iy$.

3. We may, in the expressions for x_1, y_1, substitute for θ any other real value θ_1 whatever, say the equations thus became $x_1 = p_1 + \gamma P_1$, $y_1 = q_1 + \gamma Q_1$, values which give

$$x_1 + iy_1 = p_1 + iq_1 + \gamma (P_1 + iQ_1);$$

* The investigation is taken from the memoir, Beltrami, "Delle variabili complesse sopra una superficie qualunque," *Annali di Matem.*, t. I. (1867), pp. 329—366.

the proof consists in showing that there exists for θ_1 a value, differing infinitesimally from θ and such as to reduce this equation to

$$x_1 + iy_1 = p + iq + (p' + iq') \frac{i\gamma}{C} \frac{d\theta}{dw},$$

the value assumed by $p + iq$ on substituting for w the value $w + \frac{i\gamma}{C}$: for this being so, it is obvious that we have for S' the curve in the series of curves S which corresponds to the value $\frac{\gamma}{C}$ of y in $w = x + iy$. The value of θ_1 is

$$\theta_1 = \theta - \gamma \frac{p'P + q'Q}{p'^2 + q'^2},$$

for this gives

$$p_1 = p - \gamma p' \frac{p'P + q'Q}{p'^2 + q'^2}, \quad q_1 = q - \gamma q' \frac{p'P + q'Q}{p'^2 + q'^2};$$

and since γ is infinitesimal, we may in the terms γP_1 and γQ_1 of x_1 and y_1 write $\theta_1 = \theta$, and so reduce these terms to γP and γQ respectively: we thus have

$$x_1 = p_1 + \gamma P_1 = p - \gamma p' \frac{p'P + q'Q}{p'^2 + q'^2} + \gamma P, \quad = p - \gamma q' \frac{p'Q - q'P}{p'^2 + q'^2},$$

$$y_1 = q_1 + \gamma Q_1 = q - \gamma q' \frac{p'P + q'Q}{p'^2 + q'^2} + \gamma Q, \quad = q + \gamma p' \frac{p'Q - q'P}{p'^2 + q'^2},$$

or for $\frac{p'Q - q'P}{p'^2 + q'^2}$ substituting its value $\frac{1}{C} \frac{d\theta}{dw}$, these values are

$$x_1 = p - \frac{\gamma}{C} q' \frac{d\theta}{dw}, \quad y_1 = q + \frac{\gamma}{C} p' \frac{d\theta}{dw},$$

which give the before-mentioned equation

$$x_1 + iy_1 = p + iq + (p' + iq') \frac{i\gamma}{C} \frac{d\theta}{dw},$$

and the proof is thus completed.

The following two examples are given by Beltrami.

4. First, let the curves S, S' be given confocal ellipses,

$$\frac{x_1^2}{a^2} + \frac{y_1^2}{b^2} = 1, \quad \frac{x_1^2}{a^2 - 2cy} + \frac{y_1^2}{b^2 - 2cy} = 1.$$

Here

$$p = a \cos\theta, \quad q = b \sin\theta,$$

whence

$$p'^2 + q'^2 = a^2 \sin^2\theta + b^2 \cos^2\theta;$$

$$P = -\frac{c}{a} \cos\theta, \quad Q = -\frac{c}{b} \sin\theta,$$

whence

$$p'Q - q'P = \frac{c}{ab} (a^2 \sin^2\theta + b^2 \cos^2\theta),$$

and consequently

$$Cw = \int \frac{c}{ab}\,d\theta = \frac{c}{ab}\,\theta, \text{ or taking } C = \frac{c}{ab}, \text{ then } w = \theta,$$

and for the two sets of curves we have

$$x_1 + iy_1 = a\cos(x+iy) + ib\sin(x+iy),$$

which are two sets of confocal conics. In verification, write

$$\cos x = X, \quad \sin x = X', \quad \text{whence} \quad X^2 + X'^2 = 1,$$

$$\cos iy = Y, \quad \sin iy = iY', \quad \text{,,} \quad Y^2 - Y'^2 = 1,$$

and then

$$x_1 + iy_1 = a(XY - iX'Y') + ib(YX' + iY'X),$$

that is,

$$x_1 = X(aY - bY'),$$

$$y_1 = X'(bY - aY'),$$

and consequently

$$\frac{x_1^2}{X^2} - \frac{y_1^2}{X'^2} = a^2 - b^2, \text{ which are the curves } T,$$

$$\frac{x_1^2}{(aY - bY')^2} + \frac{y_1^2}{(bY - aY')^2} = 1, \quad \text{,,} \quad \text{,,} \quad \text{,,} \quad S;$$

where observe that

$$(aY - bY')^2 - (bY - aY')^2 = (a^2 - b^2)(Y^2 - Y'^2) = a^2 - b^2,$$

so that, putting $(aY - bY')^2 = a^2 + h$, we have

$$(bY - aY')^2 = a^2 + h - (a^2 - b^2), \quad = b^2 + h,$$

and thus the curves S are the confocal ellipses

$$\frac{x^2}{a^2 + h} + \frac{y^2}{b^2 + h} = 1, \quad h > -b^2;$$

and similarly, since $X^2 + X'^2 = 1$, putting $(a^2 - b^2)X^2 = a^2 + k$, we have

$$(a^2 - b^2)X'^2 = a^2 - b^2 - (a^2 + k), \quad = -b^2 - k,$$

and thus the curves T are the confocal hyperbolas

$$\frac{x^2}{a^2 + k} + \frac{y^2}{b^2 + k} = 1, \quad -k < a^2 > b^2.$$

5. Next, let the curve S be a circle and the curve S' an interior non-concentric circle. It is convenient to take the circles, such that they have for their common chord the line $y_1 = 0$ and each meets this chord in the points $x_1 = \pm i$.

This being so, the equations of the two circles are

$$x_1^2 + y_1^2 - 2y_1 \operatorname{cosec} \mu = -1,$$

$$x_1^2 + y_1^2 - 2y_1 \operatorname{cosec} \mu (1 - \gamma \cos^2 \mu) = -1,$$

and we have

$$p = \cot\mu\cos\theta,\quad q = \cot\mu\sin\theta + \operatorname{cosec}\mu,$$

$$P = -\cot\mu\cos\theta,\quad Q = -\cot\mu\sin\theta - \cos^2\mu\operatorname{cosec}\theta,$$

giving

$$p'^2 + q'^2 = \cot^2\mu,$$

$$p'Q - q'P = \cot^2\mu(1+\cos\mu\sin\theta),$$

$$\gamma = Cdw = \frac{d\theta}{1+\cos\mu\sin\theta},\ \text{or if } C = \frac{1}{\sin\mu},\ \text{then } dw = \frac{\sin^1\mu\, d\theta}{1+\cos\mu\sin\theta}.$$

The integral of this may be written

$$\tan(\tfrac{1}{2}w - \tfrac{1}{2}iy_0) = \cot\tfrac{1}{2}\mu\tan(\tfrac{1}{4}\pi + \tfrac{1}{2}\theta);$$

and if we here assume for the constant of integration y_0 a value such that

$$\tan\tfrac{1}{2}iy_0 = i\tan\tfrac{1}{2}\mu,$$

we find for w the equation

$$\tan\tfrac{1}{2}w = \cot\mu\, e^{i\theta} + i\operatorname{cosec}\mu,$$

we have $w = x + iy$, and

$$x_1 + iy_1 = p + iq = \cot\mu\, e^{i\theta} + i\operatorname{cosec}\mu,$$

hence

$$x_1 + iy_1 = \tan\tfrac{1}{2}(x+iy);$$

or writing, as we may do, $2x$, $2y$ in place of x, y respectively, say

$$x_1 + iy_1 = \tan(x+iy),$$

for the two sets of curves S and T.

Writing $\tan x = X$, $\tan iy = iY$, this gives

$$x_1 + iy_1 = \frac{X+iY}{1-iXY} = \frac{(X+iY)(1+iXY)}{1+X^2Y^2},$$

that is,

$$x_1 = \frac{X(1-Y^2)}{1+X^2Y^2},\quad y_1 = \frac{Y(1+X^2)}{1+X^2Y^2};$$

and therefore

$$x_1^2 + y_1^2 = \frac{X^2+Y^2}{1-X^2Y^2},$$

and thence

$$x_1^2 + y_1^2 - \left(Y + \frac{1}{Y}\right)y_1 = -1,$$

$$x_1^2 + y_1^2 - \left(X - \frac{1}{X}\right)x_1 = 1,$$

for the curves S and T respectively. The value $Y = \cot\tfrac{1}{2}\mu$ gives the original S-curve

$$x_1^2 + y_1^2 - 2y_1\operatorname{cosec}\mu = -1.$$

We have thus the well-known system of orthotomic circles, where the circles of the one set pass through the points $x_1 = 0$, $y_1 = \pm 1$, and those of the other set through the antipoints $x_1 = \pm i$, $y_1 = 0$.

6. A different solution is given in the Dissertation referred to below*: I reproduce this, giving the proof in a somewhat altered form. The equations of the curve S and of the consecutive curve S' are taken to be

$$\phi(x, y) = 0, \quad \phi(x, y) - 2u\,\psi(x, y) = 0;$$

in the investigation, we introduce the conjugate variables $z = x + iy$, $\bar{z} = x - iy$, and write ϕ, ψ to denote the values assumed by the functions $\phi(x, y)$, $\psi(x, y)$ when x, y are replaced therein by their values in terms of z, $\bar{z}$, viz. $x = \frac{1}{2}(z + \bar{z})$, $y = \frac{1}{2i}(z - \bar{z})$. But in the first instance we may, without attending to the meanings of z, $\bar{z}$, consider ϕ, ψ as denoting each of them a given function of the two coordinates z, $\bar{z}$.

Suppose now that the functions $f(z)$, $f_1(\bar{z})$ are determined by the equations

$$f(z) = \int \frac{1}{\psi}\frac{d\phi}{dz}\,dz, \quad f_1(\bar{z}) = \int \frac{1}{\psi}\frac{d\phi}{d\bar{z}}\,d\bar{z},$$

where in the first integral, after the calculation of $\frac{d\phi}{dz}$, $\bar{z}$ is expressed as a function of z by means of the assumed relation $\phi = 0$, and in the second integral after the calculation of $\frac{d\phi}{d\bar{z}}$, z is expressed as a function of $\bar{z}$ by means of the same relation $\phi = 0$; and where in each integral the constant is properly determined, as presently appearing.

We have

$$f'(z)\,dz + f_1'(\bar{z})\,d\bar{z} = \frac{1}{\psi}\left(\frac{d\phi}{dz}\,dz + \frac{d\phi}{d\bar{z}}\,d\bar{z}\right),$$

where the expression on the right-hand side is $= 0$, if only $\phi = 0$; that is, ϕ being $= 0$, we have $f'(z)\,dz + f_1'(\bar{z})\,d\bar{z} = 0$; and consequently $f(z) + f_1(\bar{z}) = \text{const.}$, viz. this last equation, existing as a consequence of the equation $\phi = 0$, can be nothing else than a different form of the equation $\phi = 0$. We may, by a proper determination of the constants of integration of the two integrals, make the constant of the last-mentioned equation to be $= 0$; we thus have the relation $f(z) + f_1(\bar{z}) = 0$, equivalent to the relation $\phi = 0$.

Writing now $z + \delta z$, $\bar{z} + \delta\bar{z}$ in place of z, $\bar{z}$ respectively, where δz, $\delta\bar{z}$ are arbitrary infinitesimal increments, we have

$$f(z + \delta z) + f_1(\bar{z} + \delta\bar{z}) = 0,$$

equivalent to

$$\phi(z + \delta z, \bar{z} + \delta\bar{z}) = 0,$$

* Meyer, *Ueber die von gerade Linien und von Kegelschnitten gebildeten Schaaren von Isothermen, sowie über einige von speciellen Curven dritter Ordnung gebildete Schaaren von Isothermen:* Inaugural Dissertation der Universität zu Göttingen; 4°, Zürich, 1879. I quote the title in full as it explains the object of the paper; the Isothermals referred to are, of course, systems of orthotomic curves S and T: Plates I. to V. exhibit cases where the curves of at least one of the systems are conics, and Plates VI. to XIV. cases where the curves of at least one of the systems are cubics; the discussion of the several systems is very full and interesting.

that is,

$$f(z)+f'(z)\,\delta z+f_1(\bar z)+f_1'(\bar z)\,\delta\bar z=0,$$

equivalent to

$$\phi+\frac{d\phi}{dz}\,\delta z+\frac{d\phi}{d\bar z}\,\delta\bar z=0\,;$$

and if we here assume

$$\delta z=-u\div f'(z),$$

$$\delta\bar z=-u\div f_1'(\bar z),$$

then we have

$$f(z)+f_1(\bar z)-2u=0,$$

equivalent to

$$\phi-u\left\{\frac{1}{f'(z)}\frac{d\phi}{dz}+\frac{1}{f_1'(\bar z)}\frac{d\phi}{d\bar z}\right\}=0,$$

that is, in virtue of

$$f'(z)=\frac{1}{\psi}\frac{d\phi}{dz}\ \text{ and }\ f_1'(\bar z)=\frac{1}{\psi}\frac{d\phi}{d\bar z},$$

equivalent to

$$\phi-2u\psi=0.$$

There is the difficulty that the last-mentioned values of $f'(z)$, $f_1'(\bar z)$ do not subsist absolutely, but only when $\phi=0$; the answer to this is that the equation $\phi-2u\psi=0$, where u is infinitesimal, is in effect the equation $\phi=0$. Supposing that z, $\bar z$, instead of being connected by the equation $\phi=0$, are connected by the equation $\phi-2u\psi=0$, then the value of

$$\frac{1}{f'z}\frac{d\phi}{dz}+\frac{1}{f_1'(\bar z)}\frac{d\phi}{d\bar z},$$

instead of being actually equal to 2ϕ, will be equal to 2ϕ plus an infinitesimal value containing the factor u, and consequently on substituting this value in the expression

$$\phi-u\left\{\frac{1}{f'(z)}\frac{d\phi}{dz}+\frac{1}{f_1'(\bar z)}\frac{d\phi}{d\bar z}\right\},$$

and neglecting powers of u, we obtain as above the expression $\phi-2u\psi$.

If we now write $z=x+iy$, $\bar z=x-iy$, and assume that the equation $\phi(x, y)=0$ is a real equation, then clearly $f(z)$, $f_1(\bar z)$ as above defined will be conjugate functions of z, $\bar z$ respectively: and u being no longer infinitesimal, we shall have

$$f(z)=u+iv,$$

an equation implying the conjugate relation

$$f_1(\bar z)=u-iv.$$

Considering this equation $f(z)=u+iv$ as thus denoting the two equations, and successively eliminating the parameters v and u, we obtain first a system of curves S depending on the parameter u, and secondly a system of curves T depending on the parameter v, which are two systems of orthotomic curves. And by what precedes, it appears that the curve S belonging to the value $u=0$ of the parameter u, and the consecutive curve S' belonging to a real infinitesimal value $u = u$, are the given curves

$$\phi(x, y) = 0 \text{ and } \phi(x, y) - 2u\psi(x, y) = 0,$$

respectively.

7. For instance, let the curves S and S' be the circles

$$x^2 + y^2 - 1 = 0, \quad x^2 + y^2 - 1 - 2ux = 0;$$

here

$$\phi = z\bar{z} - 1, \quad \psi = \tfrac{1}{2}(z + \bar{z}),$$

$$\frac{1}{\psi}\frac{d\phi}{dz} = \frac{2\bar{z}}{z + \bar{z}} = \frac{2\dfrac{1}{z}}{z + \dfrac{1}{z}} = \frac{2}{z^2 + 1};$$

whence, adding $\frac{1}{2}\pi$ for the constant of integration,

$$\int \frac{1}{\psi}\frac{d\phi}{dz}\, dz = 2\tan^{-1} z + \tfrac{1}{2}\pi.$$

The two sets of curves are thus given by

$$u + iv = \tan^{-1} z = \tan^{-1}(x + iy), \quad 2\tan^{-1}(x + iy) = u - \tfrac{1}{2}\pi + iv,$$

that is,

$$x + iy = \tan(\tfrac{1}{2}u - \tfrac{1}{4}\pi + iv);$$

viz. here, if

$$\tan(\tfrac{1}{2}u - \tfrac{1}{2}\pi) = \alpha, \quad \tan\tfrac{1}{2}iv = i\beta,$$

$$x + iy = \frac{\alpha + i\beta}{1 - i\alpha\beta} = \frac{\alpha(1 - \beta^2) + i\beta(1 + \alpha^2)}{1 + \alpha^2\beta^2},$$

that is,

$$x = \frac{\alpha(1 - \beta^2)}{1 + \alpha^2\beta^2}, \quad y = \frac{\beta(1 + \alpha^2)}{1 + \alpha^2\beta^2},$$

giving

$$x^2 + y^2 = \frac{\alpha^2 + \beta^2}{1 + \alpha^2\beta^2},$$

and thence

$$x^2 + y^2 - 1 - \left(\alpha - \frac{1}{\alpha}\right)x = 0, \quad x^2 + y^2 + 1 - \left(\beta - \frac{1}{\beta}\right)y = 0,$$

for the two systems of curves. The first of these, substituting for α its value, becomes

$$x^2 + y^2 - 1 - 2x\tan u = 0,$$

and thus the curves, for $u = 0$ and u infinitesimal, are

$$x^2 + y^2 - 1 = 0 \quad \text{and} \quad x^2 + y^2 - 1 - 2ux = 0,$$

as they should be.

8. Reverting to the first-mentioned solution, the equation of the circle $x_1^2 + y_1^2 = 1$ is satisfied by $x_1 = \cos\theta$, $y_1 = \sin\theta$, and we have therefore $x_1 + iy_1 = e^{i\theta}$, and hence by what precedes, writing for convenience $X + iY$ instead of $x + iy$, we have

$$x_1 + iy_1 = e^{i\phi(X + iY)}$$

(ϕ being any real function) for a set of curves S and T, or say S_1 and T_1, wherein one of the curves S_1 is the circle $x_1^2+y_1^2=1$; in fact, writing $Y=0$, we have $x_1+iy_1=e^{i\phi(X)}$, and thence $x_1-iy_1=e^{-i\phi(X)}$, and consequently $x_1^2+y_1^2-1=0$ for one of the S_1 curves.

But similarly, if ψ be any real function, then we have $x+iy=e^{i\psi(X+iY)}$ for a set of curves S and T, wherein one of the S curves is the circle $x^2+y^2-1=0$, and thus the two equations $x_1+iy_1=e^{i\phi(X+iY)}$, $x+iy=e^{i\psi(X+iY)}$ establish a correspondence between the two circumferences $x_1^2+y_1^2-1=0$ and $x^2+y^2-1=0$. To eliminate the X, Y, we may write

$$i\psi(X+iY)=\log(x+iy),$$

or say

$$-\psi(X+iY)=i\log(x+iy);$$

then $\phi(X+iY)$ is a real function of $-\psi(X+iY)$, say it is $=f\{-\psi(X+iY)\}$, and it is thus $=f\{i\log(x+iy)\}$, or we have

$$x_1+iy_1=e^{if\{i\log(x+iy)\}}:$$

as the formula for the correspondence in question. In verification, observe that changing the sign of i, we have

$$x_1-iy_1=e^{-if\{-i\log(x-iy)\}};$$

here if $x^2+y^2-1=0$, that is, $(x+iy)(x-iy)=1$, we have

$$\log(x-iy)=-\log(x+iy),$$

and the last equation is

$$x_1-iy_1=e^{-if\{i\log(x+iy)\}},$$

so that we have $x_1^2+y_1^2-1=0$; and thus the two circumferences $x^2+y^2-1=0$ and $x_1^2+y_1^2-1=0$ correspond to each other.

9. We may write down *à priori* a formula which must be equivalent to the foregoing, viz. if $\phi(x+iy)$ be a real or imaginary function of $x+iy$, and $\bar{\phi}$ be the conjugate function (obtained by changing the sign of i in the coefficients), the formula is

$$x_1+iy_1=\frac{\phi(x+iy)}{\bar{\phi}\left(\dfrac{1}{x+iy}\right)};$$

in fact, here changing the sign of i, we have

$$x_1-iy_1=\frac{\bar{\phi}(x-iy)}{\phi\left(\dfrac{1}{x-iy}\right)},$$

which if $x^2+y^2-1=0$, that is, $(x+iy)(x-iy)=1$, becomes

$$x_1-iy_1=\frac{\bar{\phi}\left(\dfrac{1}{x+iy}\right)}{\phi(x+iy)},$$

and the two equations give $x_1^2+y_1^2-1=0$; the two circumferences thus correspond to each other. It is to be noticed that, if the numerator function contain a factor $(x+iy)^m$, say if $\phi(x+iy)=(x+iy)^m\psi(x+iy)$, then the denominator will be

$$(x+iy)^{-m}\,\bar{\psi}\left(\frac{1}{x+iy}\right),$$

and the formula thus becomes

$$x_1+iy_1=\frac{(x+iy)^{2m}\,\psi(x+iy)}{\bar{\psi}\left(\dfrac{1}{x+iy}\right)},$$

which is thus not really more general than the original form. I recur further on to this transformation between the two circumferences.

10. The curve S may be taken to be a closed curve enclosing a singly connected area F (say such a curve is a Contour), and the curve S' a like curve lying wholly inside S, and such that the normal distance between the two curves is everywhere infinitesimal. It is not in general the case, but it may happen that the successive curves S'', S''', &c. will be all of them like curves (each enclosed in the preceding and enclosing the succeeding one), of continually diminishing area, and continually approximating to a point: the curves T will then all of them pass through this point and may be called Radials. We may say that the area F is then squarewise contractible, or contracted, into a point.

11. In particular, a circle is squarewise contractible into a point, viz. if the point be the centre, then the curves S are concentric circles, and the curves T are radii; as is known and will appear, the circle is also contractible into any interior eccentric point whatever. It may be remarked that (the squares being infinitesimal) the numbers of the contours and radials are each of them infinite, but further the number of contours is infinitely great in comparison with that of the radials. Thus in the case just referred to, if to construct a figure we imagine the circumference of the circle divided into any large number n of equal parts, the radials will be the n radii drawn to the points of division: taking the radius to be $=1$, and writing for shortness $\alpha=\dfrac{2\pi}{n}$, the sides of the successive squares along any radius will be α, $\alpha(1-\alpha)$, $\alpha(1-\alpha)^2$, ..., and we require an infinite number of these to make up the entire radius; $1=\alpha\{1+(1-\alpha)+(1-\alpha)^2+\ldots\}$; viz. the number of radials being any large number n whatever, the number of contours will be actually infinite.

12. We may compare the circle as thus contracted into its centre, or rather the quadrant of a circle, with an infinite strip of finite breadth divided into squares by two sets of parallel orthotomic lines: here if x_1, y_1 refer to the positive quadrant of the circle $x_1^2+y_1^2=1$, and x, y to the infinite strip $x=0$ to $-\infty$, $y=0$ to $\frac{1}{2}\pi$, we have $x+iy=\log(x_1+iy_1)$, that is,

$$x=\log\sqrt{(x_1^2+y_1^2)},\quad y=\tan^{-1}\frac{y_1}{x_1};$$

the successive concentric quadrants correspond to equal lines parallel to the axis of y, and the successive radii to infinite lines parallel to the axis of x.

13. Considering the contour S as given, then for a consecutive interior contour S' assumed at pleasure, the area F does not contract into a point: but we have the theorem that the consecutive contour S' can be found, and that in one way only, such that the area F shall contract into a given interior point. To do this is, in effect, to make the given area to correspond say to the circle $x_1^2+y_1^2-1=0$, the contour S and the successive interior contours to the circumference of the circle and to the circumferences of the successive interior concentric circles, and finally the given interior point of F to the centre $x_1=0$, $y_1=0$ of the circle: and thus the theorem is identical with Riemann's theorem "It is possible and that in one way only to make a given singly connected area F correspond to a circle, in such wise that a given interior point of F shall correspond to the centre of the circle, and that a given boundary point of F shall correspond to a given point on the circumference of the circle." The last clause as to the given boundary point of F is necessary in order to make the correspondence a completely definite one, for without the clause it would obviously be allowable to give to the circle an arbitrary rotation about its centre.

14. The proof depends on the following lemma*.

For the given area F, it is possible to find, and that in one way only, a real function ξ of the coordinates (x, y) satisfying the following conditions:

(1) ξ is throughout the area finite and continuous, except only that in the neighbourhood of a given point thereof, taken to be the point $x=0$, $y=0$, it is $=\log\sqrt{(x^2+y^2)}$.

(2) At the boundary of the area ξ is $=0$.

(3) Throughout the area ξ satisfies the partial differential equation

$$\frac{d^2\xi}{dx^2}+\frac{d^2\xi}{dy^2}=0.$$

In fact, if ξ be thus determined, it follows from (3) that we have

$$-\frac{d\xi}{dy}dx+\frac{d\xi}{dx}dy$$

an exact differential, whence putting this $=d\eta$, or determining η by the quadrature

$$\eta=\int\left(-\frac{d\xi}{dy}dx+\frac{d\xi}{dx}dy\right),$$

we have

$$\frac{d\eta}{dx}=-\frac{d\xi}{dy},\quad \frac{d\eta}{dy}=\frac{d\xi}{dx},$$

and thence $\xi+i\eta=$ a function of $x+iy$.

* This proof is taken from the paper, Christoffel, "Sul problema delle temperature stazionarie et la rappresentazione di una data superficie," *Annali di Matem.*, t. I. (1867), pp. 89—103.

Writing then $x_1 + iy_1 = e^{\xi + i\eta} = \psi(x + iy)$ suppose, we have an orthomorphosis of the area into the circle $x_1^2 + y_1^2 - 1 = 0$, viz. for any point (x, y) of the boundary $\xi = 0$, and consequently $x_1 + iy_1 = e^{i\eta}$, thence $x_1 - iy_1 = e^{-i\eta}$, and therefore $x_1^2 + y_1^2 - 1 = 0$, or the point corresponds to a point on the circumference of the circle. Also, for a point x, y in the neighbourhood of $x = 0$, $y = 0$, we have

$$x_1 + iy_1 = e^{\log\sqrt{(x^2+y^2)} + i\eta}, \quad = \sqrt{(x^2 + y^2)}\, e^{i\eta},$$

that is, for $x = 0$, $y = 0$ we have $x_1 = 0$, $y_1 = 0$, or the given point of F corresponds to the centre of the circle.

Some further considerations as to the continuity of η would be requisite in order to show that to the series of circles $x_1^2 + y_1^2 = c^2$, as c continuously increases from 0 to 1, there correspond closed curves surrounding the assumed origin and each other, and passing continuously to the boundary of F, and thus to complete the proof that we thus obtain an orthomorphosis of F into the circle $x_1^2 + y_1^2 = 1$; but I abstain from a further discussion of the question.

15. Reverting to the lemma, this in the first place asserts the existence of a function ξ satisfying the prescribed conditions, and next that the function is completely determinate, or say that there is but one such function. As to the latter point, suppose that there is a second function ξ_1 satisfying the same conditions, then throughout the area $\xi - \xi_1$ is finite and continuous; by general principles in the theory of functions, this implies that the function has throughout the area a constant value, and since this value at the boundary is $= 0$, it must be always $= 0$, that is, $\xi - \xi_1 = 0$, or $\xi_1 = \xi$. As to the former point, it is to be remarked that we obtain $\frac{d^2\xi}{dx^2} + \frac{d^2\xi}{dy^2} = 0$ as the general condition for a minimum value of the double integral $\iint \left\{ \left(\frac{d\xi}{dx}\right)^2 + \left(\frac{d\xi}{dy}\right)^2 \right\} dx\, dy$; if then we assume that there exists a function $\xi = 0$ at the boundary of the area, and finite and continuous throughout the area, except only that in the neighbourhood of a given point thereof, say $x = 0$, $y = 0$, it becomes $= \log\sqrt{(x^2 + y^2)}$, and that for all the values which satisfy these conditions it is such that the above-mentioned double integral taken over the given area shall be a minimum, it follows that there exists a function ξ satisfying the conditions of the lemma.

16. As a simple illustration, suppose that the area F is that of the circle $x^2 + y^2 = 1$, and that the excepted point is the point $x = 0$, $y = 0$, the centre of the circle. We have here a function ξ satisfying the conditions of the lemma, viz. this is $\xi = \log\sqrt{(x^2 + y^2)}$; the resulting value of η is $\eta = \tan^{-1}\frac{y}{x}$, and we have then $\xi + i\eta$ a function of $x + iy$, viz. this is $= \log(x + iy)$. Hence $e^{\xi + i\eta} = x + iy$, and the resulting orthomorphosis of the two circles is the identical one

$$x_1 + iy_1 = x + iy,$$

viz. we have

$$x_1 = x, \quad y_1 = y.$$

17. The contraction of a circle to a given eccentric point has, in fact, been exhibited in the foregoing example in No. 5, but I resume the question from a different point of view. Writing for shortness $z_1 = x_1 + iy_1$, $z = x + iy$; using also a bar for the conjugate function, $\bar{z} = x - iy$, and so in other cases $\alpha = a + \mathrm{a}i$, $\bar{\alpha} = a - \mathrm{a}i$, $\beta = b + \mathrm{b}i$, &c., the general formula of (7) may be written in the form

$$z_1 = \frac{z-\alpha}{1-\bar{\alpha}z} \cdot \frac{z-\beta}{1-\bar{\beta}z} \ldots,$$

viz. we thus have a transformation between the circumferences $x^2 + y^2 - 1 = 0$, $x_1^2 + y_1^2 - 1 = 0$. In fact, repeating the verification, the change of sign of i gives

$$\bar{z}_1 = \frac{\bar{z}-\bar{\alpha}}{1-\alpha\bar{z}} \cdot \frac{\bar{z}-\bar{\beta}}{1-\beta\bar{z}} \ldots.$$

Here the product of the α-factors is

$$\frac{z\bar{z} - (\alpha\bar{z} + \bar{\alpha}z) + \alpha\bar{\alpha}}{1 - (\alpha\bar{z} + \bar{\alpha}z) + \alpha\bar{\alpha}z\bar{z}},$$

which, if $z\bar{z} = 1$, becomes $= 1$; the same property exists as to the β-factors, &c., and hence putting $z\bar{z} = 1$, we have $z_1\bar{z}_1 = 1$, that is, the two circumferences correspond to each other. The correspondence would have subsisted if a factor $e^{i\lambda}$ had been introduced into the expression of z_1, but the effect is merely to rotate the circle $x_1^2 + y_1^2 - 1 = 0$ about its centre, and there is no real gain of generality.

18. The most simple case is

$$z_1 = \frac{z-\alpha}{1-\bar{\alpha}z}.$$

I assume here that $\alpha_1 = a + \mathrm{a}i$, is an interior point of the circle $x^2 + y^2 - 1 = 0$ (viz. $a^2 + \mathrm{a}^2 < 1$); hence $z = \alpha$ gives $z_1 = 0$, viz. to the point $x = a$, $y = \mathrm{a}$, within the circle $x^2 + y^2 - 1 = 0$, there corresponds the centre of the circle $x_1^2 + y_1^2 - 1 = 0$; and we have in this equation the theory of the squarewise contraction of the circle to the eccentric point $x = a$, $y = \mathrm{a}$. There is no loss of generality in assuming $\mathrm{a} = 0$; doing this the formula becomes

$$z_1 = \frac{z-a}{1-az},$$

where $a^2 < 1$, or if to fix the ideas a be taken to be positive, then $a < 1$.

19. To compare with a former investigation, writing $\xi + i\eta = \log(x_1 + iy_1)$, we have

$$\xi + i\eta = \log\frac{x-a+iy}{1-ax-aiy} = \log\frac{\sqrt{\{(x-a)^2+y^2\}}}{\sqrt{\{(1-ax)^2+a^2y^2\}}} + i\tan^{-1}\left\{\frac{y}{x-a} + \frac{ay}{1-ax}\right\},$$

and consequently

$$\xi = \log\frac{\sqrt{\{(x-a)^2+y^2\}}}{\sqrt{\{(1-ax)^2+a^2y^2\}}},$$

which is $= 0$ at the boundary $x^2 + y^2 - 1 = 0$, and is finite and continuous throughout the area except in the neighbourhood of the point $x = a$, $y = 0$, for which it is

$=\log\sqrt{\{(x-a)^2+y^2\}}$; moreover, ξ satisfies the partial differential $\dfrac{d^2\xi}{dx^2}+\dfrac{d^2\xi}{dy^2}=0$, and starting from the given value of ξ we should obtain the foregoing value of η, and thence $x_1+iy_1=e^{\xi+i\eta}$, that is,

$$x_1+iy_1=\frac{x-a+iy}{1-ax-aiy},\quad \text{or}\quad z_1=\frac{z-a}{1-az},$$

as the equation for the orthomorphosis of the circle $x_1^2+y_1^2-1=0$ into the circle $x^2+y^2-1=0$, the centre $x_1=0$, $y_1=0$ corresponding to the eccentric point $x=a$, $y=0$.

20. In further development of the solution, writing

$$z_1\bar{z}_1=\frac{z\bar{z}-a(z+\bar{z})+a^2}{1-a(z+\bar{z})+a^2z\bar{z}},$$

that is,

$$x_1^2+y_1^2=\frac{x^2+y^2+a^2-2ax}{a^2(x^2+y^2)-2ax+1},$$

and then assuming $x_1^2+y_1^2=\lambda_1^2$, we have as the contour corresponding thereto

$$x^2+y^2+a^2-2ax=\lambda^2\{a^2(x^2+y^2)-2ax+1\},$$

that is,

$$(x^2+y^2)(1-a^2\lambda^2)-2a(1-\lambda^2)x+a^2-\lambda^2=0,$$

or say

$$x^2+y^2-\frac{2a(1-\lambda^2)}{1-a^2\lambda^2}x+\frac{a^2-\lambda^2}{1-a^2\lambda^2}=0,$$

or, what is the same thing,

$$\left\{x-\frac{a(1-\lambda^2)}{1-a^2\lambda^2}\right\}^2+y^2=\frac{\lambda^2(1-a^2)^2}{(1-a^2\lambda^2)^2}.$$

The equation may also be written

$$\left\{x-\tfrac{1}{2}\left(\frac{1}{a}+a\right)\right\}\left\{x-\tfrac{1}{2}\left(\frac{1}{a}+a\right)+\left(\frac{1}{a}-a\right)\frac{1+a^2\lambda^2}{1-a^2\lambda^2}\right\}+y^2+\tfrac{1}{4}\left(\frac{1}{a}-a\right)^2=0,$$

showing that the contour circles all pass through two imaginary points

$$x=\tfrac{1}{2}\left(\frac{1}{a}+a\right),\quad y=\pm\tfrac{1}{2}i\left(\frac{1}{a}-a\right).$$

21. Writing next

$$x_1+iy_1=\frac{(x-a+iy)(1-ax-aiy)}{(1-ax)^2+a^2y^2};$$

that is,

$$x_1=\frac{-a+x(1+a^2)-a(x^2+y^2)}{(1-ax)^2+a^2y^2},\quad y_1=\frac{y(1-a^2)}{(1-ax)^2+a^2y^2},$$

then to the radius $x_1-\theta y_1=0$ corresponds the radial

$$-a+x(1+a^2)-a(x^2+y^2)-\theta y(1-a^2)=0,$$

that is,

$$(x-a)\left(x-\frac{1}{a}\right)+y^2+\theta y\left(\frac{1}{a}-a\right)=0,$$

which is a circle passing through the two real points $y=0$, $x=a$, or $\frac{1}{a}$, being the antipoints of the before-mentioned pair of points.

22. Hence for the contraction of the circle $x^2+y^2-1=0$ to the interior point $x=a$, $y=0$, calling this point A, and taking A' the image hereof $\left(\text{coordinates } x=\frac{1}{a},\ y=0\right)$, we see that the contour circles are the circles all passing through the antipoints of A, A', and having thus for their common chord the line bisecting AA' at right angles, and that the radial circles are the circles all passing through the two points A, A'.

In what precedes, the distance AA' is $=\frac{1}{a}-a$, or say the half distance is $=\frac{1}{2}\left(\frac{1}{a}-a\right)$, but altering the scale so as to make this half distance $=1$, and moreover taking the origin at the middle point of AA', the equations of the contour circles and the radial circles assume the forms

$$x^2+y^2-2\alpha x+1=0,$$

and

$$x^2+y^2-2\beta y-1=0,$$

respectively.

23. It is better, interchanging x, y and altering the constants, to write these in the forms

$$x^2+y^2-\left(b+\frac{1}{b}\right)y+1=0,$$

$$x^2+y^2-\left(a-\frac{1}{a}\right)x-1=0,$$

viz. as in effect appearing in No. 5, these are the equations derived from $x+iy=\tan(u+iv)$ where $\tan u=a$, $\tan iv=ib$.

24. Passing now to the form

$$z_1=\frac{z-\alpha}{1-\bar{\alpha}z}\,\frac{z-\beta}{1-\bar{\beta}z},$$

to obtain the most simple instance, I write $a=0$, $b=\sqrt{2}$, so that the form is

$$z_1=\frac{z(z-\sqrt{2})}{1-z\sqrt{2}},$$

we thus obtain

$$z_1\bar{z}_1=\frac{z\bar{z}\{z\bar{z}-\sqrt{2}(z+\bar{z})+2\}}{1-\sqrt{2}(z+\bar{z})+2z\bar{z}},$$

that is,

$$x_1^2 + y_1^2 = \frac{(x^2+y^2)\{x^2+y^2-2\sqrt{2}\,x+2\}}{2(x^2+y^2)-2x\sqrt{2}+1},$$

and thence

$$x_1^2 + y_1^2 - 1 = \frac{(x^2+y^2-1)\{x^2+y^2-2\sqrt{2}\,x+1\}}{(x\sqrt{2}-1)^2+2y^2},$$

so that the circumference $x_1^2 + y_1^2 - 1 = 0$ corresponds to the two circumferences $x^2+y^2-1=0$, $x^2+y^2-2\sqrt{2}\,x+1=0$. The second of these is $(x-\sqrt{2})^2+y^2-1=0$; hence, putting $x+\frac{1}{\sqrt{2}}$ instead of x, the two circles are

$$\left(x+\frac{1}{\sqrt{2}}\right)^2+y^2-1=0,\quad \left(x-\frac{1}{\sqrt{2}}\right)^2+y^2-1=0,$$

(the former of these being the circle originally represented by $x^2+y^2-1=0$), and the last equation becomes

$$x_1^2+y_1^2-1=\frac{\left\{\left(x+\frac{1}{\sqrt{2}}\right)^2+y^2-1\right\}\left\{\left(x-\frac{1}{\sqrt{2}}\right)^2+y^2-1\right\}}{2(x^2+y^2)}.$$

25. Writing here $x_1^2+y_1^2-c^2=0$, we have

$$\{x^2+y^2-\tfrac{1}{2}+x\sqrt{2}\}\{x^2+y^2-\tfrac{1}{2}-x\sqrt{2}\}+2(x^2+y^2)(1-c^2)=0,$$

that is,

$$(x^2+y^2)^2-(2c^2+1)\,x^2-(2c^2-1)\,y^2+\tfrac{1}{4}=0,$$

a bicircular quartic which (or rather a branch whereof) is the contour corresponding to the circle $x_1^2+y_1^2-c^2=0$. For $c=1$, the curve breaks up into the two circles

$$\left(x+\frac{1}{\sqrt{2}}\right)^2+y^2-1=0,\quad \left(x-\frac{1}{\sqrt{2}}\right)^2+y^2-1=0;$$

and for $c=0$, it breaks up into

$$\left(x+\frac{1}{\sqrt{2}}\right)^2+y^2=0,\quad \left(x-\frac{1}{\sqrt{2}}\right)^2+y^2=0,$$

that is, into the two points $y=0$, $x=\pm\frac{1}{\sqrt{2}}$. We have to consider the curves for values of c between these two values $c=1$ and $c=0$.

26. For any such value of c, the curve consists of two symmetrical ovals, situate within the two circles respectively: we are concerned only with the oval lying within the circle $\left(x+\frac{1}{\sqrt{2}}\right)^2+y^2-1=0$; this is shown in the figure for a value of c, slightly inferior to 1, viz. it is an indented oval close approximating to the unshaded portion of the circle, say this is the contour S' consecutive to the circle which is the contour S.

As c diminishes and becomes ultimately $=0$, this oval continually diminishes, after a time losing the indentation, and approximating more and more to the form of a circle having for its centre the point $x = -\frac{1}{\sqrt{2}}$, $y = 0$, which is the centre of the circle, and ultimately reducing itself to this point.

It will be noticed that the consecutive contour S' as shown in the figure is not at an infinitesimal distance from every part of the circle which is the contour S; for a part of the circle, the consecutive contour or contour at an infinitesimal distance

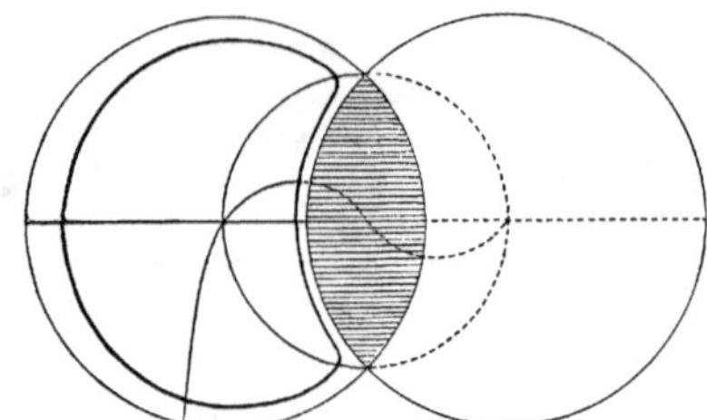

is a part of the symmetrically situated oval, lying outside of the circle. And thus in the present case we do *not* obtain a contraction of the circle into its centre: what we do obtain is a contraction of the unshaded portion of the circle into the centre; this is as it should be, for the *only* contraction of the circle into its centre is by the series of concentric circles.

27. For the radials, writing as before $x + \frac{1}{\sqrt{2}}$ in place of x, that is, $z + \frac{1}{\sqrt{2}}$ in place of z, we have

$$z_1 = x_1 + iy_1 = \frac{\left(z + \frac{1}{\sqrt{2}}\right)\left(z - \frac{1}{\sqrt{2}}\right)}{-z}, \quad = -z + \frac{\frac{1}{2}}{z} = -x - iy + \frac{\frac{1}{2}(x - iy)}{x^2 + y^2}.$$

Hence

$$x_1 = -x + \frac{\frac{1}{2}x}{x^2 + y^2}, \quad = \frac{-x(x^2 + y^2 - \frac{1}{2})}{x^2 + y^2},$$

$$y_1 = -y - \frac{\frac{1}{2}y}{x^2 + y^2}, \quad = \frac{-y(x^2 + y^2 + \frac{1}{2})}{x^2 + y^2},$$

and thus to the radius $x_1 - \theta y_1 = 0$ there corresponds the curve

$$x(x^2 + y^2 - \tfrac{1}{2}) - \theta y(x^2 + y^2 + \tfrac{1}{2}) = 0,$$

or say

$$(x - \theta y)(x^2 + y^2) - \tfrac{1}{2}(x + \theta y) = 0,$$

a circular cubic. In particular, for the line $x_1 = 0$, we have $\theta = 0$, and the curve is $x(x^2 + y^2 - \frac{1}{2}) = 0$, or say $x^2 + y^2 - \frac{1}{2} = 0$, the half circumference of which is shown in the figure: for the line $y_1 = 0$, we have $\theta = \infty$, and the curve is $y(x^2 + y^2 + \frac{1}{2}) = 0$ that is, $y = 0$.

The curve passes through the origin $x = 0$, $y = 0$, and the equation of the tangent there is $x+\theta y=0$. Moreover it passes through the centre of the circle, $x=\frac{1}{\sqrt{2}}$, $iy=0$, and the equation of the tangent there is easily found to be

$$x+\frac{1}{\sqrt{2}}-\theta y=0.$$

We may find where the curve cuts the circle $\left(x+\frac{1}{\sqrt{2}}\right)^2+y^2-1=0$, viz. this is $x^2+y^2=-x\sqrt{2}+\frac{1}{2}$, and substituting in the equation of the curve, we find

$$x^2-\theta xy+\frac{\theta}{\sqrt{2}}y=0,$$

or say

$$\left(x-\frac{1}{\sqrt{2}}\right)\left(x-\theta y+\frac{1}{\sqrt{2}}\right)+\tfrac{1}{2}=0,$$

a hyperbola which by its intersections with the circle determines the points in question. There are two real intersections, but of these only one is in the arc bounding the unshaded portion of the circle: the other intersection serves to determine by symmetry the intersection of the circular cubic with the other boundary of the unshaded portion: see the figure which exhibits the path of the circular cubic through the unshaded portion, and into and beyond the shaded portion or lens. I remark that the two systems of curves are considered by Meyer and shown in his Plate XIII.

28. The formula

$$z_1=\frac{z-\alpha}{1-\alpha z}$$

has been considered for the case of an interior point α; the case of an exterior point α might be considered in like manner, but it is obvious that we pass from one to the other, by a transformation $\frac{1}{z_1}$ for z_1; and it thus easily appears that, if α be an exterior point, we obtain a transformation between the area of the circle $x_1^2+y_1^2-1=0$ and the infinite area exterior to the circle $x^2+y^2-1=0$.

Similarly, in the formula

$$z_1=\frac{z-\alpha}{1-\bar{\alpha}z}\,\frac{z-\beta}{1-\bar{\beta}z},$$

the case considered ($\alpha=a=0$, $\beta=b=\sqrt{2}$) has been that of an interior point α, and an exterior point β. I omit the case of two exterior points, since this can be by the transformation $\frac{1}{z_1}$ for z_1 reduced to that of two interior points, but it is proper to consider this last case.

29. Considering then the case

$$z_1=\frac{z-\alpha\,.\,z-\beta}{1-\bar{\alpha}z\,.\,1-\bar{\beta}z},$$

where α, β are interior points, the most simple case is when $\alpha = a$, $\beta = -a$, positive and < 1. Here

$$z_1 = \frac{z^2 - a^2}{1 - a^2 z^2},$$

and we have

$$x_1^2 + y_1^2 = \frac{(x^2 + y^2)^2 - 2a^2(x^2 - y^2) + a^4}{1 - 2a^2(x^2 - y^2) + a^4(x^2 + y^2)^2},$$

or say

$$x_1^2 + y_1^2 - 1 = \frac{(1 - a^4)\{(x^2 + y^2)^2 - 1\}}{1 - 2a^2(x^2 - y^2) + a^4(x^2 + y^2)^2},$$

which last form shows that to the circumference $x_1^2 + y_1^2 - 1 = 0$, there corresponds the circumference $x^2 + y^2 - 1 = 0$ (and besides this, only the imaginary curve $x^2 + y^2 + 1 = 0$). Writing $x_1^2 + y_1^2 = c^2$, we have the contour

$$(c^2 a^4 - 1)(x^2 + y^2)^2 - 2a^2(c^2 - 1)(x^2 - y^2) + c^2 - a^4 = 0,$$

a bicircular quartic. For $c = 1$, we have as already mentioned the circle $x^2 + y^2 - 1 = 0$; for c less than 1, a curve lying within this circle, diminishing with c, and after a time acquiring on each side of the axis of x an indentation or assuming an hour-glass form; for the value $c^2 = a^4$, the equation becomes

$$(a^4 + 1)(x^2 + y^2)^2 - 2a^2(x^2 - y^2) = 0,$$

and the curve is a figure of eight, the two loops enclosing the points $y = 0$, $x = a$ and $x = -a$ respectively; and as c^2 diminishes to zero, the curve consists of two detached ovals lying within the two loops of the figure of eight, and ultimately reducing themselves to the two points respectively. There is no difficulty in finding the equation of the curves corresponding to a radius $x_1 - \theta y_1 = 0$, but the configuration of these curves is at once seen from that of the former system. We may in the present case say that the circle is squarewise contracted to the figure of eight; and then further that each loop of the figure of eight is squarewise contracted to a point; but we do *not* have a squarewise contraction of the circle to a point.

30. A closed curve or contour may be squarewise contracted not into a point but into a finite line: we see this in the case of a system of confocal ellipses, which gives the contraction of an ellipse into the thin ellipse which is the finite line joining the two foci. There is a like contraction of the circle; this is, in fact, given by the formula due to Schwarz, "Ueber einige Abbildungsaufgaben," *Crelle*, t. LXX. (1869), pp. 105—120 (see p. 115) for the orthomorphosis of the ellipse into a circle: this is

$$x_1 + iy_1 = \text{sn}\left\{\frac{2K}{\pi}\sin^{-1}(x + iy)\right\},$$

where, if $a^2 - b^2 = 1$ and $a = \cos\frac{i\pi K'}{4K}$, or, what is the same thing, $ib = \sin\frac{i\pi K'}{4K}$, then the ellipse $\frac{x^2}{a^2} + \frac{y^2}{a^2 - 1} = 1$ is transformed into the circle $x_1^2 + y_1^2 - \frac{1}{K} = 0$. The contours and trajectories for the ellipse are the confocal ellipses and hyperbolas respectively;

and for the circle, they are two sets of bicircular quartics, such that the portions within the circle have a configuration resembling that of the confocal ellipses and hyperbolas within the ellipse. To investigate the formulæ, it is convenient to introduce the function

$$X + iY, \quad = \sin^{-1}(x + iy);$$

we then have

$$x + iy = \sin(X + iY),$$

or say

$$x = \sin X \cos iY, \quad iy = \cos X \sin iY,$$

so that to any given value of Y there corresponds the ellipse

$$\frac{x^2}{\cos^2 iY} + \frac{y^2}{-\sin^2 iY} = 1,$$

and then

$$x_1 + iy_1 = \text{sn}\,\frac{2K}{\pi}(X + iY);$$

or if

$$s = \text{sn}\,\frac{2KX}{\pi}, \qquad is_1 = \text{sn}\,\frac{2KiY}{\pi},$$

$$c = \text{cn}\,\frac{2KX}{\pi} = \sqrt{(1 - s^2)}, \qquad c_1 = \text{cn}\,\frac{2KiY}{\pi} = \sqrt{(1 + s_1^2)},$$

$$d = \text{dn}\,\frac{2KX}{\pi} = \sqrt{(1 - k^2s^2)}, \quad d_1 = \text{dn}\,\frac{2KiY}{\pi} = \sqrt{(1 + k^2s_1^2)},$$

then

$$x_1 = \frac{sc_1d_1}{1 + k^2s^2s_1^2}, \quad y_1 = \frac{s_1cd}{1 + k^2s^2s_1^2},$$

giving

$$x_1^2 + y_1^2 = \frac{s^2 + s_1^2}{1 + k^2s^2s_1^2},$$

and therefore $x_1^2 + y_1^2 - \frac{1}{k} = 0$ if $1 - ks_1^2 = 0$, that is, if $s_1 = \frac{1}{\sqrt{k}}$, or since

$$is_1 = \text{sn}\,\frac{2KiY}{\pi}, \text{ if } \frac{2KiY}{\pi} = \tfrac{1}{2}iK', \text{ or } Y = \frac{\pi K'}{4K};$$

hence defining a as above, it appears that the elliptic periphery $\frac{x^2}{a^2} + \frac{y^2}{a^2 - 1} = 1$ corresponds to the circumference $x_1^2 + y_1^2 - \frac{1}{k} = 0$. By the introduction of $X + iY$ as above, the circle and the ellipse are each compared with a rectangle; the reduction of the circle to the rectangle, as given by the foregoing equation $x_1 + iy_1 = \text{sn}\,\frac{2K}{\pi}(X + iY)$, or what is substantially the same thing by an equation $x_1 + iy_1 = \text{sn}(X + iY)$, is more fully discussed in my paper, Cayley: "On the Binodal Quartic and the graphical representation of the Elliptic Functions," *Camb. Phil. Trans.*, t. XIV. (1889), pp. 484—494, [891], and in a paper "On some problems of orthomorphosis," *Crelle*, t. CVII. (1891), pp. 262—277, [921].

31. The whole theory, and in particular Riemann's theorem before referred to,

are intimately connected with Cauchy's theorem, "If a function $f(z)$ is holomorphic over a simply connected plane area, and if t denote any point within the area, then

$$f(t) = \frac{1}{2\pi i}\int \frac{f(z)}{z-t}\,dt,$$

where z denotes $x+iy$, and the integral is taken in the positive sense along the boundary of the area." See Briot and Bouquet, *Théorie des fonctions elliptiques* (Paris, 1875), p. 136.

Here in order to obtain by means of the theorem the value of the function $f(z)$ for a given point $t\,(=a+ib)$ within the area, we require to know the values of $f(z)$ for the several points of the boundary: viz. if z refers to a point P on the boundary, and if we represent the value $f(z)$ by a point P_1 in a second figure, then these points P_1 form a closed curve or boundary in this second figure, and we require to know not only the form of this boundary, but also the several points P_1 thereof which correspond to the several points P of the first-mentioned boundary, or say we require to know the correspondence of the two boundaries: this being known, we have by the theorem the value of $f(t)$, that is, the point a_1+ib_1 within the second area, which corresponds to the point $t=a+ib$ within the first area. The (1, 1) correspondence of the two areas is of course implied in the assertion that $f(t)$ has a determinate value, determined by means of the given values of $f(z)$ along the boundary.

921.

ON SOME PROBLEMS OF ORTHOMORPHOSIS.

[From *Crelle's Journal der Mathem.*, t. CVII. (1891), pp. 262—277.]

IN the interesting Memoir, Schwarz, "Ueber einige Abbildungsaufgaben," *Crelle*, t. LXX. (1869), pp. 105—120, the author considers the orthomorphic transformation (or, as I call it, the orthomorphosis) of a square into the infinite half-plane, or into a circle, and of a rectangle into the infinite half-plane. It is of course easy to deduce the orthomorphosis of the rectangle into a circle; and then, by giving a proper value to the modulus of the elliptic function involved in the formula, we obtain the orthomorphosis of the square into a circle, this solution (although equivalent thereto) being under a different form from that previously given for the square. But as appears from my paper "On the Binodal Quartic and the Graphical Representation of the Elliptic Functions," *Camb. Phil. Trans.* vol. XIV. (1889), pp. 484—494, [891], there is for the rectangle (and consequently also for the square) an orthomorphosis wherein the boundary of the rectangle or square corresponds, not to the circumference, but to the circumference together with two twice-repeated portions of a diameter. I propose to consider in I. these several transformations: II. relates to the orthomorphosis of a circle into a circle.

I.

1. We are concerned with the elliptic function sn for which $K' = 2K$, and also with the elliptic function snl of the lemniscate form. The modulus in the former case is

$$k = \frac{\sqrt{2}-1}{\sqrt{2}+1},\ = (\sqrt{2}-1)^2,\ = 3 - 2\sqrt{2},\ \text{ or say } \sqrt{k} = \sqrt{2}-1,\ \frac{1}{\sqrt{k}} = \sqrt{2}+1.$$

For the lemniscate function snl, the modulus is $= i$, and if (with Gauss) we write $\frac{1}{2}\varpi$ for the value of the complete function K or F_1, then we have $\frac{1}{4}\varpi = K(\sqrt{2}-1)$,

$= K\sqrt{k}$, where k has the foregoing special value: I notice the numerical values $\frac{1}{2}\varpi = 1{\cdot}311028$, $k = 3 - 2\sqrt{2} = \sin 9^\circ 52'$, $k = 1{\cdot}582548$. The relation between the lemniscate function snl, and the sn with the foregoing value of k, is easily shown to be

$$\sqrt{k}\,\mathrm{sn}\,U = \frac{(i+1)\,\mathrm{snl}\,u + i\sqrt{2}}{(i-1)\,\mathrm{snl}\,u + \sqrt{2}}, \text{ where } U - \tfrac{1}{2}iK' = \frac{1+i}{2\sqrt{k}}u;$$

and it may be added that we have

$$\mathrm{cn}\,U = \frac{\sqrt{2}}{(i-1)\,\mathrm{snl}\,u + \sqrt{2}}\sqrt{\frac{1+k}{k}}\sqrt{(1+\mathrm{snl}\,u)(1-i\,\mathrm{snl}\,u)},$$

$$\mathrm{dn}\,U = \frac{\sqrt{2}}{(i-1)\,\mathrm{snl}\,u + \sqrt{2}}\sqrt{1+k}\sqrt{(1-\mathrm{snl}\,u)(1+i\,\mathrm{snl}\,u)}.$$

2. The Schwarzian orthomorphosis of the rectangle into the infinite half-plane is given (Memoir, p. 113) by the formula $X_1 + iY_1 = \mathrm{sn}(X + iY)$, where the modulus is real, positive, and less than unity. Here, see the figures 1 (XY) and 2 (X_1Y_1),

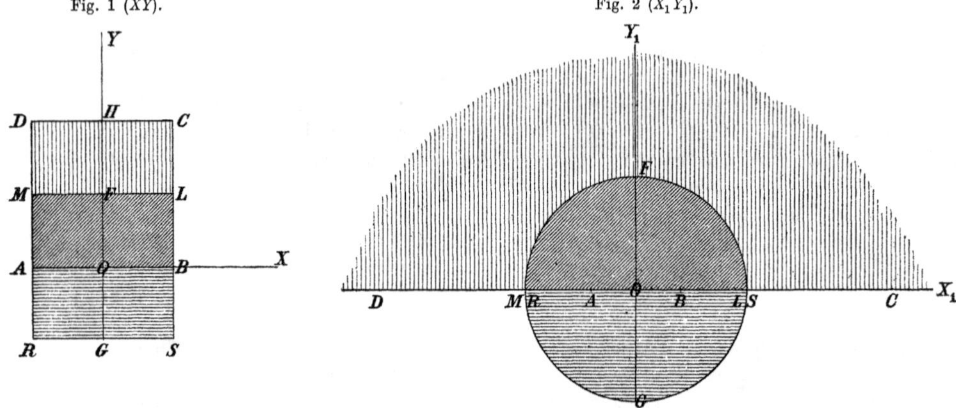

Fig. 1 (XY). Fig. 2 (X_1Y_1).

the rectangle $ABCD$, the sides of which are $AB = 2K$ and $BC = K'$, is transformed into the upper infinite half-plane $(Y_1 = +)$, the four corners of the rectangle corresponding to the points A, B, C, D on the axis of X, where $OB(=OA) = 1$, $OC(=OD) = \frac{1}{k}$.

3. We can, by a properly determined quasi-inversion (as will be explained), transform the X_1Y_1-figure into a new figure see figure 3 (X_2Y_2), the infinite X_1-axis being transformed into the circumference of a circle (the radius of which may be taken to be $= 1$) and the infinite half-plane into the area within the circle. The four points A, B, C, D are thus transformed into points on the circle, which if the quasi-inversion be a symmetrical one, will be situate, A and B symmetrically, and

also C and D symmetrically, in regard to the axis OY_2: and in the case of the before-mentioned modulus $k = 3 - 2\sqrt{2}$, for which the rectangle $ABCD$ becomes a

Fig. 3* (X_2Y_2).

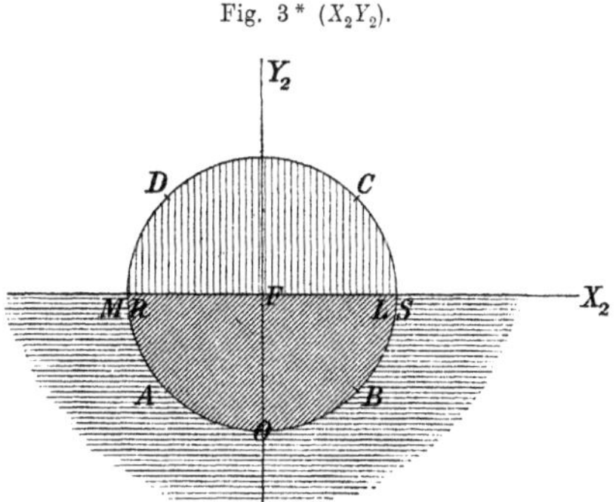

square, the quasi-inversion may be so determined that the points A, B, C, D shall be situate midway (that is, at inclinations $\pm 45°$, $\pm 135°$) in the four quadrants of the circle.

4. But if in figure 1 we take $BL = \frac{1}{2}K'$ and draw FL parallel to OX, then as shown in my memoir above referred to, the foregoing transformation $X_1 + iY_1 = \text{sn}(X + iY)$ changes the rectangle $OBLF$, the sides of which are $OB = K$ and $BL = \frac{1}{2}K'$, into the quadrant $OBLF$ of figure 2, $OB = 1$ (as already mentioned) and radius $OL = \frac{1}{\sqrt{k}}$. Hence completing the rectangle $RSLM$, the sides of which are $RS = 2K$ and $SL = K'$, (viz. this rectangle differs only in position from the first-mentioned rectangle $ABCD$), we have an orthomorphosis of the rectangle $RSLM$ into the circle of figure 2: but so nevertheless that to the boundary of the rectangle there corresponds not the circumference of the circle but the boundary $RGSBLFMAR$ composed of the circumference and the portions SB, BL and MA, AR (that is, BL, AM each twice) of a diameter of the circle. See *post* No. 12.

5. The Schwarzian orthomorphosis of the square into a circle is given (Memoir, pp. 111—113) by the formula $x_1 + iy_1 = \text{snl}(x + iy)$; viz. here—see figures 4 (xy) and 5 (x_1y_1)—the square $ABCD$, the half-diagonals whereof are each $= \frac{1}{2}\varpi$, corresponds to the circle radius $= 1$ of figure 5, the four points A, B, C, D of the circle being the quadrantal points as shown in the figure. Figure 5 is, in fact, figure 3 turned through an angle of $45°$; and it thus appears that Schwarz's lemniscate solution for the square into a circle is the quasi-inversion of his solution for the rectangle into the infinite half-plane, when by putting $k = 3 - 2\sqrt{2}$ the rectangle is made to be a square. See *post* Nos. 13 and 14.

[* The scale of this figure is double that of figures 1, 2, 4, 6, 7 in this paper.]

6. The general formula of quasi-inversion whereby the infinite X_1-axis is changed into a circumference† radius unity, is

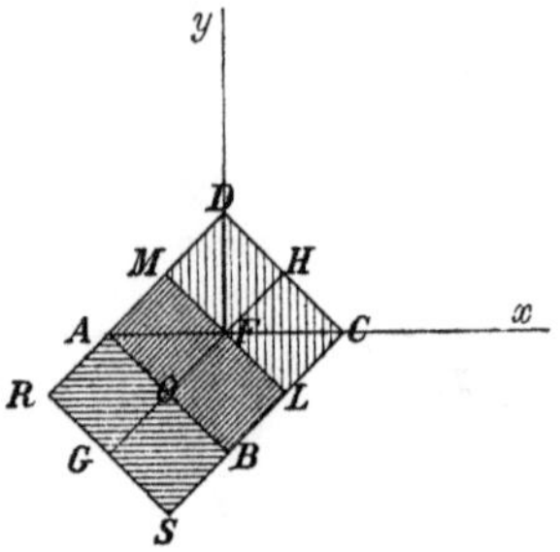

Fig. 4 (xy).

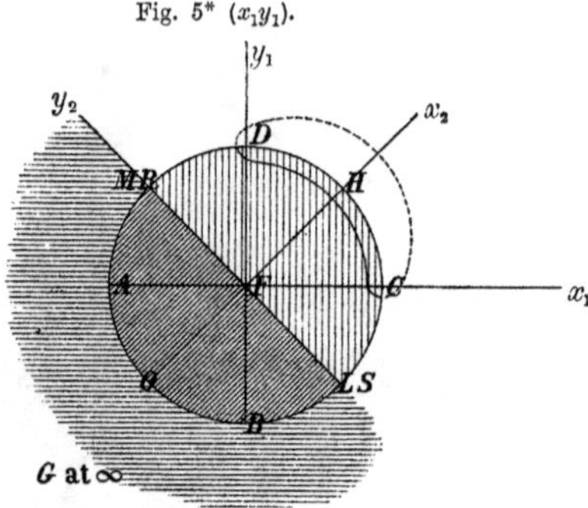

Fig. 5* (x_1y_1).

$$X_2+iY_2=\frac{1+Mi\,(X_1+iY_1)}{M\,(X_1+iY_1)+i},$$

where M is real. In fact, writing this equation in the form

$$X_2+iY_2=\frac{1+Mi\,(X_1+iY_1)}{i\,\{1-Mi\,(X_1+iY_1)\}},$$

we have

$$X_2-iY_2=\frac{1-Mi\,(X_1-iY_1)}{-i\,\{1+Mi\,(X_1-iY_1)\}},$$

and thence $X_2^2+Y_2^2-1=0$, if only $Y_1=0$; that is, the infinite X_1-axis is transformed into the circumference $X_2^2+Y_2^2-1=0$.

Writing $Y_1=0$, we have

$$X_2+iY_2=\frac{1+MiX_1}{MX_1+i}=\frac{2MX_1+i\,(M^2X_1^2-1)}{M^2X_1^2+1},$$

so that to a pair of points $(\pm X_1,\ 0)$ there corresponds a pair of points $(\pm X_2,\ Y_2)$ situate symmetrically in regard to the axis OY_2.

The equation gives

$$M(X_1+iY_1)=\frac{1-i\,(X_2+iY_2)}{(X_2+iY_2)-i}.$$

Hence, writing this in the form

$$M(X_1+iY_1)=\frac{1-i\,(X_2+iY_2)}{-i\,\{1+i\,(X_2+iY_2)\}},$$

[* The scale of this figure is double that of figures 1, 2, 4, 6, 7 in this paper.]

† I use here and elsewhere the term circumference rather than circle, to mark more clearly the distinction between the curve and the included area.

we have

$$M(X_1 - iY_1) = \frac{1 + i(X_2 - iY_2)}{i\{1 - i(X_2 - iY_2)\}},$$

and consequently $M^2(X_1^2 + Y_1^2) - 1 = 0$, if only $Y_2 = 0$; that is, the circumference $X_1^2 + Y_1^2 - \frac{1}{M^2} = 0$ is transformed into the infinite X_2-axis.

Although for the purposes of the present memoir we require the coefficient M, yet there is no real loss of generality in assuming $M = 1$, and the transformation is best studied under the form

$$X_2 + iY_2 = \frac{1 + i(X_1 + iY_1)}{X_1 + iY_1 + i};$$

see *post* No. 11.

7. As already mentioned, the coordinates (X_1, Y_1) and (X_2, Y_2) are connected by an equation of the foregoing form, and in the case of the square ($k = 3 - 2\sqrt{2}$ as before), the corresponding values for the points A, B, C, D should be

$$A, \quad X_1 + iY_1 = -1, \quad X_2 + iY_2 = \frac{-1 - i}{\sqrt{2}},$$

$$B, \quad X_1 + iY_1 = 1, \quad X_2 + iY_2 = \frac{1 - i}{\sqrt{2}},$$

$$C, \quad X_1 + iY_1 = \frac{1}{k}, \quad X_2 + iY_2 = \frac{1 + i}{\sqrt{2}},$$

$$D, \quad X_1 + iY_1 = -\frac{1}{k}, \quad X_2 + iY_2 = \frac{-1 + i}{\sqrt{2}}.$$

The proper value of M is $M = \sqrt{2} - 1, = \sqrt{k}$; the required formula thus is

$$X_2 + iY_2 = \frac{1 + i\sqrt{k}(X_1 + iY_1)}{\sqrt{k}(X_1 + iY_1) + i}$$

or say

$$\sqrt{k}(X_1 + iY_1) = \frac{1 - i(X_2 + iY_2)}{X_2 + iY_2 - i}.$$

Thus for the point B, we should have

$$\sqrt{k} = \frac{1 - i\dfrac{1 - i}{\sqrt{2}}}{\dfrac{1 - i}{\sqrt{2}} - i}, \text{ that is, } \sqrt{k} = \frac{\sqrt{2} - i - 1}{1 - i - i\sqrt{2}} = \frac{\sqrt{k} - i}{1 - \dfrac{i}{\sqrt{k}}},$$

which is right. And similarly for the point C, we should have

$$\frac{1}{\sqrt{k}} = \frac{1 - i\dfrac{1 + i}{\sqrt{2}}}{\dfrac{1 + i}{\sqrt{2}} - i}, \text{ that is, } \frac{1}{\sqrt{k}} = \frac{\sqrt{2} - i + 1}{1 + i - i\sqrt{2}} = \frac{\dfrac{1}{\sqrt{k}} - i}{1 - i\sqrt{k}},$$

which is right. And similarly for the points A and D.

8. We connect the square $ABCD$ of figure 1 with that of figure 4 by the equation

$$X + iY - \tfrac{1}{2}iK' = \frac{1+i}{2\sqrt{k}}(x+iy);$$

in fact, recollecting that $\tfrac{1}{2}K' = K = \frac{\varpi}{4\sqrt{k}}$, we have for the four points respectively

$$\begin{array}{lll} A, & X+iY=-K, & x+iy=-\tfrac{1}{2}\varpi, \\ B, & X+iY=\ \ K, & x+iy=-\tfrac{1}{2}\varpi i, \\ C, & X+iY=\ \ K+iK', & x+iy=\ \ \tfrac{1}{2}\varpi, \\ D, & X+iY=-K+iK', & x+iy=\ \ \tfrac{1}{2}\varpi i, \end{array}$$

values which satisfy the relation in question.

9. Hence writing

$$X+iY=U, \quad x+iy=u,$$

we have

$$U - \tfrac{1}{2}iK' = \frac{1+i}{2\sqrt{k}}u,$$

which is the relation in No. 1 between the arguments U, u of the elliptic functions sn, snl. We have $X_1 + iY_1 = \text{sn}\, U$, $x_1 + iy_1 = \text{snl}\, u$; and consequently

$$\sqrt{k}(X_1+iY_1) = \frac{(i+1)(x_1+iy_1)+i\sqrt{2}}{(i-1)(x_1+iy_1)+\sqrt{2}}.$$

Hence, substituting for $\sqrt{k}(X_1+iY_1)$ its value in terms of X_2+iY_2, we find

$$\frac{(i+1)(x_1+iy_1)+i\sqrt{2}}{(i-1)(x_1+iy_1)+\sqrt{2}} = \frac{1-i(X_2+iY_2)}{X_2+iY_2-i},$$

an equation which $\left(\text{multiplying on the left-hand side the numerator and the denominator each by } \frac{-i}{\sqrt{2}}\right)$ may be written

$$\frac{1 - i\dfrac{1+i}{\sqrt{2}}(x_1+iy_1)}{\dfrac{1+i}{\sqrt{2}}(x_1+iy_1)-1} = \frac{1-i(X_2+iY_2)}{X_2+iY_2-i},$$

that is, we have

$$X_2+iY_2 = \frac{1+i}{\sqrt{2}}(x_1+iy_1),$$

an equation which shows that the figure 5 is in fact the figure 3 turned through an angle of 45°. We have thus proved the conclusion stated in No. 5 as to the connexion between the lemniscate solution for the square and the solution for the rectangle.

10. It is convenient to collect here the several equations relating to the orthomorphosis of the square. We have

$$X_1 + iY_1 = \text{sn}\,(X + iY), \quad k = 3 - 2\sqrt{2};$$

$$x_1 + iy_1 = \text{snl}\,(x + iy);$$

$$\sqrt{k}\,(X_1 + iY_1) = \frac{1 - i\,(X_2 + iY_2)}{X_2 + iY_2 - i},$$

$$X + iY - \tfrac{1}{2}iK' = \frac{1+i}{2\sqrt{k}}(x + iy),$$

$$\sqrt{k}\,(X_1 + iY_1) = \frac{(i+1)(x_1 + iy_1) + i\sqrt{2}}{(i-1)(x_1 + iy_1) + \sqrt{2}},$$

$$X_2 + iY_2 = \frac{1+i}{\sqrt{2}}(x_1 + iy_1),$$

which are the equations connecting together the coordinates of the five figures.

11. I examine more in detail the above-mentioned transformation

$$X_2 + iY_2 = \frac{1 + i\,(X_1 + iY_1)}{X_1 + iY_1 + i};$$

see the foregoing figures 1 (XY) and 2 (X_1Y_1), in which we now regard the two circles as having each of them the radius unity; changing the sign of i, the equation gives

$$X_2 - iY_2 = \frac{1 - i\,(X_1 - iY_1)}{X_1 - iY_1 - i},$$

and we hence find

$$X_2^2 + Y_2^2 = \frac{X_1^2 + Y_1^2 - 2Y_1 + 1}{X_1^2 + Y_1^2 + 2Y_1 + 1};$$

consequently if $X_1^2 + Y_1^2 + 1 = 0$, then also $X_2^2 + Y_2^2 + 1 = 0$, or the transformation changes the first of these imaginary circles into the second of them: or say it changes the concentric orthotomic of the circle $X_1^2 + Y_1^2 - 1 = 0$ into the concentric orthotomic of the circle $X_2^2 + Y_2^2 - 1 = 0$.

We have moreover

$$X_2 = \frac{2X_1}{X_1^2 + Y_1^2 + 2Y_1 + 1}, \quad Y_2 = \frac{X_1^2 + Y_1^2 - 1}{X_1^2 + Y_1^2 + 2Y_1 + 1};$$

values which give the foregoing expression for $X_2^2 + Y_2^2$. But we further obtain

$$X_2^2 + Y_2^2 - 1 - \frac{2}{\mu}Y_2 = \frac{\frac{2}{\mu}(X_1^2 + Y_1^2 - 1 + 2\mu Y_1)}{X_1^2 + Y_1^2 + 2Y_1 + 1},$$

and it thus appears that the circumferences

$$X_1^2 + Y_1^2 - 1 + 2\mu Y_1 = 0, \quad X_2^2 + Y_2^2 - 1 - \frac{2}{\mu}Y_2 = 0,$$

correspond to each other. These are circles passing through the pairs of points $(Y_1=0,\ X_1=\pm 1)$, $(Y_2=0,\ X_2=\pm 1)$ respectively; or imagining them in the same figure, say they are circles which belong each to the series of circles $x^2+y^2-1+2\beta y=0$, and which moreover cut at right angles at the points $y=0$, $x=\pm 1$. But attending more carefully to the nature of the correspondence, it is to be observed that taking Y positive, and giving to X any positive value from 0 to ∞, we have in figure 2 an arc LJM lying wholly within the upper semicircle LFM; and that, corresponding hereto in figure 3, we have an arc LJM lying wholly within the lower semicircle LOM; and that as in figure 2, the arc LJM lies nearer to the semicircumference LFM or to the diameter LOM, so in figure 3 the arc LJM lies nearer to the diameter LFM or to the semicircumference LOM. Thus the upper semicircle LFM of figure 2 corresponds to the lower semicircle LOM of figure 3; but so that the semicircumference LFM of the first figure corresponds to the diameter LFM of the second figure; and the diameter LOM of the first figure to the semicircumference LOM of the second figure. And further supposing that Y_1 is still positive, but that μ has any negative value from $-\infty$ to 0, we have in figure 2 an arc in the upper half-plane lying wholly outside the semicircle; and corresponding thereto in figure 3, an arc lying wholly inside the upper semicircle LHM; that is, in figure 2 the infinite space in the upper half-plane outside the semicircle corresponds in figure 3 to the space within the upper half-circle LHM; the infinity of figure 2 corresponding to the semicircumference LHM of figure 3, and the semicircumference LFM of figure 2 to the diameter LFM of figure 3. And thus in figure 2, the upper half-plane inside and outside the semicircle corresponds in figure 3 to the lower and upper half-circles, that is, to the whole circular area $OLHM$ of figure 3.

It is to be observed, moreover, that we have

$$X_2^2+Y_2^2+1+2\lambda X_2=\frac{2(X_1^2+Y_1^2+1+2\lambda X_1)}{X_1^2+Y_1^2+2Y_1+1},$$

that is, the circles $X_1^2+Y_1^2+1+2\lambda X_1=0$ and $X_2^2+Y_2^2+1+2\lambda X_2=0$ correspond to each other. Imagining the circles as belonging to the same figure, these are one and the same circle of the series $x^2+y^2+1-2\alpha x=0$ each passing through the pair of points $(x=0,\ y=\pm i)$ which are the antipoints of the pair $(y=0,\ x=\pm 1)$; these circles thus cut at right angles those of the series $x^2+y^2-1+2\beta y=0$. We have, by means of the two series of circles, an easy construction for the correspondence between the figures 2 and 3.

12. In explanation of No. 4, observe that, starting from the equation

$$X_1+iY_1=\operatorname{sn}(X+iY),$$

and writing $\operatorname{sn}X=P$, $\operatorname{sn}iY=iQ$, we have

$$X_1+iY_1=\frac{P\sqrt{1+Q^2\,.\,1+k^2Q^2}+iQ\sqrt{1-P^2\,.\,1-k^2P^2}}{1+k^2P^2Q^2},$$

that is,

$$X_1=\frac{P\sqrt{1+Q^2\,.\,1+k^2Q^2}}{1+k^2P^2Q^2},\quad Y_1=\frac{Q\sqrt{1-P^2\,.\,1-k^2P^2}}{1+k^2P^2Q^2},$$

and thence

$$X_1^2+Y_1^2=\frac{P^2+Q^2}{1+k^2P^2Q^2}=r^2 \text{ (if } X_1^2+Y_1^2 \text{ be put } =r^2\text{);}$$

hence

$$P^2(1-k^2Q^2r^2)=r^2-Q^2,\quad Q^2(1-k^2P^2r^2)=r^2-P^2.$$

Now considering in figure 1 any line in the rectangle parallel to the axis of X, that is, taking Y constant and therefore also Q constant, and proceeding to eliminate P, we have

$$P^2=\frac{r^2-Q^2}{1-k^2Q^2r^2},\quad 1-P^2=\frac{1+Q^2-(1+k^2Q^2)\,r^2}{1-k^2Q^2r^2},$$

$$1-k^2P^2=\frac{1+k^2Q^2-k^2(1+Q^2)\,r^2}{1-k^2Q^2r^2},\quad 1+k^2P^2Q^2=\frac{1-k^2Q^4}{1-k^2Q^2r^2},$$

and consequently

$$X_1=\frac{\sqrt{1+Q^2\,.\,1+k^2Q^2}}{1-k^2Q^4}\sqrt{r^2-Q^2\,.\,1-k^2Q^2r^2},$$

$$Y_1=\frac{Q\sqrt{1+Q^2-(1+k^2Q^2)\,r^2}\sqrt{1+k^2Q^2-k^2(1+Q^2)\,r^2}}{1-k^2Q^4},$$

giving X_1, Y_1 each of them in terms of Q^2 and r^2, $=X_1^2+Y_1^2$. The former of these equations, replacing therein r^2 by its value, gives easily

$$(X_1^2+Y_1^2)^2-2AX_1^2-2BY_1^2+\frac{1}{k^2}=0,$$

where

$$2A=\frac{1+Q^2}{1+k^2Q^2}+\frac{1}{k^2}\,\frac{1+k^2Q^2}{1+Q^2},\quad 2B=Q^2+\frac{1}{k^2Q^2};$$

viz. we have thus the equation of the curve, a bicircular quartic which in figure 2 corresponds to the line parallel to the axis of X_1 in figure 1.

In particular, for the line LM of figure 1 we have $Y=\tfrac{1}{2}K'$ and thence $iQ=\text{sn}\,\tfrac{1}{2}iK'=\frac{i}{\sqrt{k}}$, that is, $Q=\frac{1}{\sqrt{k}}$, and thence $A=B=\frac{1}{k}$; the equation of the bicircular quartic is

$$(X_1^2+Y_1^2)^2-\frac{2}{k}(X_1^2+Y_1^2)+\frac{1}{k^2}=0;$$

viz. this is the circle $X_1^2+Y_1^2-\frac{1}{k}=0$ twice repeated, and we have thus this circle, or rather the half-circumference LFM of figure 2, corresponding to the line LM of figure 1. More simply,

$$X_1+iY_1=\text{sn}\,(X+\tfrac{1}{2}iK')=\frac{\frac{1+k}{\sqrt{k}}\,\text{sn}\,X+\frac{i}{\sqrt{k}}\,\text{cn}\,X\,\text{dn}\,X}{1+k^2\,.\,\frac{1}{k}\,\text{sn}^2 X}=\frac{(1+k)\,P+i\sqrt{1-P^2\,.\,1-k^2P^2}}{\sqrt{k}\,(1+kP^2)},$$

and thence

$$X_1^2+Y_1^2=\frac{(1+k)^2P^2+1-(1+k^2)P^2+k^2P^4}{k(1+2kP^2+k^2P^4)}=\frac{1}{k},$$

that is, $X_1^2+Y_1^2-\frac{1}{k}=0$ as before. It is easy to see that the points A, O, B of figure 1 correspond to the points A, O, B of figure 2, and hence that the area of the rectangle $AOBLFMA$ of figure 1 corresponds to that of the semicircle $AOBLFMA$ of figure 2.

Returning to the equation $X_1+iY_1=\text{sn}(X+iY)$, if we write herein successively

$$Y_1=\tfrac{1}{2}K'-\beta,\quad \text{sn}\, iY_1=iQ_1=\text{sn}\, i(\tfrac{1}{2}K'-\beta),$$

and

$$Y_1=\tfrac{1}{2}K'+\beta,\quad \text{sn}\, iY_1=iQ_2=\text{sn}\, i(\tfrac{1}{2}K'+\beta),$$

then we have

$$iQ_1\,.\,iQ_2=\text{sn}\, i(\tfrac{1}{2}K'-\beta)\,\text{sn}\, i(\tfrac{1}{2}K'+\beta)=-\frac{1}{k},$$

that is, $Q_1Q_2=\frac{1}{k}$: hence for q writing Q_1 or Q_2, we have in each case the same values of A and B, that is, we have the same bicircular quartic for two lines parallel to and equidistant from the line LM, but to one of these (viz. the line between LM and BA) there corresponds the half-perimeter lying within the semicircumference LFM, and to the other of them (viz. the line between LM and CD) there corresponds the half-perimeter lying without the semicircumference LFM.

It may be shown in like manner that, to any line in figure 1 parallel to the axis OY, there corresponds in figure 2 a bicircular quartic of the like form

$$(X_1^2+Y_1^2)^2-2AX_1^2-2BY_1^2+\frac{1}{k^2}=0.$$

13. Similarly in explanation of No. 5, observe that, starting from the equation $x_1+iy_1=\text{snl}(x+iy)$ and writing $\text{snl}\, x=p$, $\text{snl}\, iy=i\,\text{snl}\, y=iq$, we have

$$x_1+iy_1=\frac{p\sqrt{1-q^4}+iq\sqrt{1-p^4}}{1-p^2q^2},$$

that is,

$$x_1=\frac{p\sqrt{1-q^4}}{1-p^2q^2},\quad y_1=\frac{q\sqrt{1-p^4}}{1-p^2q^2},$$

and thence

$$x_1^2+y_1^2=\frac{p^2+q^2}{1-p^2q^2}:$$

writing $x+y=\frac{1}{2}\varpi$, we have

$$\text{sn}\, y=\frac{\text{cn}\, x}{\text{dn}\, x},\ \text{ that is, }\ q=\frac{\sqrt{1-p^2}}{\sqrt{1+p^2}},\ \text{ or }\ q^2=\frac{1-p^2}{1+p^2},\ \text{ that is, }\ \frac{p^2+q^2}{1-p^2q^2}=1;$$

and thus to the line $x+y=\frac{1}{2}\varpi$, there corresponds the circle $x_1^2+y_1^2-1=0$. More precisely, to the line CD of figure 4, there corresponds the quarter-circumference CD

of figure 5: and similarly to DA, AB and BC of figure 4 the remaining quarter-circumferences DA, AB, BC of figure 5; that is, to the whole boundary of the square in figure 4 there corresponds the whole circumference of the circle in figure 5.

14. To any line in figure 4, parallel to the axis Ox or to the axis Oy, there corresponds in figure 5 a bicircular quartic of the form $x_1^2+y_1^2-2Ax_1^2-2By_1^2-1=0$. The investigation is substantially the same as that contained in No. 12, and need not be here given. But it is remarkable that also, to any line of figure 4 parallel to a side of the square (that is, to any line $x\pm y=c$), there corresponds in figure 5 a bicircular quartic of the like form (for the sides of the square, or lines $x\pm y=\pm\frac{1}{2}\varpi$ of figure 4, this bicircular quartic becomes the twice repeated circle $x_1^2+y_1^2-1=0$ of figure 5, which is the result just obtained in No. 13). I investigate this result as follows. Writing, as in No. 13,

$$\text{snl}\, x=p,\quad \text{snl}\, iy=i\,\text{snl}\, y=iq,$$

we have, as above,

$$x_1=\frac{p\sqrt{1-q^4}}{1-p^2q^2},\quad y_1=\frac{q\sqrt{1-p^4}}{1-p^2q^2},\ \text{and thence}\ x_1^2+y_1^2=\frac{p^2+q^2}{1-p^2q^2}.$$

Now assuming between x and y the relation $x+y=C$, and writing $\text{snl}\, C=c$, this gives

$$c=\frac{p\sqrt{1-q^4}+q\sqrt{1-p^4}}{1+p^2q^2};$$

to obtain the required curve, we must between these equations (three independent equations) eliminate p and q. We have

$$x_1+y_1=\frac{p\sqrt{1-q^4}+q\sqrt{1-p^4}}{1-p^2q^2},$$

and consequently

$$(1+p^2q^2)\,c=(1-p^2q^2)\,(x_1+y_1),$$

or, writing for convenience $\Omega=1-p^2q^2$, this equation gives

$$\Omega=\frac{2c}{x_1+y_1+c}.$$

Hence $\Omega x_1=p\sqrt{1-q^4}$, $\Omega y_1=q\sqrt{1-p^4}$; these equations may be written

$$\Omega^2x_1^2=\qquad p^2-(1-\Omega)\,q^2,$$
$$\Omega^2y_1^2=-(1-\Omega)\,p^2+\qquad q^2,$$

and from these equations obtaining the expressions for p^2 and q^2, and thence the expression for p^2q^2, $=1-\Omega$, we find, after some easy reductions,

$$\{(x_1^2+y_1^2)^2-1\}+\frac{\Omega^2}{1-\Omega}\,x_1^2y_1^2=\frac{4\,(1-\Omega)}{\Omega^2}.$$

But we have

$$\frac{\Omega^2}{1-\Omega}=\frac{4c^2}{(x_1+y_1)^2-c^2},$$

or substituting this value, the equation becomes

$$(x_1^2+y_1^2)^2-1+\frac{4c^2x_1^2y_1^2}{(x_1+y_1)^2-c^2}=\frac{(x_1+y_1)^2}{c^2}-1,$$

that is,

$$\left\{(x_1^2+y_1^2)^2-\frac{(x_1+y_1)^2}{c^2}\right\}\{(x_1+y_1)^2-c^2\}+4c^2x_1^2y_1^2=0.$$

Writing for a moment $x_1^2+y_1^2=P$, $x_1y_1=Q$, this is

$$\left\{P^2-\frac{1}{c^2}(P+2Q)\right\}\{P+2Q-c^2\}+4c^2Q^2=0,$$

an equation which contains the factor $P+2Q$; throwing this out, the equation becomes, after an easy reduction,

$$(P-1)^2+(1-c^2)\left\{P-2Q-\frac{1}{c^2}(P+2Q)\right\}=0,$$

that is,

$$(x_1^2+y_1^2-1)^2+(1-c^2)\left\{(x_1-y_1)^2-\frac{1}{c^2}(x_1+y_1)^2\right\}=0,$$

the required equation. Transforming through an angle of 45° by writing

$$x_1=\frac{x_2+y_2}{\sqrt{2}},\quad y_1=\frac{x_2-y_2}{\sqrt{2}},$$

(where observe that the axis of x_2 is the line FH of figure 5), the equation becomes

$$(x_2^2+y_2^2-1)^2+(1-c^2)\left(2y_2^2-\frac{1}{c^2}2x_2^2\right)=0,$$

or writing $c=\cos\gamma$ and therefore $1-c^2=\sin^2\gamma$, this equation becomes

$$(x_2^2+y_2^2)^2-\frac{2}{\cos^2\gamma}x_2^2-2\cos^2\gamma\,.\,y_2^2+1=0,$$

a curve consisting of two indented ovals situate symmetrically in regard to the axis Fy_2 of figure 5. In fact, writing in the equation $y_2=0$, we have for x_2^2 two real positive values; but, writing $x_2=0$, we have for y_2^2 two imaginary values. For $\gamma=0$, the equation becomes

$$(x_2^2+y_2^2)^2-2(x_2^2+y_2^2)+1=0,$$

that is, we have the circle $x_2^2+y_2^2-1=0$ twice repeated. One of the ovals is shown in figure 5; the portion of it lying within the circle agrees with Schwarz's figure, p. 113, turning this round through an angle of 45°.

For the lines $x-y=C$ of figure 4, we have in figure 5 the same system of bicircular quartics turned round through an angle of 90°.

II.

15. I consider the general problem of the orthomorphosis of a circle into a circle: we can, for the transformation of the circumference of the circle $x^2+y^2-1=0$ into that of the circle $x_1^2+y_1^2-1=0$, find a formula involving an arbitrary function

or (what is the same thing) an indefinite number of arbitrary constants. In fact, writing for shortness $z = x + iy$, $z_1 = x_1 + iy_1$, and $\bar{z}$, $\bar{z}_1$ for the conjugate functions $x - iy$, $x_1 - iy_1$; also $\phi(z)$ for a function of z involving in general imaginary coefficients $a + ib$, &c., and $\bar{\phi}(z)$ for the like function with the conjugate coefficients $a - ib$, &c.; then if we assume

$$z_1 = \frac{\phi(z)}{z^m \bar{\phi}\left(\frac{1}{z}\right)},$$

where m is any positive or negative integer, this implies

$$\bar{z}_1 = \frac{\bar{\phi}(\bar{z})}{\bar{z}^m \phi\left(\frac{1}{\bar{z}}\right)};$$

consequently, if $x^2 + y^2 - 1 = 0$, that is, $z\bar{z} = 1$, or $\bar{z} = \frac{1}{z}$, we have

$$\bar{z}_1 = \frac{\bar{\phi}\left(\frac{1}{z}\right)}{\left(\frac{1}{z}\right)^m \phi(z)} = \frac{z^m \bar{\phi}\left(\frac{1}{z}\right)}{\phi(z)}, \quad = \frac{1}{z_1},$$

or $z_1 \bar{z}_1 = 1$, that is, $x_1^2 + y_1^2 - 1 = 0$.

In a slightly different form, taking α, β, &c., to denote any imaginary quantities, and $\bar{\alpha}$, $\bar{\beta}$, ... the conjugate quantities; assuming

$$\phi(z) - (z - \alpha)(z - \beta) \ldots,$$

and taking m for the number of factors, we have

$$z_1 = \frac{(z - \alpha)(z - \beta) \ldots}{(1 - \bar{\alpha}z)(1 - \bar{\beta}z) \ldots},$$

and then (repeating the demonstration) we have

$$\bar{z}_1 = \frac{(\bar{z} - \bar{\alpha})(\bar{z} - \bar{\beta}) \ldots}{(1 - \alpha\bar{z})(1 - \beta\bar{z}) \ldots},$$

which, writing therein $\bar{z} = \frac{1}{z}$, becomes

$$\bar{z}_1 = \frac{\left(\frac{1}{z} - \bar{\alpha}\right)\left(\frac{1}{z} - \bar{\beta}\right) \ldots}{\left(1 - \frac{\alpha}{z}\right)\left(1 - \frac{\beta}{z}\right) \ldots}, \quad = \frac{(1 - \bar{\alpha}z)(1 - \bar{\beta}z) \ldots}{(z - \alpha)(z - \beta) \ldots} = \frac{1}{z},$$

and consequently, if $\bar{z} = \frac{1}{z}$, then also $\bar{z}_1 = \frac{1}{z_1}$ as before.

We may in the expression for z_1 introduce a factor $\frac{A}{\bar{A}}$, or, what is the same thing, a factor A which is such that $A\bar{A} = 1$. In particular, we thus have the solution

$$z_1 = \frac{A(z - \alpha)}{(1 - \bar{\alpha}z)},$$

giving

$$\bar{z}_1 = \frac{\bar{A}(\bar{z} - \bar{\alpha})}{1 - \alpha\bar{z}},$$

which, for $z\bar{z} = 1$, becomes

$$\bar{z}_1 = \frac{\bar{A}(1 - \bar{\alpha}z)}{z - \alpha},$$

so that ($A\bar{A}$ being $= 1$) this gives $z_1\bar{z}_1 = 1$.

16. This is a solution with three arbitrary constants, viz. A (which may be put $= \cos\lambda + i\sin\lambda$) is a single arbitrary constant, and α, $= a + ib$, is two arbitrary constants; and these constants may be so determined that a given point in the interior of the one circle, and a given point on the circumference thereof, shall correspond respectively to a given point in the interior of the other circle and to a given point on the circumference thereof. According to a well-known theorem of Riemann's, any two simply connected areas included within given closed curves respectively may be made to correspond to each other, and that in one way only, under the foregoing conditions as to a pair of interior points and a pair of boundary points: and we have, in what just precedes, the solution of the problem in the case of two equal circles.

17. In the case of any other solution, we thus know that the correspondence between the two circumferences cannot and does not imply a (1, 1) correspondence between the areas of the two circles: but it is interesting to enquire what happens. I take a very particular case

$$z_1 = \frac{z(z-2)}{1-2z},$$

and therefore

$$\bar{z}_1 = \frac{\bar{z}(\bar{z}-2)}{1-2\bar{z}},$$

and consequently

$$z_1\bar{z}_1 = \frac{z\bar{z}\{z\bar{z} - 2(z+\bar{z}) + 4\}}{1 - 2(z+\bar{z}) + 4z\bar{z}},$$

that is,

$$x_1^2 + y_1^2 = \frac{(x^2+y^2)(x^2+y^2-4x+4)}{4(x^2+y^2) - 4x + 1},$$

so that, writing $x^2 + y^2 - 1 = 0$, we have $x_1^2 + y_1^2 - 1 = 0$, a correspondence of the two circumferences. But to the circumference $x_1^2 + y_1^2 - 1 = 0$ there corresponds not only the circumference $x^2 + y^2 - 1 = 0$, but another circumference. In fact, writing $x_1^2 + y_1^2 = 1$, we have

$$4(x^2+y^2) - 4x + 1 = (x^2+y^2)(x^2+y^2-4x+4),$$

that is,

$$(x^2+y^2)^2 - 4x(x^2+y^2) + 4x - 1 = 0,$$

or

$$(x^2+y^2-1)(x^2+y^2-4x+1) = 0,$$

and there is thus the other circle

$$x^2 + y^2 - 4x + 1 = 0, \text{ or say } (x-2)^2 + y^2 - 3 = 0,$$

viz. this is a circle, coordinates of centre (2, 0) and radius $=\sqrt{3}$, cutting the circle $x^2+y^2-1=0$ in two real points. Referring to the figures 6 (xy) and 7 (x_1y_1), and observing that $x_1=0$, $y_1=0$, that is, $z_1=0$, gives $z=0$, or $z=2$, that is, the points

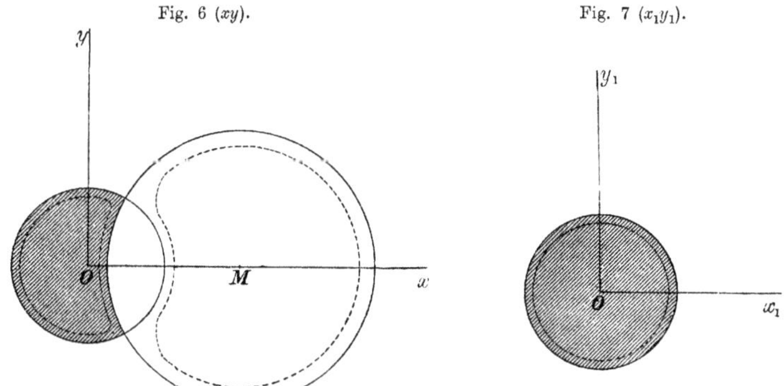

Fig. 6 (xy). Fig. 7 (x_1y_1).

$x=0$, $y=0$, and $x=2$, $y=0$, we see that to the centre O in figure 7, there correspond in figure 6 the points O, M which are the centres of the two circles. To any small closed curve, or say any small circle surrounding the point O of figure 7, there correspond in figure 6 small closed curves surrounding the points O, M respectively; and if in figure 7 the radius of the circle continually increases and becomes nearly equal to unity, the closed curves of figure 6 continually increase, changing at the same time their forms, and assume the forms shown by the dotted lines of figure 6. It thus appears that, to the whole area of the circle $x_1^2+y_1^2-1=0$ of figure 7, there correspond the two lunes ACB and ABD of figure 6; or if we attend only to the area included within the circle $x^2+y^2-1=0$ of this figure, then there corresponds not the whole area of this circle, but only the area of the lune ACB: and thus that the assumed relation $z_1=\dfrac{z(z-2)}{1-2z}$ establishes, in fact, an orthomorphosis of the circle $x_1^2+y_1^2-1=0$ into the lune ACB which lies inside the circle $x^2+y^2-1=0$ and outside the circle $(x-2)^2+y^2-3=0$. It may be added that, to the infinite area outside the circle $x_1^2+y_1^2-1=0$ of figure 7, there correspond in figure 6 first the area of the lens AB common to the two circles, and secondly the area outside the two circles: we have thus an orthomorphosis of the area outside the circle $x_1^2+y_1^2-1=0$ into these two areas respectively.

A somewhat more elegant example would have been that of the correspondence

$$z_1=\frac{z(z-\sqrt{2})}{1-\sqrt{2}z};$$

here, corresponding to the circumference $x_1^2+y_1^2-1=0$, we have the two equal circumferences $x^2+y^2-1=0$, and $(x-\sqrt{2})^2+y^2-1=0$: and to the whole area of the circle $x_1^2+y_1^2-1=0$, there correspond two equal lunes ACB and ABD.

922.

NOTE ON THE LUNAR THEORY.

[From the *Monthly Notices of the Royal Astronomical Society*, vol. LII. (1892), pp. 2—5.]

In the Lunar Theory, in whatever way worked out, the values ultimately obtained for the coordinates r, v, y should of course satisfy identically the equations of motion; and that they do so, is the ultimate verification of the correctness of the results obtained. It can hardly be hoped for that such a verification will ever be made for Delaunay's results; and yet it would seem generally that the labour of such a verification of the results to any extent, while exceeding (and possibly greatly exceeding) that of obtaining these results by any method employed for that purpose, ought still to be, so to speak, a labour of the same order. And one can, moreover, imagine the process of verification so arranged as to be a process of mere routine which could be carried out by ordinary computers. But, however this may be, I think it is not without interest to exhibit the verification to a very small extent, viz. to e, m^4.

I think there is an advantage in using capital letters for the arguments, and I accordingly write G (instead of Delaunay's g), to denote the mean anomaly.

The equations of motion are:

$$\frac{d^2r}{dt^2} - r\left\{\cos^2 y\left(\frac{dv}{dt}\right)^2\right\} + \frac{n^2a^3}{r^2} = \frac{d\Omega}{dr},$$

$$\frac{d}{dt}\left\{r^2\cos^2 y\left(\frac{dv}{dt}\right)\right\} = \frac{d\Omega}{dv},$$

$$\frac{d}{dt}\left(r^2\frac{dy}{dt}\right) + r^2\sin y\cos y\left(\frac{dv}{dt}\right)^2 = \frac{d\Omega}{dy},$$

$$\Omega = \frac{m'r^2}{r'^3}\left(\tfrac{3}{2}\cos^2 H - \tfrac{1}{2}\right), \text{ where } \cos H = \cos y\cos(v-v'),$$

or say

$$\Omega = \frac{m^2n^2a'^3}{r'^3}\, r^2\left\{\tfrac{3}{2}\cos^2 y\cos^2(v-v') - \tfrac{1}{2}\right\};$$

and thus the equations become

$$\frac{d^2r}{dt^2} - r\left\{\cos^2 y\left(\frac{dv}{dt}\right)^2 + \left(\frac{dy}{dt}\right)^2\right\} + \frac{n^2a^3}{r^2} = \frac{m^2n^2a'^3}{r'^3} r\left\{3\cos\left\{\tfrac{1}{2} + \tfrac{1}{2}\cos(2v-2v')\right\} - \tfrac{1}{2}\right\},$$

$$\frac{d}{dt} r^2\cos^2 y\left(\frac{dv}{dt}\right)^2 = \frac{m^2n^2a'^3}{r'^3} r^2\left\{-\tfrac{3}{2}\cos^2 y\sin(2v-2v')\right\},$$

$$\frac{d}{dt}\left(r^2\frac{dy}{dt}\right) + r^2\sin y\cos y\left(\frac{dv}{dt}\right)^2 = \frac{m^2n^2a'^3}{r'^3} r^2\left\{-3\sin y\cos y\,\tfrac{1}{2} + \tfrac{1}{2}\cos(2v-2v')\right\}.$$

To simplify as much as possible, take the Sun's orbit to be circular, i.e. $r' = a'$, $v' = mnt$; also neglect y^2: the first and second equations are

$$\frac{d^2r}{dt^2} - r\left(\frac{dv}{dt}\right)^2 + \frac{n^2a^3}{r^2} = m^2n^2r\left\{\tfrac{1}{2} + \tfrac{3}{2}\cos(2v-2v')\right\},$$

$$\frac{d}{dt} r^2 . \left(\frac{dv}{dt}\right) = m^2n^2r^2\left\{-\tfrac{3}{2}\sin(2v-2v')\right\};$$

and if for convenience of working we write $a = 1$, $n = 1$, then the first equation may be written

$$\frac{1}{r}\frac{d^2r}{dt^2} - \left(\frac{dv}{dt}\right)^2 + \frac{1}{r^3} = m^2\left\{\tfrac{1}{2} + \tfrac{3}{2}\cos(2v-2v')\right\},$$

and similarly the second equation is

$$\frac{1}{r^2}\frac{d}{dt}\left(r^2\frac{dv}{dt}\right) = -\tfrac{3}{2}m^2\sin(2v-2v');$$

and the third equation may be disregarded.

The two equations should be satisfied by Delaunay's values, putting therein $e' = 0$, $y = 0$; say by the values

$$\begin{aligned}
\frac{1}{r} = 1 &+ \tfrac{1}{6}m^2 - \tfrac{179}{288}m^4 \\
&+ (m^2 + \tfrac{19}{6}m^3 + \tfrac{131}{18}m^4)\cos 2D \\
&+ \qquad (\tfrac{7}{8}m^4)\cos 4D \\
&+ e\,(1 - \tfrac{7}{12}m^2) \qquad \cos G \\
&+ e\,(\tfrac{33}{16}m^2) \qquad \cos(2D + G) \\
&+ e\,(\tfrac{15}{8}m + \tfrac{187}{32}m^2) \qquad \cos(2D - G);
\end{aligned}$$

$$\begin{aligned}
v = t & \\
&+ (\tfrac{11}{8}m^2 + \tfrac{59}{12}m^3 + \tfrac{893}{72}m^4)\sin 2D \\
&+ \qquad (\tfrac{201}{256}m^4)\sin 4D \\
&+ 2e \qquad \sin G \\
&+ e\,(\tfrac{17}{8}m^2) \qquad \sin(2D + G) \\
&+ e\,(\tfrac{15}{4}m + \tfrac{263}{16}m^2) \qquad \sin(2D - G);
\end{aligned}$$

where

$$D = (1-m)\,t, \qquad G = (1 - \tfrac{3}{4}m^2)\,t, \qquad (v' = mt).$$

The verification for the first equation is

	$\frac{1}{r}\frac{d^2r}{dt^2} =$	$-\left(\frac{dv}{dt}\right)^2 =$	$\frac{1}{r^3} =$	$m^2\{-\frac{1}{2} - \frac{3}{2}(\cos 2v - 2v')\} =$	
Const.		-1	$+1$		$=0$
			$+\frac{1}{2}m^2$	$-\frac{1}{2}m^2$	$=0$
	$+2m^4$	$-\frac{121}{32}m^4$	$-\frac{9}{32}m^4$	$-\frac{33}{16}m^4$	$=0$
Cos $2D$	$+4m^2$	$-\frac{11}{2}m^2$	$+3m^2$	$-\frac{3}{2}m^2$	$=0$
	$+\frac{14}{3}m^3$	$-\frac{85}{6}m^3$	$+\frac{19}{2}m^3$		$=0$
	$+\frac{64}{9}m^4$	$-\frac{539}{18}m^4$	$+\frac{137}{6}m^4$		$=0$
Cos $4D$	$+8m^4$	$-\frac{161}{16}m^4$	$+\frac{33}{8}m^4$	$-\frac{33}{16}m^4$	$=0$
Cos G	e	$-4e$	$3e$		$=0$
	$-\frac{9}{4}em^2$	$+3em^2$	$-\frac{3}{4}em^2$		$=0$
Cos $(2D+G)$	$e\,\frac{193}{16}m^2$	$e\,(-\frac{73}{4}m^2)$	$e\,(\frac{147}{16}m^2)$	$e\,(-3m^2)$	$=0$
Cos $(2D-G)$	$em\,(\frac{15}{8})$	$em\,(-\frac{15}{2})$	$em\,(\frac{45}{8})$		$=0$
	$em^2\,(-\frac{5}{32})$	$em^2\,(-\frac{187}{8})$	$em^2\,(\frac{657}{32})$	$em^2\,(3)$	$=0.$

The verification for the second equation is

	$\frac{2}{r}\frac{dr}{dt}\cdot\frac{dv}{dt} =$	$+\frac{d^2v}{dt^2} =$	$+\frac{3}{2}m^2\sin(2v-2v') =$	
Sin $2D$	$+4m^2$	$-\frac{11}{2}m^2$	$+\frac{3}{2}m^2$	$=0$
	$+\frac{26}{3}m^3$	$-\frac{26}{3}m^3$		$=0$
	$+\frac{142}{9}m^4$	$-\frac{142}{9}m^4$		$=0$
Sin $4D$	$+\frac{21}{2}m^4$	$-\frac{201}{16}m^4$	$+\frac{33}{16}m^4$	$=0$
Sin G	$e\,.\,2$	$+e\,.-2$		$=0$
	$e\,.\,3m^2$	$+e\,.\,3m^2$		$=0$
Sin $(2D+G)$	$e\,.\,\frac{129}{8}m^2$	$e\,.-\frac{153}{8}m^2$	$e\,.\,3m^2$	$=0$
Sin $(2D-G)$	$e\,.\,\frac{15}{4}m$	$e\,.-\frac{15}{4}m$		$=0$
	$e\,.\,\frac{71}{16}m^2$	$e\,.-\frac{23}{16}m^2$	$e\,.-3m^2$	$=0;$

and the verification is thus completed.

The following intermediate results may be recorded; $\frac{1}{r} = 1 + \chi$,

	$\chi^2 =$	$-\log r =$	$\frac{dv}{dt} =$
	$\frac{19}{36}m^4$	$\frac{1}{6}m^2 - \frac{255}{288}m^4$	1
Cos $2D$	$+ \frac{1}{3}m^4$	$m^2 + \frac{19}{6}m^3 + \frac{64}{9}m^4$	$\frac{11}{4}m^2 + \frac{85}{12}m^3 + \frac{539}{36}m^4$
Cos $4D$	$+ \frac{1}{2}m^4$	$\frac{5}{8}m^4$	$\frac{201}{64}m^4$
Cos G	$+ \frac{1}{3}em^2$	$e(1 - \frac{3}{4}m^2)$	$e(2 - \frac{3}{2}m^2)$
Cos $2D + G$	$+ em^2$	$e(\frac{25}{16}m^2)$	$e(\frac{51}{8}m^2)$
Cos $2D - G$	$+ em^2$,	$e(\frac{15}{8}m + \frac{171}{32}m^2)$,	$e(\frac{15}{4}m + \frac{143}{16}m^2)$,

	$\frac{1}{r}\frac{dr}{dt} =$
Sin $2D$	$2m^3 + \frac{13}{3}m^3 + \frac{71}{9}m^4$
Sin $4D$	$\frac{5}{2}m^4$
Sin G	$e(1 - \frac{3}{2}m^2)$
Sin $(2D + G)$	$e(\frac{75}{16}m^2)$
Sin $(2D - G)$	$e(\frac{15}{8}m + \frac{51}{32}m^2)$.

These were, in fact, made use of for finding the foregoing values of r, $\frac{d^2r}{dt^2}$, &c.

923.

NOTE ON A HYPERDETERMINANT IDENTITY.

[From the *Messenger of Mathematics*, vol. XXI. (1892), pp. 131, 132.]

THE following is in effect a well-known theorem; but I am not sure whether it has been stated in a form at once so general and so precise.

If

$$\Omega = (*)(x_1, y_1)^A (x_2, y_2)^B (x_3, y_3)^C (x_4, y_4)^D \ldots$$

be a function separately homogeneous, and of the degrees $A, B, C, D, \ldots$ in the sets of variables (x_1, y_1), (x_2, y_2), (x_3, y_3), (x_4, y_4), ... respectively; and if

$$\overline{12} = \xi_1\eta_2 - \xi_2\eta_1, \; = \partial_{x_1}\partial_{y_2} - \partial_{x_2}\partial_{y_1}, \text{ \&c.,}$$

then

$$(A\,\overline{23} + B\,\overline{31} + C\,12)\,\Omega = 0,$$

when the variables (x_1, y_1), (x_2, y_2), (x_3, y_3), (x_4, y_4), ..., or only the variables (x_1, y_1), (x_2, y_2), (x_3, y_3) are therein severally replaced by (x, y).

In fact, we have

$$A\Omega = (x_1\xi_1 + y_1\eta_1)\,\Omega,\quad B\Omega = (x_2\xi_2 + y_2\eta_2),\quad C\Omega = (x_3\xi_3 + y_3\eta_3)\,\Omega;$$

thus the expression is

$$= \{(x_1\xi_1 + y_1\eta_1)\,\overline{23} + (x_2\xi_2 + y_2\eta_2)\,\overline{31} + (x_3\xi_3 + y_3\eta_3)\,\overline{12}\}\,\Omega,$$

and if we herein replace the variables (x_1, y_1), (x_2, y_2), (x_3, y_3) *in so far as they appear explicitly* by (x, y), the expression becomes

$$= \{(x\xi_1 + y\eta_1)\,\overline{23} + (x\xi_2 + y\eta_2)\,\overline{31} + (x\xi_3 + y\eta_3)\,\overline{12}\}\,\Omega,$$

where the factor in { }, substituting for $\overline{23}$, $\overline{31}$, $\overline{12}$ their values $\xi_2\eta_3 - \xi_3\eta_2$, $\xi_3\eta_1 - \xi_1\eta_3$, $\xi_1\eta_2 - \xi_2\eta_1$, becomes identically $=0$. The value of the expression is thus $=0$; and of course it remains $=0$ when, consequently upon the foregoing change (x_1, y_1), (x_2, y_2), (x_3, y_3) each into (x, y), we also change (ξ_1, η_1), (ξ_2, η_2), (ξ_3, η_3) each into (ξ, η); and if we also change (x_4, y_4), &c., into (x, y).

It is clear that Ω may denote the covariant symbol

$$\Omega = \overline{23}^{\alpha}\, \overline{31}^{\beta}\, \overline{12}^{\gamma}\, \overline{14}^{\delta}\, \overline{24}^{\epsilon}\, \overline{34}^{\zeta} \ldots U_1 V_2 W_3 T_4 \ldots,$$

where U, V, W, T, ... denote quantics

$$(a, \ldots \between x, y)^m, \; (a', \ldots \between x, y)^n, \; \ldots$$

of the degrees m, n, p, q, ... respectively, and U_1, V_2, &c., are the corresponding functions $(a, \ldots \between x_1, y_1)^m$, $(a', \ldots \between x_2, y_2)^n$, &c., the values of A, B, C being here

$$A = m - \beta - \gamma - \delta - \ldots,$$

$$B = n \; - \gamma \; - \alpha \; - \epsilon \; - \ldots,$$

$$C = p \; - \alpha \; - \beta - \zeta - \ldots;$$

the theorem expresses the property that the covariants $\overline{12}\Omega$, $\overline{23}\Omega$, $\overline{31}\Omega$, are linearly connected together; or, writing it in the form $(A\,\overline{12} - C\,\overline{13} + B\,\overline{23})\,\Omega = 0$, we have the proper linear combination $A\,\overline{12}\Omega - C\,\overline{13}\Omega$ of the two covariants $\overline{12}\Omega$ and $\overline{13}\Omega$, equal to $-B\,\overline{23}\Omega$, a determinate multiple of $\overline{23}\Omega$. Speaking roughly, we say that the *difference* of the covariants $\overline{12}\Omega$ and $\overline{13}\Omega$ is equal to $\overline{23}\Omega$.

924.

ON THE NON-EXISTENCE OF A SPECIAL GROUP OF POINTS.

[From the *Messenger of Mathematics*, vol. XXI. (1892), pp. 132, 133.]

It is well known that, taking in a plane any eight points, every cubic through these passes through a determinate ninth point. It is interesting to show that there is no system of seven points such that every cubic through these passes through a determinate eighth point.

Assuming such a system: first, no three of the points can be in a line: for, if they were, then among the cubics through the seven points we have the line through the three points and an arbitrary conic through the remaining four points, and these composite cubics have no common eighth point of intersection.

Secondly, no six of the points can be on a conic: for, if they were, then among the cubics through the seven points we have the conic through the six points and an arbitrary line through the remaining point, and these composite cubics have no common eighth point of intersection.

Taking now the points to be 1, 2, 3, 4, 5, 6, 7; among the cubics through these, we have the composite cubics (A, P), (B, Q), (C, R), where A, B, C are the lines 67, 75, 56, and P, Q, R the conics 12345, 12346, 12347 respectively; by what precedes, the points 5, 6, 7 do not lie on a line, and the points (6, 7), (7, 5) and (5, 6) neither of them lie on the conics P, Q, R respectively.

The common eighth point of intersection, if it exists, must be

$$(A, B, C);\ (A, Q, R),\ (B, R, P),\ (C, P, Q);$$
$$(P, B, C),\ (Q, C, A),\ (R, A, B);\ \text{or}\ (P, Q, R).$$

There is no point (A, B, C).

There is no point (A, Q, R): for Q, R intersect only in the points 1, 2, 3, 4, no one of which lies on A; and similarly, there is no point (B, R, P) or (C, P, Q).

B, C intersect in 5, which is a point on P; and thus 5 is the only point (P, B, C). Similarly, 6 is the only point (Q, C, A) and 7 is the only point (R, A, B).

P, Q intersect in 1, 2, 3, 4, which are each of them on R; hence 1, 2, 3, 4 are the only points (P, Q, R).

Hence the points 1, 2, 3, 4, 5, 6, 7 present themselves each once, and only once, among the intersections of the three cubics, and there is no common eighth point of intersection.

925.

ON WARING'S FORMULA FOR THE SUM OF THE mth POWERS OF THE ROOTS OF AN EQUATION.

[From the *Messenger of Mathematics*, vol. XXI. (1892), pp. 133—137.]

THE formula in question, Prob. I. of Waring's *Meditationes Algebraicæ*, Cambridge, 1782, making therein a slight change of notation, is as follows: viz. the equation being

$$x^n + bx^{n-1} + cx^{n-2} + dx^{n-3} + \ldots = 0,$$

then we have

$$(-)^m S_m =$$

$$
\begin{array}{ll}
b^m & \\
-mc \; b^{m-2} & \\
+md \; b^{m-3} & \\
\left.\begin{array}{r} -me \\ +\tfrac{1}{2}m\,.\,m-3\,.\,c^2 \end{array}\right\} b^{m-4} & \\
\left.\begin{array}{r} +mf \\ -m\,.\,m-4\,.\,cd \end{array}\right\} b^{m-5} & \\
\left.\begin{array}{r} -m\,.\,g \\ +m\,.\,m-5\,.\,ce \\ +\tfrac{1}{2}m\,.\,m-5\,.\,d^2 \\ -\tfrac{1}{6}m\,.\,m-4\,.\,m-5\,.\,c^3 \end{array}\right\} b^{m-6} & \\
\left.\begin{array}{r} +m\,.\,h \\ -m\,.\,m-6\,.\,cf \\ -m\,.\,m-6\,.\,de \\ +\tfrac{1}{2}m\,.\,m-5\,.\,m-6\,.\,c^2d \end{array}\right\} b^{m-7} & \\
\left.\begin{array}{r} +m\,.\,i \\ +m\,.\,m-7\,.\,cg \\ +m\,.\,m-7\,.\,df \\ +\tfrac{1}{2}m\,.\,m-7\,.\,e^2 \\ -\tfrac{1}{2}m\,.\,m-6\,.\,m-7\,.\,c^2e \\ -\tfrac{1}{2}m\,.\,m-6\,.\,m-7\,.\,cd^2 \\ +\tfrac{1}{24}m\,.\,m-5\,.\,m-6\,.\,m-7\,.\,c^4 \end{array}\right\} b^{m-8} & \\
+\,\&c., &
\end{array}
$$

where, reckoning the weights of b, c, d, e, ... as 1, 2, 3, 4, ..., respectively, the several terms are all the terms of the weight m, or (what is the same thing) in the coefficient of $b^{m-\theta}$ we have all the combinations of c, d, e, ..., (or say all the non-unitary combinations) of the weight θ, and where the numerical coefficient of

$$b^{m-\theta}c^{\mathrm{c}}d^{\mathrm{d}}e^{\mathrm{e}}\ldots\ (\mathrm{c}+\mathrm{d}+\mathrm{e}+\ldots=\theta),$$

is

$$=(-)^{\mathrm{c}+\mathrm{e}+\mathrm{g}+\ldots}\frac{m\,.\,m-(\theta-\delta+1)\,.\,m-(\theta-\delta+2)\ldots m-(\theta-1)}{\Pi\mathrm{c}\,.\,\Pi\mathrm{d}\,.\,\Pi\mathrm{e}\ldots}.$$

Thus for the term $b^{m-8}c^2e^1$, $\theta=8$; c, d, e $=2$, 4, 1 respectively (the other exponents each vanishing), and the coefficient is

$$(-)^3\frac{m\,.\,m-6\,.\,m-7}{1\,.\,2\,.\,1},\ =-\tfrac{1}{2}m\,.\,m-6\,.\,m-7,$$

as above; and so in other cases.

For the MacMahon form

$$1+bx+\frac{cx^2}{1\,.\,2}+\ldots=(1-\alpha x)(1-\beta x)\ldots,$$

or say

$$y^n+\frac{b}{1}y^{n-1}+\frac{c}{1\,.\,2}y^{n-2}+\ldots=(y-\alpha)(y-\beta)\ldots,$$

we must for b, c, d, ..., write b, $\frac{c}{1\,.\,2}$, $\frac{d}{1\,.\,2\,.\,3}$, ... respectively: we thus have

$$\begin{aligned}(-)^m\,.\,S_m=\ & b^m\\ & -m\frac{c}{1\,.\,2} && b^{m-2}\\ & +m\frac{d}{1\,.\,2\,.\,3} && b^{m-3}\\ & \left.\begin{aligned}&-m\frac{e}{1\,.\,2\,.\,3\,.\,4}\\ &+\tfrac{1}{2}m\,.\,m-3\left(\frac{c}{1\,.\,2}\right)^2\end{aligned}\right\} && b^{m-4}\\ & +\&c.,\end{aligned}$$

or say

$$\begin{aligned}(-)^m\,\Pi(m-1)\,S_m=\ & \Pi(m-1) && b^4\\ & -\Pi m\frac{c}{1\,.\,2} && b^{m-2}\\ & +\Pi m\frac{d}{1\,.\,2\,.\,3} && b^{m-3}\\ & \left.\begin{aligned}&-\Pi m\frac{e}{1\,.\,2\,.\,3\,.\,4}\\ &+\Pi m\frac{m-3}{2}\left(\frac{c}{1\,.\,2}\right)^2\end{aligned}\right\} && b^{m-4}\\ & +\&c.,\end{aligned}$$

the numerical coefficient of

$$b^{m-\theta}c^{\mathrm{c}}d^{\mathrm{d}}e^{\mathrm{e}}\ldots\ (\mathrm{c}+\mathrm{d}+\mathrm{e}+\ldots=\theta)$$

being

$$(-)^{\mathrm{c}+\mathrm{e}+\mathrm{g}+\ldots}\frac{\Pi m\,.\,m-(\theta-\delta+1)\,.\,m-(\theta-\delta+2)\ldots m-(\theta-1)}{\Pi\mathrm{c}\,.\,\Pi\mathrm{d}\,.\,\Pi\mathrm{e}\ldots(\Pi 2)^{\mathrm{c}}\,(\Pi 3)^{\mathrm{d}}\,(\Pi 4)^{\mathrm{e}}\ldots}.$$

It is convenient to write down the literal terms in alphabetical order (AO), calculating and affixing to each term the proper numerical coefficient; thus taking

$$1+bx+c\frac{x^2}{1.2}+\ldots=(1-\alpha x)(1-\beta x)(1-\gamma x)\ldots,$$

we find

$$\begin{array}{rll} -120S_6= & g & .\ \ \ \ \ 1 \\ & bf & .-\ \ \ 6 \\ & ce & .-\ \ 15 \\ & d^2 & .-\ \ 10 \\ & b^2e & .+\ \ 30 \\ & bcd & .+120 \\ & c^3 & .+\ \ 30 \\ & b^3d & .-120 \\ & b^2c^2 & .-270 \\ & b^4c & .+360 \\ & b^6 & .-120 \\ \hline & & \pm 541 \end{array}$$

this expression, as representing the value of the non-unitary function S_6, being in fact a seminvariant.

It is to be remarked that the foregoing expression for the sum of the mth powers of the roots of the equation

$$x^n+bx^{n-1}+cx^{n-2}+\ldots=0$$

is, in fact, the series for x^m continued so far only as the exponent of b is not negative: see as to this Note XI. of Lagrange's *Équations Numériques.* For the *à posteriori* verification, observe that we have

$$x+b+\frac{c}{x}+\frac{d}{x^2}+\ldots=0,$$

or writing for a moment $u=-b$, say this is

$$x=u+fx,$$

where

$$fx = -\frac{c}{x} - \frac{d}{x^2} - \frac{e}{x^3} - \&c.$$

Hence, by Lagrange's theorem,

$$\begin{aligned} x^m = \quad & u^m \\ & - mu^{m-1}\left(\frac{c}{u} + \frac{d}{u^2} + \frac{e}{u^3} + \ldots\right) \\ & + \left\{mu^{m-1}\left(\frac{c}{u} + \frac{d}{u^2} + \frac{e}{u^3} + \ldots\right)^2\right\}' \frac{1}{1.2} \\ & - \left\{mu^{m-1}\left(\frac{c}{u} + \frac{d}{u^2} + \frac{e}{u^3} + \ldots\right)^3\right\}'' \frac{1}{1.2.3} \\ & + \&c., \end{aligned}$$

where the accents denote differentiations in regard to u. This is

$$\begin{aligned} = \quad & u^m \\ & - m\{cu^{m-2} + du^{m-3} + eu^{m-4} + fu^{m-5} + gu^{m-6} + \ldots\} \\ & + \tfrac{1}{2}m\{(m-3)\,c^2u^{m-4} + (m-4)\,2cdu^{m-5} + (m-5)\,(d^2 + 2ce)\,u^{m-6} + \ldots\} \\ & - \tfrac{1}{6}m\{(m-4)(m-5)\,c^3u^{m-6} + \ldots\} \\ & + \&c. \end{aligned}$$

$$\begin{aligned} = \quad & u^m \\ & + u^{m-2}\,.-mc \\ & + u^{m-3}\,.-md \\ & + u^{m-4}\,.-me + \tfrac{1}{2}m\,.\,m-3\,.\,c^2 \\ & + u^{m-5}\,.-mf + \tfrac{1}{2}m\,.\,m-4\,.\,2cd \\ & + u^{m-6}\,.-mg + \tfrac{1}{2}m\,.\,m-5\,.\,(d^2 + 2ce) - \tfrac{1}{6}m\,.\,m-4\,.\,m-5\,.\,c^3 \\ & + \&c., \end{aligned}$$

which, putting therein $u = -b$ and multiplying each side by $(-)^m$, is the before-mentioned formula for $(-)^m S\alpha^m$: in that formula the series being continued only so far as the exponent of b is not negative.

I notice also that we cannot easily, by means of the known formula

$$S\alpha^m\beta^p = S\alpha^m\,.\,S\alpha^p - S\alpha^{m+p},$$

deduce an expression for $S\alpha^m\beta^p$: in fact, forming the product of the series for $S\alpha^m$, $S\alpha^p$ respectively, this product is identically equal to the series for $S\alpha^{m+p}$, or we seem to obtain $0 = S\alpha^m S\alpha^p - S\alpha^{m+p}$; to obtain the correct formula, we have to take each of the three series only so far as the exponent of b therein respectively is not negative: and it is not easy to see how the resulting formula is to be expressed.

926.

CORRECTED SEMINVARIANT TABLES FOR THE WEIGHTS 11 AND 12.

[From the *American Journal of Mathematics*, vol. XIV. (1892), pp. 195—200.]

THE tables in my paper, "Seminvariant Tables," *American Journal of Mathematics*, vol. VII. (1885), pp. 59—73, [831], are not in the best form, but the deviations present themselves only in a few columns of the tables for the weights 11 and 12, viz. in the former of these two columns, and in the latter a single column, ought to be replaced by linear combinations of other columns; there are, besides, columns which should be new named; in regard hereto, there is a point of theory which requires to be made clear. I remark that in each table the literal terms are in alphabetical order (AO); this is the proper order for the final terms, and although (as about to be explained) the proper order for the initial terms is the counter order (CO), yet as the tables cannot be at the same time arranged in the one and the other order, I adhere to the AO as the proper arrangement for the terms of the tables; we have, however, to introduce the notion that, in general, it is not the top term of a column which is to be regarded as the initial term, and in connexion herewith to consider how the columns are to be named. An instance first presents itself for the weight 11: we have, see column 5 of the table for that weight here given, a seminvariant

$$\begin{array}{ll} fg & +\ 1 \\ b^2j & \\ bci & \\ bdh & \\ beg & -\ 5 \\ bf^2 & -\ \underset{\sim\sim}{6} \\ c^2h & -\ \underset{\sim\sim}{16} \\ \vdots & \\ b^3e^2 & +\ 70\,; \end{array}$$

this is to be regarded, not as a seminvariant $fg - b^3e^2$ (it is hardly necessary to remark that, here and elsewhere the – is not a minus sign, but a mere stroke), but as a seminvariant $c^2h - b^3e^2$, viz. c^2h is a term entering into the seminvariant, and which, although it is in AO subsequent, it is in CO precedent, to the terms fg, beg and bf^2. The seminvariant contains the term $-16c^2h$ and other terms with the letter h, and it is a misnomer to call it $fg - b^3e^2$, a name implying that the highest letter thereof is g. Instead of the stroke, it would perhaps be better to write ∞, for instance $c^2h \infty b^3e^2$, where of course ∞ would be used as a mere conventional symbol.

For greater clearness, I give here the express definition of counter order, (CO), viz. whereas in AO we begin with the lowest letters, in CO we begin contrariwise with the highest letters. A term containing a higher letter or higher power of such letter precedes a term containing a lower letter or lower power of the same letter—or in the easiest form, the counter order is the alphabetical order corresponding to the reversed arrangement $z, y, \ldots, f, e, d, c, b$ of the letters.

A symbol, as $c^2h - b^3e^2$ above, may be regarded as referring to a set of terms c^2h, b^3e^2 and all the terms which are in CO subsequent to c^2h and in AO precedent to b^3e^2: as by supposition the terms are arranged in AO, the set includes no term lower than b^3e^2, or say the bottom term b^3e^2 is also the final term of the set, but it does include terms fg, beg and bf^2 higher than c^2h, and thus the top term fg is not, but c^2h is, the initial term of the set. It should be remarked that a seminvariant $ch - b^3e^2$ need not include all the terms of the set as just defined: there may very well be terms with a coefficient zero, or say accidental zeros; an instance presents itself, weight 10, where in the column $eg - bd^3$ we have $0ce^2$, no term in ce^2.

The changes actually required are very slight, viz.

Weight 11, instead of

$fg - b^3e^2$, we require $c^2h - b^3e^2$, old $fg - b^3e^2$, new named,

$c^2h - b^3d^2$, " $fg - b^3d^2$, linear combination $(fg - b^3e^2) + 8(c^2h - b^3d^2)$,

$de^2 - b^5d^2$, " $c^3f - b^5d^2$, old $de^2 - b^5d^2$, new named,

$c^3f - b^3c^4$, " $de^2 - b^3c^4$, linear combination $(de^2 - b^5d^2) + 6(c^3f - b^3c^4)$.

Weight 12, instead of

$cf^2 - bcd^3$, we require $d^2g - bcd^3$, old $cf^2 - bcd^3$, new named,

$d^2g - c^3d^2$, " $cf^2 - c^3d^2$, linear combination $(cf^2 - bcd^3) - 5(d^2g - c^3d^2)$;

but I have thought it desirable to give the complete tables for the weights in question, 11 and 12; and I have also rearranged the entire columns of the two tables so as to present in each of them the *finals* in AO. This is the case with the existing tables, except that there is a single transposition in the table weight 10.

Instead of the columns $cdf - b^4d^2$, $ce^2 - c^5$, we ought to have $ce^2 - c^5$, $cdf - b^4d^2$. The complete list up to the weight 12 is

$w=$	
2	$c - b^2$
3	$d - b^3$
4	$e - c^2$
	$c^2 - b^4$
5	$f - bc^2$
	$cd - b^5$
6	$g - d^2$
	$ce - c^3$
	$d^2 - b^2c^2$
	$c^3 - b^6$
7	$h - bd^2$
	$cf - bc^3$
	$de - b^3c^2$
	$c^2d - b^7$
8	$i - e^2$
	$cg - cd^2$
	$df - b^2d^2$
	$e^2 - c^4$
	$c^2e - b^2c^3$
	$cd^2 - b^4c^2$
	$c^4 - b^8$

$w=$	
9	$j - be^2$
	$ch - d^3$
	$dg - bcd^2$
	$ef - b^3d^2$
	$c^2f - bc^4$
	$cde - b^3c^3$
	$d^3 - b^5c^2$
	$c^3d - b^9$
10	$k - f^2$
	$ci - ce^2$
	$dh - b^2e^2$
	$eg - bd^3$
	$f^2 - c^2d^2$
	$c^2g - b^2cd^2$
	$ce^2 - c^5$
	$cdf - b^4d^2$
	$d^2e - b^2c^4$
	$c^3e - b^4c^3$
	$c^2d^2 - b^6c^2$
	$c^5 - b^{10}$

$w=$	
11	$l - bf^2$
	$cj - de^2$
	$di - bce^2$
	$eh - cd^3$
	$c^2h - b^3e^2$
	$fg - b^2d^3$
	$cef - bc^2d^2$
	$cdg - b^3cd^2$
	$d^2f - bc^5$
	$c^3f - b^5d^2$
	$de^2 - b^3c^4$
	$c^2de - b^5c^3$
	$cd^3 - b^7c^2$
	$c^3d - b^{11}$

$w=$	
12	$m - g^2$
	$ck - cf^2$
	$ei - e^3$
	$dj - b^2f^2$
	$fh - bde^2$
	$g^2 - c^2e^2$
	$ceg - d^4$
	$c^2i - b^2ce^2$
	$d^2g - bcd^3$
	$cf^2 - c^3d^2$
	$cdh - b^4e^2$
	$def - b^3d^3$
	$e^3 - b^2c^2d^2$
	$c^2e^2 - c^6$
	$c^3g - b^4cd^2$
	$cd^2e - b^2c^5$
	$c^2df - b^6d^2$
	$d^4 - b^4c^4$
	$c^4e - b^6c^3$
	$c^3d^2 - b^8c^2$
	$c^6 - b^{12}$

The two new tables are as follows: some accidental numerical errors have been corrected.

TABLE, WEIGHT 11.

	1	2	3	4	5	6	7	8	9	10	11	12	13	14
	l	cj	di	eh	c^2h	fg	cef	cdg	d^2f	c^3f	de^2	c^2de	cd^3	c^4d
l	+ 1													
bk	− 11													
cj	+ 35	+ 2												
di	− 75	− 9	+ 1											
eh	+ 90	+14	− 2	+ 1										
fg	− 42	− 7	+ 1	− 1	+ 1	+ 1								
b^2j	+ 20	− 2												
bci	− 90	+ 9	− 3											
bdh	+240	+16		− 4										
beg	−420	−63	+ 9	− 2	− 5	− 5								
bf^2	+252	+42	− 6	+ 6	− 6	− 6								
c^2h		−30	+10	+ 3	− 16									
cdg		+70	−26	− 2	+ 58	+ 2		+1						
cef		−21	+ 7	− 6	+ 5	+ 45	+1	−3						
d^2f		−56	+24	+10	−100	−100	−3	+6	+1					
de^2		+35	−15	− 5	+ 60	+ 60	+2	−4	−1	+1	+1			
b^3i			+ 2											
b^2ch			− 8		+ 32									
b^2dg			+ 8	+20	− 48	+ 8		−1						
b^2ef			− 4	−18	+ 40		−1	+3						
	1	2	3	4	5	6	7	8	9	10	11	12	13	14

TABLE, WEIGHT 11 (*continued*).

		1	2	3	4	5	6	7	8	9	10	11	12	13	14	
		l	cj	di	eh	c^2h	fg	cef	cdg	d^2f	c^3f	de^2	c^2de	cd^3	c^4d	
21	bc^2g			+ 8	− 15	− 62	− 6		− 3							21
	$bcdf$			− 16	− 24	+ 232	+ 408	+ 14	− 30	− 6						
	bce^2			+ 10	+ 45	− 205	− 405	− 11	+ 27	+ 3	− 3	− 3				
	bd^2e				− 10	+ 20	+ 20	− 1	+ 2	+ 3	− 8	− 8				
	c^3f				+ 27	− 54	− 270	− 9	+ 27	+ 4	− 12					
	c^2de				− 45	+ 90	+ 450	+ 14	− 45	− 6	+ 36	+ 6	+ 1			
	cd^3				+ 20	− 40	− 200	− 6	+ 20	+ 2	− 18		− 1	+ 1		
	b^4h					− 16										
	b^3cg					+ 56			+ 5							
30	b^3df					− 112	− 288	− 8	+ 18	+ 4						30
	b^3e^2					+ 70	+ 270	+ 9	− 23	− 2	+ 2	+ 2				
	b^2c^2f						+ 216	+ 6	− 27	− 3	+ 36					
	b^2cde						− 360	− 16	+ 51	+ 6	− 36	+ 24	− 2			
	b^2d^3						+ 160	+ 8	− 24	− 8	+ 34	+ 16	+ 1	− 1		
	bc^3e							+ 3		− 2	− 48	− 18	− 3			
	bc^2d^2							− 2		+ 6	+ 18	− 24	+ 5	− 9		
	c^4d									− 1	+ 3	+ 9	− 1	+ 4	+ 1	
	b^5g								− 2							
	b^4cf								+ 6		− 36					
40	b^4de								− 2	− 4	+ 14	− 16	+ 1			40
	b^3c^2e								− 6	+ 3	+ 72	+ 12	+ 8			
	b^3cd^2								+ 4	+ 8	− 78	− 48	− 7	+ 15		
	b^2c^3d									− 10	+ 30	+ 72	− 5	+ 11	− 4	
	bc^5									+ 3	− 9	− 27	+ 3	− 12	− 3	
	b^6f										+ 12					
	b^5ce										− 30		− 7			
	b^5d^2										+ 20	+ 32	+ 2	− 6		
	b^4c^2d											− 48	+ 10	− 39	+ 6	
	b^3c^4											+ 18	− 5	+ 29	+ 14	
50	b^7e												+ 2			50
	b^6cd												− 4	+ 32	− 4	
	b^5c^3												+ 2	− 23	− 26	
	b^8d													− 8	+ 1	
	b^7c^2													+ 6	+ 24	
	b^9c														− 11	
56	b^{11}														+ 2	56
		bf^2	de^2	bce^2	cd^3	b^3e^2	b^2d^3	bc^2d^2	b^3cd^2	bc^5	b^5d^2	b^3c^4	b^5c^3	b^7c^2	b^{11}	
		± 638	± 188	± 70	± 132	± 664	± 1640	± 57	± 170	± 43	± 278	± 192	± 35	± 98	± 48	

TABLE WEIGHT 12.

	1	2	3	4	5	6	7	8	9	10	11	12	13	14	15	16	17	18	19	20	21
	m	ck	ei	dj	fh	g^2	ceg	c^2i	d^2g	cf^2	cdh	def	e^3	c^2e^2	c^3g	cd^2e	c^2df	d^4	c^4e	c^3d^2	c^6
m	+1																				
bl	−12																				
ck	+66	+3																			
dj	−220	−15		+15																	
ei	+495	+40	+1	−40																	
fh	−792	−70	−4	+70	+25																
g^2	+462	+42	+3	−40	−24	+1															
b^2k		−3																			
bcj		+15		−45																	
bdi		−25	−4	+25																	
beh		+30	+12	−30	−125																
bfg		−14	−8	+12	+113	−12															
c^2i		−15	+3	+150				+1													
cdh		+40	−8	−400	+50			−4			+4										
ceg		−70	−22	+700	+680	−70	+1	+8			−8										
cf^2		+42	+24	−420	−675	+100	−1	−5	+2	+2	+5										
d^2g			+24		−570	+80	−1		+5												
def			−36		+925	−200	+2		−19	−4		+18									
e^3			+15		−400	+100	−1		+12	+2		−17	+1								
b^3j				+30																	
b^2ci				−135				−2													
b^2dh				+360	+200			+4			−4										
b^2eg				−630	−525	+100	−1	−8			+8										
b^2f^2				+378	+336	−64	+1	+5	−2	−2	−5										
bc^2h					−150			+4			−12										
$bcdg$					+350	−200	+2	−4	−30		+4										
$bcef$					−105	+20	−2	+2	+37	−8	−2	−54									
bd^2f					−280	+320	−2		+46	+16		−72									
bde^2					+175	−200	+2		−49	−4		+114	−12								
c^3g						+100	−1	−4	+20		+32				+1						
c^2df						+200	+2	+8	−49	−4	−64	+54			−3		+3				
c^2e^2						+125	+1	−5	−32	+28	+40	+162	−18	+1							
cd^2e							−3		+91	−44		−342	+54	−2	+4	+1					
d^4							+1		−32	+18		+135	−27	+1	−2	−1	−2	+1			
b^4i								+1													
b^3ch								−4			+20										
b^3dg								+4	+20		−4										
b^3ef								−2	−18	+12	+2	+36									
b^2c^2g								+4	−15		−60				−3						
b^2cdf								−8	−24	−24	+120	+216			+6		−6				
b^2ce^2								+5	+45	−30	−75	−360	+54	−2	+2						
b^2d^2e									−10	+20		+66	−6	+2	−6	−1					
bc^3f									+27	+12		−162			+3		−9				
bc^2de									−45	+60		+486	−180	+4	−8	−6	−15				
bcd^3									+20	−40		−252	+108	−4	+7	+8	+24	−12			
c^4e										−30		−81	+81	−2	+1	+4	+30		+1		
c^3d^2										+20		+54	−54	+2	−2	−5	−28	+8	−1	+1	
	1	2	3	4	5	6	7	8	9	10	11	12	13	14	15	16	17	18	19	20	21

TABLE, WEIGHT 12 (*continued*).

	1	2	3	4	5	6	7	8	9	10	11	12	13	14	15	16	17	18	19	20	21	
	m	ck	ei	dj	fh	g^2	ceg	c^2i	d^2g	cf^2	cdh	def	e^3	c^2e^2	c^3g	cd^2e	c^2df	d^4	c^4e	c^3d^2	c^6	
b^5h											-8											48
b^4cg											+28				+3							
b^4df											-56	-144			-3		+3					50
b^4e^2											+35	+135	-27	+1								
b^3c^2f												+108			-6		+24					
b^3cde												-180	+108	-4	+9	+10	+30					
b^3d^3												+80	-64		+1	-4	-16	+8				
b^2c^3e													-54	+2	+2	-7	-75		-4			
$b^2c^2d^2$													+36	+4	-12	-9	+6	+30	+3	-3		
bc^4d														-4	+8	+14	+48	-48	+2	-6		
c^6														+1	-2		+2	+16	-1	+4	+1	
b^6g															-1	-4	-16					
b^5cf															+3		-21					60
b^5de															-1	-4	-15					
b^4c^2e															-3	+3	+60		+6			
b^4cd^2															+2	+8		-48	-3	+3		
b^3c^3d																-10	-40	+68	-6	+22		
b^2c^5																+3	+12	-24	+3	-15	-6	
b^7f																	+6					
b^6ce																	-15		-4			
b^6d^2																	+10	+16	-23	-1		
b^5c^2d																		-24	+30	-30		
b^4c^4																		+9	-3	+21	+15	70
b^8e																			+1			
b^7cd																			+2	+14	-20	
b^6c^3																			-3	-9		
b^9d																				-4		
b^8c^2																				+3	+15	
$b^{10}c$																					-6	
b^{12}																					+1	77
	g^2	cf^2	e^3	b^2f^2	bde^2	c^2e^2	d^4	b^2ce^2	bcd^3	c^3d^2	b^4e^2	b^3d^3	$b^2c^2d^2$	c^6	b^4cd^2	b^2c^5	b^6d^2	b^4c^4	b^6c^3	b^8c^2	b^{12}	
	±1024	±212	±82	±1740	±2854	±946	±12	±46	±325	±190	±298	±1664	±442	±18	±52	±51	±258	±156	±48	±68	±32	

927.

ON CLIFFORD'S PAPER "ON SYZYGETIC RELATIONS AMONG THE POWERS OF LINEAR QUANTICS."

[From the *Proceedings of the London Mathematical Society*, vol. XXIII. (1892), pp. 99—104.]

THE paper in question, originally printed, *Proc. Lond. Math. Soc.*, t. III. (1869), pp. 9—12, is reproduced No. XIV., pp. 119—122, of the *Mathematical Papers* (8vo. Lond. 1882), where it is immediately followed by the paper No. XV., "On Syzygetic Relations connecting the Powers of Linear Quantics," pp. 123—129. The author, after referring to theorems in M. Paul Serret's *Géométrie de Direction* (Paris, 1869), proceeds as follows:—"By the use of Professor Sylvester's method of contravariant differentiation, I have arrived at certain extensions of these theorems, which I now proceed to explain," and he then states his Theorem I.: In order that a system of N points in a plane should all lie on a curve of the order n, it is sufficient that the pth powers of their distances from an arbitrary line should satisfy a linear homogeneous relation, the number N being given by the formula

$$N = \tfrac{1}{2}\alpha n (n+3) + \tfrac{1}{2}(\beta+1)(\beta+2),$$

where α is the quotient, and β the remainder of the division of p by n, so that $p = \alpha n + \beta$, and $\beta < n$. And he then gives Theorem II., a like theorem as regards points in space; and, further, two Tables, A and B, for the values of N corresponding to given values of n and p in the two cases respectively.

Theorem I. is incorrect for the first value of N in Table A, viz. if $n=1$, $p=2$, then $N=5$; the theorem here is: In order that a system of five points in a plane may lie in a line, the sufficient condition is that the squares of their distances from an arbitrary line shall satisfy a linear homogeneous relation; or, what is the same thing, if the squares of the distances of the five points satisfy a linear homogeneous relation, then the five points will lie in a line. The right conclusion is that four of the five points will lie in a line.

Before proceeding further, I slightly modify the form of the enunciation by defining (in the case of the plane figure) the power of a point to be its distance from an arbitrary line; or, what is the same thing, to be the nilfactum of the line-equation of the point; that is, for the point (x_1, y_1, z_1), the power is

$$\alpha x_1 + \beta y_1 + \gamma z_1,$$

where α, β, γ are arbitrary coefficients, or, if we please, line-coordinates. I say that $\alpha x_1 + \beta y_1 + \gamma z_1$ is the first power, $(\alpha x_1 + \beta y_1 + \gamma z_1)^2$ the second power, and so on. Clifford's Theorem I. thus is: If the pth powers of the N points satisfy a homogeneous linear relation, the N points are on a curve of the nth order.

In the case $n = 1$, $p = 2$, we have five points whose second powers satisfy a homogeneous linear relation; that is, if $(x_1, y_1, z_1), \ldots, (x_5, y_5, z_5)$ are the coordinates of the five points respectively, we have

$$\lambda_1 (\alpha x_1 + \beta y_1 + \gamma z_1)^2 + \ldots + \lambda_5 (\alpha x_5 + \beta y_5 + \gamma z_5)^2 = 0.$$

It is implicitly assumed that the points are distinct points, and we thus exclude such solutions as

$$\lambda_1 = \lambda_2 = \lambda_3 = 0, \quad (x_4, y_4, z_4) = k (x_5, y_5, z_5), \quad k^2 \lambda_4 + \lambda_5 = 0.$$

Here α, β, γ are arbitrary, and the relation is equivalent to the six equations

$$\begin{aligned}
\lambda_1 x_1^2 + \ldots + \lambda_5 x_5^2 &= 0, \\
\lambda_1 y_1^2 + \ldots \qquad\quad &= 0, \\
\lambda_1 z_1^2 + \ldots \qquad\quad &= 0, \\
\lambda_1 y_1 z_1 + \ldots \qquad\quad &= 0, \\
\lambda_1 z_1 x_1 + \ldots \qquad\quad &= 0, \\
\lambda_1 x_1 y_1 + \ldots \qquad\quad &= 0;
\end{aligned}$$

and hence, eliminating the λ's, we have the relation

$$\left\| \begin{matrix} x_1^2, & y_1^2, & z_1^2, & y_1 z_1, & z_1 x_1, & x_1 y_1 \\ x_2^2, & & & & & \\ x_3^2, & & & & & \\ x_4^2, & & & & & \\ x_5^2, & & & & & \end{matrix} \right\| = 0,$$

viz. this means that each of the determinants, obtained by selecting in any manner five out of the six columns, is $= 0$. This is a twofold relation, and thus it cannot be equivalent to the threefold relation

$$\left\| \begin{matrix} x_1, & x_2, & x_3, & x_4, & x_5 \\ y_1, & & & & \\ z_1, & & & & \end{matrix} \right\| = 0,$$

which expresses that the five points are in a line.

But we see further that the twofold relation expresses that the five points are such that we can, through them and an arbitrary sixth point (x_6, y_6, z_6), draw a determinate conic; and this is the case only if four of the five points are in a line; viz. the conic is then the line-pair composed of this line and of the line joining the remaining two points. The right conclusion therefore is that, if the above linear relation is satisfied, then four of the five points lie in a line.

Clifford's proof is rather indicated than carried out, but, from the reference to contravariant differentiation, and from the second paper mentioned above, it must have been as follows:

Starting from the linear relation considered as an identity in (α, β, γ), and operating upon it with $p\partial_\alpha + q\partial_\beta + r\partial_\gamma$, p, q, r denoting arbitrary coefficients, we obtain

$$\lambda_1 (px_1 + qy_1 + rz_1)(\alpha x_1 + \beta y_1 + \gamma z_1) + \ldots + \lambda_5 (px_5 + qy_5 + rz_5)(\alpha x_5 + \beta y_5 + \gamma z_5) = 0;$$

hence, determining the ratios of p, q, r, say by the equations

$$px_4 + qy_4 + rz_4 = 0, \quad px_5 + qy_5 + rz_5 = 0,$$

and, instead of $\lambda_1 (px_1 + qy_1 + rz_1)$, &c., writing Λ_1, &c., we have

$$\Lambda_1 (\alpha x_1 + \beta y_1 + \gamma z_1) + \Lambda_2 (\alpha x_2 + \beta y_2 + \gamma z_2) + \Lambda_3 (\alpha x_3 + \beta y_3 + \gamma z_3) = 0,$$

a homogeneous linear relation between the first powers of the three points (x_1, y_1, z_1), (x_2, y_2, z_2), (x_3, y_3, z_3). We thus see that these points, say the points 1, 2, 3, are in a line; and, similarly, by different determinations of p, q, r, that the points 1, 2, 4 are in a line; and that the points 1, 2, 5 are in a line; that is, the points 3, 4 and 5 are each of them in the line 12 joining the points 1 and 2; that is, the points 1, 2, 3, 4, 5 are in a line.

But, if we examine the reasoning more closely, it appears that the first conclusion, 1, 2, 3 are in a line, depends on the assumption that neither 1, 2, nor 3 is in the line 45. Suppose, for instance, that 3 was in the line 45, the equations

$$px_4 + qy_4 + rz_4 = 0, \quad px_5 + qy_5 + rz_5 = 0,$$

imply $px_3 + qy_3 + rz_3 = 0$, and we have

$$\lambda_1 (px_1 + qy_1 + rz_1)(\alpha x_1 + \beta y_1 + \gamma z_1) + \lambda_2 (px_2 + qy_2 + rz_2)(\alpha x_2 + \beta y_2 + \gamma z_2) = 0,$$

or say

$$\Lambda_1 (\alpha x_1 + \beta y_1 + \gamma z_1) + \Lambda_2 (\alpha x_2 + \beta y_2 + \gamma z_2) = 0,$$

satisfied by

$$(x_1, y_1, z_1) = k (x_2, y_2, z_2), \quad k\Lambda_1 + \Lambda_2 = 0,$$

that is, by making the points 1 and 2 coincident. Thus, if 3 be on the line 45 (and similarly if 1 or if 2 be on the line 45), we cannot infer that 1, 2, 3 are in a line.

I assume, therefore, that neither 1, 2, nor 3 is in the line 45; we here conclude, as above, that 1, 2, 3 are in a line. Suppose for a moment that 5 is not on this line; 4 cannot be on each of the lines 15, 25, 35, and I assume, in the first instance, that it is not on any one of these lines; thus the points 1, 2, 4 are no one of them on the line 35, and hence, by the like reasoning, 1, 2, 4 are in a line; that is, 4 is on the line 12, or, the points 1, 2, 3, 4 are in a line. If, however, 4 is on one of the lines 15, 25, 35, say it is on 35; then it is not on 25, and no one of the points 1, 3, 4 is on 25; hence, by the like reasoning, 1, 2, 4 are in a line. In this case, however, 4 being, by supposition, on the line 35, can only be the point 3; that is, 3 and 4 would be coincident, a case which need not be considered. The correct conclusion from the reasoning thus is, not that 1, 2, 3, 4, 5 are in a line, but that some four of these five points, say 1, 2, 3, 4, are in a line.

It is easy to see that, if we have on a line four points, then there exists between the second powers of these points a linear homogeneous relation. For let the distances of the points from a fixed point of the line be a, b, c, d respectively; and let the arbitrary line meet the line in a point O at a distance r from the fixed point. The distances of the four points from the point O are thus equal to $r+a$, $r+b$, $r+c$, $r+d$ respectively; and hence, writing k for the squared cosine of the inclination of the two lines, the second powers of the four points are $=k(r+a)^2$, $k(r+b)^2$, $k(r+c)^2$, $k(r+d)^2$ respectively; and we have thus, between the second powers p_1, p_2, p_3, p_4 of the four points, the homogeneous linear relation

$$\begin{vmatrix} p_1, & 1, & a, & a^2 \\ p_2, & 1, & b, & b^2 \\ p_3, & 1, & c, & c^2 \\ p_4, & 1, & d, & d^2 \end{vmatrix} = 0.$$

Moreover we can, in an infinite number of ways, form a linear combination

$$A_1 p_1 + A_2 p_2 + A_3 p_3 + A_4 p_4$$

of these second powers, which shall be independent of r, and have any given value whatever. We can therefore make this sum to be equal to the second power of any point 5 whatever (not on the line containing the four points) in relation to the arbitrary line; that is, given the four points in a line, and any other fifth point, we can establish between the second powers of these five points a linear homogeneous relation

$$A_1 p_1 + A_2 p_2 + A_3 p_3 + A_4 p_4 + A_5 p_5 = 0,$$

with non-evanescent values for each of the coefficients A. We thus see how, to the linear homogeneous relation between the second powers of the five points, there corresponds properly the non-symmetric relation: four of the five points are in a line.

The most simple cases are when $p=n$, viz. $p=n=2$, $N=6$; $p=n=3$, $N=10$, and generally $N=\frac{1}{2}(n+1)(n+2)$; and for these Clifford's theorem is easily verified. I have not examined the other cases, but it is probable that in each of them a correction is required.

928.

ON THE ANALYTICAL THEORY OF THE CONGRUENCY.

[From the *Proceedings of the London Mathematical Society*, vol. XXIII. (1892), pp. 185—188.]

IF the lines of a congruency are considered as issuing from the several points of a surface, or say as the quasi-normals of a surface, then the fundamental geometrical theory is established by an analysis closely similar to that for the theory of the curvature of a surface; viz. it is shown that each quasi-normal is intersected by two consecutive quasi-normals in two points respectively (corresponding to the centres of curvature), or say in two foci; we have on the surface two series of curves of quasi-curvature—only these do not in general intersect at right angles; the intersecting quasi-normals form two series of developable surfaces, each touching the surface of centres (or focal surface), along its cuspidal edge, &c.; and, in particular, each quasi-normal is a bitangent of the focal surface, touching it at the two foci respectively.

But the analysis assumes a very different form if we consider the congruency by itself, without thus connecting it with a surface. Regarding the congruency as determined by means of two equations,

$$U(a,\ b,\ c,\ f,\ g,\ h)=0,\quad V(a,\ b,\ c,\ f,\ g,\ h)=0,$$

between the six coordinates of a line ($af+bg+ch=0$, as usual), I take (a, b, c, f, g, h) for the coordinates of a particular line of the congruency, $(a_1, b_1, c_1, f_1, g_1, h_1)$ for those of a consecutive line. Denoting, for shortness, the derived functions

$$(\partial_a,\ \partial_b,\ \partial_c,\ \partial_f,\ \partial_g,\ \partial_h)\ U$$

by F, G, H, A, B, C, and similarly those of V by F', G', H', A', B', C', we have

$$a_1F+b_1G+c_1H+f_1A+g_1B+h_1C=0,$$
$$a_1F'+b_1G'+c_1H'+f_1A'+g_1B'+h_1C'=0;$$

viz. the consecutive line $(a_1, b_1, c_1, f_1, g_1, h_1)$ belongs to the linear congruency defined by these two equations.

Forming with these a linear combination

$$(\lambda F+\mu F')\,a_1+(\lambda G+\mu G')\,b_1+(\lambda H+\mu H')\,c_1+(\lambda A+\mu A')\,f_1+(\lambda B+\mu B')\,g_1+(\lambda C+\mu C')\,h_1=0,$$

we may determine the ratio $\lambda : \mu$ by the equation

$$(\lambda A+\mu A')(\lambda F+\mu F')+(\lambda B+\mu B')(\lambda G+\mu G')+(\lambda C+\mu C')(\lambda H+\mu H')=0,$$

that is,

$$\lambda^2(AF+BG+CH)+\lambda\mu(AF'+BG'+CH'+FA'+GB'+HC')+\mu^2(A'F'+B'G'+C'H')=0;$$

we have thus two values of $\lambda : \mu$; and, denoting the corresponding values of $\lambda F+\mu F', \ldots, \lambda C+\mu C'$ by $(f_2, g_2, h_2, a_2, b_2, c_2)$ and $(f_3, g_3, h_3, a_3, b_3, c_3)$ respectively, we have

$$a_2 f_2+b_2 g_2+c_2 h_2=0,\quad a_3 f_3+b_3 g_3+c_3 h_3=0,$$

and

$$a_1 f_2+b_1 g_2+c_1 h_2+f_1 a_2+g_1 b_2+h_1 c_2=0,$$
$$a_1 f_3+b_1 g_3+c_1 h_3+f_1 a_3+g_1 b_3+h_1 c_3=0;$$

viz. we have thus two lines $(a_2, b_2, c_2, f_2, g_2, h_2)$, $(a_3, b_3, c_3, f_3, g_3, h_3)$, not in general meeting each other, each of which is met by the line $(a_1, b_1, c_1, f_1, g_1, h_1)$; say, for shortness, the lines (a, b, c, f, g, h), $(a_1, b_1, c_1, f_1, g_1, h_1)$, $(a_2, b_2, c_2, f_2, g_2, h_2)$, $(a_3, b_3, c_3, f_3, g_3, h_3)$ are the lines 0, 1, 2, 3 respectively.

We may, in the foregoing investigation, substitute, for the coordinates of the line 1, those of the line 0; and it hence appears—what is indeed obvious—that the line 0 meets each of the lines 2 and 3. Supposing now that the lines 0 and 1 meet each other, that is, that we have

$$af_1+bg_1+ch_1+fa_1+gb_1+hc_1=0,$$

then it is clear that the line 1 must pass through the intersection of the lines 0, 2, or else through the intersection of the lines 0, 3; in fact, if 0 and 1 intersect in a point not on the line 2 or 3, then we have the line 0 as a line passing through this point and meeting each of the lines 2 and 3; and also the line 1 as a line passing through this point and meeting each of the lines 2 and 3; that is, the lines 0 and 1 would be one and the same line.

It thus appears that, considering the line 0 as given, we have two lines 2 and 3 each meeting this line, say in the points P_2 and P_3 respectively; and that, this being so, the consecutive line 1 meets the line 0 either in the point P_2 or else in the point P_3, viz. that there are two consecutive lines 1, say 1_2 and 1_3, meeting the line 0 in the points 2 and 3 respectively. These points are thus given as the intersections of the line (a, b, c, f, g, h) with the lines $(a_2, b_2, c_2, f_2, g_2, h_2)$, $(a_3, b_3, c_3, f_3, g_3, h_3)$ respectively; viz. supposing that $\lambda : \mu$ is determined by the above-mentioned quadric equation, and calling its roots $\lambda_2 : \mu_2$ and $\lambda_3 : \mu_3$, then we have

$$a_2, b_2, c_2, f_2, g_2, h_2=\lambda_2 A+\mu_2 A', \ldots, \lambda_2 H+\mu_2 H',$$
$$a_3, b_3, c_3, f_3, g_3, h_3=\lambda_3 A+\mu_3 A', \ldots, \lambda_3 H+\mu_3 H'.$$

For the point 2, we have

$$\begin{array}{llll} . & hy - gz + aw = 0, & . & h_2y - g_2z + a_2w = 0, \\ -hx & . + fz + bw = 0, & -h_2x & . + f_2z + b_2w = 0, \\ gx - fy & . + cw = 0, & g_2x - f_2y & . + c_2w = 0, \\ -ax - by - cz & . = 0, & -a_2x - b_2y - c_2z & . = 0, \end{array}$$

each set of four equations being equivalent to two equations, in virtue of the relations $af + bg + ch = 0$, $a_2f_2 + b_2g_2 + c_2h_2 = 0$ respectively. There is no completely symmetrical expression for the values of x, y, z, w; according as we derive them from the first equations, the second equations, the third equations, or the fourth equations of each set, we obtain

$$\begin{aligned} x : y : z : w = {} & \Theta_1 : ag_2 - ga_2 : ah_2 - ha_2 : gh_2 - hg_2, \\ = {} & bf_2 - fb_2 : \Theta_2 : bh_2 - hb_2 : hf_2 - fh_2, \\ = {} & cf_2 - fc_2 : cg_2 - gc_2 : \Theta_3 : fg_2 - gf_2, \\ = {} & bc_2 - cb_2 : ca_2 - ac_2 : ab_2 - ba_2 : \Theta \ ; \end{aligned}$$

where

$$\begin{aligned} \Theta_1 &= -fa_2 - bg_2 - ch_2, \quad = af_2 + gb_2 + hc_2, \\ \Theta_2 &= -af_2 - gb_2 - ch_2, \quad = fa_2 + bg_2 + hc_2, \\ \Theta_3 &= -af_2 - bg_2 - hc_2, \quad = fa_2 + gb_2 + ch_2, \\ \Theta &= -af_2 - bg_2 - ch_2, \quad = fa_2 + gb_2 + hc_2. \end{aligned}$$

For the point 3, we have, of course, the same formulæ, with the suffix 3 instead of 2.

929.

NOTE ON THE SKEW SURFACES APPLICABLE UPON A GIVEN SKEW SURFACE.

[From the *Proceedings of the London Mathematical Society*, vol. XXIII. (1892), pp. 217—225.]

THE question was considered by Bonnet, in § 7 of his "Mémoire sur la théorie générale des surfaces," *Jour. École Polyt.*, Cah. 32 (1848); I resume it here, making a greater use of the line of striction.

We may construct a skew surface, inextensible but flexible about its generating lines, as follows: Imagine a flexible extensible plane, and in it the rigid parallel lines L, L_1, L_2, L_3, &c., connected each with the following one by the rigid lines

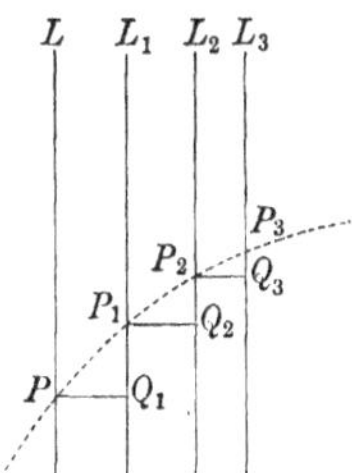

PQ_1, P_1Q_2, P_2Q_3, &c., where PQ_1 cuts L, L_1, P_1Q_2 cuts L_1, L_2, &c., at right angles; the angles LPP_1, $L_1P_1P_2$, $L_2P_2P_3$, &c., are taken to be ω, ω_1, ω_2, &c., respectively. Keeping the line L fixed, we may twist the whole plane $L_1L_2L_3$ ... round PQ_1, so that the line L_1 becomes inclined at a small angle to L, these lines now having PQ_1 for their shortest distance, and the lines L_2, L_3, &c., remaining parallel to L_1 in its new position; the foregoing twisting implies an extension (increasing with the distance on each side from P) of the strip or element between the lines L, L_1; after this twisting, we imagine the strip in question to become rigid. The amount of the twist is such that, if $d\phi$ be the inclination of the lines L, L_1, we have $d\phi = PQ_1 \div \tau$, viz. the twist $d\phi$ is in a certain proportion to the shortest distance PQ_1. Similarly, keeping the line L_1 fixed, we twist the whole plane $L_2L_3 \ldots$ round P_1Q_2, the amount of the twist or inclination $d\phi$ of the lines L_1 and L_2 being $= P_1Q_2 \div \tau_1$; after the twist, we imagine the strip or element between the two lines L_1, L_2 to become rigid. Proceeding in this manner, we have the series of rigid elements LL_1, L_1L_2, L_2L_3, &c.; putting the line L in any given position, and then keeping it fixed, we may turn the element LL_1 round this line so as to bring L_1 into a certain position; then, keeping L_1 fixed in this position, we may turn the element L_1L_2 round L_1 so as to bring L_2 into a certain position, and so on. To explain this further, imagine through a point O a series of lines K, K_1, K_2, K_3, &c., such that the inclination of K and K_1 is equal to that of L and L_1, the inclination of K_1 and K_2 is equal to that of L_1 and L_2, and so on, all these inclinations being otherwise arbitrary; or say that we have the double-triangle strips or elements KK_1, K_1K_2, K_2K_3, &c., bounded by pairs of lines, at given infinitesimal inclinations the two lines of a pair to each other, and forming a flexible double pyramid, which may be bent into any given form whatever assumed at pleasure; and this being so, we see that the system of the strips or elements LL_1, L_1L_2, L_2L_3, &c., may be so bent that the lines L, L_1, L_2, &c., shall be parallel to the lines K, K_1, K_2, &c., respectively.

Supposing the distances PQ_1, P_1Q_2, P_2Q_3, &c., to be all of them infinitesimal, we have a skew surface containing upon it a curve $P_1P_2P_3$, &c., which is the line of striction, viz. this is the locus of the point on a generating line which is the nearest point to the consecutive generating line. The line of striction cuts the several generating lines at an angle ω, variable from line to line, which is called the obliquity; and the inclination between two consecutive generating lines is in a certain ratio to the shortest distance between the two lines. Let the inclination = shortest distance $\div\, \tau$, this magnitude τ being variable from line to line: its reciprocal τ^{-1} is called the "torsion"; so that the obliquity ω and the reciprocal of the torsion τ may be regarded as functions of s, the arc of the line of striction measured from any fixed point. The skew surface is thus composed of rigid strips or elements, each included between two consecutive lines. We have further seen that the surface may be bent by turning these rigid elements about the successive generating lines, in such wise that the generating lines become parallel to the generating lines of an arbitrary cone, which is called the "asymptotic" cone (otherwise the director cone); say the surface may be bent so that it shall have a given asymptotic cone.

I consider a given skew surface; I take x, y, z for the coordinates of a point on the line of striction, and α, β, γ for the cosine-inclinations of the generating line through this point; x, y, z, α, β, γ are regarded as functions of s, the length of or distance along the line of striction measured from any fixed point thereof; and I use accents to denote differentiation in regard to s. We have

$$\alpha^2 + \beta^2 + \gamma^2 = 1,$$

and therefore

$$\alpha\alpha' + \beta\beta' + \gamma\gamma' = 0;$$

also

$$x'^2 + y'^2 + z'^2 = 1;$$

and therefore

$$x'x'' + y'y'' + z'z'' = 0;$$

x', y', z' are the cosine-inclinations of the tangent at the point x, y, z.

I remark also that, writing

$$\rho^{-1} = \sqrt{x''^2 + y''^2 + z''^2},$$

so that ρ is the radius of absolute curvature, we have $\rho x''$, $\rho y''$, $\rho z''$ for the cosine-inclinations of the binormal (or perpendicular to the osculating plane).

If from the point (x, y, z) we draw a perpendicular to the consecutive generating line through the point $(x + dx, y + dy, z + dz)$, this will also be perpendicular to the generating line through the point (x, y, z), and we thus find

$$\alpha'x' + \beta'y' + \gamma'z' = 0$$

as the condition that the point (x, y, z) may be, as it is assumed to be, a point on the line of striction.

For the proof hereof, take for the moment x_1, y_1, z_1 the coordinates of P, and α_1, β_1, γ_1 for the cosine-inclinations of L; then, considering the line PQ_1, this passes through P and cuts the line L_1; taking its equation to be

$$\frac{X - x}{A} = \frac{Y - y}{B} = \frac{Z - z}{C},$$

this meets the line

$$\frac{X - x_1}{\alpha_1} = \frac{Y - y_1}{\beta_1} = \frac{Z - z_1}{\gamma_1},$$

and we thence find

$$\begin{vmatrix} A, & B, & C \\ x_1 - x, & y_1 - y, & z_1 - z \\ \alpha_1, & \beta_1, & \gamma_1 \end{vmatrix} = 0.$$

Writing $x_1, y_1, z_1 = x + x'ds,\ y + y'ds,\ z + z'ds$, ds divides out, and we have

$$A(\beta_1 z' - \gamma_1 y') + B(\gamma_1 x' - \alpha_1 z') + C(\alpha_1 y' - \beta_1 x') = 0;$$

the line in question cuts L_1 and L; that is, we have

$$A\alpha_1 + B\beta_1 + C\gamma_1 = 0,$$

$$A\alpha + B\beta + C\gamma = 0;$$

or, eliminating A, B, C, we find

$$(\beta\gamma_1 - \beta_1\gamma)(\beta_1 z' - \gamma_1 y') + (\gamma\alpha_1 - \gamma_1\alpha)(\gamma_1 x' - \alpha_1 z') + (\alpha\beta_1 - \alpha_1\beta)(\alpha_1 y' - \beta_1 x') = 0,$$

that is,

$$(\alpha_1 x' + \beta_1 y' + \gamma_1 z')(\alpha\alpha_1 + \beta\beta_1 + \gamma\gamma_1) - (\alpha x' + \beta y' + \gamma z')(\alpha_1^2 + \beta_1^2 + \gamma_1^2) = 0.$$

But we have

$$\alpha_1^2+\beta_1^2+\gamma_1^2=1,$$

and, writing

$$\alpha_1,\ \beta_1,\ \gamma_1=\alpha+\alpha' ds,\ \beta+\beta' ds,\ \gamma+\gamma' ds,$$

we find

$$\alpha\alpha_1+\beta\beta_1+\gamma\gamma_1=\alpha^2+\beta^2+\gamma^2+(\alpha\alpha'+\beta\beta'+\gamma\gamma')\,ds=1\,;$$

the equation thus is

$$(\alpha_1 x'+\beta_1 y'+\gamma_1 z')-(\alpha x'+\beta y'+\gamma z')=0\,;$$

that is,

$$\alpha' x'+\beta' y'+\gamma' z'=0,$$

the required equation.

Calling the inclination of the generating line through the point $(x,\ y,\ z)$ to the tangent of the line of striction the "obliquity," and denoting it by ω, we have

$$\alpha x'+\beta y'+\gamma z'=\cos\omega.$$

Calling the inclination of the two consecutive generating lines divided by the shortest distance between these lines the "torsion," and denoting it by τ^{-1}, we have

$$\alpha'^2+\beta'^2+\gamma'^2=\frac{\sin^2\omega}{\tau^2}.$$

In proof hereof, if for a moment ϕ is the inclination of the lines L and L_1 to each other, then

$$\cos\phi=\alpha\alpha_1+\beta\beta_1+\gamma\gamma_1\,;$$

and therefore

$$\sin^2\phi=(\beta\gamma_1-\beta_1\gamma)^2+(\gamma\alpha_1-\gamma_1\alpha)^2+(\alpha\beta_1-\alpha_1\beta)^2,$$

viz. writing

$$\alpha_1,\ \beta_1,\ \gamma_1=\alpha+\alpha' ds,\ \beta+\beta' ds,\ \gamma+\gamma' ds,$$

this is

$$\begin{aligned}\sin^2\phi&=ds^2\,\{(\beta\gamma'-\beta'\gamma)^2+(\gamma\alpha'-\gamma'\alpha)^2+(\alpha\beta'-\alpha'\beta)^2\}\\ &=ds^2\,\{(\alpha^2+\beta^2+\gamma^2)(\alpha'^2+\beta'^2+\gamma'^2)-(\alpha\alpha'+\beta\beta'+\gamma\gamma')^2\},\\ &=ds^2\,(\alpha'^2+\beta'^2+\gamma'^2)\,;\end{aligned}$$

whence, if for a moment the shortest distance between the two lines is called δ, then we have

$$\frac{\sin\phi}{\delta}=\tau^{-1}=\frac{\sin\phi}{ds\sin\omega}\,;$$

that is,

$$\frac{\sin^2\omega}{\tau^2}=\frac{\sin^2\phi}{ds^2}=\alpha'^2+\beta'^2+\gamma'^2,$$

the required equation.

We have thus the five equations

$$\alpha^2+\beta^2+\gamma^2=1,$$

$$x'^2+y'^2+z'^2=1,$$

$$\alpha'x'+\beta'y'+\gamma'z'=0,$$

$$\alpha x'+\beta y'+\gamma z'=\cos\omega,$$

$$\alpha'^2+\beta'^2+\gamma'^2=\frac{\sin^2\omega}{\tau^2};$$

and if we herein consider ω and τ as denoting given functions of s, all the skew surfaces which satisfy these equations will be surfaces applicable one on the other.

Adding to the foregoing the derived equations

$$\alpha\alpha'+\beta\beta'+\gamma\gamma'=0,$$

$$x'x''+y'y''+z'z''=0,$$

$$\alpha x''+\beta y''+\gamma z''=-\sin\omega\,.\,\omega',$$

$$\alpha x'''+\beta y'''+\gamma z'''+\alpha'x''+\beta'y''+\gamma'z''=-\sin\omega\,.\,\omega''-\cos\omega\,.\,\omega'^2,$$

we find without difficulty

$$\alpha',\ \beta',\ \gamma'=\frac{\beta z'-\gamma y'}{\tau},\ \frac{\gamma x'-\alpha z'}{\tau},\ \frac{\alpha y'-\beta x'}{\tau},$$

$$\beta\gamma'-\beta'\gamma,\ \gamma\alpha'-\gamma'\alpha,\ \alpha\beta'-\alpha'\beta=\frac{-x'+\alpha\cos\omega}{\tau},\ \frac{-y'+\beta\cos\omega}{\tau},\ \frac{-z'+\gamma\cos\omega}{\tau}.$$

Putting, for shortness,

$$\Theta=-\sin\omega\,.\,\omega''-\cos\omega\,.\,\omega'^2-(\alpha x'''+\beta y'''+\gamma z'''),$$

$$\nabla=\begin{vmatrix}\alpha, & \beta, & \gamma\\ x', & y', & z'\\ x'', & y'', & z''\end{vmatrix},$$

and, as above,

$$\rho^{-2}=x''^2+y''^2+z''^2,$$

we find

$$\nabla^2=\sin^2\omega\left(\frac{1}{\rho^2}-\omega'^2\right),\quad \tau=\frac{\nabla}{\Theta}=\frac{\sin\omega}{\rho\Theta}\sqrt{1-\rho^2\omega'^2},$$

and to these may be joined

$$\beta z'-\gamma y'=\rho^2\{\nabla x''+\omega'\sin\omega\,(y'z''-y''z')\},$$

$$\gamma x'-\alpha z'=\rho^2\{\nabla y''+\omega'\sin\omega\,(z'x''-z''x')\},$$

$$\alpha y'-\beta x'=\rho^2\{\nabla z''+\omega'\sin\omega\,(x'y''-x''y')\}.$$

I remark that, supposing the line of striction to be given, that is, (x, y, z) to be given as functions of s, and moreover the obliquity ω to be given as a function of s, the position of the generating line through the point (x, y, z) will be given, and also the torsion τ^{-1}. In fact, among the foregoing equations, we have

$$\begin{aligned} \alpha^2 \;+\beta^2 \;+\gamma^2 \;&= 1, \\ \alpha x' \;+\beta y' \;+\gamma z' \;&= \cos\omega, \\ \alpha x'' + \beta y'' + \gamma z'' &= -\sin\omega \,.\, \omega', \end{aligned}$$

which equations determine α, β, γ, that is, the position of the generating line; the position of the consecutive generating line is then also determined; we thus have α', β', γ', and thence the torsion, which is given by the foregoing equation

$$\tau = \frac{\sin\omega}{\rho\Theta}\sqrt{1-\rho^2\omega'^2}.$$

Supposing that the line of striction is not given, but that the obliquity ω and the torsion τ^{-1} are given as functions of s, we have then the foregoing *five* equations, which are not sufficient for the determination of x, y, z, α, β, γ; but, joining to them an assumed homogeneous relation between (α, β, γ), the six quantities will be determined. The assumed homogeneous relation between (α, β, γ) is the equation of the asymptotic cone; and we have thus the theorem that, given the asymptotic cone, and also the obliquity and the torsion as functions of s (the arc of the line of striction), then this line of striction, and the skew surface the locus of the generating lines, will be determined.

We have between α, β, γ the assumed homogeneous equation, say $U=0$; and among the foregoing equations, the equations

$$\alpha^2+\beta^2+\gamma^2=1, \text{ and } \alpha'^2+\beta'^2+\gamma'^2=\frac{\sin^2\omega}{\tau^2},$$

a given function of s; these equations give, by means of an integration, α, β, γ as functions of s; and we have then the above-mentioned equations

$$x'-\alpha\cos\omega,\ y'-\beta\cos\omega,\ z'-\gamma\cos\omega=\tau(\beta\gamma'-\beta'\gamma,\ \gamma\alpha'-\gamma'\alpha,\ \alpha\beta'-\alpha'\beta),$$

which give, by integration, x, y, z as functions of s. And the skew surfaces thus obtained for an assumed asymptotic cone $U=0$ will be the required system of skew surfaces applicable the one on the other.

For example, if ω, τ are constants, and we take, for $U=0$, the equation

$$\gamma^2-c(\alpha^2+\beta^2)=0,$$

that is, if we consider the skew surfaces of constant obliquity and torsion and which have for asymptotic cone a right circular cone: it easily appears that the line of striction is a helix traced upon a right circular cylinder, and that the generating lines are at a constant inclination to the axis of the cylinder, and all of them touch the cylinder. In fact, for such a surface (a kind of helicoid), it is in the first place obvious that the helix is the line of striction, and next that the obliquity and the torsion are each of them constant.

Supposing that the obliquity and the torsion are ω, τ, and that the inclination of the generating lines to the axis of the cylinder is $90^\circ - \omega - \delta$, δ being a given constant, we find

$$\begin{aligned}
x &= \frac{\tau \cos\delta \cos(\omega+\delta)}{\sin\omega} \cos\left\{\frac{s\sin\omega}{\tau\cos(\omega+\delta)}\right\},\\
y &= \frac{\tau \cos\delta \cos(\omega+\delta)}{\sin\omega} \sin\left\{\frac{s\sin\omega}{\tau\cos(\omega+\delta)}\right\},\\
z &= s\sin\delta;\\
\alpha &= -\cos(\omega+\delta)\sin\left\{\frac{s\sin\omega}{\tau\cos(\omega+\delta)}\right\},\\
\beta &= \cos(\omega+\delta)\cos\left\{\frac{s\sin\omega}{\tau\cos(\omega+\delta)}\right\},\\
\gamma &= \sin(\omega+\delta);
\end{aligned}$$

where observe that the line of striction lies on the circular cylinder,

$$\text{radius} = \tau\cos\delta\cos(\omega+\delta) \div \sin\omega.$$

In particular, if $\delta = 0$, we have

$$\begin{aligned}
x &= \tau\cot\omega\cos\frac{s}{\tau\cot\omega}, & \alpha &= -\cos\omega\sin\frac{s}{\tau\cot\omega},\\
y &= \tau\cot\omega\sin\frac{s}{\tau\cot\omega}, & \beta &= \cos\omega\cos\frac{s}{\tau\cot\omega},\\
z &= 0, & \gamma &= \sin\omega;
\end{aligned}$$

and, for $\delta = -\omega$, we have

$$\begin{aligned}
x &= \tau\cot\omega\cos\frac{s\sin\omega}{\tau}, & \alpha &= -\sin\frac{s\sin\omega}{\tau},\\
y &= \tau\cot\omega\sin\frac{s\sin\omega}{\tau}, & \beta &= \cos\frac{s\sin\omega}{\tau},\\
z &= -s\sin\omega, & \gamma &= 0.
\end{aligned}$$

The first of these is the skew hyperboloid of revolution

$$X^2 + Y^2 - (Z^2 + \tau^2)\cot^2\omega = 0,$$

(radius of gorge $= \tau\cot\omega$); and the second of them is the helicoid (radius of cylinder $= \tau\cot\omega$), the generating lines of which are the tangents perpendicular to the axis of the cylinder; these are, in fact, surfaces found by Bonnet in his memoir above referred to. We may imagine the surface passing from the first form, in which the semi-aperture of the asymptotic cone is ω, through the series of forms belonging to the general formulæ involving δ, to the second form, in which the semi-aperture is $= 90^\circ$, i.e. in which the asymptotic cone is replaced by a plane.

930.

SUR LA SURFACE DES ONDES.

[From the *Annali di Matematica*, Ser. II., t. XX. (1892), pp. 1—18.]

1. Il y a, dans les *Annali di Matematica*, tom. II. (1859), deux Notes très intéressantes sur cette surface: Combescure, "Sur les lignes de courbure de la surface des ondes," pp. 278—285, et Brioschi, "Osservazioni sulla medesima quistione," pp. 285—287. Je me propose de réproduire et développer cette théorie, en changeant les notations et l'arrangement des recherches de la manière qui me paraît convenable.

2. Je prends a, b, c pour les carrés des semiaxes ($a > b > c$), et j'écris

$$A,\ B,\ C = a+b+c,\quad ab+ac+bc,\quad abc;$$

$$\alpha,\ \beta,\ \gamma = b-c,\quad c-a,\quad a-b,$$

($\alpha+\beta+\gamma=0$, et ainsi, $\alpha=+$, $\gamma=+$ et β, $=-\alpha-\gamma$, $=-$, et en magnitude absolue plus grand que α ou γ):

$$\xi = x^2 + y^2 + z^2,$$

$$\eta = ax^2 + by^2 + cz^2,$$

$$\zeta = a(b+c)x^2 + b(c+a)y^2 + c(a+b)z^2,$$

et de là réciproquement

$$\beta\gamma x^2 = \zeta - a\eta - bc\xi,$$

$$\gamma\alpha y^2 = \zeta - b\eta - ca\xi,$$

$$\alpha\beta z^2 = \zeta - c\eta - ab\xi.$$

3. L'équation de la surface est

$$\xi\eta - \zeta + abc = 0,$$

et de là, en écrivant $\zeta = \xi\eta - abc$, on obtient pour un point de la surface

$$\beta\gamma x^2 = (\xi - a)(\eta - bc),$$

$$\gamma\alpha y^2 = (\xi - b)(\eta - ca),$$

$$\alpha\beta z^2 = (\xi - c)(\eta - ab),$$

équations qui donnent les valeurs des coordonnées (x, y, z) du point en termes de deux paramètres ξ, η. Je remarque que ces équations donnent

$$\frac{x^2}{\xi-a}+\frac{y^2}{\xi-b}+\frac{z^2}{\xi-c}=1,$$

$$\frac{x^2}{\eta-bc}+\frac{y^2}{\eta-ca}+\frac{z^2}{\eta-ab}=0;$$

la première équation, en y considérant ξ comme dénotant $x^2+y^2+z^2$, et la seconde équation, en y considérant η comme dénotant $ax^2+by^2+cz^2$, sont équivalentes l'une et l'autre à l'équation $\xi\eta-\zeta+abc=0$ de la surface.

4. Je prends λ, μ, ν pour les cosinus des inclinations de la normale (ou, ce qui est la même chose, de la perpendiculaire par le centre sur le plan tangent) aux trois axes, et v pour le carré de la longueur de ce perpendiculaire, $v=(\lambda x+\mu y+\nu z)^2$; on a

$$\lambda=\frac{x}{D}\{a\xi+\eta-a(b+c)\},$$

$$\mu=\frac{y}{D}\{b\xi+\eta-b(c+a)\},$$

$$\nu=\frac{z}{D}\{c\xi+\eta-c(a+b)\},$$

et de là, par l'équation $\lambda^2+\mu^2+\nu^2=1$, on a

$$\begin{aligned}\alpha\beta\gamma D^2 = {} & \alpha\,(\xi-a)(\eta-bc)\{a\xi+\eta-a(b+c)\}^2\\ & +\beta\,(\xi-b)(\eta-ca)\{b\xi+\eta-b(c+a)\}^2\\ & +\gamma\,(\xi-c)(\eta-ab)\{c\xi+\eta-c(a+b)\}^2,\end{aligned}$$

équation laquelle (en réduisant à moyen des rélations $\alpha+\beta+\gamma=0$, $a\alpha+b\beta+c\gamma=0$, $a^2\alpha+b^2\beta+c^2\gamma=-\alpha\beta\gamma$, &c.) devient

$$\alpha\beta\gamma D^2=\alpha\beta\gamma\,(\xi\eta-C)(\eta-\xi^2+A\xi-B);$$

et l'on a ainsi

$$D=\sqrt{(\xi\eta-C)(\eta-\xi^2+A\xi-B)}.$$

5. On trouve de même manière

$$\lambda x+\mu y+\nu z=\frac{\xi\eta-C}{D},$$

$$a\lambda x+b\mu y+c\nu z=\frac{\eta}{D}(\eta-\xi^2+A\xi-B),$$

$$bc\lambda x+ca\mu y+ab\nu z=\frac{1}{D}\{-(\xi\eta-C)(\eta-B)-C(\eta-\xi^2+A\xi-B)\};$$

ou, en y substituant la valeur de D,

$$\lambda x+\mu y+\nu z,\quad =\sqrt{v},\quad =\sqrt{\frac{\xi\eta-C}{\eta-\xi^2+A\xi-B}}.$$

La première de ces équations donne

$$v = \frac{\xi\eta - C}{\eta - \xi^2 + A\xi - B},$$

et de là réciproquement

$$a\lambda x + b\mu y + cvz = \eta\sqrt{\frac{\eta - \xi^2 + A\xi - B}{\xi\eta - C}},$$

$$bc\lambda x + ca\mu y + ab\nu z = -(\eta - B)\sqrt{\frac{\xi\eta - C}{\eta - \xi^2 + A\xi - B}} - C\sqrt{\frac{\eta - \xi^2 + A\xi - B}{\xi\eta - C}};$$

$$\eta = \frac{b(\xi^2 - A\xi + B) - C}{v - \xi}, \quad = \xi^2 - A\xi + B + \frac{\xi - a \, . \, \xi - b \, . \, \xi - c}{v - \xi}.$$

On a ainsi les formules

$$\lambda x + \mu y + \nu z = \sqrt{v},$$

$$a\lambda x + b\mu y + c\nu z = \frac{\eta}{\sqrt{v}},$$

$$bc\lambda x + ca\mu y + ab\nu z = -\eta\sqrt{v} + B\sqrt{v} - \frac{C}{\sqrt{v}};$$

et de là aussi

$$a(b + c)\lambda x + b(c + a)\mu y + c(a + b)\nu z = \eta\sqrt{v} + \frac{C}{\sqrt{v}}.$$

6. On peut introduire dans les formules v au lieu de η; les deux paramètres seront ainsi: ξ, carré de la distance au centre; v, carré de la perpendiculaire sur le plan tangent.

On a d'abord

$$\eta - bc = \xi^2 - (a + b + c)\xi + a(b + c) + \frac{\xi - a \, . \, \xi - b \, . \, \xi - c}{v - \xi},$$

$$= (\xi - a)\left\{\xi - b - c + \frac{\xi - b \, . \, \xi - c}{v - \xi}\right\},$$

$$= \frac{\xi - a}{v - \xi}\{bc - (b + c)v + v\xi\};$$

et ainsi

$$\beta\gamma x^2 = \frac{(\xi - a)^2}{v - \xi}\{bc - (b + c)v + v\xi\},$$

et de même

$$\gamma\alpha y^2 = \frac{(\xi - b)^2}{v - \xi}\{ca - (c + a)v + v\xi\},$$

$$\alpha\beta z^2 = \frac{(\xi - c)^2}{v - \xi}\{ab - (a + b)v + v\xi\},$$

lesquelles sont les expressions des coordonnées ξ, η, ζ en termes des paramètres ξ, v.

7. On a

$$vD^2 = (\xi\eta - C)^2,$$

$$\xi\eta - C = \xi - a \,.\, \xi - b \,.\, \xi - c \,.\left\{1 + \frac{\xi}{v - \xi}\right\}, \quad = \frac{v \,.\, \xi - a \,.\, \xi - b \,.\, \xi - c}{v - \xi},$$

et de là

$$D^2 = v\left(\frac{\xi - a \,.\, \xi - b \,.\, \xi - c}{v - \xi}\right)^2.$$

On a aussi

$$a\xi + \eta - a(b + c) = \xi - b \,.\, \xi - c \,.\left(1 + \frac{\xi - a}{v - \xi}\right), \quad = \frac{v - a \,.\, \xi - b \,.\, \xi - c}{v - \xi},$$

et de là

$$\lambda = \frac{x}{D}\,\frac{v - a \,.\, \xi - b \,.\, \xi - c}{v - \xi},$$

ce qui donne

$$\beta\gamma\lambda^2 = \frac{(v - a)^2}{v \,.\, v - \xi}\{bc - (b + c)\,v + v\xi\},$$

et de même

$$\gamma\alpha\mu^2 = \frac{(v - b)^2}{v \,.\, v - \xi}\{ca - (c + a)\,v + v\xi\},$$

$$\alpha\beta\nu^2 = \frac{(v - c)^2}{v \,.\, v - \xi}\{ab - (a + b)\,v + v\xi\},$$

lesquelles sont les expressions de λ, μ, ν en termes des paramètres ξ, v.

8. On obtient

$$\begin{aligned}\alpha\beta\gamma\left\{\frac{\lambda^2}{v - a} + \frac{\mu^2}{v - b} + \frac{\nu^2}{v - c}\right\} = \frac{1}{v \,.\, v - \xi}\,.\, &\alpha\,(v - a)\,\{bc - (b + c)\,v + v\xi\}\\ &+ \beta\,(v - b)\,\{ca - (c + a)\,v + v\xi\}\\ &+ \gamma\,(v - c)\,\{ab - (a + b)\,v + v\xi\},\end{aligned}$$

ou, en réduisant comme auparavant,

$$\frac{\lambda^2}{v - a} + \frac{\mu^2}{v - b} + \frac{\nu^2}{v - c} = 0.$$

9. Je rappelle que l'équation du plan tangent est

$$\lambda x + \mu y + \nu z - \sqrt{v} = 0,$$

où v est déterminé comme fonction de λ, μ, ν par l'équation qui vient d'être donnée; et qu'en considérant λ, μ, ν, v comme des paramètres variables qui satisfont à cette équation et à l'équation $\lambda^2 + \mu^2 + \nu^2 = 1$, l'on obtient la surface comme enveloppe de ce plan tangent.

10. On a ainsi v comme l'une des racines de l'équation quadrique

$$\frac{\lambda^2}{\theta - a} + \frac{\mu^2}{\theta - b} + \frac{\nu^2}{\theta - c} = 0;$$

en dénotant par u l'autre racine, on a donc

$$\theta^2 - \{(b + c)\,\lambda^2 + (c + a)\,\mu^2 + (a + b)\,\nu^2\}\,\theta + bc\lambda^2 + ca\mu^2 + ab\nu^2 = \theta - u \,.\, \theta - v,$$

et de là

$$u+v=(b+c)\lambda^2+(c+a)\mu^2+(a+b)\nu^2,$$
$$uv=bc\lambda^2+ca\mu^2+ab\nu^2.$$

La première de ces équations peut aussi s'écrire sous la forme

$$A-u-v=a\lambda^2+b\mu^2+c\nu^2,$$

et la seconde sous la forme

$$B-uv=a(b+c)\lambda^2+b(c+a)\mu^2+c(a+b)\nu^2.$$

11. J'observe que λ, μ, ν sont les cosinus des inclinations de la perpendiculaire par le centre sur le plan tangent au point x, y, z, la longueur de cette perpendiculaire étant $\sqrt{v}$; il y a un plan tangent parallèle qui corresponde aux mêmes valeurs de λ, μ, ν, et évidemment on a alors $\sqrt{u}$ pour la longueur de la perpendiculaire sur ce plan tangent parallèle; autrement dit, les équations de deux plans tangents sont

$$\lambda X+\mu Y+\nu Z-\sqrt{v}=0,$$
$$\lambda X+\mu Y+\nu Z-\sqrt{u}=0.$$

Il convient de remarquer qu'on a pris (x, y, z) pour les coordonnées du point de la surface qui est le point de contact du premier de ces deux plans, et qu'ainsi les deux quantités v et u n'entrent pas symétriquement dans les formules.

12. En substituant dans l'expression de $u+v$ ou uv, ou ce qui est plus simple dans celle de $A-u-v$, les valeurs de λ^2, μ^2, ν^2 en termes de v, ξ, on obtient

$$\begin{aligned}\alpha\beta\gamma(A-u-v)=\frac{1}{v\,.\,v-\xi}\,.\,&\alpha a(v-a)^2\{bc-(b+c)v+v\xi\}\\ &+\beta b(v-b)^2\{ca-(c+a)v+v\xi\}\\ &+\gamma c(v-c)^2\{ab-(a+b)v+v\xi\},\end{aligned}$$

équation laquelle (en réduisant comme auparavant) devient

$$A-u-v=A-2v+\frac{1}{v\,.\,v-\xi}\{v-a\,.\,v-b\,.\,v-c\},$$

c'est-à-dire

$$(v-\xi)(v-u)=\frac{v-a\,.\,v-b\,.\,v-c}{v},$$

équation qui donne dans des formes très simples, ξ en termes de v, u, et aussi u en termes de v, ξ.

13. On a comme auparavant

$$v=\frac{\xi\eta-C}{\eta-\xi^2+A\xi-B};$$

je cherche l'expression de u en termes de ξ, η. En écrivant pour abréger

$$\eta-\xi^2+A\xi-B=M,$$

nous trouvons

$$v-a=\frac{1}{M}(\xi-a)\{a\xi+\eta-a(b+c)\},$$

$$v-b=\frac{1}{M}(\xi-b)\{b\xi+\eta-b(c+a)\},$$

$$v-c=\frac{1}{M}(\xi-c)\{c\xi+\eta-c(a+b)\},$$

$$v-\xi=\frac{1}{M}.\,\xi-a\,.\,\xi-b\,.\,\xi-c,$$

$$v\quad=\frac{1}{M}(\xi\eta-C);$$

et de là

$$v-u=\frac{v-a\,.\,v-b\,.\,v-c}{v\,.\,v-\xi}=\frac{\{a\xi+\eta-a(b+c)\}\{b\xi+\eta-b(c+a)\}\{c\xi+\eta-c(a+b)\}}{M(\xi\eta-C)},$$

ou enfin, et en restituant pour M sa valeur,

$$u=\frac{(\xi\eta-C)^2-\{a\xi+\eta-a(b+c)\}\{b\xi+\eta-c(a+b)\}\{c\xi+\eta-a(b+c)\}}{(\xi\eta-C)\{\eta-\xi^2+A\xi-B\}}.$$

14. On peut introduire dans les formules u au lieu de ξ, et ainsi exprimer les coordonnées, &c., en termes des deux paramètres v, u. On a pour cela

$$\xi=v-\frac{v-a\,.\,v-b\,.\,v-c}{v\,.\,v-u},$$

donc

$$\xi-a=(v-a)\left\{1+\frac{v-b\,.\,v-c}{v\,.\,v-u}\right\},\quad=-\frac{v-a}{v(v-u)}\{bc-(b+c)u+uv\}.$$

De plus

$$bc-(b+c)v+v\xi=(v-b)(v-c)-v(v-\xi)$$

$$=(v-b)(v-c)-(v-b)(v-c).\frac{v-a}{v-u}=v-b\,.\,v-c\,.\left(1-\frac{v-a}{v-u}\right)=-\frac{u-a\,.\,v-b\,.\,v-c}{v-u}.$$

Donc

$$\eta-bc=\frac{\xi-a}{v-\xi}\{bc-(b+c)v+v\xi\}=-\frac{\xi-a\,.\,u-a\,.\,v-b\,.\,v-c}{v-\xi\,.\,v-u}=-\frac{v\,.\,\xi-a\,.\,u-a}{v-a},$$

ou enfin, à moyen de la valeur de $\xi-a$,

$$\eta-bc=\frac{u-a}{v-u}\{bc-(b+c)u+uv\}.$$

Donc

$$-\beta\gamma x^2=\frac{u-a\,.\,v-a}{v(v-u)^2}\{bc-(b+c)u+uv\}^2,$$

et de même

$$-\gamma\alpha y^2=\frac{u-b\,.\,v-b}{v(v-u)^2}\{ca-(c+a)u+uv\}^2,$$

$$-\alpha\beta z^2=\frac{u-c\,.\,v-c}{v(v-u)^2}\{ab-(a+b)u+uv\}^2,$$

équations qui donnent les coordonnées x, y, z en termes des deux paramètres v, u.

15. De la valeur ci-dessus donnée pour $\eta - bc$ on déduit celle de η; en effet, on trouve

$$\eta(v-u) = bc(v-a) + (u-a)\{-(b+c)v + uv\}$$
$$= v\{u^2 - (a+b+c)u + ab + ac + bc\} - abc,$$

c'est-à-dire

$$\eta = \frac{v(u^2 - Au + B) - C}{v-u}, \quad = u^2 - Au + B + \frac{u-a\,.\,u-b\,.\,u-c}{v-u},$$

ainsi η est la même fonction de v, u et de v, ξ.

16. De plus

$$\beta\gamma\lambda^2 = \frac{(v-a)^2}{v\,.\,v-\xi}\{bc - (b+c)v + v\xi\}, \quad = -\frac{u-a\,.\,(v-a)^2\,.\,v-b\,.\,v-c}{v\,.\,v-\xi\,.\,v-u},$$

c'est-à-dire

$$-\beta\gamma\lambda^2 = u-a\,.\,v-a,$$

et de même

$$-\gamma\alpha\mu^2 = u-b\,.\,v-b,$$
$$-\alpha\beta\nu^2 = u-c\,.\,v-c,$$

équations qui se déduisent plus simplement des équations

$$\lambda^2 + \mu^2 + \nu^2 = 1,$$
$$\frac{\lambda^2}{v-a} + \frac{\mu^2}{v-b} + \frac{\nu^2}{v-c} = 1,$$
$$\frac{\lambda^2}{u-a} + \frac{\mu^2}{u-b} + \frac{\nu^2}{u-c} = 0.$$

Je rappelle les équations

$$\lambda x + \mu y + \nu z = \sqrt{v},$$
$$a\lambda x + b\mu y + c\nu z = \frac{\eta}{\sqrt{v}},$$
$$a(b+c)\lambda x + b(c+a)\mu y + c(a+b)\nu z = \eta\sqrt{v} + \frac{C}{\sqrt{v}};$$

et j'ajoute aussi celles-ci

$$\frac{\lambda^2}{(v-a)^2} + \frac{\mu^2}{(v-b)^2} + \frac{\nu^2}{(v-c)^2} = \frac{-1}{v\,.\,v-\xi} = -\frac{v-u}{v-a\,.\,v-b\,.\,v-c},$$

et de même

$$\frac{\lambda^2}{(u-a)^2} + \frac{\mu^2}{(u-b)^2} + \frac{\nu^2}{(u-c)^2} = \frac{v-u}{u-a\,.\,u-b\,.\,u-c}.$$

17. Formules différentielles. Nous avons

$$\lambda dx + \mu dy + \nu dz = 0, \quad x d\lambda + y d\mu + z d\nu = \tfrac{1}{2}\frac{dv}{\sqrt{v}};$$

$$2\alpha\beta\gamma(a\lambda dx + b\mu dy + c\nu dz) = \frac{a\alpha}{D}\{a\xi + \eta - a(b+c)\}\, 2\beta\gamma x dx + \&c.$$
$$= \frac{a\alpha}{D}\{a\xi + \eta - a(b+c)\}\{(\eta - bc)d\xi + (\xi - a)d\eta\}$$
$$+ \frac{b\beta}{D}\{b\xi + \eta - b(c+a)\}\{(\eta - ca)d\xi + (\xi - b)d\eta\}$$
$$+ \frac{c\gamma}{D}\{c\xi + \eta - c(a+b)\}\{(\eta - ab)d\xi + (\xi - c)d\eta\},$$

ou, en réduisant comme auparavant,

$$= \frac{\alpha\beta\gamma}{D} \{-(\xi\eta - C)\,d\xi + (\eta - \xi^2 + A\xi - B)\,d\eta\},$$

c'est-à-dire

$$2(a\lambda dx + b\mu dy + c\nu dz) = -\sqrt{\frac{\xi\eta - C}{\eta - \xi^2 + A\xi - B}}\,d\xi + \sqrt{\frac{\eta - \xi^2 + A\xi - B}{\xi\eta - C}}\,d\eta,$$

ou enfin

$$a\lambda dx + b\mu dy + c\nu dz = -\tfrac{1}{2}\sqrt{v}\,d\xi + \tfrac{1}{2}\frac{d\eta}{\sqrt{v}},$$

et de là en différentiant l'équation

$$a\lambda x + b\mu y + c\nu z = \frac{\eta}{\sqrt{v}},$$

on déduit

$$ax\,d\lambda + by\,d\mu + cz\,d\nu = \tfrac{1}{2}\sqrt{v}\,d\xi + \tfrac{1}{2}\frac{d\eta}{\sqrt{v}} - \tfrac{1}{2}\frac{\eta\,dv}{v\sqrt{v}}.$$

18. Nous avons de plus

$$2\beta\gamma x\,dx = (\eta - bc)\,d\xi + (\xi - a)\,d\eta,$$

donc

$$4\beta\gamma\,dx^2 = \frac{\{(\eta - bc)\,d\xi + (\xi - a)\,d\eta\}^2}{\xi - a\,.\,\eta - bc},$$

et de même

$$4\gamma\alpha\,dy^2 = \frac{\{(\eta - ca)\,d\xi + (\xi - b)\,d\eta\}^2}{\xi - b\,.\,\eta - ca},$$

$$4\alpha\beta\,dz^2 = \frac{\{(\eta - ab)\,d\xi + (\xi - c)\,d\eta\}^2}{\xi - c\,.\,\eta - ab};$$

et de là, après les réductions nécessaires,

$$4(\ dx^2 + \ dy^2 + \ dz^2) = \frac{\xi^2 - A\xi + B - \eta}{\xi - a\,.\,\xi - b\,.\,\xi - c}\,d\xi^2 + \frac{C - \xi\eta}{\eta - bc\,.\,\eta - ca\,.\,\eta - ab}\,d\eta^2,$$

$$4(a\,dx^2 + b\,dy^2 + c\,dz^2) = \frac{C - \xi\eta}{\xi - a\,.\,\xi - b\,.\,\xi - c}\,d\xi^2 + \frac{\eta^2 - B\eta + AC - C\xi}{\eta - bc\,.\,\eta - ca\,.\,\eta - ab}\,d\eta^2;$$

en écrivant la première de ces équations sous la forme

$$ds^2 = E\,d\xi^2 + 2F\,d\xi\,d\eta + G\,d\eta^2,$$

on a

$$E = \tfrac{1}{4}\frac{\xi^2 - A\xi + B - \eta}{\xi - a\,.\,\xi - b\,.\,\xi - c},\quad F = 0,\quad G = \tfrac{1}{4}\frac{C - \xi\eta}{\eta - bc\,.\,\eta - ca\,.\,\eta - ab}.$$

19. De plus, de l'équation $\lambda^2 + \mu^2 + \nu^2 = 1$, et des valeurs de $u + v$ et uv, on déduit

$$\lambda d\lambda + \ \mu d\mu + \ \nu d\nu = 0,$$

$$a\lambda d\lambda + \ b\mu d\mu + \ c\nu d\nu = \tfrac{1}{2}(\ du + \ dv),$$

$$bc\lambda d\lambda + ca\mu d\mu + ab\nu d\nu = \tfrac{1}{2}(v\,du + u\,dv).$$

20. Équation différentielle des courbes de courbure de la surface. En partant de l'équation

$$\begin{vmatrix} d\lambda, & d\mu, & d\nu \\ dx, & dy, & dz \\ \lambda, & \mu, & \nu \end{vmatrix} = 0,$$

ou plus simplement des équations

$$d\lambda : d\mu : d\nu = dx : dy : dz,$$

équivalentes à cette première équation, on déduit

$$(xd\lambda + yd\mu + zd\nu)(a\lambda dx + b\mu dy + c\nu dz) - (xdx + ydy + zdz)(a\lambda d\lambda + b\mu d\mu + c\nu d\nu) = 0,$$

c'est-à-dire

$$\frac{dv}{\sqrt{v}}\left(-\sqrt{v}d\xi + \frac{d\eta}{\sqrt{v}}\right) + d\xi\,(du + dv) = 0,$$

ou enfin

$$dvd\eta + vdud\xi = 0,$$

laquelle est la forme la plus simple de l'équation dont il s'agit.

21. On a ici

$$\eta = \frac{b\,(\xi^2 - A\xi + B) - C}{v - \xi}, \quad u = v - \frac{v - a\,.\,v - b\,.\,v - c}{v\,(v - \xi)},$$

et il s'agit de substituer ces valeurs dans l'équation différentielle. J'écris pour un moment

$$\eta = \frac{H}{v - \xi}, \quad u = \frac{U}{v - \xi},$$

où l'on a

$$H = v\,(\xi^2 - A\xi + B) - C,$$

$$U = -v\xi + Av - B + \frac{C}{v};$$

l'équation devient

$$-(v - \xi)\{dHdv + vd\xi dU\} + (dv - d\xi)(Hdv + vud\xi) = 0,$$

et l'on trouve

$$dH = (\xi^2 - A\xi + B)\,dv + (2\xi - A)\,vd\xi,$$

$$dU = \left(-\xi + A - \frac{C}{v^2}\right)dv - \qquad vd\xi,$$

et en substituant ces valeurs, on obtient

$$\{v^2(v - \xi) - vU\}\,d\xi^2 - \left\{(v - \xi)\left(\xi v - \frac{C}{v}\right) - vU + H\right\}dvd\xi + \{-(v - \xi)(\xi^2 - A\xi + B) + H\}\,dv^2 = 0.$$

Le coefficient de $d\xi^2$ est $v^3 - Av^2 + Bv - C$, c'est-à-dire $v - a\,.\,v - b\,.\,v - c$; de même, le coefficient de dv^2 est $\xi^3 - A\xi^2 + B\xi - C$, c'est-à-dire $\xi - a \,.\, \xi - b \,.\, \xi - c$. Le coefficient de $-dvd\xi$ est

$$\left(\xi v^2 - C - \xi^2 v + \frac{C\xi}{v}\right) + (v^2\xi - Av^2 + Bv - C) + (v\xi^2 - Av\xi + Bv - C),$$
$$= \xi\left(2v^2 - Av + \frac{C}{v}\right) - Av^2 + 2Bv - 3C;$$

l'équation différentielle est donc

$$(v^3 - Av^2 + Bv - C)\,d\xi^2 - \left\{\xi\left(2v^2 - Av + \frac{C}{v}\right) - Av^2 + 2Bv - 3C\right\} dvd\xi$$
$$+ (\xi^3 - A\xi^2 + B\xi - C)\,dv^2 = 0.$$

22. Mais on a identiquement

$$\begin{aligned} &\xi - a \,.\, v - b \,.\, v - c \\ +\, &\xi - b \,.\, v - c \,.\, v - a \\ +\, &\xi - c \,.\, v - a \,.\, v - b \\ -\, &\xi \,.\, \frac{1}{v} \,.\, v - a \,.\, v - b \,.\, v - c = \xi\left(2v^2 - Av + \frac{C}{v}\right) - Av^2 + 2Bv - 3C, \end{aligned}$$

donc en divisant par

$$v^3 - Av^2 + Bv - C, \quad = v - a \,.\, v - b \,.\, v - c,$$

l'équation devient

$$d\xi^2 - d\xi dv\left(\frac{\xi - a}{v - a} + \frac{\xi - b}{v - b} + \frac{\xi - c}{v - c} - \frac{\xi}{v}\right) + dv^2 \,.\, \frac{\xi - a \,.\, \xi - b \,.\, \xi - c}{v - a \,.\, v - b \,.\, v - c} = 0.$$

23. De même, en substituant dans l'équation $dvd\eta + vdud\xi = 0$ les valeurs

$$\eta = \frac{v\,(u^2 - Au + B)}{v - u}, \quad \xi = v - \frac{v - a \,.\, v - b \,.\, v - c}{v - u},$$

on obtient

$$(v^3 - Av^2 + Bv - C)\,du^2 - \left\{u\left(2v^2 - Av + \frac{C}{v}\right) - Av^2 + 2Bv - 3C\right\} dvdu$$
$$+ (u^3 - Au^2 + Bu - C)\,dv^2 = 0,$$

ou, ce qui est la même chose,

$$du^2 - dudv\left(\frac{u - a}{v - a} + \frac{u - b}{v - b} + \frac{u - c}{v - c} - \frac{u}{v}\right) + dv^2 \,.\, \frac{u - a \,.\, u - b \,.\, u - c}{v - a \,.\, v - b \,.\, v - c} = 0;$$

les équations différentielles entre ξ, v et entre u, v respectivement sont ainsi precisément de la même forme: on peut vérifier sans beaucoup de peine qu'en introduisant dans la première équation u au lieu de ξ par la substitution

$$\xi = v - \frac{v - a \,.\, v - b \,.\, v - c}{v - u},$$

on obtient la seconde équation: c'est là un théorème d'analyse assez remarquable.

24. L'équation différentielle en u, v est changée en entrechangeant ces deux variables: cela doit être ainsi, car autrement les deux courbes de courbure par le

point de contact du plan tangent $\lambda X+\mu Y+\nu Z-\sqrt{v}=0$, et les deux courbes de courbure par le point de contact du plan tangent parallèle $\lambda X+\mu Y+\nu Z-\sqrt{u}=0$ seraient parallèles les unes aux autres, ce qui n'est pas en général vrai. Pour que les deux équations soient identiques, on doit avoir

$$v-a\,.\,v-b\,.\,v-c\,.\left(\frac{u-a}{v-a}+\frac{u-b}{v-b}+\frac{u-c}{v-c}-\frac{u}{v}\right)$$

$$=u-a\,.\,u-b\,.\,u-c\,.\left(\frac{v-a}{u-a}+\frac{v-b}{u-b}+\frac{v-c}{u-c}-\frac{v}{u}\right),$$

c'est-à-dire

$$u\left(2v^2-Av+\frac{C}{v}\right)-Av^2+2Bv-3C=v\left(2u^2-Au+\frac{C}{u}\right)-Au^2+2Bu-3C,$$

ou enfin

$$(u-v)\{2u^2v^2-A(u+v)uv+2Buv-C(u+v)\}=0;$$

$u=v$ donne le plan à l'infinité, qui coupe la surface selon les deux coniques imaginaires $x^2+y^2+z^2=0$ et $ax^2+by^2+cz^2=0$, et aussi les plans tangents singuliers qui touchent la surface chacun selon un cercle réel ou imaginaire; je n'ai pas considéré la courbe sur la surface que l'on obtient en égalant à zéro l'autre facteur.

25. Rayons de courbure.

Les équations de la normale au point (x, y, z) sont

$$X=x-R\lambda,\quad Y=y-R\mu,\quad Z=z-R\nu,$$

et, en supposant que les coordonnées X, Y, Z se rapportent à un centre de courbure, on a

$$dx=Rd\lambda,\quad dy=Rd\mu,\quad dz=Rd\nu,$$

et de là

$$xdx+ydy+zdz=R(xd\lambda+yd\mu+zd\nu),$$

c'est-à-dire $d\xi=\dfrac{Rdv}{\sqrt{v}}$; donc $R=\sqrt{v}\dfrac{d\xi}{dv}$, $=\theta\sqrt{v}$, en posant $\theta=\dfrac{d\xi}{dv}$: θ est donc déterminé, en fonction de ξ, v, par l'équation

$$\theta^2-\left(\frac{\xi-a}{v-a}+\frac{\xi-b}{v-b}+\frac{\xi-c}{v-c}-\frac{\xi}{v}\right)\theta+\frac{\xi-a\,.\,\xi-b\,.\,\xi-c}{v-a\,.\,v-b\,.\,v-c}=0,$$

et les deux rayons de courbure sont alors donnés par la formule $R=\theta\sqrt{v}$.

26. Surface des centres de courbure. En écrivant pour R sa valeur, $=\theta\sqrt{v}$, nous avons

$$X=x-\lambda\theta\sqrt{v},\quad Y=y-\mu\theta\sqrt{v},\quad Z=z-\nu\theta\sqrt{v},$$

où X, Y, Z dénotent les coordonnées d'un point de la surface cherchée; et si nous formons les expressions Ξ, H, Z analogues à ξ, η, ζ, savoir

$$\begin{aligned}\Xi &= X^2+Y^2+Z^2,\\ \mathrm{H} &= aX^2+bY^2+cZ^2,\\ \mathrm{Z} &= a(b+c)X^2+b(c+a)Y^2+c(a+b)Z^2,\end{aligned}$$

on obtient sans peine, à moyen des valeurs trouvées pour

$$\lambda x + \mu y + \nu z, \quad a\lambda x + b\mu y + c\nu z, \quad a\lambda^2 + b\mu^2 + c\nu^2, \ \&c.,$$

$$\Xi = \xi - 2\theta v + \theta^2 v,$$

$$\mathrm{H} = \eta - 2\theta\eta + \theta^2 v (A - u - v),$$

$$\mathrm{Z} = \xi\eta + C - 2\theta(\eta v + C) + \theta^2 v (B - uv);$$

où comme auparavant

$$\theta^2 - \left(\frac{\xi - a}{v - a} + \frac{\xi - b}{v - b} + \frac{\xi - c}{v - c} - \frac{\xi}{v}\right)\theta + \frac{\xi - a \,.\, \xi - b \,.\, \xi - c}{v - a \,.\, v - b \,.\, v - c} = 0,$$

$$\eta = \frac{v(\xi^2 - A\xi + B) - C}{v - \xi},$$

$$(v - \xi)(v - u) = \frac{v - a \,.\, v - b \,.\, v - c}{v};$$

en éliminant entre ces six équations les cinq quantités ξ, η, θ, v, u, on obtient l'équation de la surface des centres en termes de Ξ, H, Z qui sont des fonctions données des coordonnées X, Y, Z d'un point de la surface.

27. Je remarque que les sections principales de la surface des ondes sont des courbes de courbure de cette surface; en particulier, pour fixer les idées, la section par le plan $z = 0$, savoir le cercle $\xi - c = 0$ et l'ellipse $\eta - bc = 0$ sont des courbes de courbure. Pour l'une ou l'autre de ces courbes, il y a une suite des courbes de courbure de l'autre espèce qui coupent le cercle ou l'ellipse à angle droit, et qui sont symétriques aux deux côtés du plan $z = 0$. En considérant par exemple l'ellipse, les normales à la surface aux points successifs de cette courbe sont situées dans le plan de xy et se coupent selon une courbe dans ce plan, la développée de l'ellipse, laquelle est une courbe sur la surface des centres; l'ordre de cette développée est $= 6$. Mais, de plus, chaque normale de la surface à un point $(x, y, 0)$ de l'ellipse est rencontrée par les normales de la surface aux points $(x, y, \pm\delta z)$ au-dessus et au-dessous du point de l'ellipse: les points d'intersection forment une courbe dans le plan de xy, laquelle est aussi une courbe sur la surface des centres, et de plus elle est une courbe cuspidale sur cette surface: nous allons voir que l'ordre de cette courbe est $= 6$. De même pour le cercle, la développée du cercle est le point $x = 0$, $y = 0$, lequel à ce que je crois doit être considéré comme cercle infiniment petit (ou deux droites imaginaires) $x^2 + y^2 = 0$; le cercle donne lieu aussi à une courbe qui est une courbe cuspidale sur la surface des centres, et nous allons voir que l'ordre de cette courbe est $= 4$. La section de la surface des centres par le plan xy est donc composée comme suit:

développée de l'ellipse,	ordre	6
courbe cuspidale, ordre 6, trois fois,	„	18
développée du cercle,	„	2
courbe cuspidale, ordre 4, trois fois,	„	12
		38,

et il paraît ainsi que l'ordre de la surface des centres de la surface des ondes doit être $= 38$.

28. Je m'arrête pour un moment pour considérer la même théorie par rapport à l'ellipsoïde $\frac{x^2}{a}+\frac{y^2}{b}+\frac{z^2}{c}=1$. La section par le plan $z=0$ est ici l'ellipse $\frac{x^2}{a}+\frac{y^2}{b}=1$; et pour la surface des centres on a dans le plan de xy, la développée d'ellipse, courbe d'ordre 6, et aussi une courbe cuspidale laquelle est une ellipse. En effet, pour trouver l'équation de cette courbe, on a pour les coordonnées X, Y du point où la normale au point (x, y, z) rencontre le plan $z=0$, les équations $x=X+\frac{\lambda}{\nu}z$, $y=Y+\frac{\mu}{\nu}z$; c'est-à-dire $X=x\left(1-\frac{c}{a}\right)$, $Y=y\left(1-\frac{c}{b}\right)$. Dans ces équations x, y se rapportent à un point de l'ellipse $\frac{x^2}{a}+\frac{y^2}{b}=1$; en éliminant entre les trois équations x, y, on obtient $\frac{aX^2}{(a-c)^2}+\frac{bY^2}{(b-c)^2}=1$, ou ce qui est la même chose $\frac{aX^2}{\beta^2}+\frac{bY^2}{\alpha^2}=1$. La section principale de la surface des centres est donc composée de cette ellipse trois fois, et de la développée de l'ellipse $\frac{x^2}{a}+\frac{y^2}{b}=1$; l'ordre de la section, et ainsi l'ordre de la surface des centres, est donc $6+3.2, =12$, comme cela doit être.

29. De même pour la surface des ondes on a $x=X+\frac{\lambda}{\nu}z$, $y=Y+\frac{\mu}{\nu}z$, c'est-à-dire

$$X=x-\frac{x\{a\xi+\eta-a(b+c)\}}{c\xi+\eta-c(a+b)},$$

$$Y=y-\frac{y\{b\xi+\eta-b(c+a)\}}{c\xi+\eta-c(a+b)},$$

où pour $z=0$ on a $(\xi-c)(\eta-ab)=0$, équation qui donne: 1°, le cercle $\xi-c=0$; 2°, l'ellipse $\eta-ab=0$.

1°. Pour le cercle $\xi=c$, en écrivant $x^2=c\theta$, $y^2=c(1-\theta)$, nous avons

$$\eta=ac\theta+bc(1-\theta),\ =c(b+\gamma\theta)$$

et de là

$$a\xi+\eta-a(b+c)=b\beta+c\gamma\theta,$$

$$b\xi+\eta-b(c+a)=b\beta+c\gamma\theta,$$

$$c\xi+\eta-c(a+b)=c(\beta+\gamma\theta),$$

valeurs qui donnent

$$X=\sqrt{c}\sqrt{\theta}\quad\left\{1-\frac{b\beta+c\gamma\theta}{c(\beta+\gamma\theta)}\right\},\quad =-\frac{\alpha\beta}{\sqrt{c}}\frac{\sqrt{\theta}}{\beta+\gamma\theta},$$

$$Y=\sqrt{c}\sqrt{1-\theta}\left\{1-\frac{b\beta+c\gamma\theta}{c(\beta+\gamma\theta)}\right\},\quad =-\frac{\alpha\beta}{\sqrt{c}}\frac{\sqrt{1-\theta}}{\beta+\gamma\theta};$$

donc

$$X^2 + Y^2 = \frac{\alpha^2\beta^2}{c}\frac{1}{(\beta+\gamma\theta)^2},$$

$$\alpha X^2 - \beta Y^2 = \frac{\alpha^2\beta^2}{c}\frac{\alpha\theta - \beta(1-\theta)}{(\beta+\gamma\theta)^2}, \quad = -\frac{\alpha^2\beta^2}{c}\frac{1}{\beta+\gamma\theta},$$

et de là

$$(\alpha X^2 - \beta Y^2)^2 = \frac{\alpha^2\beta^2}{c}(X^2+Y^2);$$

courbe de l'ordre 4, cuspidale sur la surface des centres.

2°. Pour l'ellipse $\eta = ab$, en écrivant $x^2 = b\theta$, $y^2 = a(1-\theta)$, et de là

$$\xi = b\theta + a(1-\theta), \ = a - \gamma\theta,$$

on obtient

$$a\xi + \eta - a(b+c) = -a(\beta+\gamma\theta),$$

$$b\xi + \eta - b(c+a) = -b(\beta+\gamma\theta),$$

$$c\xi + \eta - c(a+b) = -b\beta - c\gamma\theta,$$

valeurs qui donnent

$$X = \sqrt{b}\sqrt{\theta}\left\{1 - \frac{a(\beta+\gamma\theta)}{b\beta+c\gamma\theta}\right\} = -\sqrt{b}\beta\gamma\frac{\sqrt{\theta}(1-\theta)}{b\beta+c\gamma\theta},$$

$$Y = \sqrt{a}\sqrt{1-\theta}\left\{1 - \frac{b(\beta+\gamma\theta)}{b\beta+c\gamma\theta}\right\} = -\sqrt{a}\alpha\gamma\frac{\theta\sqrt{1-\theta}}{b\beta+c\gamma\theta};$$

donc

$$\frac{X^2}{b\beta^2} + \frac{Y^2}{a\alpha^2} = \gamma^2\frac{\theta(1-\theta)}{(b\beta+c\gamma\theta)^2},$$

$$\frac{X^2}{\beta} - \frac{Y^2}{\alpha} = \gamma^2\theta(1-\theta)\frac{b\beta(1-\theta) - a\alpha\theta}{(b\beta+c\gamma\theta)^2}, \quad = \gamma^2\frac{\theta(1-\theta)}{b\beta+c\gamma\theta};$$

$$\frac{X^2}{b\beta^2}\cdot\frac{Y^2}{a\alpha^2} = \gamma^4\frac{\theta^3(1-\theta)^3}{(b\beta+c\gamma\theta)^4},$$

et de là

$$\left(\frac{X^2}{b\beta^2} + \frac{Y^2}{a\alpha^2}\right)\left(\frac{X^2}{\beta} - \frac{Y^2}{\alpha}\right)^2 = \gamma^2\cdot\frac{X^2}{b\beta^2}\cdot\frac{Y^2}{a\alpha^2},$$

courbe de l'ordre 6, cuspidale sur la surface des centres.

30. On aurait pu développer la théorie des courbes et rayons de courbure à moyen de la formule ci-dessus donnée, $ds^2 = Ed\xi^2 + 2Fd\xi d\eta + Gd\eta^2$ $(F=0)$, mais pour cela il faudrait trouver plusieurs expressions qui ne sont pas encore calculées, savoir les coefficients des formules

$$d^2x = \alpha\, d\xi^2 + 2\alpha' d\xi\, d\eta + \alpha'' d\eta^2,$$

$$d^2y = \beta\, d\xi^2 + 2\beta' d\xi\, d\eta + \beta'' d\eta^2,$$

$$d^2z = \gamma\, d\xi^2 + 2\gamma' d\xi\, d\eta + \gamma'' d\eta^2,$$

et puis

$$E', F', G' = \lambda\alpha + \mu\beta + \nu\gamma, \quad \lambda\alpha' + \mu\beta' + \nu\gamma', \quad \lambda\alpha'' + \mu\beta'' + \nu\gamma''.$$

En prenant comme auparavant R pour le rayon de courbure on aurait alors, (Salmon, *Geometry of three dimensions*, Ed. 4 (1882), p. 347),

$$\left\| \begin{matrix} R, & E\,d\xi, & G\,d\eta \\ EG, & E'd\xi + F'd\eta, & F'd\xi + G'd\eta \end{matrix} \right\| = 0,$$

et

$$\left\| \begin{matrix} d\eta, & RE' - E^2G, & RF' \\ -d\xi, & RF', & RG' - EG^2 \end{matrix} \right\| = 0,$$

formules pour les courbes et rayons de courbure: en particulier, l'équation différentielle des courbes de courbure peut s'écrire sous la forme

$$\left| \begin{matrix} d\eta^2, & -d\xi\,d\eta, & d\xi^2 \\ E, & 0, & G \\ E', & F', & G' \end{matrix} \right| = 0.$$

Au reste, cette équation en $d\xi$, $d\eta$ se déduirait plus simplement de l'équation $dv\,d\eta + v\,du\,d\xi = 0$, en y introduisant les expressions de v, u en termes de ξ, η.

31. Courbes géodésiques sur la surface. L'équation différentielle du second ordre des courbes géodésiques dépend seulement des coefficients E, F, G, savoir en supposant $F = 0$, cette équation est

$$\begin{aligned} &E\,d\xi\,(-E_2 d\xi^2 + 2G_1 d\xi\,d\eta + G_2 d\eta^2) \\ &- G\,d\eta\,(\;\;E_1 d\xi^2 + 2E_2 d\xi\,d\eta - G_1 d\eta^2) \\ &+ 2EG\,(\;\;d\xi\,d^2\eta - d\eta\,d^2\xi\;\;) = 0, \end{aligned}$$

ou ce qui est la même chose

$$(-EE_2,\ 2EG_1 - GE_1,\ EG_2 - 2GE_2,\ GG_1)(d\xi,\ d\eta)^3 + 2EG\,(d\xi\,d^2\eta - d\eta\,d^2\xi) = 0,$$

où

$$E_1,\ E_2,\ G_1,\ G_2 = \frac{dE}{d\xi},\ \frac{dE}{d\eta},\ \frac{dG}{d\xi},\ \frac{dG}{d\eta}$$

respectivement, voir Cayley, "On geodesic lines, in particular those of a Quadric Surface," *Proc. London Math. Soc.*, t. IV. (1872), p. 197, [508]. Nous avons vu que pour la surface des ondes dont il s'agit les expressions de E, G sont

$$E = \tfrac{1}{4}\,\frac{\xi^2 - A\xi + B - \eta}{\xi - a\,.\,\xi - b\,.\,\xi - c}, \quad G = \tfrac{1}{4}\,\frac{C - \xi\eta}{\eta - bc\,.\,\eta - ca\,.\,\eta - ab},$$

et l'on obtiendrait de là sans peine les expressions des coefficients de la fonction cubique $(-EE_2, \ldots)(d\xi, d\eta)^3$ qui entre dans l'équation différentielle.

931.

ON SOME FORMULÆ OF CODAZZI AND WEINGARTEN IN RELATION TO THE APPLICATION OF SURFACES TO EACH OTHER.

[From the *Proceedings of the London Mathematical Society*, vol. XXIV. (1893), pp. 210—223.]

AN extremely elegant theory of the application of surfaces one upon another is developed in the memoir, Codazzi, "Mémoire relatif à l'application des surfaces les unes sur les autres," *Mém. prés. à l'Académie des Sciences*, t. XXVII. (1883), No. 6, pp. 1—47; but the notation is not presented in a form which is easily comparable with that of the Gaussian notation in the theory of surfaces. I propose to reproduce the theory in the Gaussian notation.

Codazzi considers on a given surface two systems of curves depending on the parameters t, T respectively; the curves are in the memoir taken to be orthogonal to each other, but this restriction is removed in the general formula given p. 44, *Addition au chapitre premier*. For a curve of either system, he considers the tangent, the principal normal, or normal in the osculating plane, and the binormal, or line at right angles to the osculating plane (say these are tan, prn and bin). For a curve of the one system, that in which t is variable or say a t-curve, he denotes the cosine-inclinations of these lines to the axes by the letters a, b, c, thus:

	x	y	z
tan	a_x	a_y	a_z
prn	b_x	b_y	b_z
bin	c_x	c_y	c_z

and he writes also l for the inclination of the principal normal to the normal of the surface, $\frac{dm}{dt}dt$ for the angle of contingence, or inclination of the tangent at the point (t, T) to the tangent at the point $(t+dt, T)$, and $\frac{dn}{dt}dt$ for the angle of torsion, or inclination of the osculating plane at the point (t, T) to that at the point $(t+dt, T)$, or, what is the same thing, the inclination of the binormals at these points respectively. And he gives as known formulæ

$$\frac{da}{dt}=b\frac{dm}{dt}, \quad \frac{db}{dt}=-a\frac{dm}{dt}-c\frac{dn}{dt}, \quad \frac{dc}{dt}=b\frac{dn}{dt},$$

where a, b, c denote

$$(a_x, b_x, c_x), \quad (a_y, b_y, c_y), \quad \text{or} \quad (a_z, b_z, c_z).$$

He uses the capital letters

$$A_x, A_y, A_z, B_x, B_y, B_z, C_x, C_y, C_z, L, \frac{dM}{dT}dT, \frac{dN}{dT}dT$$

with the like significations in regard to the curve for which T is variable, or say the T-curve; for greater clearness I give the accompanying figure.

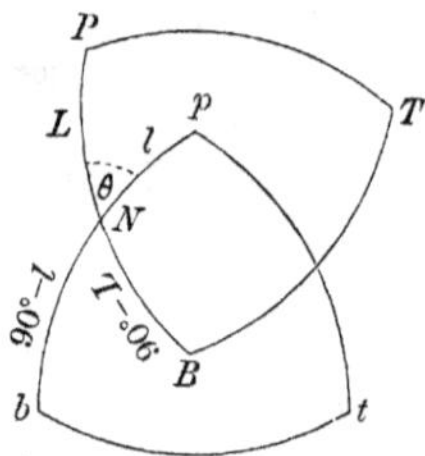

Codazzi writes further

$$\frac{dm}{dt}\cos l = u, \quad \frac{dm}{dt}\sin l = v, \quad \frac{dn}{dt}-\frac{dl}{dt}=w,$$

$$\frac{dM}{dT}\cos L = U, \quad \frac{dM}{dT}\sin L = V, \quad \frac{dN}{dT}-\frac{dL}{dT}=W;$$

also, if s, S are the arcs of the two curves respectively,

$$\frac{ds}{dt}=r, \quad \frac{dS}{dT}=R;$$

and he obtains a system of six formulæ, which in the *Addition*, p. 44, are presented in the following form—only I use therein θ, instead of his b, to denote the inclination of the two curves to each other:

$$\frac{du}{dT} = \frac{dU}{dt}\cos\theta + \frac{dW}{dt}\sin\theta + w\left(V - \frac{d\theta}{dT}\right) + (U\sin\theta - W\cos\theta)\left(v - \frac{d\theta}{dt}\right),$$

$$\frac{dU}{dt} = \frac{du}{dT}\cos\theta + \frac{dw}{dT}\sin\theta + W\left(v - \frac{d\theta}{dt}\right) + (u\sin\theta - w\cos\theta)\left(V - \frac{d\theta}{dT}\right),$$

$$\left(\frac{dv}{dT} + \frac{dV}{dt}\right)\frac{d^2\theta}{dt\,dT} + \sin\theta\,(uU - wW) - \cos\theta\,(uW + wU) = 0,$$

$$R\,(u\cos\theta + w\sin\theta) = r\,(U\cos\theta + W\sin\theta),$$

$$R\sin\theta\left(v - \frac{d\theta}{dt}\right) + \frac{dR}{dt}\cos\theta = \frac{dr}{dT},$$

$$r\sin\theta\left(V - \frac{d\theta}{dT}\right) + \frac{dr}{dT}\cos\theta = \frac{dR}{dt}.$$

In the Gaussian notation, taking p, q for the parameters, we have, with the slight variations presently referred to,

$$dx = a\,dp + a'dq + \tfrac{1}{2}\alpha dp^2 + \alpha' dp\,dq + \alpha'' dq^2,$$

$$dy = b\,dp + b'dq + \tfrac{1}{2}\beta dp^2 + \beta' dp\,dq + \beta'' dq^2,$$

$$dz = c\,dp + c'dq + \tfrac{1}{2}\gamma dp^2 + \gamma' dp\,dq + \gamma'' dq^2,$$

$$A,\ B,\ C = bc' - b'c,\ ca' - c'a,\ ab' - a'b,$$

$$E,\ F,\ G = a^2 + b^2 + c^2,\ aa' + bb' + cc',\ a'^2 + b'^2 + c'^2;$$

and therefore

$$dx^2 + dy^2 + dz^2 = Edp^2 + 2Fdp\,dq + Gdq^2;$$

$$\sqrt{EG - F^2} = V;$$

$$E',\ F',\ G' = A\alpha + B\beta + C\gamma,\ \ A\alpha' + B\beta' + C\gamma',\ \ A\alpha'' + B\beta'' + C\gamma'';$$

and I take further

$$\omega,\ \omega',\ \omega'' = a\alpha + b\beta + c\gamma,\ \ a\alpha' + b\beta' + c\gamma',\ \ a\alpha'' + b\beta'' + c\gamma'',$$

$$\varpi,\ \varpi',\ \varpi'' = a'\alpha + b'\beta + c'\gamma,\ \ a'\alpha' + b'\beta' + c'\gamma',\ \ a'\alpha'' + b'\beta'' + c'\gamma'',$$

$$\lambda,\ \lambda',\ \lambda'' = \alpha^2 + \beta^2 + \gamma^2,\ \ \alpha'^2 + \beta'^2 + \gamma'^2,\ \ \alpha''^2 + \beta''^2 + \gamma''^2,$$

$$\mu,\ \mu',\ \mu'' = \alpha'\alpha'' + \beta'\beta'' + \gamma'\gamma'',\ \ \alpha''\alpha + \beta''\beta + \gamma''\gamma,\ \ \alpha\alpha' + \beta\beta' + \gamma\gamma',$$

$$E\lambda - \omega^2 = \Delta,\quad G\lambda'' - \varpi''^2 = \Delta'',$$

where it is to be noticed that V^2, E', F', G', are written instead of Gauss's Δ, D, D', D'', and ω, ω', ω'', ϖ, ϖ', ϖ'' instead of his m, m', m'', n, n', n'': and that he gives for the last-mentioned quantities the values

$$m = \tfrac{1}{2}\frac{dE}{dp},\quad m' = \tfrac{1}{2}\frac{dE}{dq},\quad m'' = \frac{dF}{dq} - \tfrac{1}{2}\frac{dG}{dp},$$

$$n = \frac{dF}{dp} - \tfrac{1}{2}\frac{dE}{dq},\quad n' = \tfrac{1}{2}\frac{dG}{dp},\quad n'' = \tfrac{1}{2}\frac{dG}{dq},$$

or, say

$$\omega = \tfrac{1}{2}E_1,\quad \omega' = \tfrac{1}{2}E_2,\quad \omega'' = F_2 - \tfrac{1}{2}G_1,$$

$$\varpi = F_1 - \tfrac{1}{2}E_2,\quad \varpi' = \tfrac{1}{2}G_1,\quad \varpi'' = \tfrac{1}{2}G_2,$$

where the subscripts (1) and (2) denote differentiation in regard to p and q respectively.

Observing that the cosine-inclinations of the tangent to the p-curve are as a, b, c, and those of the binormal or perpendicular to the osculating plane are as $b\gamma - c\beta$, $c\alpha - a\gamma$, $a\beta - b\alpha$, we easily find

$$a_x,\ a_y,\ a_z = \frac{a}{\sqrt{E}},\quad \frac{b}{\sqrt{E}},\quad \frac{c}{\sqrt{E}},$$

$$c_x,\ c_y,\ c_z = \frac{b\gamma - c\beta}{\sqrt{\Delta}},\quad \frac{c\alpha - a\gamma}{\sqrt{\Delta}},\quad \frac{a\beta - b\alpha}{\sqrt{\Delta}},$$

$$b_x,\ b_y,\ b_z = \frac{E\alpha - a\omega}{\sqrt{E\Delta}},\quad \frac{E\beta - b\omega}{\sqrt{E\Delta}},\quad \frac{E\gamma - c\omega}{\sqrt{E\Delta}};$$

and for the cosine-inclinations of the normal of the surface

$$\Delta_x,\ \Delta_y,\ \Delta_z = \frac{A}{V},\ \frac{B}{V},\ \frac{C}{V},$$

we find

$$\cos l = \frac{A(b\gamma - c\beta) + B(c\alpha - a\gamma) + C(a\beta - b\alpha)}{V\sqrt{\Delta}}$$

$$= \frac{(a^2 + b^2 + c^2)(a'\alpha + b'\beta + c'\gamma) - (aa' + bb' + cc')(a\alpha + b\beta + c\gamma)}{V\sqrt{\Delta}}$$

$$= \frac{E\varpi - F\omega}{V\sqrt{\Delta}},$$

$$\sin l = \frac{A(E\alpha - a\omega) + B(E\beta - b\omega) + C(E\gamma - c\omega)}{V\sqrt{E\Delta}};$$

or, since $Aa + Bb + Cc = 0$, this is

$$\sin l = \frac{E(A\alpha + B\beta + C\gamma)}{V\sqrt{E\Delta}} = \frac{E'\sqrt{E}}{V\sqrt{\Delta}}.$$

We ought therefore to have

$$E'^2E + (E\varpi - F\omega)^2 = V^2\Delta,\quad = (EG - F^2)(E\lambda - \omega^2);$$

or, omitting the terms $F^2\omega^2$, which destroy each other, and throwing out a factor E, this is

$$E\varpi^2 - 2F\varpi\omega + G\omega^2 - \lambda(EG - F^2) = \lambda(EG - F^2) - E'^2;$$

viz. putting for $EG - F^2$ its value, $= A^2 + B^2 + C^2$, and for the other terms their values, this is

$$(a^2 + b^2 + c^2)(a'\alpha + b'\beta + c'\gamma)^2 - 2(aa' + bb' + cc')(a'\alpha + b'\beta + c'\gamma)(a\alpha + b\beta + c\gamma)$$
$$+ (a'^2 + b'^2 + c'^2)(\alpha^2 + \beta^2 + \gamma^2)$$
$$= (A^2 + B^2 + C^2)(\alpha^2 + \beta^2 + \gamma^2) - (A\alpha + B\beta + C\gamma)^2.$$

The right-hand side is here $= (B\gamma - C\beta)^2 + (C\alpha - A\gamma)^2 + (A\beta - B\alpha)^2$; the left-hand consists of three parts, the first whereof is

$$\{a(a'\alpha + b'\beta + c'\gamma) - a'(a\alpha + b\beta + c\gamma)\}^2 = (-B\gamma + C\beta)^2,$$

and similarly the other two parts are

$$(-C\alpha+A\gamma)^2 \text{ and } (-A\beta+B\alpha)^2;$$

the equation is thus verified.

We require Codazzi's $\frac{dm}{dt}$ and $\frac{dn}{dt}$, or say $\frac{dm}{dp}$ and $\frac{dn}{dp}$; these are to be obtained from the equations

$$\frac{d}{dp}\frac{a}{\sqrt{E}}=\frac{E\alpha-a\omega}{\sqrt{E\Delta}}\frac{dm}{dp},\quad \frac{d}{dp}\frac{b\gamma-c\beta}{\sqrt{\Delta}}=\frac{E\alpha-a\omega}{\sqrt{E\Delta}}\frac{dn}{dp},$$

and the values obtained should satisfy

$$\frac{d}{dp}\frac{E\alpha-a\omega}{\sqrt{E\Delta}}=-\frac{a}{\sqrt{E}}\frac{dm}{dp}-\frac{b\gamma-c\beta}{\sqrt{\Delta}}\frac{dn}{dp}.$$

I find

$$\frac{dm}{dp}=\frac{\sqrt{\Delta}}{E},\quad \frac{dn}{dp}=-\frac{\sqrt{E}}{\Delta}\begin{vmatrix} a, & b, & c \\ \alpha, & \beta, & \gamma \\ \alpha_1, & \beta_1, & \gamma_1 \end{vmatrix},$$

where α_1, β_1, γ_1 are the derivatives of α, β, γ in regard to p; and the equation to be verified thus is

$$\frac{d}{dp}\frac{E\alpha-a\omega}{\sqrt{E\Delta}}=-\frac{a\sqrt{\Delta}}{E\sqrt{E}}+\frac{(b\gamma-c\beta)\sqrt{E}}{\Delta\sqrt{\Delta}}\begin{vmatrix} a, & b, & c \\ \alpha, & \beta, & \gamma \\ \alpha_1, & \beta_1, & \gamma_1 \end{vmatrix}.$$

First, for $\frac{dm}{dp}$: the derivatives of a, E are α and $2(a\alpha+b\beta+c\gamma)$, $=2\omega$; we thus have

$$\frac{d}{dp}\frac{a}{\sqrt{E}},\quad =\frac{\alpha}{\sqrt{E}}-\frac{a\omega}{E\sqrt{E}}=\frac{E\alpha-a\omega}{E\sqrt{E}},$$

which is

$$=\frac{E\alpha-a\omega}{\sqrt{E\Delta}}\frac{dm}{dp};$$

viz. we have

$$\frac{dm}{dp}=\frac{\sqrt{\Delta}}{E}.$$

Next, for $\frac{dn}{dp}$: using a subscript $(_1)$ to denote derivation in regard to p, we have $\Delta=E\lambda-\omega^2$, and thence

$$\begin{aligned}\Delta_1&=E_1\lambda+E\lambda_1-2\omega\omega_1\\&=2\omega\lambda+E\,.\,2(\alpha\alpha_1+\beta\beta_1+\gamma\gamma_1)-2\omega(\lambda+a\alpha_1+b\beta_1+c\gamma_1),\\&=2\{E(\alpha\alpha_1+\beta\beta_1+\gamma\gamma_1)-\omega(a\alpha_1+b\beta_1+c\gamma_1)\},\\&=2(\alpha_1X+\beta_1Y+\gamma_1Z),\end{aligned}$$

if, for a moment,

$$X,\ Y,\ Z = E\alpha - a\omega,\ E\beta - b\omega,\ E\gamma - c\omega;$$

these values give identically

$$\alpha X + \beta Y + \gamma Z = \Delta, \quad \text{and} \quad aX + bY + cZ = 0.$$

Hence we have

$$\frac{d}{dp}\frac{b\gamma - c\beta}{\sqrt{\Delta}} = \frac{b\gamma_1 - c\beta_1}{\sqrt{\Delta}} - \frac{(b\gamma - c\beta)(\alpha_1 X + \beta_1 Y + \gamma_1 Z)}{\Delta\sqrt{\Delta}},$$

which must be

$$= \frac{X}{\sqrt{E\Delta}}\frac{dn}{dp};$$

we thus have

$$(b\gamma_1 - c\beta_1)\Delta - (b\gamma - c\beta)(\alpha_1 X + \beta_1 Y + \gamma_1 Z) = \frac{\Delta X}{\sqrt{E}}\frac{dn}{dp};$$

or, putting the left-hand side in the form

$$\begin{aligned} &(b\gamma_1 - c\beta_1)(\alpha X + \beta Y + \gamma Z) \\ -&(b\gamma - c\beta)(\alpha_1 X + \beta_1 Y + \gamma_1 Z) \\ -&(\beta\gamma_1 - \beta_1\gamma)(a X + b Y + c Z), \end{aligned}$$

this is

$$= -X \begin{vmatrix} a, & b, & c \\ \alpha, & \beta, & \gamma \\ \alpha_1, & \beta_1, & \gamma_1 \end{vmatrix},$$

and we thus find

$$\frac{dn}{dp} = -\frac{\sqrt{E}}{\Delta}\begin{vmatrix} a, & b, & c \\ \alpha, & \beta, & \gamma \\ \alpha_1, & \beta_1, & \gamma_1 \end{vmatrix}.$$

For the verification of the equation

$$\frac{d}{dp}\frac{E\alpha - a\omega}{\sqrt{E\Delta}} = -\frac{a\sqrt{\Delta}}{E\sqrt{E}} + \frac{(b\gamma - c\beta)\sqrt{E}}{\Delta\sqrt{\Delta}}\begin{vmatrix} a, & b, & c \\ \alpha, & \beta, & \gamma \\ \alpha_1, & \beta_1, & \gamma_1 \end{vmatrix},$$

we have

$$(E\alpha - a\omega)_1 = E\alpha_1 - a(a\alpha_1 + b\beta_1 + c\gamma_1) + \alpha\omega - a\lambda,$$

$$(E\Delta)_1 = 2E^2(\alpha\alpha_1 + \beta\beta_1 + \gamma\gamma_1) - 2E\omega(a\alpha_1 + b\beta_1 + c\gamma_1) + 2\Delta\omega,$$

and hence the equation is

$$-\frac{\{E^2(\alpha\alpha_1 + \beta\beta_1 + \gamma\gamma_1) - E\omega(a\alpha_1 + b\beta_1 + c\gamma_1) + \omega\}(E\alpha - a\omega)}{E\Delta\sqrt{E\Delta}} + \frac{E\alpha_1 - a(a\alpha_1 + b\beta_1 + c\gamma_1) + \alpha\omega - a\lambda}{\sqrt{E\Delta}}$$

$$= -\frac{a\sqrt{\Delta}}{E\sqrt{E}} + \frac{(b\gamma - c\beta)\sqrt{E}}{\Delta\sqrt{\Delta}}\begin{vmatrix} a, & b, & c \\ \alpha, & \beta, & \gamma \\ \alpha_1, & \beta_1, & \gamma_1 \end{vmatrix}.$$

Considering first the terms without $\alpha_1, \beta_1, \gamma_1$, these give

$$-\omega(E\alpha - a\omega) + E(\alpha\omega - a\lambda) = -a(E\lambda - \omega^2),$$

which is identically true; and then the remaining terms with α_1, β_1, γ_1 give

$$-\{E(\alpha\alpha_1+\beta\beta_1+\gamma\gamma_1)-\omega(a\alpha_1+b\beta_1+c\gamma_1)\}(E\alpha-a\omega)+\{E\alpha_1-a(a\alpha_1+b\beta_1+c\gamma_1)\}(E\lambda-\omega^2)$$

$$=(b\gamma-c\beta)E\begin{vmatrix} a, & b, & c \\ \alpha, & \beta, & \gamma \\ \alpha_1, & \beta_1, & \gamma_1 \end{vmatrix}.$$

On the left-hand side, the whole coefficient of α_1 is

$$=-(E\alpha-a\omega)^2+(b^2+c^2)(E\lambda-\omega^2),$$

which is

$$=-E^2\alpha^2+2a\alpha E\omega-a^2\omega^2+(b^2+c^2)E\lambda-(b^2+c^2)\omega^2,$$

$$=E[-E\alpha^2+2a\alpha\omega+(b^2+c^2)\lambda-\omega^2];$$

and, substituting for E, ω, and λ their values, this is found to be

$$=E\{(b^2+c^2)(\beta^2+\gamma^2)-(b\beta+c\gamma)^2\},\quad =E(b\gamma-c\beta)^2.$$

Similarly the whole coefficients of β_1 and γ_1 are found to be

$$=E(b\gamma-c\beta)(c\alpha-a\gamma),\quad\text{and}\quad E(b\gamma-c\beta)(a\beta-b\alpha),$$

respectively; and thus the left-hand side becomes

$$=(b\gamma-c\beta)E\begin{vmatrix} a, & b, & c \\ \alpha, & \beta, & \gamma \\ \alpha_1, & \beta_1, & \gamma_1 \end{vmatrix},$$

as it should do.

We have

$$\tan l=\frac{P}{Q},$$

where

$$P=E'\sqrt{E}=\sqrt{E}(A\alpha+B\beta+C\gamma),\qquad =\mathrm{a}\alpha+\mathrm{b}\beta+\mathrm{c}\gamma,\ \text{suppose},$$

$$Q=E\varpi-F\omega=(cB-bC)\alpha+(aC-cA)\beta+(bA-aB)\gamma,\quad =\mathrm{a}'\alpha+\mathrm{b}'\beta+\mathrm{c}'\gamma,\ \text{suppose},$$

$$P^2+Q^2=V^2\Delta,$$

and we have hence to find $\frac{dl}{dp}$. Using, as before, a subscript $({}_1)$ to denote derivation in regard to p, we have

$$\frac{dl}{dp}=\frac{QP_1-PQ_1}{P^2+Q^2};$$

the numerator is

$$\begin{aligned}
&= (\mathrm{a}'\alpha+\mathrm{b}'\beta+\mathrm{c}'\gamma)(\mathrm{a}\,\alpha_1+\mathrm{b}\,\beta_1+\mathrm{c}\,\gamma_1+\mathrm{a}_1\alpha+\mathrm{b}_1\beta+\mathrm{c}_1\gamma)\\
&\quad-(\mathrm{a}\,\alpha+\mathrm{b}\,\beta+\mathrm{c}\,\gamma)(\mathrm{a}'\alpha_1+\mathrm{b}'\beta_1+\mathrm{c}'\gamma_1+\mathrm{a}_1'\alpha+\mathrm{b}_1'\beta+\mathrm{c}_1'\gamma)\\
&=-[(\mathrm{bc}'-\mathrm{b}'\mathrm{c})(\beta\gamma_1-\beta_1\gamma)+(\mathrm{ca}'-\mathrm{c}'\mathrm{a})(\gamma\alpha_1-\gamma_1\alpha)+(\mathrm{ab}'-\mathrm{a}'\mathrm{b})(\alpha\beta_1-\alpha_1\beta)]\\
&\quad+Q(\mathrm{a}_1\alpha+\mathrm{b}_1\beta+\mathrm{c}_1\gamma)-P(\mathrm{a}_1'\alpha+\mathrm{b}_1'\beta+\mathrm{c}_1'\gamma).
\end{aligned}$$

For the first part hereof,

$$\text{bc}'-\text{b}'\text{c}=\sqrt{E}\,\{B\,(bA-aB)-C\,(aC-cA)\}$$
$$=\sqrt{E}\,\{A\,(aA+bB+cC)-a\,(A^2+B^2+C^2)\},\quad =-aV^2\sqrt{E},$$

since $aA+bB+cC=0$; and similarly

$$\text{ca}'-\text{c}'\text{a},\quad \text{ab}'-\text{a}'\text{b}=-bV^2\sqrt{E},\quad -cV^2\sqrt{E},$$

respectively; and thus the first part is

$$=V^2\sqrt{E}\,\{a\,(\beta\gamma_1-\beta_1\gamma)+b\,(\gamma\alpha_1-\gamma_1\alpha)+c\,(\alpha\beta_1-\alpha_1\beta)\},$$

$$=V^2\sqrt{E}\begin{vmatrix} a, & b, & c \\ \alpha, & \beta, & \gamma \\ \alpha_1, & \beta_1, & \gamma_1 \end{vmatrix}.$$

Hence, dividing by

$$P^2+Q^2,\quad =V^2\Delta,$$

the first part of $\dfrac{dl}{dp}$ is

$$=\frac{\sqrt{E}}{\Delta}\begin{vmatrix} a, & b, & c \\ \alpha, & \beta, & \gamma \\ \alpha_1, & \beta_1, & \gamma_1 \end{vmatrix}.$$

For the second part of the numerator, we require

$$\alpha\text{a}_1+\beta\text{b}_1+\gamma\text{c}_1\quad\text{and}\quad \alpha\text{a}_1'+\beta\text{b}_1'+\gamma\text{c}_1';$$

the values of a, b, c are $A\sqrt{E}$, $B\sqrt{E}$, $C\sqrt{E}$, where $E_1=2\omega$, and hence

$$\text{a}_1\alpha+\text{b}_1\beta+\text{c}_1\gamma=\left(A_1\sqrt{E}+\frac{A\omega}{\sqrt{E}}\right)\alpha+\left(B_1\sqrt{E}+\frac{B\omega}{\sqrt{E}}\right)\beta+\left(C_1\sqrt{E}+\frac{C\omega}{\sqrt{E}}\right)\gamma,$$

$$=\sqrt{E}\,(A_1\alpha+B_1\beta+C_1\gamma)+\frac{\omega}{\sqrt{E}}(A\alpha+B\beta+C\gamma).$$

From the values

$$A,\ B,\ C=bc'-b'c,\ ca'-c'a,\ ab'-a'b,$$

we have

$$A_1,\ B_1,\ C_1=\beta c'-\gamma b'+b\gamma'-c\beta',\ \gamma a'-\alpha c'+c\alpha'-a\gamma',\ \alpha b'-\beta a'+a\beta'-b\alpha',$$

and thence

$$A_1\alpha+B_1\beta+C_1\gamma=-\left[a\,(\beta\gamma'-\beta'\gamma)+b\,(\gamma\alpha'-\gamma'\alpha)+c\,(\alpha\beta'-\alpha'\beta)\right],$$

$$=-\begin{vmatrix} a; & b, & c \\ \alpha, & \beta, & \gamma \\ \alpha', & \beta', & \gamma' \end{vmatrix}.$$

Hence

$$\text{a}_1\alpha+\text{b}_1\beta+\text{c}_1\gamma=-\sqrt{E}\begin{vmatrix} a, & b, & c \\ \alpha, & \beta, & \gamma \\ \alpha', & \beta', & \gamma' \end{vmatrix}+\frac{\omega E'}{\sqrt{E}}.$$

Next, we have

$$\mathrm{a}', \mathrm{b}', \mathrm{c}' = cB - bC,\ aC - cA,\ bA - aB,$$

and thence

$$\begin{aligned}
&\alpha\mathrm{a}_1' + \beta\mathrm{b}_1' + \gamma\mathrm{c}_1' \\
&\quad = \alpha(\gamma B - \beta C + cB_1 - bC_1) + \beta(\alpha C - \gamma A + aC_1 - cA_1) + \gamma(\beta A - \alpha B + bA_1 - aB_1) \\
&\quad = (b\gamma - c\beta) A_1 + (c\alpha - a\gamma) B_1 + (a\beta - b\alpha) C_1 \\
&\quad = \quad (b\gamma - c\beta)(c'\beta - b'\gamma + b\gamma' - c\beta') \\
&\qquad + (c\alpha - a\gamma)(a'\gamma - c'\alpha + c\alpha' - a\gamma') \\
&\qquad + (a\beta - b\alpha)(b'\alpha - a'\beta + a\beta' - b\alpha');
\end{aligned}$$

the portion hereof, which is quadric in α, β, γ, is

$$\begin{aligned}
&= -(\alpha^2 + \beta^2 + \gamma^2)(aa' + bb' + cc') + (a\alpha + b\beta + c\gamma)(a'\alpha + b'\beta + c'\gamma), \\
&= -\lambda F + \omega\varpi,
\end{aligned}$$

and the remaining portion, which is lineo-linear in α, β, γ and α', β', γ', is

$$\begin{aligned}
&= (\alpha\alpha' + \beta\beta' + \gamma\gamma')(a^2 + b^2 + c^2) - (a\alpha + b\beta + c\gamma)(a\alpha' + b\beta' + c\gamma'), \\
&= E\mu'' - \omega\omega';
\end{aligned}$$

we thus have

$$\alpha\mathrm{a}_1' + \beta\mathrm{b}_1' + \gamma\mathrm{c}_1' = -\lambda F + E\mu'' + \omega(\varpi - \omega').$$

Hence the second portion of the numerator, or

$$Q(\alpha\mathrm{a}_1 + \beta\mathrm{b}_1 + \gamma\mathrm{c}_1) - P(\alpha\mathrm{a}_1' + \beta\mathrm{b}_1' + \gamma\mathrm{c}_1'),$$

is

$$= (E\varpi - F\omega)\left\{-\sqrt{E}\begin{vmatrix} a, & b, & c \\ \alpha, & \beta, & \gamma \\ \alpha', & \beta', & \gamma' \end{vmatrix} + \frac{\omega E'}{\sqrt{E}}\right\} - E'\sqrt{E}\{-\lambda F + \mu'' E + \omega(\varpi - \omega')\}.$$

There are two terms, $+E'\sqrt{E}\omega\varpi$ and $-E'\sqrt{E}\omega\varpi$, which destroy each other, and thus the whole second part of the numerator is

$$= -(E\varpi - F\omega)\begin{vmatrix} a, & b, & c \\ \alpha, & \beta, & \gamma \\ \alpha', & \beta', & \gamma' \end{vmatrix} - E'\sqrt{E}(\mu'' E - \lambda F) + \frac{E'\omega}{\sqrt{E}}(\omega' E - \omega F),$$

and, for the corresponding part of $\dfrac{dl}{dp}$, we must divide this by $P^2 + Q^2, = V^2\Delta$.

Hence, finally, we have

$$\begin{aligned}
\frac{dl}{dp} &= \frac{\sqrt{E}}{\Delta}\begin{vmatrix} a, & b, & c \\ \alpha, & \beta, & \gamma \\ \alpha_1, & \beta_1, & \gamma_1 \end{vmatrix} \\
&\quad + \frac{1}{V^2\Delta}\left\{-(\varpi E - \omega F)\begin{vmatrix} a, & b, & c \\ \alpha, & \beta, & \gamma \\ \alpha', & \beta', & \gamma' \end{vmatrix} - E'\sqrt{E}(\mu'' E - \lambda F) + \frac{E'\omega}{\sqrt{E}}(\omega' E - \omega F)\right\},
\end{aligned}$$

where I recall that the values of ω, ϖ, ω', λ, and μ'' are $a\alpha+b\beta+c\gamma$, $a'\alpha+b'\beta+c'\gamma$, $a\alpha'+b\beta'+c\gamma'$, $\alpha^2+\beta^2+\gamma^2$, and $\alpha\alpha'+\beta\beta'+\gamma\gamma'$, respectively. Observe that the first term in this expression for $\dfrac{dl}{dp}$ is

$$=-\frac{dn}{dp}.$$

We thus have

$$u=\frac{dm}{dp}\cos l=\frac{E\varpi-F\omega}{VE},$$

$$v=\frac{dm}{dp}\sin l=\frac{E'}{V\sqrt{E}},$$

$$w=\frac{dn}{dp}-\frac{dl}{dp}=-\frac{2\sqrt{E}}{\sqrt{\Delta}}\begin{vmatrix} a, & b, & c \\ \alpha, & \beta, & \gamma \\ \alpha_1, & \beta_1, & \gamma_1 \end{vmatrix}$$

$$+\frac{1}{V^2\Delta}\left\{(\varpi E-\omega F)\begin{vmatrix} a, & b, & c \\ \alpha, & \beta, & \gamma \\ \alpha', & \beta', & \gamma' \end{vmatrix}+E'\sqrt{E}(\mu''E-\lambda F)-\frac{E'\omega}{\sqrt{E}}(\omega'E-\omega F)\right\},$$

and we thence obtain at once the values of U, V, W; viz. these are

$$U=\frac{dM}{dq}\cos L=\frac{G\omega''-F\varpi''}{VG},$$

$$V=\frac{dM}{dq}\sin L=\frac{G'}{V\sqrt{G}},$$

$$W=\frac{dN}{dq}-\frac{dL}{dq}=-\frac{2\sqrt{G}}{\sqrt{\Delta''}}$$

$$+\frac{1}{V^2\Delta''}\left\{(\omega''G-\varpi''F)\begin{vmatrix} a', & b', & c' \\ \alpha'', & \beta'', & \gamma'' \\ \alpha_2'', & \beta_2'', & \gamma_2'' \end{vmatrix}+G'\sqrt{G}(\mu G-\lambda''F)-\frac{G'\varpi''}{\sqrt{G}}(\varpi'G-\varpi''F)\right\},$$

where α_2'', β_2'', γ_2'' denote the derived functions of α'', β'', γ'' in regard to q, viz. these are the third derived functions of x, y, z in regard to q.

We have, moreover,

$$dx^2+dy^2+dz^2=E\,dp^2+2F\,dp\,dq+G\,dq^2,\ =r^2\,dt^2+2rR\cos\theta\,dt\,dT+R^2\,dT^2;$$

that is,

$$r=\sqrt{E},\quad R=\sqrt{G},\quad \cos\theta=\frac{F}{\sqrt{EG}},$$

and therefore also

$$\sin\theta=\frac{\sqrt{EG-F^2}}{\sqrt{EG}},\quad =\frac{V}{\sqrt{EG}}.$$

Writing p, q in place of t, T, and for r, R substituting their values, then, with the foregoing values of u, v, w, U, V, W, Codazzi's six equations are

$$\frac{du}{dq}=\frac{dU}{dp}\cos\theta+\frac{dW}{dp}\sin\theta+w\left(V-\frac{d\theta}{dq}\right)+(U\sin\theta-W\cos\theta)\left(v-\frac{d\theta}{dp}\right),$$

$$\frac{dU}{dp}=\frac{du}{dq}\cos\theta+\frac{dW}{dq}\sin\theta+W\left(v-\frac{d\theta}{dp}\right)+(u\sin\theta-w\cos\theta)\left(V-\frac{d\theta}{dq}\right),$$

$$\left(\frac{dv}{dq}+\frac{dV}{dp}\right)\frac{d^2\theta}{dp\,dq}+\sin\theta\,(uU-wW)-\cos\theta\,(uW+wU)=0,$$

$$\sqrt{G}\,(u\cos\theta+w\sin\theta)=\sqrt{E}\,(U\cos\theta+W\sin\theta),$$

$$\sqrt{G}\sin\theta\left(v-\frac{d\theta}{dp}\right)+\frac{d\sqrt{G}}{dp}\cos\theta=\frac{d\sqrt{E}}{dq},$$

$$\sqrt{E}\sin\theta\left(V-\frac{d\theta}{dq}\right)+\frac{d\sqrt{E}}{dq}\cos\theta=\frac{d\sqrt{G}}{dp}.$$

I take the opportunity of remarking that Gauss, in § 11 of his memoir, gives the first and second of the formulæ

$$\alpha\,V^2+a\,(\varpi\,F-\omega\,G)+a'\,(\omega\,F-\varpi\,E)-AE'=0,$$

$$\alpha'\,V^2+a\,(\varpi'F-\omega'G)+a'\,(\omega'F-\varpi'E)-AF'=0,$$

$$\alpha''V^2+a\,(\varpi''F-\omega''G)+a'\,(\omega''F-\varpi''E)-AG'=0,$$

(each of them one out of a system of three like equations), where, as before,

$$V^2=EG-F^2,$$

$$E',\ F',\ G'=A\alpha+B\beta+C\gamma,\ A\alpha'+B\beta'+C\gamma',\ A\alpha''+B\beta''+C\gamma'',$$

$$\omega,\ \omega',\ \omega''=a\alpha+b\beta+c\gamma,\ a\alpha'+b\beta'+c\gamma',\ a\alpha''+b\beta''+c\gamma'',$$

$$=\tfrac{1}{2}E_1,\ \tfrac{1}{2}E_2,\ F_2-\tfrac{1}{2}G_1,$$

$$\varpi,\ \varpi',\ \varpi''=a'\alpha+b'\beta+c'\gamma,\ a'\alpha'+b'\beta'+c'\gamma',\ a'\alpha''+b'\beta''+c'\gamma'',$$

$$=F_1-\tfrac{1}{2}E_2,\ \tfrac{1}{2}G_1,\ \tfrac{1}{2}G_2.$$

These are, in fact, the formulæ (IV.),

$$\frac{d^2x}{dp^2}-\left\{\begin{matrix}11\\1\end{matrix}\right\}\frac{dx}{dp}-\left\{\begin{matrix}11\\2\end{matrix}\right\}\frac{dx}{dq}+c_{11}X=0,$$

$$\frac{d^2x}{dp\,dq}-\left\{\begin{matrix}12\\1\end{matrix}\right\}\frac{dx}{dp}-\left\{\begin{matrix}12\\2\end{matrix}\right\}\frac{dx}{dq}+c_{12}X=0,$$

$$\frac{d^2x}{dq^2}-\left\{\begin{matrix}22\\1\end{matrix}\right\}\frac{dx}{dp}-\left\{\begin{matrix}22\\2\end{matrix}\right\}\frac{dx}{dq}+c_{22}X=0,$$

of the memoir, Weingarten, "Ueber die Deformation einer biegsamen unausdehnbaren

Fläche," *Crelle*, t. c. (1887), pp. 296—310; viz. the symbols correspond to those of Weingarten, as follows:—

$$E,\ F,\ G = a_{11},\ a_{12},\ a_{22},$$

$$V^2 = EG - F^2 = a = \rho^2,$$

$$E',\ F',\ G' = -c_{11}\sqrt{a},\ -c_{12}\sqrt{a},\ -c_{22}\sqrt{a},$$

$$A,\ B,\ C = \rho X,\ \rho Y,\ \rho Z,$$

$$\omega,\ \omega',\ \omega'' = \tfrac{1}{2}\frac{da_{11}}{dp},\qquad \tfrac{1}{2}\frac{da_{11}}{dq},\qquad \frac{da_{12}}{dq} - \tfrac{1}{2}\frac{da_{22}}{dp},$$

$$\varpi,\ \varpi',\ \varpi'' = \frac{da_{12}}{dp} - \tfrac{1}{2}\frac{da_{11}}{dq},\ \tfrac{1}{2}\frac{da_{22}}{dp},\ \tfrac{1}{2}\frac{da_{22}}{dq},$$

and thus Weingarten's symbols, $\begin{Bmatrix}11\\1\end{Bmatrix}$, &c., have the values

$$\begin{Bmatrix}11\\1\end{Bmatrix} = \frac{1}{a}\left[a_{12}\left(-\frac{da_{12}}{dp} + \tfrac{1}{2}\frac{da_{11}}{dq}\right) + a_{22}\left(\tfrac{1}{2}\frac{da_{11}}{dp}\right)\right],$$

$$\begin{Bmatrix}12\\1\end{Bmatrix} = \frac{1}{a}\left[a_{12}\left(-\tfrac{1}{2}\frac{da_{22}}{dp}\right) + a_{22}\left(\tfrac{1}{2}\frac{da_{11}}{dq}\right)\right],$$

$$\begin{Bmatrix}22\\1\end{Bmatrix} = \frac{1}{a}\left[a_{12}\left(-\tfrac{1}{2}\frac{da_{22}}{dq}\right) + a_{22}\left(\frac{da_{12}}{dq} - \tfrac{1}{2}\frac{da_{22}}{dp}\right)\right],$$

$$\begin{Bmatrix}11\\2\end{Bmatrix} = \frac{1}{a}\left[a_{12}\left(-\tfrac{1}{2}\frac{da_{11}}{dp}\right) + a_{11}\left(\frac{da_{12}}{dp} - \tfrac{1}{2}\frac{da_{11}}{dq}\right)\right],$$

$$\begin{Bmatrix}12\\2\end{Bmatrix} = \frac{1}{a}\left[a_{12}\left(-\tfrac{1}{2}\frac{da_{11}}{dq}\right) + a_{11}\left(\tfrac{1}{2}\frac{da_{22}}{dp}\right)\right],$$

$$\begin{Bmatrix}22\\2\end{Bmatrix} = \frac{1}{a}\left[a_{12}\left(-\frac{da_{12}}{dq} + \tfrac{1}{2}\frac{da_{22}}{dp}\right) + a_{11}\left(\tfrac{1}{2}\frac{da_{22}}{dq}\right)\right],$$

values which give, as they should do,

$$\begin{Bmatrix}11\\1\end{Bmatrix} + \begin{Bmatrix}12\\2\end{Bmatrix} = \frac{1}{\sqrt{a}}\frac{d\sqrt{a}}{dp},\ \text{and}\ \begin{Bmatrix}12\\1\end{Bmatrix} + \begin{Bmatrix}22\\2\end{Bmatrix} = \frac{1}{\sqrt{a}}\frac{d\sqrt{a}}{dq}.$$

The foregoing comparison serves to explain the notation of Weingarten's valuable memoir.

932.

ON SYMMETRIC FUNCTIONS AND SEMINVARIANTS.

[From the *American Journal of Mathematics*, vol. xv. (1893), pp. 1—74.]

THE principal object of the present memoir is to develop further the theory of seminvariants, but in connexion therewith I was led to some investigations on symmetric functions, and I have consequently included this subject in the title. The two theories, if we adopt the MacMahon form of equation,

$$0 = 1 + bx + \frac{c}{2}x^2 + \frac{d}{6}x^3 + \ldots,$$

may be regarded as identical; but there are still two branches of the theory, viz. we may seek to obtain for the symmetric functions of the roots expressions in terms of the coefficients (which expressions, in the case of non-unitary symmetric functions, are in fact seminvariants), or we may attend to the properties of the functions of the coefficients thus obtained and which we call seminvariants. But I do not in the first instance use the MacMahon form, but retain the ordinary form of equation $0 = 1 + bx + cx^2 + dx^3 + \ldots$, and we have thus only a parallelism of the two theories, and in place of seminvariants we have functions which I call non-unitariants. In regard as well to these as to unitariant functions, I consider certain operators Θ_σ, Δ, $P - \delta b$, and $Q - 2\omega b$, which under altered forms present themselves also in the theory of seminvariants.

As regards seminvariants, I consider what I call the blunt and sharp forms respectively: the great problem is, it appears to me, that of sharp seminvariants, otherwise the I-and-F problem—viz. for any given weight we have to determine the correspondence between the initial and final terms in such wise as to obtain a system of sharp seminvariants. I obtain a "square diagram" solution, which is so far theoretically complete that for any given weight I can, without any tentative operation, determine by a laborious process the correspondence in question: but I am not thereby enabled to establish or enunciate for successive weights any general rule of correspondence; and my process is in fact, as regards practicability, far inferior to that which I call the MacMahon linkage, but of the validity of this I have not succeeded in obtaining any satisfactory proof.

I establish an umbral theory of seminvariants which will be presently again referred to, and I consider the question of the reduction of seminvariants. The final term of a seminvariant may be composite (that is, the product of two or more final terms), and that in one way only or in two or more ways, or it may be non-composite. In the case of a composite final term the seminvariant is reducible, but the converse theorem that a seminvariant with a non-composite final term is irreducible is in nowise true; the reason of this is explained. An irreducible seminvariant is a perpetuant. In regard to perpetuants, I reproduce and simplify a demonstration recently obtained by Dr Stroh as to the perpetuants for any given degree whatever: viz. the generating function for perpetuants of degree n is

$$= x^{2^{n-1}-1} \div 1 - x^2 \,.\, 1 - x^3 \ldots 1 - x^n;$$

the theorem was previously known, and more or less completely proved, for the values $n = 4$, 5, 6, and 7. Dr Stroh's investigation is conducted by an umbral representation,

$$(\alpha x + \beta y + \gamma z + \ldots)^n, \quad x + y + z + \ldots = 0,$$

of the blunt seminvariants of a given weight.

I consider in regard to seminvariants the theory of the symbols $P - \delta b$ and $Q - 2\omega b$, and the derived symbols Y and Z, each of which operating on a seminvariant gives a seminvariant. These are, in fact, connected with the derivatives (f, F) of a quantic f and any covariant thereof F; but except to point out this connexion, I do not in the present memoir consider the theory of covariants.

The Coefficients $(a, b, c, d, e, \ldots)$ *or* $(1, b, c, d, e, \ldots)$. Art. Nos. 1 to 9.

1. I consider the series $(a, b, c, d, e, \ldots)$, or putting as we most frequently do $a = 1$, say the series $(1, b, c, d, e, \ldots)$ of coefficients, the several terms whereof are taken to be of the weights 0, 1, 2, 3, 4, ... respectively. We form with these sets of isobaric terms, or say columns of the weights 0, 1, 2, 3, 4, ... respectively, for instance,

0	1	2	3	4	5	6
1	b	c	d	e	f	g
		b^2	bc	bd	be	bf
			b^3	c^2	cd	ce
				b^2c	b^2d	d^2
				b^4	bc^2	b^2e
					b^3c	bcd
					b^5	c^3
						b^3d
						b^2c^2
						b^4c
						b^6

and generally a set or column of any given weight. In each term, the letters are written in alphabetical order.

Taking the whole or any part of a column, for instance the whole column $(d,\ bc,\ b^3)$, or the part $(e,\ bd,\ c^2)$ of the next column, we may by supplying powers of a in such wise as to leave unaltered the terms of the highest degree, that is, by reading these as $(a^2d,\ abc,\ b^3)$ and $(ae,\ bd,\ c^2)$ respectively, regard them as homogeneous sets of a given degree in $(a,\ b,\ c,\ d,\ e, \ldots)$; and thus generally we may speak of the degree of a set of terms.

The terms of the several columns as above written down are in alphabetical order, AO; viz. we supply as above the proper powers of a, reading for instance col. 4 as $a^3e,\ a^2bd,\ a^2c^2,\ ab^2c,\ b^4$, where the terms are in alphabetical or dictionary order.

Each column is derived from the preceding one by Arbogast's rule, it being understood, for instance, that b^4, that is, ab^4, gives the two terms ab^3c and b^5, that is, b^3c and b^5; and so in other cases.

2. We attend in particular to the non-unitary terms, or non-unitaries, e.g. in col. 5, f, cd, which contain no b; and to the power-ending terms or power-enders, bc^2, b^5, which end in a power. It will be observed that, whenever by Arbogast's rule a term in one column gives two terms in the next column, the second of these is a power-ender; and thus in any column the excess of the number of terms above that in the preceding column is equal to the number of power-enders.

3. I consider the notion of conjugate terms: representing, for instance, the terms

$$f \qquad be \qquad cd$$

by dots in the form

$$\cdot\cdot\cdot\cdot\cdot \qquad :\cdot\cdot\cdot \qquad ::\cdot$$

and reading the number of dots in columns instead of in lines we derive the conjugate terms

$$b^5 \qquad b^3c \qquad bc^2,$$

and so in other cases. It is clear that the relation is a reciprocal one (thus the conjugates of b^5, b^3c, bc^2 are f, be, cd respectively). Moreover, a term may be its own conjugate; thus cd^2, arranging the dots in lines and reading them in columns, $\vdots\ \vdots\ :$ is again cd^2.

It is at once seen that non-unitaries and power-enders are conjugate to each other; hence in any column, the non-unitaries and the power-enders are equal in number, and a preceding result may be stated in the more complete form: in any column the excess of the number of terms above that in the preceding column is equal to the number of non-unitaries or to the number of power-enders.

4. The terms of the several columns may be arranged in counter-order CO, thus:

0	1	2	3	4	5	6
1	b	c	d	e	f	g
		b^2	bc	bd	be	bf
			b^3	c^2	cd	ce
				b^2c	b^2d	b^2e
				b^4	bc^2	d^2
					b^3c	bcd
					b^5	b^3d
						c^3
						b^2c^2
						b^4c
						b^6

viz. we arrange here according to the highest letters. The counter-order is, in fact, the alphabetical order with the reversed arrangement (..., g, f, e, d, c, b, a) of the alphabet, but in the separate terms we retain the alphabetical order, thus writing as before bf and not fb. Observe that the difference between the two arrangements, AO and CO, first presents itself in the col. 6.

In this CO arrangement, each column is derived from the next preceding one by a rule as follows: We operate on the lowest letter of each term, being a simple letter, not a power, by changing it into the next highest letter, and we further operate upon each term by multiplying it by b, the operation or (as the case may be) two operations upon any term being performed before operating upon the next term.

5. If we compare a column in AO with the same column in CO, for instance

AO	CO	AO	CO rev.
g	g	g	b^6
bf	bf	bf	b^4c
ce	ce	ce	b^2c^2
d^2	b^2e	d^2	c^3
b^2e	d^2	b^2e	b^3d
bcd	bcd	bcd	bcd
c^3	b^3d	c^3	d^2
b^3d	c^3	b^3d	b^2e
b^2c^2	b^2c^2	b^2c^2	ce
b^4c	b^4c	b^4c	bf
b^6	b^6	b^6	g

it will be seen that the terms are conjugates of each other, the first and last, the second and last but one terms, and so on; or, what is the same thing, if we reverse the order of either column, then the pairs of conjugate terms will appear each in the same line; of course, here a self-conjugate term such as *bcd* is put in evidence.

6. By writing $a, b, c, d, \ldots = a_0, a_1, a_2, a_3, \ldots$, or more simply 0, 1, 2, 3, ..., we connect the theory with that of the partition of numbers: in particular, the terms of a given weight correspond to the partitions of that weight, or number of ways in which that weight can be made up with the parts 1, 2, 3, ... It may be remarked that, in a partition, the parts are usually written in decreasing order, whereas (as remarked above) in a literal term the letters are written in alphabetical order. Thus we have 321 and *bcd*; it would be more correct to write the partition as 123.

It is frequently convenient, retaining the letters $b, c, d, \ldots$, to write for instance $q = a_\sigma$ (σ a numerical suffix), meaning thereby that q is the letter corresponding to the place σ in the series 1, 2, 3, ... If instead of the indefinite series $(1, b, c, d, \ldots)$ we consider, as is sometimes convenient, a definite series of terms $(1, b, c, \ldots, q = a_\sigma)$, then σ is said to be the "extent" of the system. The next preceding letter p will naturally be $= a_{\sigma-1}$; and if, increasing the extent by unity, we introduce a new letter r, this will be $a_{\sigma+1}$, and so in other cases, the notation being for the most part used merely as a convenient way of showing the place of a letter in the series.

7. Considering the terms of a given weight, or say a column, in *AO* or *CO*, we may represent any portion of the column by means of its initial and final terms, say *I* and *F*, by the notations *I*ao*F* and *I*co*F* respectively. But a much more important notation is *I*ca*F*; viz. this represents the series of terms of given weight which are in *CO* not superior to *I*, and in *AO* not inferior to *F* (a like notation, which however I do not employ, would be *I*ac*F*; viz. this would denote the series of terms which are in *AO* not superior to *I* and in *CO* not inferior to *F*). The definition of *I*ca*F* has been given in the above general form, but we are in fact exclusively or chiefly concerned with the case where *I* is a non-unitary and *F* a power-ender. It is to be observed that, considering the *AO* column as given, then to form from it the set or interval *I*ca*F* we may disregard altogether the terms which are in the *AO* column inferior (posterior) to *F*, for by the definition none of these enter into *I*ca*F*, but it may very well be that there are in *I*ca*F* terms which are in the *AO* column superior (anterior) to *I*. An instance of this first presents itself for the weight 11; viz. here a portion of the *AO* column is $(fg, b^2j, bci, bdh, beg, bf^2, c^2h, cdg, \ldots)$: hence in *I*ca*F*, if the initial term be c^2h, for instance in $c^2h\text{ca}b^3e^2$, we have terms fg, beg, bf^2 which are in *AO* anterior to the initial term c^2h. In order therefore to form *I*ca*F* from the *AO* column, we must first take the terms (if any) which being in *CO* posterior to *I* are in the *AO* column anterior to *I*, and then from the portion *I*ao*F* of the *AO* column reject the terms (if any) which are in *CO* anterior to *I*. In particular, starting from the *AO* column, and

arranging the non-unitaries thereof in *CO* and the power-enders in *AO*, for instance, weight 12, these are

m	g^2
ck	cf^2
dj	e^3
ei	b^2f^2
c^2i	bde^2
fh	c^2e^2
⋮	⋮

There is no difficulty in writing down the terms of the several sets or intervals

$$mcag^2,\quad mcacf^2,\quad mcae^3,\ \ldots,\ ckcag^2,\quad ckcacf^2,\ \ldots.$$

Instead of ca we may, if we please, use, and in fact I generally use the conventional symbol ∞, or write $m \infty g^2$, $m \infty cf^2$, &c. In any such set, the terms need not be arranged in *AO*; if for any purpose it is more convenient, they may be arranged in *CO*; but of course the definition of the meaning must not be departed from. The expressed initial is the highest term in *CO*, and the expressed final the lowest term in *AO*.

8. I diminish a term by replacing successively each letter thereof by the next inferior letter; for instance, if the term is cdf, then the diminished terms are $Dcdf$, $=bdf$, c^2f, cde, and so Db^2df, $=bdf$, b^2cf, b^2de (where the diminished b is a, that is, 1). Conversely, we may augment a term by replacing successively each letter thereof by the next superior letter; for instance, $Abdf$, $=b^2df$, cdf, bef, bdg, where the first augmentation b^2df is obtained from the a (which may be regarded as latent in the term operated upon). Operating upon the letters in order beginning with the lowest, the several diminutions may be called D_1, D_2, D_3, ..., and the several augmentations A_0, A_1, A_2, ... (where A_0 is in fact multiplication by b). We diminish a set by diminishing successively the several terms thereof (the diminished terms being taken without repetition; that is, each such term once only). Similarly, we may augment a set by augmenting successively the several terms thereof (the augmented terms being taken without repetition). It is to be noticed that the two operations are not reciprocal to each other; if we diminish a set, and then augment the diminished set, we obtain indeed all the terms of the original set, but in general we obtain also terms which are not included in the original set.

9. It requires some consideration to see that we have $D(I \infty F) = (D_1 I \infty D_\theta F)$, where $D_\theta F$ is the diminution performed upon the highest letter of F. Take any term M of $D(I \infty F)$, the several diminutions D_1M, D_2M, ..., $D_\phi M$ are arranged in descending order: D_1M the highest and $D_\phi M$ the lowest, as well in *CO* as in *AO*. If then D_1M is in *CO* not superior to D_1I, then all the DM's will be in *CO* not superior to D_1I; and similarly, if $D_\phi M$ is in *AO* not inferior to $D_\theta F$, then all the DM's will be in *AO* not inferior to $D_\theta F$. And this being seen, then if we take

N a term of $(D_1 I \propto D_\theta F)$, and consider the successive augmentations $A_0 N, A_1 N, \ldots, A_\phi N$ of N, then these will be in ascending order $A_0 N$ the lowest and $A_\phi N$ the highest in CO as well as in AO. It may happen that $A_\phi N$ or this and neighbouring terms are in CO higher than I, and that $A_0 N$ or this and neighbouring terms are in AO lower than F, but there will always be a term or terms which is or are in CO lower than I and in AO higher than F; and thus not only every term of $D(I \propto F)$ will be a term of $(D_1 I \propto D_\theta F)$, but conversely every term of $(D_1 I \propto D_\theta F)$ will be a term of $D(I \propto F)$, and we thus have the required relation $D(I \propto F) = (D_1 I \propto D_\theta F)$.

Symmetric Functions of the Roots. Art. Nos. 10 to 31.

10. We consider a set of roots $\alpha, \beta, \gamma, \delta, \epsilon, \ldots$ either indefinite in number, or else definite, for instance $\alpha, \beta, \gamma, \delta$. The symmetric functions (rational and integral functions) are in the first instance denoted in the usual manner

$$S\alpha = \alpha + \beta + \gamma + \delta + \ldots, \quad S\alpha\beta = \alpha\beta + \alpha\gamma + \beta\gamma + \ldots,$$
$$S\alpha^2\beta = \alpha^2\beta + \alpha\beta^2 + \alpha^2\gamma + \alpha\gamma^2 + \beta^2\gamma + \beta\gamma^2 + \ldots,$$

viz. the S refers to all the distinct combinations of like form with the combination (α, $\alpha\beta$, or $\alpha^2\beta$, as the case may be) to which it is prefixed. By omitting the S and instead of the roots considering merely their indices, these same symmetric functions would be 1, 11 (or 1^2), 21, &c., and then if instead of the numbers 1, 2, 3, &c., we introduce the symbolic capital letters $B, C, D, \ldots$, the same symmetric functions will be represented as B, B^2, BC, &c. (21, that is, 12 is written as BC, and so in other cases, the letters in alphabetical order). The letters $B, C, D, \ldots$ are considered as being of the weights 1, 2, 3, ... respectively, and thus the symmetric functions of a given degree in the roots are represented by the terms of that weight in the symbolic letters $B, C, D, \ldots$, thus the symmetric functions of the degree 4 are E, BD, C^2, B^2C, B^4; of course these terms may be arranged in AO or in CO as may be most convenient for the purpose in hand. The capital letters $B, C, D, \ldots$ are in fact umbræ, but to avoid confusion with subsequent notations I do not in general thus speak of them. A form such as $S\alpha^2$ or $S\alpha^4\beta^2$, in which there is no index 1, is said to be non-unitary; but a form $S\alpha$ or $S\alpha^2\beta$, in which there is an index $=1$ or two or more indices each $=1$, is said to be unitary: or, what is the same thing, in the symbolic representation by capital letters, the form is non-unitary or unitary according as it does not or does contain the letter B.

11. In the ordinary theory of symmetric functions, we connect the coefficients $(1, b, c, d, \ldots)$ with the roots $(\alpha, \beta, \gamma, \ldots)$ by the equation

$$1 + bx + cx^2 + dx^3 + \ldots = 1 - \alpha x \,.\, 1 - \beta x \,.\, 1 - \gamma x \ldots,$$

and we thus have

$$\begin{aligned} -b &= \alpha + \beta + \gamma + \ldots, &&= S\alpha, &&= 1, &&= B, \\ +c &= \alpha\beta + \alpha\gamma + \beta\gamma + \ldots, &&= S\alpha\beta, &&= 1^2, &&= B^2, \\ -d &= \alpha\beta\gamma + \ldots, &&= S\alpha\beta\gamma, &&= 1^3, &&= B^3, \end{aligned}$$

&c., &c.;

and it is to be remarked that, for any given number of roots, there will be this same number of coefficients: we may for instance have

$$1 + bx + cx^2 + dx^3 = 1 - \alpha x \,.\, 1 - \beta x \,.\, 1 - \gamma x,$$

that is,

$$-b = \alpha \;\; + \beta \;\; + \gamma,$$

$$+c = \alpha\beta + \alpha\gamma + \beta\gamma,$$

$$-d = \alpha\beta\gamma;$$

and similarly if the number of roots be $=4$, or any larger number.

12. The symmetric functions of a given degree, say 4, in the roots, viz.

$$S\alpha^4,\quad S\alpha^3\beta,\quad S\alpha^2\beta^2,\quad S\alpha^2\beta\gamma,\quad S\alpha\beta\gamma\delta,$$

or

$$4,\quad 31,\quad 2^2,\quad 21^2,\quad 1^4,$$

or

$$E,\quad BD,\quad C^2,\quad B^2C,\quad B^4,$$

are equal in number to the combinations of the weight 4 in the coefficients, viz.

$$e,\quad bd,\quad c^2,\quad b^2c,\quad b^4;$$

and the terms of the one set are in fact linear combinations (with mere numerical multipliers) of the terms of the other set; but more than this, we have for instance

$e = \alpha\beta\gamma\delta + \ldots$, that is, $e = B^4$:

$bd = (\alpha + \beta + \gamma + \delta \ldots)(\alpha\beta\gamma + \alpha\beta\delta + \alpha\gamma\delta + \beta\gamma\delta \ldots)$ contains only terms $\alpha^2\beta\gamma$ and $\alpha\beta\gamma\delta$, that is, bd is a linear function of B^2C and B^4:

$c^2 = (\alpha\beta + \alpha\gamma + \alpha\delta + \beta\gamma + \beta\delta + \gamma\delta \ldots)^2$ contains only terms $\alpha^2\beta^2$, $\alpha^2\beta\gamma$ and $\alpha\beta\gamma\delta$, that is, c^2 is a linear function of C^2, B^2C and B^4; and so on.

13. We have in fact the Table IV (a) which I quote from my paper "A Memoir on the Symmetric Functions of the Roots of an Equation," *Phil. Trans.* t. 147 (1857), pp. 489—496, [147],

‖	e	bd	c^2	b^2c	b^4
$S\alpha^4 \;= 4 = E$					$+1$
$S\alpha^3\beta \;= 31 = BD$				$+1$	$+4$
$S\alpha^2\beta^2 = 2^2 = C^2$			$+1$	$+2$	$+6$
$S\alpha^2\beta\gamma = 21^2 = B^2C$		$+1$	$+2$	$+5$	$+12$
$S\alpha\beta\gamma\delta = 1^4 = B^4$	$+1$	$+4$	$+6$	$+12$	$+24$

inserting on the left-hand outside margin the new symbols E, BD, &c., with their explanations: the || indicates that the table is to be read according to the columns, $e = +1B^4$, $bd = +1B^2C + 4B^4$, &c. This table gives conversely a Table IV (*b*), read according to the lines and serving to express the symmetric functions *E*, *BD*, &c., as linear functions of the combinations *e*, *bd*, c^2, b^2c, b^4 of the coefficients.

14. The (*a*) and (*b*) tables are given in the Memoir up to X (*a*) and X (*b*): it is proper to quote here the (*b*) tables up to VI (*b*) with only the change of substituting on the outside left-hand margins the literal terms such as *E*, *BD*, &c., instead of the symbols 4, 31, &c., originally used to denote these symmetric functions—it is to be observed that the left-hand symbols are in *AO*, the upper symbols in *CO*, this distinction first manifesting itself in the Table VI (*b*), so that it was necessary to go as far as this in order to put in evidence the true form of the tables.

II (*b*).

=	c	b^2
C	− 2	+ 1
B^2	+ 1	

III (*b*).

=	d	bc	b^3
D	− 3	+ 3	− 1
BC	+ 3	− 1	
B^3	− 1		

IV (*b*).

=	e	bd	c^2	b^2c	b^4
E	− 4	+ 4	+ 2	− 4	+ 1
BD	+ 4	− 1	− 2	+ 1	
C^2	+ 2	− 2	+ 1		
B^2C	− 4	+ 1			
B^4	+ 1				

V (*b*).

=	f	be	cd	b^2d	bc^2	b^3c	b^5
F	− 5	+ 5	+ 5	− 5	− 5	+ 5	− 1
BE	+ 5	− 1	− 5	+ 1	+ 3	− 1	
CD	+ 5	− 5	+ 1	+ 2	− 1		
B^2D	− 5	+ 1	+ 2	− 1			
BC^2	− 5	+ 3	− 1				
B^3C	+ 5	− 1					
B^5	− 1						

VI (b).

=	g	bf	ce	b^2e	d^2	bcd	b^3d	c^3	b^2c^2	b^4c	b^6
G	− 6	+ 6	+ 6	− 6	+ 3	− 12	+ 6	− 2	+ 9	− 6	+ 1
BF	+ 6	− 1	− 6	+ 1	− 3	+ 7	− 1	+ 2	− 4	+ 1	
CE	+ 6	− 6	+ 2	+ 2	− 3	+ 4	− 2	− 2	+ 1		
D^2	+ 3	− 3	− 3	+ 3	+ 3	− 3	0	+ 1			
B^2E	− 6	+ 1	+ 2	− 1	+ 3	− 3	+ 1				
BCD	− 12	+ 7	+ 4	− 3	− 3	+ 1					
C^3	− 2	+ 2	− 2	0	+ 1						
B^3D	+ 6	− 1	− 2	+ 1							
B^2C^2	+ 9	− 4	+ 1								
B^4C	− 6	+ 1									
B^6	+ 1										

It is hardly necessary to remark in relation to these tables that if there are only two roots, then $d=0$, &c., viz. Table II. is not affected but all the subsequent tables assume a simplified form; if there are only three roots, then $e=0$, &c., viz. Tables II. and III. are not affected but all the subsequent tables assume a simplified form; and so on.

15. We have between the differential symbols $\partial_b, \partial_c, \partial_d, \ldots$ and $\partial_\alpha, \partial_\beta, \partial_\gamma, \ldots$ certain relations which it is interesting to develop: it will be convenient to consider successively the cases, three roots, four roots, &c.

In the case of three roots, starting from

$$-b = \alpha + \beta + \gamma,$$

$$c = \alpha\beta + \alpha\gamma + \beta\gamma,$$

$$-d = \alpha\beta\gamma,$$

we have

$$\partial_\alpha = -\partial_b + (\beta+\gamma)\,\partial_c - \beta\gamma\partial_d,$$

$$\partial_\beta = -\partial_b + (\gamma+\alpha)\,\partial_c - \gamma\alpha\partial_d,$$

$$\partial_\gamma = -\partial_b + (\alpha+\beta)\,\partial_c - \alpha\beta\partial_d,$$

equations which give conversely $\partial_b, \partial_c, \partial_d$ as linear functions of $\partial_\alpha, \partial_\beta, \partial_\gamma$: I write down the three equations thus obtained together with a fourth equation which I will explain. The four equations are

$$-\partial_a+\delta'=\frac{\alpha^3}{\alpha-\beta.\alpha-\gamma}\partial_\alpha+\frac{\beta^3}{\beta-\gamma.\beta-\alpha}\partial_\beta+\frac{\gamma^3}{\gamma-\alpha.\gamma-\beta}\partial_\gamma,$$

$$-\partial_b\quad=\frac{\alpha^2}{\alpha-\beta.\alpha-\gamma}\partial_\alpha+\frac{\beta^2}{\beta-\gamma.\beta-\alpha}\partial_\beta+\frac{\gamma^2}{\gamma-\alpha.\gamma-\beta}\partial_\gamma,$$

$$-\partial_c\quad=\frac{\alpha}{\alpha-\beta.\alpha-\gamma}\partial_\alpha+\frac{\beta}{\beta-\gamma.\beta-\alpha}\partial_\beta+\frac{\gamma}{\gamma-\alpha.\gamma-\beta}\partial_\gamma,$$

$$-\partial_d\quad=\frac{1}{\alpha-\beta.\alpha-\gamma}\partial_\alpha+\frac{1}{\beta-\gamma.\beta-\alpha}\partial_\beta+\frac{1}{\gamma-\alpha.\gamma-\beta}\partial_\gamma.$$

In verification of the last three equations, observe that they give

$$-\partial_b+(\beta+\gamma)\partial_c-\beta\gamma\partial_d$$

$$=\frac{\alpha^2-\alpha(\beta+\gamma)+\beta\gamma}{\alpha-\beta.\alpha-\gamma}\partial_\alpha+\frac{\beta^2-\beta(\beta+\gamma)+\beta\gamma}{\beta-\gamma.\beta-\alpha}\partial_\beta+\frac{\gamma^2-\gamma(\beta+\gamma)+\beta\gamma}{\gamma-\alpha.\gamma-\beta}\partial_\gamma,$$

that is, $-\partial_b+(\beta+\gamma)\partial_c-\beta\gamma\partial_d=\partial_\alpha$: and similarly from the same three equations we deduce the values of ∂_β and ∂_γ; the three equations are thus equivalent to the foregoing three equations for ∂_α, ∂_β, ∂_γ.

As to the first equation, to avoid confusion with a root δ, I have written therein δ' (afterwards replaced by δ) to denote the degree of a function homogeneous in (a, b, c, d), upon which the symbols are supposed to operate; this is also the degree in the roots α, β, γ. The four equations give

$$-a(\partial_a-\delta')-b\partial_b-c\partial_c-d\partial_d=\frac{\alpha^3+b\alpha^2+c\alpha+d}{\alpha-\beta.\alpha-\gamma}\partial_\alpha+\&\text{c.},=0,$$

since

$$\alpha^3+b\alpha^2+c\alpha+d=0,\quad \beta^3+b\beta^2+c\beta+d=0,\quad \gamma^3+b\gamma^2+c\gamma+d=0.$$

The equations thus give

$$a\partial_a+b\partial_b+c\partial_c+d\partial_d=\delta',$$

which is right, and the first equation is thus verified.

16. From the last three equations for ∂_b, ∂_c, ∂_d, we deduce

$$-3\partial_b-2b\partial_c-c\partial_d=\frac{3\alpha^2+2b\alpha+c}{\alpha-\beta.\alpha-\gamma}\partial_\alpha+\frac{3\beta^2+2b\beta+c}{\beta-\gamma.\beta-\alpha}\partial_\beta+\frac{3\gamma^2+2b\gamma+c}{\gamma-\alpha.\gamma-\beta}\partial_\gamma,$$

$$=\partial_\alpha+\partial_\beta+\partial_\gamma,$$

a result more easily deducible from the first set of three equations for ∂_α, ∂_β, ∂_γ: but I have preferred to obtain it in this manner for the sake of the remark that it is a peculiarity of this combination of ∂_b, ∂_c, ∂_d that the coefficients of ∂_α, ∂_β, ∂_γ

become integral functions of the roots (in the actual case constants and $=1$): for a somewhat similar form

$$-(c\partial_b + d\partial_c), \quad = \frac{c\alpha^2 + d\alpha}{\alpha-\beta\,.\,\alpha-\gamma}\partial_\alpha + \frac{c\beta^2+d\beta}{\beta-\gamma\,.\,\beta-\alpha}\partial_\beta + \frac{c\gamma^2+d\gamma}{\gamma-\alpha\,.\,\gamma-\beta}\partial_\gamma,$$

the coefficients are fractional.

We at once have

$$\alpha\partial_\alpha + \beta\partial_\beta + \gamma\partial_\gamma = b\partial_b + 2c\partial_c + 3d\partial_d,$$

viz. these symbols operating upon a function of the roots of the degree ω, or what is the same thing, a function of the coefficients of the weight ω, are each of them equivalent to a constant factor ω.

Again, we have

$$\alpha^2\partial_\alpha + \beta^2\partial_\beta + \gamma^2\partial_\gamma = -(b^2-2c)\,\partial_b - (bc-3d)\,\partial_c - bd\partial_d,$$

$$= -b\,(b\partial_b + c\partial_c + d\partial_d) + 2c\partial_b + 3d\partial_c,$$

or since $a\partial_a + b\partial_b + c\partial_c + d\partial_d = \delta'$ (if as before δ' is the degree of the function operated upon) and therefore $b\partial_b + c\partial_c + d\partial_d = \delta' - a\partial_a$ or say $=\delta' - \partial_a$, this is

$$\alpha^2\partial_\alpha + \beta^2\partial_\beta + \gamma^2\partial_\gamma = -b\delta' + b\partial_a + 2c\partial_b + 3d\partial_c,$$

so that we have here another form $-b\delta' + b\partial_a + 2c\partial_b + 3d\partial_c$, for which the coefficients of ∂_α, ∂_β, ∂_γ are integral functions of the roots.

17. In the case of four roots, the corresponding equations are

$$\begin{aligned} -b &= \alpha + \beta + \gamma + \delta, \\ +c &= \alpha\beta + \alpha\gamma + \alpha\delta + \beta\gamma + \beta\delta + \gamma\delta, \\ -d &= \alpha\beta\gamma + \alpha\beta\delta + \alpha\gamma\delta + \beta\gamma\delta, \\ +e &= \alpha\beta\gamma\delta, \end{aligned}$$

and we then have

$$\begin{aligned} \partial_\alpha &= -\partial_b + (\beta+\gamma+\delta)\,\partial_c - (\beta\gamma + \beta\delta + \gamma\delta)\,\partial_d + \beta\gamma\delta\partial_e, \\ \partial_\beta &= -\partial_b + (\gamma+\delta+\alpha)\,\partial_c - (\gamma\delta + \gamma\alpha + \delta\alpha)\,\partial_d + \gamma\delta\alpha\partial_e, \\ \partial_\gamma &= -\partial_b + (\delta+\alpha+\beta)\,\partial_c - (\delta\alpha + \delta\beta + \alpha\beta)\,\partial_d + \delta\alpha\beta\partial_e, \\ \partial_\delta &= -\partial_b + (\alpha+\beta+\gamma)\,\partial_c - (\alpha\beta + \alpha\gamma + \beta\gamma)\,\partial_d + \alpha\beta\gamma\partial_e, \end{aligned}$$

and the converse set of equations, which for shortness I write in the form

$$-\partial_a + \delta', \; -\partial_b, \; -\partial_c, \; -\partial_d, \; -\partial_e$$

$$= \frac{\alpha^{4,\,3,\,2,\,1,\,0}}{\alpha-\beta\,.\,\alpha-\gamma\,.\,\alpha-\delta}\partial_\alpha + \frac{\beta^{4,\,3,\,2,\,1,\,0}}{\beta-\gamma\,.\,\beta-\delta\,.\,\beta-\alpha}\partial_\beta + \frac{\gamma^{4,\,3,\,2,\,1,\,0}}{\gamma-\delta\,.\,\gamma-\alpha\,.\,\gamma-\beta}\partial_\gamma + \frac{\delta^{4,\,3,\,2,\,1,\,0}}{\delta-\alpha\,.\,\delta-\beta\,.\,\delta-\gamma}\partial_\delta.$$

We have, in like manner as in the former case,

$$-4\partial_b - 3b\partial_c - 2c\partial_d - d\partial_e = \partial_\alpha + \partial_\beta + \partial_\gamma + \partial_\delta,$$
$$b\partial_b + c\partial_c + d\partial_d + e\partial_e = \alpha\partial_\alpha + \beta\partial_\beta + \gamma\partial_\gamma + \delta\partial_\delta, = \omega,$$
$$-b\delta' + b\partial_a + 2c\partial_b + 3d\partial_c + 4e\partial_d = \alpha^2\partial_\alpha + \beta^2\partial_\beta + \gamma^2\partial_\gamma + \delta^2\partial_\delta;$$

and similarly in the case of five or more roots.

18. In the case of σ' roots, I write $m = a_{\sigma'}$, and for shortness

$$\Theta_{\sigma'} = \sigma'\partial_b + (\sigma' - 1)\, b\partial_c + \ldots \quad \ldots + l\partial_m,$$
$$P = b\partial_a + \quad 2c\partial_b + 3d\partial_c + \ldots + \sigma' m\partial_l,$$

so that, besides the equation $b\partial_b + c\partial_c \ldots + m\partial_m = S\alpha\partial_\alpha = \omega$, the foregoing investigations show that we have

$$\Theta_{\sigma'} = -S\partial_\alpha,$$
$$P - b\delta = S\alpha^2\partial_\alpha.$$

The operand for these symbols is a symmetric function of the roots, which is thus also a function of the coefficients: it is of the degree ω in the roots, and consequently of the weight ω in the coefficients, and its degree in the coefficients is taken to be $= \delta$. It is sometimes convenient to represent this operand, *quà* function of the roots, by Υ and, *quà* function of the coefficients, by U, so that we have in general $\Upsilon = U$. If Υ be a non-unitary function of the roots, then we may say that Υ, $= U$, is a non-unitariant.

19. I give some illustrations of the equation $\Theta_{\sigma'} = -S\partial_\alpha$. Suppose

$$\Upsilon = U = S\alpha^4 = E = -4e + 4bd + 2c^2 - 4b^2c + b^4$$

(Table IV (b)); σ' must be $= 4$ at least and I take it to be 4 and 5 successively; we thus have

$$\Theta_4 = 4\partial_b + 3b\partial_c + 2c\partial_d + d\partial_e,$$
$$\Theta_5 = 5\partial_b + 4b\partial_c + 3c\partial_d + 2d\partial_e,$$

omitting from Θ_5 the term $e\partial_f$ which is obviously inoperative. For any number whatever of roots, we have

$$-S\partial_\alpha \,.\, S\alpha^4 = -4S\alpha^3 = -4\,(-3d + 3bc - b^3), \quad = 12d - 12bc + 4b^3,$$

and this should therefore be the value as well of $\Theta_4 E$ as of $\Theta_5 E$. The calculations may be arranged as follows:

$\Theta_4 E$

$4\,.$	$4d - 8bc + 4b^3$	$d + 16 \quad -4$	12
$3b\,.$	$4c - 4b^2$	$bc - 32 + 12 + 8$	-12
$2c\,.$	$4b$	$b^3 + 16 - 12$	$+ 4,$
$d\,. -4$			

$\Theta_5 E$

$5\ .$	$4d - 8bc + 4b^3$	$d - 20$	-8	12
$4b\ .$	$4c\ - 4b^2$	$bc - 40 + 16 + 12$		-12
$3c\ .$	$4b$	$b^3 + 20 - 16$		$+\ 4,$
$d\ .$	-4			

giving in each case the right result.

20. In the foregoing example, $S\alpha^4$ was a non-unitary function of the roots, but I take the case of a unitary function. Suppose

$$\Upsilon = U = S\alpha^3\beta = BD = 4e - bd - 2c^2 + b^2c.$$

Here $-S\partial_\alpha . S\alpha^3\beta$ is not independent of the number of the roots; in the case of 4 roots, we have

$$-S\partial_\alpha . S\alpha^3\beta = -3S\alpha^2\beta - 3S\alpha^3, \quad = -3(3d - bc) - 3(-3d + 3bc - b^3), \quad = 0d - 6bc + 3b^3;$$

and in the case of 5 roots, we have

$$-S\partial_\alpha . S\alpha^3\beta = -3S\alpha^2\beta - 4S\alpha^3, \quad = -3(3d - bc) - 4(-3d + 3bc - b^3), \quad = 3d - 9bc + 4b^3;$$

and these should therefore be the values of $\Theta_4 BD$ and $\Theta_5 BD$ respectively. The calculations are

$\Theta_4 BD$

$4\ .$	$-\ d + 2bc$	$d - 4$	$+4$	0
$3b\ .$	$-4c +\ b^2$	$bc + 8 - 12 - 2$		-6
$2c\ .$	$-\ b$	$b^3\quad +\ 3$		$+3,$
$d\ .$	$+4$			

$\Theta_5 BD$

$5\ .$	$-\ d + 2bc$	$d - 5$	$+8$	$+3$
$4b\ .$	$-4c +\ b^2$	$bc + 10 - 16 - 3$		-9
$3c\ .$	$-\ b$	$b^3\quad +\ 4$		$+4,$
$2d\ .$	$+\ b$			

giving in each case the correct result. We have $\Theta_5 - \Theta_4 = \partial_b + b\partial_c + c\partial_d + d\partial_e$, and the examples show that performing this operation on the non-unitariant $S\alpha^4$, $= E$, we obtain a result $= 0$; whereas for the unitary function $S\alpha^3\beta$, $= BD$, the result is not $= 0$.

21. Considering the question generally, I take the highest coefficient in U to be $q = a_\sigma$ (σ equal to or less than ω), or what is the same thing, the extent of U to be $= \sigma$; this implies that σ' is at least $= \sigma$; and taking it to be first $= \sigma$, and then to be any number greater than σ, we have

$$\Theta_\sigma = -S\partial_\alpha, \quad \Theta_{\sigma'} = -S\partial_\alpha,$$

where the function U operated upon by Θ_σ and $\Theta_{\sigma'}$ respectively is in each case the same function of the coefficients. It is easy to see that, if Υ is a non-unitary function of the roots, then whatever be the number of the roots we have $S\partial_\alpha . \Upsilon =$ a determinate symmetric function of the roots, and consequently = a determinate function of the coefficients. We thus have $\Theta_{\sigma'} U$ and $\Theta_\sigma U$ equal to each other; that is,

$$(\Theta_{\sigma'} - \Theta_\sigma) U = 0;$$

we may write

$$\Theta_\sigma = \sigma \partial_b + (\sigma - 1) b\partial_c + \dots + \qquad p\partial_q,$$

$$\Theta_{\sigma'} = \sigma' \partial_b + (\sigma' - 1) b\partial_c + \dots + (\sigma' - \sigma + 1) p\partial_q,$$

for the subsequent terms of $\Theta_{\sigma'}$, as involving ∂_r, ∂_s, &c., are inoperative; hence writing

$$\Delta = \partial_b + b\partial_c + c\partial_b + \dots + p\partial_q,$$

or as we may more simply express it

$$\Delta = \partial_b + b\partial_c + c\partial_b + \dots,$$

we have $\Theta_{\sigma'} - \Theta_\sigma = (\sigma' - \sigma)\Delta$, and consequently $\Delta U = 0$; Δ is thus an annihilator of any function U of the coefficients which is equal to a non-unitary function of the roots; or more shortly Δ is an annihilator of any non-unitariant.

22. Similarly, from the two equations $\Theta_\sigma = - S\partial_\alpha$, and $\Theta_{\sigma'} = S\partial_\alpha$ regarded as operating upon a non-unitary function, we deduce $\sigma'\Theta_\sigma - \sigma\Theta_{\sigma'} = (\sigma - \sigma') S\partial_\alpha$: the left-hand side is here $= (\sigma - \sigma')\Delta_1$, if

$$\Delta_1 = b\partial_c + 2c\partial_b + 3c\partial_d + \dots + (\sigma - 1) p\partial_q,$$

or say

$$\Delta_1 = b\partial_c + 2c\partial_b + 3c\partial_d + \dots,$$

viz. we have $\Delta_1 = S\partial_\alpha$; for instance, if as before

$$\Upsilon = U = S\alpha^4 = -4e + 4bd + 2c^2 - 4b^2c + b^4,$$

then

$$(b\partial_c + 2c\partial_d + 3d\partial_e)(-4e + 4bd + 2c^2 - 4b^2c + b^4) = S\partial_\alpha . S\alpha^4, \quad = 4S\alpha^3, \quad = 4(-3d + 3bc - b^3),$$

as can be at once verified. It is to be noticed, however, that $S\partial_\alpha$ operating upon a non-unitary function of the roots does not in every case give a non-unitary function; and thus successive operations with Δ_1 will not give a succession of non-unitariants.

23. I investigate the foregoing result in regard to Δ in a different manner; suppose, for instance, that $\Upsilon = U$ is the non-unitary function $S\alpha^4$ of the roots,

$$(= -4e + 4bd + 2c^2 - 4b^2c + b^4).$$

The number of roots is at least $= 4$, and I take it to be $= 4$, say the roots are α, β, γ, δ. Consider a fifth root θ, and let $\Upsilon_1 = U_1 = S\alpha^4$ be the like function for the five roots, we have $\Upsilon_1 = \Upsilon + \theta^4$, or say $U_1 = U + \theta^4$. Write $-b_1$, c_1, $-d_1$, e_1, $-f_1$

for the symmetric functions of the five roots; U_1 will not involve f_1 and it will be the same function of b_1, c_1, d_1, e_1 that U is of b, c, d, e, say we have

$$U_1 = U(b_1, c_1, d_1, e_1).$$

But we have

$$b_1 = b - \theta,\quad c_1 = c - b\theta,\quad d_1 = d - c\theta,\quad e_1 = e - d\theta;$$

and thus the foregoing equation $U_1 = U + \theta^4$ becomes

$$U(b-\theta,\ c - b\theta,\ d - c\theta,\ e - d\theta) = U + \theta^4;$$

it is in fact easy to verify that, for the foregoing value of U, the terms in θ, θ^2, θ^3 all vanish, and that the expression on the left-hand becomes $= U + \theta^4$. But attending only to the term in θ, this is $= -\theta(\partial_b + b\partial_c + c\partial_d + d\partial_e)\,U$, $= -\theta\Delta U$; viz. this term vanishing we have $\Delta U = 0$, the result which was to be proved.

In the case of a unitary function, for instance $\Upsilon = U = S\alpha^3\beta$, here introducing the new root θ we have $U_1 = U + \theta S\alpha^3 + \theta^3 S\alpha$; or there is here a term in θ, and instead of $\Delta U = 0$, we have $\Delta U = S\alpha^3$, or the unitary function is not annihilated by Δ.

The foregoing investigation is really quite general, and establishes the conclusion that Δ is an annihilator of every non-unitariant.

It is to be noticed that Θ_σ and Δ are operators, which leave each of them the degree unaltered but diminish the weight by unity: the operator $P - b\delta$, and another operator $\tfrac{1}{2}Q - b\omega$ which will be considered, increase each of them the degree by unity and also the weight by unity.

24. Coming now to the equation

$$P - b\delta = S\alpha^2\partial_\alpha,$$

it is to be remarked that, if $\sigma' = \sigma$, the expression for P ends in $q\partial_p$ where, as before, $q = a_\sigma$ is the highest coefficient in the operand; since the operand thus contains q, the next succeeding term in $r\partial_q$ would be not inoperative, and in order to include it in the expression of P we may take $\sigma' = \sigma + 1$; we thus have

$$P = b\partial_a + 2c\partial_b + 3d\partial_c + \ldots + (\sigma+1)\,r\partial_q,$$

or as we may more simply write it

$$P = b\partial_a + 2c\partial_b + 3d\partial_c + \ldots;$$

the operation thus increases the extent by unity. The symbol $S\alpha^2\partial_\alpha$ operating upon a symmetric function of the roots, gives, whatever may be the number of roots, the same symmetric function of the roots: and we see further that, operating upon a non-unitary function, it gives a non-unitary function of the roots. Hence $P - b\delta$ operating upon a non-unitariant gives a non-unitariant. I give an example.

25. Suppose, as before,

$$\Upsilon = U = S\alpha^4 = E = -4e + 4bd + 2c^2 - 4b^2c + b^4;$$

here $\delta = 4$, and therefore

$$P - b\delta = b\partial_a + 2c\partial_b + 3d\partial_c + 4e\partial_d + 5f\partial_e - 4b.$$

We have

$$S\alpha^2\partial_\alpha \,.\, S\alpha^4 = 4S\alpha^5, \quad = 4\,(-\,5f + 5be + 5cd - 5b^2d - 5bc^2 + 5b^3c - b^5),$$

and this should therefore be the result of the operation $P - b\delta$: the calculation is

$b\,.\,-12c+8bd+4c^2-4b^2c$	f					-20		-20
$2c\,.\quad 4d-8bc+4b^3$	be	-12			$+16$		$+16$	$+20$
$3d\,.\quad 4c-4b^2$	cd		$+8$	$+12$				$+20$
$4e\,.\quad 4b$	b^2d	$+8$		-12			-16	-20
$5f\,.\,-\;4$	bc^2	$+4$	-16				-8	-20
$-4b\,.\,-\;4e+4bd+2c^2-4b^2c+b^4$	b^3c	-4	$+8$				$+16$	$+20$
	b^5						-4	-4,

which is the right result.

We have seen that every non-unitariant is annihilated by Δ; it at once appears that conversely every function of the coefficients which is annihilated by Δ is a non-unitariant: it is, in fact, a symmetric function of the roots, and unless it were a non-unitary function of the roots it would not be annihilated by Δ. Non-unitariants are analogous to seminvariants; the precise relation between them will be shown further on.

26. We can, by an investigation similar to that for seminvariants, show that $P - b\delta$ operating upon a non-unitariant gives a non-unitariant. In fact, considering the two operations Δ and $P - b\delta$, we have

$$\Delta\,(P - b\delta)\dagger = \Delta\,(P - b\delta) + \Delta\,.\,(P - b\delta),$$

the meaning being that, if upon any operand U we perform first the operation $P - b\delta$ and then the operation Δ, this is equivalent to operating on U with the sum of the two operations $\Delta\,(P - b\delta)$, and $\Delta\,.\,P - b\delta$, the first of these symbols denoting the mere algebraical product of Δ and $P - b\delta$, the second of them the result of the operation Δ performed upon $P - b\delta$. We have similarly

$$(P - b\delta)\,\Delta\dagger = (P - b\delta)\,\Delta + (P - b\delta)\,.\,\Delta.$$

Hence observing that $\Delta\,(P - b\delta)$ and $(P - b\delta)\,\Delta$ are equal to each other, and subtracting, we have

$$\Delta\,(P - b\delta)\dagger - (P - b\delta)\,\Delta\dagger = \Delta\,.\,(P - b\delta) - (P - b\delta)\,.\,\Delta.$$

But from the values

$$\Delta = a\partial_b + b\partial_c + c\partial_d + \dots \qquad ,$$

and

$$P - b\delta = b\partial_a + c\partial_b + d\partial_c + \dots - b\delta,$$

we find

$$\Delta\,.\,(P-b\delta)=a\partial_a+2b\partial_b+3c\partial_c+\ldots-\delta,$$

$$(P-b\delta)\,.\,\Delta=\qquad b\partial_b+2c\partial_c+\ldots,$$

and thence

$$\Delta\,.\,(P-b\delta)-(P-b\delta)\,.\,\Delta=a\partial_a+b\partial_b+c\partial_c\ldots-\delta,\ =0,$$

since δ is the degree in the coefficients. Hence writing down the operand U,

$$\Delta\,.\,(P-b\delta)\,U-(P-b\delta)\,.\,\Delta U=0,$$

where for greater clearness I have inserted the dots, to show that Δ operates on $(P-b\delta)\,U$, and $(P-b\delta)$ on ΔU. Taking U to be a non-unitariant, we have $\Delta U=0$; and this being so, the equation gives $\Delta\,.\,(P-b\delta)\,U=0$, viz. this shows that $(P-b\delta)\,U$ is a non-unitariant.

27. There is another symbol $\tfrac{1}{2}Q-b\omega$, which is precisely analogous to $P-b\delta$, viz. operating upon a non-unitariant, it gives a non-unitariant: ω is, as before, the weight of the function operated upon, and the expression of Q is

$$\tfrac{1}{2}Q=c\partial_b+3d\partial_c+6e\partial_d+\ldots+\tfrac{1}{2}\sigma(\sigma+1)\,r\partial_q,$$

or say

$$\tfrac{1}{2}Q=c\partial_b+3d\partial_c+6e\partial_d+\ldots.$$

The proof is exactly similar, viz. we have to show that

$$\Delta\,.\,(\tfrac{1}{2}Q-b\omega)-(\tfrac{1}{2}Q-b\omega)\,.\,\Delta=0.$$

We have

$$\Delta\,.\,(\tfrac{1}{2}Q-b\omega)=b\partial_b+3c\partial_c+6d\partial_d+\ldots-\omega,$$

$$(\tfrac{1}{2}Q-b\omega)\,.\,\Delta=\qquad c\partial_c+3d\partial_d+\ldots,$$

and the difference of the two expressions is

$$b\partial_b+2c\partial_c+3d\partial_d+\ldots-\omega,\ =0,$$

since ω is the weight of the function operated upon. Hence, as before, if U be a non-unitariant and therefore $\Delta U=0$, we have $\Delta\,.\,(\tfrac{1}{2}Q-b\omega)\,U=0$, that is, $(\tfrac{1}{2}Q-b\omega)\,U$ is also a non-unitariant.

28. The symbol $\tfrac{1}{2}Q-b\omega$ has no simple expression in terms of ∂_α, ∂_β, ∂_γ, ..., and the form varies with the number of the roots: thus for 3 roots, it is

$$=-\left\{\left(\frac{c\alpha^2+3d\alpha}{\alpha-\beta\,.\,\alpha-\gamma}+b\alpha\right)\partial_\alpha+\&\text{c.}\right\},$$

for 4 roots it is

$$=-\left\{\left(\frac{c\alpha^3+3d\alpha^2+6e\alpha}{\alpha-\beta\,.\,\alpha-\gamma\,.\,\alpha-\delta}+b\alpha\right)\partial_\alpha+\&\text{c.}\right\},$$

for 5 roots it is

$$=-\left\{\left(\frac{c\alpha^4+3d\alpha^3+6e\alpha^2+10f\alpha}{\alpha-\beta\,.\,\alpha-\gamma\,.\,\alpha-\delta\,.\,\alpha-\epsilon}+b\alpha\right)\partial_\alpha+\&\text{c.}\right\},$$

and so on. It is not easy to find the effect of such a symbol upon a given symmetric function of the roots, nor in particular when the function is non-unitary is it easy to show generally that the result is non-unitary.

It is to be remarked that, if the function operated upon is of the degree δ in the roots, then we must for $\tfrac{1}{2}Q - b\omega$ take the expression with $\delta+1$ roots; for instance, if the function be of the degree 5 in the roots, then, *quà* function of the coefficients, this contains f, and it must be operated on with

$$\tfrac{1}{2}Q - b\omega, \; = c\partial_b + 3d\partial_c + 6e\partial_d + 10f\partial_e + 15g\partial_f - \omega b,$$

viz. this expression, as containing g, gives the 6-root expression for $\tfrac{1}{2}Q - b\omega$.

29. Suppose for instance the function operated upon is $F = S\alpha^5$; here taking the 6-root expression, this gives

$$-5\left\{\left(\frac{c\alpha^5 + 3d\alpha^4 + 6e\alpha^3 + 10f\alpha^2 + 15g\alpha}{\alpha-\beta\,.\,\alpha-\gamma\,.\,\alpha-\delta\,.\,\alpha-\epsilon\,.\,\alpha-\zeta} + b\alpha\right)\alpha^4 + \&\text{c.}\right\},$$

or omitting for the moment the outside factor -5, the expression in $\{\ \}$ is easily seen to be

$$= cH_4 + 3dH_3 + 6eH_2 + 10fH_1 + 15g + bS\alpha^5,$$

where H_4, H_3, H_2, H_1 denote the homogeneous functions of the degrees 4, 3, 2, 1 respectively: the values of these are obtained by adding together all the lines of the Table IV (b), all the lines of the Table III (b), &c.: the terms exclusive of $bS\alpha^5$ thus are

$$\begin{aligned} & c\,(-e + 2bd + c^2 - 3b^2c + b^4) \\ + \; & 3d\,(-d + 2bc - b^3) \\ + \; & 6e\,(-c + b^2) \\ + \; & 10f\,(-b) \\ + \; & 15g\,.\,1, \end{aligned}$$

and these are $= S\alpha^5\beta + S\alpha^4\beta^2 + S\alpha^3\beta^3$, as appears by the following calculation:

							$S\alpha^5\beta$	$S\alpha^4\beta^2$	$S\alpha^3\beta^3$	
g					+ 15	+ 15	+ 6	+ 6	+ 3	+ 15
bf				− 10		− 10	− 1	− 6	− 3	− 10
ce	− 1		− 6			− 7	− 6	+ 2	− 3	− 7
b^2e			+ 6			+ 6	+ 1	+ 2	+ 3	+ 6
d^2		− 3				− 3	− 3	− 3	+ 3	− 3
bcd	+ 2	+ 6				+ 8	+ 7	+ 4	− 3	+ 8
b^3d		− 3				− 3	− 1	− 2	0	− 3
c^3	+ 1					+ 1	+ 2	− 2	+ 1	+ 1
b^2c^2	− 3					− 3	− 4	+ 1		− 3
b^4c	+ 1					+ 1	+ 1			+ 1
b^6										

The omitted term $bS\alpha^5$, that is, $-S\alpha . S\alpha^5$, is $-S\alpha^6 - S\alpha^5\beta$; the addition hereof destroys therefore the non-unitary term $S\alpha^5\beta$, and thus the required expression, restoring the omitted factor -5, is $-5(-S\alpha^6 + S\alpha^4\beta^2 + S\alpha^3\beta^3)$, or say $= 5G - 5CE - 5D^2$, a non-unitary form: this then should be the result of the operation

$$\tfrac{1}{2}Q - b\omega, \; = c\partial_b + 3d\partial_c + 6e\partial_d + 10f\partial_e + 15g\partial_f - 5b,$$

performed upon

$$S\alpha^5 = F = -5f + 5be + 5cd - 5b^2d - 5bc^2 + 5b^3c = b^5.$$

Performing the calculation so as to omit on each side a factor 5, it is to be shown that $G - CE - D^2$ is

$$\begin{aligned} = \quad & c\,(e - 2bd - c^2 + 3b^2c - b^4) \\ + \; & 3d\,(d - 2bc + b^3) \\ + \; & 6e\,(c - b^2) \\ + \; & 10f\,(b) \\ + \; & 15g\,(-1) \\ - \; & 5b\,(-f + be + cd - b^2d - bc^2 + b^3c - \tfrac{1}{5}b^5). \end{aligned}$$

Collecting the terms, and comparing the result with the expression for $G - CE - D^2$, we have

								G	$-\ CE$	$-\ D^2$
g					-15		-15	-6	-6	-3
bf				$+10$		$+5$	$+15$	$+6$	$+6$	$+3$
ce	$+1$		$+6$				$+7$	$+6$	-2	$+3$
b^2e			-6			-5	-11	-6	-2	-3
d^2		$+3$					$+3$	$+3$	$+3$	-3
bcd	-2	-6				-5	-13	-12	-4	$+3$
b^3d		$+3$				$+5$	$+8$	$+6$	$+2$	0
c^3	-1						-1	-2	$+2$	-1
b^2c^2	$+3$					$+5$	$+8$	$+9$	-1	
b^4c	-1					-5	-6	-6		
b^6						$+1$	$+1$	$+1$		

and the two expressions are thus identical.

30. Suppose again, 6 roots as before, and that the function operated upon is $S\alpha^3\beta^2$; we find $\partial_a S\alpha^3\beta^2 = 3\alpha^2 S\alpha^2 + 2\alpha S\alpha^3 - 5\alpha^4$, and the general term is

$$\begin{aligned} & -3\left(\frac{c\alpha^5 + 3d\alpha^4 + 6e\alpha^3 + 10f\alpha^2 + 15g\alpha}{\alpha-\beta\,.\,\alpha-\gamma\,.\,\alpha-\delta\,.\,\alpha-\epsilon\,.\,\alpha-\zeta} + b\alpha\right)\alpha^2 . S\alpha^2 \\ & -2\left(\frac{c\alpha^5 + 3d\alpha^4 + 6e\alpha^3 + 10f\alpha^2 + 15g\alpha}{\alpha-\beta\,.\,\alpha-\gamma\,.\,\alpha-\delta\,.\,\alpha-\epsilon\,.\,\alpha-\zeta} + b\alpha\right)\alpha\ . S\alpha^3 \\ & +5\left(\frac{c\alpha^5 + 3d\alpha^4 + 6e\alpha^3 + 10f\alpha^2 + 15g\alpha}{\alpha-\beta\,.\,\alpha-\gamma\,.\,\alpha-\delta\,.\,\alpha-\epsilon\,.\,\alpha-\zeta} + b\alpha\right)\alpha^4. \end{aligned}$$

This gives

$$\begin{aligned}&-3\,(CH_2+3dH_1+6e+bS\alpha^3)\,S\alpha^2\\&-2\,(CH_1+3d\qquad\quad+bS\alpha^2)\,S\alpha^3\\&+5\,(CH_4+3dH_3+6eH_2+10fH_1+15g+bS\alpha^5),\end{aligned}$$

which is found to be

$$\begin{aligned}=&-3\,(BD+C^2\;+bS\alpha^3)\,S\alpha^2\\&-2\,(BC\qquad\quad+bS\alpha^2)\,S\alpha^3\\&+5\,(BF+CE+D^2+bS\alpha^5).\end{aligned}$$

Here

$$bS\alpha^3=-S\alpha S\alpha^3=-S\alpha^4-S\alpha^3\beta,\;=-E-BD,$$

$$bS\alpha^2=-S\alpha S\alpha^2=-S\alpha^3-S\alpha^2\beta,\;=-D-BC,$$

and

$$bS\alpha^5=-S\alpha S\alpha^5=-S\alpha^6-S\alpha^5\beta,\;=-G-BF;$$

the expression thus is

$$\begin{aligned}=&-3\,(-E+C^2)\,.\,C &\text{that is, }&-3\,(-S\alpha^4+S\alpha^2\beta^2)\,S\alpha^2\\&-2\,(-D\qquad)\,.\,D &&-2\,(-S\alpha^3\qquad\quad)\,S\alpha^3\\&+5\,(-G+CE+D^2), &&+5\,(-S\alpha^6+S\alpha^4\beta^2+S\alpha^3\beta^3).\end{aligned}$$

Here

$$\begin{aligned}S\alpha^2S\alpha^4\;&=S\alpha^6\;+\;S\alpha^4\beta^2, &&=G+CE,\\S\alpha^3S\alpha^3\;&-S\alpha^6\;+2S\alpha^3\beta^3, &&=G+2D^2,\\S\alpha^2S\alpha^2\beta^2&=S\alpha^4\beta^2+3S\alpha^2\beta^2\gamma^2, &&=CE+3C^3;\end{aligned}$$

and the whole is

$$\begin{aligned}&-3\,\{-G-CE+(CE+3C^2)\}\\&-2\,(-G-2D^2)\\&+5\,(-G+CE+D^2),\end{aligned}$$

which is $=5CE+9D^2-9C^3$ (a non-unitary form). This then should be the value of

$$\tfrac{1}{2}Q-b\omega,\;=c\partial_b+3d\partial_c+6e\partial_d+10f\partial_e+15g\partial_f-5b,$$

operating upon

$$S\alpha^3\beta^2,\;=CD=5f-5be+cd+2b^2d-bc^2.$$

31. There is for non-unitariants a theorem which is a much more simple form than the transformation of it afterwards obtained for seminvariants: viz. for any non-unitariant we have $\Delta U=0=(\partial_b+b\partial_c+c\partial_d+\ldots)\,U$; attending only to the portion U' of U which is of the highest degree, it is clear that we have $(b\partial_c+c\partial_d+\ldots)\,U'=0$, and if we herein diminish the letters, then $(\partial_b+b\partial_c+\ldots)\,U''=0$, where U'' is what U' becomes by a diminution of the letters; that is, U'' is a non-unitariant, viz. in any seminvariant, the terms of highest degree U' are obtained from a non-unitariant U'' by a mere augmentation of the letters: e.g. $2e-2bd+c^2$ is a non-unitariant weight 4; augmenting the letters, we have $2bf-2ce+d^2$ which with a change of sign is the portion of highest degree of the non-unitariant $2g-2bf+2ce-d^2$.

The MacMahon Form of Equation. Art. Nos. 32 to 34.

32. The equation connecting the coefficients and the roots is here taken to be

$$1+\frac{b}{1}x+\frac{c}{1.2}x^2+\frac{d}{1.2.3}x^3+\ldots=1-\alpha x.1-\beta x.1-\gamma x\ldots.$$

As to this it may be remarked that, if we had started with a form of the nth order with binomial coefficients,

$$1+\frac{n}{1}bx+\frac{n.n-1}{1.2}cx^2+\frac{n.n-1.n-2}{1.2.3}dx^3+\ldots=1-\alpha x.1-\beta x.1-\gamma x\ldots(n \text{ factors}),$$

then writing herein $\frac{x}{n}$ for x, and also $n\alpha$, $n\beta$, $n\gamma$, ..., for α, β, γ, ... and putting ultimately $n=\infty$, we have the form in question.

We pass from the ordinary form to the MacMahon form, by writing for

$$b,\ c,\ d,\ e,\ \ldots,\ \frac{b}{1},\ \frac{c}{1.2},\ \frac{d}{1.2.3},\ \frac{e}{1.2.3.4},\ \ldots \text{ or say } b,\ \frac{c}{2},\ \frac{d}{6},\ \frac{e}{24},\ \frac{f}{120},\ \frac{g}{720},\ \ldots.$$

All the results obtained for the ordinary form will, after making therein this change, apply to the new form. We thus find

$$\Theta_\sigma=\sigma\partial_b+(\sigma-1)\,2b\partial_c+(\sigma-2)\,3c\partial_d+\ldots+\qquad 1\ \sigma p\partial_q,$$

$$\Theta_{\sigma'}=\sigma'\partial_b+(\sigma'-1)\,2b\partial_c+(\sigma'-2)\,3c\partial_d+\ldots+(\sigma'-\sigma+1)\,\sigma p\partial_q,$$

$$\Theta_\sigma-\Theta_{\sigma'}=(\sigma'-\sigma)\,\Delta,$$

where

$$\Delta=\ \partial_b+2b\partial_c+3c\partial_d+\ldots+\sigma p\partial_q,$$

or say

$$=\ \partial_b+2b\partial_c+3c\partial_d+\ldots.$$

Also

$$P=b\partial_a+\ c\partial_b+\ d\partial_c+\ldots+\ r\partial_q,$$

or say

$$=b\partial_a+\ c\partial_b+\ d\partial_c+\ldots,$$

$$Q=\qquad c\partial_b+2d\partial_c+\ldots+\sigma r\partial_q,$$

or say

$$=\qquad c\partial_b+2d\partial_c+\ldots.$$

The change α, β, γ, ... into $n\alpha$, $n\beta$, $n\gamma$, ... would change $S\partial_\alpha$, $S\alpha\partial_\alpha$, $S\alpha^2\partial_\alpha$ into $n^{-1}S\partial_\alpha$, $S\alpha\partial_\alpha$, $nS\alpha^2\partial_\alpha$ respectively ($n=\infty$): but this change is, in fact, compensated for by the introduction into the formulæ of the binomial coefficients as above; it is $-S\alpha$, $S\alpha\beta$, ... not $-nS\alpha$, $n^2S\alpha\beta$, ... which are equal to b, $\frac{1}{2}c$, ...; and the conclusion is that we have to retain without alteration the symbols $S\partial_\alpha$, $S\alpha\partial_\alpha$, $S\alpha^2\partial_\alpha$: thus in the new form as in the old one, we have $\Theta_4S\alpha^4=-S\partial_\alpha.S\alpha^4=-4S\alpha^3$, see the example *ante* No. 23.

33. In the new form, a non-unitariant is annihilated by the operator

$$\Delta,\ =\partial_b+2b\partial_c+3c\partial_d+\ldots,$$

and conversely any function annihilated by Δ is a non-unitariant; comparing herewith the subsequent theory of seminvariants, this is in fact the theorem that a non-unitariant is the same thing as a seminvariant; or to state this more explicitly: for the MacMahon form of equation, a function of the coefficients which is a non-unitary symmetric function of the roots is a seminvariant.

I consider for instance the Table VI (*b*), but attend only to the non-unitary portions thereof, viz. the lines G, CE, D^2, C^3: I convert these into columns, at the same time changing the arrangement of the headings g, bf, ce, &c., from CO to AO: and then making the foregoing change b, c, d, e, f, g into $b, \frac{c}{2}, \frac{d}{6}, \frac{e}{24}, \frac{f}{120}, \frac{g}{720}$, but to avoid fractions multiplying the whole by 720, I form the table

	÷ 720 ‖	C^3	D^2	CE	G
1	g	-2	$+3$	$+6$	-6
6	bf	$+2$	-3	-6	$+6$
15	ce	-2	-3	$+2$	$+6$
20	d^2	$+1$	$+3$	-3	$+3$
30	b^2e		$+3$	$+2$	-6
60	bcd		-3	$+4$	-12
90	c^3		$+1$	-2	-2
120	b^3d			-2	$+6$
180	b^2c^2			$+1$	$+9$
360	b^4c				-6
720	b^6				$+1$
		$[d^2]$	$[c^3]$	$[b^2c^2]$	$[b^6]$

which is to be read according to the columns: and observe that the outside left-hand numbers are to be multiplied into the numbers of each column: thus the first column is to be read

$$C^3 = S\alpha^2\beta^2\gamma^2 = \frac{1}{720}(-2g + 12bf - 30ce + 20d^2),$$

the second column is to be read

$$D^2 = S\alpha^3\beta^3 = \frac{1}{720}(3g - 18bf \ldots + 90c^3),$$

and so on.

By what precedes, the columns are seminvariants,—as afterwards explained, "blunt" seminvariants; and they are named as such by the outside bottom line of symbols with a [ ]; viz.

$$[d^2] = (-2g + 12bf - 30ce + 20d^2), \quad [c^3] = (3g - 18bf \ldots + 90c^3), \text{ \&c.,}$$

where it will be observed that the symbol within the [ ] is, in fact, the power-ender which is in AO the lowest term of the column; and further that this is also the conjugate of the capital letter symbol at the head of the column.

The (b) Tables I to X, with only the change $b, c, d, e, \ldots$ into $b, \frac{c}{2}, \frac{d}{6}, \frac{e}{24}, \ldots$ are given in my paper, "Tables of the Symmetric Functions of the Roots to the degree 10, for the form

$$1 + bx + \frac{cx^2}{1.2} + \ldots = (1 - \alpha x)(1 - \beta x)(1 - \gamma x) \ldots,\text{"}$$

American Mathematical Journal, t. VII. (1885), pp. 47—56, [829].

34. By what precedes, it appears that $P - b\delta$ operating on a seminvariant gives a seminvariant, and that $Q - 2b\omega$ operating on a seminvariant gives a seminvariant: these operators will be further considered in the development of the theory of seminvariants. We see further that $\frac{1}{2}\Delta, = b\partial_c + 3c\partial_d + 6d\partial_e + \ldots$, operating on a seminvariant gives sometimes but not always a seminvariant, e.g.

$$(b\partial_c + 3c\partial_d + 6d\partial_e)(e - 4bd - 3c^2 + 12b^2c - 6b^4) = 6(d - 3bc + 2b^3).$$

Seminvariants—the I-and-F Problem, and Solution by Square Diagrams.
Art. Nos. 35 to 47.

35. Writing

$$\begin{aligned}
1 &= 1,\\
b_1 &= b + \theta,\\
c_1 &= c + 2b\theta + \theta^2,\\
d_1 &= d + 3c\theta + 3b\theta^2 + \theta^3,\\
e_1 &= e + 4d\theta + 6c\theta^2 + 4b\theta^3 + \theta^4,\\
&\&c.,
\end{aligned}$$

then there are functions of the unsuffixed letters which remain unaltered if for these we substitute the suffixed letters: any such function is termed a seminvariant. We have for instance

$$\begin{aligned}
c_1 &= c + 2b\theta + \theta^2, &&\text{i.e.,} && c_1 - b_1^2 = c - b^2,\\
-b_1^2 &= -b^2 - 2b\theta - \theta^2,\\
d_1 &= d + 3c\theta + 3b\theta^2, &&&& d_1 - 3b_1c_1 + 2b_1^3 = d - 3bc + 2b^3,\\
-3b_1c_1 &= -3bc - 6b^2\theta - 3b\theta^2,\\
&\qquad\quad - 3c\theta - 6b\theta^2 - 3\theta^3,\\
+2b_1^3 &= 2b^3 + 6b^2\theta + 6b\theta^2 + 2\theta^3,
\end{aligned}$$

and thus $c-b^2$, $d-3bc+2b^3$ are seminvariants; they are, in fact, the first and second terms of the series

$$\begin{aligned}
&c - b^2,\\
&d - 3bc + 2b^3,\\
&e - 4bd + 6b^2c - 3b^4,\\
&f - 5bc + 10b^2d - 10b^3c + 4b^5,\\
&g - 6bf + 15b^2e - 20b^3d + 15b^4c - 5b^6,\\
&\quad\vdots
\end{aligned}$$

where the law is obvious; the numbers in each line are binomial coefficients except the last number, which is the next binomial coefficient diminished by unity. The successive terms are, in fact, what c_1, d_1, e_1, f_1, g_1, ... become upon writing therein $\theta = -b$.

36. Any rational and integral function of these forms is a seminvariant, and it is to be observed that we can form functions for which (by the destruction of terms of a higher degree) there is a diminution of degree; for instance,

$$(e - 4bd + 6b^2c - 3b^4) + 3(c - b^2)^2$$

gives a seminvariant $e - 4bd + 3c^2$.

It is important to remark that a seminvariant is completely determined by its non-unitary terms; thus for $e - 4bd + 3c^2$, the non-unitary terms are $e + 3c^2$, and for this writing $e_1 + 3c_1^2$, and for e_1, c_1 substituting their above values for $\theta = -b$, we reproduce the original value $e - 4bd + 3c^2$.

37. It is at once seen that a seminvariant is reduced to zero by the operation $\Delta, = \partial_b + 2b\partial_c + 3c\partial_d + \dots$, or say that Δ is an annihilator of a seminvariant; in fact, if in any function of b, c, d, ... we write for these the suffixed letters b_1, c_1, d_1, ... then the coefficient of θ herein is at once found by operating on the function of $(b, c, d, \dots)$ with Δ, and therefore in the case of a seminvariant the result of this operation must be $=0$. And conversely, every function of $(b, c, d, \dots)$ which is reduced to zero by the operation Δ is a seminvariant.

38. For a given weight, the number of seminvariants is equal to the excess of the number of terms of that weight above the number of terms of the next preceding weight, or what is the same thing, it is equal to the number of power-enders of the given weight. More definitely, considering the terms of a seminvariant as arranged in AO, we have seminvariants the finals whereof are the several power-enders of the given weight: and we arrange the seminvariants *inter se* by taking these power-enders in AO: thus for the weight 6, we have seminvariants $[d^2]$, $[c^3]$, $[b^2c^2]$, $[b^6]$ ending in these terms respectively. We may, if we please, consider all these seminvariants as beginning with g, or say the forms may be taken to be

$g(\text{ao})\,d^2$, $g(\text{ao})\,c^3$, $g(\text{ao})\,b^2c^2$, $g(\text{ao})\,b^6$. Such forms are, in fact, furnished by the MacMahon equation: viz. up to the weight 6, we thus have for the present purpose

÷ 2	‖	C
1	c	− 2
2	b^2	+ 1
		$[b^2]$

÷ 6	‖	D
1	d	− 3
3	bc	+ 3
6	b^3	− 1
		$[b^3]$

÷ 24	‖	C^2	E
1	e	+ 2	− 4
4	bd	− 2	+ 4
6	c^2	+ 1	+ 2
12	b^2c		− 4
24	b^4		+ 1
		$[c^2]$	$[b^4]$

÷ 120	‖	CD	F
1	f	+ 5	− 5
5	be	− 5	+ 5
10	cd	+ 1	+ 5
20	b^2d	+ 2	− 5
30	bc^2	− 1	− 5
60	b^3c		+ 5
120	b^5		− 1
		$[bc^2]$	$[b^5]$

÷ 720	‖	C^3	D^2	CE	G
1	g	− 2	+ 3	+ 6	− 6
6	bf	+ 2	− 3	− 6	+ 6
15	ce	− 2	− 3	+ 2	+ 6
20	d^2	+ 1	+ 3	− 3	+ 3
30	b^2e		+ 3	+ 2	− 6
60	bcd		− 3	+ 4	− 12
90	c^3		+ 1	− 2	− 2
120	b^3d			− 2	+ 6
180	b^2c^2			+ 1	+ 9
360	b^4c				− 6
720	b^6				+ 1
		$[d^2]$	$[c^3]$	$[b^2c^2]$	$[b^6]$

read for instance

$$[d^2] = -\,2g + 12bf - 30ce + 20d^2,$$

$$[c^3] = 3g - 18bf - 45ce + 60d^2 + 90b^2e - 180bcd + 90c^3,$$

&c.

I say that $[d^2]$, $[c^3]$, $[b^2c^2]$, $[b^6]$ are "specific" when they are regarded as standing for these tabulated functions; but in general I take them to be "indefinite," that is, I regard them as denoting (as above) any seminvariants ending in d^2, c^3, b^2c^2, b^6 respectively.

39. The seminvariant $[d^2]$ is of the form $(g \propto d^2)$, including those terms which are in CO not superior to g and in AO not inferior to d^2: by a combination of $[d^2]$ and $[c^3]$, we obtain a seminvariant $(ce \propto c^3)$ containing terms which are in CO not superior to ce and in AO not inferior to c^3: similarly, from $[d^2]$, $[c^3]$, $[b^2c^2]$ we obtain a seminvariant $(d^2 \propto b^2c^2)$; and from the four forms a seminvariant $(c^3 \propto b^6)$: these four seminvariants

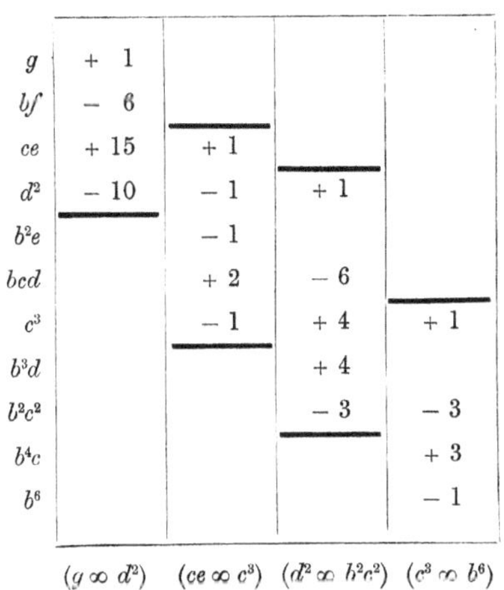

	$(g \propto d^2)$	$(ce \propto c^3)$	$(d^2 \propto b^2c^2)$	$(c^3 \propto b^6)$
g	$+ 1$			
bf	$- 6$			
ce	$+ 15$	$+ 1$		
d^2	$- 10$	$- 1$	$+ 1$	
b^2e		$- 1$		
bcd		$+ 2$	$- 6$	
c^3		$- 1$	$+ 4$	$+ 1$
b^3d			$+ 4$	
b^2c^2			$- 3$	$- 3$
b^4c				$+ 3$
b^6				$- 1$

are said to be "sharp" seminvariants: viz. considering the final as given, a sharp seminvariant is one having an initial which is in CO as low as possible; or considering the initial as given, it is one having a final which is in AO as high as possible. A seminvariant which is not sharp is said to be "blunt."

40. The sharp seminvariants are in general designated as above, $(g \propto d^2)$, &c.: but it is sometimes convenient to give the numerical coefficients of the initial and final terms respectively: as to this, it is to be noticed that the coefficient of the initial term is in most cases, but not always, $=1$,—we might of course take it to be always $=1$, but we should then in the excepted cases have fractional coefficients, and it is better to avoid this by giving a proper value to the numerical coefficient of the initial term; the numerical coefficient of the final term is in general different from ± 1, and it is not in general a multiple of the numerical coefficient of the initial term. As an instance take $dh \propto b^2e^2$, the more complete expression of which is $4dh \propto -35b^2e^2$. The sharp seminvariants up to the weight 12 are designated in this more complete form in the table *post* No. 62.

41. In the calculation of the sharp seminvariants by elimination as above, it will be noticed how unitary terms disappear: thus in combining $[d^2]$ and $[c^3]$ so as to get rid of g, the term bf disappears of itself, and we have as above the form

$(ce \infty c^3)$ beginning with the non-unitary term ce. We may, in fact, write $b=0$; we thus have

$$[d^2] = -2g - 30ce + 20d^2,$$

$$[c^3] = 3g - 45ce + 60d^2 + 90c^3,$$

giving $3[d^2] + 2[c^3] = -180(ce - d^2 - c^3)$, and then $ce - d^2 - c^3$, putting therein for c, d, e the values $c - b^2$, $d - 3bc + 2b^3$, $e - 4bd + 6b^2c - 3b^4$, gives the complete value *ut supra*, $ce - d^2 - b^2e + 2bcd - c^3$, and we thus see *à priori* that this contains no term bf, but in fact begins with ce. And in carrying out this process for any higher given weight, it is proper also to arrange the non-unitary terms not in AO but in CO, and then in each case beginning with the terms highest in CO and eliminating as many as possible of these terms we obtain the sharp seminvariant. Consider for instance the weight 12: taking the finals in AO, we have here

$$(m \infty g^2),\ (m \infty cf^2),\ (m \infty e^3),\ (m \infty b^2f^2),\ \ldots$$

the initials in CO are m, ck, dj, ei, ... and it might at first sight appear that the foregoing process of elimination would lead to the forms $(m \infty g^2)$, $(ck \infty cf^2)$, $(dj \infty e^3)$, $(ei \infty b^2f^2)$, ...; we in fact have the form $(m \infty g^2)$; and if from $(m \infty g^2)$ and $(m \infty cf^2)$ we eliminate m, we obtain the form $(ck \infty cf^2)$; but we cannot have a form $(dj \infty e^3)$ (for a form beginning with dj is of necessity of the degree 4 at least); what happens is that when from $(m \infty g^2)$, $(m \infty cf^2)$ and $(m \infty e^3)$ we eliminate m and ck, the next term dj disappears of itself, and (the following term ei not disappearing) the resulting form is $(ei \infty e^3)$: to obtain a form beginning with dj we must use the fourth form $(m \infty b^2f^2)$, and we thence obtain $(dj \infty b^2f^2)$. Arranging the initials in CO and the finals in AO, we thus have

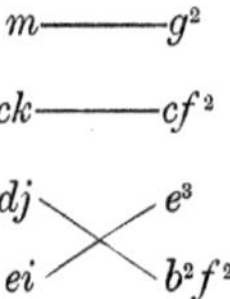

that is, we have the sharp seminvariants $m \infty g^2$, $ck \infty cf^2$, $ei \infty e^3$, $dj \infty b^2f^2$, ...; these are the results given by the MacMahon linkage as will be explained further on, but I will first approach the question from a different side.

42. It has been seen that we have $\Delta, = \partial_b + 2b\partial_c + 3c\partial_d + \ldots$, as the annihilator of a seminvariant. Considering in the first place the entire set of terms, say for the weight 6, $g(\text{ao})b^6$, we assume for a seminvariant the sum of these each multiplied by an arbitrary coefficient; the number of coefficients is equal to the number of terms of $g(\text{ao})b^6$. Operating with Δ, we obtain a function of the next inferior weight 5, containing all the terms of $Dg(\text{ao})b^6$, that is, of $f(\text{ao})b^5$, each

term multiplied by a linear function (with mere numerical factors) of the arbitrary coefficients: the expression thus obtained must be identically $=0$; and we thus find between the arbitrary coefficients a number of linear relations equal to the number of terms $f(\text{ao})\,b^5$: these relations are independent; for it is only on the supposition that they are so, that the number of coefficients which remain arbitrary will be $11-7, =4$, agreeing with the number of the seminvariants $[d^2]$, $[c^3]$, $[b^2c^2]$, $[b^6]$; whereas if the relations were not independent, there would be a larger number of seminvariants.

But if, instead of the whole set $g(\text{ao})\,b^6$, we consider a set $(g \infty d^2)$ or say $(ce \infty c^3)$ and assume for a seminvariant the sum of these terms each multiplied by an arbitrary coefficient, then operating as before with Δ we obtain between the arbitrary coefficients a number of relations equal to that of the terms $D\,(ce \infty c^3)$, and if this be less by unity than the number of the terms of $ce \infty c^3$, say if we have $(1-D)\,(ce \infty c^3)=1$, then there will be a single seminvariant $ce \infty c^3$. We, in fact, find $(1-D)$: $(g \infty d^2)$, $(ce \infty c^3)$, $(d^2 \infty b^2c^2)$, $(c^3 \infty b^6)$, each $=1$, and thus establish the existence of the foregoing seminvariants $g \infty d^2$, $ce \infty c^3$, $d^2 \infty b^2c^2$, $c^3 \infty b^6$. And similarly if in any case we have $(1-D)\,(I \infty F)=2$ or any larger number, then we have 2 or more seminvariants $I \infty F$.

43. It will be convenient to write down at once the system of square diagrams for the several weights 2 to 16; each of these may theoretically be obtained by a direct process of calculation such as I exhibit for the weight 10, but the labour would be very great indeed, and I have in fact formed the squares for the weights 11 to 16, not in this manner but by the MacMahon linkage.

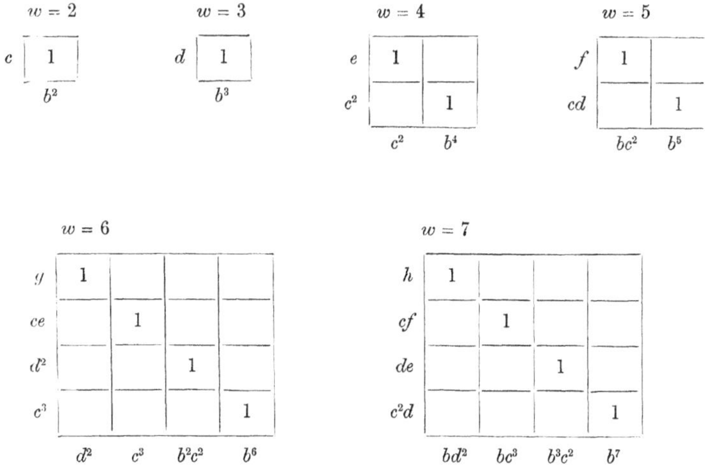

$w=2$

	b^2
c	1

$w=3$

	b^3
d	1

$w=4$

	c^2	b^4
e	1	
c^2		1

$w=5$

	bc^2	b^5
f	1	
cd		1

$w=6$

	d^2	c^3	b^2c^2	b^6
g	1			
ce		1		
d^2			1	
c^3				1

$w=7$

	bd^2	bc^3	b^3c^2	b^7
h	1			
cf		1		
de			1	
c^2d				1

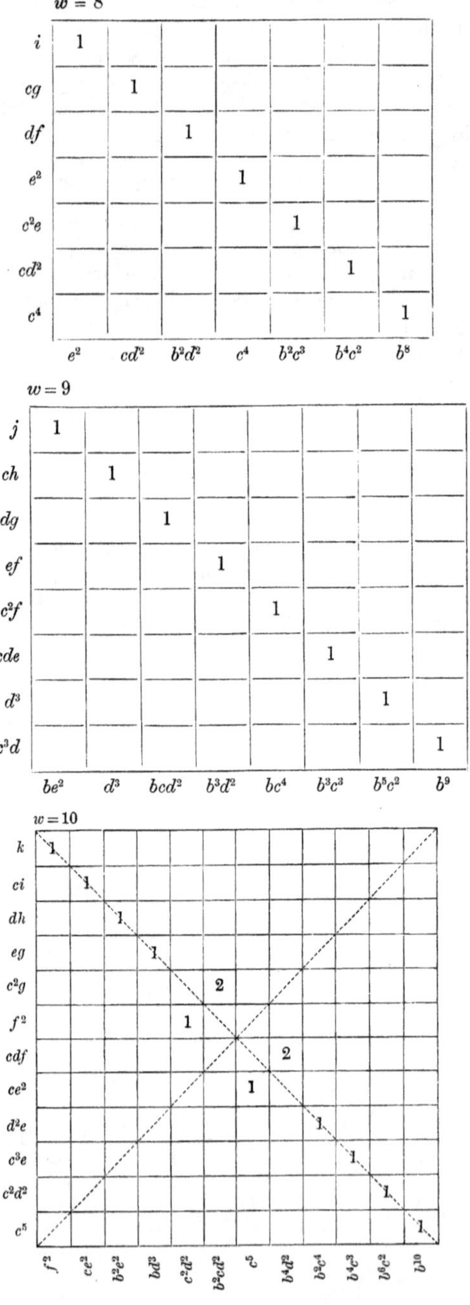

$w = 8$

	e^2	cd^2	b^2d^2	c^4	b^2c^3	b^4c^2	b^8
i	1						
cg		1					
df			1				
e^2				1			
c^2e					1		
cd^2						1	
c^4							1

$w = 9$

	be^2	d^3	bcd^2	b^3d^2	bc^4	b^3c^3	b^5c^2	b^9
j	1							
ch		1						
dg			1					
ef				1				
c^2f					1			
cde						1		
d^3							1	
c^3d								1

$w = 10$

	f^2	ce^2	b^2e^2	bd^3	c^2d^2	b^2cd^2	c^5	b^4d^2	b^2c^4	b^4c^3	b^6c^2	b^{10}
k	1											
ci		1										
dh			1									
eg				1								
c^2g						2						
f^2					1							
cdf								2				
ce^2							1					
d^2e									1			
c^3e										1		
c^2d^2											1	
c^5												1

The subsequent squares $w=11$ to 16 are, for convenience, given at the end of the present memoir (pp. 331 et seq.).

44. It is to be observed that in each square the outside left-hand terms are the non-unitaries in CO and the outside bottom terms are the power-enders in AO. I have inside each square written down only the significant numbers, but we might fill up the whole square. For instance, when $w=7$, the filled-up square would be

h	1	2	3	4
cf	0	1	2	3
de	-1	0	1	2
c^2d	0	0	0	1
	bd^2	bc^3	b^3c^2	b^7

where in the first column the numbers relate to the sets $h \infty bd^2$, $cf \infty bd^2$, $de \infty bd^2$ and $c^2d \infty bd^2$ (this last set $c^2d \infty bd^2$ is non-existent since c^2d is in AO inferior to bd^2, i.e. as well for the set as for the diminished set, number of terms is $=0$, and we have for the compartment $0-0$, $=0$). And similarly for the remaining three columns. The process of thus filling up the whole square is a direct and non-tentative one, and the conclusions to which the numbers lead are as follows: col. 1, the final being bd^2, the initial cannot be c^2d, de or cf, but taking it to be h, we have the seminvariant $h \infty bd^2$. Col. 2, the final being bc^3 the initial cannot be c^2d or de, but taking it to be cf we have the seminvariant $cf \infty bc^3$: it may be added that the top number 2 shows that there are two seminvariants $h \infty bc^3$, these are of course the foregoing ones $h \infty bd^2$ and $cf \infty bc^3$. Similarly, col. 3, the final being b^3c^2, the initial cannot be c^2d, but taking it to be de, we have the seminvariant $de \infty b^3c^2$, and col. 4, we have the seminvariant $c^2d \infty b^7$.

For the several weights up to 9, we have simply units in the dexter diagonal of each square, viz. the non-unitaries in CO correspond to the power-enders in AO, or the sharp seminvariants are $c \infty b^2$, $d \infty b^3$, &c. See *post*, Table of Reductions, No. 62, which exhibits these correspondences.

45. For the weight 10, we have deviations: the figures 1 and 2 denote as follows:

$$
\begin{array}{rlll}
1-D & k & \infty f^2 & =1 \\
 & ci & ,,\ ce^2 & ,,\ 1 \\
 & dh & ,,\ b^2e^2 & ,,\ 1 \\
 & eg & ,,\ bd^3 & ,,\ 1 \\
 & f^2 & ,,\ c^2d^2 & ,,\ 1 \\
 & c^2g & ,,\ b^2cd^2 & ,,\ 2 \\
 & ce^2 & ,,\ c^5 & ,,\ 1 \\
 & cdf & ,,\ b^4d^2 & ,,\ 2 \\
 & d^2e & ,,\ b^2c^4 & ,,\ 1 \\
 & c^3e & ,,\ b^4c^3 & ,,\ 1 \\
 & c^2d^2 & ,,\ b^6c^2 & ,,\ 1 \\
 & c^5 & ,,\ b^{10} & ,,\ 1,
\end{array}
$$

and they indicate the sharp seminvariants $k \infty f^2$, $ci \infty ce^2$, &c.: where observe that the power-enders being in AO as before, the non-unitaries are not in CO, but we have inversions (c^2g, f^2) and (cdf, ce^2).

In particular, $(1-D)(f^2 \infty c^2d^2) = 1$ indicates the seminvariant $f^2 \infty c^2d^2$;

$$(1-D)(c^2g \infty b^2cd^2) = 2,$$

means in the first instance that there are 2 seminvariants $c^2g \infty b^2cd^2$, but here the set $c^2g \infty b^2cd^2$ includes as part of itself the set $f^2 \infty c^2d^2$; so that, if $c^2g \infty b^2cd^2$ is used to denote any particular form, then the general form is $c^2g \infty b^2cd^2$ plus an arbitrary multiple of $f^2 \infty c^2d^2$, and we have thus virtually a single form $c^2g \infty b^2cd^2$. And similarly, the set $cdf \infty b^4d^2$ includes as part of itself the set $ce^2 \infty c^5$; and thus the general form $cdf \infty b^4d^2$ is = particular form plus an arbitrary multiple of $ce^2 \infty c^5$, or we have virtually a single form $cdf \infty b^4d^2$.

I remark that it would be allowable to take as a standard form of $c^2g \infty b^2cd^2$, a form not containing any term in f^2, and similarly for the standard form of $cdf \infty b^4d^2$ a form not containing any term in ce^2; but this is not done in the tables.

46. The diagram for weight 10 is constructed by the following calculation; viz. in col. 1 we calculate $(1-D)(k \infty f^2)$ and for this purpose write down the terms of $k \infty f^2$, and $D(k \infty f^2)$ in CO: in col. 2 we calculate $(1-D)(ci \infty ce^2)$, and for this purpose write down the terms of $k \infty ce^2$ and $D(k \infty ce^2)$ in CO, the terms of $ci \infty ce^2$ and $D(ci \infty ce^2)$ being thence found by rejecting the terms k, bj and the term j at the head of the two halves of the column. So in col. 3 we calculate $(1-D)(dh \infty b^2e^2)$, and for this purpose write down the terms of $(k \infty b^2e^2)$ and $D(k \infty b^2e^2)$ in CO, and for $dh \infty b^2e^2$ and $D(dh - b^2e^2)$ reject the terms k, bj, ci, b^2i and j, bi at the head of the two halves of the column. And so for the remaining columns. It is to be remarked that there is in each successive column a continually increasing number of terms to be rejected; by a properly devised variation of the algorithm it would have been possible to avoid writing down these terms at all, but for greater clearness I have inserted them.

Column	Left list	Right list	Difference	Result
1	k, bj, ci, dh, eg, f^2	j, bi, ch, dg, ef	$6-5=$	1
2	k, bj, ci, b^2i, dh, bch, eg, bdg, c^2g, f^2, bef, cdf, ce^2	j, bi, ch, b^2h, dg, bcg, ef, bdf, c^2f, be^2, cde	$11-10=$	1
3	k, bj, ci, b^2i, dh, bch, b^3h, eg, bdg, c^2g, b^2cg, f^2, bef, cdf, b^2df, ce^2, d^2e, b^2e^2	j, bi, ch, b^2h, dg, bcg, b^3g, ef, bdf, c^2f, b^2cf, be^2, cde, b^2de, d^3	$14-13=$	1
4	k, bj, ci, b^2i, dh, bch, b^3h, eg, bdg, c^2g, b^2cg, f^2, bef, cdf, b^2df, bc^2f, ce^2, b^2e^2, d^2e, $bcde$, bd^3	j, bi, ch, b^2h, dg, bcg, b^3g, ef, bdf, c^2f, b^2cf, be^2, cde, b^2de, bc^2e, d^3, bcd^2	$14-13=$	1
5	k, bj, ci, b^2i, dh, bch, b^3h, eg, bdg, c^2g, b^2cg, f^2, bef, cdf, b^2df, bc^2f, ce^2, b^2e^2, d^2e, $bcde$, c^3e, bd^3, c^2d^2	j, bi, ch, b^2h, dg, bcg, b^3g, ef, bdf, c^2f, b^2cf, be^2, cde, b^2de, bc^2e, d^3, bcd^2, c^3d	$12-11=$	1
6	k, bj, ci, b^2i, dh, bch, b^3h, eg, bdg, c^2g, b^2cg, b^4g, f^2, bef, cdf, b^2df, bc^2f, b^3cf, ce^2, b^2e^2, d^2e, $bcde$, b^3de, c^3e, b^2c^2e, bd^3, c^2d^2, b^2cd^2	j, bi, ch, b^2h, dg, bcg, b^3g, ef, bdf, c^2f, b^2cf, b^4f, be^2, cde, b^2de, bc^2e, b^3ce, d^3, bcd^2, b^3d^2, c^3d, b^2c^2d	$19-17=$	2
7	k, bj, ci, b^2i, dh, bch, b^3h, eg, bdg, c^2g, b^2cg, b^4g, f^2, bef, cdf, b^2df, bc^2f, b^3cf, ce^2, b^2e^2, d^2e, $bcde$, b^3de, c^3e, b^2c^2e, bd^3, c^2d^2, b^2cd^2, bc^3d, c^5	j, bi, ch, b^2h, dg, bcg, b^3g, ef, bdf, c^2f, b^2cf, b^4f, be^2, cde, b^2de, bc^2e, b^3ce, d^3, bcd^2, b^3d^2, c^3d, b^2c^2d, bc^4	$12-11=$	1
8	k, bj, ci, b^2i, dh, bch, b^3h, eg, bdg, c^2g, b^2cg, b^4g, f^2, bef, cdf, b^2df, bc^2f, b^3cf, b^5f, ce^2, b^2e^2, d^2e, $bcde$, b^3de, c^3e, b^2c^2e, b^4ce, bd^3, c^2d^2, b^2cd^2, b^4d^2, bc^3d, c^5	j, bi, ch, b^2h, dg, bcg, b^3g, ef, bdf, c^2f, b^2cf, b^4f, be^2, cde, b^2de, bc^2e, b^3ce, b^5e, d^3, bcd^2, b^3d^2, c^3d, b^2c^2d, b^4cd, bc^4	$19-17=$	2
9	k, bj, ci, b^2i, dh, bch, b^3h, eg, bdg, c^2g, b^2cg, b^4g, f^2, bef, cdf, b^2df, bc^2f, b^3cf, b^5f, ce^2, b^2e^2, d^2e, $bcde$, b^2de, c^3e, b^2c^2e, b^4ce, bd^3, c^2d^2, b^2cd^2, b^4d^2, bc^3d, b^3c^2d, c^5, b^2c^4	j, bi, ch, b^2h, dg, bcg, b^3g, ef, bdf, c^2f, b^2cf, b^4f, be^2, cde, b^2de, bc^2e, b^3ce, b^5e, d^3, bcd^2, b^3d^2, c^3d, b^2c^2d, b^4cd, bc^4, b^3c^3	$14-13=$	1
10	k, bj, ci, b^2i, dh, bch, b^3h, eg, bdg, c^2g, b^2cg, b^4g, f^2, bef, cdf, b^2df, bc^2f, b^3cf, b^5f, ce^2, b^2e^2, d^2e, $bcde$, b^3de, c^3e, b^2c^2e, b^4ce, b^6e, bd^3, c^2d^2, b^2cd^2, b^4d^2, bc^3d, b^2c^2d, b^5cd, c^5, b^2c^4, b^4c^3	j, bi, ch, b^2h, dg, bcg, b^3g, ef, bdf, c^2f, b^2cf, b^4f, be^2, cde, b^2de, bc^2e, b^3ce, b^5e, d^3, bcd^2, b^3d^2, c^3d, b^2c^2d, b^4cd, b^6d, bc^4, b^3c^3, b^5c^2	$14-13=$	1
11	k, bj, ci, b^2i, dh, bch, b^3h, eg, bdg, c^2g, b^2cg, b^4g, f^2, bef, cdf, b^2df, bc^2f, b^3cf, b^5f, ce^2, b^2e^2, d^2e, $bcde$, b^3de, c^3e, b^2c^2e, b^4ce, b^6e, bd^3, c^2d^2, b^2cd^2, b^4d^2, bc^3d, b^3c^2d, b^5cd, b^7d, c^5, b^2c^4, b^4c^3, b^6c^2	j, bi, ch, b^2h, dg, bcg, b^3g, ef, bdf, c^2f, b^2cf, b^4f, be^2, cde, b^2de, bc^2e, b^3ce, b^5e, d^3, bcd^2, b^3d^2, c^3d, b^2c^2d, b^4cd, b^6d, bc^4, b^3c^3, b^5c^2, b^7c	$11-10=$	1
12	k, bj, ci, b^2i, dh, bch, b^2h, eg, bdg, c^2g, b^2cg, b^4g, f^2, bef, cdf, b^2df, bc^2f, b^3cf, b^5f, ce^2, b^2e^2, d^2e, $bcde$, b^3de, c^3e, b^2c^2e, b^4ce, b^6e, bd^3, c^2d^2, b^2cd^2, b^4d^2, bc^3d, b^3c^2d, b^5cd, b^7d, c^5, b^2c^4, b^4c^3, b^6c^2, b^8c, b^6	j, bi, ch, b^2h, dg, bcg, b^2g, ef, bdf, c^2f, b^2cf, b^4f, be^2, cde, b^2de, bc^2e, b^3ce, b^5e, d^3, bcd^2, b^3d^2, c^3d, b^2c^2d, b^4cd, b^6d, bc^4, b^3c^3, b^5c^2, b^7c, b^9	$6-5=$	1

47. As to the first of the foregoing inversions c^2g, f^2, it is proper to remark, that filling up two compartments of the square we have

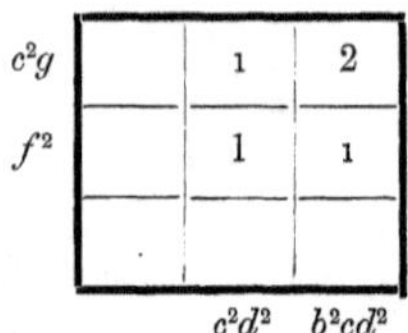

where the meaning of the numbers (1, 1) has to be considered: the first (1) seems to indicate a seminvariant $c^2g \propto c^2d^2$, but there is in fact no such form, what it really indicates is a form $0c^2g + f^2 \propto c^2d^2$, that is, $f^2 \propto c^2d^2$; and similarly, the second (1) seems to indicate a seminvariant $f^2 \propto b^2cd^2$, but there is in fact no such form, what it really indicates is $f^2 \propto c^2d^2 + 0b^2cd^2$, that is, $f^2 \propto c^2d^2$. The explanation is correct, but to make it perfectly clear some further developments would be required. The like remarks apply to the inversion cdf, ce^2.

The MacMahon Linkage. Art. Nos. 48 to 52.

48. We require the two theorems:

The first is: if a seminvariant S has q for its highest letter, then $\partial_q S$ is also a seminvariant.

The second has presented itself for unitariants (*ante* No. 31); for seminvariants the form is less simple, viz. If in any seminvariant, attending only to the terms of the highest degree, we therein change $b, c, d, e, \ldots$ into $b, 2c, 6d, 24e, \ldots$ and then diminish the letters (that is, replace each letter by the next preceding letter) and in the result so obtained change $b, c, d, e, \ldots$ into $b, \frac{c}{2}, \frac{d}{6}, \frac{e}{24}, \ldots$ we obtain a seminvariant. For instance $g - 6bf + 15ce - 10d^2$, in the terms of degree 2, making the numerical change we have $-720bf + 720ce - 360d^2$, and then diminishing the letters and making the numerical change, we obtain $-720\frac{e}{24} + 720\frac{bd}{6} - 360\frac{c^2}{4}$, that is,

$$-30(e - 4bd + 3c^2),$$

a seminvariant.

For the proof, observe that the equation $\Delta S = 0$, attending therein only to the terms of the highest degree, gives $(2b\partial_c + 3c\partial_d + \ldots)S' = 0$, if S' denote the terms of the highest degree: making the numerical change, this is $(b\partial_c + c\partial_d + \ldots)S''$, if S'' is what S' becomes thereby; diminishing the letters, this is $(\partial_b + b\partial_c + \ldots)S''' = 0$, if S'''

is the diminished value of S'', and finally making the numerical change, if T be what S''' becomes on writing therein $b, \frac{c}{2}, \frac{d}{6}, \ldots$ for $b, c, d, \ldots$, this gives

$$(\partial_b + 2b\partial_c + \ldots)\,T = 0,$$

viz. T is a seminvariant.

49. Assume that, for the weights up to a certain weight w, the forms of the sharp seminvariants are known: and for the weight w consider a seminvariant $I(\text{ca})F$: here if I be given, the first theorem establishes a limit F' such that F is in AO not higher than F'. For instance, when $w=12$, if $I=dj$, the coefficient of j as being a seminvariant can only be $d \infty b^3$, and thus the seminvariant contains a term b^3j, or the final term F must be in AO not higher than b^3j; the degree is thus $=4$ at least.

Similarly, if F be given, then the second theorem determines a limit I' such that I is in CO not lower than I'. Thus when $w=12$, as before, if $F=b^4cd^2$, then diminishing the letters we have bc^2, a term belonging to $f \infty bc^2$; the diminished form has thus terms $a^4(a^2f,\ bc^2)$, so that augmenting these the seminvariant has terms $b^4(b^2g,\ cd^2)$ and thus the initial term I is in CO not lower than b^6g.

50. A limit for I or F, when the other is given, can also in some cases be found as follows: Considering a seminvariant of the weight w as before, and denoting its extent and degree by σ and δ respectively, then we have $\sigma\delta - 2w = 0$ or positive; that is, $\sigma\delta = 2w$ at least; here given I, we have σ, and then $\delta = \frac{2w}{\sigma}$ at least; and given F we have δ, and then $\sigma = \frac{2w}{\delta}$ at least.

51. We may now explain the MacMahon linkage; for a given weight, we write down in two columns the initials or non-unitaries in CO, and the finals or power-enders in AO: by what precedes, it appears that we cannot combine the terms of the one column each with the term opposite to it in the other column; what we do is: beginning with the top of the column of initials, we combine successively each term with the highest admissible term in the column of finals: or beginning with the bottom of the column of finals, we combine successively each term with the lowest admissible term in the column of initials.

52. For the weight 12, the linkage is

read downwards.

shown by	not in AO higher than	Initial	Final	lower than not in CO	shown by
		m	g^2	bl	$k \infty f^2$
$(c \infty b^2)\, k$	b^2k	ck	cf^2	b^2k	j „ be^2
$(d$ „ $b^3)\, j$	b^3j	dj	e^3	bdi	ch „ d^3
$(e$ „ $c^2)\, i$	c^2i	ei	b^2f^2	b^3j	i „ e^2
$(c^2$ „ $b^4)\, i$	b^4i	c^2i	bde^2	b^2dh	cg „ cd^2
$(f$ „ $bc^2)\, h$	bc^2h	fh	c^2e^2	b^2eg	df „ b^2d^2
$(cd$ „ $b^5)\, h$	b^5h	cdh	d^4	b^2f^2	e^2 „ c^4
$(g$ „ $d^2)\, g$	d^2g	g^2	b^2ce^2	b^4i	h „ bd^2
$(ce$ „ $c^3)\, g$	c^3g	ceg	bcd^3	b^3dg	cf „ bc^3
$(d^2$ „ $b^2c^2)\, g$	b^2c^2g	d^2g	c^3d^2	b^3ef	de „ b^3
$(c^3$ „ $b^6)\, g$	b^6g	c^3g	b^4e^2	b^5h	g „ d^2
$(cf$ „ $bc^3)\, f$	bc^3f	cf^2	b^3d^3	b^4df	ce „ c^3
$(de$ „ $b^3c^2)\, f$	b^3c^2f	def	$b^2c^2d^2$	b^4e^2	d^2 „ b^2c^2
$(c^2d$ „ $b^7)\, f$	b^7f	c^2df	c^6	b^3d^3	c^3 „ b^6
$(e^2$ „ $c^4)\, e$	c^4e	e^3	b^4cd^2	b^6g	f „ bc^2
$(c^2e$ „ $b^2c^3)\, e$	b^2c^3e	c^2e^2	b^2c^5	b^5de	cd „ b^5
$(cd^2$ „ $b^4c^2)\, e$	b^4c^2e	cd^2e	b^6d^2	b^7f	e „ c^2
$(c^4$ „ $b^8)\, e$	b^8e	c^4e	b^4c^4	b^6d^2	c^2 „ b^4
$(d^3$ „ $b^5c^2)\, d$	b^5c^2d	d^4	b^6c^3	b^8e	d „ b^3
$(c^2d$ „ $b^9)\, d$	b^9d	c^3d^2	b^8c^2	b^9d	c „ b^2
$(c^5$ „ $b^{10})\, c$	$b^{10}c$	c^6	b^{12}		

read upwards.

Thus, beginning at the top of the column of initials, m is to be linked with g^2, that is, we have $(m \infty g^2)$; ck with cf^2, that is, we have $(ck \infty cf^2)$; dj cannot be linked with e^3, for the final must be in AO not higher than b^3j, but it is linked with the highest term b^2f^2 for which this condition is satisfied, that is, we have $(dj \infty b^2f^2)$; ei is then linked with the highest admissible term e^3, that is, we have $(ei \infty e^3)$; and so on.

Or beginning at the bottom of the column of finals, b^{12} is linked with c^6, that is, we have $(c^6 \propto b^{12})$, b^8c^2 with c^3d^2, that is, we have $(c^3d^2 \propto b^8c^2)$; b^6c^3 cannot be linked with d^4, for the initial must be in CO not lower than b^8e, but it is linked with the lowest term c^4e for which this condition is satisfied, that is, we have $(c^4e \propto b^6c^3)$; and so on.

The Umbral Notation. Stroh's Theory. Art. Nos. 53 to 56.

53. Employing the umbræ α, β, γ, δ, ..., which are such that

$$\alpha = \beta = \gamma, \ \ldots, = b; \quad \alpha^2 = \beta^2 = \gamma^2, \ \ldots, = c; \quad \alpha^3 = \beta^3 = \gamma^3, \ \ldots, = d;$$

and so on, then for instance

$$(\alpha - \beta)^2 = \alpha^2 - 2\alpha\beta + \beta^2, \quad = c - 2b^2 + c, \quad = 2(c - b^2),$$

a seminvariant;

$$(\alpha - \beta)^2 (\alpha - \gamma) = \alpha^3 - 2\alpha^2\beta + \alpha\beta^2 - \alpha^2\gamma + 2\alpha\beta\gamma - \beta^2\gamma,$$

$$= d - 2bc + bc - bc + 2b^3 - bc, \quad = d - 3bc + 2b^3,$$

a seminvariant: and so in general any rational and integral function of the differences of the umbræ developed and interpreted is a seminvariant. For the seminvariants of a given weight, e.g. $w = 6$, Dr Stroh* considers the function

$$\Omega^6 = (\alpha x + \beta y + \gamma z + \delta w + \epsilon t + \zeta u)^6,$$

where x, y, z, w, t, u are numbers the sum of which is $= 0$, or we may if we please have more than 6 such numbers: the expression is obviously a function of the differences of the umbræ and it is thus a seminvariant. To develop its value, observe that after expansion of the sixth power we have sets of similar terms, for instance $\alpha^6x^6 + \beta^6y^6 + \ldots$ which putting therein $\alpha^6 = \beta^6 = \gamma^6, \ \ldots = g$ become $= g \,.\, Sx^6$, and generally each set becomes equal to a literal term multiplied by a symmetric function of the x, y, z, w, ...; introducing capital letters to denote the elementary symmetric functions of these quantities, and recollecting that their sum is assumed to be $= 0$, say we have

$$1 + C\vartheta^2 + D\vartheta^3 + E\vartheta^4 + \ldots = 1 - x\vartheta \,.\, 1 - y\vartheta \,.\, 1 - z\vartheta \ldots,$$

(that is, $0 = Sx$, $+C = Sxy$, $-D = Sxyz$, &c.) then by aid of the Table VI (b) writing therein 0, C, D, E, F, G for b, c, d, e, f, g, we find

$$\Omega^6 = (\alpha x + \beta y + \gamma z \ldots)^6 = \alpha^6 Sx^6 + 6\alpha^5\beta Sx^5y + \text{\&c.},$$

* See the paper "Ueber die Symbolische Darstellung der Grundsyzyganten einer binären Form sechster Ordnung und eine Erweiterung der Symbolik von Clebsch," *Math. Ann.* t. XXXVI. (1890), pp. 262—303, in particular § 10, Das Formensystem einer Form unbegrenzt hoher Ordnung.

as shown in the following table:

				C^3	D^2	CE	G
1	g	Sx^6	=	-2	$+3$	$+6$	-6
$+6$	bf	Sx^5y	=	$+2$	-3	-6	$+6$
$+15$	ce	Sx^4y^2	=	-2	-3	$+2$	$+6$
$+20$	d^2	Sx^3y^3	=	$+1$	$+3$	-3	$+3$
$+30$	b^2e	Sx^4yz	=		$+3$	$+2$	-6
$+60$	bcd	Sx^3y^2z	=		-3	$+4$	-12
$+90$	c^3	$Sx^2y^2z^2$	=		$+1$	-2	-2
$+120$	b^3d	Sx^3yzw	=			-2	$+6$
$+180$	b^2c^2	Sx^2y^2zw	=			$+1$	$+9$
$+360$	b^4c	Sx^2yzwt	=				-6
$+720$	b^6	$Sxyzwtu$	=				$+1$
				$[d^2]$	$[c^3]$	$[b^2c^2]$	$[b^6]$

the numbers whereof are, it will be observed, identical with those of the foregoing table No. 33, relating to the MacMahon equation.

This is to be read

$$\Omega^6 = C^3[d^2] + D^2[c^3] + CE[b^2c^2] + G[b^6],$$

viz. Ω^6 is a linear function of C^3, D^2, CE and G, the coefficients of these, being given functions of (b, c, d, e, f, g), which given functions are the specific blunt seminvariants which have been already called $[d^2]$, $[c^3]$, $[b^2c^2]$ and $[b^6]$. And so in general, the developed value of Ω^w affords a complete definition of these specific blunt seminvariants of the weight w. Observe that $\alpha, \beta, \gamma, \delta, \ldots$ are umbræ in nowise connected with the roots $\alpha, \beta, \gamma, \delta, \ldots$ before made use of, and that $B, C, D, \ldots$ are actual quantities in nowise connected with the symbolic capitals $B, C, D, \ldots$ before made use of.

54. The capital and small letter symbols are conjugate to each other. It will be convenient to give here, in reference to subsequent investigations, a table of these conjugate forms up to the degree 6 and weight 15.

Table of Conjugates.

2	3	4	5	6	7	8	9	10	11	12	13	14	15
C b^2	D b^3	C^2 c^2	CD bc^2	C^3 d^2	C^2D bd^2	C^4 e^2	C^3D be^2	C^5 f^2	C^4D bf^2	C^6 g^2	C^5D bg^2	C^7 h^2	C^6D bh^2
		E b^4	F b^5	D^2 c^3	DE bc^3	CD^2 cd^2	D^3 d^3	C^2D^2 ce^2	CD^3 de^2	C^3D^2 cf^2	C^2D^3 df^2	C^4D^2 cg^2	C^3D^3 dg^2
				CE b^2c^2	CF b^3c^2	C^2E b^2d^2	CDE bcd^2	C^3E b^2e^2	C^2DE bce^2	D^4 e^3	C^3DE bcf^2	CD^4 ef^2	D^5 f^3
				G b^6		E^2 c^4	C^2F b^3d^2	D^2E bd^3	DE^2 cd^3	C^4E b^2f^2	D^3E be^3	C^5E b^2g^2	C^4DE bcg^2
						DF b^2c^3	EF bc^4	CE^2 c^2d^2	C^2F b^3e^2	CD^2E bde^2	CDE^2 cde^2	C^2D^2E bdf	CD^3E bef^2
						CG b^4c^2	DG b^3c^3	CDF b^2cd^2	D^2F b^2d^3	C^2E^2 c^2e^2	C^4F b^3f^2	C^3E^2 c^2f^2	C^2DE^2 cdf^2
								F^2 c^5	CEF bc^2d^2	E^3 d^4	CD^2F b^2de^2	D^2E^2 ce^3	DE^3 de^3
								C^2G b^4d^2	CDG b^3cd^2	C^2DF b^2ce^2	C^2EF bc^2e^2	CE^3 d^2e^2	C^5F b^3g^2
								EG b^2c^4	FG bc^5	DEF bcd^3	E^2F bd^4	C^3DF b^2cf^2	C^2D^2F b^2df^2
										CF^2 c^3d^2	DF^2 c^2d^3	D^3F b^2e^3	C^3EF bc^2f^2
										C^3G b^4e^2	C^2DG b^3ce^2	$CDEF$ $bcde^2$	D^2EF bce^3
										D^2G b^3d^3	DEG b^2cd^3	C^2F^2 c^3e^2	CE^2F bd^2e^2
										CEG $b^2c^2d^2$	CFG bc^3d^2	EF^2 cd^4	CDF^2 c^2de^2
										G^2 c^6		C^4G b^4f^2	F^3 d^5
												CD^2G b^3de^2	C^3DG b^3cf^2
												C^2EG $b^2c^2e^2$	D^3G b^3e^3
												E^2G b^2d^4	$CDEG$ b^2cde^2
												DFG bc^2d^3	C^2FG bc^3e^2
												CG^2 c^4d^2	EFG cd^4
													DG^2 c^3d^3

55. We can, by means of the umbral notation, write down for the blunt seminvariants of a given weight (indefinite forms, not the above-mentioned specific forms) expressions far more simple than those which are given by the foregoing theories: we can, in fact, find without difficulty *monomial* umbral expressions; and in many cases obtain also the sharp forms. To illustrate this, I consider the weight 10: I write down for convenience the symbols of the sharp forms (though the knowledge of these is in nowise required) and I form a table as follows:

Sharp forms, finals in AO.			
k	∞f^2	1	$(\alpha-\beta)^{10}$
ci	„ ce^2	2	$(\alpha-\beta)^8(\alpha-\gamma)^2$
dh	„ b^2e^2	3	$(\alpha-\beta)^8(\alpha-\gamma)(\alpha-\delta)$
eg	„ bd^3	4	$(\alpha-\beta)^6(\alpha-\gamma)^3(\alpha-\delta)$
f^2	„ c^2d^2	5	$(\alpha-\beta)^6(\alpha-\gamma)^2(\alpha-\delta)^2$
c^2g	„ b^2cd^2	6	$(\alpha-\beta)^6(\alpha-\gamma)^2(\alpha-\delta)(\alpha-\epsilon)$
ce^2	„ c^5	7	$(\alpha-\beta)^4(\alpha-\gamma)^2(\alpha-\delta)^2(\alpha-\epsilon)^2$
cdf	„ b^4d^2	8	$(\alpha-\beta)^6(\alpha-\gamma)(\alpha-\delta)(\alpha-\epsilon)(\alpha-\zeta)$
d^2e	„ b^2c^4	9	$(\alpha-\beta)^4(\alpha-\gamma)^2(\alpha-\delta)^2(\alpha-\epsilon)(\alpha-\zeta)$
c^3e	„ b^4c^3	10	$(\alpha-\beta)^4(\alpha-\gamma)^2(\alpha-\delta)(\alpha-\epsilon)(\alpha-\zeta)(\alpha-\eta)$
c^2d^2	„ b^6c^2	11	$(\alpha-\beta)^4(\alpha-\gamma)(\alpha-\delta)(\alpha-\epsilon)(\alpha-\zeta)(\alpha-\eta)(\alpha-\theta)$
c^5	„ b^{10}	12	$(\alpha-\beta)^2(\alpha-\gamma)(\alpha-\delta)(\alpha-\epsilon)(\alpha-\zeta)(\alpha-\eta)(\alpha-\theta)(\alpha-\iota)(\alpha-\kappa)$

It will be observed that all the differences used are $\alpha-\beta$, $\alpha-\gamma$, ... containing each of them an α; hence in all the forms we have α^{10}, $=k$; in $(\alpha-\beta)^{10}$, the lowest term (in AO) is $\alpha^5\beta^5$, $=f^2$; in $(\alpha-\beta)^8(\alpha-\gamma)^2$, the lowest term is $\alpha^4\beta^4.\gamma^2$, $=ce^2$; and so on, viz. in each case the lowest term is the final term of the sharp form set down in the same line.

56. The form $(\alpha-\beta)^{10}$ gives at once the sharp form $k\infty f^2$; we thus develop it:

α^{10}	$\alpha^9\beta$	$\alpha^8\beta^2$	$\alpha^7\beta^3$	$\alpha^6\beta^4$	$\alpha^5\beta^5$
β^{10}	$\alpha\beta^9$	$\alpha^2\beta^8$	$\alpha^3\beta^7$	$\alpha^4\beta^6$	
1	-10	$+45$	-120	$+210$	-252
$+1$	-10	$+45$	-120	$+210$	
$=2(k$	$-10bj$	$+45ci$	$-120dh$	$+210eg$	$-126f^2)$;

$(\alpha-\beta)^8(\alpha-\gamma)^2$ contains a term α^{10}, $=k$ and thus gives a blunt form $k\text{aoc}e^2$; if instead of it we employ the form $(\alpha-\beta)^8(\alpha-\gamma)(\beta-\gamma)$, then here as before the lowest term is $\alpha^4\beta^4.\gamma^2$, $=ce^2$, but there is no term α^{10}: there is a term $\alpha^9\beta$, $=bj$, but as this cannot appear, we must have terms of this form destroying each other. The simplest mode of effecting the development is to write

$$(\alpha-\beta)^8(\alpha-\gamma)(\beta-\gamma)=(\alpha-\beta)^8\{\alpha\beta-\gamma(\alpha+\beta)+\gamma^2\}:$$

we may herein put at once $\gamma=b$, $\gamma^2=c$, and thus the form is

$$(\alpha-\beta)^8\{\alpha\beta-b(\alpha+\beta)+c\};$$

I develop thus:

$(\alpha-\beta)^8$	1,	-8,	$+28$,	-56,	$+70$,	-56,	$+28$,	-8,	$+1$,	
		$+1$,	-8,	$+28$,	-56,	$+70$,	-56,	$+28$,	-8,	$+1$
$(\alpha-\beta)^8(\alpha+\beta)$	1,	-7,	$+20$,	-28,	$+14$,	$+14$,	-28,	$+20$,	-7,	$+1$

$$\begin{array}{l} 1bj-8ci+28dh-56eg+70f^2 \\ +1\quad -8\quad +28\quad -56 \end{array}$$

$$-b\begin{pmatrix} 1j & -7bi & +20ch & -28dg & +14ef \\ +1 & -7 & +20 & -28 & +14 \end{pmatrix}$$

$$+c\begin{pmatrix} 1i & -8bh & +28cg & -56df & +70e^2 \\ +1 & -8 & +28 & -56 & \end{pmatrix}$$

					$\div -14$
k					
bj	$+2$	-2		0	
ci	-16		$+2$	-14	$+1$
dh	$+56$			$+56$	-4
eg	-112			-112	$+8$
f^2	$+70$			$+70$	-5
b^2i		$+14$		$+14$	-1
bch		-40	-16	-56	$+4$
bdg		$+56$		$+56$	-4
bef		-28		-28	$+2$
c^2g			$+56$	$+56$	-4
cdf			-112	-112	$+8$
ce^2			$+70$	$+70$	-5
					± 23,

which, in fact, exhibits the calculation of the sharp form $ci \infty ce^2$. The disappearance of the term in bj will be noticed.

Instead of $(\alpha-\beta)^8(\alpha-\gamma)(\beta-\delta)$ which contains α^{10}, that is, k, we may take

$$(\alpha-\beta)^8(\gamma-\delta)^2,$$

that is,

$$(i-8bh+28cg-56df+35e^2)(c-b^2):$$

this is $ciaob^2e^2$, a blunt form; by subtracting from it $ci \infty ce^2$, we could obtain the next sharp form $dh \infty b^2e^2$; but this in passing; it does not appear that there is any monomial umbral expression for the last-mentioned form.

I do not stop to examine the next following forms, but pass on at once to the last of them; instead of the expression given, we may take the expression

$$(\alpha-\beta)^2(\gamma-\delta)^2(\epsilon-\zeta)^2(\eta-\theta)^2(\iota-\kappa)^2,$$

that is, $(c-b^2)^5$, which is in fact the sharp form $c^5 \infty b^{10}$.

Seminvariants of a given Degree: Generating Functions. Art. Nos. 57 to 59.

57. We may consider the seminvariants of a given degree, and arrange them according to their weights: thus in each case writing down the series of finals, and for a reason that will appear also the conjugates of these finals (see Table of Conjugates, *ante* No. 54).

For degree 2, or quadric seminvariants, we have

2	3	4	5	6 ...
C, b^2	–	C^2, c^2	–	C^3, d^2

there is here for every even weight (beginning with 2) a single form, and for every odd weight no form: the number of forms of the weight w is thus $=$ coeff. of x^w in $x^2 \div (1-x^2)$, or writing for shortness **2** to denote $1-x^2$ (and similarly **3**, **4**, ... to denote $1-x^3$, $1-x^4$, ...), say that for degree 2, Generating Function, *G.F.*, is $= x^2 \div \mathbf{2}$.

For degree 3, or cubic seminvariants, we have

3	4	5	6	7 ...
D, b^3	–	CD, bc^2	D^2, c^3	C^2D, bc^2

the counting is most easily effected by means of the conjugate forms; these contain all of them the factor D, and omitting this factor we have all the combinations of C, D which make up the weight $w-3$, viz. for weight w, we have number of ways in which $w-3$ can be made up with the parts 2, 3: that is,

for degree 3, *G.F.* is $= x^3 \div \mathbf{2}.\mathbf{3}$.

Similarly for degree **4** or quartic seminvariants, we have terms each containing E, and removing this factor, we have all the combinations of C, D, E which make up the weight $w-4$, viz.

for degree 4, *G.F.* is $= x^4 \div \mathbf{2}.\mathbf{3}.\mathbf{4}$.

Thus for degrees

$$2, \quad 3, \quad 4, \quad 5, \quad 6, \quad \ldots$$

the $G.F.$'s are

$$= x^2 \div 2, \quad x^3 \div 2.3, \quad x^4 \div 2.3.4, \quad x^5 \div 2.3.4.5, \quad x^6 \div 2.3.4.5.6, \ldots$$

58. We may analyse these results by separating the finals into classes. I use the expression $b, c, d, \ldots$ are discrete letters, meaning thereby that they are distinct letters, not of necessity consecutive but with any intervals between them. Thus deg. 3, if (b, c) are discrete letters, then the finals are b^3, and bc^2; deg. 4, if b, c, d are discrete letters, then the finals are b^4, bc^3, b^2c^2, and bcd^2; and so on, the number of classes being doubled at each step, as will presently appear for the weights 5 and 6 respectively.

I notice also a property of the conjugates of these classes; for b^3 and bc^2 themselves the conjugates are D and CD, and these occur as factors, D in the conjugate of every form of the class b^3 (for instance conjugates of c^3, d^3 are D^2, D^3), and CD in the conjugate of every form of the class bc^2 (for instance, the conjugates of bd^2, ce^2 are C^2D, C^2D^2); and the like in other cases, viz. for any class whatever the conjugate of the first or representative form occurs as a factor in the conjugates of the several other forms belonging to the same class.

59. With these explanations, the expressions for the several $G.F.$'s are obtained without difficulty, and we have

deg. 2,	class C, b^2	$G.F.$	$= x^2 \div 2$,
deg. 3,	„ D, b^3	„	$= x^3 \div 3$,
	„ CD, bc^2	„	$= x^5 \div 2.3$;

we ought here to have

$$\begin{array}{llll} x^3 \div 2.3 = & x^3 \div 3 + x^5 \div 2.3, \text{ viz. in verification} \\ \hline x^3 & = x^3.2 = x^3 - x^5 \\ & + x^5 \qquad + x^5 \\ & \qquad = x^3; \end{array}$$

deg. 4,	class E, b^4	$G.F.$	$= x^4 \div 4$,
	„ DE, bc^3	„	$= x^7 \div 3.4$,
	„ CE, b^2c^2	„	$= x^6 \div 2.4$,
	„ CDE, bcd^2	„	$= x^9 \div 2.3.4$;

we ought here to have

$$\begin{array}{lll} x^4 \div 2.3.4 = & x^4 \div 4 + x^7 \div 3.4 + x^6 \div 2.4 + x^9 \div 2.3.4, \text{ viz. in verification} \\ \hline x^4 & = x^4.2.3 = x^4 - x^6 - x^7 + x^9 \\ & + x^7.2 \qquad\qquad + x^7 - x^9 \\ & + x^6.3 \qquad + x^6 \qquad - x^9 \\ & + x^9 \qquad\qquad\qquad + x^9 \\ & \qquad\qquad = x^4; \end{array}$$

deg. 5, class	F,	b^5	$G.F. = x^5 \div 5$,
	EF,	bc^4	$x^9 \div 4.5$,
	DF,	b^2c^3	$x^8 \div 3.5$,
	CF,	b^3c^2	$x^7 \div 2.5$,
	DEF,	bcd^3	$x^{12} \div 3.4.5$,
	CEF,	bc^2d^2	$x^{11} \div 2.4.5$,
	CDF,	b^2cd^2	$x^{10} \div 2.3.5$,
	$CDEF$,	$bcde^2$	$x^{14} \div 2.3.4.5$;

and for the sum of the eight terms

$$G.F. = x^5 \div 2.3.4.5,$$

which may be verified as before.

deg. 6, class	G,	b^6	$G.F. = x^6 \div 6$,
	FG,	bc^5	$x^{11} \div 5.6$,
	EG,	b^2c^4	$x^{10} \div 4.6$,
	DG,	b^3c^3	$x^9 \div 3.6$,
	CG,	b^4c^2	$x^8 \div 2.6$,
	EFG,	bcd^4	$x^{15} \div 4.5.6$,
	DFG,	bc^2d^3	$x^{14} \div 3.5.6$,
	CFG,	bc^3d^2	$x^{13} \div 2.5.6$,
	DEG,	b^2cd^3	$x^{13} \div 3.4.6$,
	CEG,	$b^2c^2d^2$	$x^{12} \div 2.4.6$,
	CDG,	b^3cd^2	$x^{11} \div 2.3.6$,
	$DEFG$,	$bcde^3$	$x^{18} \div 3.4.5.6$,
	$CEFG$,	bcd^2e^2	$x^{17} \div 2.4.5.6$,
	$CDFG$,	bc^2de^2	$x^{16} \div 2.3.5.6$,
	$CDEG$,	b^2cde^2	$x^{15} \div 2.3.4.6$,
	$CDEFG$,	$bcdef^2$	$x^{20} \div 2.3.4.5.6$;

and for the sum of the sixteen terms

$$G.F. = x^6 \div 2.3.4.5.6,$$

which may be verified as before.

Reducible Seminvariants—Perpetuants. Art. Nos. 60 to 64.

60. Seminvariants of the degrees 2 and 3 are irreducible—or say they are perpetuants. Hence by what precedes, as regards perpetuants

$$\text{for degree } 2,\ G.F. = x^2 \div 2;$$
$$\text{for degree } 3,\ G.F. = x^3 \div 2.3.$$

For the degree 4 (if as before b, c, d denote discrete letters), then the finals are b^4, bc^3, b^2c^2 and bcd^2. For a final $b^4 = b^2.b^2$ or $b^2c^2 = b^2.c^2$, we have evidently a product of two quadric seminvariants ending in b^2 and b^2, or in b^2 and c^2, with the same final term as the quartic seminvariant; so that, considering the quartic seminvariants arranged with their finals in AO, and adding to such quartic seminvariant a proper numerical multiple of the product in question, we obtain a quartic seminvariant the final term whereof is in AO higher than the original final term b^4 or b^2c^2, and such quartic seminvariant is thus said to be reducible; a quartic seminvariant not thus reducible is a perpetuant. The quartic perpetuants are consequently those which end in bc^3 or bcd^2. The lowest form is that ending in bc^3, of the weight 7. Taking the sum of the $G.F.$'s for the forms bc^3 and bcd^2 respectively, the $G.F.$ for quartic perpetuants is

$$x^7 \div \mathbf{3.4} + x^9 \div \mathbf{2.3.4},$$

viz. this is

$$x^7(1-x^2) + x^9 \div \mathbf{2.3.4},$$

or finally

$$G.F. = x^7 \div \mathbf{2.3.4}.$$

As an instance of a reduction, we have

$$(d^2 \infty b^2c^2) - (c \infty b^2)(e \infty c^2) = (ce \infty c^3),$$

viz. this is

$$(d \infty b^2c^2) = (c - b^2)(e - 4bd + 3c^2) - (ce - d^2 - b^2e + 2bcd - c^3).$$

We have also

$$(d^2 \infty b^2c^2) = (d \infty b^3)^2 + 4(c \infty b^2)^3,$$

viz.

$$(d \infty b^2c^2) = (d - 3bc + 2b^3)^2 + 4(c - b^2)^3,$$

but this is *not* a reduction, there are on the right-hand side terms of the degree 6, which is higher than the degree of the seminvariant $d^2 \infty b^2c^2$. In general, we say that a seminvariant of any given degree is reducible when we can, by adding to it products *of its own degree* of seminvariants of inferior degrees, reduce it to a seminvariant the final of which is in AO higher than the original final.

61. For the degree 5 (taking b, c, d, e to denote discrete letters), if the final be b^5, bc^4, b^2c^3, b^3c^2, bc^2d^2 or b^2cd^2, then the seminvariant will be reducible; a perpetuant must have therefore a final bcd^3 or $bcde^2$. But it is not true that every quintic seminvariant with either of these finals is a perpetuant. To explain this, observe that the first mentioned six finals are some of them in one way only, some of them in two ways, expressible as a product of power-enders, or say they are singly, or else doubly, composite: viz. we have

$$b^5 = b^2 . b^3; \quad bc^4 = c^2 . bc^2; \quad b^2c^3 = b^2 . c^3; \quad b^3c^2 = c^2 . b^3 = b^2 . bc^2;$$

$$bc^2d^2 = c^2 . bd^2 = d^2 . bc^2; \quad b^2cd^2 = b^2 . cd^2.$$

For a doubly composite form, for instance b^3c^2, forming first the product of the quadric and cubic seminvariants ending in c^2, b^3 respectively, and secondly the product of the quadric and cubic seminvariants ending in b^2 and bc^2 respectively, we have two products each with the final b^3c^2, and forming a linear combination so as to eliminate this term b^3c^2, we have thus it may be a quintic seminvariant with a final such as bcd^3 or $bcde^2$, and the process then furnishes a reduction of such a quintic seminvariant. Or on the other hand, it may be that the finals of the degree 5 all of them disappear, and we have a relation between products of the form in question (i.e. of a quadric and a cubic seminvariant) and seminvariants of a degree inferior to 5, say this is a quintic syzygy.

In particular, a non-composite final first presents itself for the weight **12**, viz. here the finals are b^2ce^2, bcd^3, c^3d^2, the last of these is doubly composite, and it furnishes a reduction of bcd^3. For the weight 13, the finals are b^3f^2, b^2de^2, bc^2e^2, bd^4, c^2d^3 which are each of them singly or doubly composite: for the weight 14, they are b^2cf^2, b^2c^3, $bcde^2$, c^3e^2 and cd^4, and here the doubly composite form furnishes a reduction of $bcde^2$. For the weight 15, we have a final bce^3 which gives a quintic perpetuant. I have, in fact, in my paper "A Memoir on Seminvariants," *American Journal of Mathematics*, vol. VII. (1885), pp. 1—25, [828], worked out the theory of quintic syzygies and perpetuants, and subsequently connecting this with the present theory of finals, I succeeded in showing that, when the doubly composite final contains a b, then there is not a reduction but a syzygy; we thus have

$$G.F. \text{ for finals } b^3c^2,\ b^3d^2,\ \ldots = x^7 \div \mathbf{2},$$

$$\text{,, } \quad \text{,, } \quad bc^2d^2,\ \ldots = x^{11} \div \mathbf{2}\,.\,\mathbf{4},$$

whence for the two forms

$$G.F. \text{ is } x^7 \div \mathbf{2} + x^{11} \div \mathbf{2}\,.\,\mathbf{4} = \{x^7(1-x^4) + x^{11}\} \div \mathbf{2}\,.\,\mathbf{4},$$

or say for S_5, the number of quintic syzygies $G.F.$ is $= x^7 \div \mathbf{2}\,.\,\mathbf{4}$.

I further satisfied myself that the finals for the quintic perpetuants are $bc0e^3$, and $bc0ef^2$, viz. the b, c, e, f being discrete letters, the interposed 0 denotes that the c and e are not consecutive letters. The conjugates of these forms contain the factors D^2EF and CD^2EF respectively, and it hence appears that the $G.F.$'s are $= x^{15} \div \mathbf{3}\,.\,\mathbf{4}\,.\,\mathbf{5}$ and $x^{17} \div \mathbf{2}\,.\,\mathbf{3}\,.\,\mathbf{4}\,.\,\mathbf{5}$; adding these, we find

$$\text{for quintic perpetuants } G.F. \text{ is } = x^{15} \div \mathbf{2}\,.\,\mathbf{3}\,.\,\mathbf{4}\,.\,\mathbf{5},$$

which expression was given in the memoir just referred to: the result was obtained by investigating in the first instance an expression for S_5, the number of quintic syzygies of a given weight. The course of Stroh's investigation to be presently given is different; he determines directly the number of perpetuants, and we may if we please use conversely this result to obtain the number of syzygies.

62. The foregoing theory of reduction is independent of the form of the seminvariants, which may be blunt or sharp at pleasure: the actual formulæ will of course be different, and they are very much more simple for the sharp seminvariants, viz. here in many cases a seminvariant is found to be equal to a product of seminvariants of inferior degrees. I subjoin the following table of the reduction of the several sharp seminvariants up to the weight 12; the forms referred to are the tabulated forms, and to mark that this is so I write down in each case the numerical coefficients of the initial and final terms, viz. instead of $c \infty b^2$, $d \infty b^3$, &c., I write $c \infty -b^2$, $d \infty 2b^3$, &c. As appears by the table, these are for shortness denoted by C, D respectively, and so for weight 4, the forms are called E, E_2, for weight 5, F, F_2, for weight 6, G, G_2, G_3, G_4, and so on, the unsuffixed letters having thus an implied suffix, not 0 but 1. The table is

Table of Reductions.

$w=$			
2	$c \infty -b^2$	C	
3	$d \infty 2b^3$	D	
4	$e \infty 3c^2$	E	
	c^2 ,, b^4	$E_2 =$	C_2
5	$f \infty -6bc^2$	F	
	cd ,, $-2b^5$	$F_2 =$	CD
6	$g \infty -10d^2$	G	
	ce ,, $-c^3$	G_2	
	d^2 ,, $-3b^2c^2$	$G_3 =$	$CE - G_2$
	c^3 ,, $-b^6$	$G_4 =$	C^3
7	$h \infty 20bd^2$	H	
	cf ,, $3bc^3$	H_2	
	de ,, $6b^3c^2$	$H_3 =$	DE
	c^2d ,, $2b^7$	$H_4 =$	C^2D
8	$i \infty 35c^2$	I	
	cg ,, $2cd^2$	I_2	
	$3df$,, $10b^2d^2$	$I_3 =$	$CG - I_2$
	e^2 ,, $9c^4$	$I_4 =$	E^2
	c^2e ,, b^2c^3	$I_5 =$	CG_2
	cd^2 ,, $3b^4c^2$	$I_6 =$	CG_3
	c^4 ,, b^8	$I_7 =$	C_4
9	$j \infty -70be^2$	J	
	$2ch$,, $-20d^3$	J_2	
	dg ,, $-4bcd^2$	J_3	
	ef ,, $-20b^3d^2$	$J_4 =$	$\frac{1}{2}(2CH - J_2 - 7J_3)$
	$2c^2f$,, $-3bc^4$	$J_5 =$	$\frac{1}{6}(EF - J_4)$
	cde ,, $-2b^3c^3$	$J_6 =$	DG_2
	d^3 ,, $-6b^5c^2$	$J_7 =$	DG_3
	c^3d ,, $-2b^9$	$J_8 =$	C^3D
10	$k \infty -12bf^2$	K	
	ci ,, $-5ce^2$	K_2	
	$4dh$,, $-35b^2e^2$	$K_3 =$	$CI - K_2$
	$16eg$,, $-80bd^3$	K_4	
	f^2 ,, $-32c^2d^2$	$K_5 =$	$\frac{1}{15}(16EG - K_4)$
	c^2g ,, $-2b^2cd^2$	$K_6 =$	CI_2
	ce^2 ,, $-3c^5$	$K_7 =$	EG_2
	$3cdf$,, $-10b^4d^2$	$K_8 =$	CI_3
	d^2e ,, $-9b^2c^4$	$K_9 =$	EG_3
	c^3e ,, $-b^4c^3$	$K_{10} =$	C^2G_2
	c^2d^2 ,, $-3b^6c^2$	$K_{11} =$	C^2G_3
	c^5 ,, $-b^{10}$	$K_{12} =$	C_5
11	$l \infty 252bf^2$	L	
	$2cj$,, $35de^2$	L_2	
	di ,, $10bce^2$	L_3	
	eh ,, $20cd^3$	L_4	
	$16c^2h$,, $-70b^3e^2$	$L_5 =$	$-DI + L_3 + 2L_4$
	fg ,, $160b^2d^3$	$L_6 =$	$8CJ_2 - L_5$
	cef ,, $-2bc^2d^2$	$L_7 =$	$-\frac{1}{30}(FG - L_6)$
	cdg ,, $4b^3cd^2$	$L_8 =$	DI_2
	d^2f ,, $3bc^5$	$L_9 =$	$\frac{1}{2}(FG_2 - L_7)$
	$12c^3f$,, $-20b^5d^2$	$L_{10} =$	$-DI_3 + 3L_9$
	de^2 ,, $18b^3c^4$	$L_{11} =$	DE^2
	c^2de ,, $2b^5c^3$	$L_{12} =$	CDG_2
	cd^3 ,, $6b^7c^2$	$L_{13} =$	CDG_3
	c^4d ,, $2b^{11}$	$L_{14} =$	C^4D
12	$m \infty 462g^2$	M	
	$3ck$,, $42cf^2$	M_2	
	ei ,, $15e^3$	M_3	
	$15dj$,, $378b^2f^2$	$M_4 =$	$3CK - M_2$
	$25fh$,, $175bde^2$	M_5	
	g^2 ,, $125c^2e^2$	$M_6 =$	$\frac{1}{21}(25EI - 25M_3 - 4M_5)$
	ceg ,, d^4	$M_7 =$	$\frac{1}{100}(G^2 - M_6)$
	c^2i ,, $5b^2ce^2$	$M_8 =$	CK_2
	$5d^2g$,, $20bcd^3$	$M_9 =$	$-3GG_2 + 5EI_2 - 2M_7$
	cf^2 ,, $20c^3d^2$	$M_{10} =$	$GG_2 - M_7$
	$4cdh$,, $35b^4e^2$	$M_{11} =$	CK_3
	$18def$,, $80b^3d^3$	$M_{12} =$	$\frac{1}{10}(10CK_4 - 160M_7 - 32M_9 + 54M_{10})$
	e^3 ,, $36b^2c^2d^2$	$M_{13} =$	$\frac{1}{8}(9CK_5 - 9M_{10} - M_{12})$
	c^2e^2 ,, c^6	$M_{14} =$	G_2^2
	c^3g ,, $2b^4cd^2$	$M_{15} =$	C^2I_2
	cd^2e ,, $3b^2c^5$	$M_{16} =$	$EI_5 - M_{14}$
	$3c^2df$,, $10b^6d^2$	$M_{17} =$	CK_8
	d^4 ,, $9b^4c^4$	$M_{18} =$	G_2^3
	c^4e ,, $-3b^6c^3$	$M_{19} =$	C^3G_2
	c^3d^2 ,, $3b^8c^2$	$M_{20} =$	C^3G
	c^6 ,, b^{12}	M_{21}	

Where no reduction is given, the form is irreducible, i.e. it is a perpetuant.

63. As to these reductions, it may be observed that in very many cases we have the sharp seminvariant given as an actual product $E_2 = C^2$, $F_2 = CD$, $G_4 = C^3$, &c. We have next other reductions such as $G_3 = CE - G_2$, where on the right-hand side there is a single product; this has a final the same as that of the seminvariant which is to be reduced, so that, eliminating this term from the seminvariant and product in question, we have an expression which must be a linear combination (with numerical coefficients) of the preceding seminvariants of the same weight. To take a less simple example, $L_5 = -DI + L_3 + 2L_4$; here $L_5 = -fg + 16c^2h \ldots - 70b^3e^2$, and $DI = (d - 3bc + 2b^3)(i \ldots + 35e^2)$ has the final $+70b^3e^2$. The verification is

$$\begin{array}{rcl} -DI &=& -di \qquad\qquad\qquad \ldots - 70b^3e^2 \\ +L_3 &=& di - 2eh + fg \\ -2L_4 &=& \qquad 2eh - 2fg \\ \hline L_5 &=& \qquad\qquad -fg \ldots - 70b^3e^2. \end{array}$$

The only case in which we have on the right-hand side two products is $(d^2g \propto bcd^3)$, $M_9 = -3GG_2 + 5EI_2 - 2M_7$; viz. here the final of M_9 is bcd^3 which is incomposite (viz. it is not the product of two power-enders), this is in fact the first instance of a quintic seminvariant with an incomposite final and which is nevertheless reducible. For observe, the next seminvariant M_{10} has the final c^3d^2, which is a product in the two ways $c^2 . cd^2$ and $c^3 . d^2$; we have thus the two products $(e \propto c^2)(cg \propto cd^2)$ and $(ce \propto c^3)(g \propto d^2)$, that is, EI_2 and GG_2 with the same final c^3d^2, and combining them so as to eliminate this term we have an expression having the final bcd^3, and which is thus expressible in terms of M_9 and preceding seminvariants: the verification is

$$\begin{array}{rcl} -3GG_2 &=& -3ceg \qquad + 3d^2g \ldots + 60bcd^3 - 30c^3d^2 \\ +5EI_2 &=& +5ceg \qquad\qquad\qquad - 40bcd^3 + 30c^3d^2 \\ -2M_7 &=& -2ceg + 2cf^2 + 2d^2g \\ \hline M_9 &=& \qquad 2cf^2 + 5d^2g \ldots + 20bcd^3. \end{array}$$

64. I annex to this a table (taken from the square diagrams) for the initials and finals of the sharp seminvariants for the weights 13, 14, 15, and 16.

$w=$			
13	n	$\infty\ bg^2$	N
	cl	,, df^2	N_2
	dk	,, bcf^2	3
	ej	,, be^3	4
	fi	,, cde^2	5
	c^2j	,, b^3f^2	6
	gh	,, b^2de^2	7
	ceh	,, bc^2e^2	8
	d^2h	,, bd^4	9
	cfg	,, c^2d^3	10
	cdi	,, b^3ce^2	11
	deg	,, b^2cd^3	12
	df^2	,, bc^3d^2	13
	c^3h	,, b^5e^2	14
	e^2f	,, b^4d^3	15
	c^2ef	,, b^3c^2d	16
	cd^2f	,, bc^6	17
	c^2dg	,, b^5cd^2	18
	cde^2	,, b^3c^5	19
	c^4f	,, b^7d^2	20
	d^3e	,, b^5c^4	21
	c^3de	,, b^7c^3	22
	c^2d^3	,, b^9c^2	23
	c^5d	,, b^{13}	24

$w=$			
14	o	$\infty\ h^2$	O
	cm	,, cg^2	O_2
	ek	,, ef^2	3
	dl	,, b^2g^2	4
	fj	,, bdf^2	5
	gi	,, c^2f^2	6
	cei	,, ce^3	7
	h^2	,, d^2e^2	8
	c^2k	,, b^2cf^2	9
	d^2i	,, b^2e^3	10
	cfh	,, $bcde^2$	11
	cg^2	,, c^3e^2	12
	dfg	,, cd^4	13
	cdj	,, b^4f^2	14
	deh	,, b^3de^2	15
	e^2g	,, $b^2c^2e^2$	16
	c^2eg	,, b^2d^4	17
	ef^2	,, bcd^3	18
	c^2f^2	,, c^4d^2	19
	c^3i	,, b^4ce^2	20
	cd^2g	,, b^3cd^3	21
	$cdef$	,, $b^2c^3d^2$	22
	ce^3	,, c^7	23
	c^2dh	,, b^6e^2	24
	d^3f	,, b^5d^3	25
	d^2e^2	,, $b^4c^2d^2$	26
	c^3e^2	,, b^2c^6	27
	d^4g	,, b^6cd^2	28
	c^2d^2e	,, b^4c^5	29
	c^3df	,, b^8d^2	30
	cd^4	,, b^6c^4	31
	c^5e	,, b^8c^3	32
	c^4d^2	,, $b^{10}c^2$	33
	c^7	,, b^{14}	34

$w=$			
15	p	$\infty\ ch^2$	P
	cm	,, dg^2	P_2
	el	,, f^3	3
	dm	,, bcg^2	4
	fk	,, bef^2	5
	gj	,, cdf^2	6
	cej	,, de^3	7
	c^2l	,, b^3g^2	8
	d^2j	,, b^2df^2	9
	hi	,, bc^2f^2	10
	cfi	,, bce^3	11
	cgh	,, bd^2e^2	12
	dfh	,, c^2de^2	13
	e^2h	,, d^5	14
	cdk	,, b^3cf^2	15
	dei	,, b^3e^2	16
	c^2eh	,, b^2cde^2	17
	dg^2	,, bc^3e^2	18
	efg	,, bcd^4	19
	c^2fg	,, c^3d^3	20
	c^3j	,, b^5f^2	21
	e^4h	,, b^4de^2	22
	cd^2h	,, $b^3c^2e^2$	23
	$cdeg$	,, b^3d^4	24
	f^3	,, $b\ c^2d^3$	25
	cdf^2	,, bc^4d^2	26
	ce^2f	,, b^5ce^2	27
	d^3g	,, b^4cd^3	28
	d^2ef	,, $b^3c^3d^2$	29
	c^3ef	,, bc^7	30
	c^4h	,, b^7e^2	31
	c^2d^2f	,, b^6d^3	32
	dc^3	,, $b^5c^2d^2$	33
	c^2de^2	,, b^3c^6	34
	c^3dg	,, b^7cd^2	35
	cd^3e	,, b^5c^5	36
	c^5f	,, b^9d^2	37
	d^5	,, b^7c^4	38
	c^4de	,, b^9c^3	39
	c^3d^3	,, $b^{11}c^2$	40
	c^6d	,, b^{15}	41

$w=$			
16	q	$\infty\ i^2$	Q
	co	,, ch^2	Q_2
	em	,, eg^2	3
	dn	,, b^2h^2	4
	fl	,, bdg^2	5
	gk	,, bf^3	6
	cek	,, c^2g^2	7
	hj	,, cef^2	8
	i^2	,, d^2f^2	9
	cgi	,, e^4	10
	c^2m	,, b^2cg^2	11
	c^3k	,, b^2ef^2	12
	dej	,, $bcdf^2$	13
	dfi	,, bde^3	14
	e^2i	,, c^3f^2	15
	ch^2	,, c^2e^3	16
	dgh	,, cd^2e^2	17
	cdl	,, b^4g^2	18
	dej	,, b^3df^2	19
	c^2ei	,, $b^2c^2f^2$	20
	efh	,, b^2ce^3	21
	eg^2	,, $b^2d^2e^2$	22
	c^2fh	,, bc^2de^2	23
	c^2g^2	,, bd^5	24
	f^2g	,, c^4e^2	25
	ce^2g	,, c^2d^4	26
	c^3k	,, b^4cf^2	27
	cd^2i	,, b^4e^3	28
	$cdeh$	,, b^3cde^2	29
	$cdfg$	,, $b^2c^3e^2$	30
	d^2eg	,, b^2cd^4	31
	cef^2	,, bc^3d^3	32
	d^2f^2	,, c^5d^2	33
	c^2dj	,, b^6f^2	34
	d^3h	,, b^5de^2	35
	c^3ej	,, $b^4c^2e^2$	36
	c^3f^2	,, b^4d^4	37
	de^2f	,, $b^3c^2d^3$	38
	e^4	,, $b^2c^4d^2$	39
	c^2e^3	,, c^8	40
	c^4i	,, b^6ce^2	41
	c^2d^2g	,, b^5cd^3	42
	c^2def	,, $b^4c^3d^2$	43
	cd^2e^2	,, b^2c^7	44
	c^3dh	,, b^8e^2	45
	cd^3f	,, b^7d^3	46
	c^4e^2	,, $b^6c^2d^2$	47
	d^4e	,, b^4c^6	48
	c^5g	,, b^8cd^2	49
	c^3d^2e	,, b^6c^5	50
	c^4df	,, $b^{10}d^2$	51
	c^2d^4	,, b^8c^4	52
	c^6e	,, $b^{10}c^3$	53
	c^5d^2	,, $b^{12}c^2$	54
	c^8	,, b^{16}	55

It would be interesting to complete this into a table of reductions as given for the weights 2 to 12.

The Strohian Theory Resumed: Application to Perpetuants. Art. Nos. 65 to 71.

65. We can by means hereof establish, in regard to the specific blunt seminvariants, a general theory of reduction, or say a theory of the relations which exist between the seminvariants of a given degree and the powers and products of seminvariants of inferior degrees. To exhibit the form of these, it will be sufficient to take Ω a sum of two parts, $=\Omega'+\Omega''$, but the more general assumption is Ω a sum of any number of parts, $=\Omega'+\Omega''+\Omega'''+\ldots$. Taking then $\Omega=\Omega'+\Omega''$, where for the Ω' and Ω'' separately the sum of the $(x, y, z, \ldots)$ is $=0$, suppose that to the $(0, C, D, E, \ldots)$ of Ω there correspond $(0, C', D', E', \ldots)$ for Ω' and $(0, C'', D'', E'', \ldots)$ for Ω''. We have

$$\begin{aligned} C &= C'+C'',\\ D &= D'+D'',\\ E &= E'+E''+C'C'',\\ F &= F'+F''+C'D''+C''D',\\ G &= G'+G''+C'E''+C''E'+D'D'', \end{aligned}$$

the law of which is obvious.

66. We have, for instance,

$$\Omega^4=(\Omega'+\Omega'')^4, \quad =\Omega'^4+6\Omega'^2\Omega''^2+\Omega''^4, \text{ (since } \Omega'=0,\ \Omega''=0),$$

that is,

$$\begin{aligned} (C'+C'')^2 c^2 &= C'^2c^2+6C'b^2\,.\,C''b^2+C''^2c^2\\ +(E'+E''+C'C'')\,b^4 &+E'b^4 \qquad\qquad +E''b^4 \end{aligned}$$

where, and in what follows, c^2, b^4, b^2 are for shortness written instead of $[c^2]$, $[b^4]$, $[b^2]$ to denote the specific blunt seminvariants ending in c^2, b^4, b^2 respectively.

The terms in C'^2, C''^2, E', E'' are identical on each side of the equation and destroy each other: omitting these, we have only the terms in $C'C''$ which must be equivalent on the two sides of the equation, and comparing coefficients we find the relation

$$2c^2+b^4=6\,.\,b^2\,.\,b^2,$$

which of course means $2[c^2]+[b^4]=6[b^2][b^2]$, viz. this is

$$2(2e-8bd+6c^2)+(-4e+16bd+12c^2-48b^2c+24b^4)=6(-2c+2b^2)^2.$$

In like manner, for $\Omega^6=(\Omega'+\Omega'')^6$, we have

$$\begin{aligned} &(C'+C'')^3 && .\,d^2\\ &+(D'+D'')^2 && .\,c^3\\ &+(C'+C'')(E'+E''+C'C'') && .\,b^2c^2\\ &+(G'+G''+C'E''+C''E'+D'D'') && .\,b^6 \end{aligned}$$

equal to

$$\left\{\begin{matrix} C'^3 & .\,d^2 \\ +D'^2 & .\,c^3 \\ +C'E' & .\,b^2c^2 \\ +G' & .\,b^6 \end{matrix}\right\} + 15 \left\{\begin{matrix} C'^2.\,c^2 \\ +E'.\,b^4 \end{matrix}\right\} C''.\,b^2 + 20D'.\,b^3.\,D''.\,b^3 + 15C'^2.\,b^2 \left\{\begin{matrix} C''^2.\,c^2 \\ +E''.\,b^4 \end{matrix}\right\} + \left\{\begin{matrix} C''^3 & .\,d^2 \\ +D''^2 & .\,c^3 \\ +C''E'' & .\,b^2c^2 \\ +G'' & .\,b^6 \end{matrix}\right\}.$$

Here omitting the terms which destroy each other and comparing the coefficients of the remaining terms, viz. $C'^2C''+C''^2C'$, $D'D''$ and $C'E''+C''E'$, we find the relations

$$\begin{aligned} 3d^2 + b^2c^2 &= 15\,.\,c^2\,.\,b^2, \\ 2c^3 + b^6 \;\; &= 20\,.\,b^3\,.\,b^3, \\ b^2c^2 + b^6 &= 15\,.\,b^4\,.\,b^2, \end{aligned}$$

which may be easily verified. There are on the right-hand side only products of two parts, but this is on account of the special assumption $\Omega = \Omega' + \Omega''$, a sum of two parts.

67. I write now

$$\begin{aligned} &\Omega_2 = \alpha x + \beta y &&, \; S_2 x = 0, \\ &\Omega_3 = \alpha x + \beta y + \gamma z &&, \; S_3 x = 0, \\ &\Omega_4 = \alpha x + \beta y + \gamma z + \delta w &&, \; S_4 x = 0, \\ &\Omega_5 = \alpha x + \beta y + \gamma z + \delta w + \epsilon t &&, \; S_5 x = 0, \\ &\Omega_6 = \alpha x + \beta y + \gamma z + \delta w + \epsilon t + \zeta u, && \; S_6 x = 0, \end{aligned}$$

and I say that Ω_2 and Ω_3 cannot break up: but that Ω_4 breaks up if it becomes a sum of $2+2$ terms (i.e. a sum of two parts Ω_2 for each of which $S_2 x = 0$, and so in other cases): that Ω_5 breaks up if it becomes a sum of $2+3$ terms, Ω_6 breaks up if it becomes a sum of $2+4$ or $2+2+2$ terms, or if it becomes a sum of $3+3$ terms: and similarly for any higher suffix.

The condition that Ω_4 may break up is $x+y=0$, $x+z=0$, or $y+z=0$, or what is the same thing it is $\Pi_3(x+y)=0$, where $\Pi_3(x+y)$ is the product of the three sums each containing x; this is a symmetric function, we in fact have

$$\Pi_3(x+y) = x^3 + x^2(y+z+w) + x(yz+yw+zw) + yzw, \; = xyz + xyw + xzw + yzw, \quad = -D.$$

The condition in order that Ω_5 may break up is $x+y=0$, ..., or $w+t=0$, say this is $\Pi_{10}(x+y)=0$, where $\Pi_{10}(x+y)$ denotes the product of the ten sums $x+y$, ..., $w+t$. It will be shown that we have $\Pi_{10}(x+y) = -D^2E + CDF - F^2$.

The condition in order that Ω_6 may break up is, $x+y=0$, ..., or $t+u=0$, or again if $x+y+z=0$, ..., or $x+t+u=0$, viz. it is $\Pi_{15}(x+y)\,\Pi_{10}(x+y+z)=0$, where $\Pi_{15}(x+y)$ is the product of the fifteen sums $x+y$, ..., $t+u$, and $\Pi_{10}(x+y+z)$ is the product of the ten sums $x+y+z$, ..., $x+t+u$, each containing x : $\Pi_{15}(x+y)$ and

$\Pi_{10}(x+y+z)$ are symmetric functions, the expressions for which will be given further on: the weights in the capital letters are 15 and 10 respectively. And similarly for Ω with any higher suffix, we have the condition that this may break up.

I introduce the factors $\Pi_4 x = E$, $\Pi_5 x = -F$, $\Pi_6 x = G$, ... respectively and write for

$$\begin{array}{ll} \Omega_4 & M_7 = \Pi_4 x \Pi_3 \, (x+y) = -DE \text{ as above,} \\ \Omega_5 & M_{15} = \Pi_5 x \Pi_{10}(x+y) = -F(-D^2E + CDF - F^2) \text{ as above,} \\ \Omega_6 & M_{31} = \Pi_6 x \Pi_{15}(x+y)\, \Pi_{10}(x+y+z), \\ & \vdots \end{array}$$

where observe that, for the even suffixes of Ω, the last factors $\Pi_3(x+y)$, $\Pi_{10}(x+y+z)$, ... denote the products of the sums $x+y$, $x+y+z$, ... which contain x, that is in each case the products of only half the whole number of such linear factors. The suffixes of M show the weights in the capital letters C, D, E, F, G, ... viz. these are $4+3$, $=7$, $5+10$, $=15$, $6+15+10$, $=31$, and so on; the law is obvious, and for Ω_n the weight is $=2^{n-1}-1$.

68. To explain the Strohian theory of perpetuants, I assume explicitly as presently appears. For perpetuants of any given degree δ, we consider in $\Omega_\delta{}^w$ ($w=\delta$ at least) the terms containing seminvariants of the given degree: for instance when $\delta=4$, $w=12$, these are

$$\begin{array}{l} C^4E \;\;.\, b^2f^2 \\ +\, CD^2E\,.\, bde^2 \\ +\, C^2E^2\;.\, c^2e^2 \\ +\, E^3 \quad\;.\, d^4, \end{array}$$

where the capital expressions all contain as factor the letter E of the weight 4. By making Ω to break up, it is assumed that *we obtain all the reductions of the seminvariants of the degree and weight in question; and every such seminvariant, if it be reducible, will be reduced by means of the resulting formulæ.* Now there are seminvariants which are not reducible by these formulæ: in the example just considered, the seminvariant bde^2 has the coefficient CD^2E containing the factor DE, $=xyzw(x+y)(x+z)(x+w)$ which vanishes when Ω_4 breaks up; so that, supposing Ω_4 to break up, the seminvariant bde^2 disappears from the formulæ, and we have no reduction of this seminvariant. And again it is assumed *that every seminvariant which does not in this way disappear from the equation is reducible.* The irreducible seminvariants are thus the seminvariants which, when Ω breaks up into a sum of two or more parts, disappear from the formulæ; viz. the seminvariants which thus disappear are the perpetuants.

69. In the case considered of quartic seminvariants, it has just been seen that, for the weight 12, bde^2 is a perpetuant; and so in general for the weight w, every quartic seminvariant, multiplied into a product of capitals which contains the factor DE, is a perpetuant: for the weight 7 the only term is $DE\,.\,bc^3$, viz. the product

of capitals is here $=DE$; and for any higher weight w we have products which are equal to DE multiplied into products of the weight $w-7$ in C, D, E: and we thus see that the $G.F.$ for quartic perpetuants is $=x^7 \div \mathbf{2}.\mathbf{3}.\mathbf{4}$.

70. For quintic perpetuants, we consider in $\Omega_5{}^w\,(w=5$ at least) the terms which contain quintic perpetuants; for instance, when $w=15$, the terms are

$$\begin{aligned}
&C^5F\;\;\;.\,b^3g^2\\
+\;&C^2D^2F\,.\,b^2df^2\\
+\;&C^3EF\;.\,bc^2f^2\\
+\;&D^2EF\;.\,bce^3\\
+\;&CE^2F\;.\,bd^2e^2\\
+\;&CDF^2\;.\,c^2de^2\\
+\;&F^3\;\;\;\;.\,d^5,
\end{aligned}$$

where the functions of the capitals all contain the factor F; the finals b^3g^2, b^2df^2, ... are arranged in AO. Supposing Ω_5 to break up, we have an expression M, $=-D^2EF+CDF^2-F^3$, which is $=0$, and using this value of M to eliminate the term D^2EF which belongs to the seminvariant bce^3, the final whereof is highest in AO, viz. writing $D^2EF=-M+CDF^2-F^3$, the expression is

$$\begin{array}{rll}
C^5F & .\,b^3g^2 \quad\text{that is} & C^5F\;\;.\,b^3g^2\\
+\,C^2D^2F & .\,b^2df^2 & +\,C^2D^2F\,.\,b^2df^2\\
+\,C^3EF & .\,bc^2f^2 & +\,C^3EF\;.\,bc^2f^2\\
+\,(-M+CDF^2-F^3) & .\,bce^3 & -\,M\;\;\;\;.\,bce^3\\
+\,CE^2F & .\,bd^2e^2 & +\,CE^2F\;.\,bd^2e^2\\
+\,CDF^2 & .\,c^2de^2 & +\,CDF^2\;.\,(c^2de^2+bce^3)\\
+\,F^3 & .\,d^5 & +\,F^3\;\;\;.\,(d^5-bce^3);
\end{array}$$

and here when Ω_5 breaks up, we have $M=0$, that is, the seminvariant bce^3 disappears from the equation, and it is thus a perpetuant: but b^3g^2, b^2df^2, bc^2f^2 and the combinations $c^2de^2+bce^3$, and d^5-bce^3 are severally reducible.

The degree 15 is evidently the lowest degree for which there is an irreducible quintic seminvariant, and for any higher weight w the number of such seminvariants is equal to the number of capital terms which have the factor D^2EF, viz. this is equal to the number of terms weight $w-15$ which can be made up with C, D, E, F; and hence

for quintic perpetuants $G.F.=x^{15}\div \mathbf{2}.\mathbf{3}.\mathbf{4}.\mathbf{5}$.

71. For the degree 6, $M=\Pi_6 x\Pi_{15}(x+y)\,\Pi_{10}(x+y+z)$ is a function of the capitals of the weight 31, and we thence at once infer that

for sextic perpetuants $G.F.=x^{31}\div \mathbf{2}.\mathbf{3}.\mathbf{4}.\mathbf{5}.\mathbf{6}$.

But it is worth while to write down the expression for M: I do this, annexing to each term the seminvariant (i.e. final term) which belongs to it, arranging these final terms in AO; the value thus arranged is

$M =$		finals in AO
$+1$	D^4E^2FG	$bcei^3$
-2	CD^3EF^2G	$bdehi^2$
$+1$	$C^2D^2F^3G$	be^2gi^2
$+2$	D^2EF^3G	$befh^3$
-2	CDF^4G	bf^2gh^2
$+1$	F^5G	bg^5
-1	D^5EG^2	c^2di^3
$+1$	CD^4FG^2	cd^2hi^2
$+1$	$C^2D^2EFG^2$	$cdegi^2$
-4	$D^2E^2FG^2$	$cdfh^3$
-1	$C^3DF^2G^2$	ce^2fi^2
-1	$D^3F^2G^2$	ce^2h^3
$+4$	$CDEF^2G^2$	$cefgh^2$
$+1$	$C^2F^3G^2$	cf^3h^2
$+4$	EF^3G^2	cfg^4
-1	$C^2D^3G^3$	d^3gi^2
$+4$	D^3EG^2	d^2eh^3.

It thus appears that the single sextic perpetuant of the weight 31 is $bcei^3$, and generally that, for any higher weight, the sextic perpetuants are such that the conjugate capital terms contain each of them the factor D^4E^2FG.

The like reasoning shows that

$$\text{for perpetuants of degree } n,\ G.F. \text{ is } = x^{2^{n-1}-1} \div \mathbf{2}.\mathbf{3}.\mathbf{4}\ldots n.$$

Investigation of the Values of the Foregoing Functions $\Pi_{10}(x+y)$, $\Pi_{15}(x+y)$ *and* $\Pi_{10}(x+y+z)$. Art. Nos. 72 to 74.

72. If x, y, z, w, t are the roots of a quintic equation, say

$$\lambda - x\,.\,\lambda - y\,.\,\lambda - z\,.\,\lambda - w\,.\,\lambda - t = (1,\ B,\ C,\ D,\ E,\ F \between \lambda,\ 1)^5 = 0,$$

we require the product $\Pi_{10}(x+y)$ of the sum of two roots in the particular case $B=0$. But in order to the determination of the expression for $\Pi_{10}(x+y+z)$, we require the value of $\Pi_{10}(x+y)$ in the general case, B any value whatever.

Writing

$$x = -\tfrac{1}{2}(\theta + \omega),$$

$$y = -\tfrac{1}{2}(\theta - \omega),$$

and therefore

$$\theta + x + y = 0,$$

we have

$$(\theta + \omega)^5 - 2B(\theta + \omega)^4 + 4C(\theta + \omega)^3 - 8D(\theta + \omega)^2 + 16E(\theta + \omega) - 32F = 0,$$

and the like equation with $-\omega$ for ω. Hence writing $\omega^2 = M$, we have

$$(\theta^5 - 2B\theta^4 + 4C\theta^3 - 8D\theta^2 + 16E\theta - 32F) + M(10\theta^3 - 12B\theta^2 + 12C\theta - 8D) + M^2(5\theta - 2B) = 0,$$

$$(5\theta^4 - 8B\theta^3 + 12C\theta^2 - 16D\theta + 16E) + M(10\theta^2 - 8B\theta + 4C) + M^2 . 1 = 0,$$

which are of the form $A + BM + CM^2 = 0$, $A' + B'M + C'M^2 = 0$, and give therefore by elimination of M the equation

$$-(CA' - C'A)^2 + (BC' - B'C)(AB' - A'B) = 0;$$

the left-hand side is here a function of θ of the degree 10 vanishing when $\theta + x + y = 0$, and which must therefore be, save as to a numerical factor, the product $\Pi_{10}(\theta + x + y)$. And we thus find

$$\begin{aligned}\Pi_{10}&(\theta + x + y)\\ &= -\left\{24\theta^5 - 48B\theta^4 + \begin{pmatrix}56C\\+16B^2\end{pmatrix}\theta^3 + \begin{pmatrix}-72D\\-24BC\end{pmatrix}\theta^2 + \begin{pmatrix}64E\\+32BD\end{pmatrix}\theta + \begin{pmatrix}-32BE\\+32F\end{pmatrix}\right\}^2\\ &\quad + \left\{40\theta^3 - 48B\theta^2 + \begin{pmatrix}8C\\+16B^2\end{pmatrix}\theta + \begin{pmatrix}8D\\-8BC\end{pmatrix}\right\} . \left\{40\theta^7 - 112B\theta^6 + \begin{pmatrix}136C\\+80B^2\end{pmatrix}\theta^5\right.\\ &\quad \left. + \begin{pmatrix}-120D\\-200BC\end{pmatrix}\theta^4 + \begin{pmatrix}0E\\+192BD\\+128C^2\end{pmatrix}\theta^3 + \begin{pmatrix}320F\\-64BE\\-526CD\end{pmatrix}\theta^2 + \begin{pmatrix}-256BF\\+128CE\\-128D^2\end{pmatrix}\theta + \begin{pmatrix}128CF\\-128DE\end{pmatrix}\right\},\end{aligned}$$

which is

$$= 1024\theta^{10} + \ldots + 1024(-F^2 + CDF + 2BEF - BC^2F - D^2E + BCDE),$$

and which therefore for $B = 0$ gives

$$\Pi_{10}(x + y) = -F^2 + CDF - D^2E.$$

73. Suppose now x, y, z, w, t, u are the roots of a sextic equation, say

$$\lambda - x . \lambda - y . \lambda - z . \lambda - w . \lambda - t . \lambda - u = (1, B, C, D, E, F, G \between \lambda, 1)^6 = 0.$$

Considering here the product $\Pi_{20}(x + y + z)$ of the sums of 3 roots, if $B = 0$, this will be a perfect square (for each sum $x + y + z$ is equal to $-$ a sum $(w + t + u)$) say it is the square of $\Pi_{10}(x + y + z)$, where the $x + y + z$ refers to the ten sums each containing x, and we wish to find this function $\Pi_{10}(x + y + z)$. Writing for the equation whose roots are y, z, w, t, u,

$$\lambda - y . \lambda - z . \lambda - w . \lambda - t . \lambda - u = (1, B', C', D', E', F' \between \lambda, 1)^5,$$

we have by what precedes $\Pi_{10}(\theta+y+z)=$ a function $(*\between\theta, 1)^{10}$, viz. this is the above-mentioned function with B', C', D', E', F' in place of the unaccented letters. Introducing a new root x and for λ writing as we may do θ, we have

$$\theta-x.\theta-y.\theta-z.\theta-w.\theta-t.\theta-u=(\theta-x).(1, B', C', D', E', F'\between\theta, 1)^5$$
$$=\quad(1, B, C, D, E, F, G\between\theta, 1)^6;$$

that is, we have

$$\begin{array}{ll} B=B'-\theta \text{ or conversely,} & B'=B+\theta, \\ C=C'-B'\theta & C'=C+B\theta+\theta^2, \\ D=D'-C'\theta & D'=D+C\theta+B\theta^2+\theta^3, \\ E=E'-D'\theta & E'=E+D\theta+C\theta^2+B\theta^3+\theta^4, \\ F=F'-E'\theta & F'=F+E\theta+D\theta^2+C\theta^3+B\theta^4+\theta^5, \quad =-\dfrac{G}{\theta}, \\ G=\quad -F'\theta, & \end{array}$$

where I have retained B, but the value hereof is in fact $=0$. In the foregoing function $(*\between\theta, 1)^{10}$ with the accented letters, writing for these their values $B'=\theta$, $C'=C+\theta^2$, $D'=D+C\theta+\theta^3$, &c., which belong to $B=0$, we find

$$1024\Pi_{10}(\theta+y+z)=-(48\theta^5+56C\theta^3+24D\theta^2+64E\theta+32F)^2$$
$$+(16\theta^3+8C\theta+D)\{144\theta^7+264C\theta^5+72D\theta^4+128(C^2+E)\theta^3+192F\theta^2+128CE\theta$$
$$+128(CF-DE)\},$$

which equation divides by 64. Writing herein $\theta=x$, we have

$$16\Pi_{10}(x+y+z)=-(6x^5+7Cx^3+3Dx^2+8Ex+4F)^2$$
$$+(2x^3+Cx+D)\{18x^7+33Cx^5+9Dx^4+16(C^2+E)x^3+24Fx^2+16CEx+16(CF-DE)\},$$

where $x^6+Cx^4+Dx^3+Ex^2+Fx+G$ is $=0$: the value ought, in virtue of this equation, to reduce itself to a mere function of the coefficients, and we in fact find that the equation is

$$16\Pi_{10}(x+y+z)=(16C^2-64E)(x^6+Cx^4+Dx^3+Ex^2+Fx)+16CDF-16D^2E-16F^2,$$

reducing itself to

$$-(16C^2-64E)G\qquad\qquad+16CDF-16D^2E-16F^2,$$

viz. dividing each side by 16, we have

$$\Pi_{10}(x+y+z)=4EG-C^2G-F^2+CDF-D^2E,$$

which is the required result. The equation $(\theta^2-1)^3=0$, for which

$$x, y, z, w, t, u=1, 1, 1, -1, -1, -1,$$

gives a numerical verification.

74. I find also, for the same value $B=0$, the function $\Pi_{15}(x+y)$. Writing, as before,

$$x=-\tfrac{1}{2}(\theta+\omega),$$
$$y=-\tfrac{1}{2}(\theta-\omega),$$

and therefore

$$\theta + x + y = 0,$$

we have

$$(\theta+\omega)^6 + 4C(\theta+\omega)^4 - 8D(\theta+\omega)^3 + 16E(\theta+\omega)^2 - 32F(\theta+\omega) + 64G = 0,$$

and the like equation with $-\omega$ for ω. Hence writing $\omega^2 = M$, we have

$$(\theta^6 + 4C\theta^4 - 8D\theta^3 + 16E\theta^2 - 32F\theta + 64G) + M(15\theta^4 + 24C\theta^2 - 24D\theta + 16E)$$
$$+ M^2(15\theta^2 + 4C) + M^3 = 0,$$

$$(6\theta^5 + 16C\theta^3 - 24D\theta^2 + 32E\theta - 32F) + M(20\theta^3 + 16C\theta - 8D) + M^2 . 6\theta = 0,$$

say these equations are $aM^3 + bM^2 + cM + d = 0$, $pM^2 + qM + r = 0$. Eliminating M, we have

$a^2 . r^3$	$a = 1,$
$- ab . qr^2$	$b = 15\theta^2 + 4C,$
$+ ac(-2pr^2 + q^2r)$	$c = 15\theta^4 + 24C\theta^2 - 24D\theta + 16E,$
$+ b^2 . pr^2$	$d = \theta^6 + 4C\theta^4 - 8D\theta^3 + 16E\theta^2 - 32F\theta + 64G,$
$+ ad(3pqr - q^3)$	
$+ bc(-pqr)$	$p = 6\theta,$
$+ bd(-2p^2r + pq^2)$	$q = 20\theta^3 + 16C\theta - 8D,$
$+ c^2 . p^2r$	$r = 6\theta^5 + 16C\theta^3 - 24D\theta^2 + 32E\theta - 32F,$
$- cd . p^2q$	
$+ d^2 . p^3 = 0.$	

The equation, as far as I have calculated it, is

$$-32768\theta^{15} - \ldots - 32768(-D^3G + F^3 - CDF^2 + D^2EF) = 0:$$

the left-hand side is here $= -32768\Pi_{15}(x+y)$; and we have therefore

$$\Pi_{15}(x+y) = -D^3G + F^3 - CDF^2 + D^2EF,$$

the required result. It may be remarked that, writing $G = 0$ and throwing out a factor $-F$, we have $-F^2 + CDF - D^2E$, which is the expression for $\Pi_{10}(x+y)$ in the quintic equation.

We have

$$\Pi_6 x \Pi_{15}(x+y)\Pi_{10}(x+y+z)$$
$$= G\{-D^3G + (F^2 - CDF + D^2E)F\}\{(4E - C^2)G - F^2 + CDF - D^2E\},$$

the developed expression whereof is the foregoing value

$$M = D^4E^2FG - 2CD^3EF^2G + \&c., \textit{ ante } \text{No. } 71.$$

The Operators $P-\delta b$ *and* $Q-2\omega b$. Art. Nos. 75 to 84.

75. The analogous theory for non-unitariants is established, *ante* Nos. 24 *et seq.* For seminvariants, we have

$$P = b\partial_a + c\partial_b + d\partial_c + \dots,$$
$$Q = c\partial_b + 2d\partial_c + \dots,$$

or more definitely, if the seminvariant operated upon be of the degree δ, the weight ω and extent σ, say its highest letter is a_σ, $=p$, then

$$P = b\partial_a + c\partial_b + d\partial_c + \dots + q\partial_p,$$
$$Q = c\partial_b + 2d\partial_c + \dots + \sigma q\partial_p,$$

then we have

$$P-\delta b,\quad Q-2\omega b,$$

operators each of them of the deg. weight 1.1, viz. each of them operating upon a seminvariant S of the deg. weight $\delta.\omega$ gives a seminvariant S' of the deg. weight $\delta+1.\omega+1$; moreover, a new letter q is introduced, or say the extent is increased from σ to $\sigma+1$. For the proof, it is only necessary to show that $\Delta(P-b\delta)S$ and $\Delta(Q-2b\omega)$ are each $=0$, but it is unnecessary to do this, as the like proof has already been given for non-unitariants.

The two seminvariant operators were first considered in my paper "On a theorem relating to seminvariants," *Quart. Math. Journ.* t. xx. (1885), pp. 212, 213, [844].

76. We may, instead of $P-\delta b$ and $Q-2\omega b$, consider the linear combination $Y = 2\omega(P-\delta b) - \delta(Q-2\omega b)$, that is, $2\omega P - \delta Q$, which is of deg. weight 0.1, viz. it leaves the degree unaltered, while increasing as before the weight, and also the extent, each by unity. And again, the combination

$$Z = \sigma(P-\delta b) - (Q-2\omega b),$$

that is,

$$\sigma P - Q - (\sigma\delta - 2\omega)\,b,$$

where observe that $\sigma P - Q$, $= \sigma b\partial_a + (\sigma-1)\,c\partial_b + \dots + 1p\partial_o$ does not contain the new letter q; the operator Z is thus of the deg. weight 1.1 increasing the degree and also the weight each by unity, but leaving the extent unaltered.

There is a special case which it is important to attend to, we may have $\sigma\delta - 2\omega = 0$, viz. this is the case when the seminvariant operated upon is in regard to the letters comprised therein an invariant. Here the two combinations Y, Z are equivalent to each other, each of them is $=\sigma b\partial_a + (\sigma-1)\,c\partial_b + \dots + 1p\partial_o$, which is an annihilator of the seminvariant (invariant) operated upon. Hence in this case we cannot replace the original forms by the linear combinations, but must retain one (no matter which) of the original forms $P-\delta b$, $Q-2\omega b$.

77. We can, by means of the foregoing operators, starting from the quadric seminvariants $c-b^2$, &c., derive in order the seminvariants for the successive weights 3, 4, 5, ...

Thus writing down the series of finals (in AO as before),

b^2,	b^3,	c^2,	bc^2,	d^2,	bd^2,	e^2,	&c.,
		b^4	b^5	c^3	bc^3	cd^2	
				b^2c^2	b^3c^2	b^2d^2	
				b^6	b^7	c^4	
						b^2c^3	
						b^4c^2	
						b^8,	

I proceed as follows, observing, however, that when the function operated upon is an invariant seminvariant we must instead of Z write $P-\delta b$.

b^2 emerges,	$b^3=Zb^2$,	c^2 emerges,	$bc^2=Zc^2$,	d^2 emerges,	$bd^2=Zd^2$,	e^2 emerges,
		$b^4=Zb^3$	$b^5=Zb^4$	$c^3=Ybc^2$	$bc^3=Zc^3$	$cd^2=Ybd^2$
				$b^2c^2=Zbc^2$	$b^3c^2=Zb^2c^2$	$b^2d^2=Zbd^2$
				$b^6=Zb^5$	$b^7=Zb^6$	$c^4=Ybc^3$
						$b^2c^3=Zbc^2$
						$b^4c^2=Zb^3c^2$
						$b^8=Zb^7$,

viz. whenever the seminvariant to be obtained has a final containing b, it is obtained by means of the operator Z (or it may be $P-\delta b$), but when there is no b then by the operator Y.

The seminvariants operated upon may be blunt or sharp, but there is an advantage in operating on the sharp forms as these are more simple, and we thereby obtain for the next superior weight forms more nearly approximating to the sharp forms. We do not however by thus operating on a sharp form obtain directly a sharp form; to do this, the form obtained must be modified by adding thereto a numerical multiple or multiples of a preceding sharp form: and thus the theory does not determine beforehand the forms of the sharp seminvariants. But making at each step the necessary modification (if any) we have thereby, when the sharp seminvariants of the next preceding weight are known, a very convenient process for the calculation of the sharp seminvariants of any given weight, in the AO arrangement of their final terms. Thus for the weight 10; $k \propto f^2$ is taken to be known, the next two forms $ci \propto ce^2$ and $dh \propto b^2e^2$ are calculated each from $j \propto be^2$, the expression for which is

$$=j-9bi+20ch-28dg+14ef+16b^2h-56bcg+112bdf-70be^2.$$

We have for $j \propto be^2$, $\delta=3$, $\omega=9$, $\sigma=9$: and therefore

$$\tfrac{1}{3}Y=6b\partial_a+5c\partial_b+4d\partial_c+3e\partial_d+2f\partial_e+g\partial_f-i\partial_h-2j\partial_i-3k\partial_j,$$

$$Z=9b\partial_a+8c\partial_b+7d\partial_c+6e\partial_d+5f\partial_e+4g\partial_f+3h\partial_g+2i\partial_h+j\partial_i-9b.$$

78. I exhibit the calculation as follows:

$\frac{1}{3}Y(j \infty be^2)$

		1	2	3	4	5	6	7	8	9		†	÷ 70	*
k	+ 1									− 3	− 3	+ 3	0	
bj	− 10	+ 12							+ 18		+ 30	− 30	0	
ci	+ 45		− 45					− 20			− 65	+ 135	+ 70	+ 1
dh	− 120			+ 80							+ 80	− 360	− 280	− 4
eg	+ 210				− 84		+ 14				− 70	+ 630	+ 560	+ 8
f^2	− 126					+ 28					+ 28	− 378	− 350	− 5
b^2i		− 54						− 16			− 70		− 70	− 1
bch		+ 120	+ 160								+ 280		+ 280	+ 4
bdg		− 168		− 224			+ 112				− 280		− 280	− 4
bef		+ 84			+ 336	− 280					+ 140		+ 140	+ 2
c^2g			− 280								− 280		− 280	− 4
cdf			+ 560								+ 560		+ 560	+ 8
ce^2			− 350								− 350		− 350	− 5
d^2e														
b^3h														
b^2cg														
b^2df														
b^2e^2														
	± 256													± 23

$Z(j \infty be^2)$

	1	2	3	4	5	6	7	8	9	10		†	÷ − 18	*
k														
bj	+ 18								− 9	− 9	0		0	
ci		− 72						+ 40			− 32	+ 32	0	
dh			+ 140				− 84				+ 56	− 128	− 72	+ 4
eg				− 168		+ 56					− 112	+ 256	+ 144	− 8
f^2					+ 70						+ 70	− 160	− 90	+ 5
b^2i	− 81							+ 32		+ 81	+ 32	− 32	0	0
bch	+ 180	+ 256					− 168			− 180	+ 88	+ 128	+ 216	− 12
bdg	− 252		− 392			+ 448				+ 252	+ 56	− 128	− 72	+ 4
bef	+ 126			+ 672	− 700					− 126	− 28	+ 64	+ 36	− 2
c^2g		− 448									− 448	− 128	− 576	+ 32
cdf		+ 896									+ 896	+ 256	+ 1152	− 64
ce^2		− 560									− 560	− 160	− 720	+ 40
d^2e														
b^3h										− 144	− 144		− 144	+ 8
b^2cg										+ 504	+ 504		+ 504	− 28
b^2df										− 1008	− 1008		− 1008	+ 56
b^2e^2										+ 630	+ 630		+ 630	− 35
														± 149

The numbers (1, 2, ..., 9) and (1, 2, ..., 10) at the head of the columns refer to the nine terms $6b\partial_a$, $5c\partial_b$, ... of $\frac{1}{3}Y$, and the ten terms $9b\partial_a$, $8c\partial_b$, ... of Z respectively, these several operations being performed on $(j \infty be^2)$ the value of which is given above: the daggers † denote the additions which have to be made in order to obtain the proper initial term, viz. for the first † the added term is $+3(k \infty f^2)$ and for the second † the added term is $+32(ci \infty ce^2)$: the headings $\div 70$ and $\div -18$ explain themselves, and the columns headed with an asterisk * give the results, viz. the first of these is $(ci \infty ce^2)$ and the second of them is $(dh \infty b^2e^2)$. As appears above, the value of the first of these is used in the second † column for obtaining that of the second of them.

79. We may operate with $P-\delta b$ and $Q-2\omega b$ on a product (deg. weight $\delta.\omega$) ST of two seminvariants S, T, deg. weights $\delta'.\omega'$ and $\delta''.\omega''$ respectively, $\delta=\delta'+\delta''$, $\omega=\omega'+\omega''$. We have

$$(P-\delta b)ST = S.PT + T.PS - (\delta'+\delta'')bST, \quad = S(P-\delta''b)T + T(P-\delta'b)S,$$

where $(P-\delta'b)S$ and $(P-\delta''b)T$ are each of them a seminvariant. And similarly,

$$(Q-2\omega b)ST = S.QT + T.QS - 2(\omega'+\omega'')bST = S(Q-2\omega''b)T + T(Q-2\omega'b)S,$$

where $(Q-2\omega'b)S$ and $(Q-2\omega''b)T$ are each of them a seminvariant. That is, operating either with $P-\delta b$ or $Q-2\omega b$ on a product, we have a sum of products; and therefore also operating upon a sum of products (each product being of the deg. weight $\omega.\delta$) we have a sum of products, each product in such sum being of the deg. weight $\omega+1.\delta+1$, and moreover of the extent $\sigma+1$. And instead of binary products, we may, it is clear, consider ternary, quaternary, &c., products.

The like theorem applies to the derived operators Y and Z, but as to Y there is a specialty to be noticed. We have

$$\begin{aligned} Y.ST &= 2\omega(P-\delta b)ST - \delta(Q-2\omega b)ST, \\ &= 2\omega\{S(P-\delta''b)T + T(P-\delta'b)S\} - \delta\{S(Q-2\omega''b)T + T(Q-2\omega'b)S\}, \\ &= S\{2\omega(P-\delta''b)T - \delta(Q-2\omega''b)T\} + T\{2\omega(P-\delta'b)T - \delta(Q-2\omega'b)S\}, \end{aligned}$$

where the whole of the right-hand side as being equal to $Y.ST$ is of the degree δ, but except in the particular case $\left(\frac{\delta}{\omega}=\frac{\delta'}{\omega'}=\frac{\delta''}{\omega''}\right)$ the separate products $S\{\ \}$ and $T\{\ \}$ which occur on the right-hand side are each of them of the degree $\delta+1$.

It is scarcely necessary, but it may be proper, to remark that we frequently combine by addition a seminvariant S of the deg. weight $\delta.\omega$ with a seminvariant T deg. weight $\delta-\epsilon.\omega$ of the same weight but of an inferior degree, but when this is done we regard the T as standing for $a^\epsilon T$, and as being thus of the same deg. weight $\delta.\omega$. We have

$$(P-\delta b)a^\epsilon T = a^\epsilon PT + TPa^\epsilon - (\epsilon+\delta-\epsilon)ba^\epsilon T, \quad = a^\epsilon\{P-(\delta-\epsilon)b\}T + T(P-\epsilon b)a^\epsilon,$$

where $(P-\epsilon b)\,a^\epsilon=(\epsilon-\epsilon)\,b=0$, and consequently $(P-b\delta)\,a^\epsilon T=\{P-(\delta-\epsilon)\,b\}\,T$; viz. for the operation upon T, we regard $P-\delta b$ as standing for $P-(\delta-\epsilon)\,b$. As regards Q, we have $(Q-2\omega b)\,a^\epsilon T=(Q-2\omega b)\,T$; viz. the degree of T does not here present itself.

80. We may write

$$(2\omega P-\delta Q)\,S=S',$$

the new seminvariant S' being of the weight $\omega+1$; hence also

$$\{(2\omega+2)\,P-\delta Q\}\,.\,\{2\omega P-\delta Q\}\,S=S'',$$

where S'' is of the weight $\omega+2$; viz. we have an operator

$$\{(2\omega+2)\,P-\delta Q\}\,.\,\{2\omega P-\delta Q\},$$

which, operating on a seminvariant of the deg. weight $\delta\,.\,\omega$, gives a seminvariant of the deg. weight $\delta\,.\,\omega+2$. This is

$$=(4\omega^2+4\omega)(P^2+P\,.\,P)-(2\omega+2)\,\delta\,(PQ+P\,.\,Q)-2\omega\delta\,(QP+Q\,.\,P)+\delta^2\,(Q^2+Q\,.\,Q),$$

where P^2, PQ, QP and Q^2 are the mere algebraical squares and products, while $P\,.\,Q$ and $Q\,.\,P$ denote respectively P operating on Q and Q operating on P; and since $PQ=QP$, this is

$$=(4\omega^2+4\omega)(P^2+P\,.\,P)-(4\omega+2)\,\delta PQ-2\,(\omega+2)\,\delta P\,.\,Q-2\omega\delta Q\,.\,P+\delta^2\,(Q^2+Q\,.\,Q).$$

Recollecting that

$$P=b\partial_a+c\partial_b+d\partial_c+\ldots,\quad Q=c\partial_b+2d\partial_c+\ldots,$$

we have

$$\begin{aligned}
P\,.\,P&=c\partial_a+\ d\partial_b+\ e\partial_c\quad+\ldots,\\
P\,.\,Q&=\qquad\quad d\partial_b+2e\partial_c\quad+\ldots,\\
Q\,.\,P&=c\partial_a+2d\partial_b+3e\partial_c\quad+\ldots,\ =P\,.\,P+P\,.\,Q,\\
Q\,.\,Q&=\quad 1\,.\,2d\partial_b+2\,.\,3e\partial_c+\ldots,
\end{aligned}$$

and attending to the relation just obtained $Q\,.\,P=P\,.\,P+P\,.\,Q$, we find that the operator may be written

$$\begin{aligned}
&(4\omega^2+4\omega)\,\{P^2-(\delta-1)\,P\,.\,P\}\\
&-(4\omega+2)\,\delta\,\{PQ-\omega P\,.\,P-\tfrac{1}{3}(\delta-3)\,P\,.\,Q\}\\
&+\delta^2\,\{Q^2+Q\,.\,Q-\tfrac{1}{3}(4\omega+2)\,P\,.\,Q\};
\end{aligned}$$

in fact, here the terms in P^2, PQ, Q^2 are in the original form, while those in $P\,.\,P$, $P\,.\,Q$, $Q\,.\,Q$ are

$$\begin{aligned}
(4\omega^2+4\omega)(1-\delta)\,P\,.\,P+(4\omega^2+2\omega)\,\delta P\,.\,P-\tfrac{1}{3}\,(4\omega+2)\,(\delta^2-3\delta)\,P\,.\,Q+\delta^2 Q\,.\,Q\\
+\tfrac{1}{3}\,(4\omega+2)\,\delta^2\qquad P\,.\,Q,
\end{aligned}$$

which are

$$=(4\omega^2+4\omega-2\omega\delta)\,P\,.\,P-(4\omega+2)\,\delta P\,.\,Q+\delta^2 Q\,.\,Q,$$

agreeing with the original form

$$(4\omega^2+4\omega)\,P\,.\,P-(2\omega+2)\,\delta P\,.\,Q-2\omega\delta\,(P\,.\,P+P\,.\,Q)+\delta^2 Q\,.\,Q.$$

81. I find that each of the three parts is separately an operator, viz. that we have

$$P^2-(\delta-1)\,P\,.\,P,$$
$$PQ-\omega P\,.\,P-\tfrac{1}{3}(\delta-3)\,P\,.\,Q,$$
$$Q^2+Q\,.\,Q-\tfrac{1}{3}(4\omega+2)\,P\,.\,Q,$$

each of them an operator which, operating on a seminvariant of deg. weight $\delta\,.\,\omega$, gives a seminvariant of deg. weight $\delta\,.\,\omega+2$.

I verify this for the first of the three operators, say

$$\Omega=P^2-(\delta-1)\,P\,.\,P=P^2+P\,.\,P-\delta\Theta,$$

if for a moment

$$P\,.\,P,\ =c\partial_a+d\partial_b+e\partial_c+\ldots$$

is put $=\Theta$.

Here for a seminvariant S, we have

$$\Omega S=(P^2+P\,.\,P-\delta\Theta)\,S=P\,(PS)-\delta\Theta S.$$

Writing $S'=(aP-b\delta)\,S$, then S' is a seminvariant, degree $=\delta+1$, and then if $S''=(aP-b\,(\delta+1))\,S'$, S'' is a seminvariant, degree $=\delta+2$. We have $PS=a^{-1}(S'+b\delta S)$, and thence

$$\Omega S=Pa^{-1}(S'+b\delta S)-\delta\Theta S,\ =-b\,(S'+b\delta S)+P\,(S'+b\delta S)-\delta\Theta S.$$

Here

$$P\,(S'+b\delta S)=PS'+c\delta S+b\delta PS,\ =S''+b\,(\delta+1)\,S'+c\delta S+b\delta\,(S'+b\delta S),$$

and hence

$$\Omega S=S''+2b\delta S'+\{c\delta+b^2(\delta^2-\delta)\}\,S-\delta\Theta S.$$

This will be a seminvariant if $\Delta\,.\,\Omega S=0$; we have

$$\begin{aligned}\Delta\,.\,\Omega S=\Delta S''+2b\delta\Delta S'+\{c\delta+b^2(\delta^2-\delta)\}\,\Delta S-\delta\,(\Delta\Theta+\Delta\,.\,\Theta)\,S\\ +2\delta S'+\{2b\delta+2b\,(\delta^2-\delta)\}\,S,\end{aligned}$$

or, omitting the terms in $\Delta S''$, $\Delta S'$, ΔS which respectively vanish, this is

$$=2\delta S'+2b\delta^2 S-\delta\,(\Delta\Theta+\Delta\,.\,\Theta)\,S.$$

But since $PS=S'+b\delta S$, and from $\Delta S=0$ we deduce $0=(\Theta\Delta+\Theta\,.\,\Delta)\,S$, the equation becomes

$$\Delta\,.\,\Omega S=2\delta PS-\delta\,(\Delta\,.\,\Theta-\Theta\,.\,\Delta)\,S,$$

and from

$$\Delta=a\partial_b+2b\partial_c+3c\partial_d+\ldots,$$
$$\Theta=c\partial_a+d\partial_b+e\partial_c+\ldots,$$

we have

$$\Delta\,.\,\Theta=2b\partial_a+3c\partial_b+4d\partial_c+\ldots,$$
$$\Theta\,.\,\Delta=c\partial_b+2d\partial_c+\ldots,$$

and thence

$$\Delta . \Theta - \Theta . \Delta = 2b\partial_a + 2c\partial_b + 2d\partial_c + \ldots, \; = 2P,$$

and we have thus the required equation $\Delta . \Omega S = 0$.

82. If instead of P, Θ, we write B, C, so that

$$B = b\partial_a + c\partial_b + d\partial_c + \ldots,$$

$$C = B . B = c\partial_a + d\partial_b + e\partial_c + \ldots,$$

and put further

$$D = B . C = d\partial_a + e\partial_b + f\partial_c + \ldots,$$

$$E = B . D = e\partial_a + f\partial_b + g\partial_c + \ldots,$$

then the foregoing operator is $B^2 - (\delta - 1) C$, or reversing the sign, say it is $(\delta - 1) C - B^2$, which is the first of a series of operators

$$(\delta - 1) C - B^2,$$

$$(\delta - 1)(\delta - 2) D - 3 (\delta - 2) BC \qquad + 2B^3,$$

$$(\delta - 1)(\delta - 2)(\delta - 3) E - 4 (\delta - 2)(\delta - 3) BD + 6 (\delta - 3) B^2C - 3B^4,$$

$$\vdots$$

which are of the deg. weights $0 . 2$, $0 . 3$, $0 . 4$, &c., respectively, viz. operating upon a seminvariant of deg. weight $\delta . \omega$ they leave the degree unaltered, but increase the weight by 2, 3, 4,... respectively.

It is to be observed that B^2, BC, B^3, &c., denote the mere algebraical powers and products of the symbols B, C, D, &c., without any operation of one symbol on another.

As a simple illustration, take $(C - B^2)(ac - b^2)$: here,

$$C (ac - b^2) = e - 2bd + c^2$$

$$-B^2 = -(2bd\partial_a\partial_c + c^2\partial_b{}^2) (\;\; ,, \;\;) \qquad - 2bd + 2c^2$$

$$\text{Value is} \qquad e - 4bd + 3c^2;$$

and similarly for $(C - B^2)(ae - 4bd + 3c^2)$, here:

$$C (ae - 4bd + 3c^2) = g - 4bf + (6 + 1) ce - 4d^2$$

$$-B^2 = -(2bf\partial_a\partial_e + 2ce\partial_b\partial_d + d^2\partial_c{}^2) (\;\; ,, \;\;) \qquad - 2bf \quad + 8 \; ce - 6d^2$$

$$\text{Value is} \qquad g - 6bf \quad + 15 \; ce - 10d^2.$$

A direct proof may of course be obtained for any one of the foregoing operators; viz. calling it Ω, it may be shown that $\Delta\Omega S$ is $= 0$. I have not considered the like question of the derivation of series of operators from the other two forms

$$PQ - \omega P . P - \tfrac{1}{3} (\delta - 3) P . Q \text{ and } Q^2 + Q . Q - \tfrac{1}{3} (4\omega + 2) P . Q$$

respectively.

83. I do not wish in the present paper to go into the theory of covariants, but it is nevertheless proper to point out the connexion which exists between the covariant theory of derivation and the operators P and Q.

Consider a quantic $(a, b, c, \ldots, a^{\backprime} = a_{\sigma'} \between x, y)^{\sigma'}$; any covariant hereof is $(A, B, C, \ldots \between x, y)^{\mu}$, where A is a seminvariant say of degree δ and weight, $\omega = \frac{1}{2}(\sigma'\delta - \mu)$, or $\mu = \sigma'\delta - 2\omega$, reduced to zero by the operation $\Delta = a\partial_b + 2b\partial_c + \ldots + \sigma' b^{\backprime}\partial_{a^{\backprime}}$: and if we write

$$\phi_{\sigma'} = \sigma' b\partial_a + (\sigma' - 1)\, c\partial_b + \ldots + a^{\backprime}\partial_{b^{\backprime}},$$

then

$$B = \phi_{\sigma'} A,\quad C = \tfrac{1}{2}\phi_{\sigma'} B,\quad D = \tfrac{1}{3}\phi_{\sigma'} C, \ldots$$

The derivative (f, F) is

$$\begin{aligned} &= \partial_x f \,.\, \partial_y F - \partial_y f \,.\, \partial_x F \\ &= (a, b, \ldots \between x, y)^{\sigma'-1} B x^{\mu-1} + \ldots \\ &\quad - (b, c, \ldots \between x, y)^{\sigma'-1} \mu A x^{\mu-1} + \ldots \\ &= (aB - \mu bA, \ldots \between x, y)^{\sigma'+\mu-2}, \end{aligned}$$

that is, A being a seminvariant, we have $aB - \mu bA$ a seminvariant, or say

$$(\phi_{\sigma'} - \mu' b)\, A = \text{sem. } \mu' = \sigma'\delta - 2\omega;$$

and similarly

$$(\phi_{\sigma} - \mu b)\, A = \text{sem. } \mu = \sigma\delta - 2\omega.$$

Hence

$$\{\phi_\sigma - \phi_{\sigma'} - (\mu - \mu')\, b\}\, A, \text{ and } \{\sigma'\phi_\sigma - \sigma\phi_{\sigma'} - (\sigma'\mu - \sigma\mu')\, b\}\, A$$

are each of them a seminvariant: but

$$\begin{aligned} \phi_\sigma &= \sigma b\partial_a + (\sigma - 1)\, c\partial_b + \ldots, \\ \phi_{\sigma'} &= \sigma' b\partial_a + (\sigma' - 1)\, c\partial_b + \ldots, \end{aligned}$$

$$\phi_\sigma - \phi_{\sigma'} = (\sigma - \sigma')\,(b\partial_a + c\partial_b + \ldots) = (\sigma - \sigma')\, P,\quad \mu - \mu' = (\sigma - \sigma')\,\delta;$$

the first form, omitting the factor $\sigma - \sigma'$, is $= (P - \delta b)\, A$: similarly

$$\sigma'\phi_\sigma - \sigma\phi_{\sigma'} = (\sigma - \sigma')\,(c\partial_b + 2d\partial_c + \ldots) = (\sigma - \sigma')\, Q$$

and

$$\sigma'\mu - \sigma\mu' = (\sigma - \sigma')\, 2\omega,$$

and the second form is

$$= (Q - 2\omega b)\, A.$$

We thus see that the operators $P - \delta b$ and $Q - 2\omega b$ upon a seminvariant A depend on the derivation of f upon a covariant which has A for its leading coefficient: the order of f is arbitrary, and we have thus two distinct forms.

84. As an illustration, consider the quantics

$$(1, b, c, d, e, f \between x, y)^5,$$

and

$$(1, b, c, d, e, f, g \between x, y)^6:$$

each of these has a covariant the leading coefficient of which is

$$A = f - 5be + 2cd + 8b^2d - 6bc^2,$$

viz. these are

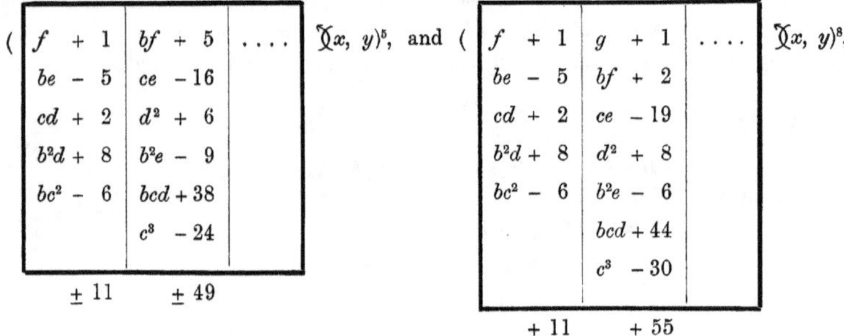

$($ … $\between x, y)^5$, and $($ … $\between x, y)^6$,

f + 1	bf + 5	....
be − 5	ce − 16	
cd + 2	d^2 + 6	
b^2d + 8	b^2e − 9	
bc^2 − 6	bcd + 38	
	c^3 − 24	
± 11	± 49	

f + 1	g + 1	....
be − 5	bf + 2	
cd + 2	ce − 19	
b^2d + 8	d^2 + 8	
bc^2 − 6	b^2e − 6	
	bcd + 44	
	c^3 − 30	
± 11	± 55	

and we find without difficulty

	$(g \propto d^2)$	$(ce \propto c^3)$	$(d^2 \propto b^2c^2)$
$(f_1, F_1) =$		-16	-10
$(f_2, F_2) =$	1	-34	-16
$(P - 3b)A =$	1	-18	-6
$(Q - 10b)A =$	5	-74	-20

and thence

$$(P - 3b)A = (f_2, F_2) - (f_1, F_1)$$

$$(Q - 10b)A = 5(f_2, F_2) - 6(f_1, F_1),$$

viz. we thus have $P - 3b$, and $Q - 10b$ upon $f \propto bc^2$ each given as a linear function of the derivatives (f_1, F_1) and (f_2, F_2), where f_1, f_2 are the quintic and the sextic function, and F_1, F_2 are like covariants of these functions respectively.

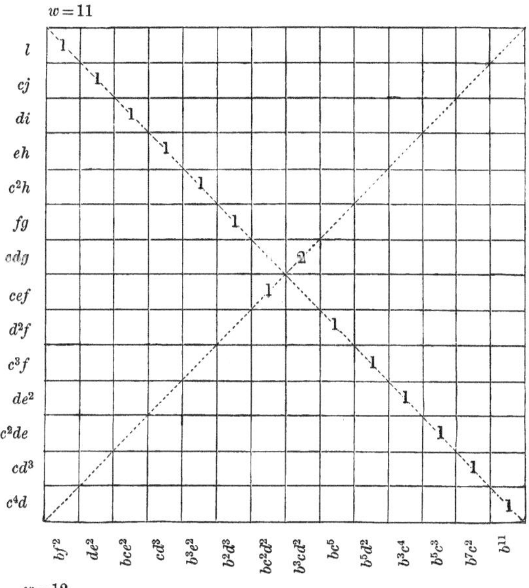

$w=11$

	bf^2	de^2	bce^2	cd^3	b^3e^2	b^2d^3	bc^2d^2	b^3cd^2	bc^5	b^5d^2	b^3c^4	b^5c^3	b^7c^2	b^{11}
l	1													
cj		1												
di			1											
eh				1										
c^2h					1									
fg						1								
cdg								2						
cef							1							
d^2f									1					
c^3f										1				
de^2											1			
c^2de												1		
cd^3													1	
c^4d														1

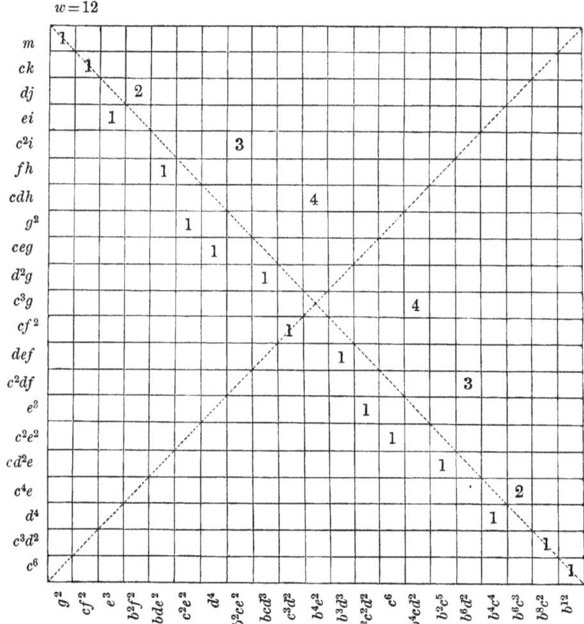

$w=12$

	g^2	cf^2	e^3	b^2f^2	bde^2	c^2e^2	d^4	b^2ce^2	bcd^3	c^3d^2	b^4e^2	b^3d^3	$b^2c^2d^2$	c^6	b^4cd^2	b^2c^5	b^6d^2	b^4c^4	b^6c^3	b^8c^2	b^{12}
m	1																				
ck		1																			
dj				2																	
ei			1																		
c^2i								3													
fh					1																
cdh											4										
g^2						1															
ceg							1														
d^2g									1												
c^3g															4						
cf^2										1											
def												1									
c^2df																	3				
e^3													1								
c^2e^2														1							
cd^2e																1					
c^4e																			2		
d^4																		1			
c^3d^2																				1	
c^6																					1

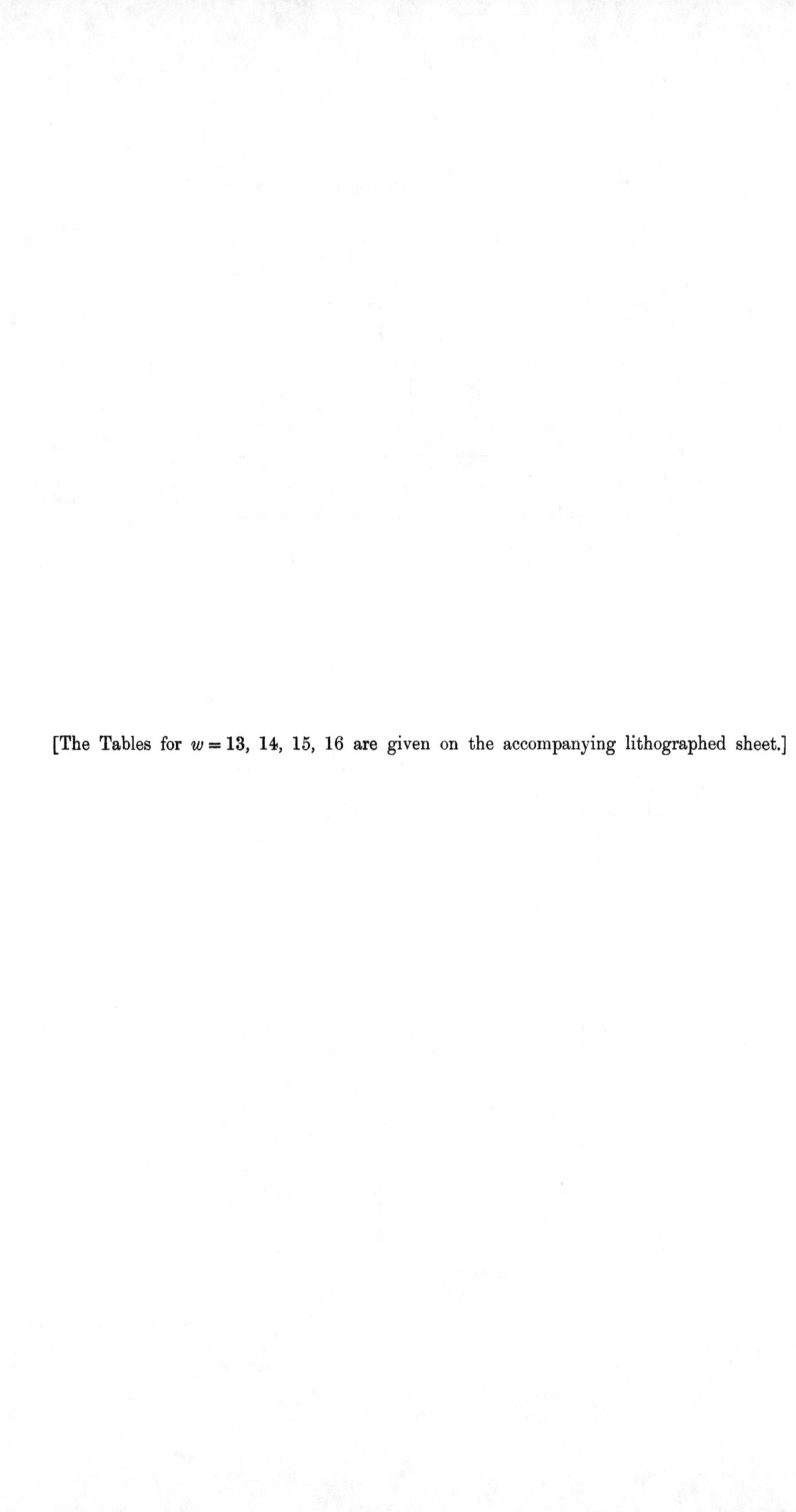

[The Tables for $w = 13$, 14, 15, 16 are given on the accompanying lithographed sheet.]

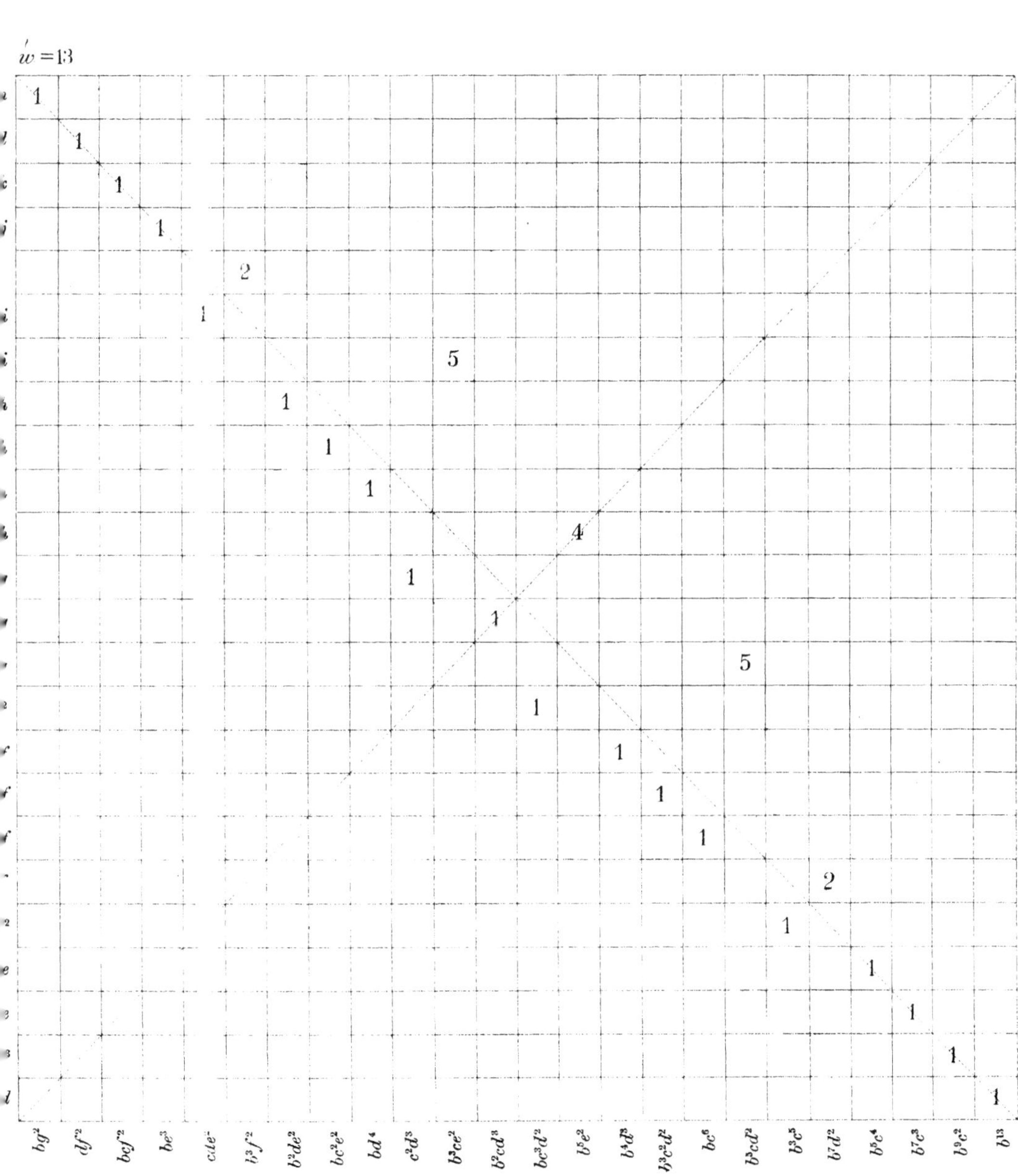

w = 13
bg²
df²
bef²
be³
cde²
b³f²
b²de²
bc²e²
bd⁴
c²d³
b³ce²
b²cd³
bc³d²
b⁵e²
b⁴d³
b³c²d²
bc⁶
b⁵cd²
b³c⁵
b⁷d²
b⁵c⁴
b⁷c³
b⁹c²
b¹³
1
1
1
1
2
1
5
1
1
1
4
1
1
5
1
1
1
1
2
1
1
1
1
1

$w = 14$

o
cm
dl
ek
c²k
fj
cdj
cei
d²i
h²
cfh
deh
c²dh
cg²
dfg
e²g
c²eg
cd²g
c⁴g
ef²
c²f²
cdef
d³f
c³df
ce³
d²e²
c³e²
c²d²e
c⁵e
cd⁴
c⁴d²
c⁷
h²
cg²
ef²
b²g²
bdf²
c²f²
ce³
d²e²
b²cf²
b²e³
bcde²
c³e²
cd⁴
b⁴f²
b³de²
b²c²e²
b²d⁴
bcd³
c⁴d²
b⁴ce²
b³cd³
b²c³d²
c⁷
b⁶e²
b⁵d³
b⁴c²d²
b²c⁶
b⁶cd²
b⁴c⁵
b⁸d²
b⁵c⁴
b⁸c³
b¹⁰c²
1
1
2
1
5
1
8
1
1
2
10
1
1
3
1
10
1
1
1
3
8
1
1
1
2
5
1
1
1
1
2
1
1

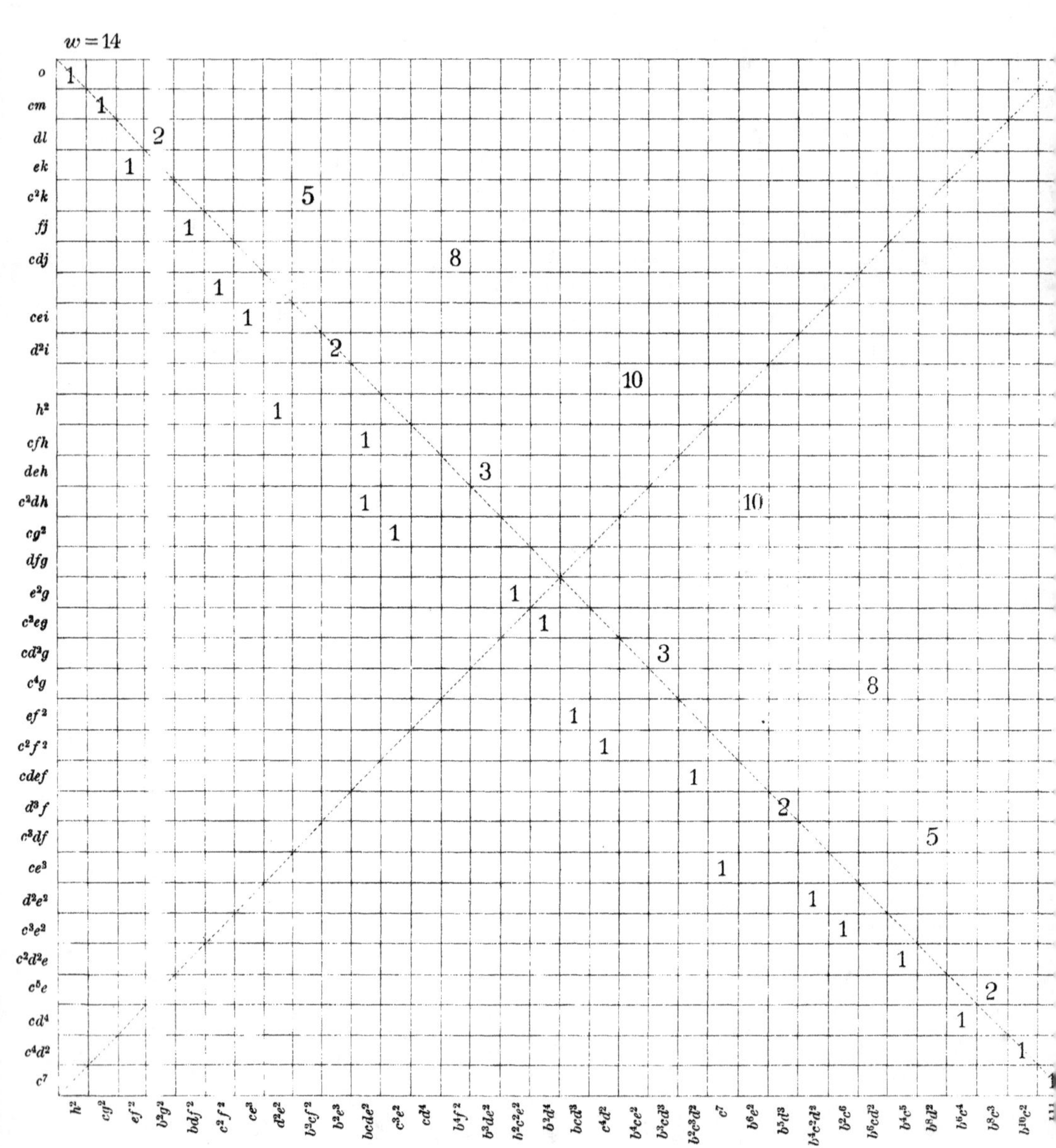

$w=15$

ch^2
dg^2
f^3
bcg^2
bef^2
cdf^2
de^3
b^3g^2
b^2df^2
bc^2f^2
bce^3
bd^2e^2
c^2de^2
d^5
b^3cf^2
b^3e^3
b^2cde^2
bc^3e^2
bcd^4
c^3d^3
b^5f^2
b^4de^2
$b^3c^2e^2$
b^3d^4
$b^2c^2d^3$
bc^4d^3
b^5ce^2
b^4cd^3
$b^3c^3d^2$
bc^7
b^7e^2
b^6d^3
$b^5c^2d^2$
b^3c^6
b^7cd^2
b^5c^5
b^8d^2
b^7c^4
b^9c^3
$b^{11}c^2$
b^{15}

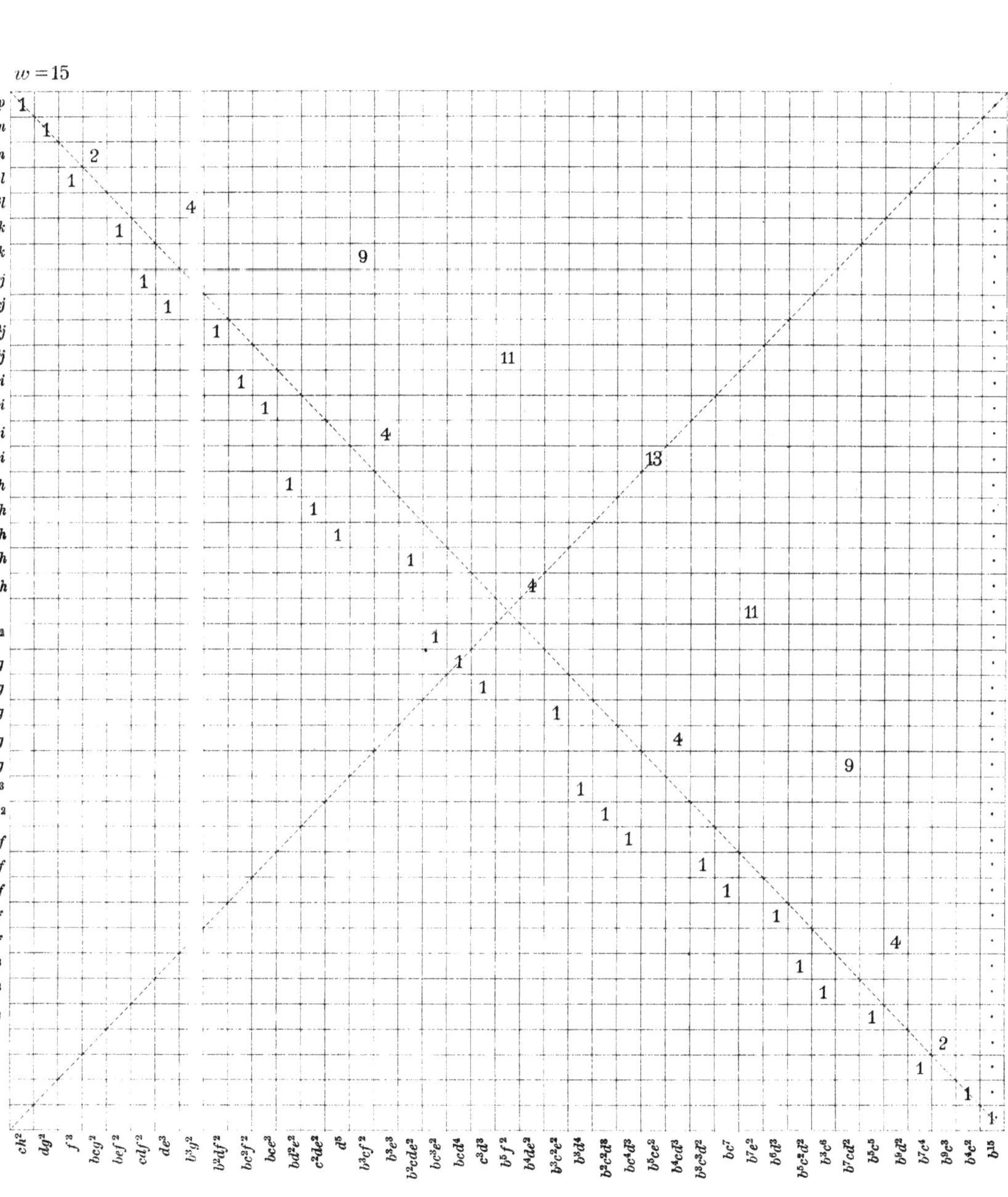

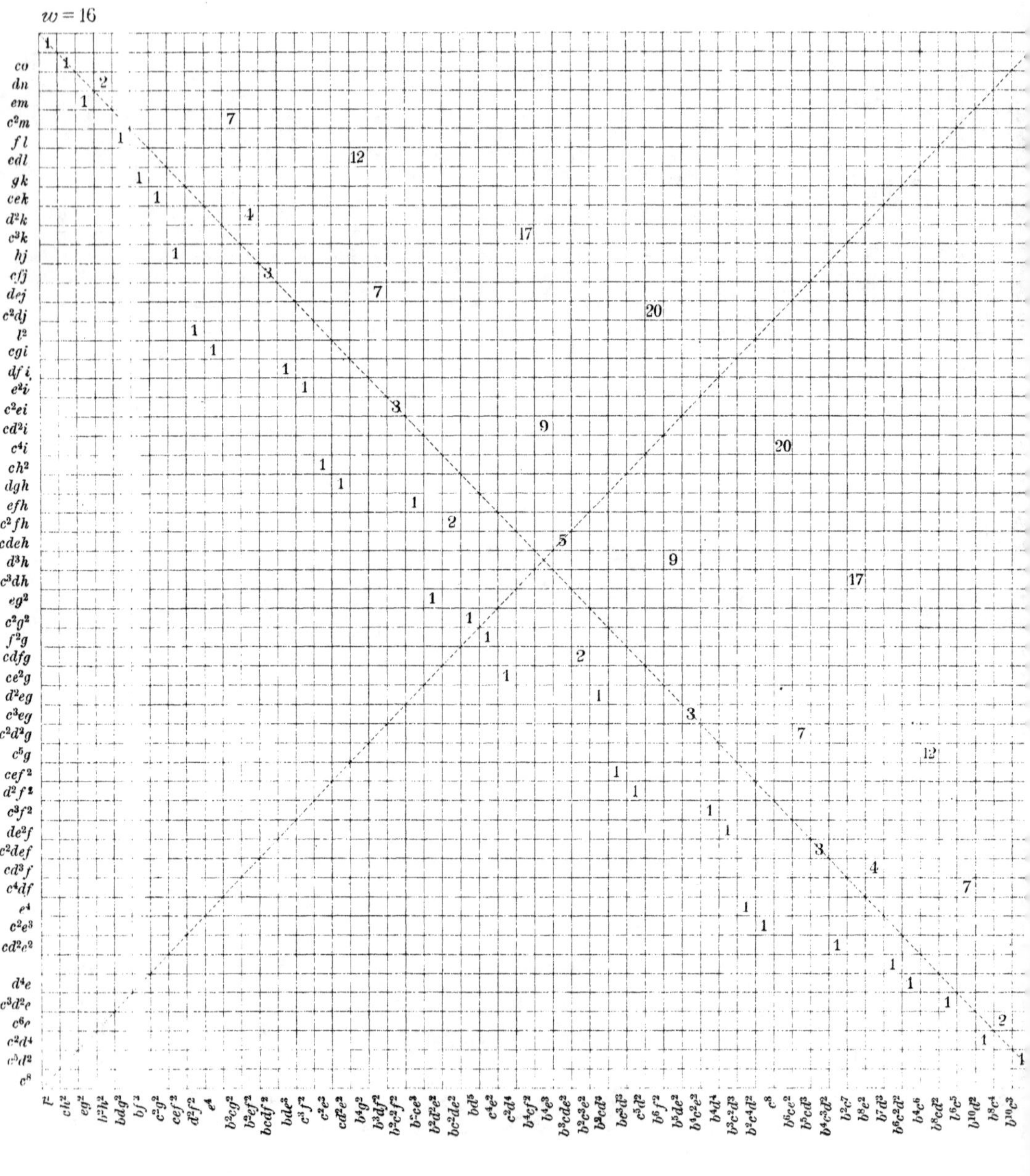

$w = 16$

933.

TABLES OF PURE RECIPROCANTS TO THE WEIGHT 8.

[From the *American Journal of Mathematics*, t. XV. (1893), pp. 75—77.]

IN the tabulation of Pure Reciprocants it is convenient to write $a=1$; we thus have for all the reciprocants of a given weight a single column of literal terms which (as in the Seminvariant Tables) I arrange in alphabetical order AO, and the several reciprocants have then each of them its own column of numerical coefficients: the form of the table is thus similar to that of the seminvariant table, the only difference being that for reciprocants the final terms are not in general power-enders: as in the seminvariant table, the columns of the table are arranged *inter se* with their final terms in AO. As remarked in my paper, "Corrected Seminvariant Tables for the Weights 11 and 12," *Amer. Math. Journ.*, t. XIV. (1892), pp. 195—200, [926], it is not in every case the top term of a column which should be regarded as the initial term; but to the extent 8, to which the reciprocant tables are here carried, this remark has no application.

I recall that the notation is the modified one employed by Halphen, and by Sylvester* in his 12th and subsequent lectures, viz. $a, b, c, d, \ldots$ denote

$$\tfrac{1}{2}\frac{d^2y}{dx^2},\ \tfrac{1}{6}\frac{d^3y}{dx^3},\ \tfrac{1}{24}\frac{d^4y}{dx^4},\ \tfrac{1}{120}\frac{d^5y}{dx^5},\ldots$$

respectively. As already noticed, a is put $=1$, but it is to be in the several terms restored in the proper powers so as to obtain for the reciprocant a homogeneous expression of a degree equal to the original degree of the final term; thus $d-3bc+2b^3$ is to be read as standing for $a^2d-3abc+2b^3$.

The ultimate verification of the expression for a pure reciprocant consists (as is known) in its annihilation by the operator

$$V = 2a^2\partial_b + 5ab\partial_c + (6ac+3b^2)\,\partial_d + (7ad+7bc)\,\partial_e + (8ae+8bd+4c^2)\,\partial_f + \text{\&c.},$$

or, say

$$V = 2\partial_b \ + 5b\partial_c \ + (6c \ + 3b^2)\,\partial_d + (7d \ + 7bc)\,\partial_e + (8e \ + 8bd + 4c^2)\,\partial_f + \text{\&c.};$$

[* *American Journal of Mathematics*, t. IX. (1887), p. 7.]

thus for the reciprocant $50e - 175bd + 28c^2 + 105b^2c$, the result obtained is

$$2(-175d + 210bc) + 5b(56c + 105b^2) + (6c + 3b^2)(-175b) + (7d + 7bc)(50),$$

or, collecting, this is

d	$-350 \qquad\qquad + 350$	$\pm\ 350$
bc	$+420 + 280 - 1050 + 350$	$\pm\ 1050$
b^3	$+525 - 525$	$\pm\ 525,$

$= 0$, as it should be.

The tables are

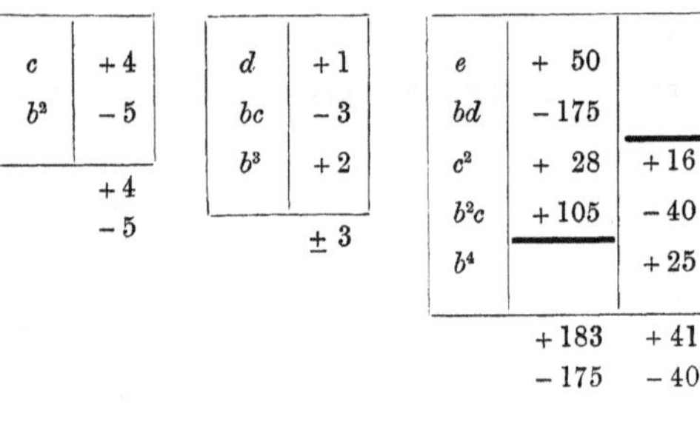

c	+4
b^2	−5
	+4
	−5

d	+1
bc	−3
b^3	+2
	±3

e	+ 50	
bd	−175	
c^2	+ 28	+16
b^2c	+105	−40
b^4		+25
	+183	+41
	−175	−40

f	+ 10	
be	− 40	
cd	− 12	+ 4
b^2d	+ 65	− 5
bc^2	+ 16	− 12
b^3c	− 39	+ 23
b^5		− 10
	± 91	± 27

g	+ 14			
bf	− 63			
ce	− 1350	+ 800		
d^2	+ 1470	− 875	+ 125	
b^2e	+ 1782	− 1000		
bcd	− 4158	+ 2450	− 750	
c^3	+ 2130	− 1344	+ 256	+ 64
b^3d			+ 500	
b^2c^2		+ 35	+ 165	− 240
b^4c			− 300	+ 300
b^6				− 125
	+ 5576	+ 3250	± 1018	+ 364
	− 5508	− 3254		− 365

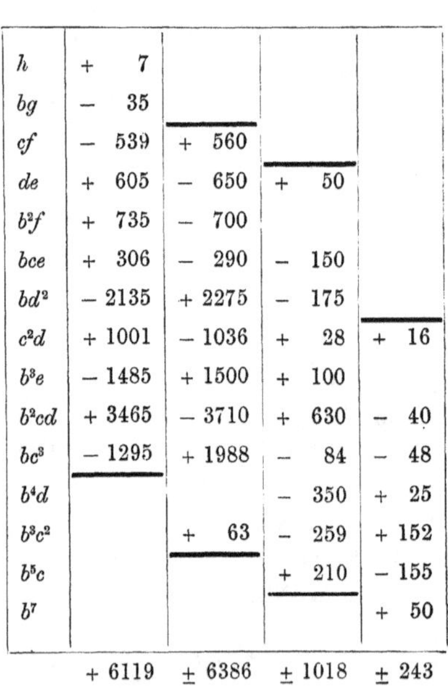

h	+ 7			
bg	− 35			
cf	− 539	+ 560		
de	+ 605	− 650	+ 50	
b^2f	+ 735	− 700		
bce	+ 306	− 290	− 150	
bd^2	− 2135	+ 2275	− 175	
c^2d	+ 1001	− 1036	+ 28	+ 16
b^3e	− 1485	+ 1500	+ 100	
b^2cd	+ 3465	− 3710	+ 630	− 40
bc^3	− 1295	+ 1988	− 84	− 48
b^4d			− 350	+ 25
b^3c^2		+ 63	− 259	+ 152
b^5c			+ 210	− 155
b^7				+ 50
	+ 6119	± 6386	± 1018	± 243
	− 5489			

i	+ 420						
bh	− 2310						
cg	− 32704	+ 1176					
df	+ 57750	− 8085	+ 20433				
e^2	− 20460	+ 7040	− 21542	+ 625			
b^2g	+ 45500	− 1470					
bcf	− 28392	+ 18963	− 61299				
bde	− 90900	− 16940	+ 69062	− 4375			
c^2e	+ 103740	− 27160	+ 80248	+ 49700	+ 3200		
cd^2	− 38320	+ 26460	− 85554	+ 55125	− 3500	+ 500	
b^3f	− 69615	− 9555	+ 40866				
b^2ce	+ 83538	+ 28098	− 106218	+ 128625	− 8000		
b^2d^2	+ 92820	+ 12740	− 54782	− 61250	+ 4375	− 625	
bc^2d	− 102102	− 52822	+ 191590	− 156800	+ 9800	− 3000	
c^4		+ 21560	− 73304	+ 84868	− 5376	+ 1024	+ 256
b^4e			− 378	− 78750	+ 5000		
b^3cd			+ 1176	+ 183750	− 12250	+ 5750	
b^2c^3				− 102165	+ 6580	− 620	− 1280
b^5d						− 2500	
b^4c^2					+ 175	− 2025	+ 2400
b^6c						+ 1500	− 2000
b^8							+ 625
	+ 383768	+ 116037	+ 403375	+ 452993	+ 29130	+ 8774	+ 3281
	− 384803	− 116032	− 403077	− 453040	− 29126	− 8750	− 3280

I remark that in the last of these tables the first column, say $i \infty bc^2d$, which ends in bc^2d, is a more simple form than Sylvester's P_8, $= i \infty c^4$, (*Amer. Math. Journ.*, t. IX. p. 35), which ends in c^4; P_8 is in fact a linear combination, first col. + 6 second col. of the first and second columns of the table: the second column, say $cg \infty c^4$ is Sylvester's (a^2cg), t. IX. p. 124.

934.

NOTE ON THE SO-CALLED QUOTIENT G/H IN THE THEORY OF GROUPS.

[From the *American Journal of Mathematics*, t. XV. (1893), pp. 387, 388.]

THE notion (see Hölder, "Zur Reduction der algebraischen Gleichungen," *Math. Ann.*, t. XXXIV. (1887), § 4, p. 31) is a very important one, and it is extensively made use of in Mr Young's paper, "On the Determination of Groups whose Order is the Power of a Prime," *American Journal of Mathematics*, t. XV. (1893), pp. 124—178; but it seems to me that the meaning is explained with hardly sufficient clearness, and that a more suitable algorithm might be adopted, viz. instead of $G_1 = G/\Gamma_1$, I would rather write $G = \Gamma_1 . QG_1$ or $QG_1 . \Gamma_1$.

We are concerned with a group G containing as part of itself a group Γ_1, such that each element of Γ_1 is commutative with each element of G. This being so, we may write

$$G = QG_1 . \Gamma_1,$$

where QG_1 is not a group but a mere array of elements, viz. if $\Gamma_1 = (1, A_2, \ldots, A_s)$, and $QG_1 = (1, B_2, \ldots, B_t)$, then the formula is

$$G = (1, B_2, \ldots, B_t)(1, A_2, \ldots, A_s),$$

where it is to be noticed that the elements B are not determinate; in fact, if A_θ be any element of Γ_1, we may, in place of an element B, write BA_θ, for

$$B(1, A_2, \ldots, A_s) \text{ and } BA_\theta(1, A_2, \ldots, A_s)$$

are, in different orders, the same elements of G.

But, G being a group, the product of any two elements of G is an element of G; viz. we thus have in general

$$B_iA_{i'} . B_jA_{j'} = B_kA_{k'};$$

that is,

$$B_i B_j = B_k A_{k'} A_{j'}^{-1} A_j^{-1} \quad (i,\ j,\ \text{unequal or equal}),$$

where the B_k is a determinate element of the series 1, $B_2, \ldots, B_t$, depending only on the elements B_i and B_j into the product of which it enters; and it is in nowise affected by the before-mentioned indeterminateness of the elements B: say B_i, B_j being any two elements of the series 1, $B_2, \ldots, B_t$, we have the last preceding equation wherein B_k is a determinate element of the same series.

We may imagine a set of elements 1, $B_2, \ldots, B_t$ for which, B_i, B_j being any two of them and B_k a third element determined as above, we have always $B_iB_j = B_k$, that is, these elements 1, $B_2, \ldots, B_t$ now form a group, say the group G_1; the original elements 1, $B_2, \ldots, B_t$ (which are subject to a different law of combination $B_iB_j = B_k A_{k'} A_{j'}^{-1} A_j^{-1}$, and do not form a group) are regarded as a mere array connected with this group, and so represented as above by QG_1; and the relation of the original group G to the group Γ_1 (consisting of elements commutative with those of G) and to the new group G_1 is expressed as above by the equation

$$G = \Gamma_1 \,.\, QG_1, \quad = QG_1 \,.\, \Gamma_1.$$

Cambridge, 2 *June*, 1893.

935.

SUR LA FONCTION MODULAIRE $\chi\omega$.

[From the *Comptes Rendus de l'Académie des Sciences de Paris*, t. CXVI. (*Janvier—Juin*, 1893), pp. 1339—1343.]

SELON les notations usitées, on a

$$\chi\omega = \sqrt[12]{kk'} = 2^{\frac{1}{6}} q^{\frac{1}{24}} \div 1 + q \,.\, 1 + q^3 \,.\, 1 + q^5 \ldots ;$$

or, d'après une transformation trouvée par M. Glaisher, on a

$$\frac{1}{1+q} + \frac{3q^2}{1+q^3} + \frac{5q^4}{1+q^5} + \ldots = \frac{1+q^2}{(1-q^2)^2} - \frac{q(1+q^4)}{(1-q^4)^2} + \frac{q^2(1+q^6)}{(1-q^6)^2} - \ldots,$$

et de là, en intégrant,

$$\log 1 + q \,.\, 1 + q^3 \,.\, 1 + q^5 \ldots = \frac{q}{1-q^2} - \tfrac{1}{2}\frac{q^2}{1-q^4} + \tfrac{1}{3}\frac{q^3}{1-q^6} - \ldots,$$

$$= -\frac{1}{q-q^{-1}} + \tfrac{1}{2}\frac{1}{q^2-q^{-2}} - \tfrac{1}{3}\frac{1}{q^3-q^{-3}} + \ldots ;$$

donc

$$\chi\omega = 2^{\frac{1}{6}} q^{\frac{1}{24}} \exp.\left(\frac{1}{q-q^{-1}} - \tfrac{1}{2}\frac{1}{q^2-q^{-2}} + \tfrac{1}{3}\frac{1}{q^3-q^{-3}} - \ldots\right),$$

ou, en écrivant $q = e^{i\pi\omega}$,

$$\chi\omega = 2^{\frac{1}{6}} \exp.\left[\frac{i\pi\omega}{24} + \frac{1}{2i}\left(\frac{1}{\sin \pi\omega} - \frac{1}{2\sin 2\pi\omega} + \frac{1}{3\sin 3\pi\omega} - \ldots\right)\right],$$

$$= 2^{\frac{1}{6}} \exp.\left[\frac{i\pi\omega}{24} - \frac{1}{2i}\Sigma_n \frac{\cos n\pi}{n \sin n\pi\omega}\right],$$

$n=1$ jusqu'à $n=\infty$, forme qui met en évidence l'équation $\chi(\omega+2) = i^{\frac{1}{6}}\chi\omega$.

Mais nous avons

$$\frac{1}{n \sin n\pi\omega} = \frac{1}{n^2\pi\omega} + \frac{1}{\pi} \Sigma_m \frac{\cos m\pi}{n(n\omega - m)},$$

$m = 1$ jusqu'à $m = +\infty$ et $m = -1$ jusqu'à $m = -\infty$.

On a, dans les parenthèses, d'abord les termes

$$\frac{1}{2i}\frac{1}{\pi\omega}\left(\frac{1}{1^2} - \frac{1}{2^2} + \frac{1}{3^2} - \dots\right), \quad = \frac{1}{2i}\frac{1}{\pi\omega}\cdot\frac{\pi^2}{12}, \quad = -\frac{\pi i}{24\omega},$$

et l'expression de $\chi\omega$ devient ainsi

$$\chi\omega = 2^{\frac{1}{6}} \exp.\left[\frac{i\pi\omega}{24} - \frac{i\pi}{24\omega} + \frac{i}{2\pi}\Sigma_n\Sigma_m \frac{\cos n\pi \cos m\pi}{n(n\omega - m)}\right],$$

ou, ce qui est plus commode,

$$\chi\omega = 2^{\frac{1}{6}} \exp.\left[\frac{i\pi\omega}{24} - \frac{i\pi}{24\omega} + \frac{i}{2\pi}(S_1 - S_2 - S_3 + S_4)\frac{1}{n(n\omega - m)}\right],$$

où les signes sommatoires $S = \Sigma_n\Sigma_m$ se rapportent: S_1 aux valeurs impaires de m et n; S_2 aux valeurs impaires de m et paires de n; S_3 aux valeurs paires de m et impaires de n; S_4 aux valeurs paires de m et n. En omettant l'expression $\frac{1}{n(n\omega - n)}$ du terme sommé, on peut considérer S_1, S_2, S_3, S_4 comme dénotant les quatre sommes dont il s'agit.

Considérons d'abord S_4; en supposant que m', n' soient l'un ou l'autre ou tous les deux impairs, on peut donner à m, n les valeurs $2m'$, $2n'$; $4m'$, $4n'$; $8m'$, $8n'$; ...; on obtient

$$S_4 = (S_1 + S_2 + S_3)\left(\frac{1}{2^2} + \frac{1}{4^2} + \frac{1}{8^2} + \dots\right) = \tfrac{1}{3}(S_1 + S_2 + S_3),$$

et le terme $S_1 - S_2 - S_3 + S_4$ dans l'expression de $\chi\omega$ devient ainsi

$$= \tfrac{4}{3}S_1 - \tfrac{2}{3}S_2 - \tfrac{2}{3}S_3.$$

Dans S_1, S_2, ou S_3, en supposant que m', n' dénotent des nombres relativement premiers, on peut donner à m, n les valeurs m', n'; $3m'$, $3n'$; $5m'$, $5n'$; ...: on obtient ainsi, pour chacune de ces sommes,

$$S = S\left(\frac{1}{1^2} + \frac{1}{3^2} + \frac{1}{5^2} + \dots\right) = \frac{\pi^2}{8}S,$$

où, au côté droit, m et n sont des nombres relativement premiers; le terme devient ainsi

$$\frac{\pi^2}{8}(\tfrac{4}{3}S_1 - \tfrac{2}{3}S_2 - \tfrac{2}{3}S_3) = \frac{\pi^2}{6}(S_1 - \tfrac{1}{2}S_2 - \tfrac{1}{2}S_3),$$

et nous avons

$$\chi\omega = 2^{\frac{1}{6}} \exp.\left[\frac{i\pi\omega}{24} - \frac{i\pi}{24\omega} + \frac{i\pi}{12}(S_1 - \tfrac{1}{2}S_2 - \tfrac{1}{2}S_3)\frac{1}{n(n\omega - m)}\right].$$

Je m'arrête pour remarquer que cette expression s'accorde parfaitement avec des résultats trouvés par M. Dedekind (*Œuvres de Riemann*, Leipzig, 1876, p. 447). En effet, en donnant à ω la valeur $\omega = \frac{m}{n} + i\alpha$, α une valeur positive très petite, M. Dedekind trouve les valeurs de $\log k$ et $\log k'$; en ajoutant ces valeurs et en ne faisant attention qu'aux termes qui deviennent infinis pour $\alpha = 0$, on a

$$\log kk' = 12 \log \chi\omega = \quad 24A,$$

m et n tous les deux impairs;

$$\log kk' = 12 \log \chi\omega = -12A,$$

m et n l'un impair, l'autre pair.

Ici $A = \frac{\pi i}{24n(n\omega - m)}$, et les valeurs de $\log \chi\omega$ sont ainsi

$$\frac{i\pi}{12}\frac{1}{n(n\omega - m)} \text{ et } -\frac{i\pi}{24}\frac{1}{n(n\omega - m)}$$

respectivement.

Dans l'expression de $\chi\omega$, il convient de réunir les termes qui correspondent aux valeurs $-m$, $+m$, c'est-à-dire au lieu de $\frac{1}{n(n\omega - m)}$, on doit écrire

$$\frac{1}{n}\left(\frac{1}{n\omega - m} + \frac{1}{n\omega + m}\right) = \frac{2\omega}{n^2\omega^2 - m^2};$$

on a ainsi finalement

$$\chi\omega = 2^{\frac{1}{6}} \exp. \left[\frac{i\pi\omega}{24} - \frac{i\pi}{24\omega} + \frac{i\pi}{6}(S_1 - \tfrac{1}{2}S_2 - \tfrac{1}{2}S_3)\frac{\omega}{n^2\omega^2 - m^2}\right],$$

où je rappelle que m et n sont des nombres positifs relativement premiers, et que les signes sommatoires S_1, S_2, S_3 se rapportent, S_1 aux valeurs impaires de m et n, S_2 aux valeurs impaires de m et paires de n, S_3 aux valeurs paires de m et impaires de n.

Écrivant $-\frac{1}{\omega}$ au lieu de ω, le terme $\frac{\omega}{n^2\omega^2 - m^2}$ auquel se rapportent les sommations se change en $\frac{\omega}{m^2\omega^2 - n^2}$; on peut échanger les lettres m et n, et l'expression de $\chi\left(-\frac{1}{\omega}\right)$ devient ainsi identiquement celle de $\chi\omega$, c'est-à-dire la forme met en évidence la relation $\chi\left(-\frac{1}{\omega}\right) = \chi\omega$.

Il y a cependant, dans cette analyse et par rapport à l'échange des lettres m et n, une difficulté qu'il convient d'écarter. Dans la formule

$$\frac{1}{\sin x} = \frac{1}{x} + \Sigma_m \frac{\cos m\pi}{x - m\pi} \quad (m = 1 \text{ jusqu'à } +\infty \text{ et } -1 \text{ jusqu'à } -\infty),$$

qui donne lieu à

$$\frac{1}{n\sin^2 n\pi\omega} = \frac{1}{n^2\pi\omega} + \frac{1}{\pi}\Sigma_m \frac{\cos m\pi}{n(n\omega - m)},$$

il est nécessaire que la variable x ait une valeur finie ou au moins infiniment petite par rapport aux valeurs extrêmes de m; ainsi, en écrivant pour x la valeur $n\pi\omega$, les valeurs extrêmes de n doivent être infiniment petites par rapport à celles de m, et la somme que nous avons dénotée simplement par $\Sigma_n\Sigma_m \frac{1}{n(n\omega - m)}$ signifie réellement $\overset{\nu}{\Sigma}_n \overset{\mu}{\Sigma}_m \frac{1}{n(n\omega - m)}$, savoir les limites pour n sont 1, ν et pour m ces limites sont -1, $-\mu$ et $+1$, $+\mu$ où μ et ν sont des nombres infiniment grands, mais $\frac{\mu}{\nu} = \infty$, ou, ce qui est la même chose, la somme est $\overset{\nu}{\Sigma}_n \overset{\infty}{\Sigma}_m \frac{1}{n(n\omega - m)}$, où ν est un très grand nombre qui devient enfin $= \infty$. En réunissant les termes pour m et $-m$, la somme à considérer est $\overset{\nu}{\Sigma}_n \overset{\infty}{\Sigma}_m \frac{1}{n^2\omega^2 - m^2}$, et il s'agit de faire voir qu'il est permis de substituer pour cela la somme

$$\overset{\nu}{\Sigma}_n \overset{\nu}{\Sigma}_m \frac{1}{n^2\omega^2 - m^2} \qquad (\nu = 1 \text{ à } \nu = \infty).$$

La différence des deux expressions est une somme double $n = 1$ jusqu'à $n = \nu$ et $m = \nu + 1$, jusqu'à $m = \infty$, savoir cette somme est égale à

$$\begin{aligned} \cos\nu\pi \Bigg\{ + &\left[\frac{1}{\omega^2 - (\nu+1)^2} - \frac{1}{\omega^2 - (\nu+2)^2} + \frac{1}{\omega^2 - (\nu+3)^2} - \dots\right] \\ - &\left[\frac{1}{4\omega^2 - (\nu+1)^2} - \frac{1}{4\omega^2 - (\nu+2)^2} + \frac{1}{4\omega^2 - (\nu+3)^2} - \dots\right] \\ &\dots\dots\dots\dots\dots\dots\dots\dots\dots\dots \\ - \cos\nu\pi &\left[\frac{1}{\nu^2\omega^2 - (\nu+1)^2} - \frac{1}{\nu^2\omega^2 - (\nu+2)^2} + \frac{1}{\nu^2\omega^2 - (\nu+3)^2} - \dots\right]\Bigg\}, \end{aligned}$$

laquelle somme $\left(\text{sauf pour une valeur imaginaire } \omega = \frac{m}{n} + i\alpha,\ \alpha \text{ positif}\right)$ devient aussi petite que l'on veut en donnant à ν une valeur suffisamment grande; c'est-à-dire qu'on peut négliger cette différence et ainsi considérer la somme $\Sigma_n\Sigma_m \frac{1}{n(n\omega - m)}$, dont je me suis servi dans l'investigation comme ayant pour n et pour les valeurs positives ou négatives de m les mêmes limites 1, ν $(\nu = \infty)$.

La forme trouvée pour $\chi\omega$ met en évidence que $\omega = 0$, $\omega = \infty$, et $\omega = \pm\frac{m}{n}$ sont des valeurs essentiellement singulières pour la fonction.

936.

NOTE ON UNIFORM CONVERGENCE.

[From the *Proceedings of the Royal Society of Edinburgh*, vol. XIX. (1893), pp. 203—207. Read December 5, 1892.]

IT appears to me that the form in which the definition or condition of uniform convergence is usually stated, is (to say the least) not easily intelligible. I call to mind the general notion: We may have a series, to fix the ideas, say of positive terms

$$(0)_x + (1)_x + (2)_x, \dots + (n)_x, \dots$$

the successive terms whereof are continuous functions of x, for all values of x from some value less than a up to and inclusive of a (or from some value greater than a down to and inclusive of a): and the series may be convergent for all such values of x, the sum of the series ϕx is thus a determinate function ϕx of x; but ϕx is not of necessity a continuous function; if it be so, then the series is said to be uniformly convergent; if not, and there is for the value $x = a$ a breach of continuity in the function ϕx, then there is for this value $x = a$ a breach of uniform convergence in the series.

Thus if the limits are say from 0 up to the critical value 1, then in the geometrical series $1 + x + x^2 + \dots$, the successive terms are each of them continuous up to and inclusive of the limit 1, but the series is only convergent up to and exclusive of this limit, viz. for $x = 1$ we have the divergent series $1 + 1 + 1 + \dots$, and this is *not* an instance; but taking, instead, the geometrical series

$$(1 - x) + (1 - x)\,x + (1 - x)\,x^2 + \dots,$$

here the terms are each of them continuous up to and inclusive of the limit 1, and the series is also convergent up to and inclusive of this limit; in fact, at the limit the series is $0+0+0+\ldots$ We have here an instance, and there is in fact a discontinuity in the sum, viz. $x<1$ the sum is

$$(1-x)(1+x+x^2+\ldots),\ =(1-x)\,.\,\frac{1}{1-x},\ =1\,;$$

whereas for the limiting value 1, the sum is $0+0+0+\ldots,\ =0$. The series is thus uniformly convergent up to and exclusive of the value $x=1$, but for this value there is a breach of uniform convergence.

I remark that Du Bois-Reymond in his paper, "Notiz über einen Cauchy'schen Satz, die Stetigkeit von Summen unendlicher Reihen betreffend," *Math. Ann.*, t. IV. (1871), pp. 135—137, shows that, when certain conditions are satisfied, the sum ϕx is a continuous function of x, but he does not use the term "uniform convergence," nor give any actual definition thereof.

M. Jordan, in his "Cours d'Analyse de l'École Polytechnique," t. I. (Paris, 1882), considers p. 116 the series $s=u_1+u_2+u_3+\ldots$, the terms of which are functions of a variable z, and after remarking that such a series is convergent for the values of z included within a certain interval, if for each of these values and for every value of the infinitely small quantity ϵ we can assign a value of n such that for every value of p,

$$\text{Mod}\,(u_{n+1}+u_{n+2}+\ldots+u_{n+p})<\text{Mod}\ \epsilon,$$

ϵ being as small as we please, proceeds:—

"Le nombre des termes qu'il est nécessaire de prendre dans la série pour arriver à ce résultat sera en général une fonction de z et de ϵ. Néanmoins on pourra très habituellement déterminer un nombre n fonction de ϵ seulement telle que la condition soit satisfaite pour toute valeur de z comprise dans l'intervalle considéré. On dira dans ce cas que la série s est *uniformément convergente* dans cet intervalle."

And similarly, Professor Chrystal in his *Algebra*, Part II. (Edinburgh, 1889), after considering, p. 130, the series

$$\frac{x}{x+1\,.\,2x+1}+\frac{x}{2x+1\,.\,3x+1}+\ldots+\frac{x}{(n-1)\,x+1\,.\,nx+1}+\ldots$$

for which the critical value is $x=0$, and in which when $x=0$ the residue R_n of the series or sum of the $(n+1)$th and following terms is $=\dfrac{1}{nx+1}$ proceeds as follows:—

Now when x has any given value, we can by making n large enough make $\dfrac{1}{nx+1}$ smaller than any given positive quantity α. But on the other hand, the smaller x is the larger must we take n in order that $\dfrac{1}{nx+1}$ may fall under α; and in general when x is variable there is no finite upper limit for n *independent of* x, say v, such that if $n>v$ then $R_n<\alpha$. When the residue has this peculiarity the series is said to be *non-uniformly convergent;* and if for a particular value of x, such as $x=0$ in the present example, the number of terms required to secure a given degree of approximation to the limit is infinite, the series is said to *converge infinitely slowly.*

And he thereupon gives the formal definition: *If for values of x within a given region in Argand's diagram we can for every value of α, however small Mod. α, assign for n an upper limit v* INDEPENDENT OF x, such that, when $n > v$, Mod. $R_n <$ Mod. α, then the series $\Sigma f(n, x)$ is said to be UNIFORMLY CONVERGENT within the region in question.

The two forms of definition (Jordan and Chrystal) appear to me equivalent, and it seems to me that construing the definition *strictly*, and applying it to the above instance

$$(1-x)+(1-x)x+(1-x)x^2+\ldots,$$

the definition does not in either case indicate a breach of uniform convergency at $x=1$, viz. the definition shows uniform convergency from $x=0$ to $x=1-\epsilon$, ϵ being a positive quantity however small, or (as I have before expressed this) uniform convergency up to and exclusive of the limit 1; and further, it shows uniform convergency at the limit 1. For at this limit, the series of terms is $0+0+0+\ldots$, the residue or sum of the $(n+1)$th and subsequent terms is thus also $0+0+0+\ldots$, and we get the value of this residue, not approximately, but exactly, by taking a single term of the series. Jordan and Chrystal calculate, each of them, the residue from the general expression thereof by writing therein for x or z the critical value; and then, comparing the value thus obtained with the values obtained for the $(n+1)$th and subsequent terms of the series on substituting therein for x or z the critical value, they seem to argue that the discrepancy between these two values indicates the breach of uniform convergency.

It may be said that the objection is a verbal one. But it seems to me that the whole notion of the residue (although very important as regards the general theory of convergence) is irrelevant to the present question of uniform convergency, and that a better method of treating the question is as follows:

Considering as before the series

$$(0)_x+(1)_x+(2)_x+\ldots+(n)_x+\ldots,$$

where the functions $(0)_x$, $(1)_x$, $(2)_x$, ... are each of them continuous up to and inclusive of the limit $x=a$, and the series has thus a definite sum ϕx, this sum is *primâ facie* a continuous function of x, and what we have to explain is the manner in which it may come to be discontinuous. Suppose that it is continuous up to and exclusive of the limit $x=a$, but that there is a breach of discontinuity at this limit: write $x=a-\epsilon$, where ϵ is a positive quantity as small as we please, and consider the two equations

$$\phi x=(0)_x+(1)_x+(2)_x+\ldots,$$

$$\phi a=(0)_a+(1)_a+(2)_a+\ldots,$$

then we have

$$\phi a-\phi x=\epsilon\left\{\frac{(0)_a-(0)_x}{a-x}+\frac{(1)_a-(1)_x}{a-x}+\ldots\right\}.$$

Hence if the sum of the series in { } is a finite magnitude M, not indefinitely large for an indefinitely small value of ϵ, we have $\phi a - \phi x = \epsilon M$, which is indefinitely small for ϵ indefinitely small, and there is no breach of continuity; the only way in which a breach of continuity can arise is by the series in { } having a value indefinitely large for ϵ indefinitely small, viz. if such a value is $\frac{N}{\epsilon}$, then $\phi a - \phi x = \epsilon \,.\, \frac{N}{\epsilon}$, $= N$, and as x changes from $a - \epsilon$ to a, the sum changes abruptly from $\phi(a-\epsilon)$ to $\phi(a-\epsilon)+N$.

The condition for a breach of uniform convergency for the value $x = a$, thus is that, writing $x = a - \epsilon$, ϵ a positive magnitude however small, the series

$$\frac{(0)_a - (0)_x}{a - x} + \frac{(1)_a - (1)_x}{(a - x)} + \ldots,$$

shall have a sum indefinitely large for ϵ indefinitely small, or say as before, a sum $= \frac{N}{\epsilon}$.

For the foregoing example, where the series is

$$(1-x) + x(1-x) + x^2(1-x) + \ldots$$

the critical value is $a = 1$: we have here $(n)_x = x^n(1-x)$, and consequently

$$\frac{(n)_a - (n)_x}{a - x} = \frac{a^n - x^n}{a - x} - \frac{a^{n+1} - x^{n+1}}{(a - x)}$$

$$= (a^{n-1} + a^{n-2}x + \ldots + x^{n-1}) - (a^n + a^{n-1}x + \ldots + x^n)$$

$$= -x^n \text{ for } a = 1.$$

The series

$$\frac{(0)_a - (0)_x}{a - x} + \frac{(1)_a - (1)_x}{a - x} + \ldots$$

thus is

$$-(1 + x + x^2 + \ldots)$$

$$= -\frac{1}{1-x}, \; = \frac{-1}{\epsilon}$$

for $x = 1 - \epsilon$, viz. we have

$$\phi 1 - \phi(1-\epsilon) = \epsilon \,.\, \frac{-1}{\epsilon}, \; = -1;$$

which is right since by what precedes

$$\phi(1-\epsilon) = 1, \;\; \phi 1 = 0.$$

937.

NOTE ON THE ORTHOTOMIC CURVE OF A SYSTEM OF LINES IN A PLANE.

[From the *Messenger of Mathematics*, vol. XXII. (1893), pp. 45, 46.]

CONSIDERING *in plano* a singly infinite system of lines, then to each point of the plane there corresponds a line (not in general a unique line), and we can therefore express in terms of the coordinates (x, y) of the point the cosine-inclinations α, β of the line to the axes. The differential equation of the orthotomic curve is then $\alpha dx + \beta dy = 0$, and it is a well-known and easily demonstrable theorem that $\alpha dx + \beta dy$ is a complete differential, say it is $= dV$; the integral equation of the orthotomic curve is therefore $V = \int (\alpha dx + \beta dy)$, $=$ const., and we see further that V is a solution of the partial differential equation $\left(\frac{dV}{dx}\right)^2 + \left(\frac{dV}{dy}\right)^2 = 1$.

If the lines are the normals of the ellipse $\frac{X^2}{a} + \frac{Y^2}{b} = 1$, then, writing the equation of the normal at the point X, Y in the form

$$\frac{a}{X}(x - X) = \frac{b}{Y}(y - Y), = \lambda,$$

suppose, we have

$$X = \frac{ax}{a + \lambda}, \quad Y = \frac{by}{b + \lambda};$$

and therefore

$$\frac{ax^2}{(a + \lambda)^2} + \frac{by^2}{(b + \lambda)^2} - 1 = 0,$$

which last equation determines λ as a function of x, y. We have α, β proportional to $\frac{X}{a}$, $\frac{Y}{b}$; or say we have

$$\alpha = M\frac{x}{a + \lambda}, \quad \beta = M\frac{y}{b + \lambda},$$

whence

$$\frac{1}{M^2} = \frac{x^2}{(a + \lambda)^2} + \frac{y^2}{(b + \lambda)^2};$$

or, writing for convenience

$$\frac{x^2}{(a+\lambda)^2}+\frac{y^2}{(b+\lambda)^2}-\frac{k^2}{\lambda^2}=0,$$

(viz. this equation defines k as a function of x, y and λ, that is, of x and y), we have

$$\alpha=\frac{\lambda x}{k(a+\lambda)},\quad \beta=\frac{\lambda y}{k(b+\lambda)};$$

and we ought therefore to have

$$\frac{\lambda}{k}\left(\frac{xdx}{a+\lambda}+\frac{ydy}{b+\lambda}\right)$$

a complete differential, $=dV$.

This is easily verified, for from the assumed value

$$k=\lambda\left(\frac{x^2}{a+\lambda}+\frac{y^2}{b+\lambda}-1\right)$$

we deduce

$$dk=2\lambda\left(\frac{xdx}{a+\lambda}+\frac{ydy}{b+\lambda}\right)+d\lambda\left(\frac{ax^2}{(a+\lambda)^2}+\frac{by^2}{(b+\lambda)^2}-1\right),\ =2\lambda\left(\frac{xdx}{a+\lambda}+\frac{ydy}{b+\lambda}\right);$$

and we have therefore

$$dV=\frac{\lambda}{k}\frac{dk}{2\lambda},\ =\tfrac{1}{2}\frac{dk}{k},$$

where k denotes a function of (x, y) defined as above; hence the equation $V=$const. gives $k=$const., or the equation of the orthotomic curve is given by the system of equations

$$\frac{ax^2}{(a+\lambda)^2}+\frac{by^2}{(b+\lambda)^2}-1=0,$$

$$\frac{x^2}{(a+\lambda)^2}+\frac{y^2}{(b+\lambda)^2}-\frac{k^2}{\lambda^2}=0,$$

where k is a constant; these equations (eliminating λ) give, in fact, the equation of the parallel curve of the ellipse, and k denotes the normal distance of a point on the curve from the ellipse. I recall that the first equation may be replaced by

$$\frac{x^2}{a+\lambda}+\frac{y^2}{b+\lambda}-\frac{k}{\lambda}-1=0,$$

and since the derived equation hereof in regard to λ is the second equation, we have the equation of the parallel curve in the known form

$$\text{Disct. }\{(\lambda+k)(\lambda+a)(\lambda+b)-(b+\lambda)x^2-(a+\lambda)y^2\}=0.$$

I notice further that, considering k a function of x, y as above, we have

$$\left(\frac{dV}{dx}\right)^2+\left(\frac{dV}{dy}\right)^2=\frac{1}{4k^2}\left\{\left(\frac{dk}{dx}\right)^2+\left(\frac{dk}{dy}\right)^2\right\},\ =\frac{\lambda^2}{k^2}\left\{\frac{x^2}{(a+\lambda)^2}+\frac{y^2}{(b+\lambda)^2}\right\},$$

that is,

$$\left(\frac{dV}{dx}\right)^2+\left(\frac{dV}{dy}\right)^2=1,$$

as it should be.

938.

ON TWO CUBIC EQUATIONS.

[From the *Messenger of Mathematics*, vol. XXII. (1893), pp. 69—71.]

STARTING from the equations

$$2 + a = b^2,$$
$$2 + b = c^2,$$
$$2 + c = a^2,$$

then eliminating b, c, we find

$$(a^4 - 4a^2 + 2)^2 - (a + 2) = 0,$$

that is,

$$a^8 - 8a^6 + 20a^4 - 16a^2 - a + 2 = 0;$$

we satisfy the equations by $a = b = c$, and thence by

$$a^2 - a - 2 = (a - 2)(a + 1) = 0:$$

there remains a sextic equation breaking up into two cubic equations; the octic equation may in fact be written

$$(a - 2)(a + 1)(a^3 + a^2 - 2a - 1)(a^3 - 3a + 1) = 0,$$

and we have thus the two cubic equations

$$x^3 + x^2 - 2x - 1 = 0, \quad x^3 - 3x + 1 = 0,$$

for each of which the roots (a, b, c) taken in a proper order are such that $2 + a = b^2$, $2 + b = c^2$, $2 + c = a^2$.

It may be remarked that starting from $y^3 + y^2 - 2y - 1 = 0$, $y^2 = x + 2$, the first equation gives $(y^3 - 2y)^2 - (y^2 - 1)^2 = 0$, that is, $y^6 - 5y^4 + 6y^2 - 1 = 0$, whence

$$(x + 2)^3 - 5(x + 2)^2 + 6(x + 2) - 1 = 0,$$

that is,

$$x^3 + x^2 - 2x - 1 = 0.$$

And similarly, starting from $y^3 - 3y + 1 = 0$, $y^2 = x + 2$, the first equation gives $(y^3 - 3y)^2 - 1 = 0$, that is, $y^6 - 6y^4 + 9y^2 - 1 = 0$, whence

$$(x + 2)^3 - 6(x + 2)^2 + 9(x + 2) - 1 = 0,$$

that is,

$$x^3 - 3x + 1 = 0.$$

To find the roots of the equation $x^3 + x^2 - 2x - 1 = 0$, taking ω an imaginary cube root of unity, and writing $\alpha = \sqrt[3]{\{7(2 + 3\omega)\}}$, $\beta = \sqrt[3]{\{7(2 + 3\omega^2)\}}$, where observe that $2 + 3\omega$, $2 + 3\omega^2$ are imaginary factors of 7, viz.

$$7 = (2 + 3\omega)(2 + 3\omega^2),$$

and therefore also $\alpha^3 + \beta^3 = 7$, $\alpha\beta = 7$, then the roots of the equation are

$$3a = -1 + \alpha + \beta,$$

$$3b = -1 + \omega\alpha + \omega^2\beta,$$

$$3c = -1 + \omega^2\alpha + \omega\beta.$$

I verify herewith the equation $a^2 = 2 + c$, viz. this gives

$$(-1 + \alpha + \beta)^2 = 18 + 3(-1 + \omega^2\alpha + \omega\beta),$$

or writing herein $2\alpha\beta = 14$, this is

$$\alpha^2 - (2 + 3\omega^2)\alpha + \beta^2 - (2 + 3\omega)\beta = 0,$$

that is,

$$\alpha^2 - \tfrac{1}{7}\beta^3\alpha \qquad + \beta^2 - \tfrac{1}{7}\alpha^3\beta \qquad = 0,$$

or finally

$$(\alpha^2 + \beta^2)(1 - \tfrac{1}{7}\alpha\beta) = 0,$$

satisfied in virtue of $\alpha\beta = 7$.

For the second equation $x^3 - 3x + 1 = 0$, ω denoting as before, the roots are

$$a = \omega^{\frac{1}{3}} + \omega^{\frac{8}{3}}, \text{ whence } a^2 = \omega^{\frac{2}{3}} + \omega^{\frac{7}{3}} + 2, = 2 + c,$$

$$b = \omega^{\frac{4}{3}} + \omega^{\frac{5}{3}}, \quad \text{,,} \quad b^2 = \omega^{\frac{8}{3}} + \omega^{\frac{1}{3}} + 2, = 2 + a,$$

$$c = \omega^{\frac{7}{3}} + \omega^{\frac{2}{3}}, \quad \text{,,} \quad c^2 = \omega^{\frac{5}{3}} + \omega^{\frac{4}{3}} + 2, = 2 + b.$$

The equation $x^3 - 5x^2 + 6x - 1 = 0$, which, writing therein $x + 2$ for x, gives

$$x^3 + x^2 - 2x - 1 = 0,$$

is considered in Hermite's *Cours d'Analyse*, Paris 1873, p. 12, and this suggested to me the foregoing investigation.

939.

ON A CASE OF THE INVOLUTION $AF+BG+CH=0$, WHERE A, B, C, F, G, H ARE TERNARY QUADRICS.

[From the *Messenger of Mathematics*, vol. XXII. (1893), pp. 182—186.]

WE have here the six conics

$$A=0,\ B=0,\ C=0,\ F=0,\ G=0,\ H=0;$$

the curves $AF=0$ and $BG=0$ are quartics intersecting in 16 points, and if 8 of these lie in a conic $H=0$, then the remaining 8 will be in a conic $C=0$. I take the first set of eight points to be 1, 2, 3, 4, 5, 6, 7, 8; the quartics $AF=0$ and $BG=0$ each pass through these eight points; and I assume for the moment

$$A=1234,\ F=5678;\ B=1256,\ G=3478,$$

viz. that $A=0$ is a conic through the points 1, 2, 3, 4, and similarly for F, G, B. Here $H=0$ is a conic through the points 1, 2, 3, 4, 5, 6, 7, 8, or attending only to the last four points it is a conic through 5, 6, 7, 8; we have therefore a linear relation between F, G, H, and supposing the implicit constant factors to be properly determined, this may be taken to be $F+G+H=0$; the identity $AF+BG+CH=0$ thus becomes $F(A-C)+G(B-C)=0$. We have thus F a numerical multiple of $B-C$, and by a proper determination of the implicit factor we may make this relation to be $F=B-C$; the last equation then gives $G=C-A$, and from the equation $F+G+H=0$, we have $H=A-B$; the six functions thus are

$$\begin{array}{lll} A,\ B-C, & \text{or if we please,} & A-D,\ B-C, \\ B,\ C-A & & B-D,\ C-A, \\ C,\ A-B & & C-D,\ A-B, \end{array}$$

where D is an arbitrary quadric function. The solution

$$(A-D)(B-C)+(B-D)(C-A)+(C-D)(A-B)=0$$

of the involution is an obvious and trivial one.

But the case which I proceed to consider is

$$A=1234,\ F=5678;\ B=1256,\ G=3478;$$

here $AF=0$, and $BG=0$, meet as before in the points 1, 2, 3, 4, 5, 6, 7, 8, and in eight other points, say that

$$\begin{array}{llllll} A=0, & B=0 & \text{meet in} & 1,\ 2 & \text{and in two other points} & \alpha,\ \beta, \\ A=0, & G=0 & ,, & 3,\ 4 & ,, \quad ,, & \gamma,\ \delta, \\ F=0, & B=0 & ,, & 5,\ 6 & ,, \quad ,, & \epsilon,\ \zeta, \\ F=0, & G=0 & ,, & 7,\ 8 & ,, \quad ,, & \eta,\ \theta; \end{array}$$

then the 8 points $\alpha, \beta, \gamma, \delta, \epsilon, \zeta, \eta, \theta$ will lie in a conic $C=0$.

I take $y^2-zx=0$ for the conic $H=0$; for any point in this conic we have $x:y:z=1:\theta:\theta^2$, and we may take $\theta_1, \theta_2, \theta_3, \theta_4, \theta_5, \theta_6, \theta_7, \theta_8$ for the parameters of the points 1, 2, 3, 4, 5, 6, 7, 8 respectively.

Write $(a, b, c, f, g, h\between x, y, z)^2=0$ for the conic A, $=1234=0$; therefore we have

$$a+b\theta^2+c\theta^4+f\theta^3+g\theta^2+h\theta=\theta-\theta_1.\theta-\theta_2.\theta-\theta_3.\theta-\theta_4;$$

or, if

$$\begin{aligned} p_{1234}&=\theta_1+\theta_2+\theta_3+\theta_4, \\ q_{1234}&=\theta_1\theta_2+\theta_1\theta_3+\theta_1\theta_4+\theta_2\theta_3+\theta_2\theta_4+\theta_3\theta_4, \\ r_{1234}&=\theta_1\theta_2\theta_3+\theta_1\theta_2\theta_4+\theta_1\theta_3\theta_4+\theta_2\theta_3\theta_4, \\ s_{1234}&=\theta_1\theta_2\theta_3\theta_4, \end{aligned}$$

then

$$c=1,\ f=-p_{1234},\ b+g=q_{1234},\ h=-r_{1234},\ a=s_{1234};$$

or, writing $g=-\lambda$, we have

$$s_{1234}x^2+q_{1234}y^2+z^2-p_{1234}yz-r_{1234}xy+\lambda(y^2-zx)=0$$

for the equation of the conic in question. We may without loss of generality put $\lambda=0$; and then if, in general,

$$\Omega=sx^2+qy^2+z^2-pyz-rxy,$$

we have $A=\Omega_{1234}=0$ for the conic $A=0$. And thus the equations of the four conics are

$$A=\Omega_{1234}=0,\ F=\Omega_{5678}=0;\ B=\Omega_{1256}=0,\ C=\Omega_{3478}=0,$$

or, as for shortness I write them,

$$A=\Omega=0,\ F=\Omega'=0;\ B=\Omega''=0,\ C=\Omega''',$$

viz. in Ω the suffixes are 1, 2, 3, 4, in Ω' they are 5, 6, 7, 8, in Ω'' they are 1, 2, 5, 6, and in Ω''' they are 3, 4, 7, 8.

I find that the implicit constant factors of AF and BG are $1, -1$, and consequently that the form of the identity is

$$\Omega\Omega'-\Omega''\Omega'''+(y^2-zx)C=0,$$

where C is a quadric function to be determined; or, what is the same thing, we have

$$(sx^2 + qy^2 + z^2 - pyz - rxy)(s'x^2 + q'y^2 + z^2 - p'yz - r'xy),$$
$$-(s''x^2 + q''y^2 + z^2 - p''yz - r''xy)(s'''x^2 + q'''y^2 + z^2 - p'''yz - r'''xy),$$
$$+(y^2 - zx)C = 0.$$

Writing for shortness

$$\theta_1 + \theta_2 = \alpha\ ,\quad \theta_1\theta_2 = \beta\ ,$$
$$\theta_3 + \theta_4 = \alpha'\ ,\quad \theta_3\theta_4 = \beta'\ ,$$
$$\theta_5 + \theta_6 = \alpha''\ ,\quad \theta_5\theta_6 = \beta''\ ,$$
$$\theta_7 + \theta_8 = \alpha''',\quad \theta_7\theta_8 = \beta''',$$

we have

$p = \alpha + \alpha'$	$p' = \alpha'' + \alpha'''$	$p'' = \alpha + \alpha''$	$p''' = \alpha' + \alpha'''$
$q = \alpha\alpha' + \beta + \beta'$	$q' = \alpha''\alpha''' + \beta'' + \beta'''$	$q'' = \alpha\alpha'' + \beta + \beta''$	$q''' = \alpha'\alpha''' + \alpha'\beta''' + \alpha'''\beta'$
$r = \alpha\beta' + \alpha'\beta$	$r' = \alpha''\beta''' + \alpha'''\beta''$	$r'' = \alpha\beta'' + \alpha''\beta$	$r''' = \alpha'\beta''' + \alpha'''\beta'$
$s = \beta\beta'$	$s' = \beta''\beta'''$	$s'' = \beta\beta''$	$s''' = \beta'\beta'''$.

In the last-mentioned equation, the first and second lines together are a quartic function of (x, y, z), say the value is

$$\begin{aligned} = \ & Ax^4 + By^4 + Cz^4, \\ & + Fy^3z + Gz^3x + Hx^3y, \\ & + Iyz^3 + Jzx^3 + Kxy^3, \\ & + Lx^2yz + Mxy^2z + Nxyz^2, \\ & + Py^2z^2 + Qz^2x^2 + Rx^2y^2, \end{aligned}$$

where after all reductions

$$\begin{aligned}
A &= ss' - s''s''' &&= 0, \\
B &= qq' - q''q''' &&= (\alpha\beta''' - \alpha'''\beta)(\alpha' - \alpha'') \\
& && \quad + (\alpha'\beta'' - \alpha''\beta')(\alpha - \alpha''') - (\beta' - \beta'')(\beta - \beta'''), \\
C &= 1 - 1 &&= 0, \\
F &= -pq' - p'q + p''q''' + p'''q'' &&= (\alpha - \alpha''')(\beta' - \beta'') + (\alpha' - \alpha'')(\beta - \beta'''), \\
G &= 0 - 0 &&= 0, \\
H &= -rs' - r's + r''s''' + r'''s'' &&= 0, \\
I &= -p - p' + p'' + p''' &&= 0, \\
J &= 0 - 0 &&= 0, \\
K &= -qr' - q'r + q''r''' + q'''r'' &&= (\alpha\beta''' - \alpha'''\beta)(\beta'' - \beta') + (\alpha'\beta'' - \alpha''\beta')(\beta''' - \beta), \\
L &= -ps' - p's + p''s''' + p'''s'' &&= (\alpha\beta''' - \alpha'''\beta)(\beta' - \beta'') + (\alpha'\beta'' - \alpha''\beta')(\beta - \beta'''), \\
M &= pr' + p'r - p''r''' - p'''r'' &&= (\alpha\beta''' - \alpha'''\beta)(\alpha'' - \alpha') + (\alpha'\beta'' - \alpha''\beta')(\alpha''' - \alpha), \\
N &= -r - r' + r'' + r''' &&= (\alpha - \alpha''')(\beta'' - \beta') + (\alpha' - \alpha'')(\beta''' - \beta), \\
P &= pp' + q + q' - p''p''' - q'' - q''' &&= 0, \\
Q &= s + s' - s'' - s''' &&= (\beta' - \beta'')(\beta - \beta'''), \\
R &= rr' + qs' + q's - r''r''' - q''s''' - q'''s'' = 0:
\end{aligned}$$

values which satisfy

$$F + N = 0,$$
$$K + L = 0,$$
$$B + M + Q = 0.$$

The quartic function is thus seen to be

$$= (y^2 - zx)(By^2 + Fyz - Qzx + Kxy) = 0,$$

viz. we have $By^2 + Fyz - Qzx + Kxy = 0$ for the equation of the conic $C = 0$.

Moreover, substituting for p, q, r, s, &c., their values, we have finally for the required involution

$$\begin{gathered}[\beta\beta' x^2 + (\alpha\alpha' + \beta + \beta')\,y^2 + z^2 - (\alpha + \alpha')\,yz - (\alpha\beta' + \alpha'\beta)\,xy] \\ \times [\beta''\beta''' x^2 + (\alpha''\alpha''' + \beta'' + \beta''')\,y^2 + z^2 - (\alpha'' + \alpha''')\,yz - (\alpha''\beta''' + \alpha'''\beta'')\,xy] \\ - [\beta\beta'' x^2 + (\alpha\alpha'' + \beta + \beta'')\,y^2 + z^2 - (\alpha + \alpha'')\,yz - (\alpha\beta'' + \alpha''\beta)\,xy] \\ \times [\beta'\beta''' x^2 + (\alpha'\alpha''' + \beta' + \beta''')\,y^2 + z^2 - (\alpha' + \alpha''')\,yz - (\alpha'\beta''' + \alpha'''\beta')\,xy],\end{gathered}$$

$$-(y^2 - zx) \times \left\{\begin{aligned} & y^2\,[(\alpha\beta''' - \alpha'''\beta)(\alpha' - \alpha'') + (\alpha'\beta'' - \alpha''\beta')(\alpha - \alpha''') - (\beta - \beta''')(\beta' - \beta'')] \\ & + yz\,[(\alpha - \alpha''')(\beta' - \beta'') + (\alpha' - \alpha'')(\beta - \beta''')] \\ & - zx\,[(\beta - \beta''')(\beta' - \beta'')] \\ & - xy\,[(\alpha\beta''' - \alpha'''\beta)(\beta' - \beta'') + (\alpha'\beta'' - \alpha''\beta')(\beta - \beta''')] \end{aligned}\right\} = 0.$$

It will be recollected that this is the solution for the case $A = 1234$, $F = 5678$; $B = 1256$, $G = 3478$: being that to which the present paper has reference.

940.

ON THE DEVELOPMENT OF $(1+n^2x)^{\frac{m}{n}}$.

[From the *Messenger of Mathematics*, vol. XXII. (1893), pp. 186—190.]

It is a known theorem that, if $\frac{m}{n}$ be any fraction in its least terms, the coefficients of the development of $(1+n^2x)^{\frac{m}{n}}$ are all of them integers, or, what is the same thing, that

$$\frac{m \,.\, m-n \ldots m-(r-1)\,n}{1 \,.\; 2 \;\ldots\; r}\, n^r$$

is an integer. The greater part, but not the whole, of this result comes out very simply from Mr Segar's very elegant theorem, *Messenger*, vol. XXII. (1893), p. 59, "the product of the differences of any r unequal numbers is divisible by $(r-1)!!$" or, as it may be stated, if $\alpha, \beta, \gamma, \ldots$ are any r unequal numbers, then $\zeta^{\frac{1}{2}}(\alpha, \beta, \gamma, \ldots)$ is divisible by $\zeta^{\frac{1}{2}}(0, 1, 2, \ldots, r-1)$.

In fact, writing $r+1$ for r and considering the numbers

$$m+n,\ n,\ 2n,\ 3n,\ \ldots\ (r-1)\,n\,;$$

then neglecting signs

$$\begin{aligned} \zeta^{\frac{1}{2}}(\alpha, \beta, \gamma, \ldots) \text{ is } = \;& m \,.\, m-n \ldots m-(r-1)\,n, \\ & \times 1n \,.\, 2n \ldots (r-1)\,n, \\ & \times 1n \,.\, 2n \ldots (r-2)\,n, \\ & \quad\vdots \\ & \times 1n \,.\, 2n, \\ & \times 1n, \end{aligned}$$

which is

$$= m \,.\, m-n \ldots m-(r-1)\,n \times n^{\frac{1}{2}r \,.\, r-1} \times \zeta^{\frac{1}{2}}(0, 1, 2, \ldots, r-1),$$

and similarly

$$\zeta^{\frac{1}{2}}(0,\ 1,\ 2,\ \ldots,\ r)=1\,.\,2\,.\,3\ldots r\times\zeta^{\frac{1}{2}}(0,\ 1,\ 2,\ \ldots,\ r-1);$$

so that, omitting the common factor $\zeta^{\frac{1}{2}}(0, 1, 2, \ldots, r-1)$, we have

$$m\,.\,m-n\ldots m-(r-1)\,n\,.\,n^{\frac{1}{2}r\,.\,r-1} \text{ divisible by } 1\,.\,2\,.\,3\ldots r.$$

It thus appears that the fraction

$$\frac{m\,.\,m-n\ldots m-(r-1)\,n}{1\ .\ \ 2\ \ \ldots\ \ \ \ r},$$

when reduced to its least terms, will contain in the denominator only products of powers of the prime factors of n; and it remains to show that multiplying this by n^r it will become integral, or what is the same thing that

$$\frac{n^r}{1\,.\,2\,.\ldots r}$$

in its least terms will not contain in the denominator any prime factor of n.

Considering in succession the prime numbers 2, 3, 5, ..., first the number 2, we see that in the product $1\,.\,2\,.\,3\,.\ldots r$, the number of terms divisible by 2 is $=\left(\frac{r}{2}\right)$, the number of terms divisible by 4 is $=\left(\frac{r}{4}\right)$, that by 8 is $=\left(\frac{r}{8}\right)$, and so on, where $\left(\frac{r}{2}\right)$ denotes the integer part of $\frac{r}{2}$, and so in other cases. Hence the product contains the factor 2, with the exponent $\left(\frac{r}{2}\right)+\left(\frac{r}{4}\right)+\left(\frac{r}{8}\right)+\ldots$, which exponent is less than

$$\frac{r}{2}+\frac{r}{4}+\frac{r}{8}+\ldots\ ad\ inf.$$

is less than r, say it is less than (r). Similarly for the number 3, the product contains the factor 3, with the exponent

$$\left(\frac{r}{3}\right)+\left(\frac{r}{9}\right)+\left(\frac{r}{27}\right)+\ldots,$$

which exponent is less than

$$\frac{r}{3}+\frac{r}{9}+\frac{r}{27}+\ldots\ ad\ inf.$$

is less than $\frac{1}{2}r$, say it is at most $=(\frac{1}{2}r)$; and so it contains the factor 5 with an exponent which is less than $\frac{1}{4}r$, say it is at most $=(\frac{1}{4}r)$, and generally the prime factor p with an exponent which is less than $\frac{1}{p-1}r$: say it is at most $=\left(\frac{1}{p-1}r\right)$.

This is

$$1\,.\,2\,.\,3\ldots r=\frac{1}{K}\,2^{(r)}\,3^{(\frac{1}{2}r)}\,5^{(\frac{1}{4}r)}\ldots,$$

where K is a whole number. Hence if $n=2^\alpha 3^\beta 5^\gamma \ldots$, we have

$$\frac{n^r}{1.2.3\ldots r}=K2^{r\alpha-(r)}.3^{r\beta-(\frac{1}{2}r)}.5^{r\gamma-(\frac{1}{4}r)}\ldots,$$

and here for every prime number 2, 3, 5, ... which is a factor of n, that is, for which the corresponding exponent $\alpha, \beta, \gamma, \ldots$ is not $=0$, the exponents $r\alpha-(r)$, $r\beta-(\frac{1}{2}r)$, $r\gamma-(\frac{1}{4}r)$, ... are all of them positive; and thus the fraction in its least terms does not contain in the denominator any prime factor of n; this is the theorem which was to be proved.

Mr Segar's theorem may without loss of generality be stated as follows: if $\beta, \gamma, \ldots$ are any $r-1$ unequal positive integers (which for convenience may be taken in order of increasing magnitude), then $\zeta^{\frac{1}{2}}(0, \beta, \gamma, \ldots)$ is divisible by $\zeta^{\frac{1}{2}}(0, 1, 2, \ldots, r-1)$. A proof, in principle the same as his, is as follows:

We have the determinant

$$\begin{vmatrix} 1, & a^\beta, & a^\gamma, \ldots \\ & b & \\ & c & \\ & \vdots & \end{vmatrix} \text{ divisible by } \begin{vmatrix} 1, & a, & a^2, \ldots \\ & b & \\ & c & \\ & \vdots & \end{vmatrix},$$

viz. the quotient is a rational and integral function of $a, b, c, \ldots$ with coefficients which are positive integers; hence putting $a=b=c, \ldots =1$, the quotient will be a positive integer number. Considering the numerator determinant, and for $a, b, c, \ldots$ writing therein $1+a$, $1+b$, $1+c$, ... respectively, where $a, b, c, \ldots$ are ultimately to be put each $=0$, the value is

$$=\begin{vmatrix} 1, & 1+\beta_1 a+\beta_2 a^2\ldots, & 1+\gamma_1 a+\gamma_2 a^2+\ldots, & \ldots \\ & b & & \\ & c & & \\ & \vdots & & \end{vmatrix},$$

where $\beta_1, \beta_2, \ldots$ denote the binomial coefficients

$$\frac{\beta}{1}, \frac{\beta.\beta-1}{1.2}, \text{\&c.}:$$

attending only to the lowest powers of $a, b, c, \ldots$ which enter into the formula, this is

$$=\begin{vmatrix} 1, & & \\ 1, & \beta_1, & \beta_2 \\ 1, & \gamma_1, & \gamma_2 \\ \vdots & & \end{vmatrix}\begin{vmatrix} 1, & a, & a^2, \ldots \\ 1, & b, & b^2, \\ 1, & c, & c^2, \\ \vdots & & \end{vmatrix},$$

or what is the same thing it is

$$= M \begin{vmatrix} 1, & & \dots \\ 1, & \beta, & \beta^2, \\ 1, & \gamma, & \gamma^2, \\ \vdots & & \end{vmatrix} \begin{vmatrix} 1, & a, & a^2 \\ 1, & b, & b^2 \\ 1, & c, & c^2 \end{vmatrix}, = M\zeta^{\frac{1}{2}}(0, \beta, \gamma, \dots) \begin{vmatrix} 1, & a, & a^2, & \dots \\ 1, & b, & b^2, & \\ 1, & c, & c^2, & \\ \vdots & & & \end{vmatrix},$$

where M is a mere number: it will be recollected that in this form, $a, b, c, \dots$ are not the original $a, b, c, \dots$. Putting herein $\beta, \gamma, \dots = 1, 2, \dots$, the denominator determinant is

$$= M\zeta^{\frac{1}{2}}(0, 1, 2, \dots) \begin{vmatrix} 1, & a, & a^2, & \dots \\ 1, & b, & b^2, & \\ 1, & c, & c^2, & \\ \vdots & & & \end{vmatrix},$$

and hence the quotient, which as already seen is an integer number, is equal to $\zeta^{\frac{1}{2}}(0, \beta, \gamma, \dots) \div \zeta^{\frac{1}{2}}(0, 1, 2, \dots)$, the theorem in question.

The original theorem as to the form of $(1+n^2x)^{\frac{m}{n}}$ is a particular case of Eisenstein's very general theorem that, in the development of any algebraical function of x, it is always possible by substituting for x a proper multiple of x, to make all the coefficients integers. It may be remarked that this would not be so if we had only

$$m \,.\, m-n \,.\dots\, m-(r-1)\,n \,.\, n^{\frac{1}{2}r\,.\,r-1}$$

divisible by $1\,.\,2\dots r$; for then, writing Nx for x, the form of the coefficient would have been

$$\frac{KN^r}{n^r\,.\,n^{\frac{1}{2}r\,.\,r-1}}, = \frac{KN^r}{n^{\frac{1}{2}r\,.\,r+1}},$$

and there would be no value (however great) of N by which the denominator factor $n^{\frac{1}{2}r\,.\,r+1}$ could be got rid of.

941.

NOTE ON THE PARTIAL DIFFERENTIAL EQUATION $Rr+Ss+Tt+U(s^2-rt)-V=0$.

[From the *Quarterly Journal of Pure and Applied Mathematics*, vol. XXVI. (1893), pp. 1—5.]

It is well known that this equation, R, S, T, U, V being any functions whatever of (x, y, z, p, q), in the case where u admits of an integral of the form $u=f(v)$ (u, v functions of x, y, z, p, q, and f an arbitrary functional symbol) can be integrated as follows; viz. taking m_1, m_2 as the roots of the quadratic equation

$$m^2-Sm+RT-UV=0,$$

(that is, writing $m_1+m_2=S$ and $m_1m_2=RT-UV$), then, m_1 denoting either root at pleasure, and m_2 the other root of the quadratic equation, if the system of ordinary differential equations

$$\begin{aligned}
m_1dx-R\,dy+U\,dq&=0,\\
-T\,dx+m_2dy+U\,dp&=0,\\
-V\,dx+m_2dq+R\,dp&=0,\\
-V\,dy+T\,dq+m_1dp&=0,\\
-p\,dx-q\,dy+\quad dz&=0,
\end{aligned}$$

(equivalent to three independent equations) admits of two integrals $u=\text{const.}$ and $v=\text{const.}$, the solution of the given partial differential equation is $u=f(v)$.

In fact, to prove this, we have

$$\begin{aligned}
du=\;&\lambda\,(\;m_1dx-R\,dy+U\,dq)\\
&+\mu\,(-T\,dx+m_2dy+U\,dp)\\
&+\nu\,(-V\,dx+m_2dq+R\,dp)\\
&+\rho\,(-V\,dy+T\,dq+m_1dp)\\
&+\sigma\,(-p\,dx-q\,dy+\quad dz),
\end{aligned}$$

that is,

$$\frac{du}{dx} = \lambda m_1 - \mu T - \nu V \qquad - \sigma p,$$

$$\frac{du}{dy} = - \lambda R + \mu m_2 \qquad - \rho V - \sigma q,$$

$$\frac{du}{dz} = \qquad \sigma,$$

$$\frac{du}{dp} = \qquad \mu U + \nu R + \rho m_1 \qquad ,$$

$$\frac{du}{dq} = \lambda U \qquad + \nu m_2 + \rho T \qquad ,$$

and thence

$$\frac{du}{dx} + \frac{du}{dz} p + \frac{du}{dp} r + \frac{du}{dq} s = \lambda (m_1 + Us) + \mu (-T + Ur) + \nu (-V + Rr + m_2 s) + \rho (m_1 r + Ts),$$

$$\frac{du}{dy} + \frac{du}{dz} q + \frac{du}{dp} s + \frac{du}{dq} t = \lambda (-R + Ut) + \mu (m_2 + Us) + \nu (Rs + m_2 t) + \rho (-V + m_1 s + Tt),$$

which equations may be represented by

$$\frac{d(u)}{dx} = A\lambda + B\mu + C\nu + D\rho,$$

$$\frac{d(u)}{dy} = A'\lambda + B'\mu + C'\nu + D'\rho;$$

and for λ, μ, ν, ρ writing λ', μ', ν', ρ', we have similarly

$$\frac{d(v)}{dx} = A\lambda' + B\mu' + C\nu' + D\rho',$$

$$\frac{d(v)}{dy} = A'\lambda' + B'\mu' + C'\nu' + D'\rho';$$

whence

$$\frac{d(u)}{dx}\frac{d(v)}{dy} - \frac{d(u)}{dy}\frac{d(v)}{dx} = \begin{vmatrix} A\lambda + B\mu + C\nu + D\rho, & A\lambda' + B\mu' + C\nu' + D\rho' \\ A'\lambda + B'\mu + C'\nu + D'\rho, & A'\lambda' + B'\mu' + C'\nu' + D'\rho' \end{vmatrix}.$$

The determinant is

$$\begin{aligned} = \; & (AD' - A'D)(\lambda\rho' - \lambda'\rho) + (BD' - B'D)(\mu\rho' - \mu'\rho) \\ & + (CD' - C'D)(\nu\rho' - \nu'\rho) + (BC' - B'C)(\mu\nu' - \mu'\nu) \\ & + (CA' - C'A)(\nu\lambda' - \nu'\lambda) + (AB' - A'B)(\lambda\mu' - \lambda'\mu). \end{aligned}$$

The determinants $AD' - A'D$, &c., each of them contain the factor

$$\Theta, = Rr + Ss + Tt + U(s^2 - rt) - V;$$

viz. we have

$$\begin{aligned} AD' - A'D &= \ \ m_1\Theta, & BC' - B'C &= -m_2\Theta, \\ BD' - B'D &= -\ T\Theta, & CA' - C'A &= -R\Theta, \\ CD' - C'D &= -\ V\Theta, & AB' - A'B &= \ \ U\Theta, \end{aligned}$$

values which give

$$(AD' - A'D)(BC' - B'C) + (BD' - B'D)(CA' - C'A) \\ + (CD' - C'D)(AB' - A'B) = \Theta^2(-m_1m_2 + TR - VU) = 0,$$

as it should be.

Hence, when the partial differential equation $\Theta = 0$ is satisfied, we have

$$\frac{d(u)}{dx}\frac{d(v)}{dy} - \frac{d(u)}{dy}\frac{d(v)}{dx} = 0;$$

and we thence have $u = f(v)$ as the integral of the partial differential equation.

It should be possible to express analytically the conditions in order that the systems of differential equations may have one or each of them two integrals.

It is interesting to remark that, if each of the two systems of ordinary differential equations has only a single integral, these two integrals do *not* lead to the solution of the partial differential equation. Consider, for instance, the case

$$R = 0,\quad S = x + y,\quad T = 0,\quad U = 0,\quad V = p + q;$$

the partial differential equation is here

$$(x + y)s - (p + q) = 0,$$

which has an integral

$$z = (x + y)\{\phi'(x) + \psi'(y)\} - 2\{\phi(x) + \psi(y)\},$$

where ϕ, ψ are arbitrary functions: the equation in m is $m^2 - m(x + y) = 0$, the roots of which are $m = 0$, and $m = x + y$.

For $m_1 = 0$, $m_2 = x + y$, the system of differential equations becomes

$$\begin{aligned} &dy = 0, \\ &-(p + q)\,dx + (x + y)\,dq = 0, \\ &-p\,dx + dz \qquad\qquad\quad = 0, \end{aligned}$$

which has only the integral $y = \text{const.}$; and similarly for $m_1 = x + y$, $m_2 = 0$, the system becomes

$$\begin{aligned} &dx = 0, \\ &-(p + q)\,dy + (x + y)\,dp = 0, \\ &-q\,dy + dz \qquad\qquad\quad = 0, \end{aligned}$$

which has only the integral $x = \text{const.}$ And these two integrals $y = \text{const.}$ and $x = \text{const.}$ do not in anywise lead to the integral of the partial differential equation.

I take the opportunity of remarking that the complete system of conditions in order that the differential

$$Adx + Bdy + Cdz + Ddw$$

may be $= MdU$ is as follows: viz. writing

$$A, B, C, D = 1, 2, 3, 4;$$

$$\frac{dB}{dz} - \frac{dC}{dy}, \frac{dC}{dx} - \frac{dA}{dz}, \frac{dA}{dy} - \frac{dB}{dx}, \frac{dA}{dw} - \frac{dD}{dx}, \frac{dB}{dw} - \frac{dD}{dy}, \frac{dC}{dw} - \frac{dD}{dz} = 23, 31, 12, 14, 24, 34,$$

where of course $12 = -21$, &c., and

$$\overline{123} = 1.23 + 2.31 + 3.12, \text{ \&c.}; \quad \overline{1234} = 1.234 - 2.341 + 3.412 - 4.123,$$

is $= 0$ identically;

$$1234 = 12.34 + 13.42 + 14.23,$$

then the conditions equivalent to three independent conditions are

$$\overline{234} = 0, \quad \overline{341} = 0, \quad \overline{412} = 0, \quad \overline{123} = 0, \quad 1234 = 0.$$

In fact, the first four equations are

$$\begin{array}{l} \quad . \quad\quad 2.34 - 3.24 + 4.23 = 0, \\ -1.34 \quad . \quad + 3.14 + 4.31 = 0, \\ \quad 1.24 - 2.14 \quad . \quad + 4.12 = 0, \\ -1.23 - 2.31 - 3.12 \quad . \quad = 0; \end{array}$$

hence, multiplying by 1, 2, 3, 4 respectively and adding, we have the identity $\overline{1234} = 0$, so that these four are equivalent to three independent equations: and multiplying by

$$\begin{array}{l} \quad . \quad\quad 12.\mu - 31.\nu + 14.\rho, \\ -12.\lambda \quad . \quad + 23.\nu + 24.\rho, \\ \quad 31.\lambda - 23.\mu \quad . \quad + 34.\rho, \\ -14.\lambda - 24.\mu - 34.\nu \quad . \end{array}$$

respectively, (where λ, μ, ν, ρ are arbitrary), we have

$$(1.\lambda + 2.\mu + 3.\nu + 4.\rho)(23.14 + 31.24 + 12.34) = 0,$$

that is,

$$23.14 + 31.24 + 12.34 = 0, \quad \text{or} \quad 1234 = 0,$$

the fifth condition.

942.

ON SEMINVARIANTS.

[From the *Quarterly Journal of Pure and Applied Mathematics*, vol. XXVI. (1893), pp. 66—69.]

I WISH to prove the following negative: a given sharp seminvariant is not in every case obtainable by mere derivation from a form of the same extent and of the next inferior degree. The meaning of the statement will be explained.

According to the general theory developed in Clebsch's *Theorie der binären algebraischen Formen*, Leipzig, 1872, the covariants of a given binary quantic f are all of them obtainable, the covariants of a given degree from those of the next inferior degree, by derivation (Ueberschiebung) of these with f; viz. if the covariants of the next inferior degree are P, Q, &c., then the covariants of the degree in question are all of them included among the forms

$$\begin{array}{lll} (f,\ P)^0\,(=fP), & (f,\ P)^1, & (f,\ P)^2, \ldots, \\ (f,\ Q)^0 \qquad\quad , & (f,\ Q)^1, & (f,\ Q)^2, \ldots, \\ \&\text{c.}, & & \end{array}$$

the index of derivation for $(f,\ P)$ being at most equal to the degree of f or to that of P, whichever of these is the smaller, and so for Q, &c. The forms thus obtained are far too numerous; but rejecting repetitions, we have a complete system of the covariants of the given degree, viz. every covariant whatever of that degree is a linear function (with numerical multipliers) of the several distinct forms thus obtained by derivation.

We can therefore, by linear combination as above, obtain all the sharp covariants of the given degree, but we may very well have a sharp covariant not included among the several distinct forms thus obtained by derivation, but only expressible as a linear combination of two or more such forms; or say we may very well have a sharp covariant not obtainable by *mere* derivation from the forms of the next inferior degree; and it is important to verify that there are sharp covariants not thus obtainable by mere derivation. I remark that the notion of a sharp covariant does not present itself in Clebsch, and that it will be presently explained.

Passing from a covariant to its leading coefficient which is a seminvariant, the statement may be applied to seminvariants; viz. it is to be verified that there are sharp seminvariants not obtainable by mere derivation from the seminvariants of the next inferior degree. But we have to introduce the notion of "extent" so as to connect the seminvariant with a quantic of some particular order, thus, if the highest letter of the seminvariant is f, we say that the extent is $=5$, and thus connect it with the quintic

$$(1,\ b,\ c,\ d,\ e,\ f\between x,\ y)^5.$$

The notion "sharp" applies to the seminvariants of a given weight. Suppose, for instance, the weight is $=8$; we have a series of initial or non-unitary terms i, cg, df, e^2, &c., and a series of final or power-ending terms e^2, cd^2, b^2d^2, c^4, &c., and we denote by $cg - c^4$ (where observe that here and in all similar cases the $-$ is not a minus sign, but is simply a stroke), the whole series of terms (including cg and c^4) which are in counter-order (CO) subsequent to cg, and in alphabetical order (AO) precedent to c^4; and so in other cases. This being so, arranging the seminvariants with their final terms in AO, we have the seminvariants

$$\begin{aligned}&i\ - e^2,\\ &cg - cd^2,\\ &df - b^2d^2,\\ &e^2\ - c^4,\\ &\&c.,\end{aligned}$$

viz. we have a seminvariant $i - e^2$ containing all or any (in fact, all) of the terms of this set as above defined; a seminvariant $cg - cd^2$ containing all or any of the terms of this set, a seminvariant $df - b^2d^2$ containing all or any of the terms of this set; and so on. These are sharp forms; a seminvariant ending in e^2, must of necessity have the leading term i, and thus belong at least to the octic

$$(1,\ b,\ c,\ d,\ e,\ f,\ g,\ h,\ i\between x,\ y)^8,$$

a seminvariant ending in cd^2 must of necessity have a leading term as high as cg, and thus belong at least to the sextic

$$(1,\ b,\ c,\ d,\ e,\ f,\ g\between x,\ y)^6.$$

Any linear combination of these would be a seminvariant $i - cd^2$, belonging to the octic, but it is not a sharp form; the final term cd^2 does not of necessity imply an initial so high as i (in fact, as we have seen, it only implies the lower initial cg): and so in other cases.

For the quintic $(1,\ b,\ c,\ d,\ e,\ f\between x,\ y)^5$, we have (for the weight 8 and degree 4) the seminvariants $df - b^2d^2$, and $e^2 - c^4$, this last belongs, of course, also to the quartic $(1,\ b,\ c,\ d,\ e\between x,\ y)^4$, it is, in fact, the squared quadrinvariant $(e - 4bd + 3c^2)^2$. I wish to show that $df-b^2d^2$ is not obtainable by mere derivation from the covariants of degree 3 of the quartic.

The quintic and its covariants up to the degree 3 are

$A = ($

1	$5b$	$10c$	$10d$	$5e$	f

$\between x, y)^5,$

say

$$(1, b, c, d, e, f \between x, y)^5;$$

$B = ($

e +1	f +1	bf +1
bd − 4	be − 3	ce − 4
c^2 + 3	cd + 2	d^2 + 3

$\between x, y)^2,$

say

$$(e-c^2,\ f-cd,\ bf-d^2 \between x, y)^2;$$

$$C = (c-b^2,\ d-bc,\ e-c^2,\ f-cd,\ bf-d^2,\ cf-de,\ df-e^2 \between x, y)^6,$$
$$D = (ce-c^3,\ cf-c^2d,\ df-cd^2,\ bdf-d^3 \between x, y)^3,$$
$$E = (f-bc^2,\ bf-c^3,\ cf-c^2d,\ df-cd^2,\ ef-d^3,\ f^2-d^2e \between x, y)^5,$$
$$F = (d-b^3,\ e-b^2c,\ f-bc^2,\ bf-c^3,\ cf-c^2d,\ df-cd^2,\ ef-d^3,\ f^2-d^2e,\ bf^2-de^2,\ cf^2-e^3 \between x, y)^9.$$

Hence all the covariants of the degree 3 are A^3, AB, AC, D, E, F, where

$$AB = (e-c^2,\ f-bc^2,\ bf-c^3,\ cf-c^2d,\ bcf-cd^2,\ bdf-d^3,\ bef-d^2e,\ bf^2-d^2f \between x, y)^7$$
$$AC = (c-b^2,\ d-bc, \ldots \between x, y)^{11};$$

and the derivatives are

	weights of leading coefficients are
$(A, A^3)^0$, (A, A^3), &c.	0, 1, 2, 3, 4, 5
$(A, AB)^0$, (A, AB), &c.	4, 5, 6, 7, $\underline{8}$, 9
$(A, AC)^0$, (A, AC), &c.	2, 3, 4, 5, $\underline{6}$, 7
$(A, D)^0$, (A, D), &c.	6, 7, $\underline{8}$, 9
$(A, E)^0$, (A, E), &c.	5, 6, 7, $\underline{8}$, 9, 10
$(A, F)^0$, (A, F), &c.	3, 4, 5, 6, 7, $\underline{8}$.

The terms giving rise to a seminvariant of weight 8 are

$$(A, AB)^4 = \begin{Bmatrix} e-c^2, & f-bc^2, & bf-c^3, & cf-c^2d, & bcf-cd^2 \\ e\ , & d\ , & c\ , & b\ , & 1 \end{Bmatrix} = df-c^4,$$

$$(A, D)^2 = \begin{Bmatrix} ce-c^3, & cf-c^2d, & df-cd^2 \\ c\ , & b\ , & 1 \end{Bmatrix} = df-c^4,$$

$$(A, E)^3 = \begin{Bmatrix} f-bc^2, & bf-c^3, & cf-c^2d, & df-cd^2 \\ d\ , & c\ , & b\ , & 1 \end{Bmatrix} = df-c^4,$$

$$(A, F)^5 = \begin{Bmatrix} d-b^3, & e-b^2c, & f-bc^2, & bf-c^3, & cf-c^2d, & df-cd^2 \\ f\ , & e\ , & d\ , & c\ , & b\ , & 1 \end{Bmatrix} = df-c^4,$$

where to explain the algorithm, I remark, that if

$$A = (A_0,\ A_1,\ A_2, \ldots \between x,\ y)^5 \text{ and } D = (D_0,\ D_1,\ D_2, \ldots \between x,\ y)^3,$$

then

$$(A,\ D)^2 = D_0 A_2 - 2D_1 A_1 + D_2 A_0,$$

represented as above by

$$(A,\ D)^2 = \begin{Bmatrix} D_0, & D_1, & D_2 \\ A_2, & A_0, & A_1 \end{Bmatrix} = \begin{Bmatrix} ce - c^3, & cf - c^2d, & df - cd^2 \\ c\ , & b\ , & 1 \end{Bmatrix}.$$

The result is in every case given as $df - c^4$; in each case there is only a single term $c \,.\, c^3$, $= c^4$, and the term in c^4 certainly presents itself. In $(A,\ AB)^4$ there is a single term $d \,.\, f$, $= df$, and in $(A,\ D)^2$ a single term df, and thus the term df certainly presents itself: in $(A,\ E)^3$ there are two terms $d \,.\, f$, $= df$ and df, and it is conceivable that, inserting the proper numerical coefficients, these might destroy each other: if this were so, the form instead of being $df - c^4$ would be $e^2 - c^4$; and similarly in $(A,\ F)^5$, there are two terms df, $= df$ and df, which it is conceivable might destroy each other, and the form would then be $e^2 - c^4$. But in every case we have the term c^4, and it thus appears that the form $df - b^2d^2$ is not obtainable by mere derivation.

The form in question is in fact obtained by a linear combination of $df - c^4$ and $e^2 - c^4$, viz. writing down the leading coefficients of the covariants B^2 and H, we have

	$3H$	$-2B^2$	$=$
df	3		3
e^2	0	-2	-2
bcf	-9		-9
bde	-15	$+16$	$+1$
c^2e	$+30$	-12	-18
cd^2	-12		-12
b^3f	$+6$		$+6$
b^2ce	-15		-15
b^2d^2	$+42$	-32	-10
bc^2d	-48	$+48$	0
c^4	$+18$	-18	0

viz. the form in question $df - b^2d^2$ is $= 3df - 2e^2 - \ldots - 10b^2d^2$.

943.

ON RECIPROCANTS AND DIFFERENTIAL INVARIANTS.

[From the *Quarterly Journal of Pure and Applied Mathematics,* vol. XXVI. (1893), pp. 169—194, 289—307.]

I USE the term Reciprocant to denote a function of an arbitrary variable or variables and its differential coefficients, not connected with any differential equation; and Differential Invariant to denote a function of the coefficients of a differential equation and the differential coefficients of these coefficients. Halphen's differential invariants are thus reciprocants, but the term reciprocant is not made use of by him. I have entitled the present paper "On Reciprocants and Differential Invariants"; in the earlier part, (except that for preserving a chronological order, I briefly refer to Sir J. Cockle's Criticoids, which are differential invariants or rather seminvariants), I attend almost exclusively to Reciprocants, reproducing and explaining, and in some parts developing, the theories of Ampère, Halphen in his first two memoirs, and Sylvester.

I.

The notion of a reciprocant first presents itself in Ampère's "Mémoire sur les avantages qu'on peut retirer dans la théorie des courbes de la considération des paraboles osculatrices, avec des reflexions sur les fonctions différentielles dont la valeur ne change pas lors de la transformation des axes," *Journ. École Polyt.* t. VII. (1808), pp. 151—191 (sent to the Institute, Dec. 1803)*. We have (p. 167) for the radius of curvature the expression

$$-\frac{(1+y'^2)^{\frac{3}{2}}}{y''},$$

* A reciprocant, the Schwarzian derivative, occurs in Lagrange's memoir, "Sur la construction des cartes géographiques," *Nouv. Mém. de Berlin,* 1779, *Œuvres,* t. IV. p. 651, but scarcely *quà* reciprocant, viz. the form in question $\frac{f'''}{f'}-\frac{3}{2}\left(\frac{f''}{f'}\right)^2$, presents itself in the equation

$$\frac{f'''(u+ti)}{f'(u+ti)}-\frac{3}{2}\left(\frac{f''(u+ti)}{f'(u+ti)}\right)^2=\frac{F'''(u-ti)}{F'(u-ti)}-\frac{3}{2}\left(\frac{F''(u-ti)}{F'(u-ti)}\right)^2,$$

and for the parameter of the osculating parabola the expression

$$-\frac{54y''^5}{\{y'''^2+3(y''^2-y'y''')^2\}^{\frac{3}{2}}};$$

and it is explicitly noticed that each of these expressions remains absolutely unaltered when the coordinates x, y are changed into any other (rectangular) coordinates whatever.

The functions here spoken of are thus orthogonal absolute reciprocants; the change which leaves them unaltered is that of x, y into X, Y, where

$$X = x\cos\theta + y\sin\theta + \alpha,\quad Y = -x\sin\theta + y\cos\theta + \beta.$$

For the more general change

$$X = ax + by + \alpha,\quad Y = cx + dy + \beta,$$

we find without difficulty

$$Y' = \frac{c+dy'}{a-by'},\quad Y'' = \frac{(ad-bc)\,y''}{(a-by')^3},$$

and thus there is *not* an identity of form in the expressions

$$\frac{(1+Y'^2)^{\frac{3}{2}}}{Y''} \text{ and } \frac{(1+y'^2)^{\frac{3}{2}}}{y''}.$$

But if we disregard factors, or (what is the same thing) attend to the equations $y''=0$ and $Y''=0$, we see that for this more general change one of these equations implies the other; each of them is, in fact, the condition for an inflexion of the curve in (x, y) or (X, Y).

II.

The differential invariants, or rather seminvariants, called "Criticoids," were considered by Sir James Cockle in his paper "On Criticoids," *Phil. Mag.* t. XXXIX. (1870), pp. 201—221; these will be more particularly considered further on, but at present it is sufficient to remark that the functions in question are connected with a linear differential equation, viz. they are functions of the coefficients of a linear differential equation, such that effecting thereon a transformation of the dependent variable, say $y=f.Y$, where f is an arbitrary function of the independent variable x, the function is equal to the like function of the coefficients of the new equation. For instance, if by the transformation in question $y=f.Y$ we transform

$$\frac{d^2y}{dx^2}+2b\frac{dy}{dx}+cy=0 \text{ into } \frac{d^2Y}{dX^2}+2B\frac{dY}{dX}+CY=0,$$

then we have

$$B = b+\frac{f'}{f},\quad C = c+2b\frac{f'}{f}+\frac{f''}{f},$$

and thence

$$C-B^2-B' = c-b^2-b',$$

where F is the conjugate function of f, an equation implying that each side thereof is equal to one and the same constant; $u-ti$ is of course not a function of $u+ti$, and thus we have here no property that the function $\frac{f'''}{f'}-\frac{3}{2}\left(\frac{f''}{f'}\right)^2$ remains unaltered when for the arguments x, y thereof we substitute X, Y, determinate functions of these arguments.

(where the accents denote differentiation in regard to x): and thus $c - b^2 - b'$ is an invariant of the form in question; say it is an α-seminvariant of the differential equation.

III.

We have four important Memoirs by Halphen, (1) "Thèse d'Analyse sur les invariants différentiels," Paris (1878), (2) "Sur les invariants différentiels des courbes gauches," *Jour. École Polyt.* Cah. XLVII. (1880), pp. 1—102; (3) "Mémoire sur la réduction des équations différentielles linéaires aux formes intégrables," *Mém. Sav. Étrang.* t. XXVIII. (1883), pp. 1—297, and (4) "Sur les invariants des équations différentielles linéaires du quatrième ordre," *Acta Math.* t. III. (1883), pp. 325—380.

In the Memoir, Halphen (1), the investigations are in the first instance presented in a geometrical form; the author considers for instance the inflexions of a plane curve, and so obtains the invariant y'', or using his notation

$$\frac{1}{1.2\ldots k}\frac{d^ky}{dx^k} = a_k,$$

say the invariant a_2 of the weight 2; similarly, the consideration of the sextactic points gives him the invariant $a_2^2a_5 - 3a_2a_3a_4 + 2a_3^3$ of the weight 9; and that of the points of nine-pointic contact a more complicated invariant of the weight 27: and with these he forms absolute invariants. But the formal analytical definition is given § 3, "Théorie des invariants différentiels jusqu'au huitième ordre exclusivement," viz. considering the general homographic transformation

$$X = \frac{\alpha x + \beta y + \gamma}{\alpha''x + \beta''y + \gamma''}, \quad Y = \frac{\alpha'x + \beta'y + \gamma'}{\alpha''x + \beta''y + \gamma''},$$

then if the equation

$$f\left(x, y, \frac{dy}{dx}, \ldots \frac{d^ny}{dx^n}\right) = 0$$

is unaltered in form by the change (x, y) into (X, Y), or what is the same thing if the function

$$f\left(x, y, \frac{dy}{dx}, \ldots \frac{d^ny}{dx^n}\right)$$

be unaltered save as to a factor, then such function is said to be a differential invariant: and the quotient of two functions having the same factor, which quotient is therefore unaltered, is said to be an absolute invariant. We may for these terms substitute reciprocant, and absolute reciprocant respectively. The theory is scarcely worked out from the definition, and indeed, as will presently appear, it is by no means easy to work out directly from the definition, even the foregoing sextactic invariant $a_2^2a_5 - 3a_2a_3a_4 + 2a_3^3$. I remark that, in what follows, instead of Halphen's $a_2, a_3, a_4, a_5, \ldots$ I write $a, b, c, d, \ldots$, viz. these are used to denote respectively

$$\tfrac{1}{2}\frac{d^2y}{dx^2}, \; \tfrac{1}{6}\frac{d^3y}{dx^3}, \; \tfrac{1}{24}\frac{d^4y}{dx^4}, \; \tfrac{1}{120}\frac{d^5y}{dx^5}, \ldots,$$

the first differential coefficient $\dfrac{dy}{dx}$ being denoted by t. The foregoing invariant, or say the sextactic reciprocant, is thus $=a^2d-3abc+2b^3$. (See *post*, XI.)

IV.

For the analytical theory, we have as above

$$X=\frac{\alpha x+\beta y+\gamma}{\alpha''x+\beta''y+\gamma''},\quad Y=\frac{\alpha'x+\beta'y+\gamma'}{\alpha''x+\beta''y+\gamma''};$$

taking h, k for the increments of x, y respectively, we have

$$k=th+ah^2+bh^3+ch^4+dh^5+\ldots,$$

and, similarly, taking H, K for the increments of X, Y respectively, we have

$$K=TH+AH^2+BH^3+CH^4+DH^5+\ldots,$$

and we require the relations between the two sets of coefficients $(t, a, b, c, d, \ldots)$ and $(T, A, B, C, D, \ldots)$: I propose to develop these up to d, D, so as to obtain the theory for the form $a^2d-3abc+2b^3$.

Writing for shortness ξ, η, ζ for

$$\alpha x+\beta y+\gamma,\quad \alpha'x+\beta'y+\gamma',\quad \alpha''x+\beta''y+\gamma''$$

respectively, we have

$$X+H=\frac{\alpha\ (x+h)+\beta\ (y+k)+\gamma}{\alpha''(x+h)+\beta''(y+k)+\gamma''},\quad =\frac{\xi+\alpha h+\beta k}{\zeta+\alpha''h+\beta''k},$$

and thence

$$H=\frac{\xi+\alpha h+\beta k}{\zeta+\alpha''h+\beta''k}-\frac{\xi}{\zeta},$$

$$=\frac{(\alpha\zeta-\alpha''\xi)\,h+(\beta\zeta-\beta''\xi)\,k}{\zeta(\zeta+\alpha''h+\beta''k)}=\frac{Ph+Qk}{\zeta(\zeta+\alpha''h+\beta''k)},$$

if

$$P=\alpha\zeta-\alpha''\xi,\quad Q=\beta\zeta-\beta''\xi,$$

or, for k substituting its value,

$$H=\frac{(P+Qt)\,h+Q\,(ah^2+bh^3+ch^4+dh^5)}{\zeta\,\{\zeta+(\alpha''+\beta''t)\,h+\beta''\,(ah^2+bh^3+ch^4+dh^5)\}},$$

$$=\frac{L\,\{h+\lambda\,(ah^2+bh^3+ch^4+dh^5)\}}{\zeta\,[1+L''\,\{h+\lambda''\,(ah^2+bh^3+ch^4+dh^5)\}]},$$

if

$$L=P+Qt\,,\quad \lambda=\frac{Q}{P+Qt},$$

$$L''=\alpha''+\beta''t,\quad \lambda''=\frac{\beta''}{\alpha''+\beta''t},$$

or, putting $\mu = \frac{\zeta}{L}$, say this is

$$\mu H = \frac{h + \lambda (ah^2 + bh^3 + ch^4 + dh^5)}{1 + L'' \{h + \lambda'' (ah^2 + bh^3 + ch^4 + dh^5)\}}.$$

Similarly, if

$$P' = \alpha'\zeta - \alpha''\eta, \quad Q' = \beta'\zeta - \beta''\eta, \quad L' = P' + Q't, \quad \lambda' = \frac{Q'}{P' + Q't}, \text{ and } \mu' = \frac{\zeta}{L'},$$

then

$$\mu' K = \frac{h + \lambda' (ah^2 + bh^3 + ch^4 + dh^5)}{1 + L'' \{h + \lambda'' (ah^2 + bh^3 + ch^4 + dh^5)\}};$$

we have thus found H and K each of them in terms of h, and the elimination of h leads to an equation between H and K expressible in the form

$$K = TH + AH^2 + BH^3 + CH^4 + DH^5.$$

Writing for convenience

$$ah^2 + bh^3 + ch^4 + dh^5 = h\Omega,$$

that is,

$$\Omega = ah + bh^2 + ch^3 + dh^4,$$

we have

$$\mu H = \frac{h(1 + \lambda\Omega)}{1 + L''h(1 + \lambda''\Omega)},$$

$$\mu' K = \frac{h(1 + \lambda'\Omega)}{1 + L''h(1 + \lambda''\Omega)},$$

and thence

$$K = \frac{\mu}{\mu'} H \frac{1 + \lambda'\Omega}{1 + \lambda\Omega}, \quad = TH \frac{1 + \lambda'\Omega}{1 + \lambda\Omega},$$

since, when h is small or say $\Omega = 0$, we have $K = TH$; viz. we thus have $T = \frac{\mu}{\mu'}$.

Developing, we have

$$K = TH \{1 + (\lambda' - \lambda)\Omega - \lambda(\lambda' - \lambda)\Omega^2 + \lambda^2(\lambda' - \lambda)\Omega^3 - \lambda^3(\lambda' - \lambda)\Omega^4\}.$$

We have

$$\Omega = ah + bh^2 + ch^3 + dh^4,$$

$$h = \frac{1}{a}\Omega - \frac{b}{a^3}\Omega^2 + \frac{1}{a^5}(2b^2 - ac)\Omega^3 - \frac{1}{a^7}(5b^3 - 5abc + a^2d)\Omega^4$$

$$= p\Omega + q\Omega^2 + r\Omega^3 - s\Omega^4,$$

suppose, and hence

$$\mu H = \frac{(p\Omega + q\Omega^2 + r\Omega^3 + s\Omega^4)(1 + \lambda\Omega)}{1 + L''(p\Omega + q\Omega^2 + r\Omega^3 + s\Omega^4)(1 + \lambda''\Omega)}$$

$$= \frac{p\Omega + (q + \lambda p)\Omega^2 + (r + \lambda q)\Omega^3 + (s + \lambda r)\Omega^4}{1 + L''p\Omega + L''(q + \lambda''p)\Omega^2 + L''(r + \lambda''q)\Omega^3 + L''(s + \lambda''r)\Omega^4},$$

which gives H in terms of Ω, and conversely Ω in terms of H. We assume

$$\Omega = XH + YH^2 + ZH^3 + WH^4;$$

K is then given as above in terms of H, Ω, that is, in terms of H, and it is assumed that we have

$$K = TH + AH^2 + BH^3 + CH^4 + DH^5.$$

In the foregoing expression for K, substituting for Ω its value

$$XH + YH^2 + ZH^3 + WH^4,$$

and comparing coefficients, we find

$$\begin{aligned}
A &= (\lambda' - \lambda) \times \; X, \\
B &= (\lambda' - \lambda) \times \; Y - \lambda X^2, \\
C &= (\lambda' - \lambda) \times \; Z - \lambda \,.\, 2XY + \lambda^2 X^3, \\
D &= (\lambda' - \lambda) \times \; W - \lambda(2XZ + Y^2) + \lambda^2 \,.\, 3X^2 Y - \lambda^3 X^4;
\end{aligned}$$

and we have then, from the relation between H, Ω, to find the values of X, Y, Z, W. We have

$$\begin{aligned}
& p\,(XH + YH^2 + ZH^3 + WH^4) \\
& \qquad - \mu H \\
& + (q + \lambda p)\,(X^2H^2 + 2XYH^3 + (2XZ + Y^2)\,H^4) \\
& \qquad - \mu H \,.\, L''p\,(XH + YH^2 + ZH^3) \\
& + (r + \lambda q)\,(X^3H^3 + 3X^2Y \,.\, H^4) \\
& \qquad - \mu H \,.\, L''(q + \lambda''p)\,(X^2H^2 + 2XYH^3) \\
& + (s + \lambda r)\,(X^4 \,.\, H^5) \\
& \qquad - \mu H \,.\, L''(r + \lambda''q)\,(X^3H^3) = 0.
\end{aligned}$$

Equating coefficients, we have

$$\begin{aligned}
& pX - \mu = 0, \\
& pY + (q + \lambda p)\,X^2 - \mu L''pX = 0, \\
& pZ + (q + \lambda p)\,2XY + (r + \lambda q)\,X^3 - \mu L''pY - \mu L''(q + \lambda''p)\,X^2 = 0, \\
& pW + (q + \lambda p)(2XZ + Y^2) + (r + \lambda q)\,3X^2Y + (s + \lambda r)\,X^4 \\
& \qquad - \mu L''pZ - \mu L''(q + \lambda''p)\,2XY - \mu L''(r + \lambda''q)\,X^3 = 0,
\end{aligned}$$

and substituting for p, q, r, s their values, we find successively after reductions, which for W are somewhat troublesome, the values

$$\begin{aligned}
X &= \mu a, \\
Y &= \mu^2 \{b + aL'' - a^2\lambda\}, \\
Z &= \mu^3 \{c + 2bL'' + aL''^2 - 3ab\lambda + 2a^3\lambda^2 - 3a^2\lambda L'' + a^2\lambda''L''\}, \\
W &= \mu^4 \{d + 3cL'' + 3bL''^2 + aL''^3 - (4ac + 2b^2)\,\lambda + 10a^2b\lambda^2 - 5a^4\lambda^3 \\
& \qquad - 12ab\lambda L'' + 3ab\lambda''L'' - 6a^2\lambda''^2 + 3a^2\lambda''L''^2 + 10a^3\lambda^2 L'' - 4a^3\lambda\lambda''L''\};
\end{aligned}$$

these values are to be substituted for X, Y, Z, W in the foregoing expressions of A, B, C, D.

I add the formulæ

$$\lambda' - \lambda = \frac{Q'}{P' + Q't} - \frac{Q}{P + Qt} = \frac{PQ' - P'Q}{(P + Qt)(P' + Q't)},$$

$$\begin{aligned}
PQ' - P'Q &= (\alpha\zeta - \alpha''\xi)(\beta'\zeta - \beta''\eta) - (\beta\zeta - \beta''\xi)(\alpha'\zeta - \alpha''\eta) \\
&= \zeta \begin{vmatrix} \alpha, & \alpha', & \alpha'' \\ \beta, & \beta', & \beta'' \\ \xi, & \eta, & \zeta \end{vmatrix} = \zeta \begin{vmatrix} \alpha, & \alpha', & \alpha'' \\ \beta, & \beta', & \beta'' \\ \gamma, & \gamma', & \gamma'' \end{vmatrix} = (\alpha''x + \beta''y + \gamma'') \begin{vmatrix} \alpha, & \beta, & \gamma \\ \alpha', & \beta', & \gamma' \\ \alpha'', & \beta'', & \gamma'' \end{vmatrix},
\end{aligned}$$

$$\begin{aligned}P + Qt &= \alpha\zeta - \alpha''\xi + (\beta\zeta - \beta''\xi)\,t\\ &= \alpha\,(\alpha''x + \beta''y + \gamma'') - \alpha''\,(\alpha x + \beta y + \gamma) + t\,\{\beta\,(\alpha''x + \beta''y + \gamma'') - \beta''\,(\alpha x + \beta y + \gamma)\},\\ P' + Q't &= \alpha'\zeta - \alpha''\eta + (\beta'\zeta - \beta''\eta)\,t\\ &= \alpha'\,(\alpha''x + \beta''y + \gamma'') - \alpha''\,(\alpha'x + \beta'y + \gamma') + t\,\{\beta'\,(\alpha''x + \beta''y + \gamma'') - \beta''\,(\alpha'x + \beta'y + \gamma')\},\end{aligned}$$

that is,

$$P + Qt = \begin{vmatrix} \alpha, & \beta, & \gamma \\ \alpha'', & \beta'', & \gamma'' \\ t, & -1, & y - tx \end{vmatrix}, \quad P' + Q't = \begin{vmatrix} \alpha', & \beta', & \gamma' \\ \alpha'', & \beta'', & \gamma'' \\ t, & -1, & y - tx \end{vmatrix},$$

whence $\lambda' - \lambda$ is known.

Also

$$\mu = \frac{\zeta}{L} = \frac{\alpha''x + \beta''y + \gamma''}{P + Qt}, \quad \mu' = \frac{\zeta}{L'} = \frac{\alpha''x + \beta''y + \gamma''}{P' + Q't};$$

and

$$T = \frac{\mu}{\mu'} = \frac{L'}{L} = \frac{P' + Q't}{P + Qt}.$$

V.

We have

$$A = (\lambda' - \lambda)\,\mu a,$$

and thus a is a reciprocant, viz. $a = 0$ is the condition of an inflexion.

I remark that $ac - b^2$ is not a reciprocant; we have

$$AC - B^2 = (\lambda - \lambda')^2\,(XZ - Y^2),$$

and then

$$XZ - Y^2 = \mu^4\,\{(ac - b^2) - a^2\lambda\,(b - a^2\lambda) + a^3\,(\lambda'' - \lambda)\,L''\},$$

and thus $AC - B^2$ is not a multiple of $ac - b^2$, even in the particular case $L'' = 0$. I notice further that we have

$$\begin{aligned}(4AC - 5B^2) &= (\lambda' - \lambda)^2\,\{4XZ - 5Y^2 + 2\lambda X^2Y - \lambda^2X^4\},\\ &= (\lambda' - \lambda)^2\,\mu^4\,\{4ac - 5b^2 - 2abL'' - a^2L''^2 + 4a^3\lambda''L''\},\end{aligned}$$

viz. in the particular case $L'' = 0$, we have

$$4AC - 5B^2 = (\lambda' - \lambda)^2\,\mu^4\,(4ac - 5b^2),$$

and thus in the particular case $L'' = 0$ (or say, if $\alpha'' = 0$, $\beta'' = 0$, that is, if X, Y are mere linear functions of x, y) we have $4ac - 5b^2 = 0$ a reciprocant. But in the general case now in hand, it is not a reciprocant.

We have now to consider the sextactic form $a^2d - 3abc + 2b^3$. We have

$$A^2D - 3ABC + 2B^3 = (\lambda' - \lambda)^3\,\{X^2W - 3XYZ + 2Y^3 + \lambda X^2\,(XZ - Y^2)\},$$

and we proceed to calculate the expression in $\{\ \}$. Writing for shortness

$$\begin{aligned}X &= \mu a,\\ Y &= \mu^2\,(b + aL'' + Y_0),\\ Z &= \mu^3\,(c + 2bL'' + aL''^2 + Z_0),\\ W &= \mu^4\,(d + 3cL'' + 3bL''^2 + aL''^3 + W_0),\end{aligned}$$

and then, omitting a factor μ^6, the expression is

$$\begin{aligned}
= \;& a^2d - 3abc + 2b^3 \\
&+ a^2W_0 - 3a(b + aL'')Z_0 + \{(-3ac + 6b^2) + 6abL'' + 3a^2L''^2\} Y_0 \\
&+ 6(b + aL'') Y_0^2 - 3aY_0Z_0 + 2Y_0^3 \\
&+ \lambda a^2 \{ac - b^2 + aZ_0 - 2(b + aL'') Y_0 - Y_0^2\},
\end{aligned}$$

where the terms after the first term are in fact $= 0$; and this being so, we have

$$A^2D - 3ABC + 2B^3 = (\lambda' - \lambda)^3 \mu^6 (a^2d - 3abc + 2b^3),$$

and thus $a^2d - 3abc + 2b^3$ is a reciprocant; but observe that the factor being here $(\lambda' - \lambda)^3 \mu^6$, and in the equation $A = (\lambda' - \lambda)\mu a$ the factor being $(\lambda' - \lambda)\mu$, we cannot with the reciprocants a and $a^2d - 3abc + 2b^3$ form an absolute reciprocant.

To verify the evanescence of the above-mentioned terms, observe that, considering first the terms independent of L'', we have

$$\begin{aligned}
a^2W_0 &= -4a^3c\lambda - 2a^2b^2\lambda + 10a^4b\lambda^2 - 5a^6\lambda^3, \\
-3abZ_0 &= +9a^2b^2\lambda - 6a^4b\lambda^2, \\
+(-3ac + 6b^2)Y_0 &= +3a^3c\lambda - 6a^2b^2\lambda, \\
-3aY_0Z_0 &= -9a^4b\lambda^2 + 6a^6\lambda^3, \\
+6bY_0^2 &= +6a^4b\lambda^2, \\
+2Y_0^3 &= -2a^6\lambda^3, \\
+\lambda a^2(ac - b^2) &= +a^3c\lambda - a^2b^2\lambda, \\
+\lambda a^3Z_0 &= -3a^4b\lambda^2 + 2a^6\lambda^3, \\
+\lambda a^2 . -2bY_0 &= +2a^4b\lambda^2, \\
+\lambda a^2 . -Y_0^2 &= -a^6\lambda^3;
\end{aligned}$$

the sum of which is in fact $= 0$. And next for the terms containing L'', we have

$$\begin{aligned}
& a^2W_0 = \\
&\quad -12a^3b\lambda L'' + 3a^3b\lambda''L'' - 6a^4\lambda L''^2 + 3a^4\lambda''L''^2 + 10a^5\lambda^2L'' - 4a^5\lambda\lambda''L'', \\
& -3abZ_0 = \\
&\quad +9a^3b\lambda L'' - 3a^3b\lambda''L'', \\
& -3a^2L''Z_0 = \\
&\quad +9a^3b\lambda L'' + 9a^4\lambda L''^2 - 3a^4\lambda''L''^2 - 6a^5\lambda^2L'', \\
& (6abL'' + 3a^2L''^2)Y_0 = \\
&\quad -6a^3b\lambda L'' - 3a^4\lambda L''^2, \\
& -3aY_0Z_0 = \\
&\quad -9a^5\lambda^2L'' + 3a^5\lambda\lambda''L'', \\
& +6(b + aL'')Y_0^2 = \\
&\quad +6a^5\lambda^2L'', \\
& +\lambda a^3Z_0 = \\
&\quad -3a^5\lambda^2L'' + a^5\lambda\lambda''L'', \\
& -2\lambda a^2(b + aL'')Y_0 = \\
&\quad +2a^5\lambda^2L'',
\end{aligned}$$

the sum of which is also $= 0$.

Before going further, it is proper to remark that $a, b, c, \ldots$ as representing the second, third, &c., differential coefficients of y are considered as being of the orders 2, 3, 4, ..., and the order of a reciprocant is taken to be the order of the highest letter contained therein: the degree means the degree in these letters $a, b, c, \ldots$, and the weight the weight in these letters, reckoning them as of the weights 2, 3, 4, ... A reciprocant is a homogeneous isobaric function of $a, b, c, \ldots$, not involving y or the first differential coefficient.

VI.

To the reciprocants thus obtained, say

$$U = a, \quad V = a^2d - 3abc + 2b^3,$$

Halphen adds another reciprocant Δ (that of nine-pointic contact) which he expresses in the form of a determinant.

Writing down in connexion therewith its developed expression, we have

$$\Delta = \begin{vmatrix} b, & c, & d, & e, & f \\ a, & b, & c, & d, & e \\ -a^2, & 0, & b^2, & 2bc, & 2bd+c^2 \\ 0, & a^2, & 2ab, & 2ac+b^2, & 2ad+2bc \\ 0, & 0, & a^2, & 3ab, & 3ac+3b^2 \end{vmatrix}, \quad = \begin{array}{rll} a^6 & df & +\ 1 \\ & e^2 & -\ 1 \\ \hline +a^5 & bcf & -\ 3 \\ & bde & +\ 3 \\ & c^2e & +\ 4 \\ & cd^2 & -\ 5 \\ \hline +a^4 & b^3f & +\ 2 \\ & b^2ce & -\ 5 \\ & b^2d^2 & -\ 1 \\ & bc^2d & +\ 14 \\ & c^4 & -\ 4 \\ \hline +a^3 & b^3cd & -\ 10 \\ & b^2c^3 & -\ 5 \\ \hline +a^2 & b^5d & +\ 4 \\ & b^4c^2 & +\ 15 \\ \hline +a^1 & b^6c & -\ 12 \\ +a^0 & b^8 & +\ 3, \end{array}$$

where observe that the whole term in the highest letter f is $a^4(a^2d - 3abc + 2b^3)f$, viz. the coefficient of f is the reciprocant $a^4(a^2d - 3abc + 2b^3)$, agreeing with a general theorem given by Halphen.

Δ is of the degree 8 and the weight 24, and it is seen without difficulty that the factor is $=(\lambda' - \lambda)^8 \mu^{16}$. He remarks that $25\Delta^3 - 27V^8$ vanishes for $a = 0$ (viz. it becomes $256 \,.\, 27b^{24} - 27 \,.\, 256b^{24} = 0$), and not only so, but that it in fact contains the factor a^4. Assuming that this is so, viz. that the terms in a^0, a, a^2, a^3 all of them vanish, I have calculated the terms in a^4, and have thus obtained an incomplete expression for the quotient $(256\Delta^3 - 27V^8) \div a^4$, viz. this is

$$\begin{aligned} H = (256\Delta^3 - 27V^8) \div a^4 = \;& 256a^{14}(df \;\; - e^2)^3 \\ & \vdots \\ & + 288b^{16}\; b^3f \; + \;\; 48 \\ & \qquad\quad b^2ce + \; 120 \\ & \qquad\quad b^2d^2 - \;\; 64 \\ & \qquad\quad bc^2d + 9288 \\ & \qquad\quad c^4 \;\; + \;\; 81; \end{aligned}$$

where I remark that the term in b^{16} presented itself in the form

$$\begin{aligned} & 256\,(54b^{19}f - 135b^{18}ce + \;\; 117b^{18}d^2 + \;\; 5346b^{17}c^2d + \;\; 9477b^{16}c^4) \\ - \; & 27\,(\qquad\qquad\qquad\qquad\quad 1792b^{18}d^2 - 48384b^{17}c^2d + 90720b^{16}c^4). \end{aligned}$$

We have thus Halphen's reciprocant H of the weight 64.

The reciprocants thus far obtained are consequently

	deg.	weight	factor
$U = a$	1	2	$(\lambda' - \lambda)\,\mu$
$V = a^2d - 3abc + 2b^3$	3	9	$(\lambda' - \lambda)^3\,\mu^6$
Δ	8	24	$(\lambda' - \lambda)^8\,\mu^{16}$
$H = (256\Delta^3 - 27V^8) \div a^4$	20	64	$(\lambda' - \lambda)^{20}\,\mu^{44}$

VII.

Reciprocants of the same degree and weight have the same factor, and may thus be combined in the way of addition, viz. R and S being reciprocants of the same degree and the same weight, then (α, β being constants) we have $\alpha R + \beta S$ a reciprocant of the same degree and the same weight. The quotient of two such reciprocants has the factor unity, or say it is an absolute invariant. Thus Δ^3 and V^8 have each of them the degree 24 and weight 72, so that $\alpha\Delta^3 + \beta V^8$ is a reciprocant; in particular, $256\Delta^3 - 27V^8$ is a reciprocant having the factor a^4, and it thus gives rise to the foregoing reciprocant $H = (256\Delta^3 - 27V^8) \div a^4$.

Moreover $\Delta^3 \div V^8$ is an absolute reciprocant.

Any two reciprocants R, S of the same degree and same weight, or, what is the same thing, any absolute reciprocant $R \div S$ gives rise to a reciprocant $RS' - R'S$, where the accents denote differentiation in regard to x ($a' = 3b$, $b' = 4c$, $c' = 5d, \ldots$); the order of the new reciprocant thus exceeds by unity the order of R or S, whichever of them is of the highest order.

From U, V, Δ, which are of the orders 2, 5, 7 respectively, it is thus possible to deduce a series of reciprocants T_8, T_9, $T_{10}, \ldots$ of the orders 8, 9, 10, ..., respectively: viz. we have from the absolute reciprocant $\Delta^3 V^{-8}$, first a reciprocant $3V\Delta' - 8V\Delta$, which, however, contains the factor U^3, and we have

$$U^3 T_8 = 3V\Delta' - 8V\Delta.$$

We then have the absolute reciprocant $U^4 T_8 \,.\, V^{-4}$, leading to

$$T_9 = UVT_8' \quad + 4(VU' - UV')\,T_8,$$

then the absolute reciprocant $U^4 T_9 \,.\, V^{-\frac{16}{3}}$, leading to

$$T_{10} = UVT_9' \quad + 4(VU' - \tfrac{4}{3}UV')\,T_9,$$

and so

$$T_{11} = UVT_{10}' \quad + 4(VU' - \tfrac{5}{3}UV')\,T_{10},$$

$$T_n = UVT'_{n-1} + 4\{VU' - \tfrac{1}{3}(n-6)\,UV'\}\,T_{n-1}.$$

Halphen considers that these are the only distinct reciprocants of the orders 2, 5, 7, 8, 9, ..., n; but remarks that we can, with the reciprocants up to any given order, form algebraical combinations, in some cases containing as factor a power of the reciprocant U or V, so that, rejecting this factor, we have a new independent reciprocant, that is, a reciprocant not expressible as a rational and integral function of inferior reciprocants; an instance hereof is the foregoing reciprocant

$$H = (256\Delta^3 - 27V^8) \div a^4.$$

Other like forms are given by Halphen, viz. writing with him T for shortness in place of T_8, we have for T the following expression in the form of a determinant

$$T = \begin{vmatrix} 3b, & 2a, & a, & 0, & 0, & 0 \\ 4c, & 3b, & b, & a, & 2a^2, & 0 \\ 5d, & 4c, & c, & 2b, & 5ab, & a^2 \\ 6e, & 5d, & d, & 3c, & 6ac + 3b^2, & 3ab \\ 7f, & 6e, & e, & 4d, & 7ad + 7bc, & 4ac + 2b^2 \\ 8g, & 7f, & f, & 6e, & 8ae + 8bd + 4c^2, & 5ad + 5bc \end{vmatrix},$$

and then

$$T_1 = (V^4 T - \tfrac{1}{6}H) \div U,$$

$$G = (U^4 T^2 + 9H) \div V^2,$$

$$\Theta = \{2U\Delta T' + T(8U'\Delta - 3U\Delta')\} \div V,$$

$$\Theta_1 = (\Theta + \tfrac{1}{4}G) \div V,$$

$$\Theta_2 = (\Theta_1 - \tfrac{45}{2}TV) \div U,$$

where Θ is given explicitly in the form of a determinant, viz. this is

$$\Theta_2 = -432 \begin{vmatrix} c, & d, & e, & f, & g, & h \\ b, & c, & d, & e, & f, & g \\ a, & b, & c, & d, & e, & f \\ a^2, & 2ab, & 2ac+b^2, & 2ad+2bc, & 2ae+2bd+c^2, & 2af+2be+2cd \\ 0, & a^2, & 2ab, & 2ac+b^2, & 2ad+2bc, & 2ae+2bd+c^2 \\ 0, & 0, & a^3, & 3a^2b, & 3a^2c+3ab^2, & 3a^2d+6abc+b^3 \end{vmatrix},$$

where $\Theta_2 = 0$ is the differential equation of the ninth order of the general cubic curve.

Halphen's S, $= UVT' + 4(VU' - UV')T$, is the above-mentioned reciprocant T_9; U is connected with Θ by the formula $2\Delta S - V^2\Theta = U^4T^2$.

Halphen considers incidentally (p. 56), what may be called a polar relation of the variables X, Y and x, y; viz. this is

$$X = \frac{dy}{dx}, \quad Y = x\frac{dy}{dx} - y,$$

whence conversely

$$x = \frac{dY}{dX}, \quad y = X\frac{dY}{dX} - Y.$$

We may here express Y and its differential coefficients in regard to X, say Y, Y_1, $Y_2, \ldots$, in terms of y and its differential coefficients in regard to x, say y, y_1, $y_2, \ldots$ (to avoid confusion with other formulæ, I purposely use these notations, abstaining from the introduction of the letters a, b, $c, \ldots$). The geometrical signification is that x, y, being point-coordinates, then X, Y are line-coordinates. We find, without difficulty,

$$Y_1 = x, \quad Y_2 = \frac{1}{y_2}, \quad Y_3 = \frac{-y_3}{y_2^3}, \quad Y_4 = \frac{-y_2y_4 - 3y_3^2}{y_2^5}, \quad Y_5 = \frac{-y_2^2y_5 + 10y_2y_3y_4 + 15y_3^3}{y_2^7}, \text{ \&c.}$$

Hence, in particular,

$$9Y_2^2Y_5 - 45Y_2Y_3Y_4 + 40Y_3^3 = -\frac{1}{y_2^9}(9y_2^2y_5 - 45y_2y_3y_4 + 40y_3^3),$$

viz. the differential equation $9y_2^2y_5 - 45y_2y_3y_4 + 40y_3^3 = 0$ (in the former notation $a^2d - 3abc + 2b^3 = 0$) of the conic in point-coordinates gives, as it should do, the equation of like form $9Y_2^2Y_5 - 45Y_2Y_3Y_4 + 40Y_3^3 = 0$ of the conic in line-coordinates.

The geometrical applications throughout the memoir are very extensive and interesting.

VIII.

In Halphen's second memoir "Sur les invariants différentiels des courbes gauches" (1880), instead of a single dependent variable y a function of x, we have two dependent variables y, z, each of them a function of x.

The relations between the original variables x, y, z, and the new variables X, Y, Z, are of the general homographic form

$$X,\ Y,\ Z = \frac{\xi}{\omega},\ \frac{\eta}{\omega},\ \frac{\zeta}{\omega},$$

where

$$\begin{aligned}\xi &= \alpha x + \beta y + \gamma z + \delta\ ,\\ \eta &= \alpha' x + \beta' y + \gamma' z + \delta'\ ,\\ \zeta &= \alpha'' x + \beta'' y + \gamma'' z + \delta''\ ,\\ \omega &= \alpha''' x + \beta''' y + \gamma''' z + \delta''',\end{aligned}$$

but he does not from these formulæ deduce the expressions of $Y_{,}$, $Y_{,,}$, ..., $Z_{,}$, $Z_{,,}$, ... in terms of y', y'', ..., z', z'', ...; the investigation is in some measure a geometrical one. The notation employed is

$$u = \tfrac{1}{12}(y''z''' - y'''z''),$$

viz. u is here the most simple reciprocant, the equation $u = 0$ is obviously the condition of a plane curve:

$$a_n = \frac{1}{4.5.6\ldots n}\,\frac{y''z^{(n)} - y^{(n)}z''}{y''z''' - y'''z''},$$

$$b_n = \frac{-1}{3.4.5\ldots n}\,\frac{y'''z^{(n)} - y^{(n)}z'''}{y''z''' - y'''z''},$$

n greater than or equal to 4; viz. these two singly infinite series of symbols a_4, a_5, a_6, ..., b_4, b_5, b_6, ..., together with u, are used for the expression of the doubly infinite series $y^{(m)}z^{(n)} - y^{(n)}z^{(m)}$, by virtue of the identity

$$\begin{aligned}&(y''z''' - y'''z'')(y^{(m)}z^{(n)} - y^{(n)}z^{(m)})\\ &\quad = (y''z^{(m)} - y^{(m)}z'')(y'''z^{(n)} - y^{(n)}z''') - (y''z^{(n)} - y^{(n)}z'')(y'''z^{(m)} - y^{(m)}z'''),\end{aligned}$$

or, as this may be written,

$$\frac{1}{1.2\ldots m.1.2\ldots n}\,\frac{y^{(m)}z^{(n)} - y^{(n)}z^{(m)}}{u} = a_n b_m - a_m b_n.$$

The most simple reciprocant (after u) is

$$v = a_6 - 2b_5 - 3a_4a_5 + 3a_4b_4 + 2a_4^3,$$

or rather u^3v, which is an integral function of the differential coefficients: the signification of the equation $v = 0$ is that the tangents of the curve belong to a linear complex. This is a property belonging to the tangents of a skew cubic, and the skew cubic thus satisfies the differential equation $u = 0$. Another reciprocant is obtained

$$w = b_6 - a_4b_5 - 4a_5b_4 + 4a_4^2b_4 - 2a_4^2a_5 + 2b_4^2 + a_5^2 + a_4^4,$$

or rather u^4w, which is an integral function of the differential coefficients: and the skew cubic satisfies the second differential equation $w = 0$. The formulæ enable the determination of the osculating skew cubic at any point of a skew curve.

The general theory is developed in a compact form in the theorems I. to VII., and it is shown that the investigation of all the reciprocants depends upon that of two reciprocants of the seventh order. The geometrical applications of the theory are very extensive and interesting.

IX.

Before going further, I remark that the homologic transformation

$$X = \alpha x + \beta y + \gamma, \quad Y = \alpha' x + \beta' y + \gamma',$$

(which is in point of generality intermediate between Halphen's homographic transformation

$$X = \frac{\alpha x + \beta y + \gamma}{\alpha'' x + \beta'' y + \gamma''}, \quad Y = \frac{\alpha' x + \beta' y + \gamma'}{\alpha'' x + \beta'' y + \gamma''},$$

and Sylvester's special transformation $X = y$, $Y = x$, and which includes as a particular case the rectangular transformation

$$X = x \cos\theta + y \sin\theta, \quad Y = -x \sin\theta + y \cos\theta),$$

does not appear to have been explicitly considered. Writing as in the general case $t, a, b, c, \ldots$, for $y', \frac{1}{2}y'', \frac{1}{6}y''', \ldots$, and $T, A, B, C, \ldots$, for $Y_{,}, \frac{1}{2}Y_{,,}, \frac{1}{6}Y_{,,,}, \ldots$; also taking h, k for the increments of x, y, and H, K for those of X, Y respectively, we have

$$\begin{aligned} H &= \alpha h + \beta k, \quad k = th + ah^2 + bh^3 + ch^4 + \ldots, \\ K &= \alpha' h + \beta' k, \end{aligned}$$

that is,

$$\begin{aligned} H &= (\alpha + \beta t)\, h + \beta\, (ah^2 + bh^3 + ch^4 + \ldots), \\ K &= (\alpha' + \beta' t)\, h + \beta'\, (ah^2 + bh^3 + ch^4 + \ldots), \end{aligned}$$

which, by the elimination of h, must lead to

$$K = TH + AH^2 + BH^3 + CH^4 + \ldots.$$

We have

$$T = \frac{\alpha' + \beta' t}{\alpha + \beta t};$$

and we write

$$\lambda = \frac{\beta}{\alpha + \beta t}, \quad \lambda' = \frac{\beta'}{\alpha' + \beta' t},$$

and thence

$$\lambda' - \lambda = \frac{\alpha\beta' - \alpha'\beta}{\alpha + \beta t \,.\, \alpha' + \beta' t};$$

moreover

$$\mu = \frac{1}{\alpha + \beta t}, \quad \mu' = \frac{1}{\alpha' + \beta' t},$$

and therefore

$$T = \frac{\mu}{\mu'}.$$

Also

$$\Omega = ah + bh^2 + ch^3 + \ldots;$$

whence, if

$$h = p\Omega + q\Omega^2 + r\Omega^3 + s\Omega^4 + \ldots,$$

then

$$p = \frac{1}{a},$$

$$q = -\frac{1}{a^3} b,$$

$$r = -\frac{1}{a^5} (ac - 2b^2),$$

$$s = -\frac{1}{a^7} (a^2d - 5abc + 5b^3),$$

$$\vdots$$

and we have

$$H = (\alpha + \beta t) h + \beta h\Omega, \ = \left(\frac{1}{\mu} + \beta\Omega\right) (p\Omega + q\Omega^2 + r\Omega^3 + \ldots),$$

giving

$$\Omega = XH + YH^2 + ZH^3 + WH^4 + \ldots,$$

where, substituting for p, q, r, ..., their values,

$$X = \mu a,$$

$$Y = \mu^2 . b - a^2\lambda,$$

$$Z = \mu^3 . c - 3ab\lambda + 2a^3\lambda^2,$$

$$W = \mu^4 . d - (4ac + 2b^2) \lambda + 10a^2b\lambda^2 - 5a^4\lambda^3,$$

$$\vdots$$

and then

$$K = HT\frac{1 + \lambda'\Omega}{1 + \lambda\Omega}, \ = HT\frac{1 + \lambda' (XH + YH^2 + \ldots)}{1 + \lambda (XH + YH^2 + \ldots)},$$

giving

$$K = TH + AH^2 + BH^3 + CH^4 + \ldots,$$

where

$$A = \lambda' - \lambda . X,$$

$$B = \lambda' - \lambda . Y - \lambda X^2,$$

$$C = \lambda' - \lambda . Z - \lambda 2XY + \lambda^2 X^3,$$

$$D = \lambda' - \lambda . W - \lambda (2XZ + Y^2) \lambda^2 . 3X^2Y - \lambda^3 X^4,$$

whence, substituting for X, Y, Z, W, ..., their values, we have $T = \frac{\mu}{\mu'}$, *ut suprà*, and

$$A = (\lambda' - \lambda) \mu . a,$$

$$B = (\lambda' - \lambda) \mu^2 . b - 2a^2\lambda,$$

$$C = (\lambda' - \lambda) \mu^3 . c - 5ab\lambda + 5a^3\lambda^2,$$

$$D = (\lambda' - \lambda) \mu^4 . d + (-6ac - 3b^2) \lambda + 21a^2b\lambda^2 - 14a^4\lambda^3,$$

$$\vdots$$

It will be observed that μ' enters only into the equation $T=\frac{\mu}{\mu'}$: there are thus no reciprocants containing t, i.e. there is here nothing analogous to Sylvester's impure reciprocants. But we have as reciprocants, all his pure reciprocants, for instance

$$4AC-5B^2=(\lambda'-\lambda)^2(4ac-5b^2), \text{ \&c.}$$

Putting

$$\alpha=0, \quad \beta=1, \quad \alpha'=1, \quad \beta'=0,$$

then

$$\lambda=\frac{1}{t}, \quad \lambda'=0, \quad \mu=\frac{1}{t}, \quad \mu'=1;$$

then $T=\frac{1}{t}$, viz. we have here no arbitrary constant entering into this equation only, and there are thus reciprocants containing t. We have, in fact, the formulæ of Sylvester's theory.

In the rectangular case,

$$\alpha=\cos\theta, \quad \beta=\sin\theta, \quad \alpha'=-\sin\theta, \quad \beta'=\cos\theta,$$

we have

$$T=\frac{\cos\theta+t\sin\theta}{-\sin\theta+t\cos\theta},$$

where θ enters into the other equations, and there are thus impure reciprocants containing t, and reciprocants pure and impure which are not reciprocants in the general homologic case; but I do not go into the question of these orthogonal reciprocants.

X.

We have next Sylvester:—"Lectures on the theory of Reciprocants" (reported by J. Hammond), *Amer. Math. Journ.*, t. VIII. (1886), pp. 196—260.

The lectures were delivered at Oxford, *Inaugural Lecture*, Dec. 1885, starting from the Schwarzian function *

$$\frac{y'''}{y'}-\tfrac{3}{2}\left(\frac{y''}{y'}\right)^2,$$

which acquires only a factor by the interchange of x and y,

$$\left[\frac{y'''}{y'}-\tfrac{3}{2}\left(\frac{y''}{y'}\right)^2=-x_1^2\left\{\frac{x_{\prime\prime\prime}}{x_\prime}-\tfrac{3}{2}\left(\frac{x_{\prime\prime}}{x_\prime}\right)^2\right\}\right].$$

In lecture 2, pp. 203 *et seq.*, Sylvester considers the general theory of the functions which remain unaltered, except as to a factor, by the interchange of x, y: viz. writing

$$t,\ a,\ b,\ c, \ldots \text{ for } y',\ y'',\ y''',\ y'''', \ldots,$$

or again

$$t,\ a_0,\ a_1,\ a_2,\ a_3, \ldots \text{ for } y',\ \tfrac{1}{2}y'',\ \tfrac{1}{6}y''',\ \tfrac{1}{24}y'''', \ldots,$$

* Schwarzian $=\frac{y'''}{y'}-\frac{3}{2}\left(\frac{y''}{y'}\right)^2=6\frac{b}{t}-\frac{3}{2}\left(\frac{2a}{t}\right)^2=\frac{6\,(bt-a^2)}{t^2}$, where the reciprocant is $=bt-a^2$.

and so

$$T,\ \alpha,\ \beta,\ \gamma, \ldots \text{ for } x_{,},\ x_{,,},\ x_{,,,},\ x_{,,,,}, \ldots,$$

or again

$$T,\ \alpha_0,\ \alpha_1,\ \alpha_2,\ \alpha_3,\ \ldots \text{ for } x_{,},\ \tfrac{1}{2}x_{,,},\ \tfrac{1}{6}x_{,,,},\ \tfrac{1}{24}x_{,,,,}, \ldots$$

he gives the equations

$$\begin{array}{llll} \alpha = -a & \div t^3, \text{ or} & \alpha_0 = -a_0 & \div t^3, \\ \beta = -bt + 3a^2 & \div t^5, & \alpha_1 = -a_1t + 2a_0^2 & \div t^5, \\ \gamma = -ct^2 + 10abt - 15a^3 \div t^7, & & \alpha_2 = -a_2t^2 + 5a_0a_1t - 5a_0^3 \div t^7, & \\ \vdots & & \vdots & \end{array}$$

and obtains with $(a, b, c, \ldots)$ reciprocants such as a, $2bt - 3a^2$, &c., viz. we have

$$a = -t^3 . \alpha,$$

$$2bt - 3a^2 = -t^6 . 2\beta T - 3\alpha^2,$$

and further on, like forms with $(a_0, a_1, a_2, \ldots)$.

It is to be remarked that it is preferable to deal with the quantities $(a_0, a_1, a_2, \ldots)$ which represent $(\tfrac{1}{2}y'', \tfrac{1}{6}y''', \tfrac{1}{24}y'''', \ldots)$ rather than with $(a, b, c, \ldots)$ which represent $(y', y'', y''', \ldots)$. I do this in the sequel, *changing the notation*, and writing $(t, a, b, c, \ldots)$ to denote $(y', \tfrac{1}{2}y'', \tfrac{1}{6}y''', \tfrac{1}{24}y'''', \ldots)$, viz. my $(t, a, b, c, \ldots)$ are not Sylvester's $(t, a, b, c, \ldots)$, but are his $(t, a_0, a_1, a_2, \ldots)$.

A reciprocant with Sylvester is thus a function of y and its differential coefficients in regard to x, which except as to a factor remain unaltered when x, y are interchanged, or say, when x, y are changed into X, Y, where $X = y$, $Y = x$. This is a much less general change than Halphen's, and thus every reciprocant (Halphen) is a reciprocant (Sylvester), but not conversely. The reciprocants (Sylvester) are far more numerous. I remark that incidentally Sylvester considers reciprocants, which remain unaltered save as to a factor, when x, y are changed into X, Y, where

$$X = \alpha x + \beta y + \gamma, \quad Y = \alpha' x + \beta' y + \gamma',$$

which again is a less general change than Halphen's—it has been in what precedes alluded to as the particular case $L'' = 0$, that is, $\alpha'' = 0$, $\beta'' = 0$. It may be remarked that the proper orthogonal substitution corresponding to Sylvester's substitution is not $X = y$, $Y = x$, but $X = y$, $Y = -x$, viz. we have here the determinant $\alpha\beta' - \alpha'\beta = +1$.

I develop Sylvester's theory of reciprocants; but I wish first to point out the resemblance in form between this theory and that of seminvariants. In the theory of seminvariants, from the set of quantities $(a, b, c, \ldots)$ in connexion with an arbitrary quantity θ, we deduce a new set $(a', b', c', \ldots)$, where

$$\begin{aligned} a' &= a, \\ b' &= b + a\theta, \\ c' &= c + 2b\theta + a\theta^2, \\ d' &= d + 3c\theta + 3b\theta^2 + a\theta^3, \\ &\vdots \end{aligned}$$

or, what is the same thing, if $\theta' = -\theta$, then

$$\begin{aligned} a &= a', \\ b &= b' + a'\theta', \\ c &= c' + 2b'\theta' + a'\theta'^2, \\ d &= d' + 3c'\theta' + 3b'\theta'^2 + a'\theta'^3, \\ &\vdots \end{aligned}$$

and this being so there exist functions which are the same for unaccented and the accented letters respectively; for instance, a', $= a$:

$$\begin{aligned} a'c' &= \quad ac + 2ab\theta + a^2\theta^2, \\ -b'^2 &= -b^2 - 2ab\theta - a^2\theta^2, \end{aligned}$$

that is, $a'c' - b'^2 = ac - b^2$; and similarly

$$a'^2d' - 3a'b'c' + 2b'^3 = a^2d - 3abc + 2b^3, \text{ \&c.};$$

these functions a, $ac - b^2$, $a^2d - 3abc + 2b^3$, &c., are called seminvariants.

Similarly, in Sylvester's theory of reciprocants, starting from y, a given function of x, we have $(t, a, b, c, \ldots)$ denoting $(y', \tfrac{1}{2}y'', \tfrac{1}{6}y''', \tfrac{1}{24}y'''')$; and conversely $(t', a', b', c', \ldots)$ denoting $(x_{,}, \tfrac{1}{2}x_{,,}, \tfrac{1}{6}x_{,,,}, \tfrac{1}{24}x_{,,,,}, \ldots)$; taking h, k for the increments of x and y respectively, we have

$$\begin{aligned} k &= th + ah^2 + bh^3 + ch^4 + dh^5 + \ldots, \\ h &= t'k + a'k^2 + b'k^3 + c'k^4 + d'k^5 + \ldots, \end{aligned}$$

which equations determine the relations between $(t, a, b, c, d, \ldots)$ and $(t', a', b', c', d', \ldots)$, viz. we have $tt' = 1$, and then

$$\begin{aligned} a' &= -a && \div t^3, \\ b' &= -bt + 2a^2 && \div t^5, \\ c' &= -ct^2 + 5abt \qquad - 5a^3 && \div t^7, \\ d' &= -dt^3 + (6ac + 3b^2)\,t^2 - 21a^2bt \qquad + 14a^4 && \div t^9, \\ e' &= -et^4 + (7ad + 7bc)\,t^3 - (28a^2c + 28ab^2)\,t^2 + 84a^3bt - 42a^5 && \div t^{11}, \\ &\vdots \end{aligned}$$

and conversely,

$$\begin{aligned} a &= -a' && \div t'^3, \\ b &= -b't' + 2a'^2 && \div t'^5, \\ c &= -c't'^2 + 5a'b't' - 5a'^3 && \div t'^7, \\ &\vdots \end{aligned}$$

and we thence deduce functions which have equal or opposite values for the accented and unaccented letters respectively; these functions may or may not contain t, t'. Thus we have

$$\begin{aligned} a't'^{-\frac{3}{2}} &= -at^{-\frac{3}{2}}, \\ (b't' - a'^2)\, t'^{-3} &= -(bt - a^2)\, t^{-3}, \\ (2c't' - 5a'b')\, t'^{-\frac{7}{2}} &= -(2ct - 5ab)\, t^{-\frac{7}{2}}, \\ (4a'c' - 5b'^2)\, t'^{-4} &= (4ac - 5b^2)\, t^{-4}, \\ \{(1 + t'^2)\, b' - 2a'^2t'\}\, t'^{-3} &= \{(1 + t^2)\, b - 2a^2t\}\, t^{-3}, \\ &\text{\&c.} \end{aligned}$$

These functions of the unaccented letters, or say the same functions omitting the exterior power of t, are called Reciprocants; thus we have the reciprocants a, $bt - a^2$ $2ct - 5ab$, $4ac - 5b^2$, $(1 + t^2)\, b - 2a^2t$, &c. Observe that, when the exterior powers of t are omitted, then the values are equal or opposite to those with the accented letters save as to a power of t, viz. the forms are

$$\begin{aligned} a &= -a't^3, \\ bt - a^2 &= -(b't' - a'^2)\, t^6, \\ 2ct - 5ab &= -(2c't' - 5a'b')\, t^7, \\ 4ac - 5b^2 &= (4a'c' - 5b'^2)\, t^8, \\ \{(1 + t^2)\, b - 2a^2t\} &= -\{(1 + t'^2)\, b' - 2a'^2t'\}\, t^6, \\ &\text{\&c.} \end{aligned}$$

A reciprocant is said to be odd or even according as the sign on the right-hand side is $-$ or $+$; the index of t on the right-hand side is said to be the weight of the reciprocant. Reciprocants may be combined in the way of addition if and only if they are each of them of the same weight and parity, viz. this being so, then if λ, λ', ..., are mere numbers, $\lambda R + \lambda' R' + \ldots$ will be a reciprocant of the same weight and parity with each of the reciprocants R, R', ...

Reciprocants may in every case be combined in the way of multiplication; viz. two or more reciprocants may be multiplied together giving a reciprocant the parity of which is the sum of the parities, and its weight the sum of the weights of the component reciprocants. Thus $bt - a^2$ and $2ct - 5ab$ being odd reciprocants of the weights 6 and 7 respectively, we have $(bt - a^2)(2ct - 5ab)$ an even reciprocant of the weight 13.

A reciprocant is pure or impure according as it does not or does contain t. The foregoing equations for a', b', c', ..., may be obtained each from the next preceding one by operating upon it with

$$\frac{1}{\lambda} t^{-1} (2a\partial_t + 3b\partial_a + 4c\partial_b + 5d\partial_c + \ldots),$$

where λ is a positive integer, $= 3 +$ weight of the letter operated upon ($3 + 3$, $= 6$ in operating on d' to obtain e', $3 + 4 = 7$ in operating on e' to obtain f', and so on).

For example,

$$e' = \tfrac{1}{6}t^{-1}(2a\partial_t + 3b\partial_a + 4c\partial_b + 5d\partial_c + 6e\partial_d)\,d'$$
$$= t^{-1}(\tfrac{1}{3}a\partial_t + \tfrac{1}{2}b\partial_a + \tfrac{2}{3}c\partial_b + \tfrac{5}{6}d\partial_c + e\partial_d)\,d',$$

viz. operating herewith on

$$d',\ = \{-dt^3 + (6ac + 3b^2)\,t^2 - 21a^2bt + 14a^4\} \div t^9,$$

we obtain

t^4,	t^3,	t^2,	t,	t^0	$\div t^{11}$,
	$2ad$,	$-14a^2c - 7ab^2$,	$56a^3b$,	$-42a^5$,	
	$3bc$,	$-21ab^2$,	$28a^3b$,		
	$4bc$,	$-14a^2c$			
	$5ad$,				
$-e$,					
$-e$,	$7ad + 7bc$,	$-28a^2c - 28ab^2$,	$84a^3b$,	$-42a^5$,	

the value given above.

The proof is at once obtained: except as to a numerical factor, the operation is in fact equivalent to the differentiation of a function of y, and its derived coefficients y', y'', ..., in regard to x.

Any pure reciprocant is reduced to zero by the operator

$$V = 4\,.\,\tfrac{1}{2}a^2\partial_b + 5ab\partial_c + 6\,(ac + \tfrac{1}{2}b^2)\,\partial_d + 7\,(ad + bc)\,\partial_e + 8\,(ae + bd + \tfrac{1}{2}c^2)\,\partial_f + \ldots,$$

or say V is an annihilator of any pure reciprocant: compare herewith the theorem $\Delta_1 = a\partial_b + 2b\partial_c + 3c\partial_d + \ldots$ is an annihilator of any seminvariant; and conversely, any homogeneous isobaric function of $a, b, c, \ldots$ reduced to zero by V is a pure reciprocant. For instance, for the pure reciprocant $5ac - 4b^2$, we have

$$(2a^2\partial_b + 5ab\partial_c)\,(4ac - 5b^2) = -20a^2b + 20a^2b,\ = 0.$$

Any rational and integral homogeneous reciprocant, operated upon with

$$G,\ = (3bt - 3a^2)\,\partial_a + (4ct - 4ab)\,\partial_b + (5dt - 5ac)\,\partial_c + \ldots,$$

produces a like reciprocant. Thus operating upon a with G, we obtain $3bt - 3a^2$, or say $bt - a^2$, and operating hereon again with G, a new reciprocant, and so on.

The series of reciprocants thus derived from a is

$$\begin{array}{l} a, \\ bt - a^2, \\ 2ct - 5ab, \\ \left.\begin{array}{r} 2dt^2 - 6ac \\ -3b^2 \end{array}\right|\ t + 7a^2b, \end{array}$$

$$\begin{array}{r|l}
2et^2 - \ \ 7ad & t^2 + \ \ 8a^2c \\
- \ \ 7bc & \ \ + \ 11ab^2,
\end{array}$$

$$\begin{array}{r|l|l}
14ft^3 - 56ae & t^2 + 103a^2d & t - \ \ 88a^3c \\
-56bd & \ \ + 199abc & \ \ - 121a^2b^2, \\
-28c^2 & \ \ + \ \ 33b^3 &
\end{array}$$

&c.

These are, in fact, connected with the terms of

$$\left(\frac{1}{\sqrt{t}}\frac{d}{dx}\right)^i \log t, \quad = -\left(\frac{1}{\sqrt{t'}}\frac{d}{dy}\right)^i \log t',$$

where observe that $\partial_x t = 2a$, $\partial_x a = 3b$, $\partial_x b = 4c$, &c. Thus

$$\left(\frac{1}{\sqrt{t}}\frac{d}{dx}\right)\log t = \frac{a}{t^{\frac{3}{2}}}; \quad \left(\frac{1}{\sqrt{t}}\frac{d}{dx}\right)^2 \log t = \frac{1}{\sqrt{t}}\frac{d}{dx}\frac{a}{t^{\frac{3}{2}}} = \frac{3(bt-a^2)}{t^3}, \text{ \&c.}$$

A like generator for pure reciprocants is

$$H, = 4(ac-b^2)\partial_b + 5(ad-bc)\partial_c + 6(ae-bd)\partial_d + \ldots;$$

thus operating herewith upon $(4ac-5b^2)$, we obtain

$$(4ac-b^2)(-10b) + 5(ad-bc)4a, \quad = 20a^2d - 60abc + 40b^3,$$

viz. we have thus the pure reciprocant $a^2d - 3abc + 2b^3$.

In repeating this process, we may reduce by means of powers and products of earlier reciprocants, and we can also in many cases throw out powers of the reciprocant a. Thus forming the expression for $H(a^2d - 3abc + 2b^3)$, this is given by column 1 of the annexed form: multiplying by 25, and adding $24(4ac-5b^2)^2$, so as to eliminate the term in b^4, we obtain the expression in col. 4: or throwing out the numerical factor 3, and also the factor a, we have the

	1,	2,	3,	4,	5,	
	H,	$25H$,	$24H^2$,	$\div 3$,		
a^3e	+ 6	+ 150		+ 150	+ 50	a^2e
a^2bd	− 21	− 525		− 525	− 175	abd
a^2c^2	− 12	− 300	+ 384	+ 84	+ 28	ac^2
ab^2c	+ 51	+ 1275	− 960	+ 315	+ 105	b^2c
b^4	− 24	− 600	+ 600			

reciprocant $+50a^2e -$&c. in col. 5, and the outside right-hand column of literal terms.

We thus obtain the series of pure reciprocants

$a + 1$	$ac + 4$	$a^2d + 1$	$a^2e + 50$	$a^3f + 10$	$a^2g + 14$	$a^3h + 7$	$a^3i + 420$	
	$b^2 - 5$	$abc - 3$	$abd - 175$	$a^2be - 40$	$abf - 63$	$a^2bg - 35$	$a^2bh - 2310$	
		$b^3 + 2$	$ac^2 + 28$	$a^2cd - 12$	$ace - 1350$	$a^2cf - 539$	$a^2cg - 25648$	$+ 1176$
			$b^2c + 105$	$ab^2d + 65$	$ad^2 + 1470$	$a^2de + 605$	$a^2df + 9240$	$- 8085$
				$abc^2 + 16$	$b^2e + 1782$	$ab^2f + 735$	$a^2e^2 + 21780$	$+ 7040$
				$b^3c - 39$	$bcd - 4158$	$abce + 306$	$ab^2g + 36680$	1470
					$c^3 + 2310$	$abd^2 - 2135$	$abcf + 85386$	$+ 18963$
						$ac^2d + 1001$	$abde - 191730$	$- 16940$
						$b^3e - 1485$	$ac^2e - 59220$	$- 27160$
						$b^2cd + 3465$	$acd^2 + 120540$	$+ 26460$
						$bc^3 - 1925$	$b^3f - 126945$	$- 9555$
							$b^2ce + 252126$	$+ 28098$
							$b^2d^2 + 169260$	$+ 12740$
							$bc^2d - 419034$	$- 52822$
							$c^4 + 129360$	$+ 21560$

I remark that a pure reciprocant not included in the foregoing series is

$a^2ce + 800$
$a^2d^2 - 875$
$ab^2e - 1000$
$abcd + 2450$
$ac^3 - 1344$
$b^2c^2 - 35$

XI.

Halphen assumes that $a^2d - 3abc + 2b^3$ is the sextactic reciprocant, or, what is the same thing, that the general conic satisfies the differential equation of the fifth order $a^2d - 3abc + 2b^3$; the proof is as follows.

The general equation of the conic may be written

$$y + Ax + B = \sqrt{k}\,\sqrt{(x - \alpha \,.\, x - \beta)},$$

which depends on the five parameters A, B, k, α, β.

We have

$$y' + A = \tfrac{1}{2}\sqrt{k}\,\frac{2x-\alpha-\beta}{\sqrt{(x-\alpha\,.\,x-\beta)}},$$

$$y'' = \tfrac{1}{4}\sqrt{k}\left\{\frac{4}{\sqrt{(x-\alpha\,.\,x-\beta)}} - \frac{(2x-\alpha-\beta)^2}{(x-\alpha\,.\,x-\beta)^{\frac{3}{2}}}\right\},$$

$$= \frac{-\tfrac{1}{4}\sqrt{k}\,(\alpha-\beta)^2}{\{x-\alpha\,.\,x-\beta\}^{\frac{3}{2}}},$$

or say

$$y''^{-\frac{2}{3}} = (\tfrac{1}{2})^{-\frac{4}{3}}\,k^{-\frac{1}{3}}\,(\alpha-\beta)^{-\frac{4}{3}}\,x-\alpha\,.\,x-\beta;$$

and we have thus the differential equation $\left(\dfrac{d}{dx}\right)^3 y''^{-\frac{2}{3}} = 0$. This gives

$$\left(\frac{d}{dx}\right)^2 y''^{-\frac{5}{3}}y''' = 0,$$

$$\frac{d}{dx}\quad y''^{-\frac{5}{3}}y'''' - \tfrac{5}{3}y''^{-\frac{8}{3}}y'''^2 = 0,$$

$$y''^{-\frac{5}{3}}y^{\text{v}} - \tfrac{5}{3}y''^{-\frac{8}{3}}y'''y''''$$
$$- \tfrac{10}{3}y''^{-\frac{8}{3}}y'''y'''' + \tfrac{40}{9}y''^{-\frac{11}{3}}y'''^3 = 0,$$

that is,

$$9y''^{-\frac{5}{3}}y^{\text{v}} - 45y''^{-\frac{8}{3}}y'''y'''' + 40y''^{-\frac{11}{3}}y'''^3 = 0,$$

or say

$$9y''^2y^{\text{v}} - 45y''y'''y'''' + 40y'''^3 = 0.$$

But y'', y''', y'''', $y^{\text{v}} = 2a$, $6b$, $24c$, $120d$, and the equation thus is $4320\,(a^2d - 3abc + 2b^3) = 0$, viz. it is $a^2d - 3abc + 2b^3 = 0$.

XII.

Sylvester's Lectures are published in the *American Mathematical Journal* as follows: Lectures 1 to 10, t. VIII. (1886), pp. 196—260; lectures 11 to 32, t. IX. (1887), viz. 11 to 16, pp. 1—37, 17 to 24, pp. 113—161, and 25 to 32, pp. 297—352; and lectures 33 and 34 (34 by Mr Hammond), t. X. (1888), pp. 1—16. In the footnote p. 7 to lecture 12, writing $y_1, y_2, y_3, y_4, \ldots$, for the derived functions of y, he adopts definitely the notation $t, a, b, c, \ldots$, to mean the reduced functions, $y_1, \dfrac{1}{1\,.\,2}y_2, \dfrac{1}{1\,.\,2\,.\,3}y_3, \dfrac{1}{1\,.\,2\,.\,3\,.\,4}y_4, \ldots$. In lecture 13, p. 20, he introduces the term Principiant—"instead of the cumbrous terms Projective Reciprocants or Differential Invariants, it is better to use the single word Principiants to denominate that crowning class or order of Reciprocants which remain to a factor *près*, unaltered for any homographic substitutions impressed on the variables"—that is, Halphen's Differential Invariant = Principiant. And in lectures 22 *et seq.*, Sylvester develops an important theory in regard to Principiants, connecting them with reciprocants and seminvariants, viz. he considers a series of seminvariants $N, A_0, A_1, A_2, \ldots$, and a series of reciprocants $M, A, B, C, \ldots$, such that the seminvariants formed with either sets of capitals are identical with each other (for instance $A_0A_2 - A_1^2 = AC - B^2$), and that any such seminvariant is a Principiant.

In what follows,

instead of $N, A_0, A_1, A_2, \ldots$, I write $N, A, B, C, \ldots$,

and instead of $M, A, B, C, \ldots$, I write $\mathfrak{M}, \mathfrak{A}, \mathfrak{B}, \mathfrak{C}, \ldots$,

so that $N, A, B, C, \ldots$, are seminvariants and $\mathfrak{M}, \mathfrak{A}, \mathfrak{B}, \mathfrak{C}, \ldots$, are reciprocants.

The expressions of these functions up to E, $\mathfrak{C}$, are

$N=$	$A=$	$B=$	$C=$	$8D=$	$6E=$
$ac + 1$	$a^2d + 1$	$a^3e + 1$	$a^4f + 1$	$a^5g + 8$	$a^6h + 6$
$b^2 - 1$	$abc - 3$	$a^2bd - 4$	$a^3be - 5$	$a^4bf - 48$	$a^5bg - 42$
	$b^3 + 2$	$c^2 - 2$	$cd - 5$	$ce - 48$	$cf - 42$
		$ab^2c + 10$	$a^2b^2d + 15$	$d^2 - 25$	$de - 45$
		$b^4 - 5$	$bc^2 + 15$	$a^3b^2e + 168$	$a^4b^2f + 168$
			$ab^3c - 35$	$bcd + 342$	$bce + 345$
			$a^0b^5 + 14$	$c^3 + 56$	$bd^2 + 180$
				$a^2b^3d - 452$	$c^2d + 174$
				$b^2c^2 - 681$	$a^3b^3e - 510$
				$ab^4c + 1020$	$b^2cd - 1578$
				$a^0b^6 - 340$	$bc^3 - 522$
					$a^2b^4d + 1299$
					$b^3c^2 + 2622$
					$ab^5c - 2877$
					$a^0b^7 + 822$
± 1	± 3	± 11	± 45	± 1594	± 5616

In terms of the fundamental seminvariants as indicated by their initial and final terms

$$ac \propto b^2 = ac - b^2, \quad a^2d \propto b^3 = a^2d - 3abc + 2b^3, \text{ \&c.},$$

the expressions of these functions are

$$N = (ac \propto b^2),$$

$$A = (a^2d \propto b^3),$$

$$B = (a^3e \propto a^2c^2) \quad - \quad 5\,(a^2c^2 \propto b^4),$$

$$C = (a^4f \propto a^2bc^2) \quad - \quad 7\,(a^3cd \propto b^5),$$

$$8D = 8\,(a^5g \propto a^4d^2) \; - 168\,(a^4ce \propto a^3c^3) \; - 113\,(a^4d^2 \propto a^2b^2c^2) + 340\,(a^3c^3 \propto b^6),$$

$$2E = 2\,(a^6h \propto a^4bd^2) - \; 32\,(a^5cf \propto a^3bc^2) - \; 37\,(a^5de \propto a^2b^3c^2) + 137\,(a^2cd^2 \propto b^7).$$

$4\mathfrak{M}=$	$\mathfrak{A}=$	$2\mathfrak{B}=$	$4\mathfrak{C}=$	$8\mathfrak{D}=$	$48\mathfrak{E}=$
$ac\ +4$	$a^2d\ +1$	$a^3e\ +2$	$a^4f\ +4$	$a^5g\ +8$	$a^6h\ +48$
$b^2\ -5$	$abc\ -3$	$a^2bd\ -7$	$a^3be\ -16$	$a^4bf\ -36$	$a^5bg\ -240$
	$b^3\ +2$	$c^2\ -4$	$cd\ -20$	$ce\ -48$	$cf\ -336$
		$a\,b^2c\ +17$	$a^2b^2d\ +45$	$d^2\ -25$	$de\ -360$
		$a^0b^4\ -8$	$bc^2\ +52$	$a^3b^2e\ +114$	$a^4b^2f\ +840$
			$a\,b^3c\ -103$	$bcd\ +282$	$bce\ +2184$
			$a^0b^5\ +38$	$c^3\ +56$	$bd^2\ +1140$
				$a^2b^3d\ -295$	$c^2d\ +1392$
				$b^2c^2\ -513$	$a^3b^3e\ -2400$
				$a\,b^4c\ +657$	$b^2cd\ -8880$
				$a^0b^6\ -200$	$bc^3\ -3504$
					$a^2b^4d\ +5955$
					$b^3c^2\ +13836$
					$a\,b^5c\ -13065$
					$a^0b^7\ +3390$
$+4-5$	± 3	± 19	± 139	± 1117	± 28785

In terms of the fundamental reciprocants of the table, *ante* p. 387, say these are

$$4ac-5b^2=P_2,\quad a^2d-3abc+2b^3=P_3,\quad 50a^2e-\ldots+105b^2c=P_4,\quad 10a^3f-\ldots-39b^3c=P_5,$$

$$14a^2g-\ldots+2310b^2c=P_6,\quad 7a^3h-\ldots-1925bc^3=P_7,^*$$

the expressions of these functions are

$$\begin{aligned}
4\mathfrak{M}&=P_2,\\
\mathfrak{A}&=P_3,\\
50\mathfrak{B}&=aP_4-8P_2^2,\\
40\mathfrak{C}&=4aP_5-38P_2P_3,\\
2800\mathfrak{D}&=200a^3P_6+1266aP_2P_4-302750P_3^3-9128P_2^3,\\
1680\mathfrak{E}&=240a^3P_7+2940aP_2P_5-3156aP_3P_4+2373P_3P_2^2.
\end{aligned}$$

Sylvester remarks that a Principiant, e.g. $a^2d-3abc+2b^3$, is at once a reciprocant, and in the theory of seminvariants (where $a, b, c, \ldots$, are the coefficients of the theory) a seminvariant; and conversely that any reciprocant which is also a seminvariant is a principiant: *quà* seminvariant, the principiant is annihilated by

$$\Omega=a\partial_b+2b\partial_c+3c\partial_d+4d\partial_e+5e\partial_f+\ldots,$$

* The eighth column $+420a^3i\ldots+129360c^4$, is *not* the proper value of P_8; the proper value is a linear combination of the eighth and ninth columns, eighth column $+6$ ninth column, viz. $P_8=420a^3i\ldots-102102bc^2d$: see as to this my paper "Tables of Pure Reciprocants to the weight 8," *Amer. Math. Jour.*, t. xv. (1893), pp. 75—77, [933].

and *quâ* reciprocant is annihilated by

$$V = 2a^2\partial_b + 5ab\partial_c + (6ac + 3b^2)\partial_d + (7ad + 7bc)\partial_e + (8ae + 8bd + 4c^2)\partial_f + \dots;$$

and any function which is annihilated by each of these operators is a principiant.

We form the foregoing series of functions $N = ac - b^2$, $A, B, C, \dots$, as follows, viz. if

$$G' = (4ac - 5b^2)\partial_b + (5ad - 7bc)\partial_c + (6ae - 9bd)\partial_d + (7af - 11be)\partial_e + \dots,$$

then

$$\begin{aligned} 5A &= G'N, \\ 6B &= G'A, \\ 7C &= G'B + NA, \\ 8D &= G'C + 2NB, \\ 9E &= G'D + 3NC; \end{aligned}$$

and similarly we form the foregoing series of functions $\mathfrak{M}, \mathfrak{A}, \mathfrak{B}, \mathfrak{C}, \dots$, as follows, viz. if

$$\mathfrak{G}' = 4(ac - b^2)\partial_b + 5(ad - bc)\partial_c + 3(ae - bd)\partial_d + 7(af - be)\partial_e + \dots,$$

($\mathfrak{G}'$ is Sylvester's G), then

$$\begin{aligned} 5\mathfrak{A} &= \mathfrak{G}'\mathfrak{M}, \\ 6\mathfrak{B} &= \mathfrak{G}'\mathfrak{A}, \\ 7\mathfrak{C} &= \mathfrak{G}'\mathfrak{B} - \mathfrak{M}\mathfrak{A}, \\ 8\mathfrak{D} &= \mathfrak{G}'\mathfrak{C} - 2\mathfrak{M}\mathfrak{B}, \\ 9\mathfrak{E} &= \mathfrak{G}'\mathfrak{D} - 3\mathfrak{M}\mathfrak{C}. \end{aligned}$$

As already mentioned, $N, A, B, C, \dots$, are seminvariants, $\mathfrak{M}, \mathfrak{A}, \mathfrak{B}, \mathfrak{C}, \dots$, are reciprocants. Putting for shortness $\tfrac{1}{2}b = \theta$, the two sets of functions are connected by

$$\begin{aligned} A &= \mathfrak{A}, & &\text{or conversely } & \mathfrak{A} &= A, \\ B &= \mathfrak{B} - \theta\mathfrak{A}, & & & \mathfrak{B} &= B + \theta A, \\ C &= \mathfrak{C} - 2\theta\mathfrak{B} + \theta^2\mathfrak{A}, & & & \mathfrak{C} &= C + 2\theta B + \theta^2 A, \\ D &= \mathfrak{D} - 3\theta\mathfrak{C} + 3\theta^2\mathfrak{B} - \theta^3\mathfrak{A}, & & & \mathfrak{D} &= D + 3\theta C + 3\theta^2 B + \theta^3 A, \\ &\vdots & & & &\vdots \end{aligned}$$

so that, one of the sets being calculated, the other set can be at once deduced therefrom. These equations give

$$AC - B^2 = \mathfrak{A}\mathfrak{C} - \mathfrak{B}^2, \quad A^2D - 3ABC + 2B^3 = \mathfrak{A}^2\mathfrak{D} - 3\mathfrak{A}\mathfrak{B}\mathfrak{C} + 2\mathfrak{B}^3, \text{ \&c.},$$

viz. as mentioned above, any seminvariant in the letters $A, B, C, \dots$, is equal to the same seminvariant in the letters $\mathfrak{A}, \mathfrak{B}, \mathfrak{C}, \dots$; or, what is the same thing, any principiant has the same expression in the letters $A, B, C, \dots$, and in the letters $\mathfrak{A}, \mathfrak{B}, \mathfrak{C}, \dots$ respectively.

We have thus the entire series of Principiants,

$$AC - B^2, \quad A^2D - 3ABC + 2B^3, \quad AE - 4BD + 3C^2, \text{ \&c.}$$

We may express Halphen's reciprocants in terms of the capitals $A, B, C, \ldots$. We have

$$U = a, \quad V = A, \quad \Delta = AC - B^2.$$

To obtain formulæ for the higher reciprocants, we require the derived functions $A', B', C', \ldots$, where the accent denotes differentiation in regard to x.

We have

$$\partial_x = 3b\partial_a + 4c\partial_b + 5d\partial_c + \ldots,$$

(whence in particular $a' = 3b$, $b' = 4b$, $c' = 5c$, ..., as is obvious). Hence, writing

$$G' = a(3b\partial_a + 4c\partial_b + 5d\partial_c + \ldots) - b(3a\partial_a + 5b\partial_b + 7c\partial_c + \ldots),$$

this may be written

$$G' = a\partial_x - bw,$$

if w be the weight of the homobaric function operated upon, reckoning the weights of $a, b, c, \ldots$, as 3, 5, 7, ... respectively, and consequently the weights of $A, B, C, \ldots$, as 15, 20, 25, ... respectively. We thus have

$$\begin{aligned}
5A &= (a\partial_x - 10b)\,N,\\
6B &= (a\partial_x - 15b)\,A,\\
7C &= (a\partial_x - 20b)\,B + NA,\\
8D &= (a\partial_x - 25b)\,C + 2NB,\\
9E &= (a\partial_x - 30b)\,D + 3NC,\\
&\vdots
\end{aligned}$$

of which the first gives only $A = a^2d - 3abc + 2b^3$. The other equations give the required formulæ

$$\begin{aligned}
aA' &= 6B + 15bA,\\
aB' &= 7C + 20bB - NA,\\
aC' &= 8D + 25bC - 2NB,\\
aD' &= 9E + 30bD - 3NC,\\
&\vdots
\end{aligned}$$

Halphen's H is given by $U^4H = 256\Delta^3 - 27V^8$, viz. we thus have

$$a^4H = 256(AC - B^2)^3 - 27A^8.$$

His T is defined by the equation $U^3T = 3V\Delta' - 8V'\Delta$, that is,

$$a^3T = 3A(AC - B^2)^2 - 8A'(AC - B^2),$$

which is

$$\begin{aligned}
&= 3A(AC' - 2BB' + CA') - 8A'(AC - B^2),\\
&= 3A^2C' - 6ABB' + (-5AC + 8B^2)\,A',
\end{aligned}$$

or substituting for A', B', C', their values we find

$$a^4T = 24(A^2D - 3ABC + 2B^3),$$

which expression for the reciprocant T was given by Sylvester.

Halphen's T_1 is $\frac{1}{U}(V^4T-\frac{1}{6}H)$, viz. we thus have

$$6a^5T_1 = 24A^4(A^2D-3ABC+2B^3)-256(AC-B^2)^3+27A^8.$$

His G is given by $V^2G=U^4T^2+9H$, viz. we have

$$A^2a^4G = 576(A^2D-3ABC+2B^3)^2+2604(AC-B^2)^3-243A^8,$$

where the whole divides by A^2; throwing out this factor, we find

$$a^4G = 576(A^2D^2-6ABCD+4AC^3+4B^3D-3B^2C^2)-243A^6.$$

From the expression for T, I find

$$a^5T' = 216A^2E-288ABD-504AC^2+576B^2C+1152(A^2D-3ABC+2B^3),$$

and I thence deduce for Halphen's Θ,

$$=\frac{1}{V}\{2U\Delta T'+T(8U'\Delta-3U\Delta')\},$$

the formula

$$a^4\Theta = 432(AC-B^2)(AE-4BD+3C^2)-576(A^2D^2-6ABCD+4AC^3+4B^3D-3B^2C^2),$$

or, what is the same thing,

$$=-144(AC-B^2)(AE-4BD+3C^2)+576A(ACE-AD^2-B^2E+2BCD-C^3).$$

Also $\Theta_1=\frac{1}{V}(\Theta+\frac{1}{4}G)$, that is, $a^4\Theta_1=\frac{1}{A}(a^4\Theta+\frac{1}{4}a^4G)$,

$$=\frac{432}{A}\{(AC-B^2)(AE-4BD+3C^2)-(A^2D^2-6ABCD+4AC^3+4B^3D-3B^2C^2)\},$$

or finally,

$$a^4\Theta_1 = 432(ACE-AD^2-B^2E+2BCD-C^3)-\tfrac{243}{4}A^5.$$

Again,

$$\Theta_2=\frac{1}{U}(\Theta_1-\tfrac{45}{2}TV)=\frac{1}{a}(\Theta_1-\tfrac{45}{2}TA),$$

whence

$$a^5\Theta_2 = a^4\Theta_1-\tfrac{45}{2}a^4TA,$$

or substituting,

$$a^5\Theta_2 = 432(ACE-AD^2-B^2E+2BCD-C^3)-540A(A^2D-3ABC+2B^3)-\tfrac{243}{4}A^5.$$

Writing

$$\begin{aligned}
&AC-B^2=\mathbf{C},\\
&A^2D-3ABC+2B^3=\mathbf{D},\\
&A^2D^2-6ABCD+4AC^3+4B^3D-3B^2C^2=\square,\\
&AE-4BD+3C^2=\mathbf{I},\\
&ACE-AD^2-B^2E+2BCD-C^3=\mathbf{J},
\end{aligned}$$

the foregoing results become $U=a$, $V=A$, $\Delta=\mathbf{C}$,

$$\begin{aligned}
a^4H \text{ (Halphen)} &= 256\mathbf{C}^3 - 27A^8,\\
a^4T &= 24\mathbf{D},\\
6a^5T_1 &= 24A^4\mathbf{D} - 256\mathbf{C}^3 + 27A^8,\\
a^4G &= 576\square - 243A^6,\\
a^4\Theta &= 144\mathbf{CI} + 576A\mathbf{J},\\
a^4\Theta_1 &= 432\mathbf{J} - \tfrac{243}{4}A^5,\\
a^5\Theta_2 &= 432\mathbf{J} - 540A\mathbf{D} - \tfrac{243}{4}A^5.
\end{aligned}$$

XIII.

The letters t, a, b, ..., of a reciprocant represent (it will be remembered) mere numerical multiples of the derived functions of a variable y, in regard to the independent variable x, and thus equating any reciprocant to zero, we have a differential equation in y. For instance, if the orthogonal reciprocant $c(1+t^2)-5abt+5a^3$ be put $=0$, (t, a, b, $c=y_1$, $\tfrac{1}{2}y_2$, $\tfrac{1}{6}y_3$, $\tfrac{1}{24}y_4$), this is the differential equation

$$y_4(1+y_1^2)-10y_1y_2y_3+15y_2^3=0$$

of the order 4. The integral hereof is expressible by the two equations

$$x=\int\frac{dt}{\sqrt{(\kappa U+\lambda V)}}+\mu,\quad y=\int\frac{tdt}{\sqrt{(\kappa U+\lambda V)}}+\nu,$$

where U, V are real functions of a parameter t, viz. $U+iV=(1+it)^6$, $U-iV=(1-it)^6$, $\{i=\sqrt{(-1)}$ as usual$\}$, so that $U=1-15t^2+15t^4-t^6$, $V=6t-20t^3+6t^5$; and κ, λ, μ, ν are the four constants of integration. In verification hereof, we have

$$dx=\frac{dt}{\sqrt{(\kappa U+\lambda V)}},\quad dy=\frac{tdt}{\sqrt{(\kappa U+\lambda V)}},$$

whence

$$\frac{dy}{dx}=y_1,\ =t,$$

hence

$$y_2=\frac{dt}{dx}=\sqrt{(\kappa U+\lambda V)},\ \text{or say}\ y_2^2=\kappa U+\lambda V,$$

and thence, using an accent to denote differentiation in regard to t, we have

$$2y_3=\kappa U'+\lambda V',$$

$$2\frac{y_4}{y_2}=\kappa U''+\lambda V'',$$

or, eliminating the κ and λ,

$$\begin{vmatrix} y_2^3, & U, & V \\ 2y_2y_3, & U', & V' \\ 2y_4, & U'', & V'' \end{vmatrix}=0.$$

Substituting for U, V either of the above-mentioned forms, there is in each case a factor in t which divides out; rejecting this factor, the equation is found to be

$$2y_4(1+t^2) - 20ty_2y_3 + 30y_2^3 = 0,$$

that is, substituting for t its value $= y_1$, and throwing out the factor 2, we have the differential equation

$$y_4(1+y_1^2) - 10y_1y_2y_3 + 15y_2^3 = 0.$$

XIV.

But when the reciprocant equated to zero is a Principiant, we have the far more important results obtained in Halphen's Memoir (1): and which are, in Sylvester's lectures, exhibited in a more complete form by expressing Halphen's reciprocants in terms of the foregoing functions $A, B, C, \ldots$. I recall that $a, b, c, \ldots$ are numerical multiples of differential coefficients of the orders 2, 3, 4, ... respectively; $A, B, C, \ldots$ contain $d, e, f, \ldots$ respectively, that is, differential coefficients of the orders 5, 6, 7, ... respectively, so that according as the highest capital letter is $A, B, C, \ldots$ respectively, the order of the differential equation is 5, 6, 7, ... respectively.

But the equation, Principiant $= 0$, may be interpreted in a different manner, viz. if instead of regarding the differential equation as an equation for the determination of y, we regard therein y as a given function of x, {that is, (x, y) as the coordinates of a point on a given curve}, then the differential equation serves for the determination of those points on the curve for which a given condition is satisfied, or say it is the condition in order to the existence of a singular point of determinate character.

Halphen's lowest reciprocants are: as already mentioned:

$$\begin{aligned} U &= a, \\ V &= a^2d - 3abc + 2b^3, \ = A, \\ \Delta & \qquad\qquad\qquad\quad , \ = AC - B^2. \end{aligned}$$

$U = 0$, that is, $a = 0$, is a differential equation of the second order, viz. it is the differential equation of a line: otherwise it is the condition for a point of inflexion.

$V = 0$, that is, $A = 0$, is a differential equation of the fifth order, viz. it is the differential of a conic: otherwise it is the condition for a sextactic point.

$\Delta = 0$, that is, $AC - B^2 = 0$, is a differential equation of the seventh order, it is the condition for what Halphen calls a point of coincidence, viz. this is a point on a given curve, such that for it the cubic of nine-pointic intersection becomes a nodal cubic having the point for node and through it one branch of eight-pointic intersection and one branch of simple intersection. Observe that this is more than the condition that the cubic of nine-pointic intersection shall be a nodal cubic, and accordingly $\Delta = 0$ is not the differential equation of a nodal cubic. In fact, a nodal cubic depends on 8 parameters and has therefore a differential equation of the order 8; this will be obtained further on. But $\Delta = 0$ is the differential equation of the order 7 of a curve such that every point thereof is a coincident point.

The differential equation of the order 7,

$$2^4 . 7^3 . \{(\lambda-2)(\lambda+1)(2\lambda-1)\}^2 . \Delta^3 - 3^3 . 5^2 (\lambda^2-\lambda+1)^3 V^8 = 0,$$

has the integral $\alpha = \beta^\lambda \gamma^{1-\lambda}$, where α, β, γ represent arbitrary linear functions $lx+my+n$: we may without loss of generality take the linear functions to be of the form $x+my+n$, introducing in this case another constant C, and writing the integral equation in the form $\alpha = C\beta^\lambda\gamma^{1-\lambda}$; the number of arbitrary constants is thus $=7$.

If $\lambda = 2$, -1 or $\frac{1}{2}$, the integral equation is $\alpha = \beta^2\gamma^{-1}$, $\alpha = \beta^{-1}\gamma^2$ or $\alpha = \beta^{\frac{1}{2}}\gamma^{\frac{1}{2}}$; or say it is $\beta^2 = \alpha\gamma$, $\gamma^2 = \alpha\beta$ or $\alpha^2 = \beta\gamma$, viz. each of these represents a conic. The differential equation is here $V=0$, viz. we have the foregoing result that $V=0$, that is, $A=0$, is the differential equation of a conic.

If $\lambda = \infty$, 0 or 1, the differential equation is

$$2^6 . 7^3 . \Delta^3 - 3^3 . 5^2 . V^8 = 0;$$

the integral equation is here of the form $\frac{\beta}{\gamma} = \log\frac{\alpha}{\gamma}$; in fact we have $\frac{\alpha}{\gamma} = \left(\frac{\beta}{\gamma}\right)^\lambda$, or for α, β, and γ writing $\lambda\alpha$, $\beta+\lambda\gamma$, and $\lambda\gamma$ respectively, this is $\frac{\alpha}{\gamma} = \left(1+\frac{\beta}{\lambda\gamma}\right)^\lambda$, which when λ is indefinitely large becomes $\frac{\alpha}{\gamma} = \exp.\frac{\beta}{\gamma}$, that is, $\frac{\beta}{\gamma} = \log\frac{\alpha}{\gamma}$. And similarly for $\lambda = 0$ and $\lambda = 1$.

If $\lambda^2 - \lambda + 1 = 0$, that is, if $\lambda = -\omega$, where ω is an imaginary cube root of unity, the differential equation is $\Delta = 0$, that is, $AC - B^2 = 0$ (of the order 7 as in the general case). The integral equation is $\alpha = \beta^{-\omega}\gamma^{-\omega^2}$, viz. this is the equation of the curve every point of which is a coincident point.

If $\lambda = 3$, the differential equation is $2^8\Delta^3 - 3^3V^8 = 0$, or say $256\Delta^3 - 27V^8 = 0$; Halphen's H is a function such that $U^4H = 256\Delta^3 - 27V^8$, so that this is

$$H_3 = U^{-4}(256\Delta^3 - 27V^8) = 0.$$

The integral equation is $\alpha = \beta^3\gamma^{-2}$, that is, $\alpha\gamma^2 = \beta^3$, or the curve is a cuspidal cubic, depending upon 7 constants; and it thus appears that the differential equation of a cuspidal cubic is the equation of the order 7

$$a^{-4}\{256(AC-B^2)^3 - 27A^8\} = 0.$$

Write

$$\Theta = 64(A^2D^2 - 6ABCD + 4AC^3 + 4B^3D - 3B^2C^2) + 144A^2(A^2D - 3ABC + 2B^3) + 81A^6$$
$$= 64\square + 144A^2\mathbf{D} + 81A^6,$$

$$\Phi = 8(A^2D - 3ABC + 2B^3) + 9A^4 = 8\mathbf{D} + 9A^4,$$

$$\Psi = 3072(AC-B^2)^3 = 3072\mathbf{C}^3,$$

where identically

$$\Phi^2 + \tfrac{1}{12}\Psi = A^2\Theta.$$

Halphen and Sylvester find that, for a cubic curve the invariants of which are S and T ($S=-l+l^4$, $T=1-20l^3-8l^6$ for the canonical form $x^3+y^3+z^3+6lxyz=0$), we have

$$4\theta^2 S = \Theta^2 - 4\Phi\Psi,$$

$$\theta^3 T = \Theta^3 - 6\Theta\Phi\Psi + 6A^2\Psi^2,$$

(θ an arbitrary multiplier); and we thence have

$$\frac{(\Theta^2-4\Phi\Psi)^3}{(\Theta^3-6\Theta\Phi\Psi+6A^2\Psi^2)^2}=\frac{64S^3}{T^2},$$

that is,

$$T^2(\Theta^2-4\Phi\Psi)^3-64S^3(\Theta^3-6\Theta\Phi\Psi+6A^2\Psi^2)^2=0,$$

or if, as with Halphen $h=\dfrac{T^2}{64S^3}$, then

$$(\Theta^3-6\Theta\Phi\Psi+6A^2\Psi^2)^2-h(\Theta^2-4\Phi\Psi)^3=0,$$

which is the differential equation of the order 8 of a cubic curve having the absolute invariant $S^3\div T^2$.

If for shortness

$$P=\Theta^2-4\Phi\Psi,$$

$$R=\Theta^3-6\Theta\Phi\Psi+6A^2\Psi^2,$$

$$Q=R^2-P^3,$$

then the foregoing equation $\dfrac{P^3}{R^2}=\dfrac{64S^3}{T^2}$ gives

$$\frac{Q}{R^2}=\frac{T^2-64S^3}{T^2}=\frac{h-1}{h},$$

and thus for a nodal cubic we have $Q=0$; we thus have

$P=0$, for the differential equation of a cubic for which $S=0$,

$R=0$, " " " " " $T=0$,

$Q=0$, " " " nodal cubic,

these being each of them of the order 8. But the equation $Q=0$ is reducible to a more simple form. We have

$$Q=R^2-P^3=\begin{array}{ll}\Theta^6 & -\ \Theta^6\\ -12\Theta^4\Phi\Psi & +12\Theta^4\Phi\Psi\\ +12\Theta^3A^2\Psi^2 & \\ +36\Theta^2\Phi^2\Psi^2 & -48\Theta^2\Phi^2\Psi^2\\ -72\Theta\Phi A^2\Psi^3 & +64\Phi^3\Psi^3\\ +36A^4\Psi^4 & \end{array}=\Psi^2\left\{\begin{array}{l}12A^2\Theta^3\\ -12\Theta^2\Phi^2\\ -72A^2\Theta\Phi\Psi\\ +64\Phi^3\Psi\\ +36A^4\Psi^2\end{array}\right\},$$

or, since in the expression in { } the first and second terms are $12\Theta^2(A^2\Theta - \Phi^2) = \Theta^2\Psi$, this is

$$Q = \Psi^3 \left\{\begin{array}{l} \Theta^2 \\ -72A^2\Theta\Phi \\ +64\Phi^3 \\ +36A^4\Psi \end{array}\right\} = \Psi^3 \left\{\begin{array}{l} \Theta^2 \\ -72A^2\Theta\Phi \\ +64\Phi^3 \\ +432A^4(A^2\Theta - \Phi^2) \end{array}\right\},$$

or finally

$$Q = \Psi^3 \{(\Theta - 36A^2\Phi + 216A^6)^2 - 64(\Phi - 9A^4)^3\}.$$

We thus have for the differential equation of the eighth order of the nodal cubic

$$(\Theta - 36A^2\Phi + 216A^6)^2 - 64(\Phi - 9A^4)^3 = 0,$$

or, since

$$\Theta = 64\Box + 144A^2\mathbf{D} + 81A^6, \text{ and } \Phi = 8\mathbf{D} + 9A^4,$$

the differential equation of the eighth order for the nodal cubic finally is

$$(64\Box - 144A^2\mathbf{D} + 27A^6)^2 + 32768\mathbf{D}^3 = 0.$$

Halphen has $\Theta_2 = 0$ for the differential equation of a cubic curve. Putting, as before,

$$\mathbf{I} = AE - 4BD + 3C^2,$$

$$\mathbf{J} = ACE - AD^2 - B^2E + 2BCD - C^3,$$

(and therefore $\mathbf{IC} - A\mathbf{J} = \Box$, $\mathbf{I}^3 - 27\mathbf{J}^2 =$ Quartic Disct.), then, by what precedes, the differential equation of the order 9 of a cubic curve is

$$a^{-5}\{432\mathbf{J} - 540A\mathbf{D} - \tfrac{243}{4}A^5\} = 0.$$

Recapitulating.

Order. The differential equation for a cubic curve:—

8. for which invt. $S = 0$, is

$$P = \Theta^2 - 12288\mathbf{C}^3(8\mathbf{D} + 9A^4) = 0,$$

8. for which invt. $T = 0$, is

$$R = \Theta^3 - 18432\mathbf{C}^3(8\mathbf{D} + 9A^4) + 55623104\mathbf{C}^9A^2 = 0,$$

8. nodal cubic

$$Q \div (3072\Psi)^3 = (64\Box - 144A^2\mathbf{D} + 27A^6)^2 + 32768\mathbf{D}^3 = 0,$$

8. for invt. $64S^3 \div T^2 = h$, is $(h-1)R^2 - hQ = 0$,

9. for general cubic is $1728\mathbf{J} - 2160A\mathbf{D} - 243A^5 = 0$,

where for shortness Θ is retained to signify its value

$$= 64\Box + 144A^2\mathbf{D} + 81A^6.$$

Halphen by a polar transformation finds that these same differential equations, changing therein the sign of **D**, apply to curves of the third class: in particular, the last equation, changing therein the sign of **D**, applies to the general curve of the third class, or sextic curve with nine cusps.

XV.

A very important notion in the theory, as well of Seminvariants as of Reciprocants, is that of MacMahon's *Multilinear Operator*, see his paper "Theory of a Multilinear Partial Differential Operator, with applications to the theories of Invariants and Reciprocants," *Proc. Lond. Math. Soc.*, t. XVIII. (1886), pp. 61—88: this operator plays so important a part in the theories to which it relates, that I venture to reproduce the definition of it in what appears to me a simplified and more easily intelligible form, and to recapitulate some of the leading properties.

I take with him the letters to be $(a_0, a_1, a_2, a_3, \ldots) = (a, b, c, d, \ldots)$, viz. the first letter is a_0 or a, the second is a_1 or b, and so on, but when we are not concerned with a term of indefinite rank, I use always $(a, b, c, d, \ldots)$ and the like in regard to any other series of letters. This being so, the definition of the *Multilinear Operator* of four elements, which is here alone in question, is

$(\mu, \nu : m, n)$, =

	$n=$ 0	1	2	3	...	n
$(\mu \quad) A$	∂_a	∂_b	∂_c	∂_d	...	∂_{a_n} ,
$+ (\mu + \nu) B$	∂_b	∂_c	∂_d	∂_e	...	$\partial_{a_{n+1}}$
$+ (\mu + 2\nu) C$	∂_c	∂_d	∂_e	∂_f	...	$\partial_{a_{n+2}}$
$+ (\mu + 3\nu) D$	∂_d	∂_e	∂_f	∂_g	...	$\partial_{a_{n+3}}$
$\vdots$						

where

	$m=$ 1	2	3	...	m
$A=$	a	$\frac{1}{2}a^2$	$\frac{1}{3}a^3$	...	$\dfrac{1}{m}a^m$
$B=$	b	ab	a^2b	...	$a^{m-1}b$
$C=$	c	$ac+\frac{1}{2}b^2$	a^2c+ab^2	...	$a^{m-1}c+\frac{1}{2}(m-1)a^{m-2}b^2$
$D=$	d	$ad+bc$	$a^2d+2abc+\frac{1}{3}b^3$	...	$a^{m-1}d+(m-1)a^{m-2}bc+\frac{1}{6}(m-1)(m-2)a^{m-3}b^3$,
$\vdots$					

where the expressions for $B, C, D, \ldots$, are obtained successively from that of $A, = \dfrac{1}{m}a^m$, by Arbogast's rule of derivation, viz. we operate on the last letter of a term, and when there is a last but one letter which in alphabetical order immediately precedes the last letter, then also upon the last but one letter, and whenever the term thus obtained ends in a power, we divide by the index of the power. Thus the term $\frac{1}{2}(m-1)a^{m-2}b^2$ in the third line gives in the fourth line

$$\tfrac{1}{2}(m-1)a^{m-2}.\,2bc+\tfrac{1}{2}(m-1)(m-2)a^{m-3}b\,.\,b^2.\,\tfrac{1}{3},$$
$$=(m-1)a^{m-2}bc+\tfrac{1}{6}(m-1)(m-2)a^{m-3}b^3.$$

Going a step further, we form in this manner the expression

$$E = a^{m-1}e + (m-1)\,a^{m-2}(bd + \tfrac{1}{2}c^2) + \tfrac{1}{6}(m-1)(m-2)\{a^{m-3}.\,3b^2c + (m-3)\,a^{m-4}b\,.\,b^3.\,\tfrac{1}{4}\},$$

that is,

$$\begin{aligned} E = {} & a^{m-1}e \\ & + a^{m-2}\ \{(m-1)\,bd + \tfrac{1}{2}(m-1)\,c^2\} \\ & + a^{m-3}.\,\tfrac{1}{2}\,(m-1)(m-2)\,b^2c \\ & + a^{m-4}.\,\tfrac{1}{24}(m-1)(m-2)(m-3)\,b^4, \end{aligned}$$

which is sufficient to explain the rule.

It may be noticed that, the operator being linear in μ, ν, we have

$$(\mu,\ \nu;\ m,\ n) = \mu\,(1,\ 0;\ m,\ n) + \nu\,(0,\ 1;\ m,\ n).$$

The Alternant of two multilinear operators.

If P, Q are the two operators, we have as usual

$$P\,.\,Q = PQ + P * Q,$$

where $P\,.\,Q$ denotes the successive operation first with Q and then with P upon any operand, PQ is the mere algebraical product of the operators, and $P * Q$ is the operator obtained by the operation of P upon Q.

Similarly,

$$Q\,.\,P = QP + Q * P,$$

and since QP is the same thing as PQ, we obtain

$$P\,.\,Q - Q\,.\,P = (P * Q) - (Q * P),$$

either of which equal expressions, or say rather the second of them, is called the alternant of P, Q and is written $[P\,.\,Q]$: viz. as the definition of the alternant, we have

$$[P\,.\,Q] = (P * Q) - (Q * P).$$

We have the remarkable theorem, that the alternant of any two operators $(\mu',\ \nu';\ m',\ n')$, $(\mu,\ \nu;\ m,\ n)$ is an operator $(\mu_1,\ \nu_1;\ m_1,\ n_1)$; where

$$\begin{aligned} \mu_1 &= (m' + m - 1)\left\{\frac{\mu'}{m'}(\mu + n'\nu) - \frac{\mu}{m}(\mu' + n\nu')\right\}, \\ \nu_1 &= (n' - n)\,\nu'\nu + \frac{m-1}{m'}\,\mu'\nu - \frac{m'-1}{m}\,\mu\nu', \\ m_1 &= m' + m - 1, \\ n_1 &= n' + n, \end{aligned}$$

or since m_1, n_1 have such simple expressions, say it is an operator $(\mu_1,\ \nu_1;\ m' + m - 1,\ n' + n)$, where μ_1, ν_1 have the values just written down.

We see at once how these values $m_1 = m' + m - 1$, $n_1 = n' + n$, arise: $Q = (\mu, \nu; m, n)$ contains the letters $(a, b, c, d, \ldots)$ in the degree m, and it contains differential symbols ∂ which, operating on any function of these letters, diminish the degree by 1; similarly $P = (\mu', \nu'; m', n')$ contains the letters $(a, b, c, d, \ldots)$ in the degree m', and it contains the differential symbols ∂ which, operating on any function of the letters, diminish the degree by 1. Hence $P * Q$ contains the letters in the degree $(m'-1)+m$, and similarly $Q * P$ contains them in the same degree $(m-1)+m'$: thus in the alternant the degree is $m'+m-1$. Again, in Q the weights of the successive functions $A, B, C, \ldots$ are $0, 1, 2, \ldots$ and these are combined with differential symbols $\partial_{a_n}, \partial_{a_{n+1}}, \ldots$ which operating on any function of the letters diminish the weight by $n, n+1, \ldots$ respectively: that is, the terms $A\partial_{a_n}, B\partial_{a_{n+1}}, \ldots$ each diminish the weight by n. So in $P = (\mu', \nu'; m', n')$ the weights of the successive terms $A, B, C, \ldots$ are $0, 1, 2, \ldots$ and these are combined with differential symbols $\partial_{a_{n'}}, \partial_{a_{n'+1}}, \ldots$ which operating on any function of the letters diminish the weights by $n', n'+1, \ldots$ respectively. We may say that Q is a sum of terms such as $\Theta_k \partial_{a_{n+k}}$, where the subscript k of Θ denotes the weight of Θ; operating hereon with P, the corresponding term of $P * Q$ is of the form $\Theta_{k-n'}\partial_{a_{n+k}}$, or (what is the same thing) the general form of the terms of $P * Q$ is $\Theta_k \partial_{a_{n'+n+k}}$; and in like manner this is the general form of the terms of $Q * P$; and it is thus also the general form of the terms of the alternant $[P.Q]$. It thus appears that, admitting the alternant to be an operator $(\mu_1, \nu_1; m_1, n_1)$, the value of n_1 is $n'+n$.

As an instance of the theorem, we may take

$$[(1, 0, 1, 1), (1, 0, 2, 1)] = (1, 0, 2, 2),$$

we have here

$$(1, 0;\ 1, 1) * (1, 0;\ 2, 1) - (1, 0;\ 2, 1) * (1, 0;\ 1, 1),$$

$$\begin{array}{r|l|l|r}
\multicolumn{1}{c}{*} & & \multicolumn{1}{c}{-} & \multicolumn{1}{c}{*} \\
a\partial_b & \tfrac{1}{2}a^2\partial_b & \tfrac{1}{2}a^2\partial_b & a\partial_b \\
+\, b\partial_c & +\, ab\partial_c & +\, ab\partial_c & +\, b\partial_c \\
+\, c\partial_d & +\, (ac+\tfrac{1}{2}b^2)\,\partial_d & +\, (ac+\tfrac{1}{2}b^2)\,\partial_d & +\, c\partial_d \\
+\, d\partial_e & +\, (ad+bc)\,\partial_e & +\, (ad+bc)\,\partial_e & +\, d\partial_e \\
+\, e\partial_f & +\, (ae+bd+\tfrac{1}{2}c^2)\,\partial_f & +\, (ae+bd+\tfrac{1}{2}c^2)\,\partial_f & +\, e\partial_f \\
\vdots & \vdots & \vdots & \vdots
\end{array}$$

$$\begin{array}{rl|lcl}
= & a\,(a\partial_c + b\partial_d + c\partial_e + d\partial_f + \ldots) - & \tfrac{1}{2}a^2\partial_c & = & \tfrac{1}{2}a^2\partial_c \\
+\, b\,(& \quad a\partial_d + b\partial_e + c\partial_f + \ldots) & +\, ab\partial_d & & +\, ab\partial_d \\
+\, c\,(& \qquad a\partial_e + b\partial_f + \ldots) & +\, (ac+\tfrac{1}{2}b^2)\,\partial_e & & +\, (ac+\tfrac{1}{2}b^2)\,\partial_e \\
+\, d\,(& \qquad\quad a\partial_f + \ldots) & +\, (ad+bc)\,\partial_f & & +\, (ad+bc)\,\partial_f, \\
\vdots & & \vdots & & \vdots
\end{array}$$

$(= 1, 0, 2, 2)$, as it should be.

For the general proof, writing for a moment

$$C_0 = \mu A_0,\quad C_1 = (\mu + \nu) A_1,\quad C_2 = (\mu + 2\nu) A_2, \text{ \&c.},$$

and similarly

$$C_0' = \mu' A_0',\quad C_1' = (\mu' + \nu') A_1',\quad C_2' = (\mu' + 2\nu') A_2', \text{ \&c.},$$

where the accented symbols refer to the values $(\mu', \nu'; m', n')$, we have

$$\begin{array}{rllll} & [(\mu', \nu'; m', n').(\mu, \nu; m, n)], & & & \\ = & C_0'\partial_{a_{n'}} \;\; * & C_0\partial_{a_n} & - C_0\partial_{a_n} \;\; * & C_0'\partial_{a_{n'}} \\ & + C_1'\partial_{a_{n'+1}} & + C_1\partial_{a_{n+1}} & + C_1\partial_{a_{n+1}} & + C_1'\partial_{a_{n'+1}} \\ & + C_2'\partial_{a_{n'+2}} & + C_2\partial_{a_{n+2}} & + C_2\partial_{a_{n+2}} & + C_2'\partial_{a_{n'+2}} \\ & \vdots & \vdots & \vdots & \vdots \\ & & + C_{n'}\partial_{a_{n+n'}} & & + C_n\partial_{a_{n'+n}} \\ & & + C_{n'+1}\partial_{a_{n+n'+1}} & & + C_{n+1}\partial_{a_{n'+n+1}}\,; \end{array}$$

or observing that in the series $C_0, C_1, C_2, \ldots$, the first term that contains $a_{n'}$ is $C_{n'}$, and the like as regards the series $C_0', C_1', C_2', \ldots$, this is

$$\begin{array}{rl} = & \{(C_0'\partial_{a_{n'}}) C_{n'} \qquad\qquad\qquad\qquad\qquad\quad - (C_0\partial_{a_n}) C_n'\}\, \partial_{a_{n+n'}} \\ & + \{(C_0'\partial_{a_{n'}} + C_1'\partial_{a_{n'+1}}) C_{n'+1} \qquad\qquad - (C_0\partial_{a_n} + C_1\partial_{a_{n+1}}) C'_{n+1}\}\, \partial_{a_{n+n'+1}} \\ & + \{(C_0'\partial_{a_{n'}} + C_1'\partial_{a_{n'+1}} + C_2'\partial_{a_{n'+2}}) C_{n'+2} - (C_0\partial_{a_n} + C_1\partial_{a_{n+1}} + C_2\partial_{a_{n+2}}) C'_{n+2}\}\, \partial_{a_{n+n'+2}} \\ & \vdots \end{array}$$

and it is thus of the form

$$C_0''\partial_{a_{n+n'}} + C_1''\partial_{a_{n+n'+1}} + C_2''\partial_{a_{n+n'+2}} + \text{\&c.},$$

i.e. the value of n_1 is $= n + n'$.

I confine myself to the comparison of the first and second coefficients: substituting for the C's their expressions in terms of the A's, we ought to have

$$\mu' A_0'\partial_{a_{n'}} (\mu + n'\nu) A_{n'} - \mu A_0 \partial_{a_n} (\mu' + n\nu') A_n' = \mu_1 A_0'',$$

$$(\mu' A_0'\partial_{a_{n'}} + (\mu' + \nu') A_1'\partial_{a_{n'+1}}) \{\mu + (n'+1)\nu\} A_{n'+1}$$

$$- (\mu A_0\partial_{a_n} + (\mu + \nu) A_1\partial_{a_{n+1}}) \{\mu' + (n+1)\nu'\} A'_{n+1} = (\mu_1 + \nu_1) A_1''.$$

Now assuming $m_1 = m' + m - 1$, and attending to the values

$$\begin{array}{lll} A_0 = \dfrac{1}{m} a_0^{m}, & A_0' = \dfrac{1}{m'} a_0^{m'}, & A_0'' = \dfrac{1}{m' + m - 1} a_0^{m'+m-1}, \\ & \quad\vdots & \quad\vdots \\ A_1 = a_0^{m-1} a_1, & & \\ A_2 = a_0^{m-1} a_2 + \tfrac{1}{2} a_0^{m-2} a_1^2, & & \\ \vdots & & \end{array}$$

we see at once that in the first equation the terms contain each of them the factor $a_0^{m'+m-1}$, and in the second equation they contain each of them the factor $a_0^{m'+m-2}a_1$; omitting these factors, we find

$$\frac{\mu'}{m'}(\mu + n'\nu) \qquad - \frac{\mu}{m}(\mu' + n\nu') \qquad = \frac{\mu_1}{m' + m - 1},$$

$$\frac{m'}{\mu'}(\mu + n'\nu + \nu)(m-1) + (\mu' + \nu')(\mu + n'\nu + \nu)$$

$$- \frac{\mu}{m}(\mu' + n\nu' + \nu')(m' - 1) - (\mu + \nu)(\mu' + n\nu' + \nu') = \mu_1 + \nu_1.$$

The first of these gives

$$\mu_1 = (m' + m - 1)\left\{\frac{\mu'}{m'}(\mu + n'\nu) - \frac{\mu}{m}(\mu' + n\nu')\right\},$$

which is the value of μ_1: substituting this in the second equation, we have

$$\nu_1 = \frac{\mu'}{m'}(\mu + n'\nu + \nu)(m-1) - \frac{\mu}{m}(\mu' + n\nu' + \nu')$$

$$+ (\mu' + \nu')(\mu + n'\nu + \nu) \qquad - (\mu + \nu)(\mu' + n\nu' + \nu')$$

$$- (m' + m - 1)\frac{\mu'}{m'}(\mu + n'\nu) - (m' + m - 1)\frac{\mu}{m}(\mu' + n\nu').$$

In the left-hand column, the terms containing $(\mu + n'\nu)$ are

$$(\mu + n'\nu)\left\{\frac{\mu'}{m'}(m-1) + \mu' + \nu' - \frac{\mu'}{m'}(m' + m - 1)\right\},$$

$$= (\mu + n'\nu)\,\nu',$$

and there are besides the terms $\frac{\mu'}{m'}\nu(m-1) + \nu(\mu' + \nu')$, hence the left-hand column is

$$= \nu\nu'(n' + 1) + \mu\nu' + \mu'\nu\left(\frac{m-1}{m'} + 1\right).$$

We have therefore

$$\nu_1 = \nu\nu'(n' + 1) + \mu\nu' + \mu'\nu\left(\frac{m-1}{m'} + 1\right) - \nu\nu'(n+1) - \mu\nu'\left(\frac{m'-1}{m} + 1\right) - \mu'\nu,$$

that is,

$$\nu_1 = \nu\nu'(n' - n) + \frac{\mu'}{m'}(m-1)\,\nu - \frac{\mu}{m}(m' - 1)\,\nu'.$$

To complete the proof, it would be of course necessary to compare the remaining terms on the two sides respectively: but in what precedes it is shown that, the form being assumed, the expression for the alternant must of necessity be $(\mu, \nu, m' + m - 1, n' + n)$, where μ, ν have their foregoing values.

Resuming the equation

$$[(\mu', \nu'; m', n')(\mu, \nu; m, n)] = (\mu_1, \nu_1; m_1, n_1),$$

where μ_1, ν_1, m_1, n_1 have the before-mentioned values, then if we herein consider μ, ν, m, n; μ_1, ν_1, m_1, n_1 as given, we can find μ', ν', m', n'.

We have $m_1 = m' + m - 1$, $n_1 = n' + n$, that is, $m' = m_1 - m + 1$, $n' = n_1 - n$, and substituting these values in the expressions of $\frac{\mu_1}{m_1}$ and ν_1, we find

$$\frac{\mu_1}{m_1} = \frac{\mu'}{m_1 - m + 1}\{\mu + (n_1 - n)\nu\} - \frac{\mu}{m}(\mu' + n\nu'),$$

$$\nu_1 = (n_1 - 2n)\nu\nu' + \frac{m-1}{m_1 - m + 1}\nu\mu' - \frac{m_1 - m}{m}\mu\nu',$$

that is,

$$\frac{\mu_1}{m_1} = \left\{\frac{\mu}{m_1 - m + 1} + \frac{\nu(n_1 - n)}{m_1 - m + 1} - \frac{\mu}{m}\right\}\mu' - \frac{\mu n}{m}\nu',$$

$$\nu_1 = \left\{\frac{(m-1)\nu}{m_1 - m + 1}\right\}\mu' + \left\{(n_1 - 2n)\nu - \frac{m_1 - m}{m}\mu\right\}\nu',$$

which are linear equations for the determination of μ' and ν'.

944.

ON PFAFF-INVARIANTS.

[From the *Quarterly Journal of Pure and Applied Mathematics*, vol. XXVI. (1893), pp. 195—205.]

1. THE functions which I propose to call Pfaff-invariants present themselves and play a leading part in the memoir, Clebsch, "Ueber das *Pfaffsche* Problem" (Zweite Abhandlung), *Crelle*, t. LXI. (1863), pp. 146—179: but it is interesting to consider them for their own sake as invariants, and in the notation which I have elsewhere used for the functions called Pfaffians. The great simplification effected by this notation is, I think, at once shown by the remark that Clebsch's expression R, which he defines by the periphrasis "Sei ferner R der rationale Ausdruck dessen Quadrat der Determinant der a_{ik} gleich ist" (*l. c.* p. 149), is nothing else than the Pfaffian $1234 \ldots 2n-1 \,.\, 2n$, and that its differential coefficients $R_{ik} = \dfrac{\partial R}{\partial a_{ik}}$ are the Pfaffians obtained from the foregoing by the mere omission of any two symbolic numbers i, k.

2. I call to mind that the symbols 12, 13, &c., made use of are throughout such that $12 = -21$, &c.; and that the definition of the successive Pfaffians 12, 1234, &c., is as follows:

$$\begin{aligned} 12 &= 12, \\ 1234 &= 12\,.\,34 + 13\,.\,42 + 14\,.\,23, \\ 123456 &= 12\,.\,3456 + 13\,.\,4562 + 14\,.\,5623 + 15\,.\,6234 + 16\,.\,2345, \end{aligned}$$

in which last expression 3456 denotes the Pfaffian $34\,.\,56 + 35\,.\,64 + 36\,.\,45$, and similarly 4562, &c.; and so on for any even number of symbols. Of course, instead of the symbolic numbers 1, 2, 3, &c., we may have any other numbers (0 is frequently used in the sequel as a symbolic number), or we may have letters or other symbols.

3. I use also a function very analogous to a Pfaffian, which is expressed in the same notation, viz. this is

$$\begin{aligned}
\phi\psi 12 &= \phi\psi 12, \\
\phi\psi 1234 &= \phi\psi 12 \,.\, 34 + \phi\psi 13 \,.\, 42 + \phi\psi 14 \,.\, 23 \\
&\quad + \phi\psi 34 \,.\, 12 + \phi\psi 42 \,.\, 13 + \phi\psi 23 \,.\, 14, \\
\phi\psi 123456 &= \phi\psi 12 \,.\, 34 \,.\, 56 + \phi\psi 34 \,.\, 12 \,.\, 56 + \phi\psi 56 \,.\, 12 \,.\, 34 + \&\text{c.},
\end{aligned}$$

viz. taking any term 12 . 34 . 56 of the Pfaffian 123456, $\phi\psi$ is connected successively with each of the binary symbols 12, 34, 56 of the term, so as to give rise to terms containing the quaternary symbols $\phi\psi 12$, &c. Such function may be called a co-Pfaffian.

4. To avoid suffixes I use different letters (x, y), (x, y, z), &c., as the case may be, associating these with the numbers (1, 2), (1, 2, 3), &c. In the case of a differential of an even number $2n$ of terms, for instance $Xdx + Ydy + Zdz + Wdw$, I consider the functions 1234, $\phi 01234$, and $\phi\psi 1234$, the first and second of which are Pfaffians, the last a co-Pfaffian, as explained above. To fix in connexion with the differential $Xdx + Ydy + Zdz + Wdw$ the meanings of these expressions, I assume

$$12 = \frac{dX}{dy} - \frac{dY}{dx}, \quad 13 = \frac{dX}{dz} - \frac{dZ}{dx}, \ \&\text{c.},$$

(of course these imply $12 = -21$, &c.),

$$01 = -10 = X, \quad 02 = -20 = Y, \ \&\text{c.};$$

ϕ is an arbitrary function of x, y, z, w, and I write

$$\phi 0 = -0\phi = 0, \quad \phi 1 = -1\phi = \frac{d\phi}{dx}, \quad \phi 2 = -2\phi = \frac{d\phi}{dy}, \ \&\text{c.};$$

ψ is also an arbitrary function of x, y, z, w, and I write

$$\phi\psi 12 = \frac{d\phi}{dx}\frac{d\psi}{dy} - \frac{d\phi}{dy}\frac{d\psi}{dx}, \quad = \frac{\partial(\phi, \psi)}{\partial(x, y)}, \ \&\text{c.}$$

(this implies $\phi\psi 21 = -\phi\psi 12$, &c.).

5. Thus, at full length, the functions are

$$\begin{aligned}
1234 &= 12 \,.\, 34 + 13 \,.\, 42 + 14 \,.\, 23 \\
&= \left(\frac{dX}{dy} - \frac{dY}{dx}\right)\left(\frac{dZ}{dw} - \frac{dW}{dz}\right) + \left(\frac{dX}{dz} - \frac{dZ}{dx}\right)\left(\frac{dW}{dy} - \frac{dY}{dw}\right) + \left(\frac{dX}{dw} - \frac{dW}{dx}\right)\left(\frac{dY}{dz} - \frac{dZ}{dy}\right),
\end{aligned}$$

$$\begin{aligned}
\phi 01234 &= \phi 0 \,.\, 1234 + \phi 1 \,.\, 2340 + \phi 2 \,.\, 3401 + \phi 3 \,.\, 4012 + \phi 4 \,.\, 0123 \\
&= \frac{d\phi}{dx}(23 \,.\, 40 + 24 \,.\, 03 + 20 \,.\, 34) \\
&\quad + \frac{d\phi}{dy}(34 \,.\, 01 + 30 \,.\, 14 + 31 \,.\, 40) \\
&\quad + \frac{d\phi}{dz}(40 \,.\, 12 + 41 \,.\, 20 + 42 \,.\, 01) \\
&\quad + \frac{d\phi}{dw}(01 \,.\, 23 + 02 \,.\, 31 + 03 \,.\, 12)
\end{aligned}$$

$$= \frac{d\phi}{dx}\left\{-W\left(\frac{dY}{dz}-\frac{dZ}{dy}\right)+Z\left(\frac{dY}{dw}-\frac{dW}{dy}\right)-Y\left(\frac{dZ}{dw}-\frac{dW}{dz}\right)\right\}$$

$$+\frac{d\phi}{dy}\left\{\quad X\left(\frac{dZ}{dw}-\frac{dW}{dz}\right)-Z\left(\frac{dX}{dw}-\frac{dW}{dx}\right)-W\left(\frac{dZ}{dx}-\frac{dX}{dz}\right)\right\}$$

$$+\frac{d\phi}{dz}\left\{-W\left(\frac{dX}{dy}-\frac{dY}{dx}\right)-Y\left(\frac{dW}{dx}-\frac{dX}{dw}\right)+X\left(\frac{dW}{dy}-\frac{dY}{dw}\right)\right\}$$

$$+\frac{d\phi}{dw}\left\{\quad X\left(\frac{dY}{dz}-\frac{dZ}{dy}\right)+Y\left(\frac{dZ}{dx}-\frac{dX}{dz}\right)+Z\left(\frac{dX}{dy}-\frac{dY}{dx}\right)\right\},$$

$$\phi\psi 1234 = \phi\psi 12\,.\,34+\phi\psi 13\,.\,24+\phi\psi 14\,.\,23$$

$$+\phi\psi 34\,.\,12+\phi\psi 24\,.\,13+\phi\psi 23\,.\,14$$

$$= \frac{\partial(\phi,\psi)}{\partial(x,y)}\left(\frac{dZ}{dw}-\frac{dW}{dz}\right)+\frac{\partial(\phi,\psi)}{\partial(x,z)}\left(\frac{dY}{dw}-\frac{dW}{dy}\right)+\frac{\partial(\phi,\psi)}{\partial(x,w)}\left(\frac{dY}{dz}-\frac{dZ}{dy}\right)$$

$$+\frac{\partial(\phi,\psi)}{\partial(z,w)}\left(\frac{dX}{dy}-\frac{dY}{dx}\right)+\frac{\partial(\phi,\psi)}{\partial(y,w)}\left(\frac{dX}{dz}-\frac{dZ}{dx}\right)+\frac{\partial(\phi,\psi)}{\partial(y,z)}\left(\frac{dX}{dw}-\frac{dW}{dx}\right).$$

6. The invariantive property of the functions consists herein, viz. if we have

$$Xdx+Ydy+Zdz+Wdw=Pdp+Qdq+Rdr+Sds,$$

so that p, q, r, s, and thence also P, Q, R, S are functions each of them of x, y, z, w, then we have

$$1234\,\partial(x,y,z,w) = (1234)'\,\partial(p,q,r,s),$$

$$\phi 01234\,\partial(x,y,z,w) = (\phi 01234)'\,\partial(p,q,r,s),$$

$$\phi\psi 1234\,\partial(x,y,z,w) = (\phi\psi 1234)'\,\partial(p,q,r,s),$$

where the accented functions refer to (p, q, r, s, P, Q, R, S), and where for greater symmetry I have separated the symbolical numerator and denominator $\partial(p, q, r, s)$ and $\partial(x, y, z, w)$; each of these equations really contains

$$\frac{\partial(p,q,r,s)}{\partial(x,y,z,w)},$$

which is the functional determinant of (p, q, r, s) in regard to (x, y, z, w): or, if we please, it contains the reciprocal hereof

$$\frac{\partial(x,y,z,w)}{\partial(p,q,r,s)},$$

which is the functional determinant of (x, y, z, w) in regard to (p, q, r, s).

7. The equations give

$$\frac{\phi 01234}{1234}=\frac{(\phi 01234)'}{(1234)'},$$

$$\frac{\phi\psi 1234}{1234}=\frac{(\phi\psi 1234)'}{(1234)'},$$

and then the expressions on the left-hand are absolute invariants in respect to the transformation of

$$Xdx + Ydy + Zdz + Wdw \quad \text{into} \quad Pdp + Qdq + Rdr + Sds.$$

They are, in fact, (for $2n = 4$) Clebsch's derivatives (ϕ) and (ϕ, ψ).

8. For the Pfaffian reduction

$$Xdx + Ydy + Zdz + Wdw = Fdf + Gdg,$$

we may write

$$P,\ Q,\ R,\ S = F,\ G,\ 0\ ,\ 0,$$

$$p\ ,\ q\ ,\ r\ ,\ s = f\ ,\ g\ ,\ F,\ G,$$

viz. we take f, g, F, G as the new independent variables; we thus have

$$01' = F,\quad 02' = G,\quad 03' = 0,\quad 04' = 0,$$

$$12' = 0\ ,\quad 13' = 1,\quad 14' = 0,\quad 23' = 0,\quad 24' = 1,\quad 34' = 0,$$

$$(1234)' = 12' \,.\, 34' + 13' \,.\, 42' + 14' \,.\, 23',\ = -1\ ;$$

and similarly

$$(\phi 01234)' = -\left\{F\frac{d\phi}{dF} \quad + G\frac{d\phi}{dG}\right\},$$

$$(\phi\psi 1234)' = -\left\{\frac{\partial(\phi,\ \psi)}{\partial(f,\ F)} + \frac{\partial(\phi,\ \psi)}{\partial(g,\ G)}\right\},$$

where, in the equations, the $-$ sign presents itself by reason that $2n$, $= 4$, is the double of an even number, or say that n is even; in the case of $2n$, the double of an odd number, that is, n odd, the sign would have been $+$.

9. We thus have

$$(\phi) = \frac{\phi 01234}{1234} \ = F\frac{d\phi}{dF} + G\frac{d\phi}{dG},$$

$$(\phi\psi) = \frac{\phi\psi 1234}{1234} = \frac{\partial(\phi,\ \psi)}{\partial(f,\ F)} + \frac{\partial(\phi,\ \psi)}{\partial(g,\ G)};$$

and in particular, by giving to ϕ and ψ the values f, g, F, G, we find

$$(f) = 0,\quad (g) = 0,\quad (F) = F,\quad (G) = G,$$

$$(f,\ g) = 0,\quad (f,\ F) = 1,\quad (f,\ G) = 0,$$

$$(F,\ G) = 0,\quad (g,\ F) = 0,\quad (g,\ G) = 1,$$

which are Clebsch's equations; in the case of $2n$ terms, the number of them is

$$n + n + \tfrac{1}{2}(n^2 - n) + \tfrac{1}{2}(n^2 - n) + n^2 = 2n + n^2 - n + n^2,$$

$= n(2n+1)$, or $\frac{1}{2}2n(2n+1)$, as it should be.

10. It may be remarked that we have

$$Fdf + Gdg = Fd\left(f + \frac{G}{F}g\right) - Fgd\frac{G}{F},$$

or writing this $= F'df' + G'dg'$, we have

$$F' = F,\quad f' = f + \frac{Gg}{F},\quad G' = +Fg,\quad g' = \frac{G}{F},$$

whence conversely

$$F = F',\quad f = f' + \frac{G'g'}{F'},\quad G = F'g',\quad g = -\frac{G'}{F'},\quad (Gg = -G'g').$$

The ten equations $(f) = 0$, $(g) = 0$, &c., ought then to lead to the corresponding ten equations $(f') = 0$, $(g') = 0$, &c., and it is easy to verify that they do so; for instance, we have

$$(f') = \left(f + \frac{Gg}{F}\right) = (f) + \frac{G}{F}(g) + g\left(\frac{G}{F}\right),$$

where

$$\left(\frac{G}{F}\right) = \frac{1}{F}(G) - \frac{G}{F^2}(F),\quad = \frac{G}{F} - \frac{G}{F^2}F,\quad = 0,$$

and thus $(f') = 0$. And again,

$$(f',\ g') = \left(f + \frac{Gg}{F},\ \frac{G}{F}\right) = \left(f,\ \frac{G}{F}\right) + \left(\frac{Gg}{F},\ \frac{G}{F}\right) = \left(f,\ \frac{G}{F}\right) + \frac{G}{F}\left(g,\ \frac{G}{F}\right) + g\left(\frac{G}{F},\ \frac{G}{F}\right),$$

where the last term vanishes; the remaining terms are

$$= \frac{1}{F}(f,\ G) - \frac{G}{F^2}(f,\ F) + \frac{G}{F^2}(g,\ G) - \frac{G}{F^3}(g,\ F),$$

$$= 0 - \frac{G}{F^2} + \frac{G}{F^2} - 0,\ \text{ that is, } (f',\ g') = 0.$$

There is, of course, the like transformation

$$Fdf + Gdg = G\left(dg + \frac{F}{G}f\right) - Gfd\frac{F}{G}.$$

11. I have, for better exhibiting the results, taken $2n = 4$, but the most simple case for an even number of terms is $2n = 2$. Here we have $Xdx + Ydy = Pdp + Qdq$, and the functions to be considered are

$$12,\ = 12 \qquad\qquad = \frac{dX}{dy} - \frac{dY}{dx},$$

$$\phi 012,\ = \phi 0 . 12 + \phi 1 . 20 + \phi 2 . 01 = -Y\frac{d\phi}{dx} + X\frac{d\phi}{dy},$$

$$\phi\psi 12,\ = \phi\psi 12 \qquad\qquad = \frac{\partial(\phi,\ \psi)}{\partial(x,\ y)}.$$

We have here

$$X = P\frac{dp}{dx} + Q\frac{dq}{dx},\quad Y = P\frac{dp}{dy} + Q\frac{dq}{dy},$$

and the invariantive properties are easily verified.

12. Thus

$$12 = \frac{dX}{dy} - \frac{dY}{dx},\quad = \left(\frac{dP}{dy}\frac{dp}{dx} - \frac{dP}{dx}\frac{dp}{dy}\right) + \left(\frac{dQ}{dy}\frac{dq}{dx} - \frac{dQ}{dx}\frac{dq}{dy}\right);$$

or, writing herein

$$\frac{dP}{dx}=\frac{dP}{dp}\frac{dp}{dx}+\frac{dP}{dq}\frac{dq}{dx},$$

and the like values for $\frac{dP}{dy}$ and for $\frac{dQ}{dx}$ and $\frac{dQ}{dy}$, we have

$$12=\left(\frac{dP}{dq}-\frac{dQ}{dp}\right)\left(\frac{dp}{dx}\frac{dq}{dy}-\frac{dp}{dy}\frac{dq}{dx}\right)=(12)'\frac{\partial(p,\ q)}{\partial(x,\ y)}.$$

Similarly, we find

$$\phi 012=-Y\frac{d\phi}{dx}+X\frac{d\phi}{dy}=\left(-Q\frac{d\phi}{dp}+P\frac{d\phi}{dq}\right)\left(\frac{dp}{dx}\frac{dq}{dy}-\frac{dp}{dy}\frac{dq}{dx}\right)=(\phi 012)'\frac{\partial(p,\ q)}{\partial(x,\ y)},$$

and

$$\phi\psi 12=\frac{d\phi}{dx}\frac{d\psi}{dy}-\frac{d\phi}{dy}\frac{d\psi}{dx}=\left(\frac{d\phi}{dp}\frac{d\psi}{dq}-\frac{d\phi}{dq}\frac{d\psi}{dp}\right)\left(\frac{dp}{dx}\frac{dq}{dy}-\frac{dp}{dy}\frac{dq}{dx}\right)=(\phi\psi 12)'\frac{\partial(p,\ q)}{\partial(x,\ y)}.$$

We thus have

$$\begin{aligned}12\,\partial(x,\ y)&=\quad(12)'\,\partial(p,\ q),\\ \phi 012\,\partial(x,\ y)&=(\phi 012)'\,\partial(p,\ q),\\ \phi\psi 12\,\partial(x,\ y)&=(\phi\psi 12)'\,\partial(p,\ q);\end{aligned}$$

or say

$$\frac{\phi 012}{12}=\frac{(\phi 012)'}{(12)'},\ \text{ and }\ \frac{\phi\psi 12}{12}=\frac{(\phi\psi 12)'}{(12)'}.$$

The proof is the same in principle for $2n=4$, or any other even value.

13. The theory is very similar in the case of an odd number $2n+1$ of terms; thus $2n+1=3$, the forms are

$$0123,\quad \phi 123,\quad\text{and}\quad \phi\psi 0123,$$

the first and second of which are Pfaffians, the third of them co-Pfaffian: the developed expression of this last is

$$\begin{aligned}\phi\psi 0123=\ &\phi\psi 01\,.\,23+\phi\psi 02\,.\,31+\phi\psi 03\,.\,12\\ &+\phi\psi 23\,.\,01+\phi\psi 31\,.\,02+\phi\psi 12\,.\,03,\end{aligned}$$

and to fix the meaning hereof we write

$$\phi\psi 01=0,\quad \phi\psi 02=0,\quad \phi\psi 03=0.$$

Hence, the differential expression being $Xdx+Ydy+Zdz$, we have

$$\begin{aligned}0123&=01\,.\,23+02\,.\,31+03\,.\,12\\ &=X\left(\frac{dY}{dz}-\frac{dZ}{dy}\right)+Y\left(\frac{dZ}{dx}-\frac{dX}{dz}\right)-Z\left(\frac{dX}{dy}-\frac{dY}{dx}\right),\\ \phi 123&=\phi 1\,.\,23+\phi 2\,.\,31+\phi 3\,.\,12\\ &=\frac{d\phi}{dx}\left(\frac{dY}{dz}-\frac{dZ}{dy}\right)+\frac{d\phi}{dy}\left(\frac{dZ}{dx}-\frac{dX}{dz}\right)+\frac{d\phi}{dz}\left(\frac{dX}{dy}-\frac{dY}{dx}\right),\\ \phi\psi 0123&=\phi\psi 23\,.\,01+\phi\psi 31\,.\,02+\phi\psi 12\,.\,03\\ &=X\frac{\partial(\phi,\ \psi)}{\partial(y,\ z)}+Y\frac{\partial(\phi,\ \psi)}{\partial(z,\ x)}+Z\frac{\partial(\phi,\ \psi)}{\partial(x,\ y)}.\end{aligned}$$

14. For the transformation

$$Xdx + Ydy + Zdz = Pdp + Qdq + Rdr,$$

we have

$$0123\,\partial(x,\ y,\ z) = (0123)'\,\partial(p,\ q,\ r),$$

$$\phi123\,\partial(x,\ y,\ z) = (\phi123)'\,\partial(p,\ q,\ r),$$

$$\phi\psi0123\,\partial(x,\ y,\ z) = (\phi\psi0123)'\,\partial(p,\ q,\ r),$$

and consequently

$$\frac{\phi123}{0123} = \frac{(\phi123)'}{(0123)'},$$

$$\frac{\phi\psi0123}{0123} = \frac{(\phi\psi0123)'}{(0123)'};$$

so that the left-hand functions are absolute invariants.

15. If in particular, $Xdx + Ydy + Zdz = df + Gdg$, then we may write

$$P,\ Q,\ R = 1,\ G,\ 0,$$

$$p,\ q,\ r\ = f,\ g,\ G.$$

Hence

$$01' = 1,\quad 02' = G,\quad 03' = 0;\quad 23' = 1,\quad 31' = 0,\quad 12' = 0,$$

and therefore

$$(0123)' = 1,\quad (\phi123)' = \frac{d\phi}{df},\quad (\phi\psi0123)' = \frac{\partial(\phi,\ \psi)}{\partial(g,\ G)} + G\frac{\partial(\phi,\ \psi)}{\partial(f,\ G)},$$

or say

$$= \frac{\partial(\phi,\ \psi)}{\partial(g,\ G)} - G\frac{\partial(\phi,\ \psi)}{\partial(f,\ G)};$$

and we thus have

$$0123\,\partial(x,\ y,\ z) = \partial(f,\ g,\ G),$$

$$\phi123\,\partial(x,\ y,\ z) = \frac{d\phi}{df}\,\partial(f,\ g,\ G),$$

$$\phi\psi0123\,\partial(x,\ y,\ z) = \left\{\frac{\partial(\phi,\ \psi)}{\partial(g,\ G)} - G\frac{\partial(\phi,\ \psi)}{\partial(f,\ G)}\right\}\partial(f,\ g,\ G),$$

and then

$$(\phi) = \frac{\phi123}{0123} = \frac{d\phi}{df},$$

$$(\phi,\ \psi) = \frac{\phi\psi0123}{0123} = \frac{\partial(\phi,\ \psi)}{\partial(g,\ G)} - G\frac{\partial(\phi,\ \psi)}{\partial(f,\ G)};$$

viz. we thus have derivatives (ϕ) and $(\phi,\ \psi)$ analogous to (but quite different in form from) those of Clebsch in the case of an even number of terms.

In particular, writing $\phi, \psi = f,\ g,\ G$, we obtain

$$(f) = 1,\quad (g) = 0,\quad (G) = 0;\quad (f,\ g) = 0,\quad (f,\ G) = -G,\quad (g,\ G) = 1,$$

which are the analogues of Clebsch's formula.

16. It is interesting to compare the formula for the two cases

$$Xdx + Ydy + Zdz + Wdw = Fdf + Gdg,$$

and

$$Xdx + Ydy + Zdz = df + Gdg.$$

In the former case f and g are symmetrically related to each other, and we may say that ($f=$ const. and $g=$ const.) is a solution of $Xdx + Ydy + Zdz + Wdw = 0$; we have $(f)=0$ and $(g)=0$. In the second case ($f=$ const. and $g=$ const.) is still a solution of $Xdx + Ydy + Zdz = 0$, but f and g are not symmetrically related to each other, and we have $(f)=1$, $(g)=0$. Moreover, in the first case $(G)=G$, but in the second case $(G)=0$, an equation of the same form as $(g)=0$; the reason is that we have here

$$Xdx + Ydy + Zdz = df + Gdg, \quad = d(f+Gg) - gdG,$$

so that, besides the solution ($f=$ const. and $g=$ const.), we have the solution

$$(f + Gg = \text{const. and } G = \text{const.}).$$

17. The remark just made may be further developed: we have

$$Xdx + Ydy + Zdz = df + Gdg, \quad = d(f+Gg) - gdG, \quad = df' + G'dg',$$

suppose, where $f' = f + Gg$, $G' = -g$, $g' = G$, and therefore also $f = f' + G'g'$, $G = g'$, $g = -G'$; the equations

$$(f)=1, \quad (g)=0, \quad (G)=0, \quad (f, g)=0, \quad (f, G)=-G, \quad (g, G)=1,$$

should lead to

$$(f')=1, \quad (g')=0, \quad (G')=0, \quad (f', g')=0, \quad (f', G')=-G', \quad (g', G')=1.$$

There is no difficulty in verifying this; thus the equations $(g)=0$, $(G)=0$, give at once $(g')=0$, $(G')=0$; and then the equation $(f)=1$ gives $(f' + G'g')=1$, that is,

$$(f') + G'(g') + g'(G') = 1, \text{ or } (f') = 1.$$

So again $(g, G)=1$ gives $(g', G')=1$; and then $(f, g)=0$ gives $(f' + G'g', G')=0$, that is,

$$(f', G') + G'(g', G') + g'(G', G') = 0, \text{ or } (f', G') = -G'.$$

And finally, $(f, G) = -G$ gives $(f' + G'g', g') + g' = 0$, that is,

$$(f', g') + G'(g', g') + g'(G', g') + g' = 0, \text{ or } (f', g') = 0.$$

I stop to give the direct verification of the equations $(f)=1$, $(g)=0$, $(G)=0$. We have

$$Xdx + Ydy + Zdz = df + Gdg,$$

that is,

$$X = \frac{df}{dx} + G\frac{dg}{dx}, \quad Y = \frac{df}{dy} + G\frac{dg}{dy}, \quad Z = \frac{df}{dz} + G\frac{dg}{dz},$$

and thence

$$23 = \frac{dY}{dz} - \frac{dZ}{dy} = \frac{dG}{dz}\frac{dg}{dy} - \frac{dG}{dy}\frac{dg}{dz},$$

$$31 = \frac{dZ}{dx} - \frac{dX}{dz} = \frac{dG}{dx}\frac{dg}{dz} - \frac{dG}{dz}\frac{dg}{dx},$$

$$12 = \frac{dX}{dy} - \frac{dY}{dx} = \frac{dG}{dy}\frac{dg}{dx} - \frac{dG}{dx}\frac{dg}{dy}.$$

Hence, multiplying first by $\frac{df}{dx}, \frac{df}{dy}, \frac{df}{dz}$, that is,

$$X - G\frac{dg}{dx}, \quad Y - G\frac{dg}{dy}, \quad Z - G\frac{dg}{dz},$$

and adding, we have

$$23\frac{df}{dx} + 31\frac{df}{dy} + 12\frac{df}{dz} = X23 + Y31 + Z12,$$

that is, $f123 = 0123$, or $(f) = 1$.

And then multiplying secondly by $\frac{dg}{dx}, \frac{dg}{dy}, \frac{dg}{dz}$ and adding, and thirdly by $\frac{dG}{dx}, \frac{dG}{dy}, \frac{dG}{dz}$ and adding, we obtain

$$23\frac{dg}{dx} + 31\frac{dg}{dy} + 12\frac{dg}{dz} = 0, \text{ that is, } (g) = 0,$$

and

$$23\frac{dG}{dx} + 31\frac{dG}{dy} + 12\frac{dG}{dz} = 0, \text{ that is, } (G) = 0.$$

To exhibit more clearly the formulæ for any odd number of terms, I take $2n + 1 = 5$,

$$Xdx + Ydy + Zdz + Wdw + Tdt = df + Gdg + Hdh.$$

We have here

$$(\phi) = \frac{\phi 12345}{012345} = \frac{d\phi}{df},$$

$$(\phi\psi) = \frac{\phi\psi 012345}{012345} + \frac{\partial(\phi, \psi)}{\partial(g, G)} + \frac{\partial(\phi, \psi)}{\partial(h, H)} - G\frac{\partial(\phi, \psi)}{\partial(f, G)} - H\frac{\partial(\phi, \psi)}{\partial(f, H)};$$

and in particular,

$$(f) = 1; \quad (g) = 0, \quad (h) = 0; \quad (G) = 0, \quad (H) = 0;$$

$$(f, g) = 0; \quad (f, h) = 0; \quad (f, G) = -G, \quad (f, H) = -H;$$

$$(g, h) = 0; \quad (G, H) = 0; \quad (g, G) = 1, \quad (g, H) = 0;$$

$$(h, G) = 0, \quad (h, H) = 1;$$

in all

$$1 + 2n + 2n + \tfrac{1}{2}(n^2 - n) + \tfrac{1}{2}(n^2 - n) + n^2,$$

$$= 1 + 4n + n^2 - n + n^2, \quad = 2n^2 + 3n + 1, \quad = \tfrac{1}{2}(2n+1)(2n+2)$$

equations.

We can, by what precedes, at once express the conditions which must be satisfied in order that a differential expression $X_1 dx_1 + X_2 dx_2 + \ldots + X_\nu dx_\nu$, may be reducible to one of the special forms df, Fdf, $df + F_1 df_1$, &c.; viz. if we have

$$\begin{array}{lll} X_1 dx_1 + X_2 dx_2 + \ldots + X_\nu dx_\nu = df, & \text{then} & 12 \quad\;\; = 0, \text{ \&c.} \\ \qquad\qquad\qquad\qquad\qquad\quad\; = Fdf, & ,, & 0123 \quad = 0, \text{ \&c.} \\ \qquad\qquad\qquad\qquad\qquad\quad\; = df + F_1 df_1, & ,, & 1234 \quad = 0, \text{ \&c.} \\ \qquad\qquad\qquad\qquad\qquad\quad\; = Fdf + F_1 df_1, & ,, & 012345 = 0, \text{ \&c.,} \\ \qquad\qquad\qquad\qquad\qquad\qquad \text{\&c.,} & & \quad \text{\&c.,} \end{array}$$

where the numbers 12, 1234, 12345, &c., represent any combinations out of the numbers 1, 2, 3, ..., ν. Of course, if ν is not sufficiently large to furnish such a combination, then there is no condition to be satisfied; thus if

$$X_1 dx_1 + X_2 dx_2 + X_3 dx_3 = df + F_1 df_1,$$

there is no condition to be satisfied.

945.

NOTE ON LACUNARY FUNCTIONS.

[From the *Quarterly Journal of Pure and Applied Mathematics*, vol. XXVI. (1893), pp. 279—281.]

THE present note is founded upon Poincaré's paper "Sur les fonctions à espaces lacunaires," *Amer. Math. Jour.*, t. XIV. (1892), pp. 201—221.

If the complex variable $z = x + iy$ is represented as usual by a point the coordinates of which are (x, y), and if $U_0, U_1, U_2, \ldots$ denote an infinite series of given functions of z, then the equation

$$fz = U_0 + U_1 + U_2 + \ldots$$

defines a function of z, but only for those values of z for which the series is convergent, or say for points within a certain region Θ; and within this region, it defines the successive derived functions $f'z, f''z, f'''z, \ldots$.

Taking $l, = h + ik$, as an increment of z, we define the function of $z + l$, by the equation

$$f(z+l) = fz + \frac{l}{1}f'z + \frac{l^2}{1\,.\,2}f''z + \ldots,$$

but only for values of l for which the series is convergent: it may very well be, and it is in general the case, that we thereby extend the definition of fz so as to make it applicable to points within a larger region Θ_1; and then considering fz as defined within this larger region Θ_1, we may pass from it to a still larger region Θ_2; and so on indefinitely, or until we cover the whole infinite plane.

For instance, the equation

$$fz = 1 + z + z^2 + z^3 + \ldots$$

defines the function $1 \div (1 - z)$ for values of z for which mod. $z < 1$, that is, for points within the circle $x^2 + y^2 - 1 = 0$; and this being so,

$$f(z+l) = \frac{1}{1-z} + \frac{l}{(1-z)^2} + \frac{l^2}{(1-z)^3} + \dots$$

extends the definition to the larger region for which this is a convergent series: the condition of convergency is

$$\text{mod.} \frac{l}{1-z} < 1, \text{ that is, mod.} \frac{h+ik}{1-x-iy} < 1, \text{ or } h^2 + k^2 < (1-x)^2 + y^2.$$

The condition is that the distance $\sqrt{(h^2+k^2)}$ must not exceed the distance $\sqrt{\{(1-x)^2+y^2\}}$ of the point $z = x + iy$, from the point $(x = 1, y = 0)$; the point z is strictly within the circle $x^2 + y^2 = 1$, but taking it on the circumference, the condition is that the point $z + l$ must lie within a circle having its centre on the circle $x^2 + y^2 - 1 = 0$ and passing through the point $(x = 1, y = 0)$. Taking $\cos\theta$ and $\sin\theta$ for the coordinates of the centre, the equation of this circle is

$$(x - \cos\theta)^2 + (y - \sin\theta)^2 = (1 - \cos\theta)^2 + \sin^2\theta,$$

that is,

$$x^2 + y^2 - 1 - 2\cos\theta\,(x-1) - 2y\sin\theta = 0;$$

and the envelope of these circles is

$$(x^2 + y^2 - 1)^2 - 4(x-1)^2 - 4y^2 = 0;$$

or, as this may be written,

$$(x^2+y^2)^2 - 6(x^2+y^2) + 8x - 3 = 0,$$

or again in the form

$$(x^2+y^2-3)^2 + 4(2x-1) = 0,$$

or in the form

$$y^4 + 2y^2(x^2-3) + (x-1)^3(x+3) = 0.$$

The curve is a cuspidal Cartesian. To put this in evidence, observe that the equation may be written

$$-4y^2 + \{(x-1)^2 + y^2\}\{(x-1)^2 + y^2 + 4x - 4\} = 0,$$

viz. writing

$$A = x + iy - 1,$$
$$B = x - iy - 1,$$
$$Z = -1,$$

then

$$A - B = 2iy,$$
$$AB \;\; = (x-1)^2 + y^2,$$
$$A + B = 2x - 2,$$

or the equation is

$$Z^2(A-B)^2 + AB\{AB - 2Z(A+B)\} = 0,$$

that is,

$$Z^2A^2 + Z^2B^2 + A^2B^2 - 2Z^2AB - 2ZA^2B - 2ZAB^2 = 0,$$

which is the equation of a bicuspidal quartic curve, having for cusps the vertices of the triangle $A = 0$, $B = 0$, $Z = 0$.

The region within which the function is $= \frac{1}{1-z}$ is thus extended to the area within the Cartesian curve, say this is the region Θ_1: starting from this curve instead of the circle (viz. by considering the envelope of the circle having its centre on the curve and passing through the point $x = 1$, $y = 0$), we obtain a second curve, a closed curve, which instead of having a cusp on the axis of x cuts this axis at right angles at a point the distance of which from the origin is greater than 1; and we thus extend the region to the area within this second curve, say this is the region Θ_2. And proceeding in this way, we ultimately extend the region to the whole of the infinite plane.

But the functions U_0, U_1, U_2, ... may be such that for every value whatever of l, for which the point $z + l$ is outside the region Θ, the series

$$fz + \frac{l}{1} f'z + \frac{l^2}{1.2} f''z + \dots$$

is divergent, and we are in this case unable to define the function fz for points outside the region Θ_1: the function then exists only for points inside the region Θ, and for points outside this region it is non-existent; a function such as this, existing only for points within a certain region and not for the whole of the infinite plane, is said to be a *lacunary* function.

946.

NOTE ON THE THEORY OF ORTHOMORPHOSIS.

[From the *Quarterly Journal of Pure and Applied Mathematics*, vol. XXVI. (1893), pp. 282—288.]

THE equation of any given curve whatever, $\Theta = 0$, may be expressed in the form

$$\phi(x+iy) + \phi(x-iy) = 0.$$

Let χ be any odd function; then since

$$\phi(x-iy) = -\phi(x+iy),$$

we have

$$\chi\phi(x-iy) = \chi\{-\phi(x+iy)\} = -\chi\phi(x+iy),$$

that is,

$$\chi\phi(x+iy) + \chi\phi(x-iy) = 0.$$

Assuming that Θ is a real function, that is, a function with real coefficients, then also $\phi(x+iy)$ will be a function with real coefficients, or say a real function of $x+iy$; the function χ may be real or imaginary, but if imaginary, then the i of the coefficients does not change its sign in the passage from $\chi\phi(x+iy)$ to $\chi\phi(x-iy)$.

In proof of the assumed theorem, imagine the equation $\Theta = 0$ expressed as an equation between $x+iy$ and $x-iy$, or, supposing it solved in regard to $x-iy$, take the form of it to be $x-iy = f(x+iy)$: let u_n be a function of n satisfying the equation of differences $u_{n+1} = fu_n$; and let $\phi(x+iy)$ be determined as a function of $x+iy$ by the elimination of n from the equations

$$x+iy = u_n, \quad \phi(x+iy) = \cos n\pi;$$

we thence have

$$x-iy = fu_n, \quad = u_{n+1},$$

and consequently

$$\phi(x-iy) = \cos(n+1)\pi,$$

that is,

$$\phi(x+iy)+\phi(x-iy)=0,$$

viz. this equation is a transformation of the equation $\Theta=0$, and thus it appears that the equation $\Theta=0$ can always be thrown into the last-mentioned form.

As an example, take the equation $y=ax+b$: which, putting for a moment $\xi=x+iy$, $\eta=x-iy$, is

$$\frac{1}{2i}(\xi-\eta)=\tfrac{1}{2}a(\xi+\eta)+b,$$

that is,

$$\eta=\frac{i+a}{i-a}\xi+\frac{2b}{i-a};$$

we have therefore

$$u_{n+1}=\frac{i+a}{i-a}u_n+\frac{2b}{i-a},$$

a solution of which is

$$u_n=\left(\frac{i+a}{i-a}\right)^n-\frac{b}{a};$$

putting this $=\xi$, we have

$$n=\frac{1}{\log\dfrac{i+a}{i-a}}\log\left(\xi+\frac{b}{a}\right),$$

and thence

$$\phi\xi=\cos\frac{\pi}{\log\dfrac{i+a}{i-a}}\log\left(\xi+\frac{b}{a}\right),$$

where observe that, writing $a+i=Re^{i\alpha}$ and therefore $a-i=Re^{-i\alpha}$, we have

$$\cos\alpha=\frac{a}{\sqrt{(a^2+1)}},\quad \sin\alpha=\frac{1}{\sqrt{(a^2+1)}},$$

or say $\cot\alpha=a$, and then

$$\frac{i+a}{i-a}=e^{2\alpha i+i\pi},\text{ or }\log\frac{i+a}{i-a}=i(2\alpha+\pi),$$

whence

$$\phi\xi=\cos\frac{\pi}{i(2\alpha+\pi)}\log\left(\xi+\frac{b}{a}\right),\ =\cosh\frac{\pi}{2\alpha+\pi}\log\left(\xi+\frac{b}{a}\right),$$

a real function of ξ.

In verification of the equation $\phi\xi+\phi\eta=0$, we have

$$\phi\eta=\cos\frac{\pi}{\log\dfrac{i+a}{i-a}}\log\left(\eta+\frac{b}{a}\right),$$

where

$$\log\left(\eta+\frac{b}{a}\right)=\log\left(\frac{i+a}{i-a}\,\xi+\frac{2b}{i-a}+\frac{b}{a}\right)=\log\frac{i+a}{i-a}\left(\xi+\frac{b}{a}\right),$$

$$=\log\frac{i+a}{i-a}+\log\left(\xi+\frac{b}{a}\right),$$

and thence

$$\phi\eta=\cos\frac{\pi}{\log\frac{i+a}{i-a}}\left\{\log\frac{i+a}{i-a}+\log\left(\xi+\frac{b}{a}\right)\right\}$$

$$=\cos\left\{\pi+\frac{\pi}{\log\frac{i+a}{i-a}}\log\left(\xi+\frac{b}{a}\right)\right\},\quad=-\cos\frac{\pi}{\log\frac{i+a}{i-a}}\log\left(\xi+\frac{b}{a}\right),$$

that is, $\phi\eta=-\phi\xi$, or $\phi\xi+\phi\eta=0$, the equation in question.

I remark, in passing, that the same equation $y=ax+b$ might have been put in the form $\phi x+\phi y=0$, viz. assuming

$$\phi x=\cos\frac{\pi}{\log a}\log\left(x-\frac{b}{1-a}\right),$$

then

$$\phi y=\cos\frac{\pi}{\log a}\log\left(ax+b-\frac{b}{1-a}\right)=\cos\frac{\pi}{\log a}\log a\left(x-\frac{b}{1-a}\right)$$

$$=\cos\frac{\pi}{\log a}\left\{\log a+\log\left(x-\frac{b}{1-a}\right)\right\}$$

$$=\cos\left\{\pi+\frac{\pi}{\log a}\log\left(x-\frac{b}{1-a}\right)\right\}$$

$$=-\cos\frac{\pi}{\log a}\log\left(x-\frac{b}{1-a}\right)=-\phi x,$$

that is, $\phi x+\phi y=0$.

If $b=0$, then

$$y=ax\ \text{ and }\ \phi x=\cos\frac{\pi\log x}{\log a};$$

in fact, repeating the proof for this particular case,

$$\phi y=\cos\pi\frac{\log ax}{\log a}=\cos\pi\left(1+\frac{\log x}{\log a}\right)=-\cos\frac{\pi\log x}{\log a},\quad=-\phi x;$$

that is,

$$\phi x+\phi y=0.$$

Considering then $(x,\ y)$ as the coordinates of a point on the curve $\Theta=0$, we have, as above,

$$\chi\phi(x+iy)+\chi\phi(x-iy)=0,$$

where ϕ is a real function determined as above, and χ is any real or imaginary odd function. This being so, assume

$$x_1+iy_1=e^{\chi\phi(x+iy)},$$

then also

$$x_1 - iy_1 = e^{\chi\phi(x-iy)},$$

and consequently

$$x_1^2 + y_1^2 = e^{\chi\phi(x+iy)+\chi\phi(x-iy)} = 1,$$

that is, we have the circumference of the circle $x_1^2 + y_1^2 - 1 = 0$ corresponding to the given curve $\Theta = 0$.

Suppose that the curve $\Theta = 0$ is a closed curve: and then writing

$$\xi + i\eta = \chi\phi(x + iy),$$

and therefore

$$\xi - i\eta = \chi\phi(x - iy),$$

we thence have

$$2\xi = \chi\phi(x + iy) + \chi\phi(x - iy),$$

a real function of (x, y).

(1) Assume that it is possible to find χ, such that ξ as defined by this last equation shall be throughout the area of the curve $\Theta = 0$ finite and continuous, except only that in the neighbourhood of a given point, taken to be the point $x = 0$, $y = 0$, it is $= \log\sqrt{(x^2 + y^2)}$.

(2) At the boundary of the area $\Theta = 0$, ξ is $= 0$.

(3) Throughout the area, ξ satisfies the partial differential equation

$$\frac{d^2\xi}{dx^2} + \frac{d^2\xi}{dy^2} = 0.$$

These conditions being satisfied, the equation

$$x_1 + iy_1 = e^{\xi + i\eta},$$

that is,

$$x_1 + iy_1 = e^{\chi\phi(x+iy)},$$

gives an orthomorphosis of the area $\Theta = 0$ into the circle $x_1^2 + y_1^2 - 1 = 0$, the point $x = 0$, $y = 0$ corresponding to the centre of the circle; (2) and (3) are satisfied as above: it remains only to satisfy (1), viz. the function χ is determined not by any equation—*but only by this condition as to finiteness and continuity;* and if it be thus determined, then the foregoing equation $x_1 + iy_1 = e^{\chi\phi(x+iy)}$ gives the required orthomorphosis.

For instance, let the curve $\Theta = 0$ be the parabola $y^2 = 4(1 - x)$, which may be regarded as a closed curve bounding the infinite parabolic area. We have $2x = \xi + \eta$, $2iy = \xi - \eta$, whence the equation is

$$-\tfrac{1}{4}(\xi - \eta)^2 = 4 - 2\xi - 2\eta,$$

that is,

$$\xi^2 - 2\xi\eta + \eta^2 - 8\xi - 8\eta + 16 = 0,$$

whence $\sqrt{\xi} + \sqrt{\eta} - 2 = 0$, or writing this in the form

$$(\sqrt{\xi} - 1) + (\sqrt{\eta} - 1) = 0,$$

we have $\phi\xi = \sqrt{\xi} - 1$, and assuming that χ can be found so that the condition as to finiteness and continuity is satisfied, then the orthomorphosis is given by

$$x_1 + iy_1 = \exp\chi(\sqrt{\xi} - 1), \quad = \exp\chi\{\sqrt{(x+iy)} - 1\}.$$

Assuming

$$\tfrac{1}{2}\chi\omega = -\tfrac{1}{2}i\pi\omega + \log\frac{1 - i\exp(\ \tfrac{1}{2}i\pi\omega)}{1 - i\exp(-\tfrac{1}{2}i\pi\omega)},$$

which is obviously an odd function, we have

$$\exp\tfrac{1}{2}\chi\omega = \frac{1}{\exp\tfrac{1}{2}i\pi\omega}\,\frac{1 - i\exp(\ \tfrac{1}{2}i\pi\omega)}{1 - i\exp(-\tfrac{1}{2}i\pi\omega)},$$

$$= \frac{1 - \exp\tfrac{1}{2}i\pi\omega}{\exp\tfrac{1}{2}i\pi\omega - 1}, \quad = \frac{i(1 - i\exp\tfrac{1}{2}i\pi\omega)}{1 + i\exp\tfrac{1}{2}i\pi\omega},$$

which is

$$= \tan\tfrac{1}{4}\pi(\omega + 1),$$

and hence, for ω writing $\sqrt{(x+iy)} - 1$, we have

$$x_1 + iy_1 = \exp\chi\{\sqrt{(x+iy)} - 1\}, \quad = \tan^2\tfrac{1}{4}\pi\sqrt{(x+iy)}.$$

This satisfies the required conditions as to finiteness and continuity; and in particular, we have

$$\xi + i\eta = \log\tan^2\tfrac{1}{4}\pi\sqrt{(x+iy)},$$

so that, x and y being small,

$$\xi + i\eta = \log\frac{\pi^2}{16}(x+iy), \quad \xi - i\eta = \log\frac{\pi^2}{16}(x-iy),$$

that is,

$$\xi = \log\frac{\pi^2}{16}\sqrt{(x^2+y^2)}.$$

Hence we have the known result: the orthomorphosis of the parabola $y^2 = 4(1-x)$ into the circle $x_1^2 + y_1^2 - 1 = 0$ is given by the equation $x_1 + iy_1 = \tan^2\tfrac{1}{4}\pi\sqrt{(x+iy)}$.

Consider the ellipse, where $a^2 - b^2 = 1$, or say

$$\frac{x^2}{\tfrac{1}{4}\left(M + \dfrac{1}{M}\right)^2} + \frac{y^2}{\tfrac{1}{4}\left(M - \dfrac{1}{M}\right)^2} = 1.$$

I show, by a less direct process, how to express this equation in the required form $\phi\xi + \phi\eta = 0$. In fact, writing

$$\xi = x + iy, \quad \eta = x - iy,$$

the equation of the ellipse is the rationalised form of

$$i\eta + \sqrt{(1-\eta^2)} = M^2\{i\xi + \sqrt{(1-\xi^2)}\}.$$

To show that this is so, call for a moment the right-hand side Ω, the equation is

$$\sqrt{(1-\eta^2)} = \Omega - i\eta,$$

hence

$$1-\eta^2 = \Omega^2 - 2\Omega i\eta - \eta^2,$$

$$2\Omega i\eta = \Omega^2 - 1,$$

or

$$2i\eta = \Omega - \frac{1}{\Omega} = M^2\{i\xi + \sqrt{(1-\xi^2)}\} + \frac{1}{M^2}\{i\xi - \sqrt{(1-\xi^2)}\},$$

$$= \left(M^2 + \frac{1}{M^2}\right) i\xi + \left(M^2 - \frac{1}{M^2}\right)\sqrt{(1-\xi^2)},$$

therefore

$$2i\eta - \left(M^2 + \frac{1}{M^2}\right) i\xi = \left(M^2 - \frac{1}{M^2}\right)\sqrt{(1-\xi^2)}$$

$$-4\eta^2 + 4\left(M^2 + \frac{1}{M^2}\right)\xi\eta + \left(M^4 + 2 + \frac{1}{M^4}\right)(-\xi)^2 = \left(M^4 - 2 + \frac{1}{M^4}\right) - \left(M^4 - 2 + \frac{1}{M^4}\right)\xi^2,$$

that is,

$$-4\eta^2 - 4\xi^2 + 4\left(M^2 + \frac{1}{M^2}\right)\xi\eta = \left(M^2 - \frac{1}{M^2}\right)^2,$$

or say

$$-\xi^2 - \eta^2 + \left(M^2 + \frac{1}{M^2}\right)\xi\eta - \tfrac{1}{4}\left(M^2 - \frac{1}{M^2}\right)^2 = 0:$$

viz. substituting for ξ, η their values, this is

$$-2(x^2 - y^2) + \left(M^2 + \frac{1}{M^2}\right)(x^2 + y^2) - \tfrac{1}{4}\left(M^2 - \frac{1}{M^2}\right)^2 = 0,$$

that is,

$$\left(M - \frac{1}{M}\right)^2 x^2 + \left(M + \frac{1}{M}\right)^2 y^2 - \tfrac{1}{4}\left(M^2 - \frac{1}{M^2}\right)^2 = 0,$$

or finally, it is

$$\frac{x^2}{\tfrac{1}{4}\left(M + \frac{1}{M}\right)^2} + \frac{y^2}{\tfrac{1}{4}\left(M - \frac{1}{M}\right)^2} - 1 = 0,$$

as it should be.

Starting then from the relation

$$i\eta + \sqrt{(1-\eta^2)} = M^2\{i\xi + \sqrt{(1-\xi^2)}\},$$

and writing

$$\phi\xi = \cos\frac{\pi}{2\log M}\log\{i\xi + \sqrt{(1-\xi^2)}\},$$

we have

$$\phi\eta = \cos\frac{\pi}{2\log M}\log M^2\{i\xi + \sqrt{(1-\xi^2)}\},$$

$$= \cos\frac{\pi}{2\log M}[\log M^2 + \log\{i\xi + \sqrt{(1-\xi^2)}\}]$$

$$= \cos\left[\pi + \frac{\pi}{2\log M}\log\{i\xi - \sqrt{(1-\xi^2)}\}\right]$$

$$= -\cos\frac{\pi}{2\log M}\log\{i\xi + \sqrt{(1-\xi^2)}\}, \quad = -\phi\xi,$$

that is, we have

$$\phi\xi + \phi\eta = 0,$$

as the required transformation of the equation of the ellipse

$$\frac{x^2}{\frac{1}{4}\left(M+\frac{1}{M}\right)^2} + \frac{y^2}{\frac{1}{4}\left(M-\frac{1}{M}\right)^2} = 1.$$

We hence derive the known formula for the orthomorphosis of the ellipse into the circle $x_1^2 + y_1^2 - 1 = 0$.

947.

ON A SYSTEM OF TWO TETRADS OF CIRCLES; AND OTHER SYSTEMS OF TWO TETRADS.

[From the *Proceedings of the Cambridge Philosophical Society*, vol. VIII. (1893), pp. 54—59.]

THE investigations of the present paper were suggested to me by Mr Orr's paper, "The Contacts of certain Systems of Circles," *Proc. Camb. Phil. Soc.*, vol. VII.

1. It is possible to find *in plano* two tetrads of circles, or say four red circles and four blue circles, such that each red circle touches each blue circle: in fact, counting the constants, a circle depends upon 3 constants, or say it has a capacity $=3$; the capacity of the eight circles is thus $=24$; and the postulation or number of conditions to be satisfied is $=16$: the resulting capacity of the system is thus *primâ facie*, $16-24=8$. It will, however, appear that, in the system considered, the true value is $=9$.

2. The *primâ facie* value of the capacity being $=8$, we are not at liberty to assume at pleasure three circles of the system. And, in fact, assuming at pleasure say 3 red circles, then touching each of these we have 8 circles, forming $\frac{1}{24}8.7.6.5, =70$, tetrads of circles: taking at random any one of these tetrads for the blue circles, the remaining red circle has to be determined so as to touch each of the four blue circles, that is, by four instead of three conditions; and there is not in general any red circle satisfying these four conditions. But the 8 tangent circles do not stand to each other in a relation of symmetry, but form in fact four pairs of circles; and it is possible out of the 70 tetrads to select (and that in 6 ways) a tetrad of blue circles, such that there exists a fourth red circle touching each of these four blue circles. We have thus a system depending upon 3 arbitrary circles, and for which, therefore, the capacity is $=9$. It is (as is known) possible, in quite a

different manner, out of the 70 tetrads to select (and that in 8 ways) a tetrad of blue circles such that there exists a fourth red circle touching each of these four blue circles—but the present paper relates exclusively to the first-mentioned 6 tetrads and not to these 8 tetrads.

3. I consider, in the first instance, a particular case in which the three red circles are not all of them arbitrary, but have a capacity $9-1, =8$; and pass from this to the general case where the capacity is $=9$. Calling the red circles 1, 2, 3 and 4; I start with the circles 1, and 2 arbitrary, and 3 a circle equal to 2: the radical axis, or common chord, of the circles 2 and 3 is thus a line bisecting at right angles the line joining the centres of the circles 2 and 3, say this is the line Ω. We have then four circles, each having its centre in the line Ω and touching the circles 1 and 2: in fact, the locus of the centre of a circle touching the circles 1 and 2 is a pair of conics, each of them having for foci the centres of these circles: the line Ω meets each of these conics in two points, and there are thus on the line Ω four points, each of them the centre of a circle touching the circles 1 and 2. But the equal circles 2 and 3 are symmetrically situate in regard to the line Ω; and it is obvious that the four circles, having their centres on the line Ω, will each of them also touch the circle 3; we have thus the four blue circles, each of them with its centre on the line Ω and touching each of the red circles 1, 2 and 3. And it is moreover clear that, taking the red circle 4 equal to 1 and situate symmetrically therewith in regard to the line Ω, then this circle 4 will touch each of the blue circles: so that we have here the four blue circles, each of them touching the four red circles. As already mentioned, the blue circles have their centre on the line Ω, that is, the line Ω is a common orthotomic of the four blue circles.

4. By inverting in regard to an arbitrary circle we pass to the general case; the line Ω becomes thus a circle Ω, orthotomic to each of the blue circles.

Starting *ab initio*, we have here at pleasure the red circles 1, 2, 3: the circle Ω is a circle having for centre a centre of symmetry of the circles 2 and 3, and passing through the points of intersection (real or imaginary) of these two circles; the circles 2 and 3 are thus the inverses (or say the images) each of the other in regard to the circle Ω. We can then find 4 circles each of them orthotomic to Ω, and touching the circles 1 and 2: but a circle orthotomic to Ω is its own inverse or image in regard to Ω; and it will thus touch the circle 3 which is the image of 2 in regard to Ω. We have thus the four blue circles each of them touching the red circles 1, 2 and 3; and then, taking the red circle 4 as the inverse or image of 1 in regard to Ω, this circle 4 will also touch each of the blue circles. Thus starting with the arbitrary red circles 1, 2, 3, we find the four blue circles and the remaining red circle 4, such that each of the blue circles touches each of the red circles. Since in the construction we group together at pleasure the two circles 2, 3 (out of the three circles 1, 2, 3) and use at pleasure either of the two centres of symmetry, it appears that the number of ways in which the figure might have been completed is $=6$.

5. The blue circles have a common orthotomic circle Ω, that is, the radical axis or common chord of each two of the blue circles passes through one and the same point, the centre of the circle Ω. The figure is symmetrical in regard to the red and blue circles respectively, and thus the red circles have a common orthotomic circle Ω', that is, the radical axis or common chord of each two of the red circles passes through one and the same point, the centre of the circle Ω'.

6. Projecting stereographically on a spherical surface, the four red circles and the four blue circles become circles of the sphere; and then making the general homographic transformation, they become plane sections of a quadric surface; we have thus the theorem that on a given quadric surface it is possible to find four red sections and four blue sections such that each blue section touches each red section; and moreover the capacity of the system is $=9$; viz. 3 of the red sections may be assumed at pleasure. But (as is well known) the theory of the tangency of plane sections of a quadric surface is far more simple than that of the tangency of circles: the condition in order that two sections may touch each other is simply the condition that the line of intersection of the two planes shall touch the quadric surface. And we construct, as follows, the sections touching each of three given sections: say the given sections are 1, 2, 3; through the sections 1 and 2 we have two quadric cones having for vertices say the points d_{12} and i_{12} (direct and inverse centres of the two sections): similarly, through the sections 1 and 3 we have two quadric cones vertices d_{13} and i_{13} respectively, and through the sections 2 and 3 we have two quadric cones vertices d_{23} and i_{23} respectively; the points d_{12}, i_{12}, d_{13}, i_{13}, d_{23}, i_{23} lie three and three in four intersecting lines or axes, viz. these are $d_{23}d_{31}d_{12}$, $d_{23}i_{31}i_{12}$, $d_{31}i_{12}i_{23}$, $d_{12}i_{23}i_{31}$ respectively. Through any one of these axes, say $d_{23}d_{31}d_{12}$, we may draw to the quadric surface two tangent planes each touching the three cones which have their vertices in the points d_{23}, d_{31}, d_{12} respectively; and the section by either tangent plane is thus a section touching each of the three given sections 1, 2, 3; we have thus the eight tangent sections of these three sections.

7. Taking as three of the red sections the arbitrary sections 1, 2, 3; and grouping together two at pleasure of these sections, say 2 and 3; we may take for the blue sections the two sections through the axis $d_{23}d_{31}d_{12}$, and those through the axis $d_{23}i_{31}i_{12}$; we have thus the four blue sections touching each of the given red sections 1, 2, 3; and this being so, there exists a remaining red section 4 touching each of the blue sections; we have thus the four blue sections touching each of the red sections 1, 2, 3 and 4. This implies that the vertices or points d_{24} and d_{34} lie on the axis $d_{23}d_{31}d_{12}$, and that the vertices or points i_{24} and i_{34} lie on the axis $d_{23}i_{21}i_{31}$; or, what is the same thing, that the four sections 1, 2, 3, 4 have in common an axis $d_{23}d_{21}d_{31}d_{24}d_{34}$ and also an axis $d_{23}i_{21}i_{31}i_{24}i_{34}$.

8. If the quadric surface be a flat surface (*surface aplatie*) or conic, then the red sections become chords of the conic; the axes are lines in the plane of the conic, and thus the tangent planes through an axis each coincide with the plane of the conic, and it would at first sight appear that any theorem as to tangency becomes nugatory. But this is not so; comparing with the last preceding paragraph, we still have the theorem: on a given conic, taking at pleasure any three chords

1, 2, 3, it is possible to find a fourth chord 4, such that the four chords have in common an axis $d_{23}d_{21}d_{31}d_{24}d_{34}$ and also an axis $d_{23}i_{21}i_{31}i_{24}i_{34}$. And the analytical theory (although somewhat complex) is extremely interesting. Considering the conic $xz - y^2 = 0$, the coordinates of a point on the conic are given by $x : y : z = 1 : \theta : \theta^2$, or say any point of the conic is determined by its parameter θ; and this being so, considering any three chords 1, 2, 3, I take for the two extremities of 1 the values ϵ, ζ; for those of 2 the values α, β; and for those of 3 the values γ, δ; the remaining chord 4 is to be determined as above, and I take for its two extremities the values E, Z.

9. Starting with the chords 1, 2, 3, we have each of the points d_{23}, &c., as the intersection of two lines, viz. these are

$$\text{for } d_{23}\begin{cases} x\alpha\delta - y(\alpha+\delta) + z = 0, \\ x\beta\gamma - y(\beta+\gamma) + z = 0,\end{cases} \quad \text{for } i_{23}\begin{cases} x\alpha\gamma - y(\alpha+\gamma) + z = 0, \\ x\beta\delta - y(\beta+\delta) + z = 0,\end{cases}$$

$$\text{,, } d_{12}\begin{cases} x\alpha\zeta - y(\alpha+\zeta) + z = 0, \\ x\beta\epsilon - y(\beta+\epsilon) + z = 0,\end{cases} \quad \text{,, } i_{12}\begin{cases} x\alpha\epsilon - y(\alpha+\epsilon) + z = 0, \\ x\beta\zeta - y(\beta+\zeta) + z = 0,\end{cases}$$

$$\text{,, } d_{13}\begin{cases} x\gamma\zeta - y(\gamma+\zeta) + z = 0, \\ x\delta\epsilon - y(\delta+\epsilon) + z = 0,\end{cases} \quad \text{,, } i_{13}\begin{cases} x\gamma\epsilon - y(\gamma+\epsilon) + z = 0, \\ x\delta\zeta - y(\delta+\zeta) + z = 0,\end{cases}$$

and we thence find without difficulty for the axis $d_{23}d_{12}d_{13}$ the equation

$$\begin{aligned} &x\{\alpha\beta\,(\delta\epsilon - \gamma\zeta) + \gamma\delta\,(\zeta\alpha - \epsilon\beta) + \epsilon\zeta\,(\beta\gamma - \alpha\delta)\} \\ &+ y\{(\beta-\alpha)(\gamma\delta - \epsilon\zeta) + (\delta-\gamma)(\epsilon\zeta - \alpha\beta) + (\zeta-\epsilon)(\alpha\beta - \gamma\delta)\} \\ &+ z\{\quad - \quad (\delta\epsilon - \gamma\zeta) - \quad (\zeta\alpha - \epsilon\beta) - \quad (\beta\gamma - \alpha\delta)\} = 0\,; \end{aligned}$$

the equation of the axis $d_{23}i_{12}i_{13}$ is obtained herefrom by the interchange of ϵ and ζ.

10. The points d_{24} and d_{34} will lie upon the first-mentioned axis if only d_{24} lies upon this axis, viz. if we have

$$\begin{aligned} &\{\quad \alpha\beta(\delta\epsilon - \gamma\zeta) + \quad \gamma\delta(\zeta\alpha - \epsilon\beta) + \quad \epsilon\zeta(\beta\gamma - \alpha\delta)\}\,(\beta + E - \alpha - Z) \\ &+ \{(\beta-\alpha)(\gamma\delta - \epsilon\zeta) + (\delta-\gamma)(\epsilon\zeta - \alpha\beta) + (\zeta-\epsilon)(\alpha\beta - \gamma\delta)\}\,(\beta E - \alpha Z) \\ &+ \{\quad - \quad (\delta\epsilon - \gamma\zeta) - \quad (\zeta\alpha - \epsilon\beta) - \quad (\beta\gamma - \alpha\delta)\}\,\{-\alpha Z(\beta + E) + \beta E(\alpha + Z)\} = 0. \end{aligned}$$

Reducing this equation, the factor $\alpha - \beta$ divides out, and we finally obtain

$$\begin{aligned} &(\gamma-\alpha)(\beta\delta + \zeta Z)(\epsilon - E) \\ &+ (\beta-\delta)(\alpha\gamma + \epsilon E)(\zeta - Z) \\ &+ (\alpha\delta - \beta\gamma)(\epsilon\zeta - EZ) \\ &+ (\alpha\beta - \gamma\delta)(\epsilon Z - \zeta E) = 0\,; \end{aligned}$$

say this is

$$A + BE + CZ + DEZ = 0,$$

where

$$\begin{aligned} A &= \quad . \quad (\gamma-\alpha)\,\beta\delta\epsilon + (\beta-\delta)\,\alpha\gamma\zeta + (\alpha\delta - \beta\gamma)\,\epsilon\zeta, \\ B &= -(\gamma-\alpha)\,\beta\delta \quad . \quad -(\alpha\beta - \gamma\delta)\,\zeta + (\beta-\delta)\,\epsilon\zeta, \\ C &= -(\beta-\delta)\,\alpha\gamma + (\alpha\beta - \gamma\delta)\,\epsilon \quad . \quad + (\gamma-\alpha)\,\epsilon\zeta, \\ D &= -(\alpha\delta - \beta\gamma) - (\beta-\delta)\,\epsilon - (\gamma-\alpha)\,\zeta \quad . \quad , \end{aligned}$$

viz. this is the condition for the existence of the axis $d_{23}d_{13}d_{12}d_{43}d_{42}$.

We interchange herein ϵ, ζ and also E, Z, and we thus obtain

$$A' + C'E + B'Z + D'EZ = 0,$$

where

$$\begin{aligned}
A' &= \quad . \quad (\beta-\delta)\,\alpha\gamma\epsilon + (\gamma-\alpha)\,\beta\delta\zeta + (\alpha\delta-\beta\gamma)\,\epsilon\zeta,\\
B' &= -(\gamma-\alpha)\,\beta\delta - (\alpha\beta-\gamma\delta)\,\epsilon \quad . \quad + \quad (\beta-\delta)\,\epsilon\zeta,\\
C' &= -(\beta-\delta)\,\alpha\gamma \quad . \quad + (\alpha\beta-\gamma\delta)\,\zeta + \quad (\gamma-\alpha)\,\epsilon\zeta,\\
D' &= -(\alpha\delta-\beta\gamma) - (\gamma-\alpha)\,\epsilon \quad - (\beta-\delta)\,\zeta \quad . \quad ,
\end{aligned}$$

viz. this is the condition for the existence of the axis $d_{23}i_{13}i_{12}i_{43}i_{42}$.

11. I remark that we have

$$A' = A + \mathrm{a}\,(\epsilon-\zeta),\quad B' = B + \mathrm{b}\,(\epsilon-\zeta),$$
$$C' = C + \mathrm{c}\,(\epsilon-\zeta),\quad D' = D + \mathrm{d}\,(\epsilon-\zeta),$$

where

$$\begin{aligned}
\mathrm{a} &= (\beta-\delta)\,\alpha\gamma - (\gamma-\alpha)\,\beta\delta = \alpha\beta\,(\gamma+\delta) - \gamma\delta\,(\alpha+\beta),\\
\mathrm{b} &= \mathrm{c} = \gamma\delta - \alpha\beta,\\
\mathrm{d} &= \alpha + \beta - \gamma - \delta,
\end{aligned}$$

and further that

$$B - C = \alpha\beta\,(\gamma+\delta) - \gamma\delta\,(\alpha+\beta) + (\gamma\delta-\alpha\beta)\,(\epsilon+\zeta) + (\alpha+\beta-\gamma-\delta)\,\epsilon\zeta,$$

say this is Π.

12. It thus appears that, for the determination of E, Z, we have

$$A + BE + CZ + DEZ = 0,$$
$$A' + C'E + B'Z + D'EZ = 0.$$

Eliminating Z, we find

$$\frac{A+BE}{A'+C'E} = \frac{C+DE}{B'+D'E},$$

that is,

$$(AC' - A'C) + (AD' - A'D + BB' - CC')\,E + (BD' - B'D)\,E^2 = 0\,;$$

upon reducing the coefficients of this equation it appears that they contain each of them the factor Π, and throwing out this factor, the equation is

$$\begin{aligned}
&\epsilon\,[\alpha\beta\,(\gamma-\delta) + \gamma\delta\,(\beta-\alpha) + (\alpha\delta-\beta\gamma)\,\zeta]\\
&+ [-\alpha\beta\,(\gamma-\delta) - \gamma\delta\,(\beta-\alpha) + (\alpha\delta-\beta\gamma)\,\epsilon + (\beta\gamma-\alpha\delta)\,\zeta + (\beta+\gamma-\alpha-\delta)\,\epsilon\zeta]\,E\\
&+ [\beta\gamma - \alpha\delta - (\beta+\gamma-\alpha-\delta)]\,E^2 = 0\,;
\end{aligned}$$

this contains obviously the factor $E-\epsilon$, or throwing out this factor, we have for E the simple equation

$$\{\alpha\beta\,(\gamma-\delta) + \gamma\delta\,(\beta-\alpha)\} + (\alpha\delta-\beta\gamma)\,(E+\zeta) + (\beta-\alpha+\gamma-\delta)\,\zeta E = 0.$$

In a similar manner it may be shown that the two equations give for Z the like simple equation

$$\{\alpha\beta\,(\gamma-\delta) + \gamma\delta\,(\beta-\alpha)\} + (\alpha\delta-\beta\gamma)\,(Z+\epsilon) + (\beta-\alpha+\gamma-\delta)\,\epsilon Z = 0,$$

viz. starting from the chords 1, 2, 3 which depend on the parameters $(\epsilon,\ \zeta)$, $(\alpha,\ \beta)$, $(\gamma,\ \delta)$ respectively, these last two equations give the parameters $(E,\ Z)$ of the chord 4.

948.

REPORT OF A COMMITTEE APPOINTED FOR THE PURPOSE OF CARRYING ON THE TABLES CONNECTED WITH THE PELLIAN EQUATION FROM THE POINT WHERE THE WORK WAS LEFT BY DEGEN IN 1817.

[From the *British Association Report*, (1893), pp. 73—120.]

We have, on the Pellian Equation, Degen's tables, the title of which is "Canon Pellianus sive Tabula simplicissimam æquationis celebratissimæ $y^2 = ax^2 + 1$ solutionem pro singulis numeri dati valoribus ab 1 usque ad 1000 in numeris rationalibus iisdemque integris exhibens." Autore Carolo Ferdinando Degen. Hafniæ, apud Gerhardum Bonnierum, MDCCCXVII., 8vo. Introductio, pp. v—xxiv. Tabula I. Solutionem æquationis $y^2 - ax^2 - 1 = 0$ exhibens, pp. 3—106. Tabula II. Solutionem æquationis $y^2 - ax^2 + 1 = 0$, quotiescunque valor ipsius a talem admiserit, exhibens, pp. 109—112.

The mode of calculation is explained in the Introduction, and illustrated by the examples of the numbers 209, 173.

As to the first of these, the entry in Table I. is

209	14, 2, 5, 3, (2) 1, 13, 5, 8, 11 3220 46551

where the first line gives the expression of $\sqrt{209}$ as a continued fraction, viz. we have

$$\sqrt{209} = 14 + \frac{1}{2+}\ \frac{1}{5+}\ \frac{1}{3+}\ \frac{1}{2+}\ \frac{1}{3+}\ \frac{1}{5+}\ \frac{1}{2+}\ \frac{1}{28+}\ \frac{1}{2+}\ \&c.,$$

the denominators being 2, 5, 3, (2), 3, 5, 2, then 28, which is the double of the integer part 14, and then again 2, 5, 3, (2), 3, 5, 2, and so on, the parentheses of the (2) being used to indicate that this is the middle term of the period.

The second row gives auxiliary numbers occurring in the calculation of the first row and having a meaning, as will presently appear. Observe that the 11 which comes under the (2) should also be printed in parentheses (11), but this is not done.

The process for the calculation of the x, y is as follows:

209			
14	1	0	+ 1
2	14	1	− 13
5	29	2	+ 5
3	159	11	− 8
(2)	506	35	+ (11)
3	1171	81	− 8
5	4019	278	+ 5
2	21266	1471	− 13
28	46551	3220	+ 1

viz. writing down as a first column the numbers of the first row, and beginning the second column with 1, 14 (14 the number at the head of the first column), and the third column with 0, 1, we calculate the numbers of the second column, $29 = 2 . 14 + 1$, $159 = 5 . 29 + 14$, $506 = 3 . 159 + 29$, &c., and the numbers of the third column in like manner, $2 = 2 . 1 + 0$, $11 = 5 . 2 + 1$, $35 = 3 . 11 + 2$, &c.; and then writing down as a fourth column the numbers of the second row with the signs +, − alternately, we have a series of equations $y^2 - ax^2 = \pm A$, viz.

$$1^2 - 209 . 0^2 = + 1,$$

$$14^2 - 209 . 1^2 = - 13,$$

$$29^2 - 209 . 2^2 = + 5,$$

$$\vdots$$

the last of them being

$$(46551)^2 - 209\,(3220)^2 = + 1,$$

this last corresponding as above to the value + 1, and the numbers 46551 and 3220 being accordingly the y and x given in the fourth and third rows of the table.

As to the second of the foregoing numbers, 173, the only difference is that the period has a double middle term, viz. the entry in the Table I. is

173	13, 6, (1, 1) 1, 4, (13, 13) 190060 2499849

The first row gives the expression of $\sqrt{173}$, viz. that is

$$\sqrt{173} = 13 + \frac{1}{6+} \quad \frac{1}{(1)+} \quad \frac{1}{(1)+} \quad \frac{1}{6+} \quad \frac{1}{26+} \text{ \&c.},$$

the denominators being 6, 1, 1, 6, then 26 (the double of the integer part 13), and then again 6, 1, 1, 6, and so on. In the second row I remark that Degen prints the parentheses (13, 13) for the double middle term.

The process for the calculation of the x, y is similar to that in the former case, viz. we have

173			
13	1	0	+ 1
6	13	1	− 4
(1	79	6	+ 13
1)	92	7	− 13
6	171	13	+ 4
26	1118	85	− 1

where the second and third columns begin 1, 13 and 0, 1 respectively, and the remaining terms are calculated $79 = 6.13 + 1$, $92 = 1.79 + 13$, &c., and $6 = 6.1 + 0$, $7 = 1.6 + 1$, &c.; and then writing down as a fourth column the terms of the second row with the signs +, − alternately, we have

$$1^2 - 173.0^2 = + 1,$$
$$13^2 - 173.1^2 = - 4,$$
$$79^2 - 173.6^2 = + 13,$$
$$\vdots$$

the last equation being

$$(1118)^2 - 173\,(85)^2 = - 1,$$

the term for the last equation being always in a case such as the present one, not $+1$, but -1. The final numbers 1118, 85 are consequently entered not in Table I., but in Table II., viz. the entry in this table is

173	85 1118

and thence we calculate the numbers y, x of Table I., viz. these are

$$2499849 = 2 \,.\, (1118)^2 + 1,$$
$$190060 = 2 \,.\, 1118 \,.\, 85.$$

Generally Table II. gives for each value of a, comprised therein, values of x, y, such that $y^2 = ax^2 - 1$, and then writing $y_1 = 2y^2 + 1$, $x_1 = 2xy$, we have

$$y_1^2 = (2ax^2 - 1)^2 = 4a^2x^4 - 4ax^2 + 1 = a \,.\, 4x^2 (ax^2 - 1) + 1 = ax_1^2 + 1,$$

so that x_1, y_1 are for the same value of a the values of x, y in Table I.

It is to be remarked that the heading of Table II. is not perfectly accurate, for it purports to give for every value of a, for which a solution exists, a solution of the equation $y^2 = ax^2 - 1$. What it really gives is the solution for each value of a for which the period has a double middle term. But if $a = \alpha^2 + 1$, then obviously we have a solution $y = \alpha$, $x = 1$, and for any such value of a the period has a single middle term, viz. the entry in Table I. is

$\alpha^2 + 1$	α, (2α) 1, 1 2α $2\alpha^2 + 1$

and we, in fact, have

α^2+1			
α	1	0	$+1$
(2α)	α	1	-1
2α	$2\alpha^2 + 1$	2α	$+1$

that is,

$$1^2 - (\alpha^2 + 1)\, 0^2 \quad = +1,$$
$$\alpha^2 - (\alpha^2 + 1)\, 1^2 \quad = -1,$$
$$(2\alpha^2 + 1)^2 - (\alpha^2 + 1)\, (2\alpha)^2 = +1.$$

The foregoing instances of the calculation of x, y in the case of the numbers 209 and 173 suggest a table which may be regarded as an extended form of Degen's tables; viz. such a table, from $a=2$ to $a=99$, is as follows:

SPECIMEN OF EXTENDED FORM OF TABLE IN REGARD TO THE PELLIAN EQUATION.

a		y	x	y^2-ax^2
2	1	1	0	+ 1
	(2)	1	1	− 1
	2	3	2	+ 1
3	1	1	0	+ 1
	(1)	1	1	− 2
	2	2	1	+ 1
5	2	1	0	+ 1
	(4)	2	1	− 1
	4	9	4	+ 1
6	2	1	0	+ 1
	(2)	2	1	− 2
	4	5	2	+ 1
7	2	1	0	+ 1
	1	2	1	− 3
	(1)	3	1	+ 2
	1	5	2	− 3
	4	8	3	+ 1
8	2	1	0	+ 1
	(1)	2	1	− 4
	4	3	1	+ 1
10	3	1	0	+ 1
	(6)	3	1	− 1
	6	19	6	+ 1
11	3	1	0	+ 1
	(3)	3	1	− 2
	6	10	3	+ 1
12	3	1	0	+ 1
	(2)	3	1	− 3
	6	7	2	+ 1

a		y	x	y^2-ax^2
13	3	1	0	+ 1
	1	3	1	− 4
	(1	4	1	+ 3
	1)	7	2	− 3
	1	11	3	+ 4
	6	18	5	− 1
14	3	1	0	+ 1
	1	3	1	− 5
	(2)	4	1	+ 2
	1	11	3	− 5
	6	15	4	+ 1
15	3	1	0	+ 1
	(1)	3	1	− 6
	6	4	1	+ 1
17	4	1	0	+ 1
	(8)	4	1	− 1
	8	33	8	+ 1
18	4	1	0	+ 1
	(4)	4	1	− 2
	8	17	4	+ 1
19	4	1	0	+ 1
	2	4	1	− 3
	1	9	2	+ 5
	(3)	13	3	− 2
	1	48	11	+ 5
	2	61	14	− 3
	8	170	39	+ 1
20	4	1	0	+ 1
	(2)	4	1	− 4
	8	9	2	+ 1

SPECIMEN OF EXTENDED FORM OF PELLIAN EQUATION TABLE—*continued.*

a		y	x	y^2-ax^2
21	4	1	0	+ 1
	1	4	1	− 5
	1	5	1	+ 4
	(2)	9	2	− 3
	1	23	5	+ 4
	1	32	7	− 5
	8	55	12	+ 1
22	4	1	0	+ 1
	1	4	1	− 6
	2	5	1	+ 3
	(4)	14	3	− 2
	2	61	13	+ 3
	1	136	29	− 6
	8	197	42	+ 1
23	4	1	0	+ 1
	1	4	1	− 7
	(3)	5	1	+ 2
	1	19	4	− 7
	8	24	5	+ 1
24	4	1	0	+ 1
	(1)	4	1	− 8
	8	5	1	+ 1
26	5	1	0	+ 1
	(10)	5	1	− 1
	10	51	10	+ 1
27	5	1	0	+ 1
	(5)	5	1	− 2
	10	26	5	+ 1
28	5	1	0	+ 1
	3	5	1	− 3
	(2)	16	3	+ 4
	3	37	7	− 3
	10	127	24	+ 1
29	5	1	0	+ 1
	2	5	1	− 6
	(1	11	2	+ 5
	1)	16	3	− 3
	2	27	5	+ 2
	10	70	13	− 1
30	5	1	0	+ 1
	(2)	5	1	− 5
	10	11	2	+ 1
31	5	1	0	+ 1
	1	5	1	− 6
	1	6	1	+ 5
	3	11	2	− 3
	(5)	39	7	+ 2
	3	206	37	− 3
	1	657	118	+ 5
	1	863	155	− 6
	10	1520	273	+ 1
32	5	1	0	+ 1
	1	5	1	− 7
	(1)	6	1	+ 4
	1	11	2	− 7
	10	17	3	+ 1
33	5	1	0	+ 1
	1	5	1	− 8
	(2)	6	1	+ 3
	1	17	3	− 8
	10	23	4	+ 1
34	5	1	0	+ 1
	1	5	1	− 9
	(4)	6	1	+ 2
	1	29	5	− 9
	10	35	6	+ 1

SPECIMEN OF EXTENDED FORM OF PELLIAN EQUATION TABLE—*continued.*

a		y	x	y^2-ax^2
35	5	1	0	+ 1
	(1)	5	1	− 10
	10	6	1	+ 1
37	6	1	0	+ 1
	(12)	6	1	− 1
	12	73	12	+ 1
38	6	1	0	+ 1
	(6)	6	1	− 2
	12	37	6	+ 1
39	6	1	0	+ 1
	(4)	6	1	− 3
	12	25	4	+ 1
40	6	1	0	+ 1
	(3)	6	1	− 4
	12	19	3	+ 1
41	6	1	0	+ 1
	(2	6	1	− 5
	2)	13	2	+ 5
	12	32	5	− 1
42	6	1	0	+ 1
	(2)	6	1	− 6
	12	13	2	+ 1
43	6	1	0	+ 1
	1	6	1	− 7
	1	7	1	+ 6
	3	13	2	− 3
	1	46	7	+ 9
	(5)	59	9	− 2
	1	341	52	+ 9
	3	400	61	− 3
	1	1541	235	+ 6
	1	1941	296	− 7
	12	3482	531	+ 1

a		y	x	y^2-ax^2
44	6	1	0	+ 1
	1	6	1	− 8
	1	7	1	+ 5
	1	13	2	− 7
	(2)	20	3	+ 4
	1	53	8	− 7
	1	73	11	+ 5
	1	126	19	− 8
	12	199	30	+ 1
45	6	1	0	+ 1
	1	6	1	− 9
	2	7	1	+ 4
	(2)	20	3	− 5
	2	47	7	+ 4
	1	114	17	− 9
	12	161	24	+ 1
46	6	1	0	+ 1
	1	6	1	− 10
	3	7	1	+ 3
	1	27	4	− 7
	1	34	5	+ 6
	2	61	9	− 5
	(6)	156	23	+ 2
	2	997	147	− 5
	1	2150	317	+ 6
	1	3147	464	− 7
	3	5297	781	+ 3
	1	19038	2807	− 10
	12	24335	3588	+ 1
47	6	1	0	+ 1
	1	6	1	− 1
	(5)	7	1	+ 2
	1	41	6	− 11
	12	48	7	+ 1
48	6	1	0	+ 1
	(1)	6	1	− 12
	12	7	1	+ 1

SPECIMEN OF EXTENDED FORM OF PELLIAN EQUATION TABLE—*continued.*

a		y	x	y^2-ax^2
50	7	1	0	+ 1
	(14)	7	1	− 1
	14	99	14	+ 1
51	7	1	0	+ 1
	(7)	7	1	− 2
	14	50	7	+ 1
52	7	1	0	+ 1
	4	7	1	− 3
	1	29	4	+ 9
	(2)	36	5	− 4
	1	101	14	+ 9
	4	137	19	− 3
	14	649	90	+ 1
53	7	1	0	+ 1
	3	7	1	− 4
	(1	22	3	+ 7
	1)	29	4	− 7
	3	51	7	+ 4
	14	182	25	− 1
54	7	1	0	+ 1
	2	7	1	− 5
	1	15	2	+ 9
	(6)	22	3	− 2
	1	147	20	+ 9
	2	169	23	− 5
	14	485	66	+ 1
55	7	1	0	+ 1
	2	7	1	− 6
	(2)	15	2	+ 5
	2	37	5	− 6
	14	89	12	+ 1
56	7	1	0	+ 1
	(2)	7	1	− 7
	14	15	2	+ 1

a		y	x	y^2-ax^2
57	7	1	0	+ 1
	1	7	1	− 8
	1	8	1	+ 7
	(4)	15	2	− 3
	1	68	9	+ 7
	1	83	11	− 8
	14	151	20	+ 1
58	7	1	0	+ 1
	1	7	1	− 9
	1	8	1	+ 6
	(1	15	2	− 7
	1)	23	3	+ 7
	1	38	5	− 6
	1	61	8	+ 9
	14	99	13	− 1
59	7	1	0	+ 1
	1	7	1	− 10
	2	8	1	+ 5
	(7)	23	3	− 2
	2	169	22	+ 5
	1	361	47	− 10
	14	530	69	+ 1
60	7	1	0	+ 1
	1	7	1	− 11
	(2)	8	1	+ 4
	1	23	3	− 11
	14	31	4	+ 1
61	7	1	0	+ 1
	1	7	1	− 12
	4	8	1	+ 3
	3	39	5	− 4
	1	125	16	+ 9
	(2	164	21	− 5
	2)	453	58	+ 5
	1	1070	137	− 9

SPECIMEN OF EXTENDED FORM OF PELLIAN EQUATION TABLE—*continued.*

a		y	x	y^2-ax^2
	3	1523	195	+ 4
	4	5639	722	− 3
	1	24079	3083	+ 12
	14	29718	3805	− 1
62	7	1	0	+ 1
	1	7	1	− 13
	(6)	8	1	+ 2
	1	55	7	− 13
	14	63	8	+ 1
63	7	1	0	+ 1
	(1)	7	1	− 14
	14	8	1	+ 1
65	8	1	0	+ 1
	(16)	8	1	− 1
	16	129	16	+ 1
66	8	1	0	+ 1
	(8)	8	1	− 2
	16	65	8	+ 1
67	8	1	0	+ 1
	5	8	1	− 3
	2	41	5	+ 6
	1	90	11	− 7
	1	131	16	+ 9
	(7)	221	27	− 2
	1	1678	205	+ 9
	1	1899	232	− 7
	2	3577	437	+ 6
	5	9053	1106	− 3
	16	48842	5967	+ 1
68	8	1	0	+ 1
	(4)	8	1	− 4
	16	33	4	+ 1
69	8	1	0	+ 1
	3	8	1	− 5

a		y	x	y^2-ax^2
	3	25	3	+ 4
	1	83	10	− 11
	(4)	108	13	+ 3
	1	515	62	− 11
	3	623	75	+ 4
	3	2384	297	− 5
	16	7775	936	+ 1
70	8	1	0	+ 1
	2	8	1	− 6
	1	17	2	+ 9
	(2)	25	3	− 5
	1	67	8	+ 9
	2	92	11	− 6
	16	251	30	+ 1
71	8	1	0	+ 1
	2	8	1	− 7
	2	17	2	+ 5
	1	42	5	− 11
	(7)	59	7	+ 2
	1	455	54	− 11
	2	514	61	+ 5
	2	1483	176	− 7
	16	3480	413	+ 1
72	8	1	0	+ 1
	(2)	8	1	− 3
	16	17	2	+ 1
73	8	1	0	+ 1
	1	8	1	− 9
	1	9	1	+ 8
	(5	17	2	− 3
	5)	94	11	+ 3
	1	487	57	− 8
	1	581	68	+ 9
	16	1068	125	− 1
74	8	1	0	+ 1
	1	8	1	− 10

SPECIMEN OF EXTENDED FORM OF PELLIAN EQUATION TABLE—*continued.*

a		y	x	y^2-ax^2
	(1	9	1	+ 7
	1)	17	2	− 7
	1	26	3	+ 10
	16	43	5	− 1
75	8	1	0	+ 1
	1	8	1	− 11
	(1)	9	1	+ 6
	1	17	2	− 11
	16	26	3	+ 1
76	8	1	0	+ 1
	1	8	1	− 12
	2	9	1	+ 5
	1	26	3	− 8
	1	35	4	+ 9
	5	61	7	− 3
	(4)	340	39	+ 4
	5	1421	163	− 3
	1	7445	854	+ 9
	1	8866	1017	− 8
	2	16311	1871	+ 5
	1	41488	4759	− 12
	16	57799	6630	+ 1
77	8	1	0	+ 1
	1	8	1	− 13
	3	9	1	+ 4
	(2)	35	4	− 7
	3	79	9	+ 4
	1	272	31	− 13
	16	351	40	+ 1
78	8	1	0	+ 1
	1	8	1	− 14
	(4)	9	1	+ 3
	1	44	5	− 14
	16	53	6	+ 1
79	8	1	0	+ 1
	1	8	1	− 15
	(7)	9	1	+ 2
	1	71	8	− 15
	16	80	9	+ 1
80	8	1	0	+ 1
	(1)	8	1	− 16
	16	9	1	+ 1
82	9	1	0	+ 1
	(18)	9	1	− 1
	18	163	18	+ 1
83	9	1	0	+ 1
	(9)	9	1	− 2
	18	82	9	+ 1
84	9	1	0	+ 1
	(6)	9	1	− 3
	18	55	6	+ 1
85	9	1	0	+ 1
	4	9	1	− 4
	(1	37	4	+ 9
	1)	46	5	− 9
	4	83	9	+ 4
	18	378	41	− 1
86	9	1	0	+ 1
	3	9	1	− 5
	1	28	3	+ 10
	1	37	4	− 7
	1	65	7	+ 11
	(8)	102	11	− 2
	1	881	95	+ 11
	1	983	106	− 7
	1	1864	201	+ 10
	3	2847	307	− 5
	18	10405	1122	+ 1

SPECIMEN OF EXTENDED FORM OF PELLIAN EQUATION TABLE—*continued.*

a		y	x	y^2-ax^2
87	9	1	0	+ 1
	(3)	9	1	− 6
	18	28	3	+ 1
88	9	1	0	+ 1
	2	9	1	− 7
	1	19	2	+ 9
	(1)	28	3	− 8
	1	47	5	+ 9
	2	75	8	− 7
	18	197	21	+ 1
89	9	1	0	+ 1
	2	9	1	− 8
	(3	19	2	+ 5
	3)	66	7	− 5
	2	217	23	+ 8
	18	500	53	− 1
90	9	1	0	+ 1
	(2)	9	1	− 9
	18	19	2	+ 1
91	9	1	0	+ 1
	1	9	1	− 10
	1	10	1	+ 9
	5	19	2	− 3
	(1)	105	11	+ 14
	5	124	13	− 3
	1	725	76	+ 9
	1	849	89	− 10
	18	1574	165	+ 1
92	9	1	0	+ 1
	1	9	1	− 11
	1	10	1	+ 3
	2	19	2	− 7
	(4)	48	5	+ 4
	2	211	22	− 7
	1	470	49	+ 3
	1	681	71	− 11
	18	1151	120	+ 1

a		y	x	y^2-ax^2
93	9	1	0	+ 1
	1	9	1	− 12
	1	10	1	+ 7
	1	19	2	− 11
	4	29	3	+ 4
	(6)	135	14	− 3
	4	839	87	+ 4
	1	3491	362	− 11
	1	4330	449	+ 7
	1	7821	811	− 12
	18	12151	1260	+ 1
94	9	1	0	+ 1
	1	9	1	− 13
	2	10	1	+ 6
	3	29	3	− 5
	1	97	10	+ 9
	1	126	13	− 10
	5	223	23	+ 3
	1	1241	128	− 15
	(8)	1464	151	+ 2
	1	12953	1336	− 15
	5	14417	1487	+ 3
	1	85038	8771	− 10
	1	99455	10258	+ 9
	3	1 84493	19029	− 5
	2	6 52934	67345	+ 6
	1	14 90361	1 53719	− 13
	18	21 43295	2 21064	+ 1
95	9	1	0	+ 1
	1	9	1	− 14
	(2)	10	1	+ 5
	1	29	3	− 14
	18	39	4	+ 1
96	9	1	0	+ 1
	1	9	1	− 15
	(3)	10	1	+ 4
	1	39	4	− 15
	18	49	5	+ 1

SPECIMEN OF EXTENDED FORM OF PELLIAN EQUATION TABLE—*continued*.

a		y	x	y^2-ax^2	a		y	x	y^2-ax^2
97	9	1	0	+ 1	98	9	1	0	+ 1
	1	9	1	− 16		1	9	1	− 17
	5	10	1	+ 3		(8)	10	1	+ 2
	1	59	6	− 11		1	89	9	− 17
	1	69	7	+ 8		18	99	10	+ 1
	(1	128	13	− 9					
	1)	197	20	+ 9	99	9	1	0	+ 1
	1	325	33	− 8		(1)	9	1	− 18
	1	522	53	+ 11		18	10	1	+ 1
	5	847	86	− 3					
	1	4757	483	+ 16					
	18	5604	569	− 1					

The meaning hardly requires explanation; for each number a, we have a series of pairs of increasing numbers, y, x, satisfying a series of equations $y^2 = ax^2 \pm b$; thus

$a = 14$

y	x	$y^2 - ax^2$
1	0	$1 - 14.0 = 1,$
3	1	$9 - 14.1 = -5,$
4	1	$16 - 14.1 = +2,$
11	3	$121 - 14.9 = -5,$
15	4	$225 - 14.16 = +1.$

The following table, calculated under the superintendence of the Committee, extends from $a = 1001$ to $a = 1500$ (square numbers omitted); it is (with slight typographical variations) nearly but not exactly in the form of Degen's Table I., the chief difference being that for a number a having a double middle term, or of the form a^2+1 (such number being further distinguished by an asterisk), the x, y entered in the table are the solutions, *not* of the equation $y^2 = ax^2 + 1$, but of the equation $y^2 = ax^2 - 1$. As remarked above, if we have $y^2 = ax^2 - 1$, then writing $y_1 = 2y^2 + 1$ and $x_1 = 2xy$, we obtain $y_1^2 = ax_1^2 + 1$.

Moreover, for each value of a, in the first line, the first term, which is the integer part of $\sqrt{a}$, is separated from the other by a semicolon, and the 1, which is the corresponding first term of the second line, is omitted.

The calculations were made by C. E. Bickmore, M.A., of New College, Oxford: his values for x and y have been revised as presently mentioned, but it has been assumed that his values for the periods and subsidiary numbers (forming the first and second lines of each division of the table) are accurate; in fact, any error therein would cause the resulting values of x and y to be wildly erroneous; but (except in a single instance which was accounted for) the errors in x and y were in every case in a single figure or two or three figures only.

The values of x and y were in every case examined by substitution in the equation ($y^2 = ax^2 + 1$, or $y^2 = ax^2 - 1$, as the case may be), which should be satisfied by them. These verifications were for the most part made by A. Graham, M.A., of the Observatory, Cambridge. As already mentioned, some errors were detected, and these have been, of course, corrected. The values of x, y given in the table thus satisfy in every case the proper equation $y^2 = ax^2 + 1$, or $y^2 = ax^2 - 1$; on the ground above referred to, it is believed that the periods and subsidiary numbers are also accurate.

It may be remarked, in regard to the verification of the equation $y^2 = ax^2 \pm 1$ for large values of x and y, it is in practice easier and safer to calculate $ax^2 \pm 1$, and then to compare the square root thereof with the given value of y, than to further calculate the value of y^2.

THE TABLE 1001 TO 1500.

1001	31; 1, 1, 1, 3, 3, 2, (4) 40, 23, 35, 16, 17, 25, (13)	33532 10 60905
1002	31; 1, 1, 1, 8, 2, 1, 1, 1, 3, (10) 41, 22, 39, 7, 23, 31, 26, 33, 17, (6)	65 35248 2068 69247
1003	31; 1, 2, (31) 42, 21, (2)	285 9026
1004	31; 1, 2, 5, 2, 2, 1, 7, 4, 1, 2, 1, 11, 1, (14) 43, 20, 11, 25, 19, 41, 8, 13, 40, 17, 44, 5, 55, (4)	85 24164 59730 2700 96330 24199
1005	31; 1, 2, 2, 1, 5, 15, 1, 2, (12) 44, 19, 20, 39, 11, 4, 41, 21, (5)	930 59568 29501 49761
1006	31; 1, 2, 1, 1, 5, 1, 3, 2, 1, 1, 1, 1, 1, 9, 1, 20, 4, 5, 1, 1, 12, 6, 1, (30) 45, 18, 29, 33, 10, 43, 15, 22, 31, 27, 30, 25, 37, 6, 55, 3, 15, 11, 30, 33, 5, 9, 53, (2)	4 45346 14025 55749 21748 141 25267 56378 02146 05455
1007	31; 1, 2, (1) 46, 17, (38)	15 476
1008	31; 1, (2) 47, (16)	4 127
1009*	31; 1, (3, 3) 48, (15, 15)	17 540
1010*	31; 1, 3, (1, 1) 49, 14, (31, 31)	41 1303
1011	31; 1, 3, 1, (9) 50, 13, 47, (6)	265 8426
1012	31; 1, 4, 3, 6, 1, 3, 8, 1, (4) 51, 12, 19, 9, 43, 16, 7, 48, (11)	1013 02110 32226 17399
1013*	31; 1, 4, 1, 4, 15, 1, 2, (2, 2) 52, 11, 44, 13, 4, 43, 19, (23, 23)	123 52985 3931 66618
1014	31; 1, 5, 2, 1, 1, 1, 1, (20) 53, 10, 23, 30, 29, 25, 38, (3)	1 46266 46 56965
1015	31; 1, 6, 10, (2) 54, 9, 6, (29)	11076 3 52871
1016	31; 1, (6) 55, (8)	8 255
1017	31; 1, 8, 7, 1, 6, 4, 1, 3, 5, 1, 1, (6) 56, 7, 8, 49, 9, 13, 41, 16, 11, 31, 32, (9)	9 09655 84992 290 09322 97217
1018*	31; 1, 9, 1, 1, 1, 6, 2, (3, 3) 57, 6, 39, 23, 38, 9, 26, (17, 17)	27 28333 870 50499
1019	31; 1, 11, 1, 3, 1, 1, 1, 3, 8, 1, 5, 2, (31) 58, 5, 47, 14, 35, 25, 34, 17, 7, 49, 10, 29, (2)	19 07764 36539 608 99233 21730
1020	31; 1, (14) 59, (4)	16 511
1021*	31; 1, 20, 3, 6, 1, 3, 2, 1, 1, 12, 5, 4, 15, 1, 2, 1, 4, 1, 1, 2, 1, 1, 1, (5, 5) 60, 3, 20, 9, 44, 15, 23, 27, 36, 5, 12, 15, 4, 45, 17, 41, 12, 33, 29, 20, 33, 25, 36, (11, 11)	98 65001 29666 69564 06909 3152 17280 37258 48825 15030

TABLE 1001 TO 1500—*continued.*

1022	31 ; 1, (30) 61, (2)	32 1023
1023	31 ; (1) (62)	1 32
1025*	32 ; (64) (1)	1 32
1026	32 ; (32) (2)	32 1025
1027	32 ; 21, 2, 1, 6, 2, (4) 3, 22, 39, 9, 27, (13)	41 54868 1331 50393
1028	32 ; (16) (4)	16 513
1029	32 ; 12, 1, 4, 2, 2, 1, 3, 15, 1, 3, 2, 1, (20) 5, 49, 12, 25, 20, 37, 17, 4, 47, 15, 20, 43, (3)	303 57068 95884 9737 94964 66615
1030	32 ; 10, 1, 2, 6, 1, 3, 1, 2, 1, 1, 2, 2, (12) 6, 41, 21, 9, 45, 14, 39, 19, 31, 30, 21, 26, (5)	2 75539 87434 88 43070 11291
1031	32 ; 9, 6, 3, 4, 1, 1, 1, 1, 1, 12, 4, 1, 1, (31) 7, 10, 19, 13, 35, 26, 31, 25, 38, 5, 14, 29, 35, (2)	2029 75370 82877 65173 74486 64200
1032	32 ; (8) (8)	8 257
1033*	32 ; 7, 7, 1, 8, 3, 3, 1, 2, 3, 2, 2, 1, 1, 1, 2, (21, 21) 9, 8, 51, 7, 19, 16, 37, 21, 17, 24, 21, 32, 27, 31, 24, (3, 3)	5 81389 30460 24093 186 86036 75961 74196
1034	32 ; 6, 2, (2) 10, 25, (22)	494 15885
1035	32 ; 5, (1) 11, (46)	35 1126
1036	32 ; 5, 2, 1, 6, (2) 12, 21, 40, 9, (28)	26322 8 47225
1037*	32 ; 4, 1, 15, (3, 3) 13, 49, 4, (19, 19)	64805 20 86882
1038	32 ; 4, 1, 1, 2, 2, 1, 2, (10) 14, 33, 29, 22, 21, 34, 23, (6)	4 78126 154 04267
1039	32 ; 4, 3, 1, 1, 5, 3, 2, 2, 21, 12, 1, 5, 1, 1, 10, 4, 1, 6, 2, 1, 3, 1, 1, 15, 17, 30, 33, 11, 18, 23, 26, 3, 5, 51, 10, 31, 33, 6, 13, 46, 9, 22, 37, 15, 34, 27, 1, 1, 1, 1, (31) 29, 30, 25, 39, (2)	1 53006 74275 15667 18643 06921 49 31946 34979 07633 93486 37520
1040	32 ; (4) (16)	4 129
1041	32 ; 3, 1, 3, 1, 1, 4, 2, 2, 7, 1, 1, 1, 12, 3, 1, (20) 17, 40, 15, 31, 32, 13, 24, 25, 8, 39, 23, 40, 5, 16, 47, (3)	2389 36879 43492 77091 86499 27575
1042*	32 ; 3, (1, 1) 18, (31, 31)	25 807

TABLE 1001 TO 1500—*continued.*

1043	32; 3, 2, 1, 1, 1, 1, (8) 19, 22, 31, 29, 26, 37, (7)	17864 5 76927
1044	32; 3, 4, 1, 1, 1, 3, 2, 1, 1, (6) 20, 13, 36, 25, 35, 16, 23, 28, 35, (9)	59 45940 1921 19201
1045	32; 3, 15, 1, (4) 21, 4, 51, (11)	14112 4 56191
1046	32; 2, 1, 12, 3, 1, 2, 1, 1, 1, 5, 1, 5, (32) 22, 41, 5, 17, 38, 19, 34, 25, 37, 16, 47, 11, (2)	12 32778 42162 398 70425 10565
1047	32; 2, 1, 3, 1, (20) 23, 37, 14, 49, (3)	4228 1 36807
1048	32; 2, 1, 2, 6, 1, 4, 1, 1, (7) 24, 33, 23, 9, 47, 12, 31, 33, (8)	32 42859 1049 80517
1049*	32; 2, 1, 1, 2, 1, 4, 3, 1, 5, 7, 1, (12, 12) 25, 29, 32, 19, 40, 13, 16, 43, 11, 8, 53, (5, 5)	4 74354 09193 153 63508 25620
1050	32; 2, 2, (10) 26, 25, (6)	270 8749
1051	32; 2, 2, 1, 1, 2, 4, 4, 10, 1, 1, 3, 12, 1, 2, 6, 7, 21, 2, 8, 1, 3, 2, 2, 27, 21, 31, 30, 23, 14, 15, 6, 35, 29, 18, 5, 42, 21, 10, 9, 3, 30, 7, 46, 15, 25, 19, 1, (31) 45, (2)	95 53827 66460 01506 59129 73250 2 94697 22410 66655 86570 86507
1052	32; 2, 3, 3, 7, 1, 4, 9, (16) 28, 17, 19, 8, 47, 13, 7, (4)	11935 79112 3 87132 00767
1053	32; 2, (4) 29, (13)	20 649
1054	32; 2, 6, 1, 2, 1, 1, 4, 2, 2, 1, 1, 1, 3, 1, 2, 3, 4, (32) 30, 9, 42, 19, 30, 33, 13, 25, 21, 33, 26, 35, 15, 38, 21, 18, 15, (2)	10335 86450 05416 3 35557 62560 56895
1055	32; 2, (12) 31, (5)	52 1689
1056	32; (2) (32)	2 65
1057	32; 1, 1, 21, 5, 1, 6, 2, 1, 1, 3, 2, 7, 1, 2, 4, 1, 1, 1, (8) 33, 32, 3, 11, 48, 9, 24, 29, 33, 16, 27, 8, 41, 21, 13, 37, 24, 39, (7)	7 37084 02001 32992 239 63733 95685 29407
1058	32; 1, 1, 8, 1, 3, 1, 3, (32) 34, 31, 7, 47, 14, 41, 17, (2)	40 53146 1318 36323
1059	32; 1, 1, 5, 2, 2, 2, 1, 2, 3, 1, (31) 35, 30, 11, 25, 23, 21, 35, 22, 15, 49, (2)	1838 68081 59834 86610
1060	32; 1, 1, 3, 1, 5, 7, (16) 36, 29, 15, 44, 11, 9, (4)	22 64856 737 38369
1061*	32; 1, 1, 2, 1, 12, 3, 5, 1, 1, 2, 15, 1, 8, 2, 1, 2, (1, 1) 37, 28, 19, 44, 5, 20, 11, 35, 28, 25, 4, 55, 7, 23, 35, 20, (31, 31)	28370 82899 91521 9 24122 86971 11530
1062	32; 1, 1, 2, 3, (32) 38, 27, 23, 19, (2)	9418 3 06917

TABLE 1001 TO 1500—*continued.*

1063	32; 1, 1, 1, 1, 10, 3, 1, 2, 1, 6, 1, 1, 21, 4, 1, (31) 39, 26, 27, 37, 6, 17, 39, 18, 43, 9, 31, 34, 3, 13, 51, (2)	5353 15274 12685 1 74532 48310 97224
1064	32; 1, 1, 1, (1) 40, 25, 31, (28)	21 685
1065	32; 1, 1, 1, 2, 1, 3, 2, 1, 5, (4) 41, 24, 35, 19, 39, 16, 21, 40, 11, (15)	25 33160 826 67999
1066*	32; 1, 1, 1, 5, 1, 6, (2, 2) 42, 23, 39, 10, 49, 9, (25, 25)	1 05205 34 34907
1067	32; 1, (1) 43, (22)	3 98
1068	32; 1, 2, 7, 1, 5, (16) 44, 21, 8, 49, 11, (4)	3 53094 115 39207
1069*	32; 1, 2, 3, 1, 1, 21, 4, 3, 5, 7, 12, 1, 15, 2, 2, 1, 3, 1, 1, 1, 4, 1, (4, 4) 45, 20, 17, 29, 36, 3, 15, 19, 12, 9, 5, 57, 4, 27, 20, 39, 15, 36, 25, 37, 12, 45, (13, 13)	186 40986 37841 77726 15285 6094 77590 16096 85726 12782
1070	32; 1, 2, 2, 5, 1, 1, (12) 46, 19, 26, 11, 31, 34, (5)	90138 29 48491
1071	32; 1, 2, 1, 1, 1, (6) 47, 18, 35, 25, 38, (9)	880 28799
1072	32; 1, 2, 1, 6, 1, 1, 8, 1, 4, 1, 1, (3) 48, 17, 44, 9, 32, 33, 7, 49, 12, 33, 31, (16)	1457 20107 47710 81927
1073*	32; 1, 3, (9, 9) 49, 16, (7, 7)	1385 45368
1074	32; 1, 3, 2, 1, 1, 2, (32) 50, 15, 23, 31, 30, 25, (2)	1 06476 34 89425
1075	32; 1, 3, 1, 2, 3, 10, 1, 1, 1, 2, 2, 6, (1) 51, 14, 39, 21, 19, 6, 39, 25, 34, 21, 26, 9, (50)	51504 12729 16 88675 74226
1076	32; 1, 4, (16) 52, 13, (4)	410 13449
1077	32; 1, 4, 2, 15, 1, (20) 53, 12, 29, 4, 59, (3)	7 16760 235 22399
1078	32; 1, (4) 54, (11)	6 197
1079	32; 1, 5, 1, 1, 2, 2, 4, 3, 1, (1) 55, 10, 35, 29, 22, 25, 14, 17, 35, (26)	54 97325 1805 76876
1080	32; 1, 6, (3) 56, 9, (20)	161 5291
1081	32; 1, 7, 4, 3, 1, 6, 1, 1, 5, 2, 3, 1, 12, 2, 1, 1, 1, 21, 3, 2, (2) 57, 8, 15, 16, 45, 9, 33, 32, 11, 27, 15, 48, 5, 24, 33, 25, 40, 3, 19, 24, (23)	8918 12221 87280 07648 2 93215 05610 17151 19615
1082*	32; 1, 8, 2, 2, 2, (1, 1) 58, 7, 26, 23, 22, (31, 31)	38369 12 62101
1083	32; 1, (9) 59, (6)	11 362

TABLE 1001 TO 1500—*continued.*

1084	32; 1, 12, 5, 2, 2, 3, 2, 6, 1, 7, 2, 1, 2, 1, 3, 1, 1, 1, 21, 3, 4, (16) 60, 5, 12, 25, 24, 17, 27, 9, 51, 8, 23, 36, 19, 40, 15, 37, 24, 41, 3, 20, 15, (4)	8 17041 27029 09269 93200 269 00393 64204 05379 19999
1085	32; 1, 15, (2) 61, 4, (31)	544 17919
1086	32; 1, (20) 62, (3)	22 725
1087	32; 1, (31) 63, (2)	33 1088
1088	32; (1) (64)	1 33
1090*	33; (66) (1)	1 33
1091	33; (33) (2)	33 1090
1092	33; (22) (3)	22 727
1093*	33; 16, (1, 1) 4, (33, 33)	545 18018
1094	33; 13, 4, 1, 1, 1, 5, 2, 1, 2, 3, 9, 6, 1, 1, (32) 5, 14, 37, 25, 38, 11, 23, 35, 22, 19, 7, 10, 31, 35, (2)	45043 80474 67914 14 89854 05815 38085
1095	33; (11) (6)	11 364
1096	33; 9, 2, 3, 1, (15) 7, 28, 15, 49, (4)	1 19595 39 59299
1097*	33; 8, 3, 1, 3, 2, 1, (1, 1) 8, 17, 41, 16, 23, 32, (29, 29)	6 34621 210 19276
1098	33; 7, 2, 1, (6) 9, 22, 41, (9)	3564 1 18097
1099	33; 6, 1, 1, 1, 1, 2, 21, 1, 2, 1, 1, 6, 1, (3) 10, 37, 27, 30, 31, 25, 3, 46, 19, 30, 35, 9, 47, (14)	4 80575 45715 159 31638 15326
1100	33; (6) (11)	6 199
1101	33; 5, 1, 1, 16, (22) 12, 31, 35, 4, (3)	7 32732 243 13015
1102	33; 5, 10, 1, 6, (2) 13, 6, 53, 9, (29)	3 42882 113 82443
1103	33; 4, 1, 2, 1, 2, 3, 1, 1, 5, 2, 9, (33) 14, 41, 19, 37, 22, 17, 31, 34, 11, 29, 7, (2)	7 06456 11145 234 62427 45024
1104	33; 4, 2, (2) 15, 25, (23)	234 7775
1105*	33; 4, (7, 7) 16, (9, 9)	857 28488

TABLE 1001 TO 1500—*continued.*

1106	33 ; 3, 1, (8) 17, 46, (7)	152 5055
1107	33 ; 3, 1, 2, 7, (33) 18, 37, 23, 9, (2)	2 18295 72 63026
1108	33 ; 3, 2, 21, 1, 3, 4, 1, 6, 1, 1, 2, 2, 1, 3, 2, 5, 9, 3, (16) 19, 28, 3, 49, 16, 13, 48, 9, 36, 29, 23, 21, 39, 16, 27, 12, 7, 21, (4)	4781 20058 69390 13510 1 59150 07379 89804 75849
1109*	33 ; 3, 3, 5, 1, 3, 13, 16, 1, 1, 2, 1, 4, (2, 2) 20, 19, 11, 44, 17, 5, 4, 37, 29, 20, 41, 13, (25, 25)	1832 35957 38617 61020 60015 42610
1110	33 ; 3, (6) 21, (10)	60 1999
1111	33 ; (3) (22)	3 100
1112	33 ; 2, 1, (7) 23, 41, (8)	75 2501
1113	33 ; 2, 1, 3, 3, 1, (8) 24, 37, 17, 16, 47, (7)	21056 7 02463
1114*	33 ; 2, 1, 1, 1, 8, 1, 10, 4, 2, 1, 3, 1, 3, 7, 6, (1, 1) 25, 33, 26, 39, 7, 55, 6, 15, 22, 39, 15, 42, 17, 9, 10, (33, 33)	18311 57471 56745 6 11178 81032 02293
1115	33 ; 2, 1, 1, 4, 5, 1, (5) 26, 29, 35, 14, 11, 49, (10)	1 36565 45 60126
1116	33 ; 2, 2, 5, 1, 2, (16) 27, 25, 11, 40, 23, (4)	1 38320 46 20799
1117*	33 ; 2, 2, 1, 2, 5, 4, 1, 21, 2, 9, 16, 1, 1, 1, 1, 6, 1, 4, 1, (2, 2) 28, 21, 36, 23, 12, 13, 52, 3, 31, 7, 4, 39, 27, 28, 37, 9, 49, 12, 41, (21, 21)	6272 59559 53855 43645 2 09639 86690 78245 41118
1118	33 ; 2, 3, (2) 29, 17, (26)	126 4213
1119	33 ; 2, 4, 1, 1, 1, 6, (22) 30, 13, 38, 25, 39, 10, (3)	9 46364 316 57255
1120	33 ; 2, 6, 1, (15) 31, 9, 55, (4)	3765 1 26001
1121	33 ; 2, 12, 1, 8, 1, 1, 1, 3, 1, 1, 7, 1, 4, 3, 1, (2) 32, 5, 56, 7, 40, 25, 37, 16, 31, 35, 8, 49, 13, 17, 40, (19)	453 93975 58620 15198 51043 64951
1122	33 ; (2) (33)	2 67
1123	33 ; 1, 1, 21, 1, 5, 7, 3, 1, 1, 2, 2, 10, 1, 3, (33) 34, 33, 3, 54, 11, 9, 18, 33, 31, 22, 27, 6, 47, 17, (2)	49257 11232 14799 16 50664 55626 32482
1124	33 ; 1, 1, 9, 13, 3, 3, 1, 1, 1, 1, 1, 2, (16) 35, 32, 7, 5, 20, 17, 35, 28, 31, 29, 32, 25, (4)	6 74757 87740 226 22006 30049
1125	33 ; 1, 1, 5, 1, 1, 2, 7, 16, 1, 1, 1, 2, 1, (6) 36, 31, 11, 36, 29, 25, 9, 4, 41, 25, 36, 19, 44, (9)	5 16002 91864 173 07264 04001
1126	33 ; 1, 1, 3, 1, (32) 37, 30, 15, 51, (2)	2718 91205

TABLE 1001 TO 1500—*continued.*

1127	33; 1, 1, 3, (33) 38, 29, 19, (2)	1645 55224
1128	33; 1, 1, 2, (2) 39, 28, 23, (24)	70 2351
1129*	33; 1, (1, 1) 40, (27, 27)	5 168
1130*	33; 1, 1, (1, 1) 41, 26, (31, 31)	13 437
1131	33; 1, 1, 1, 2, 2, 1 1, (4) 42, 25, 35, 22, 23, 30, 35, (13)	10948 3 68185
1132	33; 1, 1, 1, 4, 1, 1, 21, 1, 7, 2, 5, 7, 3, 2, 2, (16) 43, 24, 39, 13, 31, 36, 3, 57, 8, 29, 12, 9, 19, 24, 27, (4)	1 13654 94862 36362 38 23944 35058 85447
1133	33; 1, 1, 1, 16, (6) 44, 23, 43, 4, (11)	15300 5 14999
1134	33; 1, 2, 13, 7, 2, 2, (4) 45, 22, 5, 9, 26, 25, (14)	107 52040 3620 74049
1135	33; 1, 2, 4, 2, 10, 1, 3, 1, 1, 2, 1, 1, 1, 6, 1, (5) 46, 21, 14, 29, 6, 49, 15, 34, 31, 21, 35, 26, 39, 9, 51, (10)	31 50736 19505 1061 47549 56124
1136	33; 1, 2, 2, 1, 1, 2, 9, (4) 47, 20, 23, 32, 31, 25, 7, (16)	7 18620 242 20799
1137	33; 1, 2, 1, 1, 3, 2, 1, 1, 7, 1, 5, 4, (22) 48, 19, 32, 33, 17, 24, 29, 37, 8, 51, 11, 16, (3)	1 92906 26292 65 04689 34487
1138*	33; 1, 2, 1, 3, 4, (1, 1) 49, 18, 41, 17, 14, (33, 33)	10337 3 48711
1139	33; 1, (2) 50, (17)	4 135
1140	33; 1, 3, (4) 51, 16, (15)	72 2431
1141	33; 1, 3, 1, 1, 12, 1, 21, 1, 1, 2, 5, 4, 3, 7, 5, 16, 1, 2, 3, 1, 1, 1, 2, 52, 15, 31, 36, 5, 60, 3, 39, 28, 25, 12, 15, 20, 9, 13, 4, 45, 21, 17, 36, 27, 35, 20, 1, 2, 1, 4, 1, (8) 39, 19, 43, 12, 51, (7)	3 06933 85322 76565 71973 97208 103 67823 94157 22396 32371 25215
1142	33; 1, 3, 1, 5, 2, 1, 8, 1, (32) 53, 14, 47, 11, 22, 43, 7, 59, (2)	268 28010 9066 12101
1143	33; 1, 4, 4, 1, 1, 1, 2, (3) 54, 13, 14, 37, 27, 34, 23, (18)	1 39925 47 30624
1144	33; 1, 4, 1, 1, 1, 6, (1) 55, 12, 39, 25, 40, 9, (52)	16611 5 61835
1145*	33; 1, (5, 5) 56, (11, 11)	37 1252
1146	33; 1, 5, 1, 3, 1, 1, 1, (10) 57, 10, 47, 15, 38, 25, 41, (6)	1 01840 34 47551

TABLE 1001 TO 1500—*continued.*

1147	33; 1, 6, 1, 1, 5, 1, 1, 1, 1, 1, 21, (1) 58, 9, 34, 33, 11, 38, 27, 33, 26, 41, 3, (62)	2789 45403 94471 52318
1148	33; 1, 7, 2, (16) 59, 8, 31, (4)	4896 1 65887
1149	33; 1, 8, 1, 2, 3, (22) 60, 7, 44, 21, 20, (3)	2 12624 72 07295
1150	33; 1, 10, 3, 7, 4, 1, 2, (2) 61, 6, 21, 9, 14, 41, 21, (25)	349 25592 11843 84449
1151	33; 1, 12, 1, 1, 2, 2, 3, 6, 2, (33) 62, 5, 38, 29, 23, 25, 19, 10, 31, (2)	19426 07807 6 59056 71840
1152	33; 1, (15) 63, (4)	17 577
1153*	33; 1, 21, 1, 1, 1, 6, 1, 7, 1, 1, 1, 1, 1, 2, 4, 1, 5, 2, 1, 3, (1, 1) 64, 3, 43, 24, 41, 9, 53, 8, 39, 27, 32, 29, 33, 24, 13, 48, 11, 23, 39, 16, (33, 33)	3 01789 02568 75073 102 47504 00230 72656
1154	33; 1, (32) 65, (2)	34 1155
1155	33; (1) (66)	1 34
1157*	34; (68) (1)	1 34
1158	34; (34) (2)	34 1157
1159	34; 22, 1, 2, 7, 4, 2, 2, 13, 4, 1, 3, 1, (2) 3, 45, 22, 9, 15, 25, 27, 5, 14, 45, 15, 42, (19)	4902 89385 75180 1 66914 55514 24551
1160	34; (17) (4)	17 579
1161	34; 13, 1, 1, 1, 1, 2, 8, 7, 2, 4, 1, 3, (2) 5, 40, 27, 31, 32, 25, 8, 9, 29, 13, 45, 16, (27)	56 21214 40972 1915 34168 54935
1162	34; 11, 2, 1, 6, 1, (8) 6, 23, 42, 9, 54, (7)	6 63462 226 16173
1163	34; 9, 1, 2, 1, 2, 4, 1, 1, (33) 7, 46, 19, 38, 23, 14, 31, 37, (2)	369 56541 12603 21002
1164	34; 8, 1, 1, (16) 8, 33, 35, (4)	4930 1 68199
1165*	34; 7, 1, 1, 3, 16, 1, 3, 1, (1, 1) 9, 36, 31, 19, 4, 51, 15, 36, (29, 29)	867 20773 29599 61778
1166	34; 6, 1, 4, 2, 1, 1, 9, (6) 10, 49, 13, 25, 29, 38, 7, (11)	193 40870 6604 27701
1167	34; 6, 5, (11) 11, 13, (6)	10943 3 73828
1168	34; 5, 1, 2, 7, (4) 12, 41, 23, 9, (16)	66750 22 81249

TABLE 1001 TO 1500—*continued.*

1169	34; 5, 4, 13, 2, 3, 1, 1, 5, 1, 1, 1, 7, 1, (8) 13, 16, 5, 29, 17, 32, 35, 11, 40, 25, 41, 8, 55, (7)	538 84918 28064 18423 59949 83935
1170	34; 4, 1, (6) 14, 49, (9)	190 6499
1171	34; 4, 1, 1, 4, 1, 2, 2, 3, 1, 1, 1, 1, (33) 15, 33, 34, 13, 42, 21, 26, 17, 35, 30, 27, 41, (2)	15274 30263 5 22684 76130
1172	34; 4, 3, 1, 3, 1, 1, (16) 16, 17, 43, 16, 31, 37, (4)	3 59890 123 20649
1173	34; (4) (17)	4 137
1174	34; 3, 1, 3, 1, 4, 2, 13, 3, 1, 22, 11, 2, 1, 1, 1, 6, 4, 2, 2, 1, 1, 7, (34) 18, 43, 15, 46, 13, 30, 5, 17, 50, 3, 6, 25, 34, 27, 39, 10, 15, 26, 23, 30, 37, 9, (2)	1363 67209 12136 61406 74774 46724 42879 20656 93035 91365
1175	34; 3, 1, 1, 2, 5, (1) 19, 34, 31, 25, 11, (50)	12901 4 42224
1176	34; 3, 2, (2) 20, 25, (24)	140 4801
1177	34; 3, 3, 1, 22, 9, 1, 3, 7, 2, 1, 2, 1, 1, 2, 2, 2, 8, (6) 21, 16, 51, 3, 7, 48, 17, 9, 24, 37, 21, 33, 32, 23, 24, 27, 8, (11)	276 38354 60657 78460 9482 01015 39044 75351
1178	34; 3, 9, (2) 22, 7, (31)	1736 59583
1179	34; 2, 1, (33) 23, 45, (2)	300 10610
1180	34; 2, 1, 5, 1, 1, 2, 1, (2) 24, 41, 11, 36, 31, 21, 39, (20)	58950 20 24999
1181*	34; 2, 1, 2, 1, 3, 3, 5, 1, 16, 2, 1, (13, 13) 25, 37, 20, 41, 17, 20, 11, 55, 4, 23, 44, (5, 5)	35 62788 38045 1224 37647 36718
1182	34; 2, 1, 1, 1, 2, 2, 1, 2, 1, (10) 26, 33, 29, 34, 23, 22, 39, 19, 47, (6)	7 92682 272 52587
1183	34; 2, 1, 1, 7, 22, 1, (3) 27, 29, 38, 9, 3, 53, (14)	37 96401 1305 76328
1184	34; 2, 2, 3, 1, 9, (17) 28, 25, 16, 49, 7, (4)	7 95285 273 65201
1185	34; 2, 2, 1, 3, 1, 1, 2, 3, (4) 29, 21, 41, 16, 35, 31, 24, 19, (15)	12 03200 414 18751
1186	34; 2, 3, 1, 1, 4, (34) 30, 17, 33, 34, 15, (2)	1 83522 63 20195
1187	34; 2, 4, 1, 4, 9, 1, 1, 1, 2, 1, (33) 31, 13, 47, 14, 7, 41, 26, 37, 19, 49, (2)	11347 43775 3 90951 75626
1188	34; 2, 7, (6) 32, 9, (11)	1410 48599
1189*	34; 2, 13, 3, 2, 1, 2, 16, 1, 6, 1, 2, (1, 1) 33, 5, 20, 23, 36, 25, 4, 57, 9, 45, 20, (33, 33)	24 58219 58945 847 64031 17418

TABLE 1001 TO 1500—*continued.*

1190	34 ; (2) (34)	2 69
1191	34 ; 1, 1, (22) 35, 34, (3)	92 3175
1192	34 ; 1, 1, 9, 2, 1, 3, 2, 1, 1, 1, 1, 7, (17) 36, 33, 7, 24, 39, 17, 24, 33, 31, 28, 39, 9, (4)	97167 39825 33 54738 72499
1193*	34 ; 1, 1, 5, 1, 3, 2, 8, (5, 5) 37, 32, 11, 47, 16, 29, 8, (13, 13)	247 53805 8549 92268
1194	34 ; 1, 1, 4, 9, 1, 1, 1, 6, 3, 1, (10) 38, 31, 15, 7, 42, 25, 41, 10, 17, 49, (6)	6264 76750 2 16474 68749
1195	34 ; 1, 1, 3, 7, 2, 1, 1, 10, 1, (12) 39, 30, 19, 9, 26, 29, 39, 6, 59, (5)	1345 16232 46500 60959
1196	34 ; 1, 1, (2) 40, 29, (23)	12 415
1197	34 ; 1, 1, 2, 16, 1, (8) 41, 28, 27, 4, 59, (7)	74820 25 88599
1198	34 ; 1, 1, 1, 1, 2, 1, 2, 3, 2, 11, 9, 1, 4, 22, 1, 6, 1, 2, 1, 3, 3, (34) 42, 27, 31, 34, 21, 38, 23, 18, 29, 6, 7, 51, 14, 3, 58, 9, 46, 19, 42, 17, 21, (2)	1 21525 49298 36463 16460 42 06256 96350 23592 51599
1199	34 ; 1, 1, 1, 2, 9, 1, 1, 13, (3) 43, 26, 35, 25, 7, 34, 35, 5, (22)	143 75599 4977 77820
1200	34 ; 1, 1, 1, (3) 44, 25, 39, (16)	39 1351
1201*	34 ; 1, 1, 1, 9, 4, 4, 2, 1, 1, 1, 7, 13, 1, 2, 1, 2, 1, 1, 4, 22, 1, 7, 1, 45, 24, 43, 7, 16, 15, 24, 35, 27, 40, 9, 5, 48, 19, 40, 21, 32, 35, 15, 3, 59, 8, 45, 2, (2, 2) 21, (25, 25)	71 85416 85609 44230 77385 2490 13832 32746 50571 44832
1202	34 ; 1, 2, (34) 46, 23, (2)	312 10817
1203	34 ; 1, 2, 5, 1, (33) 47, 22, 11, 57, (2)	12521 4 34282
1204	34 ; 1, 2, 3, 7, 2, 2, 3, 4, 22, 1, (8) 48, 21, 20, 9, 27, 25, 19, 16, 3, 60, (7)	16 95565 54116 588 33925 37695
1205	34 ; 1, 2, 2, 16, 1, (12) 49, 20, 29, 4, 61, (5)	2 06668 71 74089
1206	34 ; 1, 2, 1, 2, (34) 50, 19, 38, 25, (2)	4202 1 45925
1207	34 ; 1, 2, 1, 6, 1, (33) 51, 18, 47, 9, 59, (2)	33387 11 59928
1208	34 ; 1, 3, 9, 1, 2, (8) 52, 17, 7, 44, 23, (8)	1 23046 42 76623
1209	34 ; 1, 3, 2, 1, 3, 2, 1, 1, (22) 53, 16, 23, 40, 17, 25, 29, 40, (3)	16 40156 570 29335

TABLE 1001 TO 1500—*continued.*

1210	34; 1, 3, 1, 1, 1, 7, 11, 2, (6) 54, 15, 39, 26, 41, 9, 6, 31, (10)	430 99524 14992 19281
1211	34; 1, (3) 55, (14)	5 174
1212	34; 1, 4, 2, 1, 2, 2, 1, (16) 56, 13, 24, 37, 23, 21, 47, (4)	3 65980 127 41151
1213*	34; 1, 4, 1, 4, 1, 1, 9, (2, 2) 57, 12, 49, 18, 33, 36, 7, (27, 27)	20 27117 706 00734
1214	34; 1, 5, 2, 1, 6, 3, 1, 1, 13, 2, 1, 2, 2, (34) 58, 11, 23, 43, 10, 19, 31, 38, 5, 25, 37, 22, 29, (2)	209 94702 60104 7315 07983 74975
1215	34; 1, (5) 59, (10)	7 244
1216	34; 1, 6, 1, 3, 4, 2, 1, 1, 4, (17) 60, 9, 48, 17, 15, 25, 31, 36, 15, (4)	1916 03685 66814 48801
1217*	34; 1, 7, 1, 2, 1, 3, 1, 1, 1, (1, 1) 61, 8, 47, 19, 43, 16, 37, 29, 32, (31, 31)	7 91969 276 28256
1218	34; 1, (8) 62, (7)	10 349
1219	34; 1, 10, 1, 1, 1, 7, 9, 1, 5, 2, 4, 5, 6, 1, 3, (1) 63, 6, 43, 25, 42, 9, 7, 54, 11, 29, 15, 13, 10, 49, 15, (46)	76791 47002 95135 26 81111 24548 55326
1220	34; 1, (12) 64, (5)	14 489
1221	34; 1, 16, (2) 65, 4, (33)	612 21385
1222	34; 1, 22, 3, 7, 2, 3, 1, 1, 1, 4, 2, 1, (4) 66, 3, 22, 9, 29, 17, 38, 27, 39, 14, 23, 42, (13)	22 12495 24122 773 42454 09307
1223	34; 1, (33) 67, (2)	35 1224
1224	34; (1) (68)	1 35
1226*	35; (70) (1)	1 35
1227	35; (35) (2)	35 1226
1228	35; 23, 2, 1, 7, 8, 1, 1, 1, 2, 2, 1, 1, 5, 3, 1, (16) 3, 24, 43, 9, 8, 41, 27, 36, 23, 24, 31, 37, 12, 17, 51, (4)	44730 59716 99506 15 67486 75428 71047
1229*	35; 17, (1, 1) 4, (35, 35)	613 21490
1230	35; (14) (5)	14 491
1231	35; 11, 1, 2, 7, 2, 4, 1, 13, 4, 1, 1, 1, 1, 6, 2, 2, 4, 3, 1, 2, 23, (35) 6, 45, 23, 9, 30, 13, 54, 5, 15, 37, 30, 29, 39, 10, 27, 26, 15, 18, 39, 25, 3, (2)	584 79034 52350 15227 51177 20517 72574 24010 98134 87200

TABLE 1001 TO 1500—*continued.*

1232	35 ; (10) (7)	10 351
1233	35 ; 8, 1, 3, 4, 7, 1, 1, 2, 6, 9, 1, (6) 8, 49, 17, 16, 9, 37, 32, 19, 11, 7, 56, (9)	210 76306 89984 7400 75542 46657
1234	35 ; 7, 1, 3, 1, 4, 4, 2, 9, 1, 1, 2, 3, 1, 1, (34) 9, 50, 15, 47, 14, 15, 30, 7, 39, 30, 25, 18, 31, 39, (2)	1669 10230 73856 58632 78690 67265
1235	35 ; (7) (10)	7 246
1236	35 ; 6, 2, 1, 1, 1, 4, 1, 3, 1, 1, 2, 1, (22) 11, 25, 35, 28, 39, 13, 47, 16, 35, 33, 20, 49, (3)	18261 82722 6 42027 25495
1237*	35 ; 5, 1, 5, 1, 1, 3, 1, 1, 2, 23, 17, 1, 1, 5, 2, 1, (7, 7) 12, 51, 11, 36, 33, 17, 36, 31, 27, 3, 4, 37, 33, 12, 23, 44, (9, 9)	30 80599 95913 30265 1083 47814 40276 61982
1238	35 ; 5, 2, 1, 1, (34) 13, 26, 29, 41, (2)	25650 9 02501
1239	35 ; (5) (14)	5 176
1240	35 ; 4, 1, 2, 7, (2) 15, 41, 24, 9, (31)	24102 8 48719
1241*	35 ; 4, 2, 1, 1, 3, 8, (1, 1) 16, 25, 32, 35, 19, 8, (35, 35)	9 65005 339 95032
1242	35 ; 4, 7, 1, 1, (2) 17, 9, 38, 31, (23)	11780 4 15151
1243	35 ; 3, 1, 9, (3) 18, 49, 7, (22)	4875 1 71874
1244	35 ; 3, 1, 2, 3, 6, 8, 1, 1, 1, 13, 2, 4, 1, (16) 19, 40, 23, 20, 11, 8, 43, 25, 44, 5, 31, 13, 55, (4)	1616 45760 60360 57013 08077 08799
1245	35 ; 3, 1, 1, 17, (14) 20, 31, 39, 4, (5)	2 13528 75 34241
1246	35 ; 3, 2, 1, 7, 6, 1, 13, 3, 1, 5, 1, 1, 1, (34) 21, 22, 45, 9, 10, 57, 5, 18, 47, 11, 42, 25, 45, (2)	1257 54312 87288 44389 66931 48735
1247	35 ; 3, 5, 9, (1) 22, 13, 7, (58)	26313 9 29188
1248	35 ; 3, (17) 23, (4)	159 5617
1249*	35 ; 2, 1, 13, 2, 7, 2, 1, 2, 3, 1, 3, 1, 1, 1, 4, 1, 3, 1, 8, 23, 2, (4, 4) 24, 45, 5, 32, 9, 25, 37, 24, 17, 45, 16, 39, 27, 40, 13, 48, 15, 57, 8, 3, 31, (15, 15)	2 61326 40028 30963 92593 92 35587 03432 90296 88240
1250*	35 ; 2, 1, 4, 2, 1, (1, 1) 25, 41, 14, 25, 34, (31, 31)	7801 2 75807
1251	35 ; 2, 1, 2, 2, 2, 5, (35) 26, 37, 23, 25, 27, 13, (2)	21 92943 775 63250
1252	35 ; 2, 1, 1, 1, 1, 5, 3, 1, 1, 4, 1, 7, 23, 2, 5, 1, (16) 27, 33, 32, 29, 39, 12, 19, 33, 36, 13, 52, 9, 3, 32, 11, 57, (4)	90969 00792 82080 32 18812 08291 34849

TABLE 1001 TO 1500—*continued.*

1253	35 ; 2, 1, 1, 17, (10) 28, 29, 41, 4, (7)	78320 27 72351
1254	35 ; 2, 2, 2, 1, (34) 29, 25, 21, 49, (2)	10234 3 62405
1255	35 ; 2, 2, 1, 7, 6, 3, 4, 1, 2, 1, 11, (14) 30, 21, 46, 9, 11, 21, 14, 45, 19, 49, 6, (5)	87 88466 34708 3113 40025 72889
1256	35 ; 2, 3, 1, 2, (17) 31, 17, 40, 25, (4)	11075 3 92499
1257	35 ; 2, 4, 1, (22) 32, 13, 56, (3)	2860 1 01399
1258*	35 ; 2, 7, 2, (1, 1) 33, 9, 26, (33, 33)	3233 1 14669
1259	35 ; 2, 13, 1, 2, 3, 2, 1, 1, 5, 1, 6, 4, (35) 34, 5, 47, 22, 19, 25, 31, 38, 11, 53, 10, 17, (2)	313 98432 81069 11140 91438 32810
1260	35 ; (2) (35)	2 71
1261*	35 ; 1, 1, 23, 5, 1, 7, 17, 1, 1, 1, 2, 5, (1, 1) 36, 35, 3, 12, 53, 9, 4, 43, 27, 36, 25, 12, (35, 35)	694 50461 60725 24662 24909 57682
1262	35 ; 1, 1, 9, 1, 1, 1, 4, 1, 4, 3, 1, (34) 37, 34, 7, 43, 26, 41, 13, 49, 14, 17, 53, (2)	17412 04708 6 18556 69263
1263	35 ; 1, 1, 5, 1, (22) 38, 33, 11, 58, (3)	4004 1 42297
1264	35 ; 1, 1, 4, (4) 39, 32, 15, (16)	360 12799
1265	35 ; 1, 1, 3, 4, (6) 40, 31, 19, 16, (11)	5820 2 06999
1266	35 ; 1, 1, 2, 1, 1, 2, 3, 1, 3, 1, (34) 41, 30, 23, 34, 33, 25, 17, 46, 15, 55, (2)	149 86708 5332 40465
1267	35 ; 1, 1, 2, 7, 1, 1, 23, (5) 42, 29, 27, 9, 34, 37, 3, (14)	175 73127 6255 14462
1268	35 ; 1, 1, 1, 1, 3, 1, 5, 1, 2, 4, 9, 1, (16) 43, 28, 31, 37, 16, 49, 11, 44, 23, 16, 7, 61, (4)	69835 99490 24 86789 07849
1269	35 ; 1, 1, 1, 1, 1, 7, 3, (2) 44, 27, 35, 28, 41, 9, 20, (27)	96264 34 29215
1270	35 ; 1, 1, 1, 3, 11, 1, 1, 1, 1, 5, 1, 7, (14) 45, 26, 39, 69, 6, 41, 29, 30, 39, 11, 54, 9, (5)	1 63267 43118 58 18371 13691
1271	35 ; 1, 1, 1, 6, (2) 46, 25, 43, 10, (31)	920 32799
1272	35 ; 1, (1) 47, (24)	3 107
1273	35 ; 1, 2, 8, 1, 1, 2, 2, 4, 23, 1, 1, 3, 1, 2, 5, 7, 1, (2) 48, 23, 8, 39, 31, 24, 27, 16, 3, 39, 32, 17, 41, 24, 13, 9, 48, (19)	5 14164 62158 51476 183 44944 08681 34807

TABLE 1001 TO 1500—*continued.*

1274	35; 1, 2, 3, 1, 6, 2, 1, 2, (2) 49, 22, 17, 49, 10, 25, 38, 23, (26)	15 15720 541 00801
1275	35; 1, 2, 2, (2) 50, 21, 26, (25)	140 4999
1276	35; 1, 2, 1, 1, 2, 2, 2, 7, 1, 1, 9, 1, 2, 13, 1, (16) 51, 20, 35, 33, 24, 25, 28, 9, 35, 36, 7, 45, 24, 5, 63, (4)	931 97427 02580 33291 18543 36799
1277*	35; 1, 2, 1, 3, 2, 5, 17, 1, 2, (6, 6) 52, 19, 44, 17, 29, 13, 4, 47, 23, (11, 11)	40427 42285 14 44679 48482
1278	35; 1, (2) 53, (18)	4 143
1279	35; 1, 3, 4, 1, 1, 13, 1, 3, 23, 1, 1, 2, 2, 1, 5, 1, 3, 1, 11, 7, 1, 6, 3, 54, 17, 15, 33, 38, 5, 51, 18, 3, 41, 30, 25, 22, 45, 11, 50, 15, 53, 6, 9, 55, 10, 19, 1, 1, 1, 1, 1, 2, (35) 37, 30, 33, 31, 34, 27, (2)	 7 89404 78828 27783 39935 75041 282 31569 87094 18598 91612 56000
1280	35; 1, 3, 2, (17) 55, 16, 31, (4)	1449 51841
1281	35; 1, 3, 1, 3, 1, 2, 13, 1, (22) 56, 15, 47, 16, 41, 25, 5, 64, (3)	220 81748 7903 29175
1282	35; 1, 4, 7, 1, 3, 9, 1, (34) 57, 14, 9, 49, 18, 7, 63, (2)	952 47138 34103 26403
1283	35; 1, 4, 1, 1, 9, 1, 2, 4, 1, 3, 2, 2, (35) 58, 13, 34, 37, 7, 46, 23, 14, 47, 17, 26, 29, (2)	5 15512 61505 184 65140 88226
1284	35; 1, (4) 59, (12)	6 215
1285*	35; 1, 5, 1, 1, 7, 2, 2, 1, 17, 4, 1, 2, 1, (1, 1) 60, 11, 35, 36, 9, 29, 21, 49, 4, 15, 44, 21, 36, (31, 31)	24 48426 94225 877 68507 57318
1286	35; 1, 6, 5, 2, 1, 2, 13, 1, (34) 61, 10, 13, 25, 37, 26, 5, 65, (2)	7072 64778 2 53631 10565
1287	35; 1, (6) 62, (9)	8 287
1288	35; 1, (7) 63, (8)	9 323
1289*	35; 1, 9, 3, 1, 2, 8, 1, 1, 1, 1, 2, 1, 1, 13, 1, 3, (1, 1) 64, 7, 19, 40, 25, 8, 41, 29, 32, 35, 23, 31, 40, 5, 53, 16, (35, 35)	835 94139 33853 30012 50804 21900
1290	35; 1, (10) 65, (6)	12 431
1291	35; 1, 13, 2, 1, 1, 2, 3, 1, 1, 1, 1, 4, 1, 11, 6, 2, 4, 3, 23, 1, 1, 1, 66, 5, 27, 33, 34, 25, 18, 37, 31, 30, 39, 13, 55, 6, 11, 30, 15, 22, 3, 45, 26, 41, 4, 7, 1, 3, 2, 1, 6, 2, (35) 15, 9, 50, 17, 23, 45, 10, 33, (2)	 1974 27745 80291 79860 94899 51267 70936 75339 57910 18687 74456 08410
1292	35; 1, (16) 67, (4)	18 647

TABLE 1001 TO 1500—*continued.*

1293	35; 1, (22) 68, (3)	24 863
1294	35; 1, (34) 69, (2)	36 1295
1295	35; (1) (70)	1 36
1297*	36; (72) (1)	1 36
1298	36; (36) (2)	36 1297
1299	36; (24) (3)	24 865
1300	36; (18) (4)	18 649
1301*	36; 14, 2, (2, 2) 5, 29, (25, 25)	6025 2 17318
1302	36; (12) (6)	12 433
1303	36; 10, 3, 2, 1, 23, 2, 1, 2, 1, 3, 3, 1, 1, 7, 2, 5, 11, 1, 5, 1, 1, 1, 4, 7, 21, 22, 49, 3, 26, 39, 21, 43, 18, 19, 33, 38, 9, 31, 13, 6, 57, 11, 42, 27, 41, 14, 1, 1, (35) 33, 39, (2)	10127 26314 51893 46887 71461 3 65564 74232 60608 71132 28808
1304	36; (9) (8)	9 325
1305	36; (8) (9)	8 289
1306*	36; 7, 4, 1, 2, 11, 1, 2, 4, 2, 9, 1, (7, 7) 10, 15, 42, 25, 6, 47, 23, 15, 31, 7, 58, (9, 9)	862 68151 05389 31176 12101 50515
1307	36; 6, 1, 1, 3, 1, 2, 1, 1, (35) 11, 37, 34, 17, 43, 22, 31, 41, (2)	54 14517 1957 48082
1308	36; (6) (12)	6 217
1309	36; 5, 1, 1, 4, 3, 1, 1, 2, 3, 17, 1, 3, 1, 7, (4) 13, 36, 35, 15, 19, 36, 33, 25, 21, 4, 55, 15, 52, 9, (17)	2219 92217 38044 80317 01740 72265
1310	36; 5, 6, 2, 1, 1, 1, 1, 1, (6) 14, 11, 26, 35, 31, 34, 29, 41, (10)	34 56382 1251 00021
1311	36; 4, 1, 4, 2, (1) 15, 49, 14, 25, (38)	5353 1 93820
1312	36; 4, 1, 1, (17) 16, 33, 39, (4)	1467 53137
1313*	36; (4, 4) (17, 17)	17 616

TABLE 1001 TO 1500—*continued.*

1314	36; (4) (18)	4 145
1315	36; 3, 1, 4, 11, 1, 7, (7) 19, 46, 15, 6, 59, 9, (10)	244 74351 8875 11646
1316	36; 3, 1, 1, 1, 1, 2, 3, (2) 20, 37, 31, 32, 35, 25, 19, (28)	65508 23 76415
1317	36; 3, 2, 3, 1, 5, 3, 1, 1, 1, 4, 1, 17, 3, 10, (24) 21, 28, 17, 49, 12, 19, 39, 28, 41, 13, 57, 4, 23, 7, (3)	1 03450 99306 81500 37 54287 70770 93751
1318	36; 3, 3, 2, 23, 1, 3, 3, 4, 1, 7, 3, 1, 9, 1, 1, 1, 1, 2, 11, 1, 2, 1, 22, 19, 31, 3, 54, 17, 21, 14, 53, 9, 18, 51, 7, 42, 29, 33, 34, 27, 6, 49, 21, 33, 1, 6, (36) 38, 11, (2)	 3138 36679 69954 08957 94554 1 13936 11468 26276 81762 76517
1319	36; 3, 6, 1, 13, 1, 1, 1, (35) 23, 10, 59, 5, 46, 25, 47, (2)	334 01011 12130 59240
1320	36; (3) (24)	3 109
1321*	36; 2, 1, 8, 2, 2, 1, 1, 3, 1, 23, 2, 4, 2, 1, 4, 6, 2, 1, 1, 7, 2, (14, 14) 25, 45, 8, 29, 24, 33, 37, 16, 55, 3, 32, 15, 24, 43, 15, 11, 27, 31, 40, 9, 33, (5, 5)	14 70631 38627 19611 69925 534 50926 71539 66127 96432
1322*	36; 2, 1, 3, 1, (1, 1) 26, 41, 17, 38, (31, 31)	821 29851
1323	36; 2, 1, 2, 7, 1, 2, (2) 27, 37, 26, 9, 47, 22, (27)	1 00360 36 50401
1324	36; 2, 1, 1, 2, 2, 3, 4, 1, 1, 3, 1, 2, 1, 2, 5, 1, 2, 3, 8, 1, 3, 1, 28, 33, 35, 24, 27, 20, 15, 36, 35, 17, 44, 21, 40, 25, 12, 45, 23, 21, 8, 53, 15, 56, 23, 2, 6, 7, 1, 13, 1, 2, 10, (18) 3. 33, 11, 9, 60, 5, 47, 24, 7, (4)	 42650 12718 55353 72450 38443 74810 15 51902 10838 17312 41819 40987 21799
1325*	36; (2, 2) (29, 29)	5 182
1326	36; 2, 2, (2) 30, 25, (26)	70 2549
1327	36; 2, 2, 1, (35) 31, 21, 51, (2)	1785 65024
1328	36; 2, 3, 1, (3) 32, 17, 47, (16)	369 13447
1329	36; 2, 5, 8, 1, 13, 1, 2, 4, 4, 1, 1, 1, 2, 2, 1, 1, 6, (24) 33, 13, 8, 61, 5, 48, 23, 16, 15, 40, 29, 37, 24, 25, 32, 39, 11, (3)	19 32883 59509 34116 704 64145 63882 01495
1330	36; 2, 7, 1, 1, 1, 1, (4) 34, 9, 41, 30, 31, 39, (14)	34182 12 46589
1331	36; 2, 14, 10, 2, 1, 4, 1, 1, 6, 1, 2, 1, (35) 35, 5, 7, 25, 43, 14, 35, 37, 10, 49, 19, 53, (2)	273 03781 10727 9961 20370 19890
1332	36; (2) (36)	2 73

TABLE 1001 TO 1500—*continued.*

1333	36; 1, 1, 23, 1, 5, 7, 1, 17, 2, 1, 1, 1, 5, (2) 37, 36, 3, 59, 12, 9, 61, 4, 27, 36, 29, 41, 12, (31)	904 20166 84140 33012 64933 32151
1334	36; 1, 1, 9, 1, 13, 1, 2, 2, 1, 1, (2) 38, 35, 7, 62, 5, 49, 22, 25, 34, 35, (23)	864 08256 31559 72095
1335	36; 1, 1, 6, (7) 39, 34, 11, (10)	1235 45124
1336	36; 1, 1, 4, 2, 1, 2, 2, 1, 4, 5, 1, 7, 3, 1, 1, (8) 40, 33, 15, 25, 39, 24, 23, 44, 15, 12, 55, 9, 20, 33, 39, (8)	174 56099 99622 6000 40737 19095
1337	36; 1, 1, 3, 2, 1, 8, 2, 4, (10) 41, 32, 19, 23, 47, 8, 31, 16, (7)	373 97272 13674 31647
1338	36; 1, 1, 2, 1, 2, 10, (12) 42, 31, 23, 39, 26, 7, (6)	4 73194 173 08813
1339	36; 1, 1, 2, 4, 1, (4) 43, 30, 27, 14, 51, (13)	4104 1 50175
1340	36; 1, 1, 1, 1, 6, (18) 44, 29, 31, 40, 11, (4)	19932 7 29631
1341	36; 1, 1, 1, 1, 1, 2, 3, 3, 1, 1, 3, 1, 2, 1, 7, 2, 2, 17, 1, 9, 1, 1, 14, (8) 45, 28, 35, 31, 36, 25, 20, 19, 35, 36, 17, 45, 20, 49, 9, 28, 29, 4, 63, 7, 36, 37, 5, (9)	36359 14354 86763 45320 13 31459 72360 20892 39201
1342	36; 1, 1, 1, (2) 46, 27, 39, (22)	30 1099
1343	36; 1, 1, 1, 4, 1, (35) 47, 26, 43, 13, 59, (2)	10591 3 88128
1344	36; 1, 1, 1, (17) 48, 25, 47, (4)	165 6049
1345	36; 1, 2, (14) 49, 24, (5)	132 4841
1346	36; 1, 2, 4, 1, 9, 1, 2, (36) 50, 23, 14, 55, 7, 46, 25, (2)	92 82362 3405 50115
1347	36; 1, 2, 2, 1, 5, 1, (35) 51, 22, 23, 46, 11, 61, (2)	1 64753 60 46682
1348	36; 1, 2, 1, 1, 23, 1, 9, 1, 1, 7, 1, 1, 1, 2, 1, 5, 2, 1, 1, 5, (18) 52, 21, 32, 41, 3, 64, 7, 37, 36, 9, 43, 28, 39, 21, 47, 12, 27, 32, 39, 13, (4)	56 21145 67828 82376 2063 81035 31297 13793
1349	36; 1, 2, 1, 2, 5, 3, 2, 17, 1, 13, 1, (2) 53, 20, 41, 25, 13, 20, 31, 4, 65, 5, 52, (19)	5 55200 34840 203 91806 65951
1350	36; 1, 2, 1, 7, 2, (2) 54, 19, 50, 9, 29, (25)	12804 4 70449
1351	36; 1, 3, (10) 55, 18, (7)	168 6175
1352	36; 1, 3, 2, 1, (17) 56, 17, 23, 49, (4)	3107 1 14243
1353	36; 1, 3, 1, 1, 1, 1, 2, 1, 8, (2) 57, 16, 39, 31, 32, 37, 21, 49, 8, (33)	11 68536 429 82433

TABLE 1001 TO 1500—*continued.*

1354*	36; 1, 3, 1, 11, 2, 6, 1, 7, 3, 4, 1, 1, (2, 2) 58, 15, 55, 6, 33, 10, 57, 9, 22, 15, 38, 33, (25, 25)	35 75033 94613 1315 49590 98165
1355	36; 1, 4, 3, 1, 2, (14) 59, 14, 19, 41, 26, (5)	49532 18 23289
1356	36; 1, 4, 1, 2, 8, 1, 5, 1, 4, (18) 60, 13, 44, 25, 8, 57, 11, 52, 15, (4)	5212 98370 1 91962 41799
1357	36; 1, 5, 6, 1, 1, 7, 1, 1, 1, 5, 2, 17, 1, 23, 1, 1, 1, 1, (2) 61, 12, 11, 36, 37, 9, 44, 27, 43, 12, 33, 4, 67, 3, 44, 29, 33, 36, (23)	9 65303 46430 64640 355 59347 94143 72351
1358	36; 1, 5, 1, 2, 2, (36) 62, 11, 47, 22, 31, (2)	81404 29 99823
1359	36; 1, 6, 2, 1, 1, 2, 2, 1, 4, 1, 1, (3) 63, 10, 27, 34, 35, 25, 23, 45, 14, 37, 35, (18)	468 78231 17281 48040
1360	36; 1, 7, 4, 1, (3) 64, 9, 15, 49, (16)	7749 2 85769
1361*	36; 1, 8, 4, 4, 2, 1, 2, 1, 1, 14, 5, 1, (1, 1) 65, 8, 17, 16, 25, 40, 23, 32, 41, 5, 13, 40, (31, 31)	35 93384 58529 1325 66186 45260
1362	36; 1, 9, 1, 1, 3, 1, 4, 2, (36) 66, 7, 39, 34, 17, 49, 14, 33, (2)	371 27048 13701 84257
1363	36; 1, 11, 3, 7, (1) 67, 6, 23, 9, (58)	93495 34 51726
1364	36; 1, 13, 1, 3, 1, 2, (6) 68, 5, 55, 16, 43, 25, (11)	2 87730 106 26551
1365	36; 1, 17, (2) 69, 4, (35)	684 25271
1366	36; 1, 23, 1, 1, 1, 7, 1, 1, 4, 2, 1, 1, 14, 5, 4, 1, 2, 1, 2, 2, 70, 3, 47, 26, 45, 9, 38, 35, 15, 27, 31, 42, 5, 14, 15, 46, 21, 42, 23, 30, 10, 7, 3, 2, 1, 1, 1, 3, 3, 1, 4, 1, 11, 2, (36) 7, 10, 21, 25, 37, 30, 39, 19, 18, 49, 13, 57, 6, 35, (2)	61 98787 91120 98468 23128 64853 64042 2291 03705 28461 89335 14238 95408 99525
1367	36; 1, (35) 71, (2)	37 1368
1368	36; (1) (72)	1 37
1370*	37; (74) (1)	1 37
1371	37; (37) (2)	37 1370
1372	37; 24, 1, 2, 7, 1, 8, 2, 1, 1, 1, 2, 2, 5, 1, 3, (18) 3, 49, 24, 9, 59, 8, 27, 36, 31, 37, 24, 29, 12, 49, 19, (4)	92062 30469 34552 34 10035 48679 27167
1373*	37; 18, (1, 1) 4, (37, 37)	685 25382
1374	37; 14, 1, 4, 2, 1, 3, 4, 1, 2, (24) 5, 57, 14, 25, 42, 19, 15, 43, 26, (3)	41422 89074 15 35443 25045

TABLE 1001 TO 1500—*continued.*

1375	37 ; 12, 2, 1, 7, 1, 1, 3, 2, 1, 2, (6) 6, 25, 46, 9, 39, 34, 19, 25, 39, 26, (11)	24090 73932 8 93308 52249
1376	37 ; 10, 1, 1, 2, 2, 3, 1, (17) 7, 41, 32, 25, 28, 17, 55, (4)	58 37205 2165 28049
1377	37 ; 9, 3, 1, 3, 1, 7, 2, 5, (4) 8, 19, 47, 16, 53, 9, 32, 13, (17)	11121 62480 4 12700 70401
1378*	37 ; 8, (4, 4) 9, (17, 17)	1153 42801
1379	37 ; 7, 2, 2, 2, 1, 1, 3, 3, (10) 10, 29, 26, 25, 34, 37, 19, 22, (7)	678 06016 25179 68895
1380	37 ; 6, 1, (2) 11, 49, (20)	182 6761
1381*	37 ; 6, 5, 1, 1, 4, 2, 2, 3, 3, 4, 14, 1, 1, 1, 2, 1, 1, 2, 1, 24, 12, 13, 37, 36, 15, 28, 27, 20, 21, 17, 5, 45, 28, 39, 23, 35, 36, 21, 52, 3, 18, 1, 1, 5, 1, 2, 7, 1, 9, 1, 2, 1, (4, 4) 4, 39, 35, 12, 45, 25, 9, 60, 7, 51, 20, 47, (15, 15)	5 76794 95050 76394 67609 26124 87905 214 34743 40923 77992 50381 97708 76282
1382	37 ; 5, 1, 2, 2, 1, 1, (36) 13, 46, 23, 26, 31, 43, (2)	3 49782 130 03237
1383	37 ; 5, 3, 2, 1, (11) 14, 21, 23, 49, (6)	34821 12 94948
1384	37 ; 4, 1, (17) 15, 57, (4)	465 17299
1385*	37 ; 4, 1, 1, 1, (3, 3) 16, 41, 29, 40, (19, 19)	2797 1 04092
1386	37 ; 4, 2, 1, (2) 17, 25, 41, (22)	572 21295
1387	37 ; 4, 8, (37) 18, 9, (2)	40557 15 10442
1388	37 ; 3, 1, 9, 1, 8, 2, 2, 5, 3, (18) 19, 52, 7, 61, 8, 29, 28, 13, 23, (4)	2 20986 97686 82 33062 52807
1389	37 ; 3, 1, 2, 2, (24) 20, 43, 23, 31, (3)	16796 6 25975
1390	37 ; 3, 1, 1, 6, 4, 1, 4, 1, 1, 11, 1, 7, 2, 1, 2, 1, (6) 21, 34, 39, 11, 15, 51, 14, 35, 39, 6, 61, 9, 26, 41, 21, 49, (10)	6165 38516 07582 2 29862 22854 32981
1391	37 ; 3, 2, 1, 1, 1, 6, 1, (4) 22, 25, 38, 29, 43, 10, 55, (13)	2 43492 90 81305
1392	37 ; 3, (4) 23, (16)	42 1567
1393	37 ; 3, (10) 24, (7)	96 3583
1394	37 ; 2, 1, (36) 25, 49, (2)	336 12545

TABLE 1001 TO 1500—*continued.*

1395	37; 2, 1, 6, (8) 26, 45, 11, (9)	3320 1 24001
1396	37; 2, 1, 3, 14, 1, 2, (18) 27, 41, 20, 5, 48, 25, (4)	45 40530 1696 48201
1397	37; 2, 1, 1, 1, 10, 18, 1, 1, 2, (6) 28, 37, 29, 44, 7, 4, 43, 31, 28, (11)	4289 66440 1 60332 48351
1398	37; 2, 1, 1, 3, 2, 1, (36) 29, 33, 38, 19, 23, 51, (2)	1 30154 48 66437
1399	37; 2, 2, 12, 14, 1, 7, 2, 1, 1, 1, 4, 2, 1, 3, 2, 7, 24, 1, 4, (37) 30, 29, 6, 5, 62, 9, 27, 37, 30, 41, 15, 25, 43, 18, 31, 10, 3, 58, 15, (2)	19 32130 14430 04420 15349 722 67866 47899 40453 09640
1400	37; 2, (2) 31, (25)	12 449
1401	37; 2, 3, 14, 1, 2, 5, 2, 2, 1, 1, 6, 4, 1, 1, 8, 1, 4, 10, 2, (24) 32, 21, 5, 49, 24, 13, 29, 25, 33, 40, 11, 16, 35, 39, 8, 55, 15, 7, 35, (3)	44183 34363 33784 97932 16 53779 66015 12296 85015
1402*	37; 2, 3, 1, 9, 1, 11, 1, 1, 2, 1, 7, 1, (1, 1) 33, 17, 54, 7, 63, 6, 41, 33, 22, 49, 9, 42, (31, 31)	2 48379 80029 93 00157 00509
1403	37; 2, 5, 3, (1) 34, 13, 19, (46)	1995 74726
1404	37; 2, 7, 1, (4) 35, 9, 56, (13)	1666 62425
1405	37; 2, (14) 36, (5)	60 2249
1406	37; (2) (37)	2 75
1407	37; 1, 1, (24) 38, 37, (3)	100 3751
1408	37; 1, 1, 10, 4, 1, 1, 2, 7, 1, (17) 39, 36, 7, 16, 39, 33, 28, 9, 63, (4)	3211 01529 1 20487 97377
1409*	37; 1, 1, 6, 3, 9, 14, 1, 9, 1, 3, 1, 3, 1, 1, 1, 1, 1, (2, 2) 40, 35, 11, 23, 8, 5, 64, 7, 55, 16, 49, 17, 40, 31, 35, 32, 37, (25, 25)	94474 40193 46217 35 46252 44206 53380
1410	37; 1, 1, (4) 41, 34, (15)	20 751
1411	37; 1, 1, 3, 2, 4, 1, 1, 3, (37) 42, 33, 19, 30, 15, 38, 35, 21, (2)	118 21953 4440 71330
1412	37; 1, 1, 2, 1, 3, 4, 6, 1, 1, 2, (18) 43, 32, 23, 44, 19, 17, 11, 41, 32, 29, (4)	2703 04352 1 01571 15393
1413	37; 1, 1, 2, 3, 1, 1, 3, 1, 6, 18, 1, 1, 1, 5, (8) 44, 31, 27, 19, 36, 37, 17, 52, 11, 4, 47, 27, 44, 13, (9)	124 23214 30740 4669 87287 31849
1414	37; 1, 1, 1, 1, 11, 1, 14, 8, 3, 2, 5, 2, 1, (4) 45, 30, 31, 43, 6, 65, 5, 9, 21, 30, 13, 25, 45, (14)	485 26483 11810 18247 50630 42299
1415	37; 1, 1, 1, 1, 1, 1, 4, 1, 3, (7) 46, 29, 35, 34, 31, 41, 14, 49, 19, (10)	5 85621 220 29004

TABLE 1001 TO 1500—*continued.*

1416	37; 1, 1, 1, 2, 2, 1, (8) 47, 28, 39, 25, 23, 49, (8)	6858 2 58065
1417*	37; 1, 1, 1, 4, 24, 1, 7, (2, 2) 48, 27, 43, 16, 3, 64, 9, (29, 29)	435 72829 16402 14636
1418*	37; 1, 1, 1, (10, 10) 49, 26, 47, (7, 7)	1033 38899
1419	37; 1, 2, (37) 50, 25, (2)	339 12770
1420	37; 1, 2, 6, 1, 1, (14) 51, 24, 11, 36, 39, (5)	25338 9 54809
1421	37; 1, 2, 3, (2) 52, 23, 20, (29)	260 9801
1422	37; 1, 2, 2, 3, 1, 3, 5, (8) 53, 22, 29, 18, 47, 19, 14, (9)	31 79792 1199 08097
1423	37; 1, 2, 1, 1, 1, 1, 6, 4, 24, 1, 9, 1, 4, 2, 12, 8, 3, 3, 3, 1, 2, (37) 54, 21, 38, 33, 31, 42, 11, 18, 3, 66, 7, 57, 14, 33, 6, 9, 22, 21, 19, 42, 27, (2)	127 24746 01715 56263 96759 4800 11423 18614 76037 04208
1424	37; 1, 2, 1, 3, 1, 2, (4) 55, 20, 47, 17, 44, 25, (16)	13250 5 00001
1425	37; 1, (2) 56, (19)	4 151
1426	37; 1, 3, 4, 1, 3, 1, 1, 1, (2) 57, 18, 15, 50, 17, 41, 30, 39, (23)	2 61132 98 60975
1427	37; 1, 3, 2, 5, 2, 1, 2, 1, 1, 2, (37) 58, 17, 31, 13, 26, 41, 23, 37, 34, 29, (2)	2481 47097 93739 18762
1428	37; 1, 3, 1, (2) 59, 16, 47, (21)	90 3401
1429*	37; 1, 4, 18, 1, (2, 2) 60, 15, 4, 51, (23, 23)	89305 33 75918
1430	37; 1, 4, 2, (2) 61, 14, 29, (26)	352 13311
1431	37; 1, 4, (1) 62, 13, (54)	35 1324
1432	37; 1, 5, 3, 8, 10, 1, 2, 4, (9) 63, 12, 23, 9, 7, 49, 24, 17, (8)	46662 66363 17 65798 05797
1433*	37; 1, 5, 1, 8, 1, 1, 1, 1, 5, 4, (1, 1) 64, 11, 59, 8, 43, 31, 32, 41, 13, 16, (37, 37)	1523 53237 57673 29724
1434	37; 1, 6, 1, 1, 2, 2, 1, 1, 1, 2, 1, 4, 3, (12) 65, 10, 41, 33, 26, 25, 38, 31, 39, 22, 47, 15, 23, (6)	46653 69200 17 66690 98751
1435	37; 1, 7, 2, 3, (7) 66, 9, 31, 21, (10)	26373 9 99046
1436	37; 1, 8, 2, (18) 67, 8, 35, (4)	6840 2 59199

TABLE 1001 TO 1500—*continued.*

1437	37; 1, 9, 1, 5, 2, 2, 4, 18, 1, 2, 1, 1, 1, (24) 68, 7, 59, 12, 29, 28, 17, 4, 53, 21, 41, 28, 47, (3)	256 83205 64040 9735 93382 10399
1438	37; 1, 11, 1, 1, 1, 7, 1, 3, 3, 24, 1, (36) 69, 6, 47, 27, 46, 9, 53, 18, 23, 3, 71, (2)	42 08945 06064 1596 07281 27743
1439	37; 1, 14, 5, 2, 1, 5, 6, 1, 2, 1, 1, 2, 2, 1, 10, 7, 2, (37) 70, 5, 14, 25, 46, 13, 11, 49, 22, 37, 35, 26, 23, 50, 7, 10, 35, (2)	338 95537 37610 21521 12857 98517 00303 64960
1440	37; 1, (17) 71, (4)	19 721
1441	37; 1, 24, 3, 8, 9, 2, 1, 2, 2, 1, 3, 1, 3, 4, 1, 3, 1, 14, 2, 1, 1, 4, 2, (6) 72, 3, 24, 9, 8, 27, 40, 25, 24, 45, 17, 48, 19, 15, 51, 16, 57, 5, 29, 33, 40, 15, 32, (11)	221 54985 95953 31104 82700 8410 14472 84306 85240 54999
1442	37; 1, (36) 73, (2)	38 1443
1443	37; (1) (74)	1 38
1445*	38; (76) (1)	1 38
1446	38; (38) (2)	38 1445
1447	38; 25, 2, 1, 7, 1, 3, 1, 1, 2, 3, 1, 5, 12, 1, 1, (37) 3, 26, 47, 9, 54, 17, 39, 34, 27, 18, 51, 13, 6, 37, 39, (2)	3 42829 14389 10093 130 41033 17574 21848
1448	38; (19) (4)	19 723
1449	38; 15, 4, 1, 2, 4, (8) 5, 16, 45, 25, 17, (9)	72 84800 2773 01249
1450*	38; 12, 1, 2, (8, 8) 6, 49, 25, (9, 9)	1 01933 38 81493
1451	38; 10, 1, 6, 1, 2, 2, 3, 1, 1, 2, 2, 14, 1, 4, 1, 1, (37) 7, 61, 10, 49, 23, 29, 19, 38, 35, 25, 31, 5, 59, 14, 35, 41, (2)	85340 31566 20707 32 50782 78443 28530
1452	38; 9, 1, 1, (18) 8, 37, 39, (4)	6878 2 62087
1453*	38; 8, 2, 5, 2, 1, 1, 5, 1, 3, 6, 10, 1, 2, 1, 2, 1, 1, 3, 18, 1, 3, 1, 9, 33, 13, 28, 33, 41, 12, 51, 19, 12, 7, 52, 21, 44, 23, 36, 37, 21, 4, 57, 17, 36, 1, 6, 2, 1, 2, (25, 25) 39, 11, 27, 39, 28, (3, 3)	37 58165 73334 54563 77143 54805 1432 54652 50389 34442 96990 99982
1454	38; 7, 1, 1, 1, 1, 2, 2, 4, 15, (38) 10, 43, 31, 34, 37, 25, 29, 17, 5, (2)	96169 61884 36 67077 58095
1455	38; 6, 1, (11) 11, 61, (6)	623 23764
1456	38; 6, 2, 1, 7, 1, (3) 12, 25, 48, 9, 55, (16)	1 29855 49 54951
1457	38; 5, 1, 6, 9, 2, 1, 1, 10, 3, 4, (2) 13, 56, 11, 8, 29, 32, 43, 7, 23, 16, (31)	18 94953 10312 723 31628 36703

TABLE 1001 TO 1500—*continued.*

1458	38; 5, 2, 3, 1, 3, 1, 2, 1, 1, 7, 1, 10 (38) 14, 31, 18, 49, 17, 46, 23, 34, 41, 9, 62, 7, (2)	78 88890 25540 3012 27540 96401
1459	38; 5, 12, 1, 1, 7, 8, 2, 1, 4, 2, 2, 2, 1, 1, 1, 5, 4, 15, 25, 2, 1, 1, (37) 15, 6, 39, 37, 10, 9, 26, 45, 15, 29, 27, 25, 39, 30, 43, 13, 18, 5, 3, 30, 31, 45, (2)	25694 24820 98124 22872 52281 9 81439 56337 99403 45032 81010
1460	38; 4, 1, 3, 4, 1, 1, (18) 16, 49, 19, 16, 35, 41, (4)	6 25898 239 15529
1461	38; 4, 2, 14, 1, 5, 2, 3, 2, 1, 3, 3, 18, 1, 4, 6, 1, 2, 1, (24) 17, 33, 5, 61, 12, 31, 20, 25, 44, 19, 23, 4, 59, 15, 11, 51, 20, 55, (3)	10897 33506 65845 02228 4 16529 16197 97289 87575
1462	38; 4, (4) 18, (17)	72 2753
1463	38; (4) (19)	4 153
1464	38; 3, 1, 4, 2, 1, (5) 20, 49, 15, 25, 47, (12)	23729 9 07925
1465*	38; 3, 1, 1, 1, 2, 1, 1, 4, 4, 1, 7, 1, 2, (3, 3) 21, 40, 31, 39, 24, 35, 39, 16, 15, 56, 9, 49, 24, (21, 21)	2 14427 21285 82 07269 84932
1466*	38; 3, 2, 7, 4, 2, 1, (2, 2) 22, 31, 10, 17, 26, 41, (25, 25)	40 05185 1533 52043
1467	38; 3, 3, 6, 1, 1, 1, (37) 23, 22, 11, 46, 27, 49, (2)	16 73045 640 80026
1468	38; 3, 5, 1, 1, 3, 2, 25, 9, 1, 1, 5, 1, 6, 8, 2, 1, 2, 1, 1, (18) 24, 13, 39, 36, 19, 33, 3, 8, 39, 37, 12, 57, 11, 9, 27, 41, 24, 33, 43, (4)	18883 73030 64922 35216 7 23520 46281 34792 85247
1469	38; 3, 18, 1, (4) 25, 4, 61, (13)	19836 7 60265
1470	38; 2, 1, (14) 26, 49, (5)	138 5291
1471	38; 2, 1, 4, 1, 4, 3, 2, 4, 12, 1, 1, 3, 1, 2, 1, 6, 1, 14, 2, 8, 25, 27, 45, 14, 53, 15, 21, 30, 17, 6, 41, 35, 18, 47, 21, 51, 10, 63, 5, 35, 9, 3, 2, 4, 1, 1, 1, 1, 1, 10, 2, 1, (37) 34, 15, 42, 31, 37, 30, 45, 7, 25, 51, (2)	 1 01712 89163 97276 23871 56715 36737 39 01057 37347 28288 54875 60183 66160
1472	38; 2, 1, (2) 28, 41, (23)	30 1151
1473	38; 2, 1, 1, 1, 2, 1, 2, 2, 9, 5, 1, 3, 1, (24) 29, 37, 32, 39, 23, 43, 24, 31, 8, 13, 53, 16, 59, (3)	6 92736 14220 265 86992 93399
1474	38; 2, 1, 1, 4, 1, 7, 1, 2, 2, 4, 1, 2, 3, 1, 10, 5, (38) 30, 33, 41, 14, 57, 9, 50, 23, 30, 15, 46, 25, 18, 55, 7, 15, (2)	1 43151 08109 79476 54 95957 60507 29745
1475	38; 2, 2, 6, 1, 1, (2) 31, 29, 11, 41, 34, (25)	14628 5 61799
1476	38; 2, 2, 1, 1, 2, 1, 3, (8) 32, 25, 36, 37, 23, 45, 20, (9)	2 18560 83 96801
1477	38; 2, 3, 6, 8, 2, 1, 1, 1, 1, 1, 5, 1, 3, 1, 2, 18, 1, 6, 25, 2, (10) 33, 21, 12, 9, 28, 37, 33, 36, 31, 43, 12, 53, 17, 44, 27, 4, 63, 11, 3, 36, (7)	2 61633 72121 47062 22240 100 55043 78017 41504 36351
1478	38; 2, 4, (38) 34, 17, (2)	3114 1 19717

TABLE 1001 TO 1500—*continued.*

1479	38 ; 2, 5, 2, 2, 1, 1, 1, (1) 35, 13, 30, 25, 38, 33, 35, (34)	1 14525 44 04376
1480	38 ; 2, 8, (19) 36, 9, (4)	5559 2 13859
1481*	38 ; 2, 14, 1, 8, 1, 2, 5, 1, 1, 2, 1, 4, 10, 1, 3, 1, 1, (1, 1) 37, 5, 65, 8, 49, 25, 13, 40, 35, 23, 47, 16, 7, 56, 17, 41, 32, (35, 35)	1 01945 97513 61405 39 23264 72655 17468
1482	38 ; (2) (38)	2 77
1483	38 ; 1, 1, 25, 5, 1, 7, 1, 2, 1, 1, 1, 1, 2, 4, 6, 1, 3, 2, 2, 1, 1, 12, 3, 1, (37) 39, 38, 3, 13, 58, 9, 51, 22, 39, 33, 34, 37, 27, 17, 11, 53, 18, 29, 26, 33, 43, 6, 19, 57, (2)	41 20928 70570 44632 38045 1586 95889 00942 30191 81226
1484	38 ; 1, 1, (10) 40, 37, (7)	44 1695
1485	38 ; 1, 1, (6) 41, 36, (11)	28 1079
1486	38 ; 1, 1, 4, 1, 1, 1, 3, (38) 42, 35, 15, 43, 30, 41, 21, (2)	4 92228 189 74735
1487	38 ; 1, 1, 3, 1, 1, 3, 1, (37) 43, 34, 19, 37, 38, 17, 59, (2)	2 05495 79 24224
1488	38 ; 1, 1, 2, 1, (5) 44, 33, 23, 49, (12)	315 12151
1489*	38 ; 1, 1, 2, 2, 1, 4, 2, 3, 1, (1, 1) 45, 32, 27, 24, 47, 15, 31, 19, 40, (33, 33)	25 78145 994 84332
1490*	38 ; 1, (1, 1) 46, (31, 31)	5 193
1491	38 ; 1, 1, 1, 1, 2, 2, (1) 47, 30, 35, 37, 26, 25, (42)	1767 68230
1492	38 ; 1, 1, 1, 2, 10, 1, 1, 1, (18) 48, 29, 39, 28, 7, 48, 27, 49, (4)	13 56270 523 87849
1493*	38 ; 1, 1, 1, 3, 2, 2, 18, 1, (10, 10) 49, 28, 43, 19, 28, 31, 4, 67, (7, 7)	1694 41225 65471 00182
1494	38 ; 1, 1, 1, 7, 15, 5, (38) 50, 27, 47, 10, 5, 25, (2)	440 05214 17009 02565
1495	38 ; 1, (1) 51, (26)	3 116
1496	38 ; 1, 2, (9) 52, 25, (8)	87 3365
1497	38 ; 1, 2, 4, 4, 1, 1, 1, 1, 6, 2, 2, 1, 8, 1, (24) 53, 24, 17, 16, 41, 33, 32, 43, 11, 31, 23, 51, 8, 67, (3)	58 19614 30932 2251 67187 51127
1498	38 ; 1, 2, 2, 1, 1, 1, 3, 1, 12, 8, 1, 1, (10) 54, 23, 26, 39, 31, 42, 17, 57, 6, 9, 38, 39, (7)	83118 18384 32 17006 59967
1499	38 ; 1, 2, 1, 1, 7, 5, 1, 4, 1, 2, 3, 1, 2, 1, 1, 2, 10, 1, 2, 15, 6, 1, (37) 55, 22, 35, 41, 10, 13, 55, 14, 47, 25, 19, 46, 23, 38, 35, 29, 7, 49, 26, 5, 11, 65, (2)	77552 05839 43067 80299 30 02576 94656 26953 60610
1500	38 ; 1, 2, 1, 2, 2, 1, 6, 2, 1, (18) 56, 21, 44, 25, 24, 49, 11, 25, 51, (4)	118 11844 4574 70751

In connexion with the subject we have a paper, "A Table of the Square Roots of Prime Numbers of the form $4m+1$ less than 10000 expanded as Periodic Continued Fractions," by C. A. Roberts, with Introduction and Explanation by Artemas Martin, the *Mathematical Magazine*, vol. II. (No. 7, for October, 1892), pp. 105—120. This extends, in fact, to numbers up to 10501, but only the denominators of the continued fractions (that is, the first lines of Degen's and the present table) are given: thus the entry for 1009 is 31; 1, (3, 3).

The paper just referred to notices errors in Degen's tables for the numbers 853 and 929. For 853 the first line should be

$$29,\ 4,\ 1,\ 5,\ 1,\ 2,\ 4,\ 1,\ 1,\ 15,\ 19,\ (2,\ 2),$$

(15 instead of Degen's 14). For 929 the first and second lines should be

$$\begin{array}{l} 30,\ 2,\ 11,\ 1,\ 2,\ 3,\ 2,\ 7,\ 5,\ (2,\ 2) \\ 1,\ 29,\ 5,\ 40,\ 19,\ 16,\ 25,\ 8,\ 11,\ (23,\ 23). \end{array}$$

The values of x, y in Table I. and those in Table II. (for the solution of $y^2 = ax^2 - 1$) are correct for each of the numbers 853 and 929.

949.

ON HALPHEN'S CHARACTERISTIC n, IN THE THEORY OF CURVES IN SPACE.

[From *Crelle's Journal d. Mathematik*, t. cxi. (1893), pp. 347—352.]

If we consider a curve in space without actual singularities, of the order (or degree) d, then this has a number h of apparent double points (adps.), viz. taking as vertex an arbitrary point in space, we have through the curve a cone of the order d, with h nodal lines; and Halphen denotes by n the order of the cone of lowest order which passes through these h lines. For a given value of d, h is at most $=\frac{1}{2}(d-1)(d-2)$, and as shown by Halphen it is at least $=[\frac{1}{4}(d-1)^2]$, if we denote in this manner the integer part of $\frac{1}{4}(d-1)^2$. For given values of d, h, it is easy to see that n must lie within certain limits, viz. if ν be the smallest number such that $\frac{1}{2}\nu(\nu+3)$ equal to or greater than h, then n is at most $=\nu$; and moreover n must have a value such that nd is at least $=2h$, or say we must have $nd=2h+\theta$, where θ is $=0$ or positive. For any given value of d, we thus have a finite number of forms (d, h, n), and we have thus *prima facie* curves in space of the several forms (d, h, n): but it may very well be, and in fact Halphen finds, that when $d=9$ or upwards, then for certain values of h, n as above, there is not any curve (d, h, n); thus $d=9$, $h=17$ the values of n are $n=4$, $n=5$, but there is not any curve $d=9$, $h=17$ for either of these values of n; or say the curves (9, 17, 4) and (9, 17, 5) are non-existent. And in the Notes and References to the papers 302, 305 in vol. v. of my *Collected Mathematical Papers*, 4to. Cambridge, 1892, see p. 615, I remarked that, in certain cases for which Halphen finds a curve (d, h, n), such curve does not exist except for special configurations of the h nodal lines not determined by the mere definition of n as the order of the cone of lowest order which passes through the h nodal lines; for instance $d=9$, $h=16$, $n=4$, for which Halphen gives a curve, I find that for the existence of the curve it is not enough that the 16 lines are situate upon a quartic cone, but they must be the 16 lines of intersection of two quartic cones.

In fact, starting from an existing curve, say the complete intersection of two given surfaces of the orders μ, ν respectively, we have, as is known,

$$d = \mu\nu, \quad 2h = \mu\nu(\mu-1)(\nu-1):$$

we find also, as will appear, $n = (\mu-1)(\nu-1)$: hence also $nd = 2h$, viz. the cone of the order n through the h nodal lines meets the cone of the order d in these lines counting each twice, and in no other lines. I remark also that h is $= \frac{1}{2}n(n+3)$ if $\mu+\nu=4$; viz. we have here two nodal lines lying in a plane; but if $\mu+\nu>4$, then h is greater than $\frac{1}{2}n(n+3)$, viz. in this case the nodal lines are not h lines taken at pleasure, but they are lines subject to the condition of lying in a cone of the order n. But this is not all; the h nodal lines lie not only in this cone of lowest order n, but also in cones of the orders $n+1$, $n+2$, ..., $n+(\mu+\nu-2)$ respectively: I do not for the moment attempt to determine the number of independent conditions which are hereby imposed upon the h lines.

The last-mentioned theorem constitutes in fact the geometrical interpretation of results contained in Jacobi's paper "De eliminatione variabilis e duabus aequationibus algebraicis," *Crelle*, t. XV. (1836), pp. 101—124, [*Ges. Werke*, t. III. pp. 295—320]. Consider the two equations

$$U = (*\between x, y, z, w)^{\mu} = 0; \quad V = (*\between x, y, z, w)^{\nu} = 0;$$

representing surfaces of the orders μ, ν respectively; since the form of the equations is quite arbitrary, we may without loss of generality assume that the point

$$(x, y, z) = (0, 0, 0)$$

is an arbitrary point in space; and this being so, we find the equation of the cone, vertex this point, which passes through the curve of intersection of the two surfaces by the mere elimination of w between the two equations. As the reasoning is exactly the same for a particular case, I write for convenience $\mu=3$, $\nu=4$, and consider the two equations

$$A_0w^3 + A_1w^2 + A_2w + A_3 = 0,$$
$$B_0w^4 + B_1w^3 + B_2w^2 + B_3w + B_4 = 0,$$

where the suffixes show the degrees in regard to (x, y, z), viz. A_0, B_0 are mere constants, A_1, B_1 are linear functions $(*\between x, y, z)^1$, A_2, B_2 quadric functions $(*\between x, y, z)^2$, and so on. Multiplying the first equation successively by 1, w, w^2, w^3, and the second equation successively by 1, w, w^2, we have 7 equations from which to eliminate 1, w, w^2, w^3, w^4, w^5, w^6, and the result is

$$\begin{vmatrix} & & & A_0, & A_1, & A_2, & A_3 \\ & & A_0, & A_1, & A_2, & A_3, & \\ & A_0, & A_1, & A_2, & A_3, & & \\ A_0, & A_1, & A_2, & A_3, & & & \\ & & B_0, & B_1, & B_2, & B_3, & B_4 \\ & B_0, & B_1, & B_2, & B_3, & B_4, & \\ B_0, & B_1, & B_2, & B_3, & B_4, & & \end{vmatrix} = 0,$$

viz. this is an equation

$$(*\between x, y, z)^{12}=0,$$

of the cone of the order $d=12$, through the curve of intersection of the two surfaces: say this equation is $\Omega=0$.

But if we only multiply the first equation by 1, w, w^2 successively and the second equation by 1, w successively, then we have 5 equations serving to determine the ratios of w^5, w^4, w^3, w^2, w, 1, viz. we have these quantities proportional to the six determinants which can be formed out of the matrix

$$\left|\begin{array}{cccccc} & & A_0, & A_1, & A_2, & A_3 \\ & A_0, & A_1, & A_2, & A_3, & \\ A_0, & A_1, & A_2, & A_3, & & \\ & B_0, & B_1, & B_2, & B_3, & B_4 \\ B_0, & B_1, & B_2, & B_3, & B_4, & \end{array}\right|;$$

say we have

$$\begin{aligned} &w^5 : w^4 : w^3 : w^2 : w : 1 \\ &= L : M : N : P : Q : R, \end{aligned}$$

where L, M, N, P, Q, R represent homogeneous functions $(*\between x, y, z)^\theta$, of the degrees 11, 10, 9, 8, 7, 6 respectively. We may if we please write

$$w=\frac{L}{M}=\frac{M}{N}=\frac{N}{P}=\frac{P}{Q}=\frac{Q}{R};$$

or eliminating w, we have the series of equations which may be written

$$\left|\begin{array}{ccccc} L, & M, & N, & P, & Q \\ M, & N, & P, & Q, & R \end{array}\right|=0,$$

viz. we thus denote that the determinants formed with any two columns of this matrix are severally $=0$. This of course implies that each of the determinants in question is the product of Ω and a factor which is a homogeneous function of the proper degree in (x, y, z), so that the several equations are all of them satisfied if only $\Omega=0$. We have for instance $PR-Q^2=A\Omega$, where A is a quadric function $(*\between x, y, z)^2$; similarly $NR-PQ=B\Omega$, where B is a cubic function $(*\between x, y, z)^3$; and the like as regards the other determinants.

If the ratios $x : y : z$ have any given values such that we have for these $\Omega=0$, then w has a determinate value, that is, on each line of the cone $\Omega=0$, there is a single point of the curve of intersection of the two surfaces: the only exceptions are when, for the given values of $x : y : z$, the expressions for w assume an indeterminate form, viz. w has then two values, and there are upon the line two points of the curve, or what is the same thing, the line is a nodal line of the cone: the conditions for a nodal line thus are $L=0$, $M=0$, $N=0$, $P=0$, $Q=0$, $R=0$, viz. each of these equations is that of a cone passing through the nodal lines of the cone $\Omega=0$; the cone of lowest order is $R=0$, a cone of the order 6 meeting the cone $\Omega=0$ of the order 12 in 36 lines each twice, which lines are consequently the nodal lines of the cone $\Omega=0$. The mere condition of the 36 lines lying upon a cone of the order 6 shows that the 36 lines are not arbitrary; and we have moreover, through the 36 lines, cones of the orders 7, 8, 9, 10 and 11 respectively. Obviously the foregoing reasoning is quite general, and for the surfaces of the orders μ, ν we have (as stated above) the cone Ω of the order $\mu\nu$, with $h=\frac{1}{2}\mu\nu(\mu-1)(\nu-1)$ nodal lines, the intersections (each counting twice) of this cone with a cone of the order $n=(\mu-1)(\nu-1)$; and moreover the h nodal lines lie also in cones of the orders $n+1$, $n+2$, ..., $n+\mu+\nu-2$ respectively.

To examine the meaning of the theorem, I form the table

μ, ν	d	h	n, ..., $n+\mu+\nu-2$
2, 2	4	2	1, 2, 3
2, 3	6	6	2, 3, 4, 5
2, 4	8	12	3, 4, 5, 6, 7
2, 5	10	20	4, 5, 6, 7, 8, 9
⋮			
3, 3	9	18	4, 5, 6, 7, 8
3, 4	12	36	6, 7, 8, 9, 10, 11
3, 5	15	60	8, 9, 10, 11, 12, 13, 14
⋮			
4, 4	16	72	9, 10, 11, 12, 13, 14, 15
4, 5	20	120	12, 13, 14, 15, 16, 17, 18, 19
⋮			

Here μ, $\nu=2$, 2, there are 2 nodal lines, which are arbitrary, and of course lie on cones of the orders 1, 2, 3 respectively. So μ, $\nu=2$, 3; there are 6 nodal lines, which are not arbitrary, inasmuch as they lie on a cone of the order 2; but regarding them as arbitrary lines on such a cone, we can through them draw a cone of the order 3 or any higher order, and it is thus no specialisation to say that they lie upon cones of the orders 3, 4, and 5. But going a step further $\mu=2$, $\nu=4$: here we have 12 nodal lines which, inasmuch as they lie on a cone of the order 3, are not arbitrary: and they are not arbitrary lines upon this cone, for they lie on a cone of the order 4, and such a cone can be drawn through at most 11 arbitrary lines on a cubic cone. In fact, upon a cone of the order θ, taking at pleasure N lines, the condition that it may be possible through these to draw a cone of the order $\theta+1$ is $\frac{1}{2}(\theta+1)(\theta+4)=N+3$ at least; for if this number were $=N+2$, then through the N points we have only the improper cone $(x+\beta y+\gamma z)\,U_\theta=0$, if $U_\theta=0$ is the cone of the order θ. It thus appears that the 12 nodal lines are not arbitrary lines on a cubic cone, but that they constitute the complete intersection of a cubic cone and a quartic cone. But through such 12 lines we may draw cones of the 5th and higher orders, and it is thus no further condition that the 12 lines lie on cones of the orders 5, 6 and 7 respectively.

So again $\mu=3$, $\nu=3$; we have here 18 nodal lines which, inasmuch as they lie on a cone of the order 4, are not arbitrary: and they are not arbitrary lines on this cone inasmuch as they lie also on a cone of the order 5, and such a cone can be drawn through at most 17 arbitrary lines on the quartic cone: it thus appears that the 18 nodal lines are 18 out of the 20 lines of intersection of a quartic cone and a quintic cone. But there is no further condition, for through such lines we can draw a cone of the order 6 or any higher order and thus the lines lie on cones of the orders 6, 7 and 8 respectively. It appears probable however that, for higher values of μ, ν, it would be necessary to take account not only (as in these examples) of the cones of the orders n and $n+1$, but of those of higher orders $n+2$, &c.; and thus that it is *not* the true form of the theorem to say that the h nodal lines must be h out of the $n(n+1)$ lines of intersection of two cones of the orders n and $n+1$ respectively.

It appears, by what precedes, that the h, $=\frac{1}{2}\mu\nu(\mu-1)(\nu-1)$, lines which are the nodal lines of the cone of arbitrary vertex which passes through the curve of intersection of two surfaces of the orders μ, ν respectively, form a remarkable special system of lines, which well deserve further study. I remark also that, without having proved the negative, it seems to me clear that given the values of d, h, n it is only in the cases where the h lines form some such special system (and not in the general case where the h nodal lines are any lines whatever on a cone of the order n) that there exists a curve (d, h, n); and thus that the question for further investigation is, for given values of (d, h, n) to determine the conditions necessary for the existence of a curve in space with these characteristics (d, h, n).

950.

ON THE SEXTIC RESOLVENT EQUATIONS OF JACOBI AND KRONECKER.

[From *Crelle's Journal d. Mathematatik*, t. CXIII. (1894), pp. 42—49.]

THE equations referred to are: the first of them, that given by Jacobi in the paper "Observatiunculae ad theoriam aequationum pertinentes," *Crelle*, t. XIII. (1835), pp. 340—352, [*Ges. Werke*, t. III., pp. 269—284], under the heading "Observatio de aequatione sexti gradus ad quam aequationes sexti gradus revocari possunt," and the second, that of Kronecker in the note "Sur la résolution de l'équation du cinquième degré," *Comptes Rendus*, t. XLVI. (1858), pp. 1150—1152. Jacobi's equation is closely connected with that obtained by Malfatti in 1771, see Brioschi's paper "Sulla resolvente di Malfatti per l'equazione del quinto grado," *Mem. R. Ist. Lomb.*, t. IX. (1863); but the characteristic property first presents itself in Jacobi's form, and I think the equation is properly described as Jacobi's resolvent equation. The other equation has been always known as Kronecker's resolvent equation; it belongs to the class of equations for the multiplier of an elliptic function considered by Jacobi in the paper "Suite des notices sur les fonctions elliptiques," *Crelle*, t. III. (1828), pp. 303—310, see p. 308, [*Ges. Werke*, t. I., pp. 255—263, see p. 261]: say Kronecker's equation belongs to the class of Jacobi's Multiplier Equations. We have in regard to it the paper by Brioschi, "Sul metodo di Kronecker per la resoluzione delle equazioni di quinto grado," *Atti Ist. Lomb.*, t. I. (1858), pp. 275—282, and see also the "Appendice terza" to his translation of my *Elliptic Functions* (Milan, 1880): it seems to me however that the theory of Kronecker's equation has not hitherto been exhibited in the clearest form.

I consider the forms

12345	13524
13254	12435
24315	23541
35421	34152
41532	45213
52143	51324

which, if in the first instance the figures are regarded as points, represent the twelve pentagons which can be formed with the points 1, 2, 3, 4, 5; each form in the right-hand column is derived from the corresponding form in the left-hand column by *stellation*, say we have $13524 = S12345$, and so in other cases.

A pentagon is in general reversible, but we sometimes consider it as irreversible (viz. we distinguish between the pentagons 12345 and 15432); when this is so, we write $15432 = R12345$, and we have thus twelve new forms, in all twenty-four forms. The symbols R, S are such that $R^2 = 1$, $S^2 = R$, $RS = SR$, $S^4 = 1$. But for a reversible pentagon, there is no occasion to use the symbol R, and we have simply $S^2 = 1$.

In a somewhat different point of view, we may for an irreversible pentagon write

12345 to denote any one of the forms 12345, 23451, 34512, 45123, 51234,

R12345 " " " " 15432, 54321, 43215, 32154, 21543,

and for a reversible pentagon 12345 to denote any one of these same ten forms.

Each pentagon gives thus ten forms, viz. we have in all 120 forms which all are the different arrangements of the five figures. But further regarding the arrangement 12345 as positive, then the forms in the left-hand column are each of them positive, and the ten forms derived from any one of these are each of them positive, that is, the forms in the left-hand column give all the 60 positive arrangements of the five figures: and similarly the forms in the right-hand column give all the 60 negative arrangements of the five figures.

Taking 1, 2, 3, 4, 5 to denote any quantities, or say the five roots x_1, x_2, x_3, x_4, x_5 of a quintic equation, we regard 12345, ..., as denoting functions of these roots: in particular, 12345 may denote a cyclic reversible function, the analogue of the reversible pentagon, or it may denote a cyclic irreversible function, the analogue of the irreversible pentagon.

Jacobi's resolvent equation.

The most simple course is to take 12345 a cyclic reversible function of the roots x; a root of the resolvent equation is then $12345 - 13524$, $= (1 - S)\,12345$, and the six roots are

$$z_1 = (1 - S)\,12345,$$
$$z_2 = (1 - S)\,13254,$$
$$z_3 = (1 - S)\,24315,$$
$$z_4 = (1 - S)\,35421,$$
$$z_5 = (1 - S)\,41532,$$
$$z_6 = (1 - S)\,52143.$$

Here effecting on the roots x any positive substitution whatever, we permute *inter se* the roots z; but effecting on the roots x any negative substitution whatever, then reversing the signs of all the roots z, we permute *inter se* these reversed values. Thus effecting on the root z_1 the negative substitution 12, it becomes $(1-S)\,21345$, which is

$$=(1-S)\,S23514=(S-S^2)\text{ (that is, } S-1 \text{ or } -(1-S))\ 23514=-(1-S)\,41532=-z_5;$$

and similarly for the effect of the same substitution 12 upon any other of the roots z.

It follows that any rational symmetrical function of the roots z is a two-valued function of the coefficients of the quintic equation, viz. it is a function of the form $P+Q\sqrt{\Delta}$, where Δ is the discriminant and P, Q are rational functions of the coefficients of the quintic equation.

In particular, if 12345, ..., are rational and integral functions of the roots x, then for any rational and integral function of the roots z, we have P and Q rational and integral functions of the coefficients, and any rational and integral homogeneous function of the roots z, according as it is of an even or an odd degree in these roots, will be of the form P or of the form $Q\sqrt{\Delta}$; the resolvent equation is thus of the form

$$(1,\ B\sqrt{\Delta},\ C,\ D\sqrt{\Delta},\ E,\ F\sqrt{\Delta},\ G\between z,\ 1)^6=0,$$

where Δ is the discriminant, and B, C, D, E, F, G are rational and integral functions of the coefficients of the quintic equation.

The most simple form of the function 12345 is that employed by Jacobi and for which $12345=12+23+34+45+51$, where 12, ..., denote x_1x_2, ..., respectively.

For comparison with Kronecker's equation, it is proper to take 12345 a cyclic *irreversible* function of the roots x; we have then

$$12345+15432=(1+R)\,12345,$$

a cyclic reversible function, and the roots of Jacobi's equation will be in the first (or since Kronecker writes x_0, x_1, x_2, x_3, x_4 instead of x_1, x_2, x_3, x_4, x_5, in the second) of the following two forms, say

$z_1=(1+R)(1-S)\,12345,$	01234,
$z_2=(1+R)(1-S)\,13254,$	02143,
$z_3=(1+R)(1-S)\,24315,$	13204,
$z_4=(1+R)(1-S)\,35421,$	24310,
$z_5=(1+R)(1-S)\,41532,$	30421,
$z_6=(1+R)(1-S)\,52143;$	41032,

viz. in the first form the terms are 12345, ..., and in the second they are 01234, ..., but the theory is in no wise altered by this change of form.

Kronecker's resolvent equation.

Kronecker writes x_m to denote x_0, x_1, x_2, x_3, x_4 according as the residue of m (mod. 5) is $=0$, 1, 2, 3 or 4: then putting

$$x_m x^2_{m+n} x^2_{m+2n} + v x^3_m x_{m+n} x_{m+2n} = m \,.\, m+n \,.\, m+2n,$$

a root is

$$\begin{aligned} f = \;& (012+123+234+340+412)\sin\frac{2\pi}{5} \\ & + (024+130+241+302+413)\sin\frac{4\pi}{5} \\ & + (031+142+203+314+420)\sin\frac{6\pi}{5} \\ & + (043+104+210+321+432)\sin\frac{8\pi}{5}, \end{aligned}$$

and the other roots are deduced from this by changing 01234 into 03412, 14023, 20134, 31240, 42301 respectively.

Taking ϵ an imaginary fifth root of unity, say $\epsilon = \cos\frac{2\pi}{5} + i\sin\frac{2\pi}{5}$, so that $\sin\frac{2\pi}{5}$, $\sin\frac{4\pi}{5}$, $\sin\frac{6\pi}{5}$, $\sin\frac{8\pi}{5}$ are as $\epsilon-\epsilon^4$, $\epsilon^2-\epsilon^3$, $\epsilon^3-\epsilon^2$, $\epsilon^4-\epsilon$; also writing

$$01234 = 012+123+234+340+412, \;\ldots,$$

so that 01234, ..., are cyclic irreversible functions of the roots x, then the expression for the root f is

$$\begin{aligned} f = \;& (\epsilon-\epsilon^4)\,01234 + (\epsilon^2-\epsilon^3)\,02413 \\ & - (\epsilon-\epsilon^4)\,04321 - (\epsilon^2-\epsilon^3)\,03142, \end{aligned}$$

or, as this may be written,

$$f = \{(\epsilon-\epsilon^4)(1-R) + (\epsilon^2-\epsilon^3)(1-R)S\}\,01234,$$

and the expressions for the six roots are

$$\begin{aligned} f\; &= \{(\epsilon-\epsilon^4)(1-R) + (\epsilon^2-\epsilon^3)(1-R)S\}\,01234, \\ f_0 &= \{(\epsilon-\epsilon^4)(1-R) + (\epsilon^2-\epsilon^3)(1-R)S\}\,03412, \\ f_1 &= \{(\epsilon-\epsilon^4)(1-R) + (\epsilon^2-\epsilon^3)(1-R)S\}\,14023, \\ f_2 &= \{(\epsilon-\epsilon^4)(1-R) + (\epsilon^2-\epsilon^3)(1-R)S\}\,20134, \\ f_3 &= \{(\epsilon-\epsilon^4)(1-R) + (\epsilon^2-\epsilon^3)(1-R)S\}\,31240, \\ f_4 &= \{(\epsilon-\epsilon^4)(1-R) + (\epsilon^2-\epsilon^3)(1-R)S\}\,42301. \end{aligned}$$

I write down the analogous functions

$$\begin{aligned} F\; &= \{(\epsilon-\epsilon^4)(1-R) + (\epsilon^2-\epsilon^3)(1-R)S\}\,03142, \quad (=RS01234), \\ F_0 &= \{(\epsilon-\epsilon^4)(1-R) + (\epsilon^2-\epsilon^3)(1-R)S\}\,01324, \quad (=RS03412), \\ F_1 &= \{(\epsilon-\epsilon^4)(1-R) + (\epsilon^2-\epsilon^3)(1-R)S\}\,12430, \quad (=RS14023), \\ F_2 &= \{(\epsilon-\epsilon^4)(1-R) + (\epsilon^2-\epsilon^3)(1-R)S\}\,23041, \quad (=RS20134), \\ F_3 &= \{(\epsilon-\epsilon^4)(1-R) + (\epsilon^2-\epsilon^3)(1-R)S\}\,34102, \quad (=RS31240), \\ F_4 &= \{(\epsilon-\epsilon^4)(1-R) + (\epsilon^2-\epsilon^3)(1-R)S\}\,40213, \quad (=RS42301). \end{aligned}$$

This being so, I say that any positive substitution on the roots x permutes *inter se* the roots f, reversing in some cases the signs; and that any negative substitution on the roots x changes the roots f into the roots F, permuting these roots *inter se* and reversing in some cases the signs. And similarly as to the effect on the roots F of a positive substitution and a negative substitution respectively.

Thus the positive substitution 123 changes

$$\begin{array}{lll} f \text{ into } & \{(\epsilon-\epsilon^4)(1-R)+(\epsilon^2-\epsilon^3)(1-R)S\}\,02314 & = f_1, \\ f_0 \quad ,, & \{(\epsilon-\epsilon^4)(1-R)+(\epsilon^2-\epsilon^3)(1-R)S\}\,01423 & = f_4, \\ f_1 \quad ,, & \{(\epsilon-\epsilon^4)(1-R)+(\epsilon^2-\epsilon^3)(1-R)S\}\,24031 & = f_3, \\ f_2 \quad ,, & \{(\epsilon-\epsilon^4)(1-R)+(\epsilon^2-\epsilon^3)(1-R)S\}\,30214 = R03412 & = -f_0, \\ f_3 \quad ,, & \{(\epsilon-\epsilon^4)(1-R)+(\epsilon^2-\epsilon^3)(1-R)S\}\,12340 & = f, \\ f_4 \quad ,, & \{(\epsilon-\epsilon^4)(1-R)+(\epsilon^2-\epsilon^3)(1-R)S\}\,43102 = R20134 & = -f_2; \end{array}$$

viz. $f, f_0, f_1, f_2, f_3, f_4$ are changed into $f_1, f_4, f_3, -f_0, f, -f_2$.

And similarly the negative substitution 12 changes

$$\begin{array}{lll} f \text{ into } & \{(\epsilon-\epsilon^4)(1-R)+(\epsilon^2-\epsilon^3)(1-R)S\}\,02134 & = F_4, \\ f_0 \quad ,, & \{(\epsilon-\epsilon^4)(1-R)+(\epsilon^2-\epsilon^3)(1-R)S\}\,03421 = R12430 & = -F_1, \\ f_1 \quad ,, & \{(\epsilon-\epsilon^4)(1-R)+(\epsilon^2-\epsilon^3)(1-R)S\}\,24013 & = F_0, \\ f_2 \quad ,, & \{(\epsilon-\epsilon^4)(1-R)+(\epsilon^2-\epsilon^3)(1-R)S\}\,10234 & = F_3, \\ f_3 \quad ,, & \{(\epsilon-\epsilon^4)(1-R)+(\epsilon^2-\epsilon^3)(1-R)S\}\,32140 = R23041 & = -F_2, \\ f_4 \quad ,, & \{(\epsilon-\epsilon^4)(1-R)+(\epsilon^2-\epsilon^3)(1-R)S\}\,41302 = R03142 & = -F, \end{array}$$

viz. $f, f_0, f_1, f_2, f_3, f_4$ are changed into $F_4, -F_1, F_0, F_3, -F_2, -F$.

Hence considering the equation the roots of which are $f^2, f_0^2, f_1^2, f_2^2, f_3^2, f_4^2$, this is an equation of the form

$$(1, b, c, d, e, f, g \between f^2, 1)^6 = 0,$$

and similarly the equation the roots of which are $F^2, F_0^2, F_1^2, F_2^2, F_3^2, F_4^2$ is an equation of the form

$$(1, B, C, D, E, F, G \between F^2, 1)^6 = 0,$$

where b and B are conjugate values $\beta+\beta'\sqrt{\Delta}$, $\beta-\beta'\sqrt{\Delta}$, c and C are conjugate values $\gamma+\gamma'\sqrt{\Delta}$, $\gamma-\gamma'\sqrt{\Delta}$, ..., of two-valued functions of the form $P \pm Q\sqrt{\Delta}$, Δ being the discriminant equation, and P and Q rational functions of the coefficients of the quintic equation.

Each term $x_0x_1^2x_2^2$ and $x_0^3x_1x_2$ of Kronecker's function $x_0x_1^2x_2^2+vx_0^3x_1x_2$ is of the form 0.12 $(=0.21)$, a function of x_0 multiplied by a symmetrical function of x_1, x_2; and it is by reason hereof that the roots $f, f_0, f_1, f_2, f_3, f_4$ are connected by linear relations such that the equation belongs to the class of Jacobi's multiplier equations. Thus in the expressions for these roots, attending first to the terms multiplied by $\epsilon-\epsilon^4$, we have

$$\begin{aligned}
f\; &= (\epsilon-\epsilon^4).0(12-34)+1(23-04)+2(34-01)+3(04-12)+4(01-23),\\
\hline
f_0 &= (\epsilon-\epsilon^4).0(34-12)+1(02-34)+2(03-14)+3(14-02)+4(12-03),\\
f_1 &= (\epsilon-\epsilon^4).0(23-14)+1(04-23)+2(13-04)+3(14-02)+4(02-13),\\
f_2 &= (\epsilon-\epsilon^4).0(13-24)+1(34-02)+2(01-34)+3(24-01)+4(02-13),\\
f_3 &= (\epsilon-\epsilon^4).0(13-24)+1(24-03)+2(04-13)+3(12-04)+4(03-12),\\
f_4 &= (\epsilon-\epsilon^4).0(14-23)+1(24-03)+2(03-14)+3(01-24)+4(23-01);
\end{aligned}$$

and next to the terms multiplied by $\epsilon^2-\epsilon^3$, we have

$$\begin{aligned}
f\; &= (\epsilon^2-\epsilon^3).0(24-13)+1(03-24)+2(14-03)+3(02-14)+4(13-02),\\
\hline
f_0 &= (\epsilon^2-\epsilon^3).0(24-13)+1(04-23)+2(13-04)+3(01-24)+4(23-01),\\
f_1 &= (\epsilon^2-\epsilon^3).0(34-12)+1(03-24)+2(01-34)+3(24-01)+4(12-03),\\
f_2 &= (\epsilon^2-\epsilon^3).0(23-14)+1(04-23)+2(14-03)+3(12-04)+4(03-12),\\
f_3 &= (\epsilon^2-\epsilon^3).0(14-23)+1(34-02)+2(01-34)+3(02-14)+4(23-01),\\
f_4 &= (\epsilon^2-\epsilon^3).0(34-12)+1(02-34)+2(04-13)+3(12-04)+4(13-02);
\end{aligned}$$

we have thus

$$\begin{aligned}
f_0+f_1+f_2+f_3+f_4 = (\epsilon-\epsilon^4).\,&0\{34-12+2(13-24)\}+(\epsilon^2-\epsilon^3).\,0\{24-13+2(34-12)\}\\
&+1\{04-23+2(24-03)\}+\qquad 1\{03-24+2(04-23)\}\\
&+2\{01-34+2(03-14)\}+\qquad 2\{14-03+2(10-24)\}\\
&+3\{12-04+2(14-02)\}+\qquad 3\{20-14+2(12-04)\}\\
&+4\{23-01+2(02-13)\}+\qquad 4\{13-02+2(23-01)\}\\
=-\{(\epsilon-\epsilon^4)+2(\epsilon^2-\epsilon^3)\}&(f)_1-\{2(\epsilon-\epsilon^4)-(\epsilon^2-\epsilon^3)\}(f)_2,
\end{aligned}$$

if for a moment $(f)_1$ denotes the terms of f which are multiplied by $\epsilon-\epsilon^4$ and $(f)_2$ the terms of f which are multiplied by $\epsilon^2-\epsilon^3$.

But we have

$$\sqrt{5}=\epsilon+\epsilon^4-\epsilon^2-\epsilon^3,$$

whence

$$\begin{aligned}
\sqrt{5}(\epsilon-\epsilon^4) &= \epsilon-\epsilon^4+2(\epsilon^2-\epsilon^3),\\
\sqrt{5}(\epsilon^2-\epsilon^3) &= 2(\epsilon-\epsilon^4)+\epsilon^2-\epsilon^3;
\end{aligned}$$

and hence the equation just obtained is $f_0+f_1+f_2+f_3+f_4=-\sqrt{5}f$, viz. this equation, being satisfied separately by the terms such as $x_0x_1^2x_2^2$ and $x_0^3x_1x_2$, will be satisfied for $x_0x_1^2x_2^2+vx_0^3x_1x_2$: and so for the like equations which follow. (I notice that Brioschi has $+\sqrt{5}f$; the difference is quite immaterial, since the formulæ would coincide by reversing the sign of f, or those of f_0, f_1, f_2, f_3, f_4.)

We show further that $f_0+\epsilon^2f_1+\epsilon^4f_2+\epsilon f_3+\epsilon^3f_4=0$; to verify this, observe that in this expression the terms multiplied by $0\ (=x_0)$ are

$$\begin{aligned}
&(\epsilon-\epsilon^4)\{34-12+\epsilon^2(23-14)+\epsilon^4(13-24)+\epsilon(13-24)+\epsilon^3(14-23)\}\\
&+(\epsilon^2-\epsilon^3)\{24-13+\epsilon^2(34-12)+\epsilon^4(23-14)+\epsilon(14-23)+\epsilon^3(34-12)\},
\end{aligned}$$

where the terms containing 12, 13, 14, 23, 24, 34 respectively are each $=0$, viz. the coefficient of 12 is $(-\epsilon+\epsilon^4)+(-\epsilon^4+1)+(-1+\epsilon)=0$: that of 13 is

$$1-\epsilon^3+(\epsilon^2-1)+(-\epsilon^2+\epsilon^3)=0:$$

and so for the other coefficients. In like manner it appears that the terms multiplied by 1, 2, 3, 4 ($=x_1, x_2, x_3, x_4$) respectively are each $=0$, and thus the equation in question is verified. And in like manner it is shown that

$$f_0+\epsilon^3 f_1+\epsilon f_2+\epsilon^4 f_3+\epsilon^2 f_4=0.$$

The roots f thus satisfy the relations

$$\begin{aligned} f_0+\ f_1+\ f_2+\ f_3+\ f_4&=-f\sqrt{5},\\ f_0+\epsilon^2 f_1+\epsilon^4 f_2+\ \epsilon f_3+\epsilon^3 f_4&=0,\\ f_0+\epsilon^3 f_1+\ \epsilon f_2+\epsilon^4 f_3+\epsilon^2 f_4&=0,\end{aligned}$$

or the equation for f^2 belongs to the class of Jacobi's multiplier equations. Hence (see Brioschi's "Appendice terza" before referred to) the form of the equation is

$$(f^2-a)^6-4a(f^2-a)^5+10b(f^2-a)^3-4c(f^2-a)+5b^2-4ac=0,$$

or determining the arbitrary coefficient v so that a may be $=0$, the form is

$$f^{12}+10bf^6-4cf^2+5b^2=0,$$

which is Kronecker's equation

$$f^{12}-10\phi f^6+5\psi^2=\psi f^2.$$

As to the meaning of the coefficients a, b, c, I recall that, in virtue of the foregoing linear relations between the roots, these may be expressed in terms of three arbitrary quantities a_0, a_1, a_2 as follows:

$$\begin{aligned} f\ &=a_0\sqrt{5},\\ f_0&=a_0+\ a_1+\ a_2,\\ f_1&=a_0+\ \epsilon a_1+\epsilon^4 a_2,\\ f_2&=a_0+\epsilon^2 a_1+\epsilon^3 a_2,\\ f_3&=a_0+\epsilon^3 a_1+\epsilon^2 a_2,\\ f_4&=a_0+\epsilon^4 a_1+\ \epsilon a_2,\end{aligned}$$

and a, b, c are then determinate functions of a_0, a_1, a_2, viz. we have

$$\begin{aligned} a&=a_0^2+a_1a_2,\\ b&=8a_0^4a_1a_2-2a_0^2a_1^2a_2^2+a_1^3a_2^3-a_0(a_1^5+a_2^5),\\ c&=80a_0^6a_1^2a_2^2-40a_0^4a_1^3a_2^3+5a_0^2a_1^4a_2^4+a_1^5a_2^5\\ &\quad -a_0(32a_0^4-20a_0^2a_1a_2+5a_1^2a_2^2)(a_1^5+a_2^5)\\ &\quad +\tfrac{1}{4}(a_1^5+a_2^5)^2;\end{aligned}$$

so that, for $a=0$ and therefore $a_0=\sqrt{-a_1a_2}$, we have

$$\begin{aligned} b&=11a_1^3a_2^3-a_0(a_1^5+a_2^5),\\ c&=-44a_1^5a_2^5-57a_0a_1^2a_2^2(a_1^5+a_2^5)+\tfrac{1}{4}(a_1^5+a_2^5)^2,\end{aligned}$$

but I do not know that for Kronecker's form the actual values of a_0, a_1, a_2 in terms of the coefficients of the quintic equation have been calculated.

951.

NON-EUCLIDIAN GEOMETRY.

[From the *Transactions of the Cambridge Philosophical Society*, vol. XV. (1894), pp. 37—61. Read January 27, 1890.]

I CONSIDER ordinary three-dimensional space, and use the words point, line, plane, &c., in their ordinary acceptations; only the notion of distance is altered, viz. instead of taking the Absolute to be the circle at infinity, I take it to be a quadric surface: in the analytical developments this is taken to be the imaginary surface $x^2+y^2+z^2+w^2=0$, and the formulæ arrived at are those belonging to the so-called Elliptic Space. The object of the Memoir is to set out, in a somewhat more systematic form than has been hitherto done, the general theory; and in particular, to further develop the analytical formulæ in regard to the perpendiculars of two given lines. It is to be remarked that not only all purely descriptive theorems of Euclidian geometry hold good in the new theory; but that this is the case also (only we in nowise attend to them) with theorems relating to parallelism and perpendicularity, in the Euclidian sense of the words. In Euclidian geometry, infinity is a special plane, the plane of the circle at infinity, and we consider (for instance) parallel lines, that is, lines which meet in a point of this plane: in the new theory, infinity is a plane in nowise distinguishable from any other plane, and there is no occasion to consider (although they exist) lines meeting in a point of this plane, that is, parallel lines in the Euclidian sense. So again, given any two lines, there exists always, in the Euclidian sense, a single line perpendicular to each of the given lines, but this is not in the new sense a perpendicular line; there is nothing to distinguish it from any other line cutting the two given lines, and consequently no occasion to consider it: we do consider the lines—there are, in fact, two such lines—which in the new sense of the word are perpendicular to each of the given lines.

It should be observed that the term distance is used to include inclination: we have, say, a linear distance between two points; an angular distance between two lines which meet; and a dihedral distance between two planes. But all these are distances of the same kind, having a common unit, the quadrant, represented by $\frac{1}{2}\pi$; and in fact, any distance may be considered indifferently as a linear, an angular, or a dihedral distance: the word, perpendicular, usually represented by $\perp$, refers of course to a distance $=\frac{1}{2}\pi$. We have moreover the distance of a point from a plane, that of a point from a line, and that of a plane from a line. Two lines which do not meet may be $\perp$, and in particular they may be reciprocal: in general, they have two distances; and they have also a "moment" and "comoment," the values of which serve to express those of the two distances. Lines may be, in several distinct senses, as will be explained, parallel; and for this reason the word parallel is never used simpliciter; the notion of parallelism does not apply to planes, nor to points.

Elliptic space has been considered and the theory developed in connexion with the imaginaries called by Clifford biquaternions, and as applied to Mechanics: I refer to the names, Ball, Buchheim, Clifford, Cox, Gravelius, Heath, Klein, and Lindemann: in particular, much of the purely geometrical theory is due to Clifford. Memoirs by Buchheim and Heath are referred to further on.

Geometrical Notions. Art. Nos. 1 to 16.

1. The Absolute is a general quadric surface: it has therefore lines of two kinds, which it is convenient to distinguish as directrices and generatrices: through each point of the surface there is a directrix and a generatrix, and the plane through these two lines is the tangent plane at the point. A line meets the surface in two points, say A, C; the generatrix at A meets the directrix at C; and the

Fig. 1.

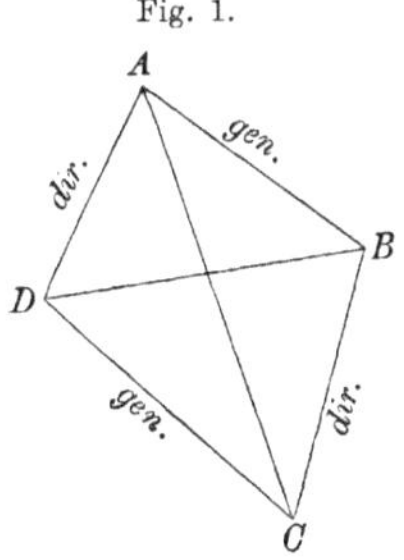

directrix at A meets the generatrix at C; and we have thus on the surface two new points B, D; joining these we have a line BD, which is the reciprocal of AC; viz. BD is the intersection of the planes BAD, BCD which are the tangent planes at A, C respectively, and similarly AC is the intersection of the planes ABC, ADC which are the tangent planes at B, D respectively.

According to what follows, reciprocal lines are $\perp$, but $\perp$ lines are not in general reciprocal; thus the two epithets are not convertible, and there will be occasion throughout to speak of reciprocal *lines*.

2. Two points may be harmonic; that is, the two points and the intersections of their line of junction with the Absolute may form a harmonic range: the two points are in this case said to be $\perp$.

Two planes may be harmonic: that is, the two planes and the tangent planes of the Absolute through their line of intersection may form a harmonic plane-pencil: the two planes are said to be $\perp$.

Two lines which meet may be harmonic: that is, the two lines and the tangents from their point of intersection to the section of the Absolute by their common plane may form a harmonic pencil: the two lines are said to be $\perp$.

The locus of all the points $\perp$ to a given point is a plane, the reciprocal or polar plane of the given point; and similarly the envelope of all the planes $\perp$ to a given plane is a point, the pole of the given plane: a point and plane reciprocal to each other, or say a pole and polar plane, are said to be $\perp$.

3. If a point is situate anywhere in a given line, the $\perp$ plane passes always through the reciprocal line: each point of the reciprocal line is thus a point of the $\perp$ plane, i.e. it is $\perp$ to the given point: that is, considering two reciprocal lines, any point on the one line and any point on the other line are $\perp$. Similarly any plane through the one line and any plane through the other line are $\perp$.

A line and plane may be harmonic; that is, they may be reciprocal in regard to the cone, vertex their point of intersection, circumscribed to the Absolute; the line and plane are said to be $\perp$. The $\perp$ plane passes through the reciprocal line, and conversely every plane through the reciprocal line is a $\perp$ plane. It may be added that the line passes through the $\perp$ point of the plane; and conversely, that every line through the $\perp$ point of a plane is $\perp$ to the plane. Moreover if a line and plane be $\perp$, the line is $\perp$ to every line in the plane and through the point of intersection.

A line and point may be harmonic; that is, they may be reciprocal in regard to the section of the Absolute by their common plane: the line and point are said to be $\perp$. The $\perp$ point lies in the reciprocal line, and conversely every point of the reciprocal line is a $\perp$ point. It may be added that the line lies in the $\perp$ plane of the point: and conversely that every line in the $\perp$ plane of a point is $\perp$ to the point. Moreover if a line and point be $\perp$, the line is $\perp$ to every line through the point and in the plane of junction.

4. We may have a triangle ABC composed of three lines BC, CA, AB in the same plane: the six parts hereof are the linear distances B, C; C, A; A, B of the angular points, and the angular distances of the sides CA, AB; AB, BC; BC, CA. Similarly we may have a trihedral composed of three lines meeting in a point, say the planes through the several pairs of lines are A, B, C respectively: the six parts hereof are the angular distances CA, AB; AB, BC; BC, CA of the three lines, and the dihedral distances B, C; C, A; A, B of the three planes. According to the definitions of distance hereinafter adopted, the relation of the six parts is that of the sides and angles of a spherical triangle: in particular, if two sides are each

$=\frac{1}{2}\pi$, then the opposite angles are each $=\frac{1}{2}\pi$, and the included angle and the opposite side have a common value; and so also if two angles are each $=\frac{1}{2}\pi$, then the opposite sides are each $=\frac{1}{2}\pi$, and the included side and the opposite angle have a common value.

5. Let A, C be points on a line, and B, D points on the reciprocal line; by what precedes, each of the lines AB, AD, CB, CD is $=\frac{1}{2}\pi$: also each of the angles ACD, ACB, CAB, CAD is $=\frac{1}{2}\pi$. The line AC is $\perp$ to the plane BCD and to the lines BC, CD, in that plane; it is also $\perp$ to the plane BAD and to the lines BA, AD in that plane; and similarly for the line BD. From the trihedral of the planes which meet in C, distance of planes ACB, ACD = distance of lines BC, CD, viz. the dihedral distance of two planes through the line AC is equal to the angular distance of their intersections with the $\perp$ plane BCD; and it is therefore equal also to the

Fig. 2.

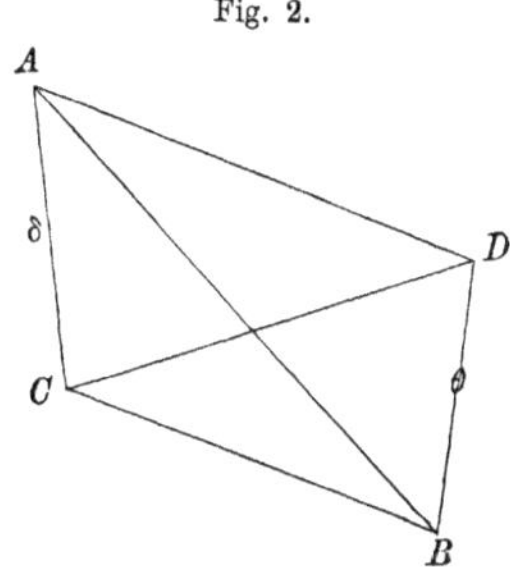

linear distance of their intersections with the other $\perp$ plane BAD: and so from the triangle BCD, where BC, CD are each $=\frac{1}{2}\pi$, the angular distance BCD is equal to the linear distance BD; that is, the distance of the planes ACB, ACD, that of the lines BC, CD, that of the lines BA, AD, and that of the points B, D, are all of them equal; say the value of each of them is $=\theta$. And in like manner the distance of the planes ABD, CBD, that of the lines AB, BC, that of the lines AD, DC, and that of the points A, C, are all of them equal: say the value of each of them is $=\delta$.

The theorem may be stated as follows: all the planes $\perp$ to a given line intersect in the reciprocal line: and if we have through the given line any two planes, the distance of these two planes, the distance between their lines of intersection with any one of the $\perp$ planes, and the distance between their points of intersection with the reciprocal line, are all of them equal.

And it thus appears also that a distance may be represented indifferently as a linear distance, an angular distance, or a dihedral distance.

6. Consider a point and a plane: we may through the point draw a line $\perp$ to the plane, and intersecting it in a point called the "foot": the distance of the point and plane is then (as a definition) taken to be equal to that of the point and foot. It may be added that the $\perp$ line is, in fact, the line joining the point with the $\perp$

point of the plane; and that the distance of the point and plane is equal to the complement of the distance of the point and the ⊥ point. Or again, we may in the plane draw a line ⊥ to the point, and determining with it a plane called the roof: and then (as an equivalent definition) the distance of the plane and point is equal to the distance of the plane and roof. It may be added that the ⊥ line is, in fact, the intersection of the plane with the ⊥ plane of the point, and that the distance of the point and plane is also equal to the complement of the distance of the plane and the ⊥ plane of the point.

7. Consider a point and line: we have through the point a line ⊥ to the line and cutting it in a point called the foot; the distance of the point and line is then (as a definition) equal to the distance of the point and foot. It may be added that the foot is the intersection with the line of a plane ⊥ thereto through the point.

Again, consider a plane and line: we have in the plane a line ⊥ to the line and determining with it a plane called the roof: the distance of the plane and line is then (as a definition) equal to the distance of the plane and roof. It may be added that the roof is the plane determined by the line and a point ⊥ thereto in the plane.

8. If two lines intersect, then their reciprocals also intersect. Say the intersecting lines are X, Y; and their reciprocals X', Y' respectively; then K, the point of intersection of X, Y, has for its reciprocal the plane of the lines X', Y'; and similarly K', the point of intersection of the lines X', Y', has for its reciprocal the plane of the lines X, Y: hence KK' has for its reciprocal the line of intersection of the planes XY and $X'Y'$; say this is the line Λ, meeting X, Y, X', Y', in the

Fig. 3.

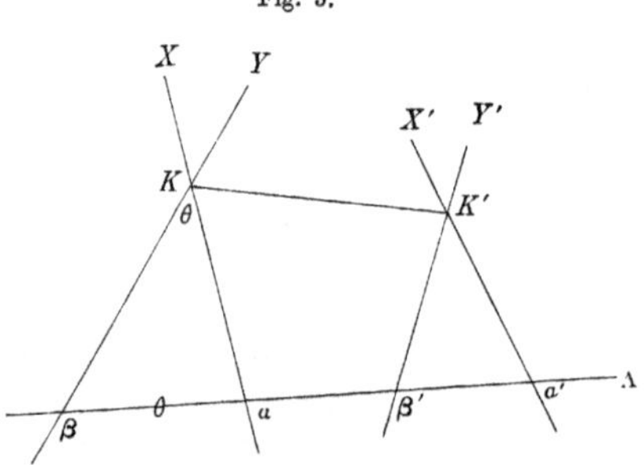

points α, β, α', β' respectively. Since K, K' are points in the reciprocal lines X, X' (or in the reciprocal lines Y, Y'), the distance KK' is $=\frac{1}{2}\pi$; and since the plane XY passes through the line Λ which is the reciprocal of KK', the line KK' is ⊥ to the plane XY and also to each of the lines X, Y: (it is also ⊥ to the plane $X'Y'$ and to each of the lines X', Y'). Again, since the lines KK' and Λ are reciprocal, each of the distances $K\alpha$, $K\beta$ is $=\frac{1}{2}\pi$; that is, the line Λ is ⊥ to each of the lines X and Y, (and similarly it is ⊥ to each of the lines X' and Y'). Moreover the angle at K or distance of the lines X and Y (which is equal to the distance of the planes $K'KX$ and $K'KY$) is equal to the distance $\alpha\beta$ of the intersections of Λ with the lines X and Y respectively. We have thus for the two intersecting lines X and Y, the two lines KK' and Λ each of them $\perp$ to the two lines: where observe that KK' is the line of junction of the point of intersection of the two given lines with the point of intersection of the reciprocal lines; and that Λ is the line of intersection of the plane of the two given lines with the plane of the reciprocal lines. The linear distance along KK' between the two lines is $=0$; the dihedral distance between the planes, which KK' determines with the two lines respectively, is equal to the angular distance between the two lines. The linear distance along Λ is equal to the angular distance between the two lines; the dihedral distance between the two planes, which Λ determines with the two lines respectively, is $=0$.

9. If two lines are such that the first of them intersects the reciprocal of the second of them, then also the second will intersect the reciprocal of the first; the two lines are in this case said to be contrasecting lines; or more simply, to contrasect: and contrasecting lines are said to be $\perp$. Supposing that the two lines are X, Y and their reciprocals X', Y' respectively, we have here X, Y' intersecting in a point K, and X', Y intersecting in a point K': and the planes XY', $X'Y$ intersect in a line Λ which meets the lines X, Y, X', Y' in the points α, β, α', β' respectively. As before, the lines KK' and Λ are reciprocal: the distance KK' is $=\frac{1}{2}\pi$; and KK' is $\perp$ to the plane XY', that is, to each of the lines X, Y'; and also to the plane $X'Y$, that is, to each of the lines X', Y; it is thus $\perp$ to each of the lines X and Y. Again each of the angles at α, β, α', β' is $=\frac{1}{2}\pi$; that is,

Fig. 4.

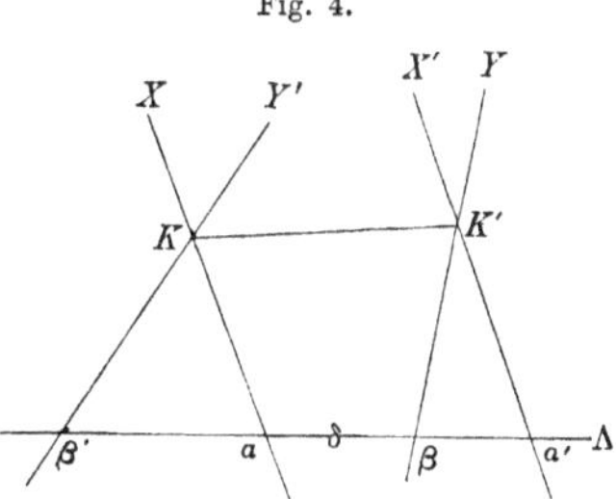

the line Λ is $\perp$ to each of the lines X, Y', X', Y, or say to each of the lines X and Y. Moreover the angle at K, or say the angular distance of the intersecting lines X and Y', is equal to the distance $\alpha\beta'$; and similarly the angle at K', or say the angular distance of the intersecting lines X' and Y, is equal to the distance $\alpha'\beta$: but the distances $\alpha\alpha'$, $\beta\beta'$ are each equal to $\frac{1}{2}\pi$; and hence the distances $\alpha\beta'$, $\alpha'\beta$ are equal to each other and each of them is equal to the complement of the distance $\alpha\beta$. Thus in the case of two contrasecting lines we have the lines KK' and Λ each of them $\perp$ to the two given lines; where observe that KK' is the line joining the point of intersection of X with the reciprocal of Y and the point of intersection of Y with the reciprocal of X; and that Λ is the line of intersection of the plane through X and the reciprocal of Y with the plane through Y and the reciprocal of X. The linear distance KK' between the two lines along the first of these lines is thus $=\frac{1}{2}\pi$.

10. We have KK' and Λ reciprocal lines; on the first of these, we have the points K, K' which are $\perp$ points: hence also the planes ΛK and $\Lambda K'$ are $\perp$; but the plane ΛK is the plane $\Lambda XY'$ or say the plane ΛX, and the plane $\Lambda K'$ is the plane $\Lambda X'Y$ or say the plane ΛY; hence the planes ΛX and ΛY are $\perp$. Similarly the line Λ cuts the two lines in the points α, β; and the line KK' determines with these two points respectively the plane $KK'\alpha$, that is $KK'X$, and $KK'\beta$, that is $KK'Y$; and thus the linear distance between the two points α, β is equal to the dihedral distance between the two planes $KK'X$ and $KK'Y$. Thus the $\perp$ line Λ cuts the two lines in two points α, β the linear distance of which is, say, δ: and it determines with them two planes the dihedral distance of which is $=\frac{1}{2}\pi$. And the other $\perp$ line KK' cuts the two lines in the points K, K' the linear distance of which is $=\frac{1}{2}\pi$, and it determines with them two planes the dihedral distance of which is $=\delta$.

11. Consider a line X and its reciprocal X': a line intersecting each of these also contrasects each of them and is thus $\perp$ to each of them: and similarly if Y be any other line and Y' its reciprocal, a line intersecting Y and Y' also contrasects each of them and is thus $\perp$ to each of them. Hence a line which meets each of the four lines X, X', Y, Y' is also $\perp$ to each of them, or attending only to the lines X, Y, say it is a $\perp$ of these lines: there are two $\perp$s; and clearly these are reciprocal to each other, for if a line meets X, Y, X', Y', then its reciprocal meets X', Y', X, Y, that is, the same four lines. Looking back to figure 2, we may take AB, CD for the given lines, and AC, BD for the two $\perp$s; as just remarked, these are reciprocal to each other. The $\perp$ AC cuts the two lines respectively in the two points A and C the linear distance of which is say $=\delta$; and it determines with them two planes ACB, ACD, the dihedral distance of which is say $=\theta$. Similarly the other $\perp$ BD meets the two lines respectively in the two points B and D the linear distance of which is $=\theta$, and it determines with them two planes BDA, BDC the dihedral distance of which is $=\delta$. In the plane triangles which are the faces of the tetrahedron $ABCD$, there is in each triangle an angle opposite to AC or BD and which, or say the angular distance of the two including sides, is thus $=\delta$ or θ. Except as aforesaid, the sides, angles, and dihedral angles, or say the linear, angular, and dihedral distances, of the tetrahedron are each of them $=\frac{1}{2}\pi$.

12. Considering the lines X and Y as given, the distances δ and θ depend upon two functions called the Moment and the Comoment: viz. moment $=0$ is the condition in order that the two lines may intersect (or, what is the same thing, in order that their reciprocals may intersect): comoment $=0$ is the condition in order that the two lines may contrasect, that is, each line meet the reciprocal of the other one. It may be convenient to mention here that the actual relations are

$$\sin\delta\sin\theta = \text{Moment},\quad \cos\delta\cos\theta = \text{Comoment}.$$

In particular, if moment $=0$, then the lines intersect; we have, say $\delta=0$, and therefore $\cos\theta=$ comoment; if comoment $=0$, then the lines contrasect, that is they are $\perp$: we have, say $\theta=\frac{1}{2}\pi$, that is, $\sin\delta=$ moment. These are the two particular cases which have been considered above.

13. Consider as above the two lines X, Y met by the $\perp$ δ in the two points A and C respectively. Consider at A a line I $\perp$ to the lines X, δ; and take Π the plane of the lines (X, δ) and Ω the plane of the lines (X, I). Similarly consider at C a line K $\perp$ to the lines Y, δ, and take Π_1 the plane of the lines (Y, δ) and Ω_1 the plane of the lines (Y, K): we have thus through A two planes Π, Ω meeting in the line X; and through C two planes Π_1, Ω_1, meeting in the line Y. It requires only a little reflection to see that the distances of these planes are

$$(\Pi, \Pi_1)=\theta, \quad (\Omega, \Omega_1)=\delta;$$

$$(\Pi, \Omega)=\tfrac{1}{2}\pi, \ (\Pi_1, \Omega_1)=\tfrac{1}{2}\pi; \ (\Pi, \Omega_1)=\tfrac{1}{2}\pi, \ (\Pi_1, \Omega)=\tfrac{1}{2}\pi.$$

Fig. 5.

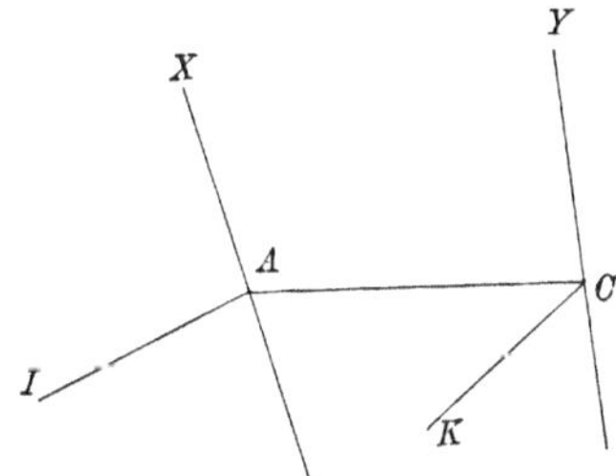

In fact, Π, Π_1 are the before-mentioned planes ACB, ACD the distance of which was $=\theta$: Ω, Ω_1 are planes having the common $\perp$ AC, which is the line through the poles of these planes, and such that the distance AC is equal to the distance of the two poles, that is, the distance of the two planes. Moreover from the definitions, the distances (Π, Ω) and (Π_1, Ω_1) are each $=\frac{1}{2}\pi$: the plane Π passes through the $\perp$ at C to the plane Ω_1 that is, $(\Pi, \Omega_1)=\frac{1}{2}\pi$; and similarly the plane Π_1 passes through the $\perp$ at A to the plane Ω, that is, $(\Pi_1, \Omega)=\frac{1}{2}\pi$; and we have thus the relations in question.

The consideration of these planes leads, (see *post* 31 and 32), to the before-mentioned equation, $\cos\delta\cos\theta=$ comoment; if instead of one of the lines, say Y, we consider the reciprocal line Y', then the angles δ, θ are changed each of them into its complement, and we deduce immediately the other equation, $\sin\delta\sin\theta=$ Moment.

14. It may happen that, instead of the determinate number 2, we have a singly infinite system of $\perp$s: viz. this will be so if the lines X, X', Y, Y' are generating lines (of the same kind) of a hyperboloid. They will be so if the lines X and Y each of them meet the same two lines (of the same kind) of the Absolute, say if X, Y each meet two directrices D_1, D_2, or two generatrices G_1, G_2; but it seems

less easy to prove conversely that the lines X and Y must satisfy one of these two conditions. Suppose first that X, Y each meet the two directrices D_1, D_2; say X meets them in α_1, α_2, and Y in β_1, β_2 respectively. We have at α_1 a generatrix which meets D_2, suppose in α_2', and at α_2 a generatrix which meets D_1, suppose in α_1'; joining α_1', α_2', we have the line X' which is the reciprocal of X; viz. X' meets each of the lines D_1, D_2: similarly the generatrices at β_1, β_2 meet D_2, D_1 in the points β_2', β_1' respectively, and joining these, we have the line Y' which is the reciprocal of Y: thus Y' meets each of the lines D_1 and D_2: the line D_1 meets the four generatrices in the points α_1, α_1', β_1, β_1' respectively, and the line D_2 meets the same four generatrices in the points α_2', α_2, β_2', β_2: thus

$$AH(\alpha_1, \alpha_1', \beta_1, \beta_1') = AH(\alpha_2', \alpha_2, \beta_2', \beta_2),$$

AH denoting anharmonic ratio as usual. But

$$AH(\alpha_2', \alpha_2, \beta_2', \beta_2) = AH(\alpha_2, \alpha_2', \beta_2, \beta_2'),$$

and thus the equation may be written

$$AH(\alpha_1, \alpha_1', \beta_1, \beta_1') = AH(\alpha_2, \alpha_2', \beta_2, \beta_2');$$

viz. the lines X, X', Y, Y', cut D_1, D_2 homographically; and there is thus a singly infinite system of lines cutting D_1, D_2 homographically: that is, X, X', Y, Y', are lines (of the same kind) of a hyperboloid. And similarly if X, Y each cut the same two generating lines G_1, G_2, then will X', Y' also cut these lines and X, X', Y, Y' will cut them homographically, that is, X, X', Y, Y' will be lines (of the same kind) of a hyperboloid.

Fig. 6.

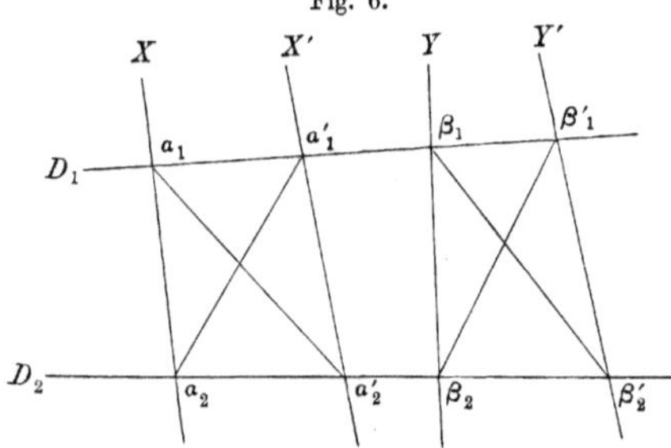

The condition may be otherwise stated; if the lines X, Y have for $\perp$s any two directrices D_1, D_2 or any two generatrices G_1, G_2 of the Absolute, then in either case there will be a singly infinite series of $\perp$s: the $\perp$ distances are all of them equal; say we have $\theta = \delta$, and therefore $\sin^2\delta =$ moment, $\cos^2\delta =$ comoment; and therefore moment + comoment = 1; or as the equation is more properly written, $\pm$ moment $\pm$ comoment = 1.

15. Two lines X, Y, each of them meeting the same two directrices D_1, D_2, are said to be "right parallels"; and similarly two lines X, Y each meeting the same two generatrices G_1, G_2, are said to be "left parallels": the selection as to which set

of lines of the Absolute shall be called directrices and which shall be called generatrices will be made further on, (see *post* 35). We have just seen that, if two lines are right parallels, or are left parallels, then in either case there is a singly infinite series of $\perp$s. It may be remarked that reciprocal lines are at once right parallels and left parallels; and that in this case there is a doubly infinite series of $\perp$s, viz. every line cutting the two lines is a $\perp$.

Observe that right parallels do not meet, and left parallels do not meet: their doing so would imply in the one case the meeting of two directrices, and in the other case the meeting of two generatrices.

16. If instead of the foregoing definitions by means of two directrices or two generatrices, we consider a directrix and a generatrix of the Absolute, and define parallel lines by reference thereto, then it is at once seen that there are 3 chief forms, and several subforms; the directrix and generatrix meet in a point, or say an ineunt, of the Absolute, and lie in a plane which is a tangent plane of the Absolute: we may have two lines X, Y which

1°. Each pass through the ineunt, neither of them lying in the tangent plane;

2°. Each lie in the tangent plane, neither of them passing through the ineunt;

3°. One passes through the ineunt, but does not lie in the tangent plane: the other lies in the tangent plane, but does not pass through the ineunt.

Observe that in the cases 1° and 2° the lines X and Y intersect, but in the case 3° they do not intersect. The lines in the case 3° are I believe what Buchheim has termed β-parallels, his α-parallels being the foregoing right or left parallels*. The subforms arise by omitting in 1°, 2°, or 3°, as the case may be, the negative condition in regard to the two lines or to one of them; as the question is not here further pursued, I do not attempt to give names to these several kinds of parallel lines.

Point-, line-, and plane-coordinates: General formulæ. Art. Nos. 17 to 20.

17. We consider point-coordinates (x, y, z, w): line-coordinates (a, b, c, f, g, h), where $af+bg+ch=0$: and plane-coordinates $(\xi, \eta, \zeta, \omega)$; if we have a line which is at once through two points and in two planes, then the line-coordinates are given by

$$\begin{aligned} &a : b : c : f : g : h \\ &= y_1z_2 - y_2z_1 : z_1x_2 - z_2x_1 : x_1y_2 - x_2y_1 : x_1w_2 - x_2w_1 : y_1w_2 - y_2w_1 : z_1w_2 - z_2w_1 \\ &= \xi_1\omega_2 - \xi_2\omega_1 : \eta_1\omega_2 - \eta_2\omega_1 : \zeta_1\omega_2 - \zeta_2\omega_1 : \eta_1\zeta_2 - \eta_2\zeta_1 : \zeta_1\xi_2 - \zeta_2\xi_1 : \xi_1\eta_2 - \xi_2\eta_1. \end{aligned}$$

Similarly, if a plane be determined by three points thereof, then the coordinates of the plane are given by

$$\xi : \eta : \zeta : \omega = \left|\begin{matrix} 1 & & & \\ x_1, & y_1, & z_1, & w_1 \\ x_2, & y_2, & z_2, & w_2 \\ x_3, & y_3, & z_3, & w_3 \end{matrix}\right| : \left|\begin{matrix} & 1 & & \\ x_1, & y_1, & z_1, & w_1 \\ x_2, & y_2, & z_2, & w_2 \\ x_3, & y_3, & z_3, & w_3 \end{matrix}\right| : \left|\begin{matrix} & & 1 & \\ x_1, & y_1, & z_1, & w_1 \\ x_2, & y_2, & z_2, & w_2 \\ x_3, & y_3, & z_3, & w_3 \end{matrix}\right| : \left|\begin{matrix} & & & 1 \\ x_1, & y_1, & z_1, & w_1 \\ x_2, & y_2, & z_2, & w_2 \\ x_3, & y_3, & z_3, & w_3 \end{matrix}\right|;$$

* See Buchheim, "A Memoir on Biquaternions," *Amer. Math. Jour.* t. VII. (1885), pp. 293—326.

and if a point be given as the intersection of three planes, the coordinates of the point are

$$x : y : z : w = \begin{vmatrix} 1 & & & \\ \xi_1, & \eta_1, & \zeta_1, & \omega_1 \\ \xi_2, & \eta_2, & \zeta_2, & \omega_2 \\ \xi_3, & \eta_3, & \zeta_3, & \omega_3 \end{vmatrix} : \begin{vmatrix} & 1 & & \\ \xi_1, & \eta_1, & \zeta_1, & \omega_1 \\ \xi_2, & \eta_2, & \zeta_2, & \omega_2 \\ \xi_3, & \eta_3, & \zeta_3, & \omega_3 \end{vmatrix} : \begin{vmatrix} & & 1 & \\ \xi_1, & \eta_1, & \zeta_1, & \omega_1 \\ \xi_2, & \eta_2, & \zeta_2, & \omega_2 \\ \xi_3, & \eta_3, & \zeta_3, & \omega_3 \end{vmatrix} : \begin{vmatrix} & & & 1 \\ \xi_1, & \eta_1, & \zeta_1, & \omega_1 \\ \xi_2, & \eta_2, & \zeta_2, & \omega_2 \\ \xi_3, & \eta_3, & \zeta_3, & \omega_3 \end{vmatrix}.$$

18. The conditions in order that a point (x, y, z, w) may be situate on a line (a, b, c, f, g, h) are

$$\begin{array}{rrrrl} . & hy & -gz & +aw & = 0, \\ -hx & . & +fz & +bw & = 0, \\ gx & -fy & . & +cw & = 0, \\ -ax & -by & -cz & . & = 0, \end{array}$$

viz. these constitute a twofold relation.

Similarly, the conditions in order that the plane $(\xi, \eta, \zeta, \omega)$ may contain the line (a, b, c, f, g, h) are

$$\begin{array}{rrrrl} . & c\eta & -b\zeta & +f\omega & = 0, \\ -c\xi & . & +a\zeta & +g\omega & = 0, \\ b\xi & -a\eta & . & +h\omega & = 0, \\ -f\xi & -g\eta & -h\zeta & . & = 0, \end{array}$$

viz. these constitute a twofold relation.

19. The condition in order that two lines (a, b, c, f, g, h), (A, B, C, F, G, H) may meet is

$$Af + Bg + Ch + Fa + Gb + Hc = 0.$$

Supposing that the two lines meet, we have at the point of intersection

$$\begin{array}{rrrrl} . & hy & -gz & +aw & = 0, \\ -hx & . & +fz & +bw & = 0, \\ gx & -fy & . & +cw & = 0, \\ -ax & -by & -cz & . & = 0, \end{array} \qquad \begin{array}{rrrrl} . & Hy & -Gz & +Aw & = 0, \\ -Hx & . & +Fz & +Bw & = 0, \\ Gx & -Fy & . & +Cw & = 0, \\ -Ax & -By & -Cz & . & = 0; \end{array}$$

and from these equations we can find the coordinates x, y, z, w of the point of intersection in a fourfold form, viz. we may write

$$\begin{array}{rlllll} x : y : z : w = & fA + bG + cH & : gA - aG & : hA - aH & : hG - gH \\ = & fB - bF & : gB + cH + aF & : hB - bH & : fH - hF \\ = & fC - cF & : gC - cG & : hC + aF + bG & : gF - fG \\ = & bC - cB & : cA - aC & : aB - bA & : fA + gB + hC. \end{array}$$

There is no real advantage in any one over any other of these forms, but it is convenient to work with the last of them

$$x : y : z : w = bC - cB : cA - aC : aB - bA : fA + gB + hC.$$

20. In like manner if two lines intersect, the plane which contains each of them is given by

$$\begin{array}{lllll} \xi : \eta : \zeta : \omega = & aF + gB + hC & : bF - fB & : cF - fC & : cB - bC \\ = & aG - gA & : bG + hC + fA & : cG - gC & : aC - cA \\ = & aH - hA & : bH - hB & : cH + fA + gB & : bA - aB \\ = & gH - hG & : hF - fH & : fG - gF & : aF + bG + cH; \end{array}$$

or say we have

$$\xi : \eta : \zeta : \omega = gH - hG : hF - fH : fG - gF : aF + bG + cH.$$

The Absolute. Art. Nos. 21 to 27.

21. The equation is

in point coordinates $x^2 + y^2 + z^2 + w^2 = 0$,

in plane coordinates $\xi^2 + \eta^2 + \zeta^2 + \omega^2 = 0$,

in line coordinates $a^2 + b^2 + c^2 + f^2 + g^2 + h^2 = 0$.

Hence

$\perp$ of plane $(\xi, \eta, \zeta, \omega)$ is point $(\xi, \eta, \zeta, \omega)$,

$\perp$ of point (x, y, z, w) is plane (x, y, z, w).

Reciprocal of line (a, b, c, f, g, h) is line (f, g, h, a, b, c);

Points (x, y, z, w), (x', y', z', w') are $\perp$ if $xx' + yy' + zz' + ww' = 0$;

Planes $(\xi, \eta, \zeta, \omega)$, $(\xi', \eta', \zeta', \omega')$ are $\perp$ if $\xi\xi' + \eta\eta' + \zeta\zeta' + \omega\omega' = 0$.

22. A line (a, b, c, f, g, h) and plane $(\xi, \eta, \zeta, \omega)$ are $\perp$ when the line passes through the $\perp$ point of the plane, that is, the point $(\xi, \eta, \zeta, \omega)$: the conditions (equivalent to two equations) are

$$\begin{array}{rrrrl} . & h\eta & - g\zeta & + a\omega & = 0, \\ -h\xi & . & + f\zeta & + b\omega & = 0, \\ g\xi & - f\eta & . & + c\omega & = 0, \\ -a\xi & - b\eta & - c\zeta & . & = 0. \end{array}$$

A line (a, b, c, f, g, h) and point (x, y, z, w) are $\perp$ when the line lies in the $\perp$ plane of the point, that is, in the plane (x, y, z, w): the conditions (equivalent to two equations) are

$$\begin{array}{rrrrl} . & cy & - bz & + fw & = 0, \\ -cx & . & + az & + gw & = 0, \\ bx & - ay & . & + hw & = 0, \\ -fx & - gy & - hz & . & = 0. \end{array}$$

Two lines (a, b, c, f, g, h), (a', b', c', f', g', h') which meet, that is, for which $af' + bg' + ch' + a'f + b'g + c'h = 0$, are $\perp$ if

$$aa' + bb' + cc' + ff' + gg' + hh' = 0.$$

23. There will be occasion to consider the pair of tangent planes drawn through the line (a, b, c, f, g, h) to the Absolute. Writing for shortness

$$\begin{aligned} P &= \quad . \quad hy - gz + aw, \\ Q &= -hx \quad . \quad + fz + bw, \\ R &= \quad gx - fy \quad . \quad + cw, \\ S &= -ax - by - cz \quad . \quad , \end{aligned}$$

it may be shown that the equation of the pair of planes is

$$P^2 + Q^2 + R^2 + S^2 = 0.$$

In fact, writing for a moment $(\xi, \eta, \zeta, \omega)$ and $(\xi', \eta', \zeta', \omega')$ to denote the coefficients of (x, y, z, w) in P and Q respectively, so that

$$(\xi, \eta, \zeta, \omega) = (0, h, -g, a), \quad (\xi', \eta', \zeta', \omega') = (-h, 0, f, b),$$

then the equation of the planes is

$$(\xi' P - \xi Q)^2 + (\eta' P - \eta Q)^2 + (\zeta' P - \zeta Q)^2 + (\omega' P - \omega Q)^2 = 0,$$

that is,

$$(\xi'^2 + \eta'^2 + \zeta'^2 + \omega'^2) P^2 - 2 (\xi\xi' + \eta\eta' + \zeta\zeta' + \omega\omega') PQ + (\xi^2 + \eta^2 + \zeta^2 + \omega^2) Q^2 = 0,$$

viz. this equation is

$$(b^2 + h^2 + f^2) P^2 + 2 (fg - ab) PQ + (a^2 + g^2 + h^2) Q^2 = 0.$$

But P, Q, R, S are connected by the identical equations

$$\begin{aligned} . \quad cQ - bR + fS &= 0, \\ -cP \quad . \quad + aR + gS &= 0, \\ bP - aQ \quad . \quad + hS &= 0, \\ -fP - gQ - hR \quad . \quad &= 0; \end{aligned}$$

using these equations to express R, S in terms of P, Q, viz. writing

$$R = -\frac{1}{h}(fP + gQ), \quad S = -\frac{1}{h}(bP - aQ),$$

we see that the last preceding equation is equivalent to $P^2 + Q^2 + R^2 + S^2 = 0$.

24. Similarly, if

$$\begin{aligned} P_1 &= \quad . \quad cy - bz + fw, \\ Q_1 &= -cx \quad . \quad + az + gw, \\ R_1 &= \quad bx - ay \quad . \quad + hw, \\ S_1 &= -fx - gy - hz \quad . \quad , \end{aligned}$$

functions which are connected by the identical relations

$$\begin{aligned} . \quad hQ_1 - gR_1 + aS_1 &= 0, \\ -hP_1 \quad . \quad + fR_1 + bS_1 &= 0, \\ gP_1 - fQ_1 \quad . \quad + cS_1 &= 0, \\ -aP_1 - bQ_1 - cR_1 \quad . \quad &= 0; \end{aligned}$$

then in like manner we have

$$P_1^2 + Q_1^2 + R_1^2 + S_1^2 = 0,$$

for the equation of the pair of tangent planes from the reciprocal line (f, g, h, a, b, c) to the Absolute. And we may remark the identity

$$(P^2 + Q^2 + R^2 + S^2) + (P_1^2 + Q_1^2 + R_1^2 + S_1^2) = (a^2 + b^2 + c^2 + f^2 + g^2 + h^2)(x^2 + y^2 + z^2 + w^2).$$

We, in fact, have

$P^2 + Q^2 + R^2 + S^2$	x	y	z	w
x	$a^2 + g^2 + h^2$	$ab - fg$	$ac - hf$	$cg - bh$
y	$ab - fg$	$b^2 + h^2 + f^2$	$bc - gh$	$ah - cf$
z	$ac - hf$	$bc - gh$	$c^2 + f^2 + g^2$	$bf - ag$
w	$cg - bh$	$ah - cf$	$bf - ag$	$a^2 + b^2 + c^2$

;

and in like manner

$P_1^2 + Q_1^2 + R_1^2 + S_1^2 =$	x	y	z	w
x	$b^2 + c^2 + f^2$	$-ab + fg$	$-ac + hf$	$-cg + bh$
y	$-ab + fg$	$c^2 + a^2 + g^2$	$-bc + gh$	$-ah + cf$
z	$-ac + hf$	$-bc + gh$	$a^2 + b^2 + h^2$	$-bf + ag$
w	$-cg + bh$	$-ah + cf$	$-bf + ag$	$f^2 + g^2 + h^2$

.

25. For the distance of two points (x, y, z, w) and (x', y', z', w'), we have

$$\cos\delta = \frac{xx' + yy' + zz' + ww'}{\sqrt{x^2 + y^2 + z^2 + w^2}\,\sqrt{x'^2 + y'^2 + z'^2 + w'^2}},$$

whence also

$$\sin\delta = \frac{\sqrt{a^2 + b^2 + c^2 + f^2 + g^2 + h^2}}{\sqrt{x^2 + y^2 + z^2 + w^2}\,\sqrt{x'^2 + y'^2 + z'^2 + w'^2}},$$

where, in the numerator, (a, b, c, f, g, h) stand for the coordinates of the line of junction of the two points, taken to be equal to

$$yz' - y'z,\ zx' - z'x,\ xy' - x'y,\ xw' - x'w,\ yw' - y'w,\ zw' - z'w$$

respectively.

Similarly, for the distance of two planes $(\xi, \eta, \zeta, \omega)$ and $(\xi', \eta', \zeta', \omega')$, we have

$$\cos\delta = \frac{\xi\xi' + \eta\eta' + \zeta\zeta' + \omega\omega'}{\sqrt{\xi^2 + \eta^2 + \zeta^2 + \omega^2}\,\sqrt{\xi'^2 + \eta'^2 + \zeta'^2 + \omega'^2}},$$

whence also

$$\sin\delta = \frac{\sqrt{a^2 + b^2 + c^2 + f^2 + g^2 + h^2}}{\sqrt{\xi^2 + \eta^2 + \zeta^2 + \omega^2}\,\sqrt{\xi'^2 + \eta'^2 + \zeta'^2 + \omega'^2}},$$

where, in the numerator, (a, b, c, f, g, h) stand for the coordinates of the line of intersection of the two planes, taken to be equal to

$$\xi\omega' - \xi'\omega,\ \eta\omega' - \eta'\omega,\ \zeta\omega' - \zeta'\omega,\ \eta\zeta' - \eta'\zeta,\ \zeta\xi' - \zeta'\xi,\ \xi\eta' - \xi'\eta$$

respectively.

The distance of a point (x, y, z, w) and plane $(\xi', \eta', \zeta', \omega')$ is the complement of the distance of the point (x, y, z, w) and the point $(\xi', \eta', \zeta', \omega')$ which is the ⊥ point of the plane; viz. we have

$$\sin\delta = \frac{x\xi' + y\eta' + z\zeta' + w\omega'}{\sqrt{x^2+y^2+z^2+w^2}\sqrt{\xi'^2+\eta'^2+\zeta'^2+\omega'^2}},$$

$$\cos\delta = \frac{\sqrt{a^2+b^2+c^2+f^2+g^2+h^2}}{\sqrt{x^2+y^2+z^2+w^2}\sqrt{\xi'^2+\eta'^2+\zeta'^2+\omega'^2}},$$

where, in the numerator, (a, b, c, f, g, h) stand for the coordinates of the line of junction of the two points. Of course the same result might have been equally well derived from the formula for the distance of two planes.

26. If we now consider a plane triangle ABC, and write

(x_1, y_1, z_1, w_1) for the coordinates of A,

(x_2, y_2, z_2, w_2) „ „ B,

(x_3, y_3, z_3, w_3) „ „ C,

then the coordinates

$$a,\quad b,\quad c,\quad f,\quad g,\quad h,$$

of the line BC will be

$$y_2z_3 - y_3z_2,\ z_3x_2 - z_2x_3,\ x_2y_3 - x_3y_2,\ x_2w_3 - x_3w_2,\ y_2w_3 - y_3w_2,\ z_2w_3 - z_3w_2,$$

and similarly for the coordinates of the lines AB, CA; the equations

$$a_1f_2 + b_1g_2 + c_1h_2 + a_2f_1 + b_2g_1 + c_2h_1 = 0, \text{ \&c.},$$

which express that these lines meet in pairs in the points A, B, C respectively, are of course satisfied identically; and we then have for the sides and angles (linear and angular distances) of the triangle

$$\cos a = \frac{x_2x_3 + y_2y_3 + z_2z_3 + w_2w_3}{\sqrt{x_2^2+y_2^2+z_2^2+w_2^2}\sqrt{x_3^2+y_3^2+z_3^2+w_3^2}},$$

$$\sin a = \frac{\sqrt{a_1^2+b_1^2+c_1^2+f_1^2+g_1^2+h_1^2}}{\sqrt{x_2^2+y_2^2+z_2^2+w_2^2}\sqrt{x_3^2+y_3^2+z_3^2+w_3^2}},$$

$$\cos A = \frac{a_2a_3 + b_2b_3 + c_2c_3 + f_2f_3 + g_2g_3 + h_2h_3}{\sqrt{a_2^2+b_2^2+c_2^2+f_2^2+g_2^2+h_2^2}\sqrt{a_3^2+b_3^2+c_3^2+f_3^2+g_3^2+h_3^2}}, \text{ \&c.};$$

and this being so, if with the values of $\cos a$, $\cos b$, $\cos c$, we form the expression for $\cos a - \cos b \cos c$, then reducing to a common denominator, the expression for the numerator is at once found to be

$$= a_2a_3 + b_2b_3 + c_2c_3 + f_2f_3 + g_2g_3 + h_2h_3,$$

and thence easily

$$\cos A = \frac{\cos a - \cos b \cos c}{\sin b \sin c},$$

viz. the expressions for the angles in terms of the sides are those of ordinary spherical trigonometry.

27. Hence also

$$\sin A = \frac{\sqrt{1 - \cos^2 a - \cos^2 b - \cos^2 c + 2 \cos a \cos b \cos c}}{\sin b \sin c};$$

whence

$$\sin A : \sin B : \sin C = \sin a : \sin b : \sin c,$$

and

$$\cos A + \cos B \cos C = \frac{\cos a\,(1 - \cos^2 a - \cos^2 b - \cos^2 c + 2 \cos a \cos b \cos c)}{\sin^2 a \sin b \sin c},$$

and consequently

$$\cos a = \frac{\cos A + \cos B \cos C}{\sin B \sin C},$$

which completes the system of formulæ.

And similarly for a trihedral, that is, if we have three planes A, B, C (meeting of course in a point, O) then the dihedral distances BC, CA, AB and the angular distances CA, AB; AB, BC; BC, CA are related to each other in the same way as the angles and sides of an ordinary spherical triangle.

Distance of a point and line. Art. Nos. 28, 29.

28. The point is taken to be (x_1, y_1, z_1, w_1), the line (a, b, c, f, g, h). Drawing through the point a $\perp$ plane, say $(\xi, \eta, \zeta, \omega)$ meeting the line in the foot, and taking the coordinates hereof to be (x_2, y_2, z_2, w_2), then $\xi x_1 + \eta y_1 + \zeta z_1 + \omega w_1 = 0$ and

$$\begin{array}{rrrrl} \cdot & h\eta & -g\zeta & +a\omega & = 0, \\ -h\xi & \cdot & +f\zeta & +b\omega & = 0, \\ g\xi & -f\eta & \cdot & +c\omega & = 0, \\ -a\xi & -b\eta & -c\zeta & \cdot & = 0, \end{array}$$

giving, say,

$$\begin{array}{rrrrl} \xi = & \cdot & cy_1 & -bz_1 & +fw_1, \\ \eta = & -cx_1 & \cdot & +az_1 & +gw_1, \\ \zeta = & bx_1 & -ay_1 & \cdot & +hw_1, \\ \omega = & -fx_1 & -gy_1 & -hz_1 & \cdot\ \ . \end{array}$$

We have here

$$\xi^2 + \eta^2 + \zeta^2 + \omega^2 = (b^2 + c^2 + f^2) x_1^2 + \&c.,$$

where $(b^2 + c^2 + f^2) x_1^2 + \&c.$ denotes the before-mentioned quadric function of (x_1, y_1, z_1, w_1), which, equated to zero and regarding therein (x_1, y_1, z_1, w_1) as current coordinates, represents the pair of tangent-planes from the reciprocal line (f, g, h, a, b, c) to the Absolute.

Resuming the question in hand, we have then

$$\xi x_2 + \eta y_2 + \zeta z_2 + \omega w_2 = 0,$$

which, with

$$\begin{aligned} . \quad\quad hy_2 - gz_2 + aw_2 &= 0, \\ -hx_2 \quad . \quad + fz_2 + bw_2 &= 0, \\ gx_2 - fy_2 \quad . \quad + cw_2 &= 0, \\ -ax_2 - by_2 - cz_2 \quad . \quad &= 0, \end{aligned}$$

gives, say,

$$\begin{aligned} -x_2 &= \quad . \quad c\eta - b\zeta + f\omega, \\ -y_2 &= -c\xi \quad . \quad + a\zeta + g\omega, \\ -z_2 &= -b\xi - a\eta \quad . \quad + h\omega, \\ -w_2 &= -f\xi - g\eta - h\zeta \quad . \quad , \end{aligned}$$

that is,

$$\begin{aligned} x_2 &= (b^2 + c^2 + f^2) x_1 + (-ab + fg) y_1 + (-ac + hf) z_1 + (-cg + bh) w_1, \\ y_2 &= (-ab + fg) x_1 + (c^2 + a^2 + g^2) y_1 + (-bc + gh) z_1 + (-ah + cf) w_1, \\ z_2 &= (-ca + hf) x_1 + (-bc + gh) y_1 + (a^2 + b^2 + h^2) z_1 + (-bf + ag) w_1, \\ w_2 &= (-cg + bh) x_1 + (-ah + cf) y_1 + (-bf + ag) z_1 + (f^2 + g^2 + h^2) w_1. \end{aligned}$$

We have therefore

$$x_1x_2 + y_1y_2 + z_1z_2 + w_1w_2 = (b^2 + c^2 + f^2) x_1^2 + \&c.,$$

and

$$x_2^2 + y_2^2 + z_2^2 + w_2^2 = (a^2 + b^2 + c^2 + f^2 + g^2 + h^2) \{(b^2 + c^2 + f^2) x_1^2 + \&c.\},$$

where $(b^2 + c^2 + f^2) x_1^2 + \&c.$ denotes in each case the above-mentioned quadric function of (x_1, y_1, z_1, w_1).

In verification of the expression for $x_2^2 + y_2^2 + z_2^2 + w_2^2$, it is to be remarked that we have identically

$$\begin{aligned} &\xi^2 + \eta^2 + \zeta^2 + \omega^2 + (af + bg + ch)^2 (x_1^2 + y_1^2 + z_1^2 + w_1^2) \\ &\quad\quad = (a^2 + b^2 + c^2 + f^2 + g^2 + h^2) \{(b^2 + c^2 + f^2) x_1^2 + \&c.\}; \end{aligned}$$

here on the left-hand side the whole coefficient of x_1^2 is

$$(b^2 + c^2 + f^2)^2 + (ab - fg)^2 + (ca - hf)^2 + (cg - bh)^2 + (af + bg + ch)^2,$$

where the last four terms are together $= (b^2 + c^2 + f^2)(a^2 + g^2 + h^2)$, and thus the whole coefficient is (as it should be) $= (b^2 + c^2 + f^2)(a^2 + b^2 + c^2 + f^2 + g^2 + h^2)$: and similarly for the coefficients of the remaining terms.

29. Writing then δ for the required distance, we have

$$\cos\delta = \frac{x_1x_2 + y_1y_2 + z_1z_2 + w_1w_2}{\sqrt{x_1^2 + y_1^2 + z_1^2 + w_1^2}\sqrt{x_2^2 + y_2^2 + z_2^2 + w_2^2}},$$

that is,

$$\cos\delta = \frac{\sqrt{(b^2 + c^2 + f^2)\,x_1^2 + \&c.}}{\sqrt{x_1^2 + y_1^2 + z_1^2 + w_1^2}\sqrt{a^2 + b^2 + c^2 + f^2 + g^2 + h^2}},$$

where $(b^2 + c^2 + f^2)\,x_1^2 + \&c.$ is the above-mentioned quadric function

	x_1	y_1	z_1	w_1
x_1	$b^2+c^2+f^2$	$-ab+fg$	$-ac+hf$	$-cg+bh$
y_1	$-ab+fg$	$c^2+a^2+g^2$	$-bc+gh$	$-ah+cf$
z_1	$-ac+hf$	$-bc+gh$	$a^2+b^2+h^2$	$-bf+ag$
w_1	$-cg+bh$	$-ah+cf$	$-bf+ag$	$f^2+g^2+h^2$

Distance of a plane and line. Art. No. 30.

30. This may be deduced from the last preceding result: the formula, as written down, gives the distance of the ⊥ plane (x_1, y_1, z_1, w_1) from the reciprocal line (f, g, h, a, b, c): hence writing $(\xi, \eta, \zeta, \omega)$ for (x_1, y_1, z_1, w_1) and (a, b, c, f, g, h) for (f, g, h, a, b, c), we have for the distance of plane $(\xi, \eta, \zeta, \omega)$ and line (a, b, c, f, g, h) the expression

$$\cos\delta = \frac{\sqrt{(a^2 + g^2 + h^2)\,\xi^2 + \&c.}}{\sqrt{\xi^2 + \eta^2 + \zeta^2 + \omega^2}\sqrt{a^2 + b^2 + c^2 + f^2 + g^2 + h^2}},$$

where $(a^2 + g^2 + h^2)\,\xi^2 + \&c.$ denotes the quadric function

	ξ	η	ζ	ω
ξ	$a^2+g^2+h^2$	$ab-fg$	$ac-hf$	$cg-bh$
η	$ab-fg$	$b^2+h^2+f^2$	$bc-gh$	$ah-cf$
ζ	$ac-hf$	$bc-gh$	$c^2+f^2+g^2$	$bf-ag$
ω	$cg-bh$	$ah-cf$	$bf-ag$	$a^2+b^2+c^2$

The theory of two lines. Art. Nos. 31 to 38.

31. Considering any two lines X, Y, it has been seen that these have two ⊥s, viz. each ⊥ is a line cutting as well the two lines X, Y as the reciprocal lines X', Y', say that one of them cuts the lines X, Y in the points A, C respectively, and the other of them cuts the two lines in the points B, D respectively: and take, as before, the distances AC and BD to be $=\delta$ and θ respectively.

The coordinates of the lines X, Y are

$$(a,\ b,\ c,\ f,\ g,\ h) \text{ and } (a_1,\ b_1,\ c_1,\ f_1,\ g_1,\ h_1)$$

respectively; and if we consider, as before, the planes Π, Ω, Π_1, Ω_1 the coordinates of which are (l, m, n, p), $(\lambda, \mu, \nu, \varpi)$, (l_1, m_1, n_1, p_1), $(\lambda_1, \mu_1, \nu_1, \varpi_1)$ respectively, then X is the intersection of the planes Π, Ω, and we have

$$\begin{aligned} &a : b : c : f : g : h \\ &= l\varpi - \lambda p : m\varpi - \mu p : n\varpi - \nu p : m\nu - n\mu : n\lambda - l\nu : l\mu - m\lambda; \end{aligned}$$

and similarly Y is the intersection of the planes Π_1, Ω_1, and we have

$$\begin{aligned} &a_1 : b_1 : c_1 : f_1 : g_1 : h_1 \\ &= l_1\varpi_1 - \lambda_1 p_1 : m_1\varpi_1 - \mu_1 p_1 : n_1\varpi_1 - \nu_1 p_1 : m_1\nu_1 - n_1\mu_1 : n_1\lambda_1 - l_1\nu_1 : l_1\mu_1 - m_1\lambda_1. \end{aligned}$$

Also the planes (Π, Ω), (Π_1, Ω_1), (Π, Ω_1), (Π_1, Ω) being naturally $\perp$, we have

$$\begin{aligned} l\lambda + m\mu + n\nu + p\varpi &= 0, \\ l_1\lambda_1 + m_1\mu_1 + n_1\nu_1 + p_1\varpi_1 &= 0, \\ l\lambda_1 + m\mu_1 + n\nu_1 + p\varpi_1 &= 0, \\ l_1\lambda + m_1\mu + n_1\nu + p_1\varpi &= 0; \end{aligned}$$

and for the inclinations to each other of the planes (Π, Π_1) and (Ω, Ω_1), we have

$$\cos\delta = \frac{\lambda\lambda_1 + \mu\mu_1 + \nu\nu_1 + \varpi\varpi_1}{\sqrt{\lambda^2 + \&c.}\ \sqrt{\lambda_1{}^2 + \&c.}},$$

$$\cos\theta = \frac{ll_1 + mm_1 + nn_1 + pp_1}{\sqrt{l^2 + \&c.}\ \sqrt{l_1{}^2 + \&c.}}.$$

32. The expressions for the coordinates of the two lines give

$$\begin{aligned} aa_1 + bb_1 + cc_1 + ff_1 + gg_1 + hh_1 &= (ll_1 + mm_1 + nn_1 + pp_1)(\lambda\lambda_1 + \mu\mu_1 + \nu\nu_1 + \varpi\varpi_1) \\ &\quad - (l\lambda_1 + m\mu_1 + n\nu_1 + p\varpi_1)(l_1\lambda + m_1\mu + n_1\nu + p_1\varpi) \\ &= (ll_1 + mm_1 + nn_1 + pp_1)(\lambda\lambda_1 + \mu\mu_1 + \nu\nu_1 + \varpi\varpi_1) \\ &= \sqrt{l^2 + \&c.}\ \sqrt{l_1{}^2 + \&c.}\ \sqrt{\lambda^2 + \&c.}\ \sqrt{\lambda_1{}^2 + \&c.}\ \cos\delta\cos\theta. \end{aligned}$$

But we have

$$\begin{aligned} a^2 + b^2 + c^2 + f^2 + g^2 + h^2 &= (l^2 + m^2 + n^2 + p^2)(\lambda^2 + \mu^2 + \nu^2 + \varpi^2) - (l\lambda + m\mu + n\nu + p\varpi)^2 \\ &= (l^2 + \&c.)(\lambda^2 + \&c.); \end{aligned}$$

and similarly

$$\begin{aligned} a_1{}^2 + b_1{}^2 + c_1{}^2 + f_1{}^2 + g_1{}^2 + h_1{}^2 &= (l_1{}^2 + m_1{}^2 + n_1{}^2 + p_1{}^2)(\lambda_1{}^2 + \mu_1{}^2 + \nu_1{}^2 + \varpi_1{}^2) - (l_1\lambda_1 + m_1\mu_1 + n_1\nu_1 + p_1\varpi_1)^2 \\ &= (l_1{}^2 + \&c.)(\lambda_1{}^2 + \&c.). \end{aligned}$$

Hence the last result gives

$$\frac{aa_1 + bb_1 + cc_1 + ff_1 + gg_1 + hh_1}{\sqrt{a^2 + \&c.}\ \sqrt{a_1{}^2 + \&c.}} = \cos\delta\cos\theta;$$

or calling the expression on the left-hand side the comoment of the two lines, and denoting it by M_1, the equation just obtained is

$$\cos\delta\cos\theta = \text{comoment}, \ = M_1.$$

And if for either of the lines we substitute its reciprocal, then for δ, θ we have $\frac{1}{2}\pi - \delta$, $\frac{1}{2}\pi - \theta$ respectively, and consequently

$$\frac{af_1 + bg_1 + ch_1 + a_1 f + b_1 g + c_1 h}{\sqrt{a^2 + \&c.}\,\sqrt{a_1^2 + \&c.}} = \sin\delta\sin\theta\,;$$

or calling the expression on the left-hand side the moment of the two lines and denoting it by M, the equation is

$$\sin\delta\sin\theta = \text{moment}, \ = M,$$

where observe that $M=0$ is the condition for the intersection of the two lines, $M_1 = 0$ the condition for their contrasection*.

33. But to determine the coordinates (A, B, C, F, G, H) of the $\perp$ line AC or BD, and the coordinates of the points A and C or B and D of the points in which it meets the lines X and Y respectively, I employ a different method.

We consider the lines

$$(a,\ b,\ c,\ f,\ g,\ h),\quad (a_1,\ b_1,\ c_1,\ f_1,\ g_1,\ h_1),$$

and their reciprocals

$$(f,\ g,\ h,\ a,\ b,\ c),\quad (f_1,\ g_1,\ h_1,\ a_1,\ b_1,\ c_1).$$

A line (A, B, C, F, G, H) meeting each of these four lines is said to be a perpendicular. We have

$$\begin{aligned}
(A,\ B,\ C,\ F,\ G,\ H)\,&(a,\ b,\ c,\ f,\ g,\ h) = 0,\\
\text{,,}\qquad &(f,\ g,\ h,\ a,\ b,\ c) = 0,\\
\text{,,}\qquad &(a_1,\ b_1,\ c_1,\ f_1,\ g_1,\ h_1) = 0,\\
\text{,,}\qquad &(f_1,\ g_1,\ h_1,\ a_1,\ b_1,\ c_1) = 0,
\end{aligned}$$

equations which determine say A, B, C, F in terms of G, H, and then substituting in $AF + BG + CH = 0$ we have two values of $G : H$; i.e. there are two systems of values (A, B, C, F, G, H), that is, two perpendiculars.

The equations may be written

$$\begin{aligned}
(A+F)(a+f) + (B+G)(b+g) + (C+H)(c+h) &= 0,\\
(A+F)(a_1+f_1) + (B+G)(b_1+g_1) + (C+H)(c_1+h_1) &= 0,\\
(A-F)(a-f) + (B-G)(b-g) + (C-H)(c-h) &= 0,\\
(A-F)(a_1-f_1) + (B-G)(b_1-g_1) + (C-H)(c_1-h_1) &= 0.
\end{aligned}$$

* The foregoing demonstration of the fundamental formulæ $\cos\delta\cos\theta = M_1$, $\sin\delta\sin\theta = M$, is, in effect, that given by Heath in his Memoir "On the Dynamics of a Rigid Body in Elliptic Space," *Phil. Trans.* t. 175 (for 1884), pp. 281—324.

Hence we have

$$\begin{array}{ccccc} A+F & : & B+G & : & C+H, = \\ (b+g)(c_1+h_1)-(b_1+g_1)(c+h) & : & (c+h)(a_1+f_1)-(a+f)(c_1+h_1) & : & (a+f)(b_1+g_1)-(a_1+f_1)(b+g), = \\ \mathfrak{A}+\alpha & : & \mathfrak{B}+\beta & : & \mathfrak{C}+\gamma; \\ A-F & : & B-G & : & C-H, = \\ (b-g)(c_1-h_1)-(b_1-g_1)(c-h) & : & (c-h)(a_1-f_1)-(a-f)(c_1-h_1) & : & (a-f)(b_1-g_1)-(a_1-f_1)(b-g), = \\ \mathfrak{A}-\alpha & : & \mathfrak{B}-\beta & : & \mathfrak{C}-\gamma; \end{array}$$

equations which may be written

$$A+F,\ B+G,\ C+H = 2\lambda\,(\mathfrak{A}+\alpha,\ \mathfrak{B}+\beta,\ \mathfrak{C}+\gamma),$$

$$A-F,\ B-G,\ C-H = 2\mu\,(\mathfrak{A}-\alpha,\ \mathfrak{B}-\beta,\ \mathfrak{C}-\gamma),$$

where

$$\mathfrak{A} = bc_1 - b_1c + gh_1 - g_1h, \quad \alpha = bh_1 - b_1h - (cg_1 - c_1g),$$

$$\mathfrak{B} = ca_1 - c_1a + hf_1 - h_1f, \quad \beta = cf_1 - c_1f - (ah_1 - a_1h),$$

$$\mathfrak{C} = ab_1 - a_1b + fg_1 - f_1g, \quad \gamma = ag_1 - a_1g - (bf_1 - b_1f).$$

34. We have

$$(\mathfrak{A}+\alpha)^2+(\mathfrak{B}+\beta)^2+(\mathfrak{C}+\gamma)^2 = \{(a+f)^2+(b+g)^2+(c+h)^2\}\{(a_1+f_1)^2+(b_1+g_1)^2+(c_1+h_1)^2\}$$
$$- \{(a+f)(a_1+f_1)+(b+g)(b_1+g_1)+(c+h)(c_1+h_1)\}^2,$$

$$(\mathfrak{A}-\alpha)^2+(\mathfrak{B}-\beta)^2+(\mathfrak{C}-\gamma)^2 = \{(a-f)^2+(b-g)^2+(c-h)^2\}\{(a_1-f_1)^2+(b_1-g_1)^2+(c_1-h_1)^2\}$$
$$- \{(a-f)(a_1-f_1)+(b-g)(b_1-g_1)+(c-h)(c_1-h_1)\}^2;$$

or putting

$$\rho^2 = a^2+b^2+c^2+f^2+g^2+h^2,$$

$$\rho_1^2 = a_1^2+b_1^2+c_1^2+f_1^2+g_1^2+h_1^2,$$

$$\sigma_1 = aa_1+bb_1+cc_1+ff_1+gg_1+hh_1,$$

$$\sigma = af_1+bg_1+ch_1+a_1f+b_1g+c_1h,$$

the foregoing values are

$$= \rho^2\rho_1^2 - (\sigma+\sigma_1)^2, \quad \rho^2\rho_1^2 - (\sigma-\sigma_1)^2.$$

Hence

$$A^2+B^2+C^2+F^2+G^2+H^2 = 4\lambda^2\{\rho^2\rho_1^2-(\sigma+\sigma_1)^2\} = 4\mu^2\{\rho^2\rho_1^2-(\sigma-\sigma_1)^2\};$$

or we may write

$$\lambda^2 = \rho^2\rho_1^2 - (\sigma-\sigma_1)^2, \quad \text{or say} \quad \lambda = \sqrt{\rho^2\rho_1^2-(\sigma-\sigma_1)^2},$$

$$\mu^2 = \rho^2\rho_1^2 - (\sigma+\sigma_1)^2, \qquad \mu = -\sqrt{\rho^2\rho_1^2-(\sigma+\sigma_1)^2}.$$

Making a slight change of notation, if we put

$$M = \frac{af_1+bg_1+ch_1+a_1f+b_1g+c_1h}{\sqrt{a^2+\&c.}\,\sqrt{a_1^2+\&c.}} = \frac{\sigma}{\rho\rho_1},$$

$$M_1 = \frac{aa_1+bb_1+cc_1+ff_1+gg_1+hh_1}{\sqrt{a^2+\&c.}\,\sqrt{a_1^2+\&c.}} = \frac{\sigma_1}{\rho\rho_1},$$

then the values are

$$\lambda = rr_1 \sqrt{1-(M-M_1)^2}, \quad \mu = -rr_1 \sqrt{1-(M+M_1)^2}.$$

And, this being so, the two systems of values of A, B, C, F, G, H, are

$$\begin{array}{l|l} \lambda(\mathfrak{A}+\alpha)+\mu(\mathfrak{A}-\alpha), & \lambda(\mathfrak{A}+\alpha)-\mu(\mathfrak{A}-\alpha), \\ \lambda(\mathfrak{B}+\beta)+\mu(\mathfrak{B}-\beta), & \lambda(\mathfrak{B}+\beta)-\mu(\mathfrak{B}-\beta), \\ \lambda(\mathfrak{C}+\gamma)+\mu(\mathfrak{C}-\gamma), & \lambda(\mathfrak{C}+\gamma)-\mu(\mathfrak{C}-\gamma), \\ \lambda(\mathfrak{A}+\alpha)-\mu(\mathfrak{A}-\alpha), & \lambda(\mathfrak{A}+\alpha)+\mu(\mathfrak{A}-\alpha), \\ \lambda(\mathfrak{B}+\beta)-\mu(\mathfrak{B}-\beta), & \lambda(\mathfrak{B}+\beta)+\mu(\mathfrak{B}-\beta), \\ \lambda(\mathfrak{C}+\gamma)-\mu(\mathfrak{C}-\gamma), & \lambda(\mathfrak{C}+\gamma)+\mu(\mathfrak{C}-\gamma); \end{array}$$

viz. the two perpendiculars are reciprocals each of the other.

35. Before going further I notice that, if

$$\frac{a_1+f_1}{a+f} = \frac{b_1+g_1}{b+g} = \frac{c_1+h_1}{c+h} \quad \text{or} \quad \frac{a_1-f_1}{a-f} = \frac{b_1-g_1}{b-g} = \frac{c_1-h_1}{c-h},$$

then the four equations for (A, B, C, F, G, H) reduce themselves to three equations only: and thus instead of two perpendiculars we have a singly infinite series of perpendiculars (see *ante* 15).

To explain the meaning of the equations, I observe that a line (a, b, c, f, g, h) will be a generating line of the one kind, or say a "generatrix," of the Absolute if

$$a+f=0, \quad b+g=0, \quad c+h=0:$$

and it will be a generating line of the other kind, or say a "directrix," of the Absolute if $a-f=0$, $b-g=0$, $c-h=0$. Or what is the same thing, we have

$(\mathrm{a}, \mathrm{b}, \mathrm{c}, -\mathrm{a}, -\mathrm{b}, -\mathrm{c})$, where $\mathrm{a}^2+\mathrm{b}^2+\mathrm{c}^2=0$ for a generatrix,

and

$(\mathrm{a}, \mathrm{b}, \mathrm{c}, \mathrm{a}, \mathrm{b}, \mathrm{c})$, where $\mathrm{a}^2+\mathrm{b}^2+\mathrm{c}^2=0$ for a directrix, of the Absolute.

Consider now two directrices $(\mathrm{a}, \mathrm{b}, \mathrm{c}, \mathrm{a}, \mathrm{b}, \mathrm{c})$ and $(\mathrm{a}_1, \mathrm{b}_1, \mathrm{c}_1, \mathrm{a}_1, \mathrm{b}_1, \mathrm{c}_1)$: if a line (a, b, c, f, g, h) meets each of these, then

$$(a+f)\,\mathrm{a} + (b+g)\,\mathrm{b} + (c+h)\,\mathrm{c} = 0,$$
$$(a+f)\,\mathrm{a}_1 + (b+g)\,\mathrm{b}_1 + (c+h)\,\mathrm{c}_1 = 0,$$

and consequently

$$a+f : b+g : c+h = \mathrm{bc}_1-\mathrm{b}_1\mathrm{c} : \mathrm{ca}_1-\mathrm{c}_1\mathrm{a} : \mathrm{ab}_1-\mathrm{a}_1\mathrm{b};$$

and similarly if $(a_1, b_1, c_1, f_1, g_1, h_1)$ meets each of the two directrices, then

$$a_1+f_1 : b_1+g_1 : c_1+h_1 = \mathrm{bc}_1-\mathrm{b}_1\mathrm{c} : \mathrm{ca}_1-\mathrm{c}_1\mathrm{a} : \mathrm{ab}_1-\mathrm{a}_1\mathrm{b},$$

that is, if the lines each of them meet the same two directrices of the Absolute, then

$$\frac{a_1+f_1}{a+f} = \frac{b_1+g_1}{b+g} = \frac{c_1+h_1}{c+h}.$$

Conversely, if these relations are satisfied, then the lines each of them meet two directrices of the Absolute.

In like manner, if the lines each meet two generatrices of the Absolute, then

$$\frac{a_1-f_1}{a-f}=\frac{b_1-g_1}{b-g}=\frac{c_1-h_1}{c-h};$$

and conversely, if these relations are satisfied, then the lines each of them meet the same two generatrices of the Absolute. In the former case, the lines are said to be "right parallels": in the latter case, "left parallels."

A line (a, b, c, f, g, h) meets the Absolute in two points, and through each of these we have a directrix and a generatrix: that is, the line meets two directrices and two generatrices.

Through a given point we may draw, meeting the two directrices, or meeting the two generatrices, a line: that is, through a given point we may draw a line

$$(a_1, b_1, c_1, f_1, g_1, h_1)$$

which is a right parallel, and a line which is a left parallel, to a given line. That is, regarding as given the first line, and also a point of the second line, there are two positions of the second line such that for each of them, the $\perp$s of the pair of lines, instead of being two determinate lines, are a singly infinite series of lines.

36. Reverting to the general case, we have found (A, B, C, F, G, H) the coordinates of either of the lines $\perp$ to the given lines (a, b, c, f, g, h) and $(a_1, b_1, c_1, f_1, g_1, h_1)$: supposing that the $\perp$ intersects the first of these lines in the point the coordinates of which are (x, y, z, w), and the second in the point the coordinates of which are

$$(x_1, y_1, z_1, w_1),$$

then we have for each set of coordinates a fourfold expression; the choice of the form is indifferent, and I write

$$x : y : z : w = cB-bC : aC-cA : bA-aB : fA+gB+hC,$$

$$x_1 : y_1 : z_1 : w_1 = c_1B-b_1C : a_1C-c_1A : b_1A-a_1B : f_1A+g_1B+h_1C.$$

We have then, for the distance of these two points,

$$\cos\phi=\frac{xx_1+yy_1+zz_1+ww_1}{\sqrt{x^2+y^2+z^2+w^2}\sqrt{x_1^2+y_1^2+z_1^2+w_1^2}},\quad \sin\phi=\frac{\sqrt{(yz_1-y_1z)^2+\&c.}}{\sqrt{x^2+y^2+z^2+w^2}\sqrt{x_1^2+y_1^2+z_1^2+w_1^2}};$$

where $\phi=\delta$ or θ, according to the sign of the radical $\lambda:\mu$ contained in the expressions for A, B, C, F, G, H.

I have not succeeded in obtaining in this manner the final formulæ for the determination of the distances: these in fact are, by what precedes, given by the equations

$$\sin\delta\sin\theta=M,\quad \cos\delta\cos\theta=M_1.$$

For then, writing ϕ to denote either of the distances δ, θ, at pleasure, we have

$$\frac{M^2}{\sin^2\phi}+\frac{M_1^2}{\cos^2\phi}=1,$$

that is,

$$\cos^4\phi + \cos^2\phi\,(M_1^2 - M^2 + 1) + M_1^2 = 0,$$

or

$$\cos^2\phi = \tfrac{1}{2}\{M_1^2 - M^2 + 1 \pm \sqrt{M_1^4 + M^4 - 2M_1^2M^2 - 2M_1^2 - 2M^2 + 1}\},$$

which is the expression for the cosine of the distance.

In the case where the two lines intersect $M = 0$, and if δ be the $\perp$ distance which vanishes, then $\delta = 0$, and consequently $\cos\theta = M_1$: the last-mentioned formula, putting therein $M = 0$ and taking the radical to be $= M_1^2 - 1$, gives $\cos^2\phi = M_1^2$, that is, $\phi = \theta$, and $\cos^2\theta = M_1^2$, as it should be.

37. I verify as follows, in the case in question of two *intersecting* lines,

$$(af_1 + bg_1 + ch_1 + a_1f + b_1g + c_1h = 0),$$

the formula

$$\cos\theta = \frac{xx_1 + yy_1 + zz_1 + ww_1}{\sqrt{x^2 + y^2 + z^2 + w^2}\,\sqrt{x_1^2 + y_1^2 + z_1^2 + w_1^2}}.$$

We have here

$$\begin{aligned}
A &= \mathfrak{A} = bc_1 - b_1c + gh_1 - g_1h,\\
B &= \mathfrak{B} = ca_1 - c_1a + hf_1 - h_1f,\\
C &= \mathfrak{C} = ab_1 - a_1b + fg_1 - f_1g,\\
F &= \alpha = bh_1 - b_1h - cg_1 + c_1g,\\
G &= \beta = cf_1 - c_1f - ah_1 + a_1h,\\
H &= \gamma = ag_1 - a_1g - bf_1 + b_1f.
\end{aligned}$$

I stop to notice that these formulæ may be obtained in a different and somewhat more simple manner: the two lines (a, b, c, f, g, h) and $(a_1, b_1, c_1, f_1, g_1, h_1)$ intersect; hence their reciprocals also intersect: the equation of the plane through the two lines and that of the plane through the two reciprocal lines are respectively

$$\begin{aligned}
(gh_1 - g_1h)\,x + (hf_1 - h_1f)\,y + (fg_1 - f_1g)\,z + (af_1 + bg_1 + ch_1)\,w &= 0,\\
(bc_1 - b_1c)\,x + (ca_1 - c_1a)\,y + (ab_1 - a_1b)\,z + (fa_1 + gb_1 + hc_1)\,w &= 0:
\end{aligned}$$

the line (A, B, C, F, G, H) is thus the line of intersection of these two planes, and it is thence easy to obtain the foregoing values.

From the values of A, B, C, F, G, H, we have to find x, y, z, w and x_1, y_1, z_1, w_1 by the formulæ given above. We have

$$\begin{aligned}
x = cB - bC &= c^2a_1 - cc_1a + chf_1 - ch_1f\\
&\quad - abb_1 + a_1b^2 - bfg_1 + bgf_1\\
&= (bg + ch)\,f_1 + a_1(b^2 + c^2) - b_1ab - c_1ac - bfg_1 - cfh_1\\
&= -f(af_1 + bg_1 + ch_1) + a_1(b^2 + c^2) - b_1ab - c_1ac;
\end{aligned}$$

or writing here $a_1f + b_1g + c_1h$ in place of $-(af_1 + bg_1 + ch_1)$, this is a linear function of a_1, b_1, c_1; and similarly finding the values of y, z, w, we have

$$\begin{aligned}
x &= a_1(b^2 + c^2 + f^2) + b_1(fg - ab) + c_1(hf - ca),\\
y &= a_1(fg - ab) + b_1(c^2 + a^2 + g^2) + c_1(gh - bc),\\
z &= a_1(hf - ca) + b_1(gh - bc) + c_1(a^2 + b^2 + h^2),\\
w &= a_1(bh - cg) + b_1(cf - ah) + c_1(ag - bf).
\end{aligned}$$

And in like manner, (I introduce for convenience the sign $-$, as is allowable),

$$\begin{aligned}
-x_1 &= a(b_1^2+c_1^2+f_1^2) + b(f_1g_1-a_1b_1) + c(h_1f_1-c_1a_1),\\
-y_1 &= a(f_1g_1-a_1b_1) + b(c_1^2+a_1^2+g_1^2) + c(g_1h_1-b_1c_1),\\
-z_1 &= a(h_1f_1-c_1a_1) + b(g_1h_1-b_1c_1) + c(a_1^2+b_1^2+h_1^2),\\
-w_1 &= a(b_1h_1-c_1g_1) + b(c_1f_1-a_1h_1) + c(a_1g_1-b_1f_1).
\end{aligned}$$

38. Write for shortness

$$p = a^2+b^2+c^2, \quad p_1 = f^2+g^2+h^2, \quad \omega = a_1f+b_1g+c_1h,$$

and therefore

$$q = aa_1+bb_1+cc_1, \quad q_1 = ff_1+gg_1+hh_1, \quad -\omega = af_1+bg_1+ch_1,$$

$$r = a_1^2+b_1^2+c_1^2, \quad r_1 = f_1^2+g_1^2+h_1^2.$$

We have

$$\begin{aligned}
x &= a_1p - aq + f\omega, & x_1 &= -ar + a_1q + f_1\omega,\\
y &= b_1p - bq + g\omega, & y_1 &= -br + b_1q + g_1\omega,\\
z &= c_1p - cq + h\omega, & z_1 &= -cr + c_1q + h_1\omega,
\end{aligned}$$

$$w = -\begin{vmatrix} f, & g, & h \\ a, & b, & c \\ a_1, & b_1, & c_1 \end{vmatrix}, \quad w_1 = -\begin{vmatrix} f_1, & g_1, & h_1 \\ a, & b, & c \\ a_1, & b_1, & c_1 \end{vmatrix};$$

from which we easily obtain

$$x^2+y^2+z^2 = p(pr-q^2) + (p_1+2p)\,\omega^2,$$

and by expressing w^2 in the form of a determinant

$$w^2 = p_1(pr-q^2) - p\omega^2,$$

we obtain

$$x^2+y^2+z^2+w^2 = (p+p_1)(pr-q^2+\omega^2);$$

and in like manner

$$x_1^2+y_1^2+z_1^2+w_1^2 = (r+r_1)(pr-q^2+\omega^2).$$

And again

$$xx_1+yy_1+zz_1 = q(pr-q^2) + (q_1+2q)\,\omega^2,$$

and by expressing ww_1 in the form of a determinant

$$ww_1 = q_1(pr-q^2) - q\omega^2,$$

we find

$$xx_1+yy_1+zz_1+ww_1 = (q+q_1)(pr-q^2+\omega^2).$$

Hence substituting in

$$\cos\theta = \frac{xx_1+yy_1+zz_1+ww_1}{\sqrt{x^2+y^2+z^2+w^2}\,\sqrt{x_1^2+y_1^2+z_1^2+w_1^2}},$$

the factor $pr-q^2+\omega^2$ disappears, and we have

$$\cos\theta = \frac{q+q_1}{\sqrt{p+p_1}\,\sqrt{r+r_1}} = M_1,$$

the required result.

952.

ON THE KINEMATICS OF A PLANE, AND IN PARTICULAR ON THREE-BAR MOTION: AND ON A CURVE-TRACING MECHANISM.

[From the *Transactions of the Cambridge Philosophical Society*, vol. XV. (1894), pp. 391—402.]

THE first part of the present paper, On the Kinematics of a Plane, and on Three-bar Motion, is purely theoretical: the second part contains a brief description of a Curve-tracing Mechanism, which at my suggestion has been constructed by Prof. Ewing in the workshops of the Engineering Laboratory, Cambridge.

PART I.

1. The theory of the motion of a plane, when two given points thereof describe given curves, has been considered by Mr S. Roberts in his paper, "On the motion of a plane under given conditions," *Proc. Lond. Math. Soc.* t. III. (1871), pp. 286—318, and he has shown that, if for the given curves the order, class, number of nodes, and of cusps, are (m, n, δ, κ) and $(m', n', \delta', \kappa')$ respectively $(n = m^2 - m - 2\delta - 3\kappa,\ n' = m'^2 - m' - 2\delta' - 3\kappa')$, then for the curve described by any fixed point of the plane:

$$\begin{aligned}
&\text{order} &&= 2mm',\\
&\text{class} &&= 2\,(mm' + mn' + nm'),\\
&\text{number of nodes} &&= mm'\,(2mm' - m - m') + 2\,(m\delta' + m'\delta),\\
&\text{number of cusps} &&= 2\,(m\kappa' + m'\kappa);
\end{aligned}$$

but he remarks that these formulæ require modification when the directrices or either of them pass through the circular points at infinity. And he has considered the case where the two directrices become one and the same curve.

2. It will be convenient to speak of the line joining the two given points as the link; the two given points, say B and D, are then the extremities of the link; and I take the length of the link to be $= c$, and the two directrices to be b and d; we have thus the link $c = BD$ moving in suchwise that its extremity B describes the curve b of the order m, and its extremity D the curve d of the order m': in Mr Roberts' problem, the locus is that described by a point P rigidly connected with the link, or say by a point P the vertex of the triangle PBD.

3. The points B, D describe of course the directrices b, d respectively: taking on b a point B_1 at pleasure, then if B be at B_1, the corresponding positions of D are the intersections of d by the circle centre B_1 and radius c, viz. there are thus $2m'$ positions of D: and similarly taking on d a point D_1 at pleasure, then if D be at D_1, the corresponding positions of B are the intersections of b by the circle centre D_1 and radius c, viz. there are thus $2m$ positions of B. The motion thus establishes a $(2m, 2m')$ correspondence between the points of the directrices b and d, viz. to a given point on b there correspond $2m'$ points on d, and to a given point on d there correspond $2m$ points on b. Of course, for a given point on either directrix, the corresponding points on the other directrix may be any or all of them imaginary; and thus it may very well be that for either directrix not the whole curve but only a part or detached parts thereof will be actually described in the course of the motion. In saying that a part is described, we mean described by a continuous motion; say that the point B (the point D remaining always on a part of d) is capable of describing continuously a part of b; it may very well happen that the point B (the point D remaining always on a different part of d) is capable of describing continuously a different part of b, but that it is not possible for B to pass from the one to the other of these parts of b without removing D from the one part and placing it on the other part of d, and thus that we have on b detached parts each of them continuously described by B; and similarly, we may have on d detached parts each of them continuously described by D.

4. But dropping for the moment the question of reality, to a given position of B on b there correspond as was mentioned $2m'$ positions of D on d, or say $2m'$ positions of the link c: in the entire motion of the link it must assume each of these $2m'$ positions, and for each of them the point B comes to assume the position in question on b; the directrix b is thus described $2m'$ times, that is, the locus described by B will be the directrix b repeated $2m'$ times, or say a curve of the order $m \times 2m'$, $= 2mm'$. Similarly, the locus described by D will be the directrix d repeated $2m$ times, or say a curve of the order $m' \times 2m$, $= 2mm'$.

5. In general, if B_1D_1 be any position of the link and if B moves from B_1 along b in a determinate sense, then D will move from D_1 along d in a determinate sense; and if B moves from B_1 along b in the opposite sense, then also D will move from D_1 along d in the opposite sense. Or what is the same thing, we may have B moving in a determinate sense through B_1, and D moving in a determinate sense through D_1; and reversing the sense of B's motion, we reverse also the sense of D's motion. But there are certain critical positions of the link, viz. we have a critical position when

the link is a normal at B_1 to the directrix b, or a normal at D_1 to the directrix d. Say first the link is a normal at B_1 to the directrix b. The infinitesimal element at B_1 may be regarded as a straight line at right angles to the link; hence if for a moment D_1 is regarded as a fixed point, the link may rotate in either direction round D_1, that is, B may move from B_1 along b in either of the two opposite senses, say B_1 is a "two-way point." But if on d we take on opposite sides of D_1 the consecutive points D_1' and D_1'', say $D_1'D_1$ cuts D_1B_1 at an acute angle and $D_1''D_1$ cuts it at an

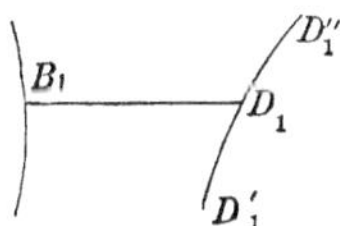

obtuse angle, then D_1' will be nearer to b than was D_1, and thus the circle centre D_1' and radius c will cut b in two real points B_1' and B_1'' near to and on opposite sides of B_1; or as D moves to D_1', B will move from B_1 indifferently to B_1' or B_1''. Contrariwise, D_1'' is further from b than was D_1, and thus the circle centre D_1'' and radius c, will not meet b in any real point near to B_1, and hence D is incapable of moving from D_1 in the sense D_1D_1''. Or what is the same thing, the described portion of d, which includes a point D_1', will terminate at D_1, or say D_1 is a "summit" on the directrix d. We have thus a summit on d, corresponding to the two-way point on b. And of course in like manner, if the link is a normal at D_1 to the directrix d, then D_1 is a two-way point on d, and the corresponding point B_1 is a summit on b.

6. If the link is at the same time a normal at B_1 to b and at D_1 to d, then each of the points B_1, D_1 is a two-way point and also a summit; or more accurately, each of them is a two-way point and also a pair of coincident summits.

But the case requires further investigation. Considering the position B_1D_1 as given, we may take the axis of x coincident with this line, and the origin O in suchwise

that OB_1, OD_1 are each positive and $OD_1 > OB_1$; say we have $OD_1 = \delta$, $OB_1 = \beta$, and therefore $\delta - \beta = c$. The equation of the curve b in the neighbourhood of B_1 is $y^2 = 2\rho(x - \beta)$, where ρ is the radius of curvature at B_1, assumed to be positive when the curve is convex to O, or what is the same thing when the centre of curvature R lies to the right of B_1 $(OR - OB_1 = +)$; and similarly the equation of d in the neighbourhood of D_1 is $y^2 = 2\sigma(x - \delta)$, where σ is the radius of curvature at D_1 assumed to be positive when the curve is convex to O, or what is the same thing when the centre of curvature S lies to the right of D_1 $(OS - OD_1 = +)$.

Consider now (x_1, y_1) the coordinates of a point on b in the neighbourhood of B_1, $y_1^2 = 2\rho(x_1 - \beta)$, and taking B at this point, let (x_2, y_2) be the coordinates of the corresponding point D on d in the neighbourhood of D_1, $y_2^2 = 2\sigma(x_2 - \delta)$. We have

$$c^2 = (x_1 - x_2)^2 + (y_1 - y_2)^2,$$

and here

$$x_1 = \beta + \frac{y_1^2}{2\rho}, \quad x_2 = \delta + \frac{y_2^2}{2\sigma},$$

whence

$$x_1^2 = \beta^2 + \frac{\beta y_1^2}{\rho}, \quad x_1x_2 = \beta\delta + \tfrac{1}{2}\frac{\delta y_1^2}{\rho} + \tfrac{1}{2}\frac{\beta y_2^2}{\sigma}, \quad x_2^2 = \delta^2 + \frac{\delta y_2^2}{\sigma}.$$

The equation thus becomes

$$(\delta - \beta)^2 + \frac{y_1^2}{\rho}(\beta - \delta) + \frac{y_2^2}{\sigma}(\delta - \beta) + (y_1 - y_2)^2 = c^2,$$

that is,

$$y_1^2\left(1 + \frac{\beta - \delta}{\rho}\right) - 2y_1y_2 + y_2^2\left(1 + \frac{\delta - \beta}{\sigma}\right) = 0,$$

a quadric equation between y_1 and y_2. Evidently if we had taken D a point on d, coordinates (x_2, y_2), in the neighbourhood of D_1 and had sought for the coordinates (x_1, y_1) of the corresponding point B on b in the neighbourhood of B_1, we should have found the same equation between y_1 and y_2.

7. The equation will have real roots if

$$1 > \left(1 + \frac{\beta - \delta}{\rho}\right)\left(1 + \frac{\delta - \beta}{\sigma}\right);$$

viz. ρ, σ having the same sign, this is

$$\rho\sigma > (\rho + \beta - \delta)(\sigma + \delta - \beta):$$

but ρ, σ having opposite signs, then

$$\rho\sigma < (\rho + \beta - \delta)(\sigma + \delta - \beta).$$

These conditions may be written

$$(OR - OB_1)(OS - OD_1) - (OS - OB_1)(OR - OD_1) > \text{ or } < 0,$$

that is,

$$(OS - OR)(OD_1 - OB_1) > \text{ or } < 0.$$

But we have $OD_1 - OB_1 = +$, and therefore, ρ, σ having the same sign, the condition of reality is $OS > OR$, i.e. S to the right of R; but ρ, σ having opposite signs, the condition of reality is $OS < OR$, i.e. S to the left of R. Observe that, S lying to the left of R, we cannot have $\rho = -$, $\sigma = +$, and that the second alternative thus is $\rho = +$, $\sigma = -$, then $OS < OR$, or S lies to the left of R.

The condition was investigated as above in order to exhibit more clearly the geometrical signification: but of course the original form, or say the equation

$$1 - \left(1 + \frac{\beta - \delta}{\rho}\right)\left(1 + \frac{\delta - \beta}{\sigma}\right) > 0,$$

gives at once

$$\frac{\delta - \beta}{\rho\sigma}(\delta + \sigma - \beta - \rho) > 0.$$

8. Writing the quadric equation in the form

$$y_1^2\left(1-\frac{c}{\rho}\right)-2y_1y_2+\left(1+\frac{c}{\sigma}\right)y_2^2=0,$$

we have

$$\left(1-\frac{c}{\rho}\right)y_1=\left\{1\pm\sqrt{\frac{c}{\rho\sigma}(c+\sigma-\rho)}\right\}y_2;$$

the two values of $y_1 : y_2$ will have the same sign or opposite signs according as $1-\frac{c}{\rho}$ and $1+\frac{c}{\sigma}$ have the same sign or opposite signs, and in the case where these have the same sign, then this is also the sign of each of the two values of $y_1 : y_2$. Or what is the same thing, if $1-\frac{c}{\rho}$ and $1+\frac{c}{\sigma}$ are each of them positive, then the two values of $y_1 : y_2$ are each of them positive; if $1-\frac{c}{\rho}$ and $1+\frac{c}{\sigma}$ are each of them negative, then the two values of $y_1 : y_2$ are each of them negative; and if $1-\frac{c}{\rho}$ and $1+\frac{c}{\sigma}$ have opposite signs, then the two values of $y_1 : y_2$ have opposite signs. Considering the different cases ρ, $\sigma = ++$, $+-$, $--$, we find

ρ, $\sigma = ++$, then values of $y_1 : y_2$ are $++$ or $--$, according as DR, BS are $++$ or $--$,
ρ, $\sigma = +-$ " " " " " DR, SB " "
ρ, $\sigma = --$ " " " " " RD, SB " "

and in each case the values of $y_1 : y_2$ are $+-$, if the two distances referred to have opposite signs: $DR = +$ means that R is to the right of, or beyond, D, and so in other cases.

9. The different cases, two real roots as above, are

O R D R S
ρ, $\sigma = ++$ { R S } $y_1 : y_2 = ++$, " $\pm$,
S R
ρ, $\sigma = +-$ { S R; S R; S R } $y_1 : y_2 = ++$, " $\pm$, " $\pm$, " $--$,
R S
ρ, $\sigma = --$ { R S } $y_1 : y_2 = ++$, " $\pm$.
B D

Obviously the cases ρ, $\sigma = --$, correspond exactly to the cases ρ, $\sigma = +, +$; the only difference is that the concavities, instead of the convexities, of the two curves are turned towards the point O.

10. If the two roots of the quadratic equation are imaginary, then B_1D_1 is a conjugate or isolated position of the link, and B_1, D_1 are isolated points on the curves b and d respectively.

11. If the roots are real, then the three cases $y_1 : y_2 = ++$, $--$ and $+-$, may be delineated as in the annexed figures, viz. taking in each case y_1 as positive, that is, imagining B to move upwards from B_1 through an infinitesimal arc of b, then D moves from D_1 through either of two infinitesimal arcs of d, both upwards, both downwards, or the one upwards and the other downwards, as shown in the figures

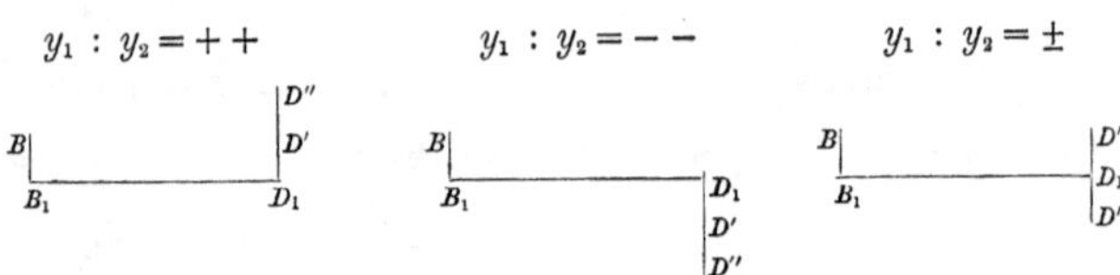

and where it is to be observed that, reversing the sense of the motion of B from B_1, we reverse also the senses of the motion of D from D_1: moreover that, considering D as moving through an infinitesimal arc of d from D_1, we have the like relations thereto of the two infinitesimal arcs of b described by B from B_1. Thus the points B_1 and D_1 are singular points of like character.

If $y_1 : y_2 = ++$, we may say that B_1 (or D_1) is a for-forwards point; if $y_1 : y_2 = --$, then that B_1 (or D_1) is a back-backwards point; and if $y_1 : y_2 = \pm$, then that B_1 (or D_1) is a back-forwards point.

12. The separating case between two imaginary roots and two real roots is that of two equal real roots: the condition for this is $\delta + \sigma = \beta + \rho$, that is, $OS = OR$, or the two centres of curvature are coincident; the characters of the points B_1 and D_1 would in this case depend on the aberrancies of curvature of the curves b and d at these points respectively. If each of the curves is a circle, then the curves are concentric circles, and the link BD moves in suchwise that its direction passes always through the common centre of the two circles—or say so that BD is always a radius of the annulus formed by the two circles—and for any position of BD, the two extremities B, D are related to each other in like manner with the points B_1 and D_1. Thus, in this case, there are no singular points B_1 and D_1 to be considered.

13. In the case where the curves b, d are circles, we have three-bar motion: say

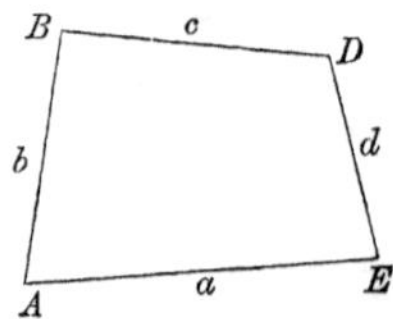

the figure is as here shown; I take in it b, d for the radii of the two circles respectively and a for the distance of their centres; viz. we have the link $BD = c$, pivoted at its extremities to the arms or radii $AB = b$, and $ED = d$, which rotate about the fixed centres A, E at a distance from each other $= a$. Here a, b, c, d are each of them positive; a, b, d may have any values, but then c is at most $= a + b + d$, and if $a > b + d$ then c is at least $= a - b - d$; but if $a =$ or $< b + d$, then c may be $= 0$, viz. it may have any value from 0 to $a + b + d$. And in either case there will be critical values of c. The cases are very numerous. To make an exhaustive enumeration, we may assume d at most $= b$, and in each of the two cases $d < b$ and $d = b$, considering the centre of the circle d as moving from the right of the centre of the circle b towards this centre, we may in the first instance divide as follows:

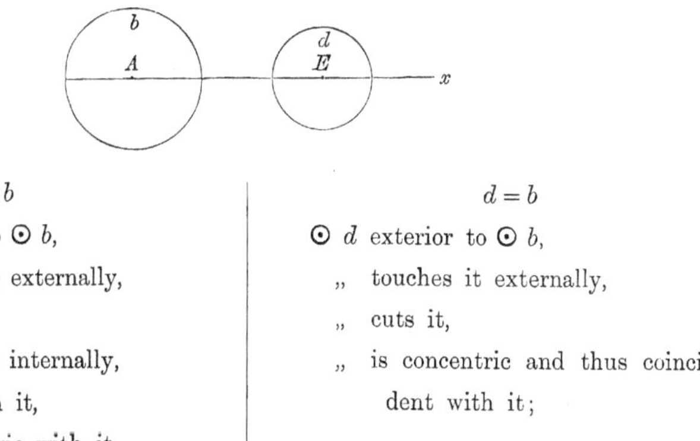

$d < b$	$d = b$
⊙ d exterior to ⊙ b,	⊙ d exterior to ⊙ b,
„ touches it externally,	„ touches it externally,
„ cuts it,	„ cuts it,
„ touches it internally,	„ is concentric and thus coincident with it;
„ lies within it,	
„ is concentric with it,	

and then, in each of these cases, give to the length c of the link its different admissible values.

14. Considering the case $d < b$, then we have (see Plate I., at p. 516), exterior series, the figures 1, 1—2, 2, 2—3, 3, 3—4, 4, viz.

fig. 1,	$c = a - b - d$,
1—2,	„ intermediate,
2,	$c = a - b + d$,
2—3,	„ intermediate,
3,	$c = a + b - d$,
3—4,	„ intermediate,
4,	$c = a + b + d$.

15. In figure 1, the curves described by the extremities B and D respectively are each of them a mere point.

In figure 1—2, we have $a + d > b + c$ and $a + b > d + c$. Hence in the course of the motion the arms b, c come into a right line, giving a position B_1D_1' of the link, where B_1 is a two-way point on b and D_1' a summit on d; or rather, there are two such positions symmetrically situate on opposite sides of the axis Ax. And again, in the course of the motion, the arms d, c come into a right line, giving a position $B_1'D_1$, where D_1 is a two-way point on d and B_1' a summit on b; or rather, there are two such positions symmetrically situate on opposite sides of the axis Ax. Only an arc of the circle b is described, viz. the arc adjacent to d included between the two summits B_1' on b; and in like manner, only an arc of the circle d is described, viz. the arc adjacent to b included between the two summits D_1' on d. The described portions on b and d respectively are to be regarded each of them as a double line or indefinitely thin bent oval: and it is to be observed that for a given position of B (or D) there are two positions of the link BD, each of these positions being assumed by the link in the course of its motion.

16. In figure 2, the two positions B_1D_1' of the link come to coincide together in a single axial position BD, but we still have the other two positions $B_1'D_1$ of the link, where B_1' is a summit on b, and D_1 a two-way point on d. As regards BD, this is the configuration ρ, $\sigma = --$, R, B, S, D; $y_1 : y_2 = \pm$, and thus each of the axial points B, D is a back-and-forwards point. Thus only the arc $B'B_1'$ of the circle b is described by the point B, but the whole circumference of the circle d is described by the point D. If we further examine the motion it will appear that, as B moves from the axial point B say to the upper summit B_1' and returns to B, then D starting from the axial point D may describe (*and that in either sense*, viz. $y_1 = +$, then we have $y_2 = \pm$) the entire circumference of d, returning to the axial point D; and similarly, as B moves from the axial point B to the lower summit B_1' and returns to B, then D starting as before from the axial point D may describe (*and that in either sense*, viz. $y_1 = -$, then we have $y_2 = \pm$) the entire circumference of d, returning to the axial point D. It is thus not the entire arc $B_1'B_1'$ but each of the half-arcs BB_1' which corresponds, and that in either of two ways, to the circumference of d.

17. In figure 2—3, there are four critical positions $B_1'D_1$ (forming two pairs, those of the same pair situate symmetrically on opposite sides of the axis Ax) where, as before, B_1' is a summit on b, and D_1 a two-way point on d. The described portions of b are the detached arcs $B_1'B_1'$ between the two upper summits, and $B_1'B_1'$ between the two lower summits: the described portion of d is the whole circumference. In fact, attending to one of the arcs on b, say the upper arc $B_1'B_1'$, as B moves from one of the summits, say the left-hand summit B_1', and then returns to the left-hand summit B_1', then D, starting from the corresponding two-way point D_1, may describe, *and that in either sense*, the entire circumference of d, returning to the same point D_1; and similarly, as B describes the lower arc $B_1'B_1'$, starting from and returning to a summit, then D, starting from the corresponding two-way point D_1, may describe, *and that in either sense*, the entire circumference of d, returning to the same two-way point D_1.

18. In figure 3, two of the positions $B_1'D_1$ have come to coincide together in the axial position BD: but we still have the other two positions $B_1'D_1$, where B_1' is a summit on b, and D_1 a two-way point on d. As regards the axial points B, D, this is the configuration ρ, $\sigma = ++$; B, R, D, S; $y_1 : y_2 = \pm$, viz. each of the points B, D is a back-and-forwards point. The two detached arcs $B_1'B_1'$ of b have united themselves into a single arc $B_1'B_1'$, which is the described portion of b; the described portion of

d is, as before, the entire circumference. It is to be observed (as in fig. 2) that properly it is not the entire arc $B_1'B_1'$, but each of the half-arcs BB_1', which corresponds to the entire circumference of d.

19. The figure 3—4 closely corresponds to fig. 1—2, the only difference being that the arcs $B_1'B_1'$ and $D_1'D_1'$, which are the described portions of b and d respectively, (instead of being the nearer portions, or those with their convexities facing each other) are the further portions, or those with their concavities facing each other, of the two circles respectively.

Finally, in fig. 4, the described portions of the two circles reduce themselves to the axial points B and D respectively.

20. Still assuming $d < b$, and passing over the case of external contact, we come to that in which the circles intersect each other; but this case has to be subdivided. Since the circles intersect, we have $b + d > a$, consistently herewith we may have:—

b, d each $< a$,	A, E each outside the lens common to the two circles,
$b = a$, $d < a$,	A outside, E on boundary of the lens,
$b > a$, $d < a$,	A outside, E inside the lens,
$b > a$, $d = a$,	A on boundary of, E inside the lens,
b, d each $> a$,	A, E, each inside the lens;

and in each case we have to consider the different admissible values of c. I omit the discussion of all these cases.

21. Still assuming $d < b$, and passing over the case of internal contact, we come to that of the circle d included within the circle b: we have here again a subdivision of cases; viz. we may have $d > a$, that is, A inside d; $d = a$, that is, A on the circumference of d; or $d < a$, that is, A outside d. The critical values of c, arranged in order of increasing magnitude in these three cases respectively, are:—

$d > a$	$d = a$	$d < a$
$b - d - a$,	$b - 2d$,	$b - d - a$,
$b - d + a$,	b,	$b + d - a$,
$b + d - a$,	b,	$b - d + a$,
$b + d + a$,	$b + 2d$,	$b + d + a$.

I attend only to the first case; we have here (see Plate II., at p. 516), interior series, the figures 1, 1—2, 2, 2—3, 3, 3—4, 4, viz.

fig. 1	$c = b - d - a$,
1—2	„ intermediate,
2	$c = b - d + a$,
2—3	„ intermediate,
3	$c = b + d - a$,
3—4	„ intermediate,
4	$c = b + d + a$.

22. In figure 1, the curves described by the points B_1D are each of them a mere point. In figure 1—2, we have two critical positions $B_1'D_1$ situate symmetrically on opposite sides of the axis, B_1' being a summit on b, and D_1 a two-way point on d, and moreover two critical positions B_1D_1' situate symmetrically on opposite sides of the axis, B_1 being a two-way point on b, and D_1' a summit on d. The described portion of b is the arc $B_1'B_1'$, and the described portion of d is the arc $D_1'D_1'$, these two arcs being thus the nearer portions of the two circles respectively.

23. In figure 2, the four critical positions coalesce all of them in the axial position BD; the described portions are thus the entire circumferences of the two circles respectively. This is a remarkable case. The configuration is $\rho, \sigma = + +$; B, D, R, S; $y_1 : y_2 = + +$. Imagine D to move from the axial point D in a given sense round the circle d, say with uniform velocity, then B moves from the axial point B in the same sense *but with either of two velocities* round the circle b; one of these velocities is at first small but ultimately increases rapidly, the other is at first large but ultimately decreases rapidly, so that the two revolutions of B from the axial point B round the entire circumference to the axial point B correspond each of them to the revolution of D from the axial point D round the entire circumference to the axial point D. And similarly, if we imagine B to move in a given sense from the axial point B round the circle b, say with uniform velocity, then D moves from the axial point D in the same sense but with either of two velocities round the circle d: one of these velocities is at first small but ultimately increases rapidly, the other is at first large but ultimately decreases rapidly, so that the two revolutions from the axial point D round the entire circumference of d to the axial point D correspond each of them to the revolution from the axial point B round the entire circumference of b to the axial point B.

24. In figure 2—3, there are no critical positions; the described portions of the circles b, d are the entire circumferences of the two circles respectively, these being described in the same sense, by the points B and D respectively. It is to be observed that, to a given position of B on b, there correspond two positions of D on d, or say two positions of the link, but the link does not in the course of its motion pass from one of these positions to the other; the motions are separate from each other, and may be regarded as belonging to different configurations of the system. And of course in like manner, to a given position of D on d, there correspond two positions of B on b, or say two positions of the link: we have thus the same two separate motions.

25. In figure 3, the critical axial position BD of the link makes its appearance: the described portions are still the entire circumferences of the two circles respectively. As the point D is here to the left of the point B, we must take the origin O to the right of B, and reverse the direction of the axis Ox; the configuration is thus $\rho, \sigma = + -$, B, S, R, D; $y_1 : y_2 = - -$. Everything is the same as in fig. 2 except (the signs of $y_1 : y_2$ being, as just mentioned, $- -$) that the motions in the circles b and d instead of being in the same sense are in opposite sense, viz. as D moves from the axial point D in a given sense round the circle d to the axial point D say with uniform velocity, then B moves from the axial point B round the circle b in the opposite sense, *and with either of two velocities;* and similarly, as B moves from the

axial point B in a given sense round the circle b say with uniform velocity, then D moves from the axial point D round the circle d in the opposite sense, *and with either of two velocities.*

26. In figure 3—4, we have again the two critical positions $B_1'D_1$ symmetrically situate on opposite sides of the axis, B_1' a summit on b, D_1 a two-way point on d: and also the two critical positions B_1D_1' symmetrically situate on opposite sides of the axis, B_1 a two-way point on b, D_1' a summit on d. The described portion of b is the arc $B_1'B_1'$, and the described portion of d the arc $D_1'D_1'$, these arcs being thus the further portions of the two circles respectively.

Finally, in figure 4, the described portions reduce themselves to the two points B, D respectively.

27. The several forms for $d = b$ can be at once obtained from those for $d < b$; the only difference is that several intermediate forms disappear, and the entire series of divisions is thus not quite so numerous.

Part II.

1. The curve-tracing mechanism was devised with special reference to the curves of three-bar motion, viz. the object proposed was that of tracing the curve described by a point K of the link BD, the extremities whereof B and D describe given circles respectively, or more generally by a point K, the vertex of a triangle KBD, whereof the other vertices B and D describe given circles respectively, and that in suchwise that the points B and D might be free to describe the two entire circumferences respectively: but the principle applies to other motions, and I explain it in a general way as follows.

2. Imagine the cranked link BD, composed of the bars $B\beta$ and $D\delta$, rigidly attached $B\beta$ to the top and $D\delta$ to the bottom of the cylindrical disk K (this same letter K is used to denote the axis of the disk), and where $B\beta$ and $D\delta$ may be either parallel or inclined to each other at any given angle, so that, referring the points B, K, D to a horizontal plane, BKD is either a right line, or else K is the vertex of a triangle the other vertices whereof are B and D. The disk K, with the attached

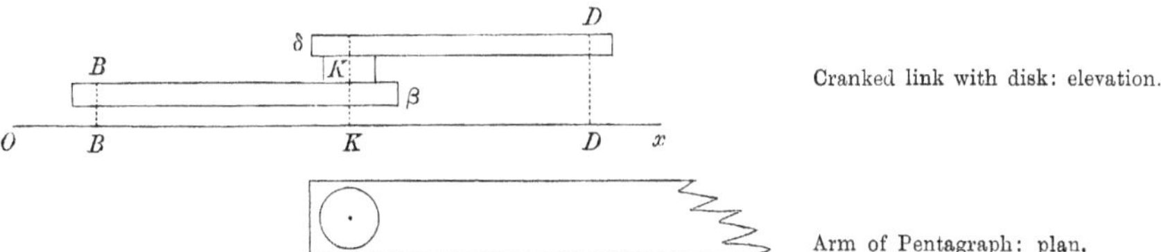

Cranked link with disk: elevation.

Arm of Pentagraph: plan.

bars $B\beta$ and $D\delta$, moves in a horizontal plane: and if the motion of the point B be regulated in any manner by a mechanism lying wholly below B and supported by the bed of the entire mechanism, and similarly if the motion of the point D be regulated in any manner by a mechanism lying wholly above D and supported by a bridge of sufficient length (resting on the bed of the entire mechanism), then the disk K moves in its own horizontal plane unimpeded by other parts of the mechanism: and if we fit the disk K so as to move smoothly within a circular aperture in the arm of a pentagraph, then the pencil of the pentagraph will trace out on a sheet of paper the curve described by the point K on the axis of the disk, or say by the point K of the beam BKD. Of course for the three-bar motion, all that is required is that the point B shall describe a circle, viz. it must be pivoted on to an arm AB, which is itself pivoted at A to the bed: and that the point D shall describe a circle, viz. it must be pivoted on to an arm DE, which is itself pivoted at E to the bridge. Special arrangements are required to enable the variation of the several lengths AB, BK, KD, DE and ED, and the mechanism thus unavoidably assumes a form which appears complicated for the object intended to be thereby effected.

3. The form of Pentagraph which I use consists of a parallelogram $ABCD$, pivoted together at the points A, B, C, D, the bars AB and DC being above AD and BC. There is a cradle G, rotating about a fixed centre, and which carries between guides the arm AD, which has a sliding motion, so that the lengths GD and GA may be

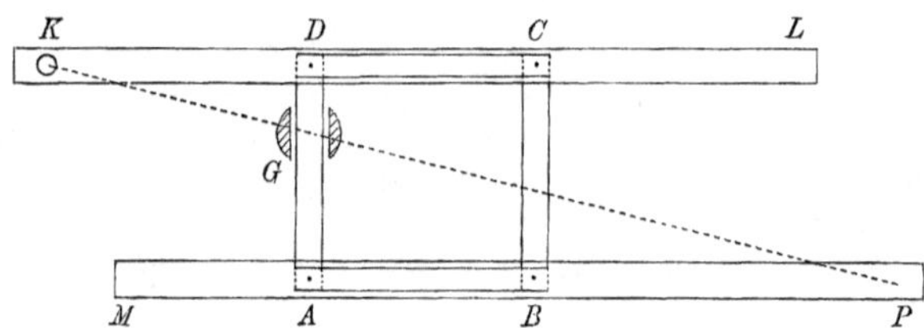

made to have any given ratio to each other. Above the bar DC and sliding along it, we have the arm KL (where K is the circular aperture which fits on to the disk K of the cranked link): and above AB and sliding along it, we have the arm MP which carries the pencil P: of course, in order that the pentagraph may be in adjustment, the points K, G, P must be *in lined*.

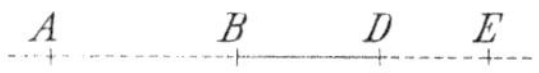

Fig 1. $c=a-b-d$.

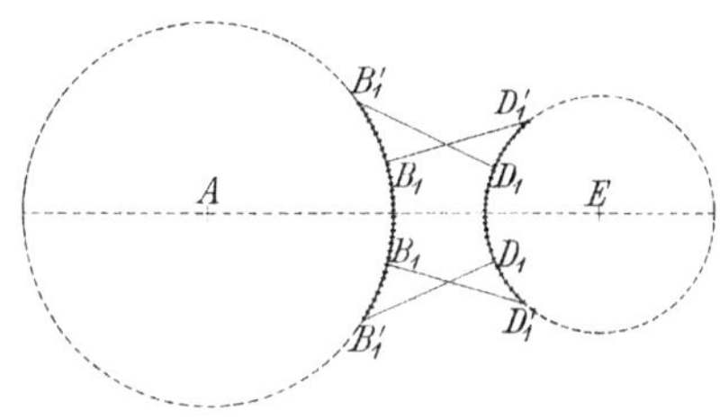

Fig. 1-2.

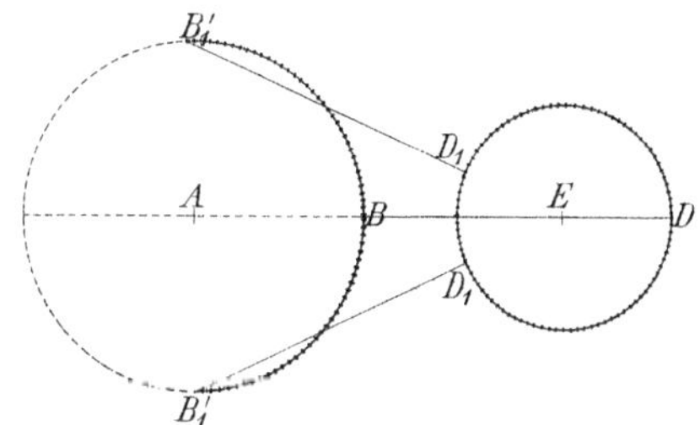

Fig. 2. $c=a-b+d$

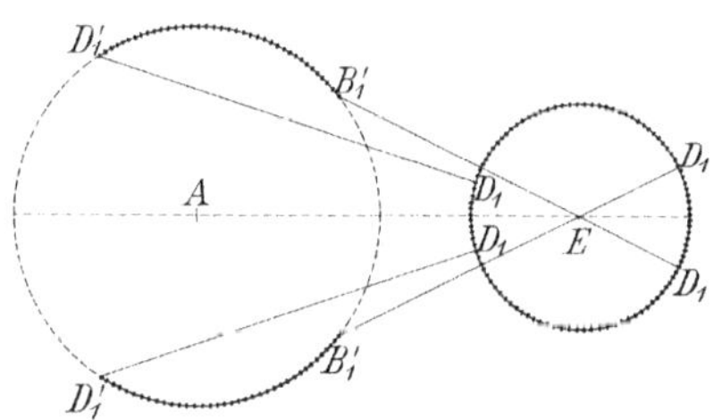

Fig. 2-3.

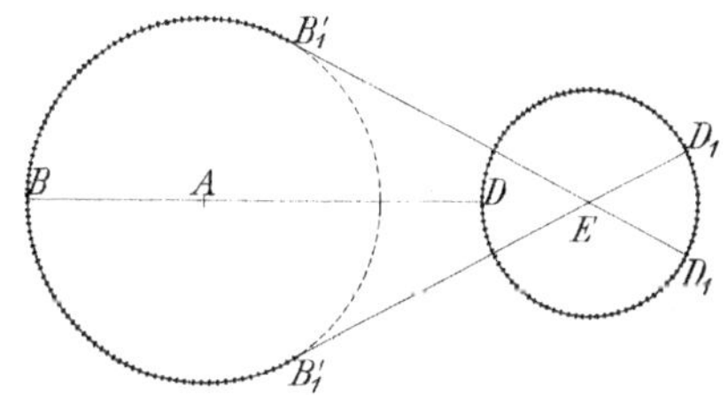

Fig. 3. $c=a+b-d$

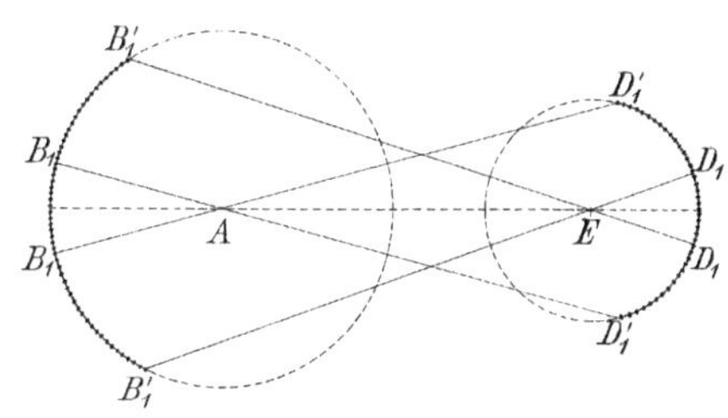

Fig. 3-4.

Fig. 4. $c=a+b+d$

Exterior Series.

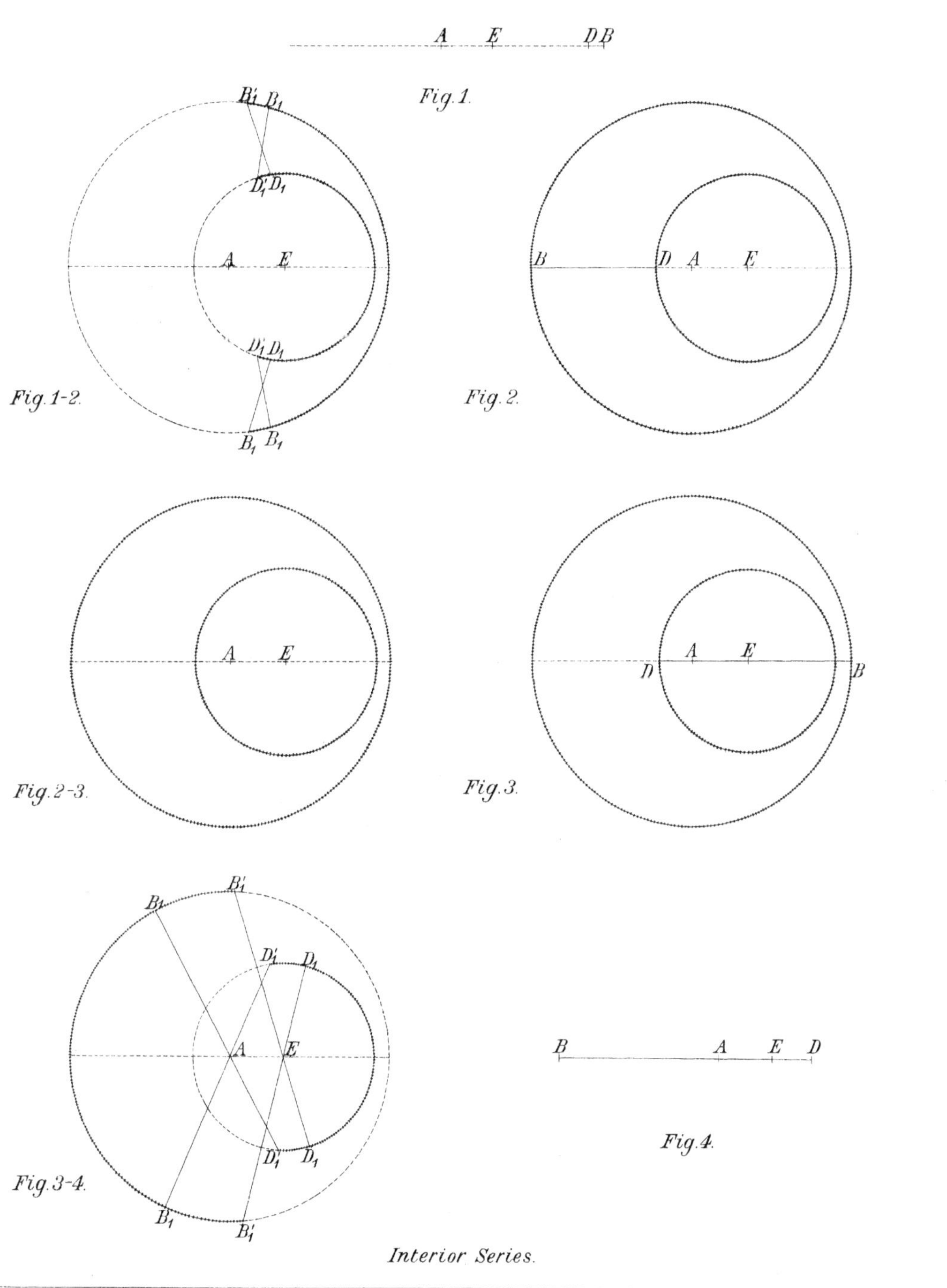

Interior Series.

953.

ON THE NINE-POINTS CIRCLE.

[From the *Messenger of Mathematics*, vol. XXIII. (1894), pp. 23—25.]

IF from the angles A, B, C of a triangle we draw tangents to a conic Ω, meeting the opposite sides in the points α, α'; β, β'; γ, γ' respectively, then it is known that these six points lie in a conic. In particular, if the conic Ω reduce itself to a point-pair OO', then we have the theorem that, if from the angles A, B, C, we draw to the point O lines meeting the opposite sides in the points α, β, γ respectively; and to the point O' lines meeting the opposite sides in the points α', β', γ' respectively, then the six points α, α'; β, β'; γ, γ' lie in a conic. We may enquire the conditions under which this conic becomes a circle. It may be remarked that one of the points say O' remains arbitrary: for if through the points α', β', γ', we draw a conic (or in particular a circle) meeting the three sides respectively in the remaining points α, β, γ, then (by a converse of the general theorem) the lines $A\alpha$, $B\beta$, $C\gamma$ will meet in a point O.

Using trilinear coordinates (x, y, z) and writing $x : y : z = a : b : c$ for the point O, and $x : y : z = a' : b' : c'$ for the point O', it is at once seen that the equation of the conic through the six points is

$$aa'x^2 + bb'y^2 + cc'z^2 - (bc' + b'c)\,yz - (ca' + c'a)\,zx - (ab' + a'b)\,xy = 0\,;$$

in fact, writing herein successively $x = 0$, $y = 0$, $z = 0$, we see that the equation is satisfied by $x = 0$, $(by - cz)(b'y - c'z) = 0$; by $y = 0$, $(cz - ax)(c'z - a'x) = 0$; and by $z = 0$, $(ax - by)(a'x - b'y) = 0$ respectively. And it is to be observed that the equation may also be written

$$(aa'x + bb'y + cc'z)(x + y + z) - (b + c)(b' + c')\,yz - (c + a)(c' + a')\,zx - (a + b)(a' + b')\,xy = 0.$$

Suppose now that x, y, z represent areal coordinates, viz. that (x, y, z) are proportional to the perpendicular distances of the point from the sides, each divided by the perpendicular distance of the opposite angle from the same side; or, what is the same thing, coordinates such that the equation of the line infinity is $x+y+z=0$. Then if A, B, C denote the angles of the triangle, the general equation of a circle is

$$(yz\sin^2 A + zx\sin^2 B + xy\sin^2 C) + (\lambda x + \mu y + \nu z)(x+y+z) = 0,$$

where λ, μ, ν are arbitrary coefficients.

Hence, putting this

$$= \Theta\{-(b+c)(b'+c')\,yz - (c+a)(c'+a')\,zx$$
$$-(a+b)(a'+b')\,xy + (aa'x + bb'y + cc'z)(x+y+z)\},$$

we must have

$$\Theta(b+c)(b'+c') = -\sin^2 A,$$
$$\Theta(c+a)(c'+a') = -\sin^2 B,$$
$$\Theta(a+b)(a'+b') = -\sin^2 C;$$

and then

$$\Theta aa' = \lambda,\quad \Theta bb' = \mu,\quad \Theta cc' = \nu,$$

which last equations determine the values of λ, μ, ν.

Taking a', b', c' at pleasure, we have

$$2a = \frac{1}{\Theta}\left(\ \frac{\sin^2 A}{b'+c'} - \frac{\sin^2 B}{c'+a'} - \frac{\sin^2 C}{a'+b'}\right),$$
$$2b = \frac{1}{\Theta}\left(-\frac{\sin^2 A}{b'+c'} + \frac{\sin^2 B}{c'+a'} - \frac{\sin^2 C}{a'+b'}\right),$$
$$2c = \frac{1}{\Theta}\left(-\frac{\sin^2 A}{b'+c'} - \frac{\sin^2 B}{c'+a'} + \frac{\sin^2 C}{a'+b'}\right),$$

viz. a, b, c having these values, the conic through the six points α, β, γ, α', β', γ' is the circle having for its equation

$$yz\sin^2 A + zx\sin^2 B + xy\sin^2 C + \Theta(aa'x + bb'y + cc'z)(x+y+z) = 0;$$

and we may obviously without loss of generality give to Θ any specific value, say $\Theta = 1$.

If $a' = b' = c'$, $= 1$, then we have

$$-4a = \frac{1}{\Theta}(-\sin^2 A + \sin^2 B + \sin^2 C),$$
$$-4b = \frac{1}{\Theta}(\ \ \sin^2 A - \sin^2 B + \sin^2 C),$$
$$-4c = \frac{1}{\Theta}(\ \ \sin^2 A + \sin^2 B - \sin^2 C),$$

or writing for convenience $\Theta = -\frac{1}{2}$, the values of a, b, c are

$$\tfrac{1}{2}(-\sin^2 A + \sin^2 B + \sin^2 C),\quad \tfrac{1}{2}(\sin^2 A - \sin^2 B + \sin^2 C),\quad \tfrac{1}{2}(\sin^2 A + \sin^2 B - \sin^2 C)$$

respectively. But we have

$$A + B + C = \pi,$$

and thence

$$\begin{aligned} &\sin^2 A + \sin^2 B - \sin^2 C, \\ = &\sin^2 A + \sin^2 B - \sin^2 (A + B) \\ = &\ 2 \sin A \sin B (\sin A \sin B - \cos A \cos B), \\ = &- 2 \sin A \sin B \cos (A + B), \\ = &\ 2 \sin A \sin B \cos C, \end{aligned}$$

and we thus have

$$a,\ b,\ c = \sin B \sin C \cos A,\ \ \sin C \sin A \cos B,\ \ \sin A \sin B \cos C,$$

(or, what is the same thing, $a : b : c = \cot A : \cot B : \cot C$), and the equation of the circle is

$$\begin{aligned} &yz \sin^2 A + zx \sin^2 B + xy \sin^2 C \\ &\qquad - \tfrac{1}{2} (x \sin B \sin C \cos A + y \sin C \sin A \cos B + z \sin A \sin B \cos C)(x + y + z) = 0. \end{aligned}$$

We thus have $x : y : z = 1 : 1 : 1$ for the point O', and $x : y : z = \cot A : \cot B : \cot C$ for the point O; viz. O' is the point of intersection of the lines from the angles to the mid-points of the opposite sides respectively; and O is the point of intersection of the perpendiculars from the angles on the opposite sides respectively: and the foregoing equation is consequently that of the Nine-points Circle.

954.

ON THE NINE-POINTS CIRCLE OF A PLANE TRIANGLE.

[From the *Messenger of Mathematics*, vol. XXIII. (1894), pp. 25—27.]

I CONSIDER the circle which meets the sides of a triangle ABC in the points F, L; G, M; H, N respectively, where ultimately F, G, H are the feet of the perpendiculars let fall from the angles on the opposite sides, and L, M, N are the mid-points of the sides: but in the first instance, they are taken to be arbitrary points. Taking the radius of the circle to be unity, the coordinates of the point F may be taken to be $\cos F$, $\sin F$, and these may be expressed rationally in terms of the tangent of the half-angle, $f = \tan \frac{1}{2}F$; and similarly for the other points, viz. we may determine the six points by the parameters f, g, h, l, m, n respectively. The sides of the triangle are the lines joining the points L, F; M, G; N, H respectively: thus the equations of the sides are

$$\begin{aligned}
&\text{for } BC: \quad x(1-lf) + y(l+f) - (1+lf) = 0, \quad \text{say } U = 0,\\
&\;\;,, \;\; CA: \quad x(1-mg) + y(m+g) - (1+mg) = 0, \quad ,, \;\; V = 0,\\
&\;\;,, \;\; AB: \quad x(1-nh) + y(n+h) - (1+nh) = 0, \quad ,, \;\; W = 0.
\end{aligned}$$

We have AF, a line through the intersections of BC and CA; its equation is therefore of the form $BV - CW = 0$, and to determine B, C we have $BV_0 - CW_0 = 0$, if V_0, W_0 are the values of V, W belonging to the point F, the coordinates of which are

$$\frac{1-f^2}{1+f^2}, \quad \frac{2f}{1+f^2};$$

we find

$$\begin{aligned}
V_0 &= -2(f-g)(f-m) \div (1+f^2);\\
W_0 &= \;\;\, 2(h-f)(f-n) \div (1+f^2),
\end{aligned}$$

and then $B \div C = W_0 \div V_0$: we thus find the following equations:

$$\begin{aligned}
&\text{that of } AF \text{ is } BV - CW = 0,\\
&\quad ,, \quad BG \;\; ,, \;\; C'W - A'U = 0,\\
&\quad ,, \quad CH \;\; ,, \;\; A''U - B''V = 0,
\end{aligned}$$

where

$$\begin{aligned} B\ &: C\ = -(h-f)(f-n) : (f-g)(f-m),\\ C'\ &: A' = -(f-g)(g-l) : (g-h)(g-n),\\ A''\ &: B'' = -(g-h)(h-m) : (h-f)(h-l). \end{aligned}$$

The condition in order that the three lines may meet in a point is $BC'A'' = CA'B''$, viz. this is

$$(f-n)(g-l)(h-m) + (f-m)(g-n)(h-l) = 0,$$

or, as this may also be written,

$$2fgh - gh(m+n) - hf(n+l) - fg(l+m) + mn(g+h) + nl(h+f) + lm(f+g) - 2lmn = 0.$$

Similarly, the equation of

$$\begin{aligned} AL\ &\text{is}\ \mathfrak{B}V - \mathfrak{C}W = 0,\\ BM\ &\ ,,\ \ \mathfrak{C}'W - \mathfrak{A}'U = 0,\\ CN\ &\ ,,\ \ \mathfrak{A}''U - \mathfrak{B}''V = 0, \end{aligned}$$

where

$$\begin{aligned} \mathfrak{B}\ &: \mathfrak{C}\ = -(n-l)(h-l) : (l-m)(g-l),\\ \mathfrak{C}'\ &: \mathfrak{A}'\ = -(l-m)(f-m) : (m-n)(h-m),\\ \mathfrak{A}''\ &: \mathfrak{B}'' = -(m-n)(g-n) : (n-l)(f-n). \end{aligned}$$

The condition in order that the three lines may meet in a point is $\mathfrak{B}\mathfrak{C}'\mathfrak{A}'' = \mathfrak{C}\mathfrak{A}'\mathfrak{B}''$, viz. this is the same condition as before; that is, if the lines AF, BG, CH meet in a point, then also the lines AL, BM, CN will meet in a point.

In the case of the nine-points circle, we have MN, NL, LM parallel to LF, MG, NH, respectively: the equation of MN is

$$x(l-mn) + y(m+n) - (l+mn) = 0,$$

and this is parallel to LF, if

$$\frac{m+n}{1-mn} = \frac{l+f}{1-lf},\ \text{that is,}\ L+F = M+N.$$

Hence, for the nine-points circle, we have

$$L+F = M+N,\quad M+G = N+L,\quad N+H = L+M,$$

or, as these equations may be written,

$$2L = G+H,\quad 2M = H+F,\quad 2N = F+G,$$

viz. it thus appears that the radii to the points L, M, N respectively, or say the radii L, M, N, bisect the angles made by the radii G and H, H and F, F and G respectively.

It may be added that we have

$$\begin{aligned} m+n-l+lmn &= f\{1-mn+l(m+n)\},\\ n+l-m+lmn &= g\{1-nl+m(n+l)\},\\ l+m-n+lmn &= h\{1-lm+n(l+m)\}, \end{aligned}$$

viz. f, g, h are expressible each of them as a rational function of l, m, n.

955.

THE NUMERICAL VALUE OF Πi, $=\Gamma(1+i)$.

[From the *Messenger of Mathematics*, vol. XXIII. (1894), pp. 36—38.]

I DO not know whether the numerical value of Πx for an imaginary value of x has ever been calculated; and I wish to calculate it for a simple case $x=i$.

We have

$$\begin{aligned}\frac{1}{\Pi z}=&\left(1+\frac{z}{1}\right)\\ &\left(1+\frac{z}{2}\right)e^{z\,\mathrm{hl}\,\frac{1}{2}}\\ &\left(1+\frac{z}{3}\right)e^{z\,\mathrm{hl}\,\frac{2}{3}}\\ &\quad\vdots\\ &\left(1+\frac{z}{s}\right)e^{z\,\mathrm{hl}\,\frac{s-1}{s}}\\ &\quad\vdots\end{aligned}$$

where hl denotes the hyperbolic logarithm. Hence, in particular, when $z=i$, we have

$$\begin{aligned}\frac{1}{\Pi i}=&\,1+\frac{i}{1}\\ &\,1+\frac{i}{2}\,.\cos \mathrm{hl}\,\tfrac{1}{2}+i\sin \mathrm{hl}\,\tfrac{1}{2}.\\ &\,1+\frac{i}{3}\,.\cos \mathrm{hl}\,\tfrac{2}{3}+i\sin \mathrm{hl}\,\tfrac{2}{3}.\\ &\,1+\frac{i}{4}\,.\cos \mathrm{hl}\,\tfrac{3}{4}+i\sin \mathrm{hl}\,\tfrac{3}{4}.\\ &\quad\vdots\\ =&\,\sqrt{(1+1)}\,.\cos\theta_1+i\sin\theta_1\,.\cos\phi_1-i\sin\phi_1.\\ &\,\sqrt{(1+\tfrac{1}{4})}\,.\cos\theta_2+i\sin\theta_2\,.\cos\phi_2-i\sin\phi_2.\\ &\,\sqrt{(1+\tfrac{1}{9})}\,.\cos\theta_3+i\sin\theta_3\,.\cos\phi_3-i\sin\phi_3.\\ &\quad\vdots\end{aligned}$$

($\phi_1=0$, and in the subsequent terms the imaginary part is taken with a negative sign in order to obtain positive values for ϕ_2, ϕ_3, &c.), $=\Omega(\cos\Theta+i\sin\Theta)$, if Ω be the modulus and Θ the sum

$$(\theta_1-\phi_1)+(\theta_2-\phi_2)+(\theta_3-\phi_3)+\ldots.$$

We have

$$\Omega_1=\sqrt{(1+1)}\,.\,\sqrt{(1+\tfrac{1}{4})}\,.\,\sqrt{(1+\tfrac{1}{9})}\ldots,$$

which may be calculated directly. The value of Ω admits, however, of a finite expression, viz. we have

$$\Omega^2=\frac{1}{\Pi i\,\Pi(-i)}=\frac{\sin\pi i}{\pi i}=\frac{e^{\pi}-e^{-\pi}}{2\pi},$$

the approximate numerical value is $\Omega=1{\cdot}9173$, viz. we have

$$e^{\pi}-e^{-\pi}=23{\cdot}141-{\cdot}043=23{\cdot}098:\ \log=1{\cdot}3635744,\quad -\log 2\pi=\bar{1}{\cdot}201819,$$

whence

$$\log\Omega^2={\cdot}5653935,\quad \log\Omega={\cdot}2826967,\ \text{or}\ \ \Omega=1{\cdot}9173.$$

We have

$$\tan\theta_1=1,\quad \tan\theta_2=\tfrac{1}{2},\quad \tan\theta_3=\tfrac{1}{3},\ \&\text{c.},$$

also

$$\phi_1=0,\quad \phi_2=\frac{180^\circ}{M\pi}\log\tfrac{1}{2},\quad \phi_3=\frac{180^\circ}{M\pi}\log\tfrac{2}{3},\ \&\text{c.},$$

where M is the modulus for the Briggian logarithms,

$$M=\ {\cdot}4342944\ \log\ =\bar{1}{\cdot}6377843,$$
$$\pi=3{\cdot}1415926\ \text{,,}\ =\ {\cdot}4971499,$$
$$180\ \ \text{,,}\ =2{\cdot}2552755,$$

whence

$$\log\frac{180}{M\pi}=2{\cdot}1203383,\quad \frac{180^\circ}{M\pi}=131^\circ{\cdot}9284.$$

We hence calculate the succession of values of θ and ϕ as follows:

θ	tan	arc
1	1	45°
2	·5	26 34′
3	·3333333	18 26
4	·25	14 2
5	·2	11 19
6	·1666666	9 28
7	·1428571	8 8
8	·125	7 8
9	·1111111	6 20
10	·1	5 43

ϕ	$131^\circ{\cdot}93\ \times$				$\theta - \phi =$		
1		$= 0$			1	45°	
2	log 1/2 =	·3010300 =	39°	$43'$	2	-13°	$9'$
3	2/3	·1760913	23	14	3	4	48
4	3/4	·1249387	16	29	4	2	27
5	4/5	·0969100	12	47	5	1	28
6	5/6	·0791813	10	26	6	0	58
7	6/7	·0669467	8	50	7	0	42
8	7/8	·0579920	7	39	8	0	31
9	8/9	·0511525	6	44	9	0	24
10	9/10	·0457575	6	2	10	0	19

The sum of all the negative arcs $\theta_2 - \phi_2$, $\theta_3 - \phi_3$, ... as far as calculated, that is, up to $\theta_{10} - \phi_{10}$ is $= 24^\circ\, 46'$, or, writing x for the sum of the remaining arcs $\theta_{11} - \phi_{11}$ to infinity, we have

$$\frac{1}{\Pi i} = 1{\cdot}9173\,(\cos\Theta + i\sin\Theta),$$

where

$$\Theta = 45^\circ - 24^\circ\, 46' - x, \quad = 20^\circ\, 14' - x.$$

It would not be difficult to calculate a limit to the value of x.

956.

ON RICHELOT'S INTEGRAL OF THE DIFFERENTIAL EQUATION $\frac{dx}{\sqrt{X}} + \frac{dy}{\sqrt{Y}} = 0$.

[From the *Messenger of Mathematics*, vol. XXIII. (1894), pp. 42—47.]

IN the Memoir "Einige neue Integralgleichungen des Jacobischen Systems Differentialgleichungen," *Crelle*, t. XXV. (1843), pp. 97—118, Richelot, working with the more general problem of a system of $n-1$ differential equations between n variables, obtains a result which in the particular case $n = 2$ (that is, for the differential equation

$$\frac{dx}{\sqrt{X}} + \frac{dy}{\sqrt{Y}} = 0, \quad X = a + bx + cx^2 + dx^3 + ex^4,$$

and Y the same function of y), is in effect as follows: an integral is

$$\left\{\frac{\sqrt{X}(\theta - y) - \sqrt{Y}(\theta - x)}{x - y}\right\}^2 = \square\,(\theta - x)(\theta - y) + \Theta + e(\theta - x)^2(\theta - y)^2,$$

where $\square$, θ are arbitrary constants, and Θ denotes the quartic function

$$a + b\theta + c\theta^2 + d\theta^3 + e\theta^4;$$

viz. this is theorem 3, p. 107 (*l. c.*), taking therein $n = 2$, and writing θ, $\square$ for Richelot's α and const.

The peculiarity is that the integral contains apparently *two* arbitrary constants, and it is very interesting to show how these really reduce themselves to a single arbitrary constant.

Observe that, on the right-hand side, there are terms in θ^4, θ^3 whereas no such terms present themselves on the left-hand side. But by changing the constant $\square$, we can get rid of these terms, and so bring each side to contain only terms in θ^2, θ, 1; viz. writing

$$\square = -2e\theta^2 - d\theta - c + C,$$

where C is a new arbitrary constant, the equation becomes

$$\begin{aligned}\left\{\frac{\sqrt{X}(\theta-y)-\sqrt{Y}(\theta-x)}{x-y}\right\}^2 = \ &\theta^2 [\quad e(x+y)^2 \quad + d(x+y) + C \quad]\\ &+\theta\,[-2e\,xy(x+y) - d\,xy \quad -(C-c)(x+y)+b]\\ &+\quad [\quad e\,x^2y^2 \qquad\qquad +(C-c)\,xy \quad +a],\end{aligned}$$

which still contains the two arbitrary constants θ, C.

But this gives the three equations

$$\frac{(\sqrt{X}-\sqrt{Y})^2}{(x-y)^2} = e(x+y)^2 + d(x+y) + C,$$

$$-2\frac{(\sqrt{X}-\sqrt{Y})(y\sqrt{X}-x\sqrt{Y})}{(x-y)^2} = -2e\,xy(x+y) - d\,xy - (C-c)(x+y) + b,$$

$$\frac{(y\sqrt{X}-x\sqrt{Y})^2}{(x-y)^2} = e\,x^2y^2 + (C-c)\,xy + a.$$

The first of these is Lagrange's integral containing the arbitrary constant C; and it is necessary that the three equations shall be one and the same equation; viz. the second and third equations must be each of them a mere transformation of the first, equation.

It is easy to verify that this is so. Starting from the first equation, we require, first the value of

$$-2\frac{(\sqrt{X}-\sqrt{Y})(y\sqrt{X}-x\sqrt{Y})}{(x-y)^2}, \ = \Omega,$$

for a moment.

We form a rational combination, or combination without any term in $\sqrt{XY}$; this is

$$(x+y)\frac{(\sqrt{X}-\sqrt{Y})^2}{(x-y)^2} - 2\frac{(\sqrt{X}-\sqrt{Y})(y\sqrt{X}-x\sqrt{Y})}{(x-y)^2} = e(x+y)^3 + d(x+y)^2 + C(x+y) + \Omega,$$

where the left-hand side is

$$\frac{(x-y)(X-Y)}{(x-y)^2}, \ = \frac{X-Y}{x-y},$$

which is

$$= e(x^3 + x^2y + xy^2 + y^3) + d(x^2 + xy + y^2) + c(x+y) + b,$$

and we thence have for

$$\Omega, \ = -2\frac{(\sqrt{X}-\sqrt{Y})(y\sqrt{X}-x\sqrt{Y})}{(x-y)^2},$$

the value given by the second equation.

Secondly, starting again from the first equation, and proceeding in like manner to find the value of

$$\frac{(y\surd X - x\surd Y)^2}{(x-y)^2}, \ =\Omega,$$

for a moment, we form a rational combination

$$-xy\frac{(\surd X-\surd Y)^2}{(x-y)^2}+\frac{(y\surd X - x\surd Y)^2}{(x-y)^2} = -exy(x+y)^2 - dxy(x+y) - Cxy + \Omega,$$

where the left-hand side is

$$\frac{(x-y)(-yX+xY)}{(x-y)^2}, \ = \frac{-yX+xY}{x-y},$$

which is

$$= -exy(x^2+xy+y^2) - dxy(x+y) - cxy + a;$$

and we thence have for

$$\Omega, \ = \frac{(y\surd X - x\surd Y)^2}{(x-y)^2},$$

the value given by the third equation.

In conclusion, I give what is in effect the process by which Richelot obtained his integral. The integral is $v = \square$, where

$$v = \frac{-\Theta}{\theta - x\,.\,\theta - y} - e(\theta - x\,.\,\theta - y) + (\theta - x\,.\,\theta - y)\,\Omega^2,$$

if, for shortness,

$$\Omega = \frac{\surd X}{\theta - x\,.\,x - y} + \frac{\surd Y}{\theta - y\,.\,y - x},$$

and it is required thence to show that

$$\frac{dx}{\surd X}+\frac{dy}{\surd Y}=0,$$

or, what is the same thing, to show that v satisfies the partial differential equation

$$\surd X\frac{dv}{dx} - \surd Y\frac{dv}{dy} = 0.$$

We have

$$\frac{dv}{dx} = \frac{-\Theta}{(\theta-x)^2(\theta-y)} + e(\theta-y) - (\theta-y)\,\Omega^2 + 2(\theta-x)(\theta-y)\,\Omega\frac{d\Omega}{dx},$$

$$\frac{dv}{dy} = \frac{-\Theta}{(\theta-x)(\theta-y)^2} + e(\theta-x) - (\theta-x)\,\Omega^2 + 2(\theta-x)(\theta-y)\,\Omega\frac{d\Omega}{dy},$$

and thence, attending to the value of Ω,

$$\sqrt{X}\frac{dv}{dx}-\sqrt{Y}\frac{dv}{dy}=\frac{-\Theta}{\theta-x\,.\,\theta-y}(x-y)\,\Omega+(e-\Omega^2)(\theta-x)(\theta-y)(x-y)\,\Omega$$
$$+2(\theta-x)(\theta-y)\,\Omega\left(\sqrt{X}\frac{d\Omega}{dx}-\sqrt{Y}\frac{d\Omega}{dy}\right),$$

or say

$$-\frac{\left(\sqrt{X}\frac{dv}{dx}-\sqrt{Y}\frac{dv}{dy}\right)}{(\theta-x)(\theta-y)(x-y)\,\Omega}=\frac{\Theta}{(\theta-x)^2(\theta-y)^2}-e+\Omega^2-\frac{2}{x-y}\left(\sqrt{X}\frac{d\Omega}{dx}-\sqrt{Y}\frac{d\Omega}{dy}\right);$$

and it is consequently to be shown that the function on the right-hand side is $=0$. We have

$$\sqrt{X}\frac{d\Omega}{dx}=\frac{\frac{1}{2}X'}{(\theta-x)(x-y)}+\frac{X}{(\theta-x)^2(x-y)}-\frac{X}{(\theta-x)(x-y)^2}+\frac{\sqrt{(XY)}}{(\theta-y)(x-y)^2},$$

$$\sqrt{Y}\frac{d\Omega}{dy}=\frac{\frac{1}{2}Y'}{(\theta-y)(y-x)}+\frac{Y}{(\theta-y)^2(y-x)}-\frac{Y}{(\theta-y)(x-y)^2}+\frac{\sqrt{(XY)}}{(\theta-x)(x-y)^2},$$

and thence

$$\sqrt{X}\frac{d\Omega}{dx}-\sqrt{Y}\frac{d\Omega}{dy}=\frac{\frac{1}{2}X'}{(\theta-x)(x-y)}-\frac{\frac{1}{2}Y'}{(\theta-y)(y-x)}$$
$$+\left\{\frac{X}{(\theta-x)^2}+\frac{Y}{(\theta-y)^2}\right\}\frac{1}{x-y}$$
$$-\left(\frac{X}{\theta-x}-\frac{Y}{\theta-y}\right)\frac{1}{(x-y)^2}$$
$$-\frac{\sqrt{(XY)}}{(\theta-x)(\theta-y)(x-y)}.$$

Multiplying by $\frac{2}{x-y}$, we may put the result in the form

$$\frac{2}{x-y}\left(\sqrt{X}\frac{d\Omega}{dx}-\sqrt{Y}\frac{d\Omega}{dy}\right)=\frac{1}{\theta-x}\frac{d}{dx}\frac{X}{(x-y)^2}+\frac{1}{\theta-y}\frac{d}{dy}\frac{Y}{(x-y)^2}$$
$$+\frac{2X}{(\theta-x)^2(x-y)^2}+\frac{2Y}{(\theta-x)^2(x-y)^2}-\frac{2\sqrt{(XY)}}{(\theta-x)(\theta-y)(x-y)^2};$$

and the equation to be verified thus is

$$0=\frac{\Theta}{(\theta-x)^2(\theta-y)^2}-e+\Omega^2$$
$$-\frac{1}{\theta-x}\frac{d}{dx}\frac{X}{(x-y)^2}-\frac{2X}{(\theta-x)^2(x-y)^2}$$
$$-\frac{1}{\theta-y}\frac{d}{dy}\frac{Y}{(x-y)^2}-\frac{2Y}{(\theta-x)^2(x-y)^2}$$
$$+\frac{2\sqrt{(XY)}}{(\theta-x)(\theta-y)(x-y)^2}.$$

But decomposing the first term into simple fractions, we have

$$\frac{\Theta}{(\theta-x)^2(\theta-y)^2}=+e$$
$$+\frac{1}{\theta-x}\frac{d}{dx}\frac{X}{(x-y)^2}+\frac{X}{(\theta-x)^2(x-y)^2}$$
$$+\frac{1}{\theta-y}\frac{d}{dy}\frac{Y}{(x-y)^2}+\frac{Y}{(\theta-y)^2(x-y)^2}.$$

Also for the third term, we have

$$\Omega^2=\quad\frac{X}{(\theta-x)^2(x-y)^2}$$
$$+\frac{Y}{(\theta-y)^2(x-y)^2}$$
$$-\frac{2\sqrt{(XY)}}{(\theta-x)(\theta-y)(x-y)^2},$$

and substituting these values the several terms destroy each other, so that the right-hand side is $=0$, as it should be.

957.

ILLUSTRATIONS OF SYLOW'S THEOREMS ON GROUPS.

[From the *Messenger of Mathematics*, vol. XXIII. (1894), pp. 59—62.]

THE theorems 1, 2, and 3 in the paper Sylow, "Théorèmes sur les groupes de Substitutions," *Math. Ann.* t. V. (1872), pp. 584—594, apply to groups in general, and not only to groups of substitutions. They are as follows:

THEOREM 1. If n^α be the highest power of the prime number n which divides the order of a group G, this group contains a group g of the order n^α: if, moreover, $n^\alpha\nu$ is the order of the highest group contained in G, the operations whereof are permutable with the group g, then the order of G is of the form $n^\alpha\nu\,(nk+1)$. (I write k for Sylow's p, since it is convenient to have p to denote a prime number; and for Sylow's "Substitutions" I write "Operations.")

THEOREM 2. Everything being as in the preceding theorem, the group G contains precisely $nk+1$ distinct groups of the order n^α; and these are obtained by transforming any one of them by the operations of G, each group being given by $n^\alpha\nu$ distinct transformations.

THEOREM 3. If the order of a group is n^α, n being prime, then any operation ϑ whatever of the group may be expressed by the formula

$$\vartheta = \theta_0{}^i\theta_1{}^k\theta_2{}^l \ldots \theta^r{}_{\alpha-1},$$

where

$$\begin{aligned}
\theta_0{}^n &= 1,\\
\theta_1{}^n &= \theta_0{}^a,\\
\theta_2{}^n &= \theta_0{}^b\theta_1{}^c,\\
\theta_3{}^n &= \theta_0{}^d\theta_1{}^e\theta_2{}^f,\\
&\vdots
\end{aligned}$$

and where

$$\vartheta^{-1}\theta_0\vartheta = 1,$$
$$\vartheta^{-1}\theta_1\vartheta = \theta_0^{\beta}\theta_1,$$
$$\vartheta^{-1}\theta_2\vartheta = \theta_0^{\gamma}\theta_1^{\delta}\theta_2,$$
$$\vartheta^{-1}\theta_3\vartheta = \theta_0^{\epsilon}\theta_1^{\zeta}\theta_2^{\eta}\theta_3.$$
$$\vdots$$

But at present I attend only to the theorems 1 and 2.

For instance, consider the group G of the order $n=6$,

$$1,\ \beta,\ \beta^2,\ \alpha,\ \alpha\beta,\ \alpha\beta^2\ (\alpha^2=1,\ \beta^3=1,\ \alpha\beta^2=\beta\alpha,\ \alpha\beta=\beta^2\alpha).$$

Here $n=2$ or 3: if $n=2$, we have $N=n^a\nu(nk+1)=2.1(2+1)$; if $n=3$, we have $N=n^a\nu(nk+1)=3.2.1$.

First, $n=2$; we should have a group g of the order 2; one such group is $(1, \alpha)$, and the only group the substitutions whereof are permutable with $(1, \alpha)$ is the group $(1, \alpha)$ itself: for, taking any other operation of the group, for instance β, it is not true that $\beta(\gamma, \alpha)=(1, \alpha)\beta$; in fact, the left-hand is $(\beta, \beta\alpha)$ and the right-hand is $(\beta, \alpha\beta)$ or $(\beta, \beta^2\alpha)$: hence $n^a\nu$, $=2\nu$, $=2$, or ν is $=1$.

Hence also, by theorem 2, there should be 3 groups of the order 2 such as $(1, \alpha)$, viz. these are $(1, \alpha)$, $(1, \alpha\beta)$, $(1, \alpha\beta^2)$, derived from $(1, \alpha)$ as follows:

$$\begin{array}{ll} 1\ \ (1, \alpha)\,1^{-1} & =(1, \alpha), \\ \alpha\ \ (1, \alpha)\,\alpha^{-1} & =(1, \alpha), \\ \beta\ \ (1, \alpha)\,\beta^{-1} & =(1, \alpha\beta), \\ \beta^2\ (1, \alpha)\,\beta^{-2} & =(1, \alpha\beta^2), \\ \alpha\beta\ (1, \alpha)(\alpha\beta)^{-1} & =(1, \alpha\beta^2), \\ \alpha\beta^2(1, \alpha)(\alpha\beta^2)^{-2} & =(1, \alpha\beta), \end{array}$$

$$\begin{array}{llllll} \text{since} & \beta^{-1} & =\beta^2, & \text{and therefore the second term is} & \beta\alpha\beta^2 & =\alpha\beta^2.\beta^2=\alpha\beta, \\ \text{,,} & \beta^{-2} & =\beta, & \text{,,}\qquad\text{,,} & \beta^2\alpha\beta & =\alpha\beta\,.\,\beta\ =\alpha\beta^2, \\ \text{,,} & (\alpha\beta)^{-1} & =\alpha\beta, & \text{,,}\qquad\text{,,} & \alpha\beta\alpha\alpha\beta & =\alpha\beta\,.\,\beta\ =\alpha\beta^2, \\ \text{,,} & (\alpha\beta^2)^{-1} & =\alpha\beta^2, & \text{,,}\qquad\text{,,} & \alpha\beta^2\alpha\alpha\beta^2 & =\alpha\beta^2.\beta^2=\alpha\beta\,; \end{array}$$

viz. the derivatives are $(1, \alpha)$, $(1, \alpha\beta)$, $(1, \alpha\beta^2)$, each twice.

Secondly, $n=3$; there should be here a group of the order 3, viz. this is $(1, \beta, \beta^2)$. The group, the substitutions whereof are permutable with $(1, \beta, \beta^2)$, is the entire group $(1, \beta, \beta^2, \alpha, \alpha\beta, \alpha\beta^2)$; in fact, taking any substitution hereof, for instance α, we have $\alpha(1, \beta, \beta^2)=(1, \beta, \beta^2)\alpha$, viz. the left-hand side is $(\alpha, \alpha\beta, \alpha\beta^2)$, and the right-hand side is $(\alpha, \beta\alpha, \beta^2\alpha)$, $=(\alpha, \alpha\beta^2, \alpha\beta)$, which is the left-hand side, *the change of order being immaterial*; this is the meaning of the expression used, "the operations whereof are permutable with the group g." Hence, we have $n^a\nu$, $=3\nu$, $=6$, or $\nu=2$; and thence also $nk+1$, $=3k+1$, $=1$, viz. $k=0$. There is thus only a single group of the order 3, viz. the group $(1, \beta, \beta^2)$.

As another instance, I take the group of the order 12 formed by the positive substitutions of four letters, viz. these are

$$\begin{array}{lll} 1, & ab \,.\, cd, & abc, \\ & ac \,.\, bd, & acb, \\ & ad \,.\, bc, & abd, \\ & & adb, \\ & & acd, \\ & & adc, \\ & & bcd, \\ & & bdc. \end{array}$$

Here, taking $n=2$, we have $N=n^{\alpha}\nu\,(nk+1)=2^2 \,.\, 3 \,.\, 1$; there is a group g of the order 4, viz. this is

$$(1,\ ab \,.\, cd,\ ac \,.\, bd,\ ad \,.\, bc),$$

and the greatest group, the substitutions whereof are permutable with this group g, is the entire group of the order 12; thus, considering any substitution thereof, for instance abc, we have

$$abc \begin{Bmatrix} 1 \\ ab \,.\, cd \\ ac \,.\, bd \\ ad \,.\, bc \end{Bmatrix} = \begin{Bmatrix} 1 \\ ab \,.\, cd \\ ac \,.\, bd \\ ad \,.\, bc \end{Bmatrix} abc,$$

viz. the left-hand is $\begin{Bmatrix} abc \\ acd \\ bdc \\ adb \end{Bmatrix}$, the right-hand is $\begin{Bmatrix} abc \\ bdc \\ adb \\ acd \end{Bmatrix}$;

hence $n^{\alpha}\nu$, $=4\nu$, $=12$ or $\nu=3$; whence also $nk+1$, $=2k+1$, $=1$: and thus the foregoing group g is the only group of the order 4.

Similarly, taking $\nu=3$, we have $N=n^{\alpha}\nu\,(nk+1)$, $=3 \,.\, 1 \,.\, 4$. There is a group g of the order 3, say $(1,\ abc,\ acb)$; the greatest group, the substitutions whereof are permutable with g, is the group g itself, viz. we have $n^{\alpha}\nu$, $=3\nu$, $=3$, or $\nu=1$; and then $nk+1$, $=3k+1$, $=4$: there are thus 4 groups of the order 3, viz. these are

$$(1,\ abc,\ acb),\ (1,\ abd,\ adb),\ (1,\ acd,\ adc),\ (1,\ bcd,\ bdc).$$

Reverting to the before-mentioned group of the order 6, this not only contains each of the groups $(1, \alpha)$, $(1, \alpha\beta)$, $(1, \alpha\beta^2)$ of order 2, and the group $(1, \beta, \beta^2)$ of order 3; but it is the permutable product of a group of order 2 by a group of order 3, say it is

$$G = (1, \alpha)(1, \beta, \beta^2), \; = (1, \beta, \beta^2)(1, \alpha).$$

A group, which is thus a permutable product of two factors, is said to be a true product; and when it cannot be thus expressed as a permutable product of two factors, it is prime or simple. A group, the order of which is equal to a prime number p (the cyclical group of the order p) is simple; but the order may be a composite number and yet the group be simple—it was remarked by Galois, *Liouville*, t. XI. (1865), p. 409, that the order of the lowest simple group of composite order is 60, $= 2^2 . 3 . 5$, and it has been recently shown, Hölder, "Die einfachen Gruppen im ersten und zweiten Hundert der Ordnungszahlen," *Math. Ann.* t. XL. (1892), pp. 55—88, that the only other composite order of a simple group in the first 200 numbers is 168. Moreover, in the paper Cole, "Simple groups from order 201 to order 500," *Amer. Math. Jour.* t. XIV. (1892), pp. 378—388, it is shown that within these limits the only numbers which can give a simple group or groups are 360 and 432. I take the opportunity of referring to two other important papers, Young, "On the determination of groups whose order is a power of a prime," *Amer. Math. Jour.* t. XV. (1893), pp. 124—178, and Cole and Glover, "On groups whose orders are products of three prime factors," *ib.* pp. 191—220.

958.

ON THE SURFACE OF THE ORDER n WHICH PASSES THROUGH A GIVEN CUBIC CURVE.

[From the *Messenger of Mathematics*, vol. XXIII. (1894), pp. 79, 80.]

It is natural to assume that, taking A, B, C to denote the general functions $(x, y, z, w)^{n-2}$ of the order $n-2$, the general surface of the order n which passes through the curve

$$\begin{Bmatrix} x, & y, & z \\ y, & z, & w \end{Bmatrix} = 0,$$

(or, what is the same thing, the curve $x : y : z : w = 1 : \theta : \theta^2 : \theta^3$), has for its equation

$$\begin{vmatrix} A, & B, & C \\ x, & y, & z \\ y, & z, & w \end{vmatrix} = 0;$$

but the formal proof is not immediate. Writing the equation in the form

$$U = Sax^\alpha y^\beta z^\gamma w^\delta, \; = 0, \; \alpha+\beta+\gamma+\delta = n,$$

then U must vanish on writing therein $x : y : z : w = 1 : \theta : \theta^2 : \theta^3$; a term $ax^\alpha y^\beta z^\gamma w^\delta$ becomes $= a\theta^p$, where $p = \beta + 2\gamma + 3\delta$ is the weight of the term reckoning the weights of x, y, z, w as 0, 1, 2, 3 respectively; and hence the condition is that, for each given weight p, the sum Sa of the coefficients of the several terms of this weight shall be $= 0$. Using any such equation to determine one of the coefficients thereof in terms of the others, the function U is reduced to a sum of duads $a(x^\alpha y^\beta z^\gamma w^\delta - x^{\alpha'} y^{\beta'} z^{\gamma'} w^{\delta'})$, where in each duad the two terms are of the same degree and of the same weight, and where a is an arbitrary coefficient; it ought therefore to be true that each such duad $x^\alpha y^\beta z^\gamma w^\delta - x^{\alpha'} y^{\beta'} z^{\gamma'} w^{\delta'}$ has the property in question—or writing P, Q, $R = yw - z^2$, $zy - xw$, $xz - y^2$, say that each such duad is of the form $AP + BQ + CR$.

Suppose for a moment that α' is greater than α, but that β', γ', δ' are each less than β, γ, δ respectively: the duad is $x^{\alpha'} y^{\beta} z^{\gamma} w^{\delta} (x^\lambda - y^\mu z^\nu w^\rho)$, where λ, μ, ν, ρ are each positive, and hence $x^\lambda - y^\mu z^\nu w^\rho$ is a duad having the property in question, or changing the notation say $x^\alpha - y^\beta z^\gamma w^\delta$ has the property in question; and in like manner, by considering the several cases that may happen, we have to show that each of the duads

$$x^\alpha - y^\beta z^\gamma w^\delta,\quad y^\beta - x^\alpha z^\gamma w^\delta,\quad z^\gamma - x^\alpha y^\beta w^\delta,\quad w^\delta - x^\alpha y^\beta z^\gamma,$$
$$x^\alpha y^\beta - z^\gamma w^\delta,\quad x^\alpha z^\gamma - y^\beta w^\delta,\quad x^\alpha w^\delta - y^\beta z^\gamma,$$

has the property in question; it being of course understood that, in each of these duads, the two terms have the same degree and the same weight. The first form cannot exist; for we must have therein $\alpha = \beta + \gamma + \delta$ and $0 = \beta + 2\gamma + 3\delta$, which is inconsistent with α, β, γ, δ each of them positive. For the second form $\beta = \alpha + \gamma + \delta$, $\beta = 2\gamma + 3\delta$: this is $\alpha = \gamma + 2\delta$ or the duad is $y^{2\gamma+3\delta} - x^{\gamma+2\delta} z^\gamma w^\delta$, $= (y^2)^\gamma\, y^{3\delta} - (xz)^\gamma\, (x^2w)^\delta$. Writing $y^2 = xz - R$, we have terms containing the factor R, and a residual term $(xz)^\gamma\, \{y^{3\delta} - (x^2w)^\delta\}$, and writing herein

$$xw = yz - Q \text{ or } x^2w = xyz - Q,$$

we have terms containing Q as a factor and a residual term

$$(xz)^\gamma\, \{y^{3\delta} - (xyz)^\delta\},\ = (xz)^\gamma\, y^\delta\, \{(y^2)^\delta - (xz)^\delta\},$$

and again writing herein $y^2 = xz - R$, we see that this term contains the factor R: hence the duad in question consists of terms having the factor R or the factor Q. Similarly for the other cases: either α, β, γ, δ can be expressed as positive numbers, and then the duad consists of terms each divisible by P, Q, or R; or else α, β, γ, δ cannot be expressed as positive numbers, and then the duad does not exist: thus for the third form $z^\gamma - x^\alpha y^\beta w^\delta$, here $\gamma = \alpha + \beta + \delta$, $2\gamma = \beta + 3\delta$, or say $\gamma = 3\alpha + 2\beta$, $\delta = 2\alpha + \beta$, and the duad is $z^{3\alpha+2\beta} - x^\alpha y^\beta w^{2\alpha+\beta}$, $= z^{3\alpha}\,(z^2)^{2\beta} - (xw^2)^\alpha\,(yw)^\beta$, which can be reduced to the required form. But for the duad $x^\alpha y^\beta - z^\gamma w^\delta$, we have $\alpha + \beta = \gamma + \delta$, $\beta = 2\gamma + 3\delta$, which cannot be satisfied by positive values of α, β, γ, δ, and thus the duad does not exist.

A surface of the order n which passes through $3n+1$ points of a cubic curve contains the curve: hence the number of constants, or say the capacity of a surface of the order n, through the curve $P = 0$, $Q = 0$, $R = 0$, is

$$\tfrac{1}{6}(n+1)(n+2)(n+3) - 1 - (3n+1),\ = \tfrac{1}{6}(n^3 + 6n^2 - 7n - 6).$$

Primâ facie the capacity of the surface $AP + BQ + CR = 0$, A, B, C being general functions of the order $n-2$, is

$$3 \,.\, \tfrac{1}{6}(n-1)\, n\, (n+1) - 1,\ = \tfrac{1}{2}(n^3 - n - 2),$$

but there is a reduction on account of the identical equations

$$xP + yQ + zR = 0,\quad yP + zQ + wR = 0,$$

which connect the functions P, Q, R: for $n = 2$, the formulæ give each of them as it should do, Capacity $= 2$; viz. the quadric surface through the curve is

$$aP + bQ + cR = 0.$$

959.

NOTE ON PLÜCKER'S EQUATIONS.

[From the *Messenger of Mathematics*, vol. XXIV. (1895), pp. 23, 24.]

It is well known that if

$$A,\ B,\ C,\ D = 2,\ 3,\ 6,\ 8,$$

then the equations

$$\begin{aligned} n &= m^2 - m - A\delta - B\kappa, \\ \iota &= 3m^2 - 6m - C\delta - D\kappa, \\ m &= n^2 - n - A\tau - B\iota, \\ \kappa &= 3n^2 - 6n - C\tau - D\iota, \end{aligned}$$

are equivalent to three independent equations giving n, τ, ι in terms of m, δ, κ. It is easy to show that the *necessary* conditions in order that this may be so, are

$$C = 3A, \quad \text{and} \quad D = 3B - 1,$$

that is,

$$A,\ B,\ C,\ D = A,\ B,\ 3A,\ 3B-1,$$

where A and B are arbitrary.

In fact, from the last two equations eliminating τ, and for n, ι substituting their values, we have

$$\begin{aligned} Cm - A\kappa = &\ (C-3A)(m^2 - m - A\delta - B\kappa)^2, \\ &- (C-6A)(m^2 - m - A\delta - B\kappa), \\ &+ (AD - BC)(3m^2 - 6m - C\delta - D\kappa), \end{aligned}$$

which must therefore be an identity. In order that the term in m^4 may vanish we must have $C = 3A$; and then substituting this value for C, we must have

$$3Am - A\kappa = 3A(m^2 - m - A\delta - B\kappa) + (AD - 3AD)(3m^2 - 6m - 3A\delta - D\kappa).$$

Here the coefficient of m^2 must vanish, that is,

$$0 = 3A + 3AD - 9AB, \quad \text{or} \quad D = 3B - 1,$$

and, substituting this value, the equation is

$$\begin{aligned} 3Am - A\kappa = &\ 3A(-m - A\delta - B\kappa) \\ &- A\{-6m - 3A\delta - (3B-1)\kappa\}, \end{aligned}$$

that is,

$$\begin{aligned} 3m - \kappa = &- 3m - 3A\delta - 3B\kappa \\ &+ 6m + 3A\delta + (3B-1)\kappa, \end{aligned}$$

an identity.

960.

ON THE CIRCLE OF CURVATURE AT ANY POINT OF AN ELLIPSE.

[From the *Messenger of Mathematics*, vol. XXIV. (1895), pp. 47, 48.]

LET
$$u = \frac{x}{a}, \quad v = \frac{y}{b}, \quad u^2 + v^2 = 1.$$

The equation of the circle of curvature at the point (x, y) of the ellipse $\frac{x^2}{a^2} + \frac{y^2}{b^2} = 1$ is
$$X^2 + Y^2 + 2\frac{a^2 - b^2}{ab}(-bXu^3 + aYv^3) + u^2(a^2 - 2b^2) + v^2(b^2 - 2a^2) = 0.$$

Write $X = a\xi$, $Y = b\eta$, then this becomes
$$a^2\xi^2 + b^2\eta^2 + 2(a^2 - b^2)(-\xi u^3 + \eta v^3) + u^2(a^2 - 2b^2) + v^2(b^2 - 2a^2) = 0.$$

To find where this meets the ellipse, we must write $\xi^2 + \eta^2 = 1$; eliminating η, we have
$$a^2\xi^2 + b^2(1 - \xi^2) - 2(a^2 - b^2)\xi u^3 + u^2(a^2 - 2b^2) + v^2(b^2 - 2a^2) + 2(a^2 - b^2)v^3\sqrt{(1 - \xi^2)} = 0,$$
or putting for shortness
$$a^2 - b^2 = A, \quad u^2(a^2 - 2b^2) + v^2(b^2 - 2a^2) = B,$$
the equation for ξ is
$$A\xi^2 - 2A\xi u^3 + b^2 + B + 2Av^3\sqrt{(1 - \xi^2)} = 0,$$
but
$$b^2 + B = b^2(u^2 + v^2) + u^2(a^2 - 2b^2) + v^2(b^2 - 2a^2) = u^2(a^2 - b^2) + v^2(2b^2 - 2a^2) = A(u^2 - 2v^2),$$
viz.
$$\xi^2 - 2\xi u^3 + u^2 - 2v^2 + 2v^3\sqrt{(1 - \xi^2)} = 0,$$
that is,
$$(\xi^2 - 2u^3\xi + u^2 - 2v^2)^2 - 4v^6(1 - \xi^2) = 0,$$
which is without difficulty reduced to the form
$$(\xi - u)^3\{\xi - (u^3 - 3uv^2)\} = 0,$$
that is,
$$\xi = u^3 - 3uv^2,$$
and hence
$$\eta = v^3 - 3vu^2,$$
viz. writing $u, v = \cos\theta, \sin\theta$, then we have
$$\xi = \cos^3\theta - 3\cos\theta\sin^2\theta = \cos 3\theta,$$
$$\eta = \sin^3\theta - 3\sin\theta\cos^2\theta = -\sin 3\theta,$$
or the circle of curvature at $(a\cos\theta, b\sin\theta)$ cuts the ellipse in $(a\cos 3\theta, -b\sin 3\theta)$, as is known.

961.

A TRIGONOMETRICAL IDENTITY.

[From the *Messenger of Mathematics*, vol. XXIV. (1895), pp. 49—51.]

THE following was proposed as a Senate House Problem: Given the equations

$$a\cos(\beta+\gamma)+b\cos(\beta-\gamma)+c=0,$$
$$a\cos(\gamma+\alpha)+b\cos(\gamma-\alpha)+c=0,$$
$$a\cos(\alpha+\beta)+b\cos(\alpha-\beta)+c=0,$$

it is to be shown that $a^2+2bc-b^2=0$.

Assume

$$\cos\alpha+i\sin\alpha,\ \cos\beta+i\sin\beta,\ \cos\gamma+i\sin\gamma=x,\ y,\ z,$$

then the equations are

$$a\left(yz+\frac{1}{yz}\right)+b\left(\frac{y}{z}+\frac{z}{y}\right)+2c=0,$$
$$a\left(zx+\frac{1}{zx}\right)+b\left(\frac{z}{x}+\frac{x}{z}\right)+2c=0,$$
$$a\left(xy+\frac{1}{xy}\right)+b\left(\frac{x}{y}+\frac{y}{x}\right)+2c=0,$$

that is,

$$a(1+y^2z^2)+b(y^2+z^2)+2cyz=0,$$
$$a(1+z^2x^2)+b(z^2+x^2)+2czx=0,$$
$$a(1+x^2y^2)+b(x^2+y^2)+2cxy=0,$$

the second and third equations give

$$a:b:2c=x(x^2-yz):x(-1+x^2yz):(1-x^4)(y+z);$$

or, say a, b, $2c$ are equal to these values; and then, substituting in the first equation, we have

$$x(1+y^2z^2)(x^2-yz)+x(y^2+z^2)(-1+x^2yz)+(1-x^4)(y^2z+yz^2)=0,$$

which is

$$(x-y)(x-z)\{x+y+z-(yz+zx+xy)(x+y+z)\}=0,$$

viz. our relation between x, y, z is

$$x+y+z-(yz+zx+xy)\,xyz=0,$$

or, what is the same thing,

$$\frac{1}{yz}+\frac{1}{zx}+\frac{1}{xy}-(yz+zx+xy)=0.$$

Then

$$\begin{aligned} a+b &= x(-1+x^2)(1+yz),\\ a-b &= x(\ \ 1+x^2)(1-yz),\\ 2c\ \ &= \ (\ \ 1-x^4)(y+z), \end{aligned}$$

$$a^2-b^2=x^2(1-x^4)(y^2z^2-1),\quad 2bc=x(-1+x^2yz)(1-x^4)(y+z).$$

The equation to be verified is

$$x(y^2z^2-1)=(1-x^2yz)(y+z),$$

that is,

$$x+y+z-(yz+zx+xy)\,xyz=0,$$

as it should be.

A somewhat different form of the proof is as follows:—We have *identically*

$$\begin{vmatrix} \cos(\beta+\gamma), & \cos(\beta-\gamma), & 1\\ \cos(\gamma+\alpha), & \cos(\gamma-\alpha), & 1\\ \cos(\alpha+\beta), & \cos(\alpha-\beta), & 1 \end{vmatrix}$$

$$=4\sin\tfrac{1}{2}(\beta-\gamma)\sin\tfrac{1}{2}(\gamma-\alpha)\sin\tfrac{1}{2}(\alpha-\beta)\{\sin(\beta+\gamma)+\sin(\gamma+\alpha)+\sin(\alpha+\beta)\},$$

and therefore the relation between the angles is

$$\sin(\beta+\gamma)+\sin(\gamma+\alpha)+\sin(\alpha+\beta)=0.$$

From the second and third equations,

$$a:b:c=\sin\{\tfrac{1}{2}(\beta+\gamma)-\alpha\}\ :\ -\sin\{\tfrac{1}{2}(\beta+\gamma)+\alpha\}\ :\ 2\sin\alpha\cos\alpha\cos\tfrac{1}{2}(\beta-\gamma),$$

or say

$$\begin{aligned} a &= \ \ \sin\tfrac{1}{2}(\beta+\gamma)\cos\alpha-\cos\tfrac{1}{2}(\beta+\gamma)\sin\alpha,\\ b &= -\sin\tfrac{1}{2}(\beta+\gamma)\cos\alpha-\cos\tfrac{1}{2}(\beta+\gamma)\sin\alpha,\\ c &= \ \ 2\sin\alpha\cos\alpha\cos\tfrac{1}{2}(\beta-\gamma), \end{aligned}$$

therefore

$$a^2-b^2=-4\sin\alpha\cos\alpha\sin\tfrac{1}{2}(\beta+\gamma)\cos(\beta+\gamma)=-2\sin\alpha\cos\alpha\sin(\beta+\gamma),$$

$$\begin{aligned} bc &= 2\sin\alpha\cos\alpha\{-\cos\tfrac{1}{2}(\beta-\gamma)\sin\tfrac{1}{2}(\beta+\gamma)\cos\alpha-\cos\tfrac{1}{2}(\beta-\gamma)\cos\tfrac{1}{2}(\beta+\gamma)\sin\alpha\},\\ &= 2\,.\,\tfrac{1}{2}\sin\alpha\cos\alpha\{-(\sin\beta+\sin\gamma)\cos\alpha-(\cos\beta+\cos\gamma)\sin\alpha\},\\ &= \sin\alpha\cos\alpha\{-\sin(\gamma+\alpha)-\sin(\alpha+\beta)\}=\sin\alpha\cos\alpha\sin(\beta+\gamma), \end{aligned}$$

whence therefore

$$a^2 - b^2 + 2bc = 0,$$

which is the required relation.

The equation to be proved may also be written

$$\begin{vmatrix} \cos(\beta+\gamma), & \cos(\beta-\gamma), & 1 \\ \cos(\gamma+\alpha), & \cos(\gamma-\alpha), & 1 \\ \cos(\alpha+\beta), & \cos(\alpha-\beta), & 1 \end{vmatrix}$$

$$= 4 \sin \tfrac{1}{2}(\beta-\gamma) \sin \tfrac{1}{2}(\gamma-\alpha) \sin \tfrac{1}{2}(\alpha-\beta) \{\sin(\beta+\gamma) + \sin(\gamma+\alpha) + \sin(\alpha+\beta)\},$$

or putting

$$\begin{aligned} \beta+\gamma &= a, & b-c &= \gamma-\beta, \\ \gamma+\alpha &= b, & c-a &= \alpha-\gamma, \\ \alpha+\beta &= c, & a-b &= \beta-\alpha, \end{aligned}$$

this becomes

$$\begin{vmatrix} \cos a, & \cos(b-c), & 1 \\ \cos b, & \cos(c-a), & 1 \\ \cos c, & \cos(a-b), & 1 \end{vmatrix}$$

$$= -4 \sin \tfrac{1}{2}(b-c) \sin \tfrac{1}{2}(c-a) \sin \tfrac{1}{2}(a-b) (\sin a + \sin b + \sin c),$$

an identity which may be proved without difficulty.

962.

COORDINATES VERSUS QUATERNIONS.

[From the *Proceedings of the Royal Society of Edinburgh*, vol. XX. (1895), pp. 271—275. Read July 2, 1894.]

It is contended that Quaternions (as a method) are more comprehensive and less artificial than—and, in fact, in every way far superior to—Coordinates. Thus Professor Tait, in the Preface to his *Elementary Treatise on Quaternions* (1867), reproduced in the second and third editions (1873 and 1890), writes—"It must always be remembered that Cartesian methods are mere particular cases of quaternions where most of the distinctive features have disappeared; and that when, in the treatment of any particular question, scalars have to be adopted, the quaternion solution becomes identical with the Cartesian one. Nothing, therefore, is ever lost, though much is generally gained, by employing quaternions in place of ordinary methods. In fact, even when quaternions degrade to scalars, they give the solution of the most general statement of the problem they are applied to, quite independent of any limitations as to choice of particular coordinate axes." And he goes on to speak of "such elegant trifles as trilinear coordinates," and would, I presume, think as lightly of quadriplanar coordinates. It is right to notice that the claims of quaternions are chiefly insisted upon in regard to their applications to the physical sciences; and I would here refer to his paper, "On the Importance of Quaternions in Physics" (*Phil. Mag.*, Jan. 1890), being an abstract of an address to the Physical Society of the University of Edinburgh, Nov. 1889; but these claims certainly extend to and include the science of geometry.

I wish to examine into these claims on behalf of quaternions. My own view is that quaternions are merely a particular method, or say a theory, in coordinates. I have the highest admiration for the notion of a quaternion; but (I am not sure whether I did or did not use the illustration many years ago in conversation with Professor Tait), as I consider the full moon far more beautiful than any moonlit view, so I regard the notion of a quaternion as far more beautiful than any of its applications. As another illustration which I gave him, I compare a quaternion formula to a pocket-map—a capital thing to put in one's pocket, but which for use must be unfolded: the formula, to be understood, must be translated into coordinates.

I remark that the imaginary of ordinary algebra—for distinction call this θ—has no relation whatever to the quaternion symbols i, j, k; in fact, in the general point of view, all the quantities which present themselves are, or may be, complex values $a+\theta b$, or, in other words, say that a scalar quantity is in general of the form $a+\theta b$. Thus quaternions do not properly present themselves in plane or two-dimensional geometry at all—although, as will presently appear, we may use them in plane geometry; but they belong essentially to solid or three-dimensional geometry, and they are most naturally applicable to the class of problems which in coordinates are dealt with by means of the three rectangular coordinates x, y, z.

In plane geometry, considering an origin O, and through it two rectangular axes Ox, Oy, then in coordinates we determine the position of a point by means of its coordinates x, y; or, writing x, y, z to denote given linear functions of the original rectangular coordinates x, y, we may, if we please, determine it by trilinear coordinates, or say by the ratios $x : y : z$. The advantage is, that we thereby deal with the line infinity as with any other line, whereas with the rectangular coordinates x, y the line infinity presents itself as a line *sui generis*, and that we thereby bring the theory into connexion with that of the homogeneous functions $(*\between x, y, z)^n$.

In quaternions, the position of a point is determined in reference to the fixed point O, by its vector α, which is in fact $=ix+jy$, where i, j are the quaternion imaginaries ($i^2=-1$, $j^2=-1$, $ij=-ji$), but the idea is to use as little as possible the foregoing equation $\alpha=ix+jy$, and thus to conduct the investigations independently, as far as may be, of the particular positions of the axes Ox, Oy.

As the most simple example, take the theorem that the lines joining the extremities of equal and parallel lines in a plane are themselves equal and parallel, viz. (writing $\sim$ to denote equal and parallel), if $AB \sim CD$, then $AC \sim BD$.

Coordinates.

A, B, C, D are determined by their coordinates

$$(x_1, y_1),\ (x_2, y_2),\ (x_3, y_3),\ (x_4, y_4).$$

$$AB \sim CD$$

gives

$$\left.\begin{aligned} x_2 - x_1 &= x_4 - x_3 \\ y_2 - y_1 &= y_4 - y_3 \end{aligned}\right\},$$

whence

$$\left.\begin{aligned} x_3 - x_1 &= x_4 - x_2 \\ y_3 - y_1 &= y_4 - y_2 \end{aligned}\right\},$$

that is,

$$AC \sim BD.$$

Quaternions.

AB, CD are determined by their vectors α, β, and then writing γ for the vector AD,

$$AB \sim CD$$

gives

$$\alpha = \beta,$$

whence

$$\gamma - \beta = -\alpha + \gamma,$$

that is,

$$AC \sim BD.$$

And for the comparison of the two solutions, we have

$$\alpha = i(x_2 - x_1) + j(y_2 - y_1), \quad \beta = i(x_4 - x_3) + j(y_4 - y_3).$$

But this example of a plane theorem is a trivial one, given only for the sake of completeness.

Passing to solid geometry, we have—

Coordinates.—Considering a fixed point O, and through it the rectangular axes Ox, Oy, Oz, the position of a point is determined by its coordinates x, y, z. But we may, in place of these, consider the quadriplanar coordinates (x, y, z, w) linear functions of the original rectangular coordinates x, y, z.

Quaternions.—The position of a point in reference to the fixed origin O is determined by its vector α, which is in fact $= ix + jy + kz$, where i, j, k are the Hamiltonian symbols $(i^2 = j^2 = k^2 = -1,\ jk = -kj = i,\ ki = -ik = j,\ i = -ji = k)$; but the idea is to use as little as possible the foregoing equation $\alpha = ix + jy + kz$, and thus to conduct the investigations independently, as far as may be, of the particular positions of the axes Ox, Oy, Oz.

I consider the problem to determine the line OC at right angles to the plane of the lines OA, OB.

Coordinates.

Taking O as origin, the coordinates of A, B, C are taken to be

$$(x_1, y_1, z_1),\quad (x_2, y_2, z_2),\quad (x, y, z)$$

respectively. Then

$$xx_1 + yy_1 + zz_1 = 0,$$

$$xx_2 + yy_2 + zz_2 = 0;$$

whence

$$x : y : z = y_1z_2 - y_2z_1 : z_1x_2 - z_2x_1 : x_1y_2 - x_2y_1.$$

Quaternions.

Points A, B, C are determined by their vectors α, β, γ. Then

$$S\alpha\gamma = 0,\quad S\beta\gamma = 0;$$

whence

$$m\gamma = V\alpha\beta,$$

m being an arbitrary scalar.

Here to compare the two solutions, observe that the two equations $S\alpha\gamma = 0$, $S\beta\gamma = 0$ are in fact the equations $xx_1 + yy_1 + zz_1 = 0$, $xx_2 + yy_2 + zz_2 = 0$; and so also $m\gamma = V\alpha\beta$ denotes the relations $x : y : z = y_1z_2 - y_2z_1 : z_1x_2 - z_2x_1 : x_1y_2 - x_2y_1$. But a quaternionist says that $m\gamma = V\alpha\beta$ *is* the compendious and elegant solution of the problem as opposed to the artificial and clumsy one $x : y : z = y_1z_2 - y_2z_1 : z_1x_2 - z_2x_1 : x_1y_2 - x_2y_1$. And it is upon this that I join issue; $m\gamma = V\alpha\beta$ is a very pretty formula, like the folded-up pocket-map, but, to be intelligible, I consider that it requires to be developed into the other form. Of course, the example is as simple a one as could have been selected; and, in the case of a more complicated example, the mere abbreviation of the quaternion formula would be very much greater, but just for this reason there is the more occasion for the developed coordinate formula. To take another example, the condition, in order that the vectors α, β, γ may be coplanar, is $S\alpha\beta\gamma=0$, and Professor Tait contrasts this with the prolixity of the corresponding coordinate formula

$$\begin{vmatrix} x, & y, & z \\ x_1, & y_1, & z_1 \\ x_2, & y_2, & z_2 \end{vmatrix} = 0.$$

I remark that, when all the components of a determinant have to be expressed, nothing can be shorter than this, the ordinary determinant notation, which simply expresses the several components in their line-and-column relation to each other. But as a mere abbreviation, it would be allowable to write Δ, $=(ABC)$, to denote the determinant formed by the coordinates of the three points.

In conclusion, I would say that while coordinates are applicable to the whole science of geometry, and are the natural and appropriate basis and method in the science, quaternions seem to me a particular and very artificial method for treating such parts of the science of three-dimensional geometry as are most naturally discussed by means of the rectangular coordinates x, y, z.

963.

NOTE ON DR MUIR'S PAPER, "A PROBLEM OF SYLVESTER'S IN ELIMINATION."

[From the *Proceedings of the Royal Society of Edinburgh*, vol. XX. (1895), pp. 306—308. Received November 6, 1894.]

I IN part reproduce this very interesting paper for the sake of a remark which appears to me important. I write (a, b, c, f, g, h) in place of Muir's (A, B, C, A', B', C'), and take as usual (A, B, C, F, G, H) and K to denote

$$(bc - f^2,\ ca - g^2,\ ab - h^2,\ gh - af,\ hf - bg,\ fg - ch)$$

and the discriminant $abc - af^2 - bg^2 - ch^2 + 2fgh$

I then write

$$\begin{aligned} U &= bz^2 - 2fyz + cy^2, & P &= fx^2 + ayz - hzx - gxy, & L &= bcx^2 + afyz - bgzx - chxy, \\ V &= cx^2 - 2gzx + az^2, & Q &= gy^2 - hyz + bzx - fxy, & M &= cay^2 - afyz + bgzx - chxy, \\ W &= ay^2 - 2hxy + bx^2, & R &= hz^2 - gyz - fzx + cxy, & N &= abz^2 - afyz - bgzx + chxy. \end{aligned}$$

The equations $U = 0$, $V = 0$, $W = 0$, imply $P = 0$, $Q = 0$, $R = 0$, but observe that P, Q, R are not the sums of mere numerical multiples of U, V, W; we, in fact, have identically

$$\begin{aligned} 2yzP &= -x^2U + y^2V + z^2W, \\ 2zxQ &= x^2U - y^2V + z^2W, \\ 2xyR &= x^2U + y^2V - z^2W. \end{aligned}$$

If then $U = 0$, $V = 0$, $W = 0$, we have also $P = 0$, $Q = 0$, $R = 0$, and we can from the six equations dialytically eliminate x^2, y^2, z^2, yz, zx, xy, thus obtaining a result, Determinant = 0, which is $K^2 = 0$; this is, in fact, Sylvester's process for the elimination.

But L, M, N are sums of mere numerical multiples of U, V, W, viz. we have

$$\begin{aligned} 2L &= -aU + bV + cW, \\ 2M &= aU - bV + cW, \\ 2N &= aU + bV - cW, \end{aligned}$$

so that the original equations $U=0$, $V=0$, $W=0$ are equivalent to and may be replaced by $L=0$, $M=0$, $N=0$.

Muir shows that we have identically

$$\begin{aligned} L - fP &= x\,(Ax + Hy + Gz), \\ M - gQ &= y\,(Hx + By + Fz), \\ N - hR &= z\,(Gx + Fy + Cz), \end{aligned}$$

where observe that the first of these equations is

$$\left.\begin{aligned} &(fx^2 - ayz)\,(bz^2 - 2fyz + cy^2) \\ -&(fy^2 - byz)\,(cx^2 - 2gzx + az^2) \\ -&(fz^2 - cyz)\,(ay^2 - 2hxy + bx^2) \end{aligned}\right\} = 2xyz\,(Ax + Hy + Gz);$$

and similarly for the second and third equations.

He thence infers that the elimination may be performed by eliminating x, y, z from the equations

$$\begin{aligned} Ax + Hy + Gz &= 0, \\ Hx + By + Fz &= 0, \\ Gx + Fy + Cz &= 0, \end{aligned}$$

viz. that the result is

$$\begin{vmatrix} A, & H, & G \\ H, & B, & F \\ G, & F, & C \end{vmatrix} = 0,$$

that is, $K^2=0$ as before.

The natural inference is that K being $=0$, the three linear equations in (x, y, z) are equivalent to two equations giving for the ratios $x : y : z$ rational values which should satisfy the original equations $U=0$, $V=0$, $W=0$: the fact is that there are no such values, but that, K being $=0$, the three equations are equivalent to a single equation: for observe that, combining for instance the first and second equations, these will be equivalent to each other if only

$$\frac{A}{H} = \frac{H}{B} = \frac{G}{F},$$

that is,

$$AB - H^2 = 0, \quad GH - AF = 0, \quad HF - BG = 0,$$

which are $cK=0$, $fK=0$, $gK=0$, all satisfied by $K=0$; and similarly for the first and third, and the second and third equations. It will be remembered that the true form of the result is not $K=0$ but $K^2=0$, and this seems to be an indication that the three equations should be, as they have been found to be, equivalent to a single equation.

The problem may be further illustrated as follows: instead of the original equations $U=0$, $V=0$, $W=0$, consider the like equations with the inverse coefficients (A, B, C, F, G, H), viz.

$$Bz^2 - 2Fyz + Cy^2 = 0,$$

$$Cx^2 - 2Gzx + Az^2 = 0,$$

$$Ay^2 - 2Hxy + Bx^2 = 0,$$

so that the result of the elimination should be

$$(ABC - AF^2 - BG^2 - CH^2 + 2FGH)^2 = 0.$$

Here considering in connexion with the triangle $x=0$, $y=0$, $z=0$ (say the vertices hereof are the points A, B, C) the conic

$$(a, b, c, f, g, h)(x, y, z)^2 = 0,$$

the first equation represents the pair of tangents from the point A to the conic, the second the pair of tangents from the point B to the conic, and the third the pair of tangents from the point C to the conic. The first and second pairs of tangents intersect in four points, and if one of the third pair of tangents passes through one of the four points, then it is at once seen that the conic must touch one of the sides $x=0$, $y=0$, $z=0$ of the triangle, viz. we must have $bc-f^2=0$, $ca-g^2=0$, or $ab-h^2=0$. But we have $a=BC-F^2$, &c., or writing

$$K_1 = ABC - AF^2 - BG^2 - CH^2 + 2FGH,$$

then these equations are $K_1A=0$, $K_1B=0$, $K_1C=0$, all satisfied by $K_1=0$. We may regard $K_1=0$ as the condition in order that the conic $(a, b, c, f, g, h)(x, y, z)^2=0$ may be a *point-pair*: the analytical reason for this is not clear, but we see at once that, if the conic be a point-pair, then the three pairs of tangents are the lines drawn from the points A, B, C respectively to the two points of the point-pair, so that the three pairs of tangents have in common these two points. Regarding $K_1=0$ as the condition in order to the existence of a single common point, and recollecting that the true result of the elimination is $K_1^2=0$, the form perhaps indicates what we have just seen is the case, that there are in fact two common points of intersection: but at any rate the foregoing geometrical considerations lead to $K_1=0$, as the condition for the coexistence of the three equations.

I remark in conclusion that I do not know that there is any general theory of the case where a result of elimination presents itself in the form $\Omega^2=0$, as distinguished from the ordinary form $\Omega=0$.

964.

ON THE NINE-POINTS CIRCLE OF A SPHERICAL TRIANGLE.

[From the *Quarterly Journal of Pure and Applied Mathematics*, vol. XXVII. (1895), pp. 35—39.]

THE definition is in effect given in Hart's paper, "Extension of Terquem's theorem respecting the circle which bisects three sides of a triangle," *Quarterly Mathematical Journal*, t. IV. (1861), pp. 260, 261, viz. if we have a spherical triangle ABC, then we have a circle (i.e. a small circle of the sphere), say the nine-points circle, meeting the sides BC, CA, AB in the points F, L; G, M; H, N respectively, where

$$\tan \tfrac{1}{2} BF = \frac{\cos \frac{1}{2} b - \cos \frac{1}{2} c \cos \frac{1}{2} a}{\cos \frac{1}{2} c \sin \frac{1}{2} a},$$

$$\tan \tfrac{1}{2} CF = \frac{\cos \frac{1}{2} c - \cos \frac{1}{2} a \cos \frac{1}{2} b}{\cos \frac{1}{2} b \sin \frac{1}{2} a},$$

(which equations agree with $BF + FC = BC$), and

$$\tan \tfrac{1}{2} BL = \frac{\cos \frac{1}{2} c \sin \frac{1}{2} a}{\cos \frac{1}{2} b + \cos \frac{1}{2} c \cos \frac{1}{2} a},$$

$$\tan \tfrac{1}{2} CL = \frac{\cos \frac{1}{2} b \sin \frac{1}{2} a}{\cos \frac{1}{2} c + \cos \frac{1}{2} a \cos \frac{1}{2} b},$$

(which equations agree with $BL + CL = BC$); and with the like formulæ for the points G, M: and H, N: respectively.

If, as usual, the sides of the triangle are called a, b, c, and for shortness we write

$$(\cos \tfrac{1}{2} a,\ \cos \tfrac{1}{2} b,\ \cos \tfrac{1}{2} c;\ \sin \tfrac{1}{2} a,\ \sin \tfrac{1}{2} b,\ \sin \tfrac{1}{2} c) = (p,\ q,\ r;\ p_1,\ q_1,\ r_1),$$

then the formulæ are

$$\tan\tfrac{1}{2}BF = \frac{q-rp}{rp_1},\quad \tan\tfrac{1}{2}CF = \frac{r-pq}{qp_1},$$

$$\tan\tfrac{1}{2}CG = \frac{r-pq}{pq_1},\quad \tan\tfrac{1}{2}AG = \frac{p-qr}{rq_1},$$

$$\tan\tfrac{1}{2}AH = \frac{p-qr}{qr_1},\quad \tan\tfrac{1}{2}BH = \frac{q-rp}{pr_1},$$

and

$$\tan\tfrac{1}{2}BL = \frac{rp_1}{q+rp},\quad \tan\tfrac{1}{2}CL = \frac{qp_1}{r+pq},$$

$$\tan\tfrac{1}{2}CM = \frac{pq_1}{r+pq},\quad \tan\tfrac{1}{2}AM = \frac{rq_1}{p+qr},$$

$$\tan\tfrac{1}{2}AN = \frac{qr_1}{p+qr},\quad \tan\tfrac{1}{2}BN = \frac{pr_1}{q+rp}.$$

Before going further, it may be remarked that for a, b, c, each of them small, we have

$$\tfrac{1}{2}BF = \frac{(1-\tfrac{1}{8}b^2)-(1-\tfrac{1}{8}c^2)(1-\tfrac{1}{8}a^2)}{\tfrac{1}{2}a},$$

that is, $BF = \dfrac{a^2+c^2-b^2}{2a} = c\cos B$, and similarly $CF = a\cos C$, if A, B, C are the angles of the plane triangle; that is, in the plane triangle F, G, H are the foot of the perpendiculars let fall from the angles on the opposite sides. Moreover, $\tan\tfrac{1}{2}BL = \dfrac{\tfrac{1}{2}a}{1+1.1}$, that is, $BL = \tfrac{1}{2}a$, and similarly $CL = \tfrac{1}{2}a$; that is, L, M, N are the median points of the three sides respectively.

In the general case of the spherical triangle ABC, the construction is effected by means of a triangle $A'B'C'$, the sides whereof are respectively the halves of those

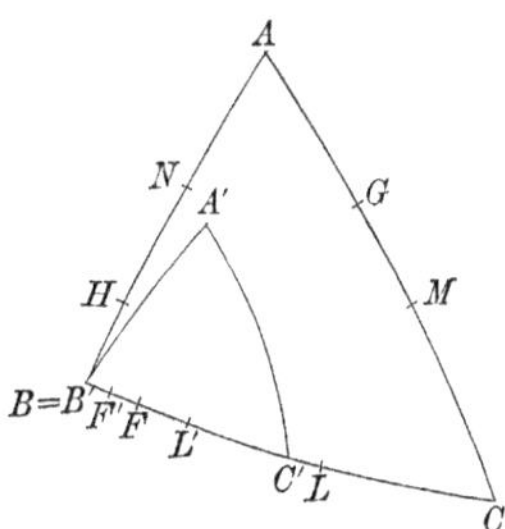

of the original triangle: viz. for this triangle $A'B'C'$, we construct the points F', G', H' and L', M', N', and then on the sides of the triangle ABC taking $BF = 2BF'$, $BL = 2BL'$, &c., we have the points F, G, H, L, M, N.

Thus p, q, r, p_1, q_1, r_1 denoting as above the cosines and sines of the half-sides of the triangle ABC, that is, the cosines and sines of the sides of the triangle $A'B'C'$, we have

$$\tan B'F' = \frac{q - rp}{rp_1}, \quad \tan C'F' = \frac{r - pq}{qp_1},$$
$$\vdots$$
$$\tan B'L' = \frac{rp_1}{q + rp}, \quad \tan C'L' = \frac{qp_1}{r + pq}.$$
$$\vdots$$

First, for the points F', G', H', we have $A'F'$, $B'G'$, $C'H'$, the perpendiculars from the angles on the opposite sides, meeting in a point O', the orthocentre of the triangle $A'B'C'$: in fact, from the triangle $A'B'F'$, right-angled at F', we have

$$\tan B'F' = \tan A'B' \cos B' = \tan \tfrac{1}{2}c \, \frac{\cos \frac{1}{2}b - \cos \frac{1}{2}c \cos \frac{1}{2}a}{\sin \frac{1}{2}c \sin \frac{1}{2}a},$$
$$= \frac{r_1}{r} \frac{q - rp}{r_1 p_1} = \frac{q - rp}{rp_1},$$

as above, and similarly for the points G' and H'.

I notice that we have

$$\sin B'F' = \frac{q - rp}{\sqrt{\{(q - rp)^2 + r^2p_1^2\}}}, \quad = \frac{q - rp}{\sqrt{(q^2 + r^2 - 2pqr)}},$$

and thus

$$\sin B'F' = \frac{q - rp}{\sqrt{(q^2 + r^2 - 2pqr)}}, \quad \sin C'F' = \frac{r - pq}{\sqrt{(q^2 + r^2 - 2pqr)}},$$

$$\sin C'G' = \frac{r - pq}{\sqrt{(r^2 + p^2 - 2pqr)}}, \quad \sin A'G' = \frac{p - qr}{\sqrt{(r^2 + p^2 - 2pqr)}},$$

$$\sin A'H' = \frac{p - qr}{\sqrt{(p^2 + q^2 - 2pqr)}}, \quad \sin B'H' = \frac{q - rp}{\sqrt{(p^2 + q^2 - 2pqr)}};$$

hence

$$\sin B'F' \,.\, \sin C'G' \,.\, \sin A'H' = \sin C'F' \,.\, \sin A'G' \,.\, \sin B'H',$$

which (as is well known) is the condition for the intersection of the arcs $A'F'$, $B'G'$, $C'H'$ in the orthocentre O'.

But I say further that we have

$$\sin 2B'F' \; (= \sin BF) = \frac{2\,(q - rp)\, rp_1}{q^2 + r^2 - 2pqr},$$

$$\sin 2C'F' \; (= \sin CF) = \frac{2\,(r - pq)\, qp_1}{q^2 + r^2 - 2pqr},$$

and thence

$$\sin BF \,.\, \sin CG \,.\, \sin AH = \sin CF \,.\, \sin AG \,.\, \sin BH,$$

and thus the arcs AF, BG, CH meet in a point which is obviously *not* the orthocentre of the triangle ABC.

Secondly, for the points L', M', N', we have

$$\sin B'L' = \frac{rp_1}{\sqrt{(q^2+r^2+2pqr)}}, \quad \sin C'L' = \frac{qp_1}{\sqrt{(q^2+r^2+2pqr)}},$$

that is,

$$\sin B'L' : \sin C'L' = r : q, \quad = \cos B'A' : \cos C'A';$$

and similarly

$$\sin C'M' : \sin A'M' = p : r, \quad = \cos C'B' : \cos A'B',$$

$$\sin A'N' : \sin B'N' = q : p, \quad = \cos A'C' : \cos B'C',$$

viz. the sides $B'C'$, $C'A'$, $A'B'$ are by the points L', M', N' divided each into two parts such that for any side the sines of the two parts are proportional to the cosines of the other two sides. We have

$$\sin B'L' \,.\, \sin C'M' \,.\, \sin A'N' = \sin C'L' \,.\, \sin A'M' \,.\, \sin B'N',$$

viz. the arcs $A'L'$, $B'M'$, $C'N'$ meet in a point K' which may be called the cos-centre of the triangle $A'B'C'$ (where observe that, for a, b, c indefinitely small, i.e. for a plane triangle, the points L', M', N' are the mid-points of the sides, and the centre K' is the C. G. or median point of the triangle).

But further, we have

$$\sin 2B'L' \,(=\sin BL) = \frac{2rp_1(q+rp)}{q^2+r^2+2pqr},$$
$$\vdots$$
$$\sin 2C'L' \,(=\sin CL) = \frac{2qp_1(r+pq)}{q^2+r^2+2pqr},$$
$$\vdots$$

and thence

$$\sin BL \,.\, \sin CM \,.\, \sin AN = \sin CL \,.\, \sin AM \,.\, \sin BN,$$

viz. the arcs AL, BM, CN meet in a point, which is obviously *not* the cos-centre of the triangle ABC.

We have thus the construction of the nine-points circle as a six-points circle, by means of the points F, G, H, L, M, N; and by way of recapitulation we may say that the nine-points circle meets the sides BC, CA, AB in the points F, L; G, M; H, N respectively, where the points F, G, H depend on the ortho-centre of the semi-triangle, and the points L, M, N depend on the cos-centre of the semi-triangle.

The triangle ABC has an inscribed circle and three escribed circles, and we have (as is known) the theorem that the nine-points circle touches each of these four circles. The circles BC, CA, AB and the nine-points circle form a tetrad of circles, and the inscribed circle and the three escribed circles a tetrad of circles, or say the eight circles form a bitetrad, such that each circle of the one tetrad touches each circle of the other tetrad.

965.

ON THE SIXTY ICOSAHEDRAL SUBSTITUTIONS.

[From the *Quarterly Journal of Pure and Applied Mathematics*, vol. XXVII. (1895), pp. 236—242.]

THE Sixty Icosahedral Substitutions were obtained in an elegant form by Gordan in the paper "Ueber endliche Gruppen linearer Transformationen einer Veränderlichen," *Math. Ann.* t. XII. (1877), pp. 23—46, see p. 45, where the group is exhibited in the canonical form

$$\epsilon^\mu \eta, \quad \frac{-\epsilon^\mu}{\eta}, \quad \epsilon^\mu \frac{(\epsilon^2+\epsilon^4)\,\eta+\epsilon^\rho}{\epsilon^{-\rho}\eta-(\epsilon+\epsilon^3)}, \quad \epsilon^\mu \frac{-\epsilon^{-\rho}\eta+(\epsilon+\epsilon^3)}{(\epsilon^2+\epsilon^4)\,\eta+\epsilon^\rho}$$

(ϵ an imaginary fifth root of unity), agreeing with the developed form given in my paper, "On the Schwarzian Derivative and the Polyhedral Functions," *Camb. Phil. Trans.*, t. XIII. Part 1 (1881), [745], pp. 5—68, see pp. 55—56, [this Collection, vol. XI., pp. 204—205].

But a slight change of form is desirable. I write $\begin{Bmatrix} a, & b \\ c, & d \end{Bmatrix}$ to denote the unimodular matrix

$$\begin{Bmatrix} \dfrac{a}{\sqrt{(ad-bc)}}, & \dfrac{b}{\sqrt{(ad-bc)}} \\[2ex] \dfrac{c}{\sqrt{(ad-bc)}}, & \dfrac{d}{\sqrt{(ad-bc)}} \end{Bmatrix},$$

so that the value is independent of the absolute magnitudes of a, b, c, d, but depends only on their ratios; and observe that $m\begin{Bmatrix} a, & b \\ c, & d \end{Bmatrix}$ denotes $\begin{Bmatrix} ma, & mb \\ c\;, & d \end{Bmatrix}$, the factor m applying to the top-line only. As before, ϵ denotes an imaginary fifth root of unity, and I write $A=\epsilon+\epsilon^4$, $B=\epsilon^2+\epsilon^3$, $A'=\epsilon-\epsilon^4$, $B'=\epsilon^2-\epsilon^3$, but, in fact, I scarcely need more than the first of these symbols.

The substitutions are

$$(x_1, y_1) = \begin{pmatrix} a, & b \\ c, & d \end{pmatrix}(x, y),$$

that is, $x_1 = ax + by$, $y_1 = cx + dy$; and we then have sixty matrices $\begin{Bmatrix} a, & b \\ c, & d \end{Bmatrix}$, forming the group in question.

Giving to ρ the values 0, 1, 2, 3, 4, I write

$$K_\rho = \begin{Bmatrix} \epsilon^\rho, & 0 \\ 0, & \epsilon^{-\rho} \end{Bmatrix},$$

so that K_0 is the matrix unity $\begin{Bmatrix} 1, & 0 \\ 0, & 1 \end{Bmatrix}$,

$$L_\rho = \begin{Bmatrix} 0, & -\epsilon^\rho \\ \epsilon^{-\rho}, & 0 \end{Bmatrix},$$

$$M_\rho = \begin{Bmatrix} A, & \epsilon^\rho \\ \epsilon^{-\rho}, & -A \end{Bmatrix}, \quad \text{mod.} \begin{vmatrix} A, & \epsilon^\rho \\ \epsilon^{-\rho}, & A \end{vmatrix} \text{ is } = -A^2 - 1, \ = B'^2,$$

$$N_\rho = \begin{Bmatrix} -\epsilon^\rho, & A \\ A, & \epsilon^{-\rho} \end{Bmatrix}, \quad \text{mod.} \begin{vmatrix} -\epsilon^\rho, & A \\ A, & \epsilon^{-\rho} \end{vmatrix} \text{ is } = -A^2 - 1, \ = B'^2;$$

then the matrices are

K_0, K_1, K_2, K_3, K_4

L_0, L_1, L_2, L_3, L_4

M_0, M_1, M_2, M_3, M_4

1		read $M_0, M_1, \ldots,$
ϵ		$\epsilon M_0,$
ϵ^2		$\vdots$
ϵ^3		
ϵ^4		

N_0, N_1, N_2, N_3, N_4

1		,, $N_0, N_1, \ldots,$
ϵ		$\epsilon N_0,$
ϵ^2		$\vdots$
ϵ^3		
ϵ^4		

viz. $5 + 5 + 25 + 25, = 60$ matrices in all, including the matrix unity.

It is to be observed that the system must remain unaltered if we change ϵ into ϵ^2, ϵ^3, or ϵ^4: this is at once obvious for the change ϵ into ϵ^4, but less immediately so for the other two changes. Suppose ϵ changed into ϵ^2, then M_ρ becomes

$$= \begin{Bmatrix} B, & \epsilon^{2\rho} \\ \epsilon^{-2\rho}, & -B \end{Bmatrix}, \ = \begin{Bmatrix} AB, & A\epsilon^{2\rho} \\ A\epsilon^{-2\rho}, & -AB \end{Bmatrix} = \begin{Bmatrix} 1, & A\epsilon^{2\rho} \\ A\epsilon^{-2\rho}, & -1 \end{Bmatrix} \text{ (since } AB = 1) = \epsilon^{4\rho} \begin{Bmatrix} -\epsilon^{-2\rho}, & A \\ A, & \epsilon^{2\rho} \end{Bmatrix},$$

or changing ρ into $-\frac{1}{2}\rho$, which has the same values, this is $= \epsilon^{-2\rho}\begin{Bmatrix}-\epsilon^{\rho}, & A\\ A, & \epsilon^{-\rho}\end{Bmatrix}$, viz. M_ρ is equal to a power of ϵ multiplied by N_ρ: and similarly N_ρ becomes equal to a power of ϵ multiplied by M_ρ. And the like as regards the change ϵ into ϵ^3.

I find that the sixty matrices may be coordinated with the sixty positive substitutions of the letters *abcde* as follows:

1	$K_0 \begin{Bmatrix}1, & 0\\ 0, & 1\end{Bmatrix}$	1	17	$\epsilon^2 M_0,\ \epsilon^2 \begin{Bmatrix}A, & 1\\ 1 & -A\end{Bmatrix}$	*ced*
2	$L_0 \begin{Bmatrix}0, & -1\\ 1, & 0\end{Bmatrix}$	*ad . bc*	18	$\epsilon^2 M_1,\ \epsilon^2 \begin{Bmatrix}A, & \epsilon\\ \epsilon^4 & -A\end{Bmatrix}$	*adc*
3	$L_1 \begin{Bmatrix}0, & -\epsilon\\ \epsilon^4, & 0\end{Bmatrix}$	*ab . de*	19	$\epsilon^2 M_2,\ \epsilon^2 \begin{Bmatrix}A, & \epsilon^2\\ \epsilon^3 & -A\end{Bmatrix}$	*adb*
4	$L_2 \begin{Bmatrix}0, & -\epsilon^2\\ \epsilon^3, & 0\end{Bmatrix}$	*ac . be*	20	$\epsilon^2 M_3,\ \epsilon^2 \begin{Bmatrix}A, & \epsilon^3\\ \epsilon^2 & -A\end{Bmatrix}$	*acb*
5	$L_3 \begin{Bmatrix}0, & -\epsilon^3\\ \epsilon^2, & 0\end{Bmatrix}$	*bd . ce*	21	$\epsilon^2 M_4,\ \epsilon^2 \begin{Bmatrix}A, & \epsilon^4\\ \epsilon & -A\end{Bmatrix}$	*bce*
6	$L_4 \begin{Bmatrix}0, & -\epsilon^4\\ \epsilon, & 0\end{Bmatrix}$	*ae . cd*	22	$\epsilon^3 M_0,\ \epsilon^3 \begin{Bmatrix}A, & 1\\ 1 & -A\end{Bmatrix}$	*abc*
7	$M_0 \begin{Bmatrix}A, & 1\\ 1 & -A\end{Bmatrix}$	*ac . bd*	23	$\epsilon^3 M_1,\ \epsilon^3 \begin{Bmatrix}A, & \epsilon\\ \epsilon^4 & -A\end{Bmatrix}$	*bec*
8	$M_1 \begin{Bmatrix}A, & \epsilon\\ \epsilon^4 & -A\end{Bmatrix}$	*ae . bd*	24	$\epsilon^3 M_2,\ \epsilon^3 \begin{Bmatrix}A, & \epsilon^2\\ \epsilon^3 & -A\end{Bmatrix}$	*cde*
9	$M_2 \begin{Bmatrix}A, & \epsilon^2\\ \epsilon^3 & -A\end{Bmatrix}$	*ae . bc*	25	$\epsilon^3 M_3,\ \epsilon^3 \begin{Bmatrix}A, & \epsilon^3\\ \epsilon^2 & -A\end{Bmatrix}$	*acd*
10	$M_3 \begin{Bmatrix}A, & \epsilon^3\\ \epsilon^2 & -A\end{Bmatrix}$	*bc . de*	26	$\epsilon^3 M_4,\ \epsilon^3 \begin{Bmatrix}A, & \epsilon^4\\ \epsilon & -A\end{Bmatrix}$	*abd*
11	$M_4 \begin{Bmatrix}A, & \epsilon^4\\ \epsilon & -A\end{Bmatrix}$	*ac . de*	27	$\epsilon N_0,\ \epsilon \begin{Bmatrix}-1, & A\\ A, & 1\end{Bmatrix}$	*aec*
12	$N_0 \begin{Bmatrix}-1, & A\\ A, & 1\end{Bmatrix}$	*ab . cd*	28	$\epsilon^4 N_1,\ \epsilon^4 \begin{Bmatrix}-\epsilon, & A\\ A, & \epsilon^4\end{Bmatrix}$	*bcd*
13	$N_1 \begin{Bmatrix}-1, & A\epsilon\\ A\epsilon^4, & 1\end{Bmatrix}$	*ad . be*	29	$\epsilon^2 N_2,\ \epsilon^2 \begin{Bmatrix}-\epsilon^2, & A\\ A, & \epsilon^3\end{Bmatrix}$	*abc*
14	$N_2 \begin{Bmatrix}-1, & A\epsilon^2\\ A\epsilon^3, & 1\end{Bmatrix}$	*ab . ce*	30	$N_3,\ \begin{Bmatrix}-\epsilon^3, & A\\ A, & \epsilon^2\end{Bmatrix}$	*aed*
15	$N_3 \begin{Bmatrix}-1, & A\epsilon^3\\ A\epsilon^2, & 1\end{Bmatrix}$	*be . cd*	31	$\epsilon^3 N_4,\ \epsilon^3 \begin{Bmatrix}-\epsilon^4, & A\\ A, & \epsilon\end{Bmatrix}$	*bed*
16	$N_4 \begin{Bmatrix}-1, & A\epsilon^4\\ A\epsilon, & 1\end{Bmatrix}$	*ad . ce*			

32	$\epsilon^4 N_0,\ \epsilon^4 \begin{Bmatrix} -1, & A \\ A, & 1 \end{Bmatrix}$	*bde*
33	$\epsilon^2 N_1,\ \epsilon^2 \begin{Bmatrix} -\epsilon, & A \\ A, & \epsilon^4 \end{Bmatrix}$	*acb*
34	$N_2,\ \begin{Bmatrix} -\epsilon^2, & A \\ A, & \epsilon^3 \end{Bmatrix}$	*ade*
35	$\epsilon^3 N_3,\ \epsilon^3 \begin{Bmatrix} -\epsilon^3, & A \\ A, & \epsilon^2 \end{Bmatrix}$	*bde*
36	$\epsilon N_4,\ \epsilon \begin{Bmatrix} -\epsilon^4, & A \\ A, & \epsilon \end{Bmatrix}$	*ace*
37	$K_1,\ \begin{Bmatrix} \epsilon, & 0 \\ 0, & \epsilon^4 \end{Bmatrix}$	*aedbc*
38	$K_2,\ \begin{Bmatrix} \epsilon^2, & 0 \\ 0, & \epsilon^3 \end{Bmatrix}$	*adceb*
39	$K_3,\ \begin{Bmatrix} \epsilon^3, & 0 \\ 0, & \epsilon^2 \end{Bmatrix}$	*abecd*
40	$K_4,\ \begin{Bmatrix} \epsilon^4, & 0 \\ 0, & \epsilon \end{Bmatrix}$	*acbde*
41	$\epsilon M_0,\ \epsilon \begin{Bmatrix} A, & 1 \\ 1 - A \end{Bmatrix}$	*adecb*
42	$\epsilon M_1,\ \epsilon \begin{Bmatrix} A, & \epsilon \\ \epsilon^4 - A \end{Bmatrix}$	*acdeb*
43	$\epsilon M_2,\ \epsilon \begin{Bmatrix} A, & \epsilon^2 \\ \epsilon^3 - A \end{Bmatrix}$	*acebd*
44	$\epsilon M_3,\ \epsilon \begin{Bmatrix} A, & \epsilon^3 \\ \epsilon^2 - A \end{Bmatrix}$	*abdce*
45	$\epsilon M_4,\ \epsilon \begin{Bmatrix} A, & \epsilon^4 \\ \epsilon - A \end{Bmatrix}$	*adcbe*
46	$\epsilon^4 M_0,\ \epsilon^4 \begin{Bmatrix} A, & 1 \\ 1 - A \end{Bmatrix}$	*aebcd*
47	$\epsilon^4 M_1,\ \epsilon^4 \begin{Bmatrix} A, & \epsilon \\ \epsilon^4 - A \end{Bmatrix}$	*abced*
48	$\epsilon^4 M_2,\ \epsilon^4 \begin{Bmatrix} A, & \epsilon^2 \\ \epsilon^3 - A \end{Bmatrix}$	*abedc*
49	$\epsilon^4 M_3,\ \epsilon^4 \begin{Bmatrix} A, & \epsilon^3 \\ \epsilon^2 - A \end{Bmatrix}$	*adbec*
50	$\epsilon^4 M_4,\ \epsilon^4 \begin{Bmatrix} A, & \epsilon^4 \\ \epsilon - A \end{Bmatrix}$	*aecdb*
51	$\epsilon^2 N_0,\ \epsilon^2 \begin{Bmatrix} -1, & A \\ A, & 1 \end{Bmatrix}$	*acbed*
52	$N_1,\ \begin{Bmatrix} -\epsilon, & A \\ A, & \epsilon^4 \end{Bmatrix}$	*abcde*
53	$\epsilon^3 N_2,\ \epsilon^3 \begin{Bmatrix} -\epsilon^2, & A \\ A, & \epsilon^3 \end{Bmatrix}$	*aecbd*
54	$\epsilon N_3,\ \epsilon \begin{Bmatrix} -\epsilon^3, & A \\ A, & \epsilon^2 \end{Bmatrix}$	*acdbe*
55	$\epsilon^4 N_4,\ \epsilon^4 \begin{Bmatrix} -\epsilon^4, & A \\ A, & \epsilon \end{Bmatrix}$	*abdce*
56	$\epsilon^3 N_0,\ \epsilon^3 \begin{Bmatrix} -1, & A \\ A, & 1 \end{Bmatrix}$	*adbce*
57	$\epsilon N_1,\ \epsilon \begin{Bmatrix} -\epsilon, & A \\ A, & \epsilon^4 \end{Bmatrix}$	*aebdc*
58	$\epsilon^4 N_2,\ \epsilon^4 \begin{Bmatrix} -\epsilon^2, & A \\ A, & \epsilon^3 \end{Bmatrix}$	*acedb*
59	$\epsilon^2 N_3,\ \epsilon^2 \begin{Bmatrix} -\epsilon^3, & A \\ A, & \epsilon^2 \end{Bmatrix}$	*adebc*
60	$N_4,\ \begin{Bmatrix} -\epsilon^4, & A \\ A, & \epsilon \end{Bmatrix}$	*aedcb*

As an example of the composition of the forms, take 56, 57: we have

$$56 = adbce\,.\,abdec,$$
$$57 = aebdc\,.\,edacb,$$
$$abcde,$$

whence

$$56\,.\,57 = cde, \quad = 24,$$

and we have

$$56\,.\,57 = \begin{matrix} -\epsilon^3, & A\epsilon^3 \\ A, & 1 \end{matrix}\;\left|\;\begin{matrix}(-\epsilon^2,\ A),\ (\epsilon A,\ \epsilon^4) \\ \hline \\ \end{matrix}\right. = \left|\begin{matrix} 1+A^2\epsilon^3, & -A\epsilon^4+A\epsilon^2 \\ -A\epsilon^2+A, & A^2\epsilon+\epsilon^4 \end{matrix}\right|;\quad 24 = \begin{matrix} A\epsilon^3, & 1, \\ \epsilon^3, & -A, \end{matrix}$$

viz. taking out the factor $A(\epsilon^2-\epsilon^4)$, we ought to have

$$\left|\begin{matrix} 1+A^2\epsilon^3, & A(\epsilon^2-\epsilon^4) \\ A(1-\epsilon^2), & A(\epsilon^2-\epsilon^4) \end{matrix}\right| = \left|\begin{matrix} A\epsilon^3, & 1 \\ \epsilon^3, & -A \end{matrix}\right|\begin{matrix} A(\epsilon^2-\epsilon^4) \\ A(\epsilon^2-\epsilon^4) \end{matrix}:$$

a relation which will be identically true if only

$$1+A^2\epsilon^3 = A^2(1-\epsilon^2),$$
$$\epsilon^4+A^2\epsilon = A^2(-\epsilon^2+\epsilon^4).$$

These are

$$1 = A^2(\ \ 1-\epsilon^2-\epsilon^3),$$
$$\epsilon^4 = A^2(-\epsilon-\epsilon^2+\epsilon^4),$$

the second of which is obviously equivalent to the first: and this first equation is

$$1 = A^2(A+2),$$

that is,

$$A^3+2A^2-1=0, \text{ or } (A+1)(A^2+A-1)=0,$$

which is true in virtue of $A^2+A-1=0$; we have thus the required equation

$$56\,.\,57 = 24,$$

or say

$$\begin{Bmatrix} -\epsilon^3, & A\epsilon^3 \\ A, & 1 \end{Bmatrix}\begin{Bmatrix} -\epsilon^2, & A\epsilon \\ A, & \epsilon^4 \end{Bmatrix} = \begin{Bmatrix} A, & \epsilon^2 \\ \epsilon^3, & -A \end{Bmatrix}.$$

In like manner it may be shown that the product of any two of the sixty matrices is a matrix of the group.

There is an interesting case of linear transformation in Gordan's paper "Ueber die Auflösung der Gleichungen vom fünften Grade," *Math. Ann.*, t. XIII. (1878), pp. 375—404, see p. 379. Writing as before $A=\epsilon+\epsilon^4$, $B=\epsilon^2+\epsilon^3$, $A'=\epsilon-\epsilon^4$, $B'=\epsilon^2-\epsilon^3$, then for y_1, y_2 we substitute Ay_1+y_2, y_1-Ay_2, each divided by B'; and for x_1, x_2 we make the like substitution, ϵ^2 instead of ϵ, viz. this is for x_1, x_2, Bx_1+x_2, x_1-Bx_2, each divided by $-A'$.

This being so, the bipartite function

$$f = (y_1^3,\ y_1^2y_2,\ y_1y_2^2,\ y_2^3)(x_1^2x_2,\ x_2^3,\ x_1^3,\ -x_1x_2^2),$$

or, what is the same thing,

	x_1^3,	$x_1^2x_2$,	$x_1x_2^2$,	x_2^3
$=\ y_1^3$		1		
$y_1^2y_2$				1
$y_1y_2^2$	1			
y_2^3			-1	

remains unaltered by the substitutions. Observe that we have $AB = -1$, so that, omitting common factors, the substitutions are

$$y_1, y_2 \text{ multiplied by } \quad Ay_1 + y_2,\ y_1 - Ay_2,$$
$$x_1, x_2 \text{ multiplied by } -(x_1 - Ax_2),\ Ax_1 + x_2,$$

where it is to be noticed that (x_1, x_2) are neither cogredient nor contragredient with (y_1, y_2); but they are what may be termed socio-gredient, viz. the substitution for (y_1, y_2) determines uniquely that for (x_1, x_2).

The verification of the invariance of f might be effected rather more simply by means of the last-mentioned forms, but it is interesting to use the original forms

$$y_1, y_2 \text{ multiplied by } Ay_1 + y_2,\ y_1 - Ay_2,$$
$$x_1, x_2 \text{ multiplied by } Bx_1 + x_2,\ x_1 - Bx_2.$$

Making the substitution, the whole coefficient of y_1^3 is found to be

$$\begin{aligned}&(A^3B^2 + A^2 + AB^3 - B)\, x_1^3\\ &+ (2A^3B - A^3B^3 - 3A^2B + 3AB^2 + 2B^2 - 1)\, x_1^2x_2\\ &+ (A^3 - 2A^3B^2 + 3A^2B^2 + 3AB - B^3 + 2B)\, x_1x_2^2\\ &+ (-A^3B - A^2B^3 + A - B^2)\, x_2^3,\end{aligned}$$

where, reducing by $AB = -1$, the several coefficients are

$$\begin{aligned}A + A^2 - B^2 - B, &\quad = 0,\\ -2A^2 + 3A + 2B^2 - 3B, &\quad = 5(A - B),\\ A^3 - 2A - B^3 + 2B, &\quad = 0,\\ A^2 - B + A - B^2, &\quad = 0,\end{aligned}$$

in virtue of the relations $A^2 + A = 1$, $B^2 + B = 1$. There are similar reductions for the coefficients of $y_1^2y_2$, $y_1y_2^2$ and y_2^3; and the whole thus becomes $= 5(A - B)$, multiplied by

	x_1^3	$x_1^2x_2$	$x_1x_2^2$	x_2^3
y_1^3		1		
$y_1^2y_2$				1
$y_1y_2^2$	1			
y_2^3			-1	

;

but there is a denominator $-A'^3B'^3$, and we have $A'B' = B - A$, or this denominator is $(A - B)^3$, $= (A - B)(A - B)^2$, where

$$(A - B)^2 = A^2 - 2AB + B^2, \ = 1 - A + 2 + 1 - B, \ = 4 - A - B, \ = 5,$$

viz. the denominator is $= 5(A - B)$. Thus the new value of f is equal to its original value, which is the theorem in question.

966.

NOTE ON A MEMOIR IN SMITH'S COLLECTED PAPERS.

[From the *Bulletin of the American Mathematical Society*, Ser. 2, vol. I. (1895), pp. 94—96.]

AMONG the most noticeable papers in the *Collected Mathematical Papers* of H. J. S. Smith we have the hitherto unpublished "Memoir on the Theta and Omega Functions," XLIII. (vol. II. pp. 415—623), written in connexion with Dr Glaisher's Tables of the Theta-Functions and originally intended as an Introduction thereto. It appears that in 1873 or 1874 Dr Glaisher asked him, as a member of the British Association committee for the calculation of the Tables, whether he would contribute an Introduction. His reply was that he did not see his way to writing anything appropriate to the tables themselves, but that he "could say something with respect to the constants at the head of the pages." These constants were K, K', E, J, J', &c., the numerical values whereof were given for every minute of the modular angle. The memoir grew in extent, and it was finally decided that it should follow these yet unpublished tables with the before-mentioned title, "Memoir on the Theta and Omega Functions," but fortunately it has at length appeared in the Collected Papers as above.

In explanation of the title and scope of the memoir, it will be remembered that the Theta-Functions are functions of two arguments, x and q; so that giving to x the value zero or any numerical value, or any value depending on that of q, we obtain a series of functions containing the single argument q, or writing as usual $q=e^{i\pi\omega}$, say the single argument ω; and in the memoir, the attention is directed chiefly but not exclusively to these functions of a single argument which are termed *Omega Functions*. The functions chiefly considered under this designation are Hermite's functions $\phi\omega$, $\psi\omega$, $\chi\omega$, which represent the values of $\sqrt{k}$, $\sqrt[4]{k'}$, and $\sqrt[12]{kk'}$ considered as functions of $q=e^{i\pi\omega}$. To fix the ideas, it may be mentioned that the actual values (in one of their very numerous forms) are

$$\phi\omega=\sqrt{2}q_{\frac{1}{8}}\prod_1^\infty\frac{1+q^{2m}}{1+q^{2m-1}},$$

$$\psi\omega=\prod_1^\infty\frac{1-q^{2m-1}}{1+q^{2m-1}},$$

$$\chi\omega=\sqrt[6]{2}q_{\frac{1}{24}}\prod_1^\infty\frac{1}{1+q^{2m-1}},$$

each of them a one-valued function of ω: any rational and integral function of $\phi\omega$, $\psi\omega$, $\chi\omega$ is termed a *Modular* Function. It is right to add that the definitions extend only to those values of q for which the series are convergent, or (what is the same thing) ω regarded as a point must be situate within a certain region of the upper infinite half-plane $y = +$.

The requisite formulæ for the Theta-Functions are obtained from Jacobi's fundamental formula for the multiplication of four Theta-Functions; and the Elliptic Functions are introduced by means of their definitions in terms of the Theta-Functions: and the whole theory of Elliptic Functions is thus brought into connexion with the Theta and Omega Functions. The theory of Transformation depends in a great measure on the arithmetical and geometrical theory of binary matrices, of which the constituents are integral numbers; this theory plays an extensive part throughout the memoir.

An abstract of the contents of the memoir is as follows:

Arts. 1—14. Definitions and Elementary Properties of the Theta, Omega, and Elliptic Functions.

15—23. Arithmetical Theory of Binary Matrices.

24—34. The Transformation of the Theta and Omega Functions.

35—45. Geometrical Representation of Binary Quadratic Forms.

46—51. Geometrical Representation of the Modular Functions

$$\Phi\omega \text{ and } \Psi\omega \; (= \phi^8\omega, \; \psi^8\omega).$$

52—58. The Modular Equation.

59—62. The Equation of the Multiplier.

63—73. The Modular Curves.

74—82. Theory of the Modular Functions $\phi\omega$ and $\psi\omega$.

83—88. Theory of the Modular Function

$$T\omega = (1 - \chi^{24}\omega)^3 \div \chi^{48}\omega.$$

89—90. The Differential Equation of the Modular Equations and Curves (this last section somewhat incomplete).

A good deal of the same ground is gone over in Weber's *Elliptische Functionen und Algebraische Zahlen* (8vo. Brunswick, 1891), a work which exhibits in a very compendious form the higher parts of the theory of Elliptic Functions, and which well deserves to be carefully studied.

967.

AN ELEMENTARY TREATISE ON ELLIPTIC FUNCTIONS.

The first edition was published in 1876; the second edition was in the press at the time of Professor Cayley's death in 1895, and it was published in the course of that year.

END OF VOL. XIII.

CAMBRIDGE: PRINTED BY J. AND C. F. CLAY, AT THE UNIVERSITY PRESS.

www.ingramcontent.com/pod-product-compliance
Lightning Source LLC
LaVergne TN
LVHW011249110826
845149LV00001B/85

* 9 7 8 1 4 1 8 1 8 5 5 8 9 *